THE
DEVELOPMENT
OF
MODERN
CHEMISTRY

Aaron J. Ihde

THE UNIVERSITY OF WISCONSIN

DOVER PUBLICATIONS, INC.

NEW YORK

Published in Canada by General Publishing Company, Ltd.,
30 Lesmill Road, Don Mills, Toronto, Ontario.
Published in the United Kingdom by Constable and Company, Ltd., 10 Orange Street, London WC2H 7EG.

This Dover edition, first published in 1984, is an unabridged,
slightly corrected republication of the third printing (1970) of
the work originally published by Harper & Row, Publishers,
Inc., in 1964. Appendix I (the list of the chemical elements)
and Appendix IV (the list of Nobel Prize winners in Chemistry, Physics, Medicine and Physiology) have been updated.

Manufactured in the United States of America
Dover Publications, Inc., 180 Varick Street, New York, N.Y.
10014

Library of Congress Cataloging in Publication Data

Ihde, Aaron John, 1909–
The development of modern chemistry.

Reprint. Originally published: New York : Harper & Row,
1970 printing. With updated appendix.
Includes bibliographical references and indexes.
1. Chemistry—History. I. Title.
[QD11.I44 1984] 540'.9 82-18245
ISBN 0-486-64235-6

Frontispiece: Antoine Laurent Lavoisier with his wife, painted
from life in 1788 by Jacques Louis David. (Original is in the
Rockefeller Institute. Courtesy of the Rockefeller Institute.)

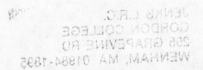

To Olive

CONTENTS

III

THE GROWTH OF SPECIALIZATION

IV

THE CENTURY OF THE ELECTRON

PREFACE

From small beginnings associated with the study of gases and minerals, chemistry has grown in two centuries to hold a central position among the modern sciences. The story of its unfolding holds a particular fascination because it impinges on so many intellectual developments. Further interest lies in the influence which the growth of chemistry has had on practical areas during the past century, this facet of the subject serving as an object lesson of the role of pure science in the development of technology, agriculture, and medicine.

It must not be assumed, of course, that chemistry is a development of only the past two centuries. The fundamentals of the discipline have accumulated from the day that primitive man mastered fire. Numerous authors, however, have extensively treated the material philosophies of the Greeks and the long period of alchemical activity. I have therefore given only a brief treatment of the incidents in the growth of chemistry before Joseph Black.

On the other end of the time span, most historians of chemistry have been content to stop at the beginning of the twentieth century. This has been true despite the fact that more advances in chemistry have taken place since 1900 than in all previous centuries combined. Although there are obvious pitfalls in dealing with the history of recent events, these must no longer serve as excuses for avoiding this century which is so rich in chemical discoveries.

The objective of this book, therefore, has been to portray the flow of events which brought chemistry from its primitive unspectacular state in 1750 to its dramatic vigor in the present day. I have sought to give proper attention to the part played by individuals without making the account a series of biographical sketches. At the same time I have attempted to place chemistry in the framework of the times. It has influenced human life in major ways, particularly in the nature of industrial and agricultural activity. At the same time, the growth of chemistry has been influenced by human affairs—political, economic, and social. These interactions I have sought to reveal.

Many individuals have influenced the writing of this book. Nearly two decades ago, with the generous encouragement of J. Howard Mathews, I had the opportunity to revive a then defunct course in history of chemistry. During the years since then, interested students have been a constant source of inspiration, while at the same time bringing me to realize the inadequacies in my understanding of the subject. I owe much to these students whose questions and interest forced me to probe ever more deeply into the subject.

The encouragement of a few individuals is particularly worthy of note. Three remarkable teachers developed in me a deep appreciation for the role of history in the understanding of chemistry. Louis Kahlenberg brought me into my first contact with history of chemistry. Henry A. Schuette, in his courses in food chemistry and organic analysis, utilized historical background whenever practical to develop better understanding of the subject. Karl Paul Link missed no opportunity to enliven his teaching with reference to historical developments.

The late Georg Urdang, historian of pharmacy, was a source of constant inspiration during the few years I was privileged to know him. The late George Sarton, too, set a notable example and provided understanding and encouragement. Numerous other friends and associates provided stimulating opportunities for the deepening of understanding and the testing of ideas. I am particularly grateful to Farrington Daniels for his encouragement, his understanding, and his patience.

To those colleagues who gave their time to read one or more chapters in the manuscript I owe particular thanks. They are Robert Alberty, Walter Blaedel, Marshall Clagett, I. B. Cohen, Erwin Hiebert, Edwin M. Larsen, Robert Stauffer, Robert West, and John Willard. I derived great benefit from their suggestions and realize that they saved the book from some serious shortcomings.

Finally, I express my gratitude to the members of my family who were forced to live with the discomforts of authorship in their midst. Not only were they gracious in accepting this as a way of life but all of them made positive contributions in many ways. Gretchen and Hal Serrie read much of the manuscript from the viewpoint of the nonchemist and made valuable suggestions. My son John read the material from the vantage point of a science student. My wife Olive contributed in ways too numerous to mention—critical, mechanical, inspirational. Her interest and encouragement and her indispensable collaboration had much to do with making the work possible.

AARON J. IHDE

May, 1964

I

THE FOUNDATIONS
OF CHEMISTRY

ALTHOUGH THE RISE OF CHEMISTRY
as a modern science becomes evident in the events of the last half of the
eighteenth century, the foundations of the science were being clearly
laid down during the preceding three millennia. Certain empirical advances
of a chemical nature reach back into an even earlier period.

Alchemy is commonly looked upon as the precursor of chemistry. This
represents an oversimplification, for chemistry inherited fully as much
from medicine and the technological arts as it did from alchemy. The
latter discipline actually left the subject matter of chemistry encumbered
with a system of erroneous beliefs which had to be cleared away before
fruitful pursuit of chemical knowledge could proceed.

These beliefs were part of the Greek philosophy which became incorporated into natural philosophy and hindered the rise of modern science
in the seventeenth century. They were perhaps even more deeply ingrained
in the subject matter of chemistry as the consequence of one and a half
millennia of alchemical activity, than they were in astronomy and
physics. As a result, these latter sciences were able to rid themselves of
their confining handicaps a century earlier than chemistry.

Although chemistry as a modern science may be considered to have
originated in the work of the pneumatic chemists following the discoveries
concerning gases after 1750, it is necessary to understand earlier ideas
about matter if the problems of the Chemical Revolution are to be properly
comprehended. Hence, it is necessary to examine the growth in material
knowledge, both practical and theoretical, which preceded 1750.

1

CHAPTER 1

PRELUDE
TO CHEMISTRY

ANCIENT KNOWLEDGE OF MATTER

Throughout man's study of matter, two divergent facets are readily apparent: the technical or factual, and the philosophical or theoretical. Advances in factual knowledge soon lead to reflection about why and how; advances in theory inevitably have an effect on practical knowledge—fruitful if the theoretical concept is sound, retarding if it is not sound.

Our knowledge of theoretical ideas regarding matter really originates with the Greeks. They seem to have been the first to ponder extensively about the nature of matter and they left a sizable literature regarding their speculations. Our knowledge of the chemical arts in antiquity goes back much further, partly through written records but primarily through archeological remains.

Technological Knowledge

It may be argued that man has practiced chemistry from the time he learned to control fire. *Sinanthropus pekinensis*, huddling in his caves near Choukoutien a half million years ago, left an accumulation of ashes, charcoal, and charred bones testifying to his use of fire. It would be foolhardy to consider him a chemist in any more than the vaguest sense, for he undoubtedly used fire more for warmth and protection from wild animals than for transformation of matter. He was a fire-tender rather than a fire-maker. Only with the latter advance did primitive man reach the point where he was truly able to control fire and to go on to use it effectively in cooking, pottery-making, and metal smelting, all of them involved in early chemical arts. The significance of fire in man's rise to a dominant species is well summarized by Coon:

> The use of fire is the only open-and-shut difference between man and all other animals. Fire was the first source of power which man found out how to use which did not come from the conversion of food and air into energy inside his own body. In Early Pleistocene times he made beautiful tools and brought up his children without it. In Middle Pleistocene he used it only to warm his knuckles in the mouth of a cave. In Late Pleistocene times it made him a more efficient animal, and during the last eight thousand years he has found increasing uses for it, and burned ever greater quantities of fuel. Fire has been the key to his rapid rise in mastering the forces of nature, his conquest and partial destruction of the earth, and his current problems.[1]

Although paleolithic man became a fire-tender, his mastery of thermal changes of materials extended little beyond the roasting of meats. During the neolithic revolution, however, this knowledge was extended to include the baking of pottery and the smelting of the less active metals like copper. In addition, such chemical operations were mastered as the making of beer and wine, the tanning of leather, the scouring of wool, the dyeing of textiles, and the making of glass. Red dyestuffs were prepared from kermes, arachil, madder, and henna; yellows from safflower, saffron, turmeric, and pomegranate; blues from indigo and woad; purple from certain Mediterranean snails. Mordants containing iron, aluminum, and copper were in use. Butter and cheese were prepared from milk by pastoral people. A variety of drug materials were in use, mostly prepared from plants.

The mastery of metal smelting was an important step forward and it had profound cultural effects. Copper was perhaps the first metal to be isolated from its ores—at first from the colored basic carbonates, later from the sulfides. By 1200 B.C. lead, tin, and iron were being produced, although iron was still a monopoly of the Hittites. Gold, silver, and mercury were also known, but their use goes back to precopper times because their low chemical activity permits their occurrence as the native metal.

Thus, in moving from the Stone Age successively through the Copper, Bronze, and Iron Ages, man acquired mastery over a variety of chemical processes. This mastery was gained in part by accident, in part by shrewd observation. Once attained, it was passed from generation to generation by a sort of apprenticeship procedure. Understanding of the chemical processes remained on a highly empirical level.

Philosophical Concepts of Matter

Although technicians continued to work in purely empirical fashion for many centuries, the Greek philosophers sought to formulate a theory of

[1] Carleton S. Coon, *The Story of Man*, Knopf, New York, 1954, p. 53.

matter which would be intellectually satisfying. As early as the sixth century B.C., Thales of Miletus sought to create a system in which water was the basis of all things. Anaximander of Miletus (c. 610–545 B.C.) believed in a primary substance, unidentified with any known substance, but eternal, boundless, and containing within itself all the contraries such as hot and cold, wet and dry. Eternal motion brought about separation of qualities and resulted in the materials of the known world. Anaximines of Miletus (c. 585–524 B.C.) modified the ideas of his predecessors by making air the primary substance and suggesting its transformation into other materials by thinning (fire) or thickening (winds, clouds, rain, hail, earth, rock).

The Pythagorean contemporaries of the younger Miletans preferred to derive their universe from number and geometric relationships. They

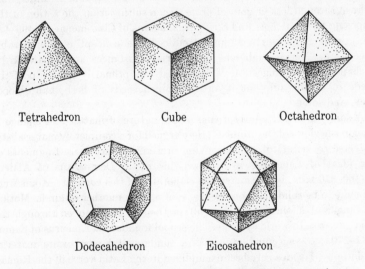

Tetrahedron   Cube   Octahedron

Dodecahedron   Eicosahedron

Fig. 1.1. THE FIVE REGULAR SOLIDS (PLATONIC BODIES).

knew the relationship between pitch and length of vibrating strings, and between triangular numbers, square numbers, pyramidal numbers, and cubic numbers; and they were aware of the five regular solids—the tetrahedron, cube, octahedron, dodecahedron, and eicosahedron—later referred to as the Platonic bodies. The urge to design a geometric universe based upon numerical relations was irresistible to them and has presented a continuing temptation to philosophers ever since.

Heraclitus of Ephesus (c. 544–484 B.C.) developed a dynamic concept in which change was the only reality. To the extent that he accepted a primary substance it was fire, not fire as a material but fire as the representative of change. Heraclitus' philosophy also placed much importance on the role of opposites in the creation of strife.

The early philosophies, and particularly the stress on change, met strong opposition from the Eleatic school which, through logical argument, found it necessary to deny the possibility of change. Parmenides used the concepts of *Being* and *Not-being* to argue that only *Being* could exist. Since *Being* was eternal and unchanging, change could take place only through dilution with *Not-being*. But the latter did not exist; hence change was impossible.

In order to overcome such arguments, which conflicted with the evidence of the senses, the pluralist philosophers abandoned unitary philosophies for a material world composed of at least several fundamental substances. Empedocles of Agrigentum (c. 495–435 B.C.) considered fire, air, earth, and water to be elemental substances, capable of undergoing change through loss or gain of properties resulting from the action of the opposing forces of *Love* and *Strife*. Anaxagoras of Clazomenae (c. 499–428 B.C.) went much further; he introduced the concept of innumerable "seeds" or "germs" inherent in material substances and separable by such processes as digestion. Thus, bread was primarily made up of the seeds of bread-stuff, but it also contained seeds of flesh, hair, blood, rock, and so on.

A somewhat different approach to the nature of matter and change is seen in the views of the atomists; they argued for a completely materialistic universe consisting of atoms moving in a void. Since mere fragments of the ideas of Leucippus are known, his pupil, Democritus of Abdera (c. 460–370 B.C.), is considered the elaborator of this concept. Atoms were thought to be solid and homogeneous, of a finite number of kinds. Motion was inherent in them, and their interactions as they moved through the void gave rise to world systems. The school founded by Epicurus of Samos (341–270 B.C.) accepted atoms as a fundamental part of its material philosophy. Its ideas are best exemplified in the Latin verse of the Roman poet Lucretius (96–55 B.C.). His *De rerum natura* describes, more completely and clearly than the extant fragments by Epicurus and Democritus, the physics, psychology, and ethics of this philosophical group.

Despite the enthusiasm and speculative logic of the Epicureans, atomism was soon pushed aside and received little further consideration until the seventeenth century. The speculative nature of atoms made their rejection easy, particularly in a world dominated by the Roman Stoics and soon to see the rise of powerful spiritual movements. Furthermore, the great authority of Aristotle, who found atoms untenable on mathematical grounds,[2] removed them from serious consideration until the decline of Scholasticism.

[2] Aristotle, *De caelo*, 303 a 21.

The major Greek impact on natural philosophy derives from the thought of Socrates, Plato, and Aristotle. Socrates (470–399 B.C.) actually directed attention away from the nature of the universe because he felt that such studies were of little consequence in contrast to studies of the nature of man and his relationship to society. Except for mathematics, Socrates found natural philosophy too speculative for serious consideration. He thereby halted a type of philosophy which was really going nowhere. At the same time he left philosophy with positive positions which were important for the future growth of science.[3] His insistence on unambiguous definitions and classifications, on logical argument, on respect for order, and on rational skepticism benefited every intellectual discipline, including science.

Plato (428–347 B.C.), the pupil of Socrates and the author of the dialogues which most clearly show us his master's ideas, was not content to ignore the realm of nature. Late in life he began a projected trilogy—the *Timaeus*, *Critias*, and *Hemocrates*—which sought to deal with the nature of the universe. Only the *Timaeus* was ever finished. The *Critias* was started but was abandoned in the middle of a sentence. The *Laws*, Plato's last work, was perhaps an expansion of the political ideas originally intended to accompany the *Timaeus*.

Besides Socrates' influence, Plato's works show the influence of Pythagoras, Heraclitus, and Parmenides. From Pythagoras comes the immense respect for mathematics and the mystical religious tone of his philosophy; from Heraclitus, the feeling of lack of permanence in the world of the senses; from Parmenides the notion that, on logical grounds, change is an illusion. Although the two latter ideas appear to be mutually contradictory, they lead logically to Plato's Theory of Ideas, according to which an observed object is only a reflection of the *Idea*, only the latter being real, perfect, and eternal. Thus, the senses are not to be trusted and knowledge is to be gained by deductive reasoning. The Platonic approach to learning places little emphasis on direct observation, much less on experimentation.

The *Timaeus*, which today is little known, beautifully reflects the Platonic approach to knowledge. The central figure, Timaeus, is a Pythagorean astronomer. The dialogue deals principally with the nature of the heavens and is outside the scope of the present book. However, the theory of matter developed in the *Timaeus* is briefly considered in order to gain insight into the nature of the thought which long influenced natural philosophy.

Plato adopted the four elements—earth, water, air, and fire—as primary and hinted at a fifth substance, an ether or quintessence which was left undeveloped but came to be associated with the material of the heavens. The four elements were bodies associated with the four regular solids; the

[3] George Sarton, *A History of Science*, vol. 1, *Ancient Science Through the Golden Age of Greece*, Harvard Univ. Press, Cambridge, 1952, pp. 271–272.

cube was assigned to earth, the tetrahedron to fire, the octahedron to air, and the eicosahedron to water. The dodecahedron was assigned to the heavenly element.

The sides of the four regular solids are resolvable into smaller triangles —the cube into isosceles right triangles; the solids with four, eight, and twenty sides into scalene triangles having angles of 30, 60, and 90 degrees —and these triangles permit various-sized squares to be built from the

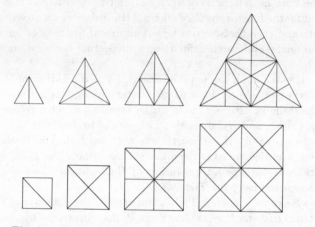

Fig. 1.2.   PLATO'S USE OF TRIANGLES FOR THE CONSTRUC-
TION OF SQUARES AND EQUILATERAL TRIANGLES.

isoceles right triangles, and various-sized equilateral triangles from the scalene. The squares may then be joined together to form the cubes of earth. The equilateral triangles may be joined to form the tetrahedra of fire, the octahedra of air, and the eicosahedra of water. Since the latter three are composed of the same triangles, they are considered inter-convertible. The fact that particle size may vary accounts for different kinds of fire, air, water, and earth. However, this particular kind of geometric atomism is not a projection of Democritean atomism because Plato firmly rejected the void. Any spaces between atoms must be filled by smaller atoms.

Although these logical ideas have no place in today's science, they were influential for a long time, inasmuch as the *Timaeus*, alone of Plato's works, was translated into Latin during the ancient period and exerted a continuing influence on the development of scientific thought in the western world.

Aristotle of Stagira (384–322 B.C.), although strongly influenced by his teacher Plato in many areas, differed somewhat in his ideas on matter. He held a more sympathetic point of view toward observation since he rejected the notion that only the idea or form was real. However, the results of his observational approach are apparent primarily in his biological works which were written late in life. His books on the physical sciences like *De caelo, De generatione et corruptione*, and the *Meteorologica* are from an earlier period when he was still strongly influenced by Platonism. It must also be realized that the subject matter of physical science was less amenable to observation in those times than that of biology. Chemistry in particular could hardly progress very far until an experimental tradition had developed from some of the other sciences.

Aristotle fully examined the ideas of his predecessors before formulating his own system. He was deeply concerned with the causes of things and with their purpose. Hence, his physical world is intensely logical but is based on premises open to grave question. He accepted a limited number of elements, but placed more importance on their qualities than on their substance. His definition of elements is quite modern:

> Let us then define the element in bodies as that into which other bodies may be analysed, which is present in them either potentially or actually . . . , and which cannot itself be analysed into constituents differing in kind.[4]

However, complete analysis of his works suggests that he believed in some sort of formless, undefined *prima materia* which became elemental only when suitable qualities were impressed upon it. To the Empedoclean elements—fire, air, water, and earth—Aristotle added the ether, or quintessence. The latter was perfect; it was the material of the heavens, starting with the sphere of the moon and extending to the periphery of his finite and spherical universe. Its nature was unchanging and its motion was perfect, that is, circular.

The four terrestrial elements were imperfect and moved up and down. The natural motion of earth was downward since it was absolutely heavy; fire moved upward since it was absolutely light. The respective motions of water and air were in accord with their relative heaviness and lightness.

The four qualities, hot and cold, wet and dry, were associated with the four elements and gave each its character. A particular element, in its most perfect form, contained two of the qualities in maximum degree. We may represent this in the diagram which later appeared so frequently in the alchemical literature. The Aristotelian elements were not unchanging, however, but were idealizations unlike the substances of everyday experience. Their qualities might be changed, thereby transforming them into one of the other elements. Thus, earth possessed dryness and coldness in the ultimate degree. By abstracting coldness and substituting hotness it might be transmuted into fire. This kind of thinking was

[4] Aristotle, *De caelo*, 302 a 16.

consistent with the practice of alchemy which became prominent early in the Christian Era.

Aristotle's immediate successors, Theophrastus and Strato, both wrote on physical subjects, but most of their works have been lost. Intellectual activity then shifted to the museum at Alexandria where scientific thought

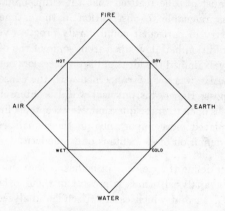

Fig. 1.3.　RELATION OF THE FOUR ELEMENTS
AND THE FOUR QUALITIES.

was actively pursued for several centuries. Although important steps were taken in mathematics, physics, medicine, geography, and astronomy during this period, there is no extant evidence of real activity in the field of chemistry. Concepts remained Platonic or Aristotelian. Factual knowledge likewise was perhaps little changed from that of a millennium before, although the literary sources are considerably more abundant during the period of Roman domination.

The Romans were not a scientific people and made almost no scientific contributions of their own; to the extent that they utilized scientific ideas they were content to accept those of the Greeks. The Latin writers on science, Varro, Vitruvius, Seneca, and Pliny the Elder, as well as the Greek physician Dioscorides, were compilers rather than innovators. Nevertheless, we owe a great deal to these writers, particularly to Pliny, for factual information about minerals, technological materials, and biological substances.

THE ALCHEMICAL HERITAGE

Alchemy began to be a formal discipline toward the end of the first century of the Christian Era. It arose among the Greek scholars of Alexandria and spread through the lands of the eastern Mediterranean, particularly Syria, before being taken over by the Arabs following the rise of Islam. Later, interest in the subject was transmitted to the Latin West with the transfer of Arabic scholarship to Spain and Italy in the twelfth century.

Because of its ambiguous literature, the origins of alchemy are shrouded in mystery. Besides the Egyptian development, there are claims for Chinese, Hindu, Mesopotamian, and Hebrew origins. There is no question that activities of an alchemical sort were taking place in China and India by the time alchemy was developing in Alexandria. However, scholarly studies to date have not established sound evidence for either contact or lack of contact between these cultures which would shed light on the origin and transmission of alchemy. Claims for Hebrew origin largely derive from reference to Biblical names in alchemical literature, a practice common in the later periods. Claims for the early presence of alchemy in Mesopotamia are largely based on confusion between chemical technology and alchemy.

Their difference is significant. Alchemy represents an amalgamation of certain chemical technologies and philosophical speculations. This amalgamation took place in Egypt where the Greek philosophers could see the activities of metalworkers; these artisans could enhance the appearance of ornaments fashioned from less precious metals and stones. Although these workmen doubtless realized that they were fabricating imitation jewelry when they alloyed and gilded, the philosophers convinced themselves that genuine silver and gold were being produced. The Platonic idea and the stress on perfection, coupled with the Aristotelian notions regarding alteration of elemental materials by adjustment of forms and qualities, were attractive concepts which made the transmutations of base metals into gold appear not unreasonable.

Original literature from the period of Greek alchemy is nonexistent[5] except for two tomb papyri dating from the end of the third century. Known as the *Leyden Papyrus X* and the *Stockholm Papyrus*, they are, respectively, recipe collections for preparing false silver and gold, and false gems.[6] There is none of the mysterious speculative philosophy associated with the alleged works of Democritos (first or second century),

[5] The lack of original Egyptian works has been attributed to an edict of Diocletian, A.D. 296, condemning alchemical works to the flames since they were believed to be a threat to the state (M. Berthelot, *Les origines de l'alchimie*, Steinheil, Paris, 1885, pp. 26, 72). Modern scholarship is inclined to question the issuance of such an edict; instead it attributes the lack of an Egyptian literature to the ravages of time and the secretive nature of alchemy. The earliest extant source that recounts the Diocletian legend is a writing by John of Antioch (fl. A.D. 620) referring to an Egyptian monk, Panodorus, who lived a whole century later than Diocletian.

[6] For general information, together with an English translation of the papyri, see E. R. Caley, *J. Chem. Educ.*, **3**, 1149 (1926), and **4**, 979 (1927).

Fig. 1.4. GREEK ALCHEMICAL MANUSCRIPT
PAGE SHOWING TEXT AND OUROBOROUS
SERPENT.
(From MS. Parisianus Greek 2327 by permission
of the *Journal of Hellenic Studies*.)

Zosimos (late third century), and Synesios (c. 370–415) as copied in Greek
manuscripts of the tenth century and later.[7]

Interpretation of existing materials makes it evident that transmutation
was associated with certain purifications and regenerations connected
with color changes.[8] *Melanosis* (blackening) was the first step in which
base metals were fused to produce a black mass, supposedly without
character, which could be subjected successively to *leukosis* (whitening),
xanthosis (yellowing), and *iosis* (purpling). The last step supposedly
produced an iridescent gold, the ultimate in perfection. Greek alchemy

---

[7] The oldest Greek manuscript dealing with Egyptian alchemy is *Marcianus 299*
in the library of St. Mark's in Venice, dating from the tenth or eleventh century.
Two other important Greek manuscripts are *Parisianus 2325* and *P. 2327* in the
Bibliothèque Nationale; these date from the thirteenth and fifteenth centuries
respectively.

[8] Arthur J. Hopkins, *Alchemy, Child of Greek Philosophy*, Columbia Univ. Press,
New York, 1934.

already carried the concept of the *ferment*, the catalyst which could generate silver and gold. Later alchemy placed great stress on the importance of the ferment, variously known as the *philosopher's stone* and the *elixir of life*.

Despite the philosophical influences, further mixed with religious ideas, animism, and astrology, Greek alchemy developed apparatus, procedures, and practical knowledge of chemical substances which persisted down to the modern period. Heat was the only useful energy source for the alchemist, but it was used in varying degrees—ranging from the warmth of a fermenting dung heap, through the steady heat of a water bath, to the more intense temperatures of an ash or sand bath, to the direct heat of the coals in a furnace.

Distillation became important early in alchemical history. Taylor[9] reasons that the still was evolved from Dioscorides' observation of the condensation of mercury on the under side of a pot lid. The alembic, as the still-head came to be called, was not a very efficient condenser and the distillation process was incapable of making good separations during the Greek and Arabic periods. Such apparatus as pans, beakers, flasks, mortars and pestles, funnels, sieves, tongs, and crucibles all were in use at this time. In addition to the seven known metals, zinc, antimony, and arsenic probably entered into the operations, but in the form of compounds. Sulfur was well known, but the name was frequently associated with other flammable materials. Knowledge of acids was limited to vinegar, fruit juices, and hydrolyzed salts. A number of salts like sodium carbonate, alum, iron sulfate, and common salt were available in a fairly pure form, but the alchemical writers often confused them with related substances.

The spread of alchemy to Islam accompanied the translation of Greek works on philosophy, mathematics, and medicine in the ninth century. A tradition which places the origin of Arabic alchemy early in the preceding century has been largely dissipated as a result of the investigations by Ruska.[10] Modern scholarship even casts doubt on the authenticity of Jabir ibn Hayyan, the best-known name in Arabic alchemy. Although Jabir supposedly dates from 722 to 815, Kraus[11] showed that the Jabirian works he examined could not date from the time of Harun-al-Raschid but were the work of a religious sect in the tenth century.

The Jabirian corpus, especially the *Seventy Books* and the *Books of the Balances*, developed the four Aristotelian elements into a numerological system that purported to reveal the balance in the qualities in different

[9] F. Sherwood Taylor, *Ann Sci.*, **5**, 201 (1945); see also Taylor's *The Alchemists*, Schuman, New York, 1949, pp. 38 ff. Martin Levey *Centaurus*, **4**, 23 (1955), basing his opinion on Akkadian perfumery texts of the second millennium B.C. and the shape of certain Mesopotamian pots, believes that distillation was practiced long before it became an alchemical operation.

[10] J. Ruska, *Arabische Alchemisten*, Winter, Heidelberg, 1924.

[11] Paul Kraus, "Jabir ibn Hayyan," *Mém. l'Institut d'Egypte*, vol. 44, Cairo, 1943.

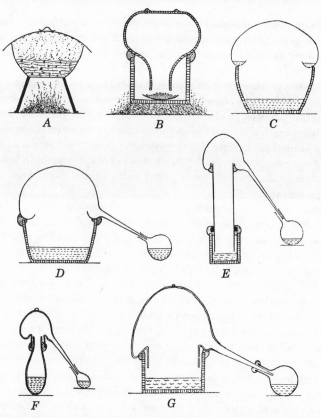

Fig. 1.5. EVOLUTION OF THE "STILL." *A* Collection of
condensate under the pot lid. *B* Apparatus for sublimation.
C Edges of the lid turned under to collect condensate.
D Delivery spout sealed into trough of lid. *E, F, G,* Variant
forms of stills.

(According to the conjecture of F. Sherwood Taylor in *Annals
of Science.* Used with permission of the publisher.)

metals and thereby suggested what changes were necessary in order to
achieve transmutations. The corpus also dealt with the role of alchemical
sulfur and mercury in the nature of metals, a concept perhaps traceable
to Aristotle's theory of smoky and watery exhalations in the formation
of minerals. Since metals differed in the proportion and purity of the
sulfur and mercury they contained, suitable treatment with elixirs should
bring about adjustments that would make the inferior metals more
perfect. Alchemy thereafter placed great stress upon the preparation of
elixirs, particularly from animal and plant materials. Eggs were especially
favored since they appeared to be endowed with a regenerative power.

Historical evidence for Abu-Bakr Muhammed ibn Zakariya al-Razi
(c. 865–923), a Persian known in the Latin world as Bubachar or Rhazes,
is much sounder. His medical works were authoritative in both Islam and
Christendom, and alchemical works attributed to him were widely circulated. His classification of chemical substances into mineral, vegetable,
animal, and derivative, and of the mineral into the subclasses of spirits,

Fig. 1.6. ARABIC ALCHEMICAL MANUSCRIPT PAGE FROM
THE *Sharḥ dī wān al-shudur.*
(Courtesy of the National Library of Medicine, Washington, D.C.)

bodies (metals), stones, vitriols, boraces, and salts, shows greater sophistication than the Jabirian classification into spirits, metals, and bodies that
may be powdered.

Abu Ali al-Hussin ibn Abdallah ibn Sina (980–1037; his Latin name is
Avicenna), the great Persian authority on medicine, also left works
dealing with alchemy. His ideas on the nature of metals were Jabirian,
but he was skeptical about the possibility of transmutation.[12]

---

[12] E. J. Holmyard and D. C. Mandeville, *Avicennae de congelatione et conglutinatione
lapidum, being sections of the* Kitab al-Shifa. *The Latin and Arabic texts edited with an
English translation of the latter and with critical notes*, Guethner, Paris 1927, pp. 41–42.

At the time that Arabic scholarship was being transmitted to western Europe (around the twelfth century), alchemical activity began to appear there. At first, translations of Arabic works were the principal sources of knowledge, but by the close of the thirteenth century an indigenous Latin literature began to appear. These books revealed strong Arabic overtones and, particularly the Geber works, were generally considered to be translations from the Arabic of Jabir. However, modern studies by Kopp, Hoefer, Berthelot, and Lippmann have led to the conclusion that although the Geber works reflect Arabic alchemy, they are compilations of a Latin alchemist around 1300, not direct translations of Jabirian works. Hence, they reflect the state of knowledge in the early phase of European alchemy, not of Arabic knowledge around 800.

In subsequent years the Latin corpus was expanded, frequently with authorship attributed to respected scholars like Roger Bacon, Albertus Magnus, Thomas Aquinas, Ramon Lull, and Arnold of Villanova. Actually, none of these men was clearly identified with the pursuit of alchemy and the works are almost entirely spurious. During the European period few first-rank scholars did more than dabble in alchemy and important additions to theory were negligible, although there were important steps forward in distillation procedures and knowledge of chemical substances.

The traditional alembic was never an efficient distillation vessel. Being made of glass or pottery, it was a poor conductor of heat and after the

Fig. 1.7. DISTILLATION APPARATUS WITH ROSENHÜTTE.
(From Brunschwygk, *The vertuose Boke of Distyllacyon of the Waters of all manner of Herbes* . . . [1527].)

Fig. 1.8. ALCHEMICAL LABORATORY WITH A MOOR'S HEAD
AT THE LEFT. After the painting by David Teniers the
Younger (1610–1690). The seated figure at the extreme
right, holding the book and flask, is believed to be a self-
portrait of the artist.

(Courtesy of the Edgar Fahs Smith Collection.)

initial warming its condensing effectiveness diminished greatly. Two
developments—the *Rosenhütte* and the *Moor's head*—which took place
about the time that knowledge of alchemy reached the West changed
this greatly. The *Rosenhütte* was a conical alembic made of metal which,
because it exposed a large surface of heat-conducting metal to the air,
was superior in condensing power. The Moor's head was a glass or pottery
vessel so designed that water might be placed around the alembic.

These inventions made it possible to prepare alcohol distillates of high
concentration. *Aqua ardens*, so called because it was flammable, also called

aqua vitae, quickly took its place in medicine as a powerful remedy. At the same time the production of brandy and whiskey became a widespread commercial enterprise. Up to this time the public had had to get along with such fermented beverages as wine and beer.

Improved stills also enabled the production of such mineral acids as *oil of vitriol* (H_2SO_4), *aqua fortis* (HNO_3), and *aqua regia* ($HNO_3 + HCl$). Oil of vitriol was first produced by the distillation of vitriols such as green vitriol ($FeSO_4 \cdot 7H_2O$). Its action on saltpeter (KNO_3) gave aqua fortis; on saltpeter and common salt, aqua regia. Probably because of its volatile nature, muriatic acid (HCl) was not clearly recognized or consciously prepared until about 1600.

Late European alchemy was frequently a tool of charlatans on the one hand, of occultists on the other. Although many alchemists were sincere and honest practitioners of the art, there were enough of the other variety to give alchemy a bad name. The failure of sincere alchemists to achieve transmutations, coupled with the outright fraud of such practitioners as attached themselves to the court of the Holy Roman Emperor, Rudolf II, in Prague, and the vague occult works of such mystics as Heinrich Khunrath, Michael Maier, and Thomas Vaughan, caused many scholars, clergymen, and rulers to take a dim view of the art. Alchemy attracted the interest of writers (Dante, Petrarch, Chaucer, Ben Jonson, Balzac) and artists (Pieter Brueghel the Elder, David Teniers the Younger, Thomas Rowlandson), but its treatment at their hands seldom created a favorable impression of the alchemist.

THE MEDICAL HERITAGE

Of the three pillars[13] of modern chemistry, the medical and technological perhaps contributed much more than the alchemical, although it is not always possible to distinguish the three clearly, since medical men frequently practiced alchemy and alchemy dealt with the transformation of metals. Through the Middle Ages medicine was characterized by adherence to concepts systematized by Galen in the second century after Christ, together with various appendages from astrology, magic, religion, and folklore. During the sixteenth century a change began which ultimately altered the nature of medical education and practice. The medical works of Vesalius, Paré, and Harvey played the same role in the reform of biology and medicine that the works of Copernicus, Kepler, Galileo, and Newton did in the reform of the physical sciences.

The improved methods of distillation brought new substances into medicine. As early as the thirteenth century Thaddeus of Florence

[13] A. J. Ihde, *J. Chem. Educ.*, **33**, 107 (1956).

described the medical value of alcohol in his *De virtutibus aquae vitae*,[14] From this time forward, alcoholic distillates continued to be used. Distillation also resulted in a variety of essential oils which were tried as medicines. Evidence of the interest in distillates is seen in the large number of books on distillation which appeared in print during the century after the introduction of printing in Europe. Particularly important were the two by Hieronymus Brunschwygk (c. 1452–1512) which appeared in 1500 (Small Book) and in 1512 (Big Book). Both dealt with the construction of furnaces and stills, the herbs suitable for distillation, and the medical use of the distillates. These books—both text and woodcuts—were frequently reprinted, translated, plagiarized, adapted, and emended.

Both these works reveal the contemporary idea that distillation was important in the preparation of essences or spirits from the perishable crude raw herbs. A double purpose was served by distillation. The essence was separated from inert material, thereby becoming more efficacious; and it was separated from the plant portions which quickly deteriorated, wherefore its keeping quality was enhanced.

Stills underwent steady changes during this period, becoming more efficient. The use of a continuous countercurrent flow of water led to better condensation. Steam distillation by injection of live steam is described in a book by Claude Dariot (1533–1594).[15]

Although chemical substances had been associated with medicine for some time, this association was given major emphasis and was developed into the system ultimately known as *iatrochemistry* by the Swiss physician Theophrastus Bombast von Hohenheim (1493–1541), generally known by his scholarly name of (Philippus Aureolus) Paracelsus. His irregular education was followed by service as an army surgeon and by extensive travels during which he became acquainted with local medical lore and established a considerable reputation as a physician. In 1527 he became town physician of Basel, but he quickly antagonized the local medical faculty by giving his surgical lectures in German, and by his open contempt for such medical authorities as Galen and Avicenna, and his brashness. Within a year he was forced to flee Basel. For the remainder of his life he wandered from one Germanic city to another, for his enemies refused to let him settle down to a profitable practice. His time was spent in drinking heavily and in dictating voluminous works on medicine, chemistry, and theology. His writings seldom found a publisher but were left among various of his diminishing group of friends.

These works were eagerly sought out by his followers in the century after his death. He was glorified by his students and friends and violently condemned by his enemies. Opinion about him even today is seldom balanced. His extensive writings reveal the remarkable insight of a

[14] E. O. von Lippmann, *Arch. Geschichte Med.*, **7**, 379 (1914).
[15] R. J. Forbes, *A Short History of the Art of Distillation*, Brill, Leyden, 1948, p. 150.

Fig. 1.9. Paracelsus.
(Courtesy of the Chester Fisher Collection.)

forward-looking physician and scientist, and at the same time are full of contradictions and reactionary viewpoints.

Paracelsus' position in chemistry has been greatly overemphasized. There is no evidence that he made any original contributions to empirical knowledge in the field, and his theoretical contributions merely introduced new and useless concepts without discarding existing ones.

In the maze of their inconsistencies and inaccuracies, the Paracelsian writings reveal certain ideas which are characteristic of his thought. Paracelsus believed alchemy to be one foundation of medicine, along with philosophy, astrology, and occult virtue. The real purpose of alchemy should be to produce medicines rather than to transmute metals. At the same time, he had a good understanding of the chemistry of metals and believed in transmutation.

He continued to retain his belief in the four elements while he popularized the concept of the three principles—sulfur, mercury, and salt. This *tria prima* derived from Arabic alchemy and was connected with the behavior of matter toward fire, sulfur being associated with combustion, mercury with fluidity and volatility, and salt with inertness. These three principles had only a superficial connection with the substances whose names they bore, but there was a philosophical connotation. The three principles in the body must be in balance or various illnesses would result.

Paracelsus believed in a *tartarus* as the cause of such illnesses as gout, and kidney and gall stones. He looked upon the body as a chemical system whose functioning was dominated by vital spirits. Food, which he considered as being composed of both poisonous and wholesome parts, was digested under the supervision of an *Archaeus* which separated the poisonous from the wholesome. When the *Archaeus* was indisposed, the poisonous, not being properly separated and eliminated, caused illness. The function of medicine was to restore the *Archaeus* to its normal state.

He also believed that diseases were cured by *arcana*, essences prepared by the alchemist by distillation, percolation, extraction, and similar operations. A certain *arcanum* should be effective against a certain disease. This point of view led his followers to experiment with a wide variety of chemical substances, including inorganic poisons like the vitriols, corrosive sublimate ($HgCl_2$), sugar of lead (lead acetate), antimony compounds, and potable gold. Such Paracelsians as Oswald Croll, Quercetanus, Mynsicht, and Turquet de Mayerne were influential in bringing about the introduction of mineral drugs into seventeenth-century pharmacopoeias.

Critics, particularly at the medical school in Paris, found much to condemn in the drugs introduced by some of the Paracelsians. However, such moderates as Daniel Sennert, Andreas Libavius, Angelo Sala, J. B. van Helmont, and Johann Glauber made progress in treating diseases with chemical medicines. Along with such figures as Sylvius and Tachenius later in the seventeenth century, these men sparked an amalgamation of chemistry and medicine which benefited both fields.

Franciscus Sylvius (Franz de la Boë, 1614–1672) was responsible for the systematization of Paracelsian medicine into iatrochemistry. He looked upon the human body as a chemical system that was delicately balanced between acid and alkaline *acrimonies*. Diseases were due to lack of balance and treatment consisted of restoration of balance. Like other theoretical medical systems before it, iatrochemistry failed because it represented an attempt to oversimplify biology. Nevertheless, it stimulated both medicine and chemistry through the renewal of interest and an increase in experimentation.

THE TECHNOLOGICAL HERITAGE

The technological arts, particularly metallurgy, made significant contributions to the advance of chemistry. However, the technological artisans were largely unlettered and left no literary works. But in the sixteenth century, when mining had become an important economic activity in Europe, there suddenly appeared a number of books dealing with the subject. These works are valuable in showing the state of the technical arts at the time.

The books vary from the *Büchlein* or "do it yourself" type to comprehensive treatises by Biringuccio, Agricola, and Ercker. All are characterized by a matter-of-fact quality not found in contemporary alchemical books. They were written by practical men for practical men; hence sterile theory was kept at a minimum.

Ein nützlich Bergbüchlein (Little Book of Ores) by an unknown author appeared very early in the century and was frequently reprinted. It dealt with the location of ores. Of greater significance to chemistry were the *Probierbüchlein*, concerned with the assaying of metals. The earliest edition was brought out around 1520 and it also was frequently reprinted. These books discuss balances and weights; sampling; the touchstone; furnaces and crucibles; cupellation; the separation of gold, silver, and

Fig. 1.10. Franciscus (de la Boë) Sylvius.
(Courtesy of the Edgar Fahs Smith Collection.)

Fig. 1.11. Agricola.
(Courtesy of the Edgar Fahs Smith Collection.)

copper; cementation; recovery of precious metals from scrap; coining and coinage; and miscellaneous operations on silver and gold.

A comprehensive treatise on mining and metallurgy was published in 1540 under the authorship of Vannoccio Biringuccio (1480–1539), a mine superintendent in northern Italy. His *De la pirotechnia* deals with the ores of commercial metals, economically valuable minerals, metallurgy, assaying, the separation of metals, alchemy (here skepticism regarding transmutation is expressed), distillation and sublimation, and pyrotechnics. His attitude is empirical and the book is relatively free of the sort of mystery found in works on alchemy of that time.

A more voluminous writer on geological and metallurgical subjects was Agricola (Georg Bauer, 1494–1555), a physician of Saxony whose *De re metallica* was published posthumously in 1556. Like Biringuccio, Agricola dealt factually with mining and smelting problems. His book is beautifully illustrated with woodcuts which reveal the technical nature of mining, smelting, and assaying operations. In earlier books Agricola discussed weights and measures, classification of minerals, origin of ores, and related subjects.

A third major work of high quality was the *Beschreibung Allerfürnemisten mineralischen Ertzt und Bergkwerksarten* by Lazarus Ercker, which was published in Prague in 1574. Ercker (d. 1593), who was inspector general of mines in the Holy Roman Empire, covered the same subject matter as Agricola but gave more emphasis to assay methods. In all these books the principal emphasis was on the precious metals and their common impurities like lead and copper. Iron was inadequately treated in such works until Réaumur published *L'art de convertir le fer forgé en acier* in 1722. New books on metallurgy appeared during the next century but none of them made substantial advances over the classical works. The art was undergoing little change because chemistry had not yet developed to the point where it could make an impact.

Assay methods described in these books indicate a groping toward quantitative analysis. They imitated smelting procedures and their objective was to determine the amount of gold and silver that could be extracted from the ore rather than the exact amount of metal present. Nevertheless, they introduced a quantitative aspect which was almost entirely lacking in alchemy. Their purpose was entirely practical—to determine the amount of extractable metal in ores and to find out whether coins and the metal used in jewelry were pure.

Assaying equipment consisted mainly of muffle furnaces and balances, with subsidiary items like weights, cupels, crucibles, flasks, alembics, and tongs. The best balances were well-made instruments that were sensitive to 0.1 mg. The most sensitive balances were encased and some had beam-lifting devices to protect knife-edges. Weights were not of standard sizes but represented fractions of the amounts used in large-scale operations.

In the common cupellation process, metallic contaminants of silver and gold were converted into oxides which were absorbed by the bone ash cupel, leaving a button of the precious metal which could then be weighed. If the button contained both silver and gold, the two had to be separated by means of nitric acid. Other assay procedures included using the touchstone and measuring density, but neither gave highly accurate results. In 1659 Glauber[16] referred to the qualitative importance of the color of flames and fumes. He also used glass beads for identification purposes.

This period also saw arsenic, antimony, bismuth, and zinc recognized as metals distinct from the seven known in antiquity. All four of these

---

[16] J. R. Glauber, *Tractatus de signatura salium, metallorum, et planetarum*, Amsterdam, 1659.

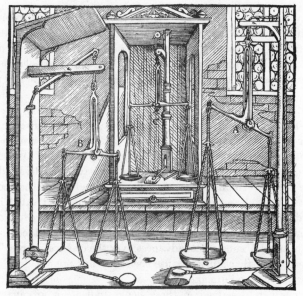

Fig. 1.12. ASSAY BALANCES.
(From Agricola, *De re metallica* [1556].)

had been recognized earlier but had been confused with compounds and with metals like lead and mercury.

Other inorganic technologies treated in the metallurgical books included the making of glass, pottery, gunpowder, salts, and acids. Some of these arts were being reexamined empirically in line with the changes being introduced. Bernard Palissy (1510–1589) experimented with pottery enamels and developed a distinctive French pottery that for a time had a unique appeal. Christoph Schürer of Saxony introduced the use of cobalt compounds in the production of blue glass in 1540. Cobalt itself was not discovered for another two centuries, although it was causing trouble for miners who encountered refractory ores that failed to yield copper when treated according to customary procedures. The superstitious miners believed that the ores had been bewitched by the gnomes or *Kobalds* of the mountains. Nickel caused similar trouble, wherefore it was called "Old Nick's Copper," or *Kupfernickel*.

Gunpowder, which was now important in warfare, was manufactured in large quantities following well-standardized recipes. Recipes for it are given in both Biringuccio and Agricola. Its quality was extremely variable, however, on account of the uncertain quality of saltpeter. The preparation of various salts like common salt, soda, alum, and vitriols was also discussed by the metallurgical writers.

Fig. 1.13. ASSAY FURNACE AND OTHER ASSAY EQUIPMENT.
(From Agricola, *De re metallica* [1556].)

THE BEGINNING OF HOUSECLEANING

Before chemistry could become a flourishing science, a period of house-cleaning was necessary during which the accumulated concepts of previous centuries would be critically examined and either kept or discarded. The seventeenth and eighteenth centuries were such a period. Unfortunately the housecleaning was slow, since there was a tendency to set the rubbish to one side rather than discard it. Inadequate or erroneous concepts in modified form continued to handicap chemistry's progress toward becoming a vigorous discipline.

The experimental approach to science, which had been so fruitful in connection with physics and medicine, was available for use in chemistry but was hardly likely to produce striking results until the inadequate concepts of the past were swept aside, as had already been done in physics and medicine. Slowly, the rejection of the ancient cosmology of the four elements in the astronomical field led to their rejection in the chemical field as well.

The changing attitude toward nature in the seventeenth century was sparked not only by physicists and physiologists but by philosophers. Francis Bacon (1561–1626) pioneered in leading philosophy away from the Aristotelian syllogism to an inductive position. Although he overstated the case for induction, he was a powerful factor in diverting science from an unfruitful approach that depended too heavily on logical argumentation from premises which should have been questioned many centuries earlier. In practice, the Baconian emphasis on fact-collecting proved too unwieldy to be practical. Bacon's contempt for mathematics was fortunately not taken seriously in most quarters and his hostility to deduction was often overlooked. At the same time, his belief that the role of science was "the endowment of human life with new inventions and riches"[17] was taken up with great enthusiasm by the English founders of the Royal Society and the French compilers of the *Encyclopédie*.

In a somewhat different fashion René Descartes (1596–1650) left a philosophical impact on science. In his *Discours de la méthode* (1637) he introduced the principle of systematic doubt—accept nothing which cannot be clearly established to be true. From certain reality he would then proceed to deduce consequences by dividing a problem into as many parts as possible and moving from the simple to the complex. When different consequences were deducible from the same principles, experimentation was necessary to establish the correct consequences. He examined scientific matters extensively, utilizing mechanical arguments. Motion and extension, along with mass and time, were fundamental. Descartes recognized a sort of atomism; but since he did not believe in action at a distance, his universe had no void and his atoms were expansible.

During the seventeenth century atomism was reexamined, mostly in connection with physical properties of matter, particularly gases. Gassendi, Boyle, and Newton all speculated on the role of corpuscles in the expansion and contraction of gases. Other men speculated about atoms in connection with chemical change. Angelo Sala (1575–1640) considered fermentation as a regrouping of elementary particles to form new substances. Daniel Sennert (1572–1627) postulated four kinds of atoms corresponding to the four elements, and suggested that substances of the second order were formed by the combination of primary atoms. Joachim Jung, Sebastian Basso, and David van Goorle also published books dealing with atomic speculations.

The principal problem that had to be faced was chemical identity. Different substances had been confused with one another since antiquity, and identical substances were known under a variety of names and considered to be different. There was no clear recognition of the fact that some substances were building blocks of which other substances were composed, or that subtle replacements could take place in chemical reactions. Some men like Paracelsus and Libavius believed that the fact that metallic copper appeared when a piece of iron was placed in a blue vitriol solution was an example of a transmutation. This was denied by Sala, Sennert, and van Helmont. Sala sought to demonstrate that the

[17] F. Bacon, *Novum organum*, 81st aphorism.

copper formed because there was copper in the vitriol. Sennert showed that gold could be recovered from acids in which it had been dissolved, presumably because gold atoms retained their fundamental character while in solution.

Johann Baptista van Helmont (1577–1644), despite his Paracelsian leanings, made progress toward understanding chemical entities. He rejected the four ancient elements and the three Paracelsian principles; instead, he maintained that air and water were the only primitive elements. Since he believed that air underwent no chemical changes, water must be the basis of all things.

Van Helmont's most significant work was done with chemically produced gases, which he distinguished from ordinary air. He recognized the identity of the gas produced during the burning of combustible substances like charcoal or alcohol—he called it *gas sylvestre*—with such gases as are given off (1) during the fermentation of grape juice or grain mash, (2) by the action of vinegar on sea shells, and (3) by the reactions which take place in certain caves. Unfortunately, his *gas sylvestre* also included other gases such as that given off during the reaction of silver with nitric acid. He also distinguished a flammable gas produced by the heating of organic matter, or by putrefaction, as in the intestines. This, named *gas pingue*, was a mixture, primarily of hydrocarbons. Nevertheless, van Helmont shrewdly recognized differences between gases and realized that a unique gas might be produced in several ways from different sources.

Johann Rudolf Glauber (1604–1670), a younger medical contemporary of van Helmont, acquired extensive knowledge about acids and salts and their interrelationships. He improved the current methods of preparing several mineral acids and was perhaps the first to understand metathesis, or double decomposition reactions. Otto Tachenius (c. 1620–1690), a former student of Sylvius, recognized that salts are the product of the reaction of an acid and a base. His use of spot tests (e.g., nutgall extract for detecting iron compounds) helped lay the foundations of qualitative analysis.

Perhaps the most important of the work along this line was that of Robert Boyle (1627–1691), the rich Englishman whose Baconian enthusiasm for the value of science stimulated the formation of the Royal Society. By 1662, with the aid of Robert Hooke (1635–1703), he had carried out the experiments with the air pump which led to recognition of the pressure-volume relationship in gases which bears Boyle's name.

In 1661 Boyle published *The Sceptical Chymist* in which he set out to demolish the four elements and the three principles. He showed that there was no sound basis for regarding these substances as elemental and pointed out that the clearest-speaking chemists looked upon elements as ". . . certain primitive and simple, or perfectly unmingled bodies; which not being made of any other bodies, or of one another, are the ingredients

Fig. 1.14. JOHANN BAPTISTA VAN HELMONT AND HIS SON FRANCISCUS MERCURIUS.

(From Helmont, *Oriatrike, or Physik Refined* [1662].)

Fig. 1.15. ROBERT BOYLE. The first air pump is shown in the background.

(From Boyle, *Appara veria* [1680].)

of which all those called perfectly mixt bodies are immediately compounded, and into which they are ultimately resolved."[18] It is doubtful if Boyle considered any known substance as elemental. Examination of his works suggests that he believed in a primary matter out of which all bodies were formed.[19] As a corpuscularian he sought to explain properties on the basis of increasing degrees of aggregation of primary corpuscles.[20]

Boyle's works are an excellent source as regards the chemical knowledge of his time. Along with Tachenius, he laid the foundations of qualitative analysis. He utilized flame colors, spot tests, fumes, precipitates, specific gravity, and solvent action as analytical tools. His work with indicators like syrup of violets led to the association of various acidic and alkaline substances.

---

[18] R. Boyle, *The Sceptical Chymist*, Caldwell & Crooke, London, 1661, p. 350. In the more readily available Everyman's Edition, the statement is on p. 187.

[19] T. S. Kuhn, *Isis*, **43**, 12 (1952).

[20] R. Boyle, *Origin of Forms and Qualities*, in T. Birch (ed.), *Works of the Honourable Robert Boyle*, 5 vols., 1774, vol. 2, p. 470.

Another important problem facing chemists concerned the nature of combustion. Fire had always been the principal energy source utilized by chemical philosophers, but explanations of it were inclined to be purely speculative. Boyle and Hooke studied the behavior of combustible substances under various conditions, including evacuated vessels. They observed that combustion stops when air is withdrawn, and that sulfur fumes when heated in a vacuum but does not ignite. But gunpowder burned under water! They realized that air was involved in the combustion of most substances, but that saltpeter might serve as a substitute for air in gunpowder.

Boyle also observed that metals gained in weight when heated in air but he reasoned that the gain was due to the absorption of igneous particles which passed even through glass. Boyle believed that air consisted of three kinds of particles; only one kind was really air particles, the other two being exhalations from the earth and from celestial bodies. The role of the air in combustion, calcination, and respiration was due to these extraneous particles which were present in only minor amounts.

Hooke independently suggested that the air acted as a solvent for "sulphureous bodies," the union producing heat. John Mayow (c. 1641–1679) published a book in 1674 which emphasized the role in combustion of nitro-aerial particles both in the air and in saltpeter. He also believed that a sulfureous substance must be present in the combustible material.

The studies of these London chemists were moving toward recognition of the part air played in combustion. However, their work must in no way be construed as anticipating the discovery of oxygen. Their ideas were still highly speculative because their experiments were of necessity limited in character since there was as yet no convenient way of preparing and manipulating gases. Consequently these ideas could not resist the rise of the phlogiston theory introduced by Johann Joachim Becher (1635–1682) and extended by his disciple, Georg Ernst Stahl (1660–1734).

In his *Physicae subterraneae* (1667) Becher concluded that bodies were composed of three earths—*terra lapidea* (vitreous), *terra mercurialis* (mercurial), and *terra pinguis* (fatty). He considered combustible substances to be rich in *terra pinguis*, which was lost during burning. Even metals that were calcinable contained some of the fatty earth. Stahl later developed these ideas into an elaborate chemical system in which the term "phlogiston" replaced *terra pinguis*.

The potential comprehensiveness of the phlogiston theory proved amazingly good in a world in which chemistry still held a qualitative attitude toward matter. Not only did the theory explain combustion and calcination as being caused by the loss of phlogiston, it explained the smelting of ores equally well. Since the ore (calx) of a metal is converted into metal by heating with charcoal, a substance rich in phlogiston, Stahl argued that phlogiston was transferred from the charcoal to the calx, converting the latter into the metal. Respiration, as well as many other chemical changes, was explained in terms of phlogistic concepts.

This theory tended to direct chemical thought toward mineral and pneumatic studies, in contrast to the predominantly medical interests which had previously characterized the science. The theory was readily accepted by German chemists like Marggraf, Pott, Juncker, and Neumann,

Fig. 1.16. Johann Joachim Becher.
(From *Chemischen Glucks Hafen* [1726].)

Fig. 1.17. Georg Ernst Stahl.
(From Stahl, *Opusculum chymico-physico-medicum* [1715].)

and the Swedes, Bergman and Scheele. The French and British took a more casual attitude toward it. A few chemists, particularly Friedrich Hoffman and Herman Boerhaave, were critical of parts of it.

CONCLUSION

As the middle of the eighteenth century approached, chemistry was still in the grip of a speculative conceptual scheme, albeit a comprehensive one. The ancient ideas of elements based on qualities were beginning to give way, largely because of their inadequacy, but the somewhat similar phlogiston theory still misdirected chemical investigations. To a certain extent the theory stimulated thought and experimentation and chemistry benefited thereby. Frequently the experimental results exposed inconsistencies which were covered up by patchwork extensions and alterations of the theory.

Despite these theoretical handicaps, empirical chemistry had made considerable progress since antiquity. Fourteen elements were known, although they were not recognized as elemental substances. The important mineral acids were known and oil of vitriol was produced commercially. Both weak and strong alkalies (carbonates and hydroxides respectively) were known. Numerous salts were available for medical and commercial use as well as for laboratory experimentation. Few organic compounds were known, although the medical chemists had been distilling enthusiastically for over two centuries. There was some grasp of the importance of identity, some understanding of the nature of the gross changes that occurred in simple reactions. The available knowledge was sufficient to permit further progress, particularly as the unsuitability of current concepts began to be recognized.

CHAPTER 2

PNEUMATIC CHEMISTRY

Although such investigators as van Helmont and Boyle made a certain amount of progress in the study of gases, they failed to establish a continuing tradition for several reasons. One was that they were largely dominated by the idea that air (or gas) was an element. When confronted with gases that had unique properties, the natural philosopher explained their difference from air as being due to impurities, not as being due to the fact of their being uniquely different substances.

The studies by van Helmont, Mayow, and Hooke all suggested the existence of different kinds of air, but were not convincing. Moreover, the rise of the phlogiston theory coincided with a decline in the study of gases. Boyle's studies that resulted in his famous law also undoubtedly convinced skeptics regarding the uniformity of air, since gases in general followed the inverse pressure-volume relationship. Such convictions naturally did not stimulate studies of the chemical characteristics of gases, and there was no significant research on the subject between that of Mayow and of Black, except for the work done by Hales.

There was another factor which discouraged chemical studies of air. This was the difficulty of collecting gases in a reasonably pure state. Because so many of his vessels shattered when he prepared gases by chemical reactions, van Helmont used the term "gas," after the Greek word for chaos, for such substances. The evolution of gases during these reactions could be observed, but collecting a gas uncontaminated by air remained an unsolved problem until Hales' work.

STEPHEN HALES

The English biologist Stephen Hales (1677–1761) was an Anglican minister by profession. He was educated at Cambridge, where he came under the school's strong Newtonian influences, and he determined to apply to the living world the quantitative approaches which had been so fruitful in mechanics. His two books, *Vegetable Staticks* (1727) and *Haemastaticks* (1733), described his experiments and their results.

Fig. 2.1. STEPHEN HALES.
(Courtesy of the Edgar Fahs Smith Collection.)

Hales, interested in the renewal of substances in plants, observed that plants absorb air through their leaves. He knew that the absorbed air must be "fixed," that is, converted into solid material. Many experiments involving collecting and measuring the fixed air in solid substances were done. These experiments were made possible by the invention of the pneumatic trough, an extraordinarily simple device for collecting gases over water. Although Mayow had experimented with air confined in vessels inverted over water, the collection of gases over water awaited Hales; however, he may have been anticipated by the little-known Moitrel d'Element who is said to have had the idea of separating the generating vessel from the receiving vessel.

Hales used a bent gun barrel as a delivery tube. The closed end containing the solid material was placed in the fire; the other end opened into a vessel of water inverted in a pan of water. In this manner he heated a variety of

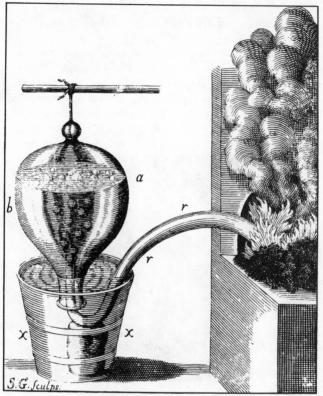

Fig. 2.2. HALES' PNEUMATIC TROUGH FOR THE COLLECTION
OF GASES.
(From Hales, *Vegetable Statiks* [1727].)

biological materials such as blood, tallow, horn, oyster shells, wood, seeds,
honey, beeswax, sugar, coal, tartar, and urinary calculi, and collected the
gases driven out by the heat. He also collected gases from fermentation
and putrefaction, as well as those given off when chalk, pyrites, and saltpeter were heated. Unfortunately, Hales was concerned with quantitative
rather than qualitative aspects. Thus he measured the volume of gases
collected over water without appreciating their difference in solubility
or recognizing that the gases from various sources differed from one another. Although he reported that the air distilled from peas "flashed"
and that that from Newcastle coal killed a sparrow, he believed that all
these gases were "true air" and thus missed the opportunity of discovering
a variety of gases. In short, he was interested in the quantity of air fixed
in solid substances, not in possible differences in the air itself.

In other studies Hales measured the water consumption of plants, the pressure of sap rising in stems, the blood pressure of animals, and the speed at which the blood circulates in veins and capillaries. His work brought him great renown in scientific circles, not only in Britain but on the Continent as well.[1]

Fig. 2.3. JOSEPH BLACK.
(Courtesy of the Edgar Fahs Smith Collection.)

JOSEPH BLACK

Pneumatic chemistry achieved great prominence among British chemists in the last half of the century. Joseph Black's studies of fixed air were particularly important in establishing the foundations of pneumatic chemistry because he recognized the uniqueness of the gas formed during experiments on a variety of substances.

Joseph Black (1728–1799) was born in Bordeaux where his father, a Scot-Irish wine merchant, was living at the time. His student years were spent in Scotland where he studied at the universities of Glasgow and Edinburgh. At Glasgow he became a student assistant to William Cullen

[1] Henry Guerlac, *Archiv. int. hist. sci.*, **15**, 393 (1951).

(c. 1710–1790), who was teaching chemistry not as a mere adjunct to medicine, but as a science entitled to recognition on its own merits. About 1751 Black transferred to Edinburgh where he took his degree as doctor of medicine in 1754. Two years later he succeeded Cullen as lecturer in chemistry at Glasgow when the latter accepted a similar position at Edinburgh. When Cullen discontinued his chemistry course in 1766, Black again succeeded him and spent the remainder of his life in Edinburgh.

Black took great pains with his teaching and became very popular as a lecturer. He placed little stress on theories, except when they were well founded on experiments, and he was willing to change his position if there were sound reasons for doing so. He was interested in the application of science to agriculture and industry, and frequently offered advice to practical operators. Although he published very little, he exerted a significant influence on the development of science through his many students and friends.

His most famous research had its beginnings in his doctoral dissertation, *The Acid Humours Arising from Food, and Magnesia Alba.*[2] His immediate purpose was to study stomach acidity and the effectiveness of alkaline substances in alleviating its distress. Actually, he was interested in the use of these substances for dissolving calculi (gall and kidney stones) in patients; however, he had little success in this problem.[3] He had come across some interesting information with respect to magnesia alba ($MgCO_3$), but he found that this substance had little effectiveness in dissolving calculi. Then, as so frequently happens when graduation draws near, he related his studies on magnesia alba to the medical problem of stomach acidity. During the time still available, he wrote up the experiments and included prefatory remarks that would suitably orient the professors who must read his dissertation. The next year Black extended his experiments on carbonates and read before the Physical, Literary and Philosophical Society of Edinburgh his classic paper in which he described fully his experiments on the carbonates of magnesium and calcium.[4]

According to Friedrich Hoffman, magnesia alba, the basic carbonate of magnesium, was prepared from the brine remaining after crystallization of saltpeter, or from the bitter saline solution that remained after the separation of common salt. The compound was used in medicine as an antacid. Black prepared magnesia alba by the reaction of Epsom salts ($MgSO_4$) with potash, and thoroughly studied its properties. He observed that it lost weight when heated, and that this loss was due to the escape of air. He realized that this air was not identical with ordinary air, but

[2] Joseph Black, *De humore acido a cibis orto et magnesia alba*, Univ. of Edinburgh Dissertation, 1754. An English translation by Alexander Crum Brown was made available by Leonard Dobbin to *J. Chem. Educ.*, **12**, 225, 268 (1935).

[3] Henry Guerlac, *Isis*, **48**, 137 ff. (1957).

[4] Published in the Society's *Essays and Observations, Philosophical and Literary* **2**, 157 (1756). Most readily available as Alembic Club Reprint No. 1, Edinburgh, 1898.

had unique characteristics. The same air was obtained by calcining lime-stone and by treating magnesia alba, calcareous earths, and mild alkalies with acids. He called it *fixed air*.

He further observed that the magnesia (MgO) and the quicklime (CaO) produced during the calcination of magnesia alba and chalk, respectively, reacted with solutions of mild alkalies and formed magnesia alba and chalk once more. (*Mild alkali* was the term used for the carbonates of potassium, sodium, and ammonia.) When treated with slaked lime, produced by the reaction of quicklime and water, the mild alkalies became caustic and chalk was formed. Black correctly reasoned that limestone became caustic when it lost its fixed air, and regained its mildness when it recombined with fixed air. He observed that limewater clouded when in contact with air containing fixed air, and that caustic alkalies of soda and potash became mild on standing in contact with air. Since all the mild alkalies effervesced with acid, he reasoned that they must be a combination of caustic alkali and fixed air.

When slaked lime and mild alkali were mixed, the fixed air in the mild alkali was transferred to the lime which precipitated as chalk; the residual solution took on the causticity of alkali deprived of fixed air.

Modern equations are as follows:

$$\overset{\text{heat}}{CaCO_3 \rightarrow} \underset{\text{Quicklime}}{CaO} + \underset{\text{Fixed air}}{CO_2}$$
$$\underset{\text{Chalk}}{\vphantom{CaCO_3}}$$

$$CaO + H_2O \rightarrow \underset{\text{Slaked lime}}{Ca(OH)_2}$$

$$Ca(OH)_2 + \underset{\text{Potash}}{K_2CO_3} \rightarrow \underset{\text{Chalk}}{CaCO_3} + \underset{\text{Caustic potash}}{2KOH}$$

On standing in air:

$$Ca(OH)_2 + CO_2 \rightarrow CaCO_3 + H_2O$$

On merely standing:

$$CaO + CO_2 \rightarrow CaCO_3$$

Caustic alkalies on standing became mild as a result of the absorption of fixed air from the atmosphere.

$$2KOH + CO_2 \rightarrow K_2CO_3 + H_2O$$

Black further observed that no gas was liberated when quicklime or magnesia was treated with acids.

Since his experiments were carried out in quantitative terms, they revealed the role of fixed air in the cycle of reactions he studied. His work also was influential in finally dispelling the notion that all gases were the element "air." He showed that fixed air was a trace component of the atmosphere but was not identical with atmospheric air. In his lectures he demonstrated that fixed air was a component of exhaled air by blowing through a tube into limewater, which became turbid. In similar fashion, he demonstrated the formation of fixed air during combustion by passing air over glowing charcoal. He also studied the effect of fixed air on animals.

Black's researches on fixed air were soon extended by others. David Macbride, a Dublin surgeon, dealt with gases produced by fermentation processes in his *Experimental Essays* published in 1764. Macbride recognized van Helmont's *gas sylvestre* as fixed air and associated that gas with the air reported by other investigators. He measured the carbon dioxide content of various gas mixtures, using an apparatus credited to Black. The gas to be analyzed was generated in a glass bottle by fermentation or chemical action, then bubbled through ammonia water in a second bottle. When the reaction was completed, limewater was added to the ammonia water to precipitate the carbonate as chalk. The chalk was then acidified and the carbon dioxide was collected and measured. Macbride also investigated the carbon dioxide in human blood; he concluded that it was carried in association with the red blood corpuscles.

Bergman, Cavendish, and Priestley likewise studied fixed air. Cavendish made careful determinations of its specific gravity, whereas Bergman and Priestley were interested in its weakly acid character.

One of Black's students, Daniel Rutherford (1749–1819), is generally credited with discovering nitrogen, although the gas was found independently about the same time by Priestley, Cavendish, and Scheele. Black had observed that a residual gas remained after carbonaceous matter was burned in atmospheric air and the fixed air was absorbed by caustic potash. Rutherford studied this gas and reported his results in his medical dissertation at Edinburgh in 1772.[5] Although the gas failed to support combustion or respiration, it differed from fixed air in that it was not absorbed by caustic substances. It was a *noxious air*, that is, one not suitable for burning combustibles or supporting life. Rutherford clearly distinguished it from fixed air, the primary subject of his dissertation, because of its insolubility in alkalies. However, he failed to recognize it as a distinct chemical species; instead, he considered it to be atmospheric air saturated with phlogiston.

HENRY CAVENDISH

Hydrogen, known for some years as *inflammable air*, was discovered in 1766 by Henry Cavendish. Several times in earlier years (i.e., Boyle,

[5] For an English translation of Rutherford's dissertation, see Leonard Dobbin *J. Chem. Educ.*, **12**, 370 (1935).

Fig. 2.4. HENRY CAVENDISH.
(Courtesy of the Edgar Fahs Smith Collection.)

c. 1670) it had been noted that a flammable gas was evolved when certain metals like iron were treated with acids, but it was Cavendish who collected the gas and subjected it to systematic study. His findings were read before the Royal Society in the first of a series of three papers on *Factitious Airs*, that is, gases contained in bodies in an inelastic state and liberated by chemical art. The second paper dealt with fixed air; the third, with gases produced by fermentation and putrefaction, an extension of Macbride's studies.

Flammable air had been known from Boyle's time, but there was much confusion regarding it because not only hydrogen but carbon monoxide and hydrocarbon gases had been encountered. Cavendish generated his "inflammable air" from dilute sulfuric or hydrochloric acid, using zinc, iron, or tin. He observed that it could not be produced from nitric acid or from concentrated sulfuric acid. The flammable air produced from different acids by different metals was found to be identical and the density was far less than that of atmospheric air. He interpreted his experiments in terms of phlogiston.

> It seems likely from hence that either of the above-mentioned metallic substances [zinc, iron, tin] are dissolved in spirit of salt, or the diluted vitriolic acid, their phlogiston flies off, without having its nature changed by the acid, and forms the inflammable air; but that when they are dissolved in the nitrous acid, or united by heat to the vitriolic acid, their phlogiston unites to part of the acid used for their solution, and flies off with it in fumes, the phlogiston losing its inflammable property by the union.[6]

The suggestion that "inflammable air" was phlogiston was quickly adopted by Scheele, Kirwan, and others.

In experiments with fixed air, using bladders for weighing, Cavendish estimated the specific gravity to be 1.57 times that of atmospheric air. He obtained the proportion of the gas in marble and other carbonates, and studied the uptake of carbon dioxide by concentrated potassium carbonate solutions. In the latter case, in which he was pursuing one of Black's observations, he found that the crystals thus formed were distinctly richer in fixed air than common potash. (The crystals were the bicarbonate.)

Other of his experiments showed that the gases produced by fermentation of sugar solutions or apple juice were composed entirely of fixed air. Gas from putrefying gravy or meat, however, contained "inflammable air" and a heavy inert gas, in addition to fixed air.

In unpublished notes, Cavendish also recorded experiments on nitrogen, prepared by passing atmospheric air repeatedly over hot charcoal and absorbing the fixed air with caustic potash.

In 1783 Cavendish published results of experiments which proved that water was formed when "inflammable air" (H_2) was burned in Priestley's dephlogisticated air (O_2), and soon thereafter he showed that nitric acid was produced by passing electric sparks through atmospheric air. The importance of these experiments will be examined further in the next chapter.

JOSEPH PRIESTLEY

In only a few years an English clergyman, Joseph Priestley, isolated and studied more new gases than any person before or since. When it is realized that he had no formal training as a scientist, his achievement becomes even more remarkable.

Priestley (1733–1804), the son of a humble cloth-finisher, was born near Leeds. His mother died when he was seven and his upbringing was placed

[6] H. Cavendish, *Phil. Trans.* **56**, 141 (1766).

in the hands of an aunt who wished him to study for the dissenting ministry. Upon leaving the theological academy at Daventry he became assistant to the Presbyterian minister at Needham market. The poor salary and the congregation's dissatisfaction with his liberal religious views caused him to leave the position. He spent the next years teaching school and fulfilling occasional ministerial duties. In 1761 he became instructor in classical languages at the Warrington Academy. Here he came in contact with Matthew Turner, a Liverpool physician who lectured on chemistry at the Academy. Though he may have been attracted to chemistry at this time, he did no work in it, his activities being concentrated on education,

Fig. 2.5. JOSEPH PRIESTLEY.
(Courtesy of the Edgar Fahs Smith Collection.)

history, and biography. His tabulated *Chart of Biography* won him an honorary LL.D. from Edinburgh. His *Essay on the First Principles of Government* (1768) influenced Jeremy Bentham in developing the concept of "the greatest happiness of the greatest number" as a criterion of moral goodness.

During trips to London Priestley came in contact with eminent persons, among them Benjamin Franklin, who stimulated his interest in scientific matters and encouraged him to write on scientific subjects. His *History of the Present State of Electricity* (1767) was largely a compilation from papers published in the *Philosophical Transactions of the Royal Society*, but it entailed in addition a great deal of correspondence and some

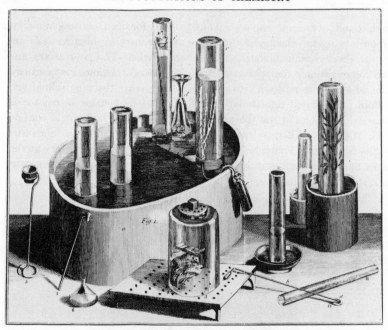

Fig. 2.6. APPARATUS USED BY PRIESTLEY FOR STUDYING GASES. *Fig. 1* An earthenware trough *a* about 8 in. deep, containing flat stones *bb* under water at one end. Gases are confined in cylindrical vessels *ccc* about 10 in. tall. A tall beer glass *d* is used to confine a mouse in gas over water. The phial *e* is used for generation of gases. The cylinder *f* shows a stand for supporting a container in the gas over water. *Fig. 2* Additional cylinders containing gases over water, one of them with a sprig of mint in it. *Fig. 3* Jar and perforated metal plates for holding mice to be used in experiments. *Fig. 4* Wire thrust through a cork—for use in closing vessels which are to be introduced through water into a cylinder (as in *Fig. 1f*). *Fig. 5* Wire stand for supporting objects above water in the cylinders (as in *Fig. 1f*). *Fig. 6* Funnel for pouring gas from vessel with a wide mouth into one with a narrow mouth. *Fig. 11* Cylinder, open at both ends, for testing gases with a candle. *Fig. 12* Candle *a* at the end of a wire *b*, for insertion into the top the cylinder of gas. The other end of the wire has a candle *c* which can be inserted into a container from the bottom.

(From Priestley, *Experiments and Observations on Different Kinds of Air*, vol. 1 [1774].)

experimentation. The success of the history caused him to plan a whole series of books on the history of science; but because of new activities and interests, the *History of Optics* (1772) was the only one to be published.

In 1767 Priestley accepted a Unitarian ministry at Leeds, where his scientific activities began in earnest with his studies on gases. Living next door to a brewery, he set out to experiment with the fixed air which appeared on the surface of the fermenting mash. His experiments led him to conclude that perhaps the medicinal qualities of the waters of Spa and Pyrmont were due to the dissolved fixed air. When he dissolved this gas in water, a pleasant taste and effervescence were produced. His invention of "soda water" received much publicity. Priestley advocated it as a cure for scurvy, but it proved unsuccessful.

In 1772 his reputation brought him an invitation to become librarian and companion to Sir William Petty (1737–1805), second Earl of Shelburne. Lord Shelburne had recently retired as Secretary of State because of his sympathy toward the American colonies. He was able to offer Priestley a tempting salary, with sufficient leisure to pursue his literary and scientific interests. The six years Priestley spent with Lord Shelburne were highly productive; his earlier work with gases bore fruit during this period in his investigation of nitric oxide, nitrous oxide, hydrogen chloride, ammonia, sulfur dioxide, silicon tetrafluoride, and oxygen. Evidence for each of these had been encountered elsewhere, but it was Priestley who systematically collected and studied these gases. By substituting mercury for water in the pneumatic trough, he succeeded in collecting the gases that are appreciably soluble in water.

Priestley continued to write voluminously. At first his pneumatic experiments were published in the *Philosophical Transactions*; but because of his great output, Sir Joseph Banks, Secretary of the Royal Society, soon suggested that he seek some other publisher. The result was a series of six volumes entitled *Experiments and Observations on Different Kinds of Air*, published at irregular intervals between 1774 and 1786.[7] He also found time for religious writings; his *Institutes of Natural and Revealed Religion* (1772–1774) aroused much antagonism among the orthodox. In 1780 he left the employ of Lord Shelburne, who continued paying a generous annuity during the rest of Priestley's life.

He then became junior minister at the New Meeting House in Birmingham, and also a member of the Lunar Society, so named because its meetings were held on nights when the moon was full so that the members could have some light on their way home. The society had been founded around 1766 by Erasmus Darwin, grandfather of Charles and author of *The Botanic Garden*; Matthew Boulton, the industrialist associated with James Watt in developing the steam engine; and a few other men with philosophical leanings.

At this time the last two volumes of Priestley's *Experiments and Observations* were published, but his scientific activity never again rose

---

[7] Priestley's work on gases was actually published as three volumes under the title *Experiments and Observations on Different Kinds of Air*, Johnson, London, 1774, 1775, 1777; and three volumes under the title *Experiments and Observations Relating to Various Branches of Natural Philosophy; with a Continuation of the Observations on Air*, vol. 1, London, 1779; vols. 2 and 3, Birmingham, 1781, 1786.

Fig. 2.7. APPARATUS USED BY PRIESTLEY IN EXPERIMENTAL STUDIES.
7 Hearth with gun barrel for driving gas out of solid substances. A pipe stem
delivers gas to an inverted cylinder filled with mercury. *8* Apparatus for collecting water-soluble gases over mercury: *a* a bowl of mercury; *b* glass cylinder
originally filled with mercury; *c* tube containing the ingredients from which
the gas is to be produced; *d* trap to intercept any ingredients which splash out
of *c*. *9* Bladder for storage and transfer of gases. This has a funnel at the bottom
and a glass at the top. *10* Apparatus for impregnating a liquid with a gas:
a bottle containing the liquid to be impregnated; *b* bowl of the same liquid;
c flask for generating gas; *d* bladder and flexible leather tube. *13* Glass tube
for use as a siphon. *14* Vessel to be exhausted by an air pump, after which a
dry gas can be admitted. This apparatus was used for studying behavior of
dry substances with various gases. *15* Apparatus for testing minute quantities
of gases. Its use is explained on pages 20–21 of Priestley's book. *16–18* Apparatus for subjecting gases to a spark. For its use, see pages 21 ff. of Priestley's
book. *19* Apparatus for passing electricity through air trapped in the top of
the bent tube. There is mercury in each glass and in each arm of the tube up
to *a*. Part *ab* of the tube contains water colored blue with vegetable juice.
Part *bb* contains common air. After electricity was passed through the system
the volume of air contracted and the water turned red.
(From Priestley, *Experiments and Observations on Different Kinds of Air*,
vol. 1 [1774].)

to its earlier heights. He became steadily more involved in liberal religious and political controversy. His *Corruption of Christianity* (1782) brought condemnation from the Calvinists and Lutherans, nor did his four-volume *History of Early Opinion Concerning Jesus Christ* have a pacifying effect on his critics. He was sympathetic to the revolutionists in France as he had been to the American colonists. Late in 1790 he drew the animosity of Edmund Burke by his criticism of the latter's *Reflections on the French Revolution.* Feelings against him in certain quarters were sufficiently violent to make his home a primary target in the Birmingham riots of July 14, 1791. That date, the second anniversary of the fall of the Bastille in Paris, occasioned an intemperate uprising in Birmingham which the magistrates made no serious attempt to quell. In the extensive vandalism that occurred at the homes of known religious and political liberals, Priestley's meetinghouse was set afire, his home was raided, and his books, scientific apparatus, and furniture were thrown from the windows. The Priestley family fled, thus avoiding personal harm, but found it expedient

Fig. 2.8. PRIESTLEY'S HOUSE AND LABORATORY AT FAIR HILL, DESTROYED IN THE BIRMINGHAM RIOTS.
(Courtesy of the Edgar Fahs Smith Collection.)

not to return to Birmingham. Priestley went to London where he encountered great hostility, even among members of the Royal Society. His efforts to carry on his work were frustrated. Finally, in 1794, he took his family to America. He declined a Unitarian ministry in New York City and a professorship of chemistry at the University of Pennsylvania, preferring to continue writing in Northumberland, Pennsylvania, during the last decade of his life.

Priestley approached chemistry in the true spirit of an amateur. He was gifted with a great deal of manipulative skill and ingenuity and he had an intense curiosity and enthusiasm. At the same time he was completely naïve and his experiments were conducted largely without plan. In spite of his rather casual experimentation, his studies of gases led to systematic results. He examined their solubility in water, ability to support or extinguish a flame, respirability, behavior toward hydrogen chloride and ammonia gases and toward nitrous air (NO) and hydrogen, the effect of an electric spark, their ability to transmit sound, and their density. Since Priestley had little patience in using a balance, his density determinations were unreliable, in contrast to the careful measurements made by Cavendish. His reports of his experiments are lengthy and detailed, and reveal with unusual candor his thoughts as well as his observations.

Throughout his life he was a confirmed phlogistonist. His experiments were regularly explained with reference to phlogiston, even at the expense of consistency. Though he was advanced as far as politics and religion were concerned, he remained scientifically a conservative, despite discoveries which were vitally important in bringing about the abandonment of the concept of air as an element and of phlogiston as an essential entity in explaining chemical phenomena.

Priestley's early studies of gases were primarily repetitions and extensions of even earlier studies of fixed and flammable airs, but during this time he developed the techniques that were so fruitful later.

Of special importance in his work on gases was his discovery of "nitrous air," now known as nitric oxide. Hales had noted the formation of a red gas when air came in contact with a gas generated from pyrites by spirit of niter (nitric acid). Cavendish suggested to Priestley that spirit of niter might be the source of the reddish gas and that metals might serve as well as pyrites in its production. Priestley investigated the action of a number of metals on nitric acid and succeeded in isolating the colorless, water-insoluble nitrous air. In studying its properties he noted its ability to diminish the volume of common air with the simultaneous formation of brown fumes (nitrogen dioxide) which were soluble in water. He further observed that the diminution of common air amounted to about one-fifth.

In extending these observations he noticed that the diminution was apparently proportional to the fitness of common air for respiration; this provided a better test for its fitness than that in which a mouse was used. He went on to study the most suitable proportions of nitrous and common air to achieve maximum diminution in a test for the "goodness of air." Apparently the best results were obtained when 1 volume of nitrous air was added to 2 volumes of common air confined over water. After the brown fumes dissolved in the water, the residual gas occupied about 1.8 volumes if the common air was "good." If the latter had been "spoiled" by burning or breathing, the residual volume increased with the degree of "spoilage."[8] Thus, eudiometry—the measurement of the purity of air— was born. The term soon acquired a broader meaning as it came to refer to gas analysis in general. The eudiometer was further developed as a practical instrument by Felix Fontana in Italy and by Cavendish, who obtained some excellent results with it.

Priestley found that when nitrous air stood over iron a change occurred that resulted in the production of an air in which a candle burned brilliantly. This air, which was nitrous oxide (N_2O), was named "dephlogisticated nitrous air" because it attracted phlogiston from the burning candle better than common air did. Priestley reasoned that the nitrous air, coming as it did from niter, was rich in phlogiston. This was withdrawn by the iron, leaving an air depleted in the principle of fire.

"Marine acid air" (hydrogen chloride gas) was prepared when spirit of salt (hydrochloric acid solution) was heated, with provision made for collecting the water-soluble gas over mercury. Later Priestley found that it was best prepared by the action of sulfuric acid on common salt. When dissolved in water, the gas caused iron to dissolve rapidly, with the evolution of flammable air.

"Alkaline air" (ammonia) was similarly produced by heating "volatile alkali" (ammonium hydroxide solution) and collecting the gas over mercury. This procedure was improved by treating the ammonium hydroxide with quicklime. In an attempt to learn whether a "neutral air" might also exist, Priestly mixed ammonia and hydrogen chloride gases but obtained a white cloud which proved to be sal ammoniac (NH_4Cl). When ammonia was brought in contact with an electric spark, he was surprised to find that mephitic air (N_2) and "inflammable air" (H_2) were produced. The density of ammonia was found to be less than that of common air.

On heating sulfuric acid with olive oil he obtained "vitriolic acid air" (SO_2). The same gas was obtained when the acid was heated with charcoal; but when he heated sulfuric acid to see if this acid air could be obtained directly, he observed no change until a laboratory accident caused mercury

---

[8] The 2 volumes of air would contain 2/5 volume of oxygen which would unite with 4/5 volume of NO to form 4/5 volumes of NO_2. Since all the NO_2 would dissolve in the water, the original total of 3 volumes of mixed gases would diminish 6/5 volumes, leaving 1⅘ volumes of residual gas. If part of the oxygen were replaced by carbon dioxide, as in spoiled air, the residual volume after completion of the reaction would be greater.

to be sucked back into the retort that contained the hot acid. The vessel was shattered and sulfurous fumes escaped. In a deliberate test, he learned that metals such as mercury and copper liberated the "vitriolic acid air" from the acid. He reasoned that the metals imparted their phlogiston to the hot acid, thus converting it into a permanent elastic air.

In attempting to secure a "vegetable acid air" Priestley heated concentrated acetic acid; however, the gas he obtained was not a new one but was vitriolic acid air, resulting from the fact, as he later learned, that the vinegar was badly adulterated with sulfuric acid.

Silicon tetrafluoride was encountered in experiments on "fluoro acid air" which had been discovered by Scheele in Sweden. Priestley tried to prepare hydrogen fluoride by heating fluorspar (CaF_2) with sulfuric acid in glass vessels. He found that the glass was seriously "corroded," even to holes being etched through the vessels. When water was present, a stony crust formed. He collected over mercury the gas produced by the action of the acid air on glass. When water was admitted, a reaction occurred in which a stony film formed. (The SiF_4 was hydrolyzed to silica.)

In other pneumatic studies, Priestley isolated nitrogen (already studied by Rutherford), oxygen, and hydrogen sulfide. He also prepared chlorine and phosphine, but failed to recognize them. In early experiments in which he prepared fixed air by heating limestone in a gun barrel, he noted the presence of a flammable air which burned with a blue flame—clearly carbon monoxide formed by reduction of fixed air in the hot iron gun barrel. However, he failed to isolate the gas; and therefore its discovery must be attributed to William Cruikshank. Priestley worked with hydrogen, carbon monoxide, and impure methane as flammable airs, but he failed to distinguish clearly between them and consequently was frequently misled in his interpretations.

He found that common air confined over a moist paste of sulfur and iron filings lost one-fifth of its volume, and he knew that the residual gas differed from fixed air because it failed to react with limewater and was lighter than common air, though equally noxious to animals. He considered noxious gases, such as nitrogen and carbon dioxide, to be common air which was heavily phlogisticated, that is, air which had taken up from burning substances or respiring animals the phlogiston being given off.

Air that had been spoiled because of animal respiration or putrefaction or burning could be restored to good air by being shaken with water, or better, with alkali. However, the number of times that restoration could be effected was limited because the air became permanently noxious (as the oxygen was used up). Priestley discovered that vegetation could be relied upon to restore the goodness, regardless of how often the air had been vitiated by burning candles in it.

His most famous studies concern oxygen. On August 1, 1774, he prepared this gas by heating *mercurius calcinatus per se* (mercuric oxide prepared by heating mercury in air).[9] He had obtained a 12-in. lens with which to heat various substances to high temperatures by focusing the sunlight. Placing the substance that was to be heated over mercury inside a mercury-filled vessel which was itself inverted in a pneumatic trough filled with mercury made it possible to collect any gases that were formed. On testing the gas from the red calx of mercury, he found that it was apparently insoluble in water; to his surprise, a candle burned brilliantly in it, and a red-hot piece of charcoal sparkled. Since this behavior was similar to that of his dephlogisticated nitrous air, he suspected that the gas was derived from niter, especially when he found that the same gas could be prepared from *red precipitate* (mercuric oxide prepared by dissolving mercury in nitric acid, evaporating to dryness, and heating the mercuric nitrate). Because he suspected that the gas might be dephlogisticated nitrous air, deriving ultimately from nitric acid, he naturally believed that his *mercurius calcinatus per se* was really *red precipitate* derived from niter and misrepresented to him by the pharmacist from whom he had bought it. However, genuine samples of *mercurius calcinatus per se* obtained from a friend named Warltire and from the Parisian pharmacist, Cadet, gave identical results. This led to the conclusion, particularly after he had prepared the same gas from red lead, that calcination imparted to the metal the property of yielding this kind of air from the atmosphere. He finally concluded that common air consisted of nitrous air and earth with as much phlogiston as was necessary to produce elasticity.

During later studies, solubility experiments revealed that the gas prepared from red mercury calx behaved differently than dephlogisticated nitrous air. Finding that the gas behaved like good common air, he concluded that, like common air, it was fit for respiration. Furthermore, on testing with nitrous air it left a residual volume similar to that left by good common air.[10]

The next day he was surprised to find that when a lighted candle was placed in the residual gas that remained after the test, it burned brilliantly. Ordinarily this gas would have extinguished a candle (since it would be primarily nitrogen). Priestley wrote:

> I cannot, at this distance of time, recollect what it was that I had in view in making this experiment; but I know I had no expectation of the real issue of it. Having acquired a considerable degree of readiness

[9] The gas had also been prepared by Pierre Bayen in February, 1774. Bayen reported the decomposition of red calx of mercury with the liberation of a gas which he erroneously identified as fixed air (Rozier, *Observations sur la physique*, **5**, 154, 1775). Priestley had isolated oxygen as early as 1771 when he heated saltpeter, but he failed to study the gas at that time.

[10] The volumes Priestley used for the test—2 of the test gas to 1 of NO—would give a residual volume of 1.5 with oxygen. The difference between 1.8 for good air and 1.5 for oxygen is small enough to have been overlooked, even if probable impurities in the gases he used are not considered.

in making experiments of this kind, a very slight and evanescent motive would be sufficient to induce me to do it. If, however, I had not happened for some other purpose, to have had a lighted candle before me, I should probably never have made the trial; and the whole train of my future experiments relating to this kind of air might have been prevented.[11]

A week later he placed a mouse in a vessel containing the new gas. The animal remained conscious fully half an hour—twice as long as it would have in good common air—and was revived when taken out and held near the fire. He reasoned that since individual animals vary, it might be possible for one mouse to live twice as long as another in common air. A day later, he decided to make the nitrous air test on the air in which the mouse had been placed and was surprised to find that, far from being noxious, it gave a better test than common air. On testing with a second portion of nitrous air there was further diminution in volume. Only then did Priestley realize that he was dealing with a gas like common air, but whose respirable and combustion-supporting properties were considerably enhanced. Further experiments with mice and with nitrous air confirmed this. His phlogiston philosophy, however, kept him from realizing the unique importance of the new gas which he named "dephlogisticated air." He considered it to be common air that had been deprived of a large part of its phlogiston, and hence was capable of readily absorbing phlogiston from respiring animals, burning fuels, and calcining metals.

CARL WILHELM SCHEELE

The gas oxygen was isolated and studied even earlier in Sweden by Scheele. But his book, *Chemische Abhandlung von der Luft und Feuer*, was not published until 1777, so the gas was well known by the time the book appeared.

Carl Wilhelm Scheele (1742–1786) was born in Stralsund, the chief town of what was then Swedish Pomerania, where his father was a merchant. Because the family was large, an academic education was impossible and young Carl was apprenticed at age fourteen to an apothecary in Göteborg and later to one in Malmö. Sympathetic masters gave him ample opportunity to study chemistry and pharmacy, and to experiment with the substances in their shops. He had access to the texts of Neumann, Lemery, Boerhaave, Kunckel, and Stahl. Following this training, he worked in apothecary shops in Stockholm, Uppsala, and Köping. He began his original investigations in Stockholm. At Uppsala he became a close friend of Bergman, who encouraged his work and suggested

[11] Priestley, *Experiments and Observations*, 1775, vol. ii, p. 43.

the researches on *magnesia nigra* (MnO₂) which resulted in the discovery of baryta, chlorine, and oxygen, and paved the way for Gahn's discovery of manganese.

During his short, poverty-stricken lifetime he accomplished a great deal in practical chemistry. His technique for isolating the calcium salts led to the discovery of such organic acids as tartaric, oxalic, gallic, pyrogallic, uric, mucic, lactic, and citric. His research on inorganic compounds

Fig. 2.9. CARL WILHELM SCHEELE.
(Courtesy of the American Institute for the History of Pharmacy.)

resulted in the discovery of hydrofluoric, hydrocyanic, nitrosulfonic, molybdic, tungstic, fluosilicic, and arsenic acids. He isolated glycerin from olive oil after saponification with litharge in the classic pharmaceutical procedure for preparing lead plaster. He also isolated lactose from milk. He ascertained the nature of hydrogen sulfide, borax, microcosmic salt, and Prussian blue, and demonstrated that plumbago contains no lead but is simply carbon associated with traces of iron. He prepared arsine and copper arsenite, the green pigment still known as Scheele's green, and devised new methods for the preparation of numerous other substances.

In the field of analytical chemistry he developed a method for separating iron and manganese, fused silicates with alkaline carbonates, and discovered ferrous ammonium sulfate. In studying the influence of various portions of the spectrum on silver chloride he noted that blackening was enhanced by the violet portion and was actually extended beyond into the dark region now known as the ultraviolet.

Scheele's book on air and fire had as its purpose the understanding of fire. Since combustion takes place in air, he believed that the constitution of air had to be determined before the nature of fire could be understood. Being a phlogistonist, he set out to study the evolution of phlogiston from various substances in a confined volume of air, and to ascertain the effect of phlogiston on the volume and properties of the air. He knew that air was reduced in volume when exposed to liver of sulfur (K_2S), sulfur fumes, turpentine, iron vitriol precipitated by caustic alkali ($Fe(OH)_2$), and phosphorus. He believed that the air combined with the phlogiston escaping from the combustible substance and was reduced in volume as a result. Since he knew that the foul air (*verdorbene Luft*) remaining after combustion was less dense than common air, he reasoned that the volume decrease could not be due to the contraction of air when it combined with phlogiston. Hence he assumed that phlogiston combined with a component of common air and escaped as heat. This component he called "fire air" (*Feuer Luft*), and he set out to isolate it.

He decided that fire air could be isolated from heat, provided the phlogiston could be transferred to a better receptor. He reasoned that nitric acid should absorb phlogiston from heat, because it was so effective in absorbing it from metals.

$$\text{Metal} \ + \ \text{Nitric acid} \rightarrow \text{Salt} \ + \ \text{Red fumes}$$
$$(\text{Calx} + \phi) \qquad\qquad\qquad (\text{Calx}) \quad (\text{Nitric acid} + \phi)$$

Therefore nitric acid should liberate the fire air from heat.

$$\text{Heat} \ + \ \text{Nitric acid} \rightarrow \text{Red fumes} \ + \ \text{Fire air}$$
$$(\text{Fire air} + \phi) \qquad\qquad (\text{Nitric acid} + \phi)$$

He fixed the nitric acid with potash and heated the niter in a retort that led into a bladder containing slaked lime to absorb the red fumes. The bladder slowly filled with a gas which proved to have the characteristics he expected. A taper burned brilliantly in it; it was completely absorbed by liver of sulfur and by phosphorus. When mixed with foul air, the mixture had the properties of common air. He also succeeded in preparing fire air from the nitrates of silver and mercury, the oxides of mercury and manganese, and the carbonates of silver and mercury (the carbon dioxide was absorbed by alkali in the case of the carbonates).

Apparently it never occurred to Scheele that the concept of fire air enabled him to explain combustion without recourse to phlogiston. In line with the thinking of his time, he took phlogiston for granted and incorporated it into all phenomena involving heat and light. The red fumes given off in reactions with nitric acid he considered were the vapors of the acid combined with phlogiston. He pointed out, for example, that

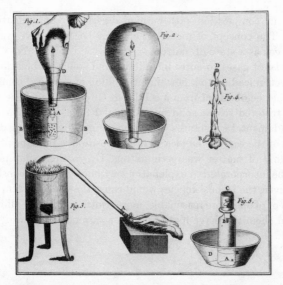

Fig. 2.10. EXPERIMENTS DESCRIBED BY SCHEELE. *Fig. 1*
The burning of hydrogen prepared by the action of acid on
metal. Water rises to *D* in the flask, showing loss of volume
of air. *Fig. 2* Burning of candle in flask. On being unsealed
under limewater, the solution rises in the flask. *Fig. 3*
Furnace, retort, and bladder for generating and collecting
gases. *Fig. 4* Bladder. *Fig. 5* Bees in *C* were proved to give
off "fixed air" by the fact that limewater rose in *A*.
(From Scheele, *Chemical Observations and Experiments on Air
and Fire* [1780].)

these fumes were obtained on dissolving copper (calx $+ \phi$) but not on
dissolving copper calx because the latter contained no phlogiston.

Light had an equal role with heat in Scheele's theorization. When he
found that light causes silver chloride to blacken, he maintained that the
substance acquired phlogiston from the light. The blackened residue gave
red fumes with nitric acid; the unilluminated silver chloride did not.

CONCLUSION

The studies on pneumatic chemistry were significant in the rise of
chemistry for several reasons. They brought to the attention of scientists
a group of gaseous substances which were clearly unique from one another. Although they might still be called "air," it was hard to avoid the
conclusion that they differed from one another because of fundamental

differences in composition rather than superficial contamination with impurities. The concept of air as an element began to give way to the concept of gas as a state of matter.

The ability of gases to enter into liquid and solid combinations had a further effect in leading to a new understanding of chemical combination. If a gas could be obtained from a solid, put through a sequence of operations, and restored to the solid in its original form, then there must be certain fundamental substances of a primary nature which made up the substances of the earth. The idea that composition is fundamental to the understanding of matter was germinating. Up to this point phlogiston could still be incorporated in explanations without glaring inconsistencies. Soon, however, the inconsistencies would cause the whole system to break down in favor of one more reasonable, more fruitful, and more all-inclusive. Thus would chemistry have its beginnings as a modern science.

II

THE PERIOD
OF FUNDAMENTAL
THEORIES

T HE PERIOD FROM 1750 TO 1860 SAW
chemistry achieve maturity as a science. After Boyle's death there was
a growing disinterest in the alchemical vestiges which had so long handicapped philosophers in their study of matter. These had by no means been
discarded by 1775, but an increasing unrest was evident. The genius of a
Lavoisier was essential to set chemistry off on a sound path. Chemical
techniques and factual knowledge had now advanced to the point where
his insight could be fruitful.

Once the break with tradition was made, the next necessary step—
the introduction of chemical atomism—followed rapidly. Dalton had the
insight to recognize the virtues of an atomic concept for chemistry, but
not the genius to follow through. The problems connected with chemical
atomism were solved only slowly during the next half century. The work
of Berzelius, Gay-Lussac, Davy, Dumas, Liebig, Laurent, Gerhardt,
Kekulé, and numerous others was vital in obtaining a clear understanding
of chemical elements and their behavior. The problem of atomic weights
was so formidable that it taxed the ingenuity of the best chemists of
Europe for half a century. This period of chaos did not end until the body
of chemical knowledge was sufficient to permit the necessary pieces of the
puzzle to be drawn together from a number of sources. This state was
reached by 1860.

CHAPTER 3

LAVOISIER AND THE
CHEMICAL REVOLUTION

THE ATTACK ON THE PHLOGISTON THEORY

The phlogiston theory had become generally established in the chemical world by the middle of the eighteenth century. Nonetheless, there were some who were lukewarm or even frankly critical. In spite of the apparent logic of the system which had been built around this hypothetical substance, some facts continued to embarrass the enthusiasts.

It was repeatedly observed that air was essential to combustion; in the absence of air, burning failed to take place. Moreover, combustion continued only for a limited time, in an enclosed vessel, and the air remaining in the vessel not only failed to support combustion in a freshly introduced burning substance, but did not sustain the life of animals that were placed in the vessel. Supporters of the theory pointed out that air attracted phlogiston. If no air was present, or if the air that was present was saturated with phlogiston, the phlogiston in the combustible substance could not leave it.

A second criticism was based on the common observation that metals actually gained weight on calcination. This seemed not in accord with a theory which postulated a loss of phlogiston on calcination. The sixteenth-century metallurgists knew about this increase in weight; in fact, as early as 1630, Jean Rey sought to explain it by postulating the absorption of air by the calces that had been made dense by heating. In his *Essays . . . of Effluviums*, Boyle doubted that a constituent of the metal was expelled on heating and that only its earth remained behind. He repeatedly noted a gain in weight when metals were heated; this was true even when they

were heated in sealed containers. (He always weighed the calx after unsealing the vessel, but he failed to recognize the true significance of the
inrush of air on unsealing!) "For it *does not* appear by our Tryals," he
observed, "that any proportion, worth regarding, of moist and fugitive
parts was expell'd in the Calcination; but it *does* appear very plainly,
that by this Operation the Metals gain'd more weight than they lost."[1]
He suggested that the fire introduced a ponderable effluvium which left
the calx heavier than the original metal. This fire effluvium could pass
through the sealed glass vessel.

The reaction of the phlogistonists toward such facts varied. The majority
were inclined to dismiss as unimportant any facts based on using the
balance. Some attributed to phlogiston a *levity* or a *negative weight*; hence
its escape should leave the residue heavier. Others confused weight and
specific gravity. Such eminent chemists as Macquer and Guyton de
Morveau made this error. This argument held that the calx was lighter
for its volume than the original metal, so that weight per unit volume
rather than absolute weight must be considered. Still others altered the
hypothesis by suggesting that the calcining metal gained weight from the
fire while the phlogiston was escaping, the gain being more than sufficient
to offset the weight lost when the phlogiston escaped—a return to Boyle's
notion.[2] This was the chaos in theories that confronted Lavoisier at the
beginning of his scientific career.

Antoine Laurent Lavoisier (1743-1794) was the son of a wealthy Parisian
lawyer. His mother died when he was five, and his upbringing was left
to a maiden aunt. His father was able to provide the best possible education at the Collège Mazarin. Though the family hoped for a legal career,
young Antoine found his greatest interest in the sciences, and made marked
progress under Abbé de la Caille, a professor of mathematics and astronomy.

An interest in chemistry developed as a consequence of his being taught
by Guillaume François Rouelle (1703–1770) at the latter's apothecary
shop.[3] Rouelle also served as demonstrator to Professor Louis-Claude
Bourdelain (1696–1777) at the Jardin du Roi between 1742 and 1768.
The fashionable audience at the Jardin customarily heard Bourdelain
lecture ponderously on chemical theory, after which he left the room.
Thereupon, Rouelle proceeded to demonstrate the facts of chemistry
experimentally, often at variance with the statements made earlier by
the lecturer. Even when not much interested in science, the audience loved
Rouelle's absent-minded enthusiasm. His influence on French chemistry
was notable. In addition to his fashionable audience such scientists as

---

[1] R. Boyle, *Essays . . . of Effluviums*, part iv, sect. ii (discussion of the pervious-
ness of glass), Pitt, London, 1673, pp. 76–77.

[2] For an excellent study of the problem of weight gain on calcination, see J. R.
Partington and D. McKie, *Ann. Sci.*, **2**, 361 (1937); **3**, 1, 337 (1938); **4**, 113 (1939).

[3] Henry Guerlac, *Isis*, **45**, 51· (1954).

Macquer, Bayen, and Cadet sat at his feet. Though a follower of the phlo-giston theory he was unenthusiastic. His real importance in chemistry, besides his teaching, was his recognition of salts as the product of the reaction of acids and bases. He even distinguished acid salts and, in a confused fashion, basic salts.

Lavoisier's enthusiasm for science was further whetted when, in 1766, he had the opportunity of accompanying the geologist, Jean Étienne Guettard (1715–1786), on a mineral survey of Alsace and Lorraine. Also in 1766 he received an Académie prize for the best essay on the problem of lighting a large city. In it he demonstrated his ability to get to the heart of a problem by the way he studied the nature of illuminants and the effi-ciency of illuminating devices.

The Académie Royale des Sciences elected him to a joint membership in 1768 at the age of twenty-five. He was made a regular member a year later on the death of the metallurgist Jars. The academicians unques-tionably recognized talent in this young amateur who was sufficiently wealthy to finance scientific investigations. During the remainder of his life he was called upon repeatedly to report on problems confronting the Académie.

His full energies were not devoted to scientific endeavors, however, for in 1768 he purchased a partial membership in the Ferme Générale, a tax-collecting firm. Taxes were not collected by the government directly; instead, the privilege was sold to the Ferme, a group of independent businessmen organized to make collections on a profit-and-loss basis. After some years Lavoisier acquired full membership in the organization.

These business connections were also responsible for his marriage. One of the members of the Ferme, Jacques Paulze—his home was a mecca for the philosophical and political leaders of Paris—had a fourteen-year-old daughter named Marie-Anne Pierrette, who became Lavoisier's wife in December, 1771. She worked closely with her husband and complemented him in the areas where he was deficient. Because she was an accomplished linguist, the content of books by English chemists was available to him as soon as they appeared. She was a competent artist and not only drew the sketches for his textbook, but left two sketches of the respiration experi-ments. She regularly kept his notes.

In 1775 Lavoisier was appointed *régisseur des poudres* and moved to the Arsenal where he set up his laboratory. Here he studied the quality of the saltpeter used in French gunpowder, and rapidly brought about improve-ments. Within a few years French gunpowder, up to then the poorest in Europe, became the best. Here also he carried out his later experiments on combustion and his respiration studies. Not only was this laboratory a center of research activity, but it became famous as the meeting place of the scientific leaders of Europe and even America (Franklin and Jefferson).

Lavoisier's investigations are characterized by a search for measurable fact uninfluenced by tradition. The balance was an important tool in his laboratory. Although he was not the first, as some have intimated, to make serious use of the balance, he was unquestionably a leader in quantitative chemistry. In line with the concept established by Black, there was a notion that quantities must be the same after as before an experiment. In his chemical studies, Lavoisier proceeded in accordance with what later came to be known as the law of conservation of matter.

This was evident in his early experiments which proved that water was not transmuted into earth by heating. A belief common among chemists of the time held that prolonged distillation of water caused it to change into earth in accordance with the ancient idea of the four elements. Lavoisier placed distilled water in a weighed glass retort called a pelican, heated it until the air was largely expelled, sealed the pelican, then weighed it with the water. There followed a heating period of 101 days, during which time flecks of suspended solid material appeared in the water. The pelican and its contents were then cooled and weighed. Lavoisier found that there was no change in weight; i.e., no ponderable material had been introduced from the fire. The pelican was next emptied, dried, and weighed. It showed a loss of 17.38 *grains* (0.92 g.), which was accounted for by the flecks of solid material, plus the residue left by the evaporation of the water.[4]

In his next important experiments he studied the behavior of diamonds on being heated. This subject was attracting considerable interest at the time and Rozier's journal published a steady stream of papers dealing with the work and opinions of various men. From the time Boyle was experimenting, it was known that diamonds are destroyed by strong heat; but whether this was due to evaporation, combustion, or fragmentation was not known. Maillard, the prominent Parisian jeweler, held that the diamonds would not be destroyed if they were properly protected from air, and he offered to furnish stones for experiments if he himself could supervise their protection from air. In collaboration with Macquer and Cadet, Lavoisier heated diamonds which Maillard had sealed in a clay pipe; there was no detectable loss of weight. It was evident that destruction occurred only in the air. In later experiments the three scientists, together with Brisson, studied the decomposition of diamonds in the heat obtained by means of a large Tschirnhaus-type lens. When heated in air enclosed in a bell jar over water, the diamond lost weight, the volume of air diminished about 12 per cent, and a precipitate formed on the addition of limewater to the confining liquid. The results were the same over mercury, except that the volume of air decreased only after the addition

---

[4] A. L. Lavoisier, *Mémoires de l'Académie Royale des Sciences*, 1770, pp. 73, 90 (published in 1773). The paper describing the research was read before the Académie on November 14, 1770, and was published anonymously in Rozier's *Introduction aux observations sur la physique*, **1**, 78 (1771).

of limewater over the mercury. Thus it was clear that the destruction of the diamond involved a form of combustion, with fixed air as a product.[5]

During the autumn of 1772 Lavoisier was becoming interested in the weight gain shown by phosphorus during combustion; this is revealed by three notes written at this time.[6] In the first, a notebook entry on September 10, 1772, recording the initial experiment with phosphorus, Lavoisier indicated his desire to learn whether air is absorbed by phosphorus during combustion. In a note to the Académie on October 20, 1772, he reported the absorption of air when phosphorus was burned under a bell jar. A sealed note deposited with the Académie on November 1, 1772, and read May 5, 1773, reported:

> . . . Sulphur, in burning far from losing weight, on the contrary gains it; that is to say that from a *livre* of sulphur one can obtain much more than a *livre* of vitriolic acid, making allowance for the humidity of the air; it is the same with phosphorus; this increase of weight arises from a prodigious quantity of air that is fixed during the combustion and combines with the vapors. This discovery . . . has led me to think that what is observed . . . may well take place in the case of all substances that gain in weight by combustion and calcination.[7]

A notebook entry of February 20, 1772,[8] contains remarks on fixed and ordinary air and plans for experiments to determine its nature. Lavoisier's characteristic self-confidence is reflected in the memorandum:

> The importance of the end in view prompted me to undertake all this work, which seemed to me destined to bring about a revolution in physics and chemistry. I have felt bound to look upon all that has been done before me as suggestive. I have proposed to repeat it all with new safeguards, in order to link our knowledge of the air that goes into combination or that is liberated from substances, with other acquired knowledge, and to form a theory.[9]

[5] The diamond experiments were given preliminary publication in Rozier, *op. cit.*, **2**, 108, 612 (1772). More complete accounts were published in *Mémoires de l'Académie*, 1772, vol. ii, pp. 564, 591. Since the *Mémoires* for 1772 were not printed until 1776, there was time for additional work; hence the papers do not entirely represent Lavoisier's thought in 1772.

[6] Andrew Meldrum, *Archeion*, **14**, 15 (1932); Max Speter, *Angew, Chem.*, **45**, 104 (1932).

[7] Lavoisier, *Mémoires de chimie*, **2**, 85 (1805); *Oeuvres*, **2**, 103 (1862). Translation from D. McKie, *Antoine Lavoisier, the Father of Modern Chemistry*, Lippincott, Philadelphia, 1935, p. 117.

[8] The notebook date is actually 1772; but I agree with Grimaux, Guerlac, and Meldrum and disagree with Berthelot and Daumas in believing that the 2 is the absent-minded type of error so frequent early in a new year and should actually read 1773.

[9] Reported from Lavoisier's notes by M. Berthelot, *La révolution chimique Lavoisier*, Alcan, Paris, 1890, p. 48. The translation is that by Andrew Meldrum, *The Eighteenth Century Revolution in Science—The First Phase*, Longmans, Green, New York, 1930, p. 9.

In January, 1774, the *Opuscules physiques et chymiques* appeared in print. Here Lavoisier recorded his repetition of Black's experiments with fixed air, and his own experiments on the reduction of minium with charcoal to obtain fixed air and the absorption of air during calcination of tin and lead. He was ready to conclude, on the basis of rather doubtful experimental evidence, that an elastic fluid combined with metals on calcination, but that not all the air was suitable for fixing during calcination. He showed that the elastic fluid formed during the reduction of red lead with charcoal was the same as that liberated by the action of acid on chalk. Both failed to support respiration in living animals or combustion of candles, and both precipitated when limewater was added. When phosphorus was ignited over water or mercury by means of a lens the volume decreased as much as one-fifth, and the weight of the phosphorus increased significantly. When phosphorus or sulfur was heated in a vacuum by means of a lens, each sublimed without combustion.

The *Opuscules* reveals that Lavoisier was still confused regarding the nature of combustion. He was certain that the gain in weight accompanying combustion and calcination was due to combination with a component of the air. But he inclined toward the belief that this component was Black's fixed air and that combustion somewhat resembled what took place when fixed air combined with quicklime.

During 1774 he reported on the results of heating lead and tin in sealed vessels.[10] The weight was the same before and after heating the vessels; but when he unsealed them, there was an inrush of air and a resultant increase in weight of the total. It was apparent that the calx was formed without loss of material and without gain in weight due to absorption of fire particles; rather, there was combination with the air. However, the real clue to the nature of burning and calcination was not yet apparent.

In October, 1774, Lord Shelburne, accompanied by Priestley, spent some time in Paris. During their stay, Priestley met the scientific leaders of France and spoke about his experiments on gases. In later writings he recalled telling Lavoisier about the gas prepared by heating the calx of mercury without charcoal. The significance of Priestley's visit in relation to Lavoisier's subsequent thinking has been the subject of much debate. Some authors maintain that this information was the clue Lavoisier needed, but others are not so sure. French concludes that their meeting had no significance.[11] At the same time Lavoisier must have been familiar

[10] Lavoisier, in Rozier's *Observations sur la physique*, **4**, 446 (1774). A revised version including data for two experiments with tin was published in *Mémoires de l'Académie*, 1774, p. 351 (published in 1778).

[11] S. J. French, *J. Chem. Educ.*, **27**, 83 (1950). A formerly lost letter from Scheele to Lavoisier, written in September, 1774, reveals that the French chemist learned of Scheele's work on oxygen at about the same time. See Uno Boklund *Lychnos*, **17**, 1 (1957–1958).

with Bayen's work, though he never gave this Parisian chemist recognition. Bayen published two papers in Rozier's journal early in 1774.[12] The first described the reduction, with weight loss, of mercury calx by heating. The second reported that the elastic fluid accompanying the reduction was fixed air, but this was soon shown to be in error. Whatever the stimulus, Lavoisier made experiments with mercury calx in November, though his crucial studies were not made until the following March. On April 26 he spoke before the Académie; the paper was published in the May issue of Rozier's *Observations*.[13] This paper, which is often referred to as the Easter Memoir, was also published in the *Mémoires de l'Académie* for 1775, but because of the tardy publication schedule did not actually appear until 1778.[14] In the interval Lavoisier's ideas developed considerably, as is evident on comparing the original paper as it appeared in Rozier's journal and the revised version in the official Académie memoirs. Failure to take into consideration the delay in publication of the official memoirs has led several early historians of chemistry to conclude that Lavoisier formulated his fundamental ideas earlier than was actually the case. The problem has now been clarified by Meldrum,[15] who showed that the papers as Lavoisier read them were frequently first published in Rozier's *Observations*, and that a revised form appeared several years later in the *Mémoires de l'Académie*. It is the *Mémoires* versions that are included in the collected works usually consulted by scholars.

Lavoisier's notebook for March, 1775, reveals that he now knew that the gas liberated on heating mercury calx was not fixed air, but a gas closely resembling common air. In a series of experiments he disposed of the argument that the red compound of mercury was not a true calx, as had been suggested, by heating the compound with charcoal in a small retort whose stem was bent so that the liberated gas could be collected over water in a bell jar. During the heating, the red calx and charcoal disappeared, and mercury and a gas were formed. Tests revealed that the gas was soluble in water, extinguished the flame of a candle, asphyxiated animals, formed a precipitate with limewater, and combined with alkalies. Clearly the gas was fixed air and the red calx was a true calx, since, like the calces of other metals, it could be reduced with charcoal to form fixed air.

When the red calx was heated in the same fashion without charcoal, however, a different gas was produced—one that was insoluble in water, made a candle flame brighter, did not asphyxiate a bird, and showed a decrease in volume like common air when a third of nitrous air (NO) was added. The gas behaved like specially pure common air.

[12] P. Bayen, in *Observations sur la physique*, **3**, 127, 278 (1774).
[13] A. L. Lavoisier, in *ibid.*, **5**, 429 (1775).
[14] A. L. Lavoisier, *Mémoires de 'l'Académie*, 1775, p. 520 (published in 1778).
[15] Andrew Meldrum, *Isis*, **20**, 396 (1934).

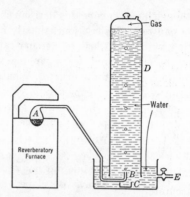

Fig. 3.1. LAVOISIER'S APPARATUS FOR DE-
COMPOSING MERCURIC OXIDE.
(From James B. Conant (Ed.), *The Overthrow of
the Phlogiston Theory*, Harvard Univ. Press,
Cambridge, 1950, copyright by the President
and Fellows of Harvard College, with permission.)

The original version of the Easter Memoir, "On the Nature of the Principle Which Combines with Metals During Calcination and Increases
Their Weight," showed clearly that Lavoisier had given up the idea that
metals and combustibles united with fixed air. Instead, he supposed that
they combined with common air during calcination and combustion. On
reduction with charcoal the metal gave up its common air to the charcoal,
and fixed air was the resultant product. With a calx such as that of
mercury, in which decomposition took place without charcoal, the common
air was liberated in a more respirable and more combustible form.

Priestley, who read the paper soon after its publication, pointed out in
his *Experiments and Observations* published that winter, that the gas
driven out by heating mercury calx was more than common air; it was
dephlogisticated air. Had Lavoisier tested the residual gas from the
nitrous air test with more nitrous air, it would have been obvious to him,
as it had been to Priestley that March day, that the gas was several times
purer than common air.[16]

This time Priestley's experiments were a definite factor in further
clarifying Lavoisier's ideas. His notebooks for 1776 state that he prepared
l'air déphlogistique de M. Prisley (sic) and tested it with candle and nitrous
air.[17] In subsequent experiments he studied the air remaining after
calcination of mercury. Although these last experiments were not

---

[16] J. Priestley, *Experiments and Observations on Different Kinds of Air*, Johnson,
London, 1775, vol 2, pp. 31 ff.
[17] M. Berthelot, *op. cit.*, p. 271

adequately described at the time, a description is available in the *Traité élémentaire de chimie*, published in 1789. It is as follows:

I took a matrass . . . of about 36 cubical inches capacity, having a long neck B C D E, of six or seven lines internal diameter, and having bent the neck as in [Fig. 3.2]. Fig. 2. so as to allow of its being placed in the furnace M M N N, in such a manner that the extremity of its neck E might be inserted under a bell-glass F G, placed in a trough of quicksilver R R S S; I introduced four ounces of pure mercury into the matrass, and, by means of a syphon, exhausted the air in the receiver F G, so as to raise the quicksilver to L L, and I carefully marked the height at which it stood by pasting on a slip of paper. Having accurately noted the height of the thermometer and barometer, I lighted a fire in the furnace M M N N, which I kept up almost continually during twelve days, so as to keep the quicksilver always almost at its boiling point. Nothing remarkable took place during the first day: The Mercury, though not boiling, was continually evaporating, and covered the interior surface of the vessels with small drops, at first very minute, which gradually augmenting to a sufficient size, fell back into the mass at the bottom of the vessel. On the

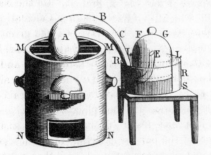

Fig. 3.2. APPARATUS USED BY LAVOISIER
IN DEMONSTRATING THE FORMATION OF
MERCURY CALX FROM MERCURY AND AIR.
(From *Œuvres de Lavoisier*, J. B. Dumas, Ed.
[1862], vol. 1, Plate I.)

second day, small red particles began to appear on the surface of the mercury, which, during the four or five following days, gradually increased in size and number; after which they ceased to increase in either respect. At the end of twelve days, seeing that the calcination of the mercury did not at all increase, I extinguished the fire, and allowed the vessels to cool. The bulk of air in the body and neck of the matrass, and in the bell-glass, reduced to a medium of 28 inches

of the barometer and 10° (54.5°) of the thermometer, at the commence-
ment of the experiment was about 50 cubical inches. At the end of
the experiment the remaining air, reduced to the same medium
pressure and temperature, was only between 42 and 43 cubical inches;
consequently it had lost about $\frac{1}{6}$ of its bulk. Afterwards, having collected all the red particles, formed during the experiment, from the
running mercury in which they floated, I found these to amount to
45 grains.

I was obliged to repeat this experiment several times, as it is difficult
in one experiment both to preserve the whole air upon which we operate, and to collect the whole of the red particles, or calx of mercury,
which is formed during the calcination. It will often happen in the
sequel, that I shall, in this manner, give in one detail the results of
two or three experiments of the same nature.

The air which remained after the calcination of the mercury in this
experiment, and which was reduced to $\frac{5}{6}$ of its former bulk, was no
longer fit either for respiration or for combustion; animals being
introduced into it were suffocated in a few seconds, and when a taper
was plunged into it, it was extinguished as if it had been immersed
into water.

In the next place, I took the 45 grains of red matter formed during
this experiment, which I put into a small glass retort, having a proper
apparatus for receiving such liquid, or gaseous product, as might be
extracted: Having applied a fire to the retort in a furnace, I observed
that, in proportion as the red matter became heated, the intensity of
its color augmented. When the retort was almost red hot, the red
matter began gradually to decrease in bulk, and in a few minutes after
it disappeared altogether; at the same time $41\frac{1}{2}$ grains of running
mercury were collected in the recipient and 7 or 8 cubical inches of
elastic fluid, greatly more capable of supporting both respiration and
combustion than atmospherical air, were collected in the bell-glass.

A part of this air being put into a glass tube of about an inch
diameter, showed the following properties: A taper burned in it with
a dazzling splendour, and charcoal, instead of consuming quietly as it
does in common air, burnt with a flame, attended with a decrepi-
tating noise, like phosphorus and threw out such a brilliant light that
the eyes could hardly endure it. This species of air was discovered
almost at the same time by Mr. Priestley, Mr. Scheele, and myself.
Mr. Priestley gave it the name of *dephlogisicated air*, Mr. Scheele called
it *empyreal air*. At first I named it *highly respirable air*, to which
has since been substituted the term of *vital air*. We shall presently
see what we ought to think of these denominations.

In reflecting upon the circumstances of this experiment, we readily
perceive, that the mercury, · during its calcination, absorbs the

salubrious and respirable part of the air, or, to speak more strictly, the base of this respirable part; that the remaining air is a species of mephitic, incapable of supporting combustion or respiration; and consequently that atmospheric air is composed of two elastic fluids of different and opposite qualities. As a proof of this important truth, if we recombine these two elastic fluids, which we have separately obtained in the above experiment, viz. the 42 cubical inches of mephitic, with the 8 cubical inches of respirable air, we reproduce an air precisely similar to that of the atmosphere, and possessing nearly the same power of supporting combustion and respiration, and of contributing to the calcination of metals.[18]

The revised Easter Memoir published in 1778 differed from the original only in changes in a few words and phrases. For example, the statement in the original that the "principle which unites with metals during calcination, which increases their weight, and which is a constituent of the calx is

neither one of the constituent parts of the air, nor a particular acid distributed in the atmosphere, that it is the air itself entire without alteration, without decomposition even to the point that if one sets. it free after it has been so combined it comes out more pure, more respirable, if this expression may be permitted, than the air of the atmosphere and is more suitable to support ignition and combustion"[19]

appeared in the 1778 version as:

"nothing else than the healthiest and purest part of air; so that if air, after entering into combination with a metal, is set free again, it emerges in an eminently respirable condition, more suited than atmospheric air to support ignition and combustion."[20]

In describing the gas obtained when the mercury calx was heated, Lavoisier made no significant changes in his statement regarding identifying tests until he came to the last one. The original statement, "It was diminished like common air by an addition of a third of common air,"[21] was deleted entirely in the revision. This looks like a face-saving device, for Priestley had pointed out[22] that Lavoisier had been misled into thinking the gas was common air because he had noted an approximately normal diminution in volume without testing the residue with further additions

[18] A. L. Lavoisier, *Elements of Chemistry*, transl. by Robert Kerr, Creech, Edinburgh, 1790, pp. 33–37.
[19] James B. Conant has made available an English translation of parts of the Easter Memoir, comparing the original version with the final revision in parallel columns. See his *The Overthrow of the Phlogiston Theory*, Harvard Case Histories in Experimental Science, No. 2, Harvard Univ. Press, Cambridge, 1950, pp. 22–23.
[20] *Ibid.*, pp. 22–23.
[21] *Ibid.*, p. 26.
[22] J. Priestley, *op. cit.*, vol. 2, pp. 320–323.

of nitrous air. Lavoisier is known to have made such tests after learning of Priestley's criticism; therefore the omission seems strange.

The uncertainty revealed in the original is replaced by certainty in the revision. In 1775 the nature of the gas that combined with metals and with sulfur and phosphorus was still questionable. It was not fixed air. It seemed to be common air, but when it was recovered from mercury calx it apparently was superior to common air. Hence the hedging statement that it was "not only common air but that it was more respirable, more combustible." After reading Priestley's experiments and criticisms, however, Lavoisier performed further experiments which enabled him to speak with conviction of the "eminently respirable part of the air" in the revision. His real debt to Priestley at this stage was acknowledged, albeit somewhat grudgingly.

For Lavoisier, the years 1776 and 1777 were a period of experimentation along lines indicated by Priestley's work. He wrote eight memoirs during these years, though they did not appear in the *Mémoires de l'Académie* until 1779 and 1781. These papers reveal that his ideas on combustion were clarified before the end of 1777. In the first of these memoirs[23] he concluded that all acids contain a portion of the "purest part of the air." When he dissolved mercury in nitric acid and heated the crystals he obtained nitrous air (NO), air in which a candle burned better, moisture, and the amount of mercury he started with. This memoir also contains a statement that some of the experiments are not Lavoisier's own; actually there are none for which Priestley could not claim to have had the idea first.

A following memoir on phosphorus reported that this substance gained $2\frac{1}{2}$ times in weight upon combustion and that this increase corresponded to the weight of air absorbed. The density of the remaining air, called *mofette atmosphérique*, was stated to be slightly less than that of common air. Adding the "eminently respirable" air to the *mofette* made the latter common air.

Particularly important was the memoir on the respiration of animals. In it Lavoisier reported that air was decreased in volume more during calcination than during respiration. Air after respiration precipitated limewater, but this was not true of air in which calcination had taken place. Lavoisier concluded that the "eminently respirable part" of the air either was converted into fixed air in the lungs or was absorbed in the lungs and replaced by fixed air. He favored the latter theory, since Priestley had shown that blood remained red only while in contact with atmospheric air or dephlogisticated air. Priestley found that blood blackened in the presence of fixed air, Cavendish's "inflammable air," nitrous air, and *in vacuo*, but the redness was restored in dephlogisticated air; the latter change was accompanied by a decrease in volume of the gas.

[23] A. L. Lavoisier, *Mémoires de l'Académie*, 1776, p. 671 (published in 1779).

Lavoisier noted that when the red calx of mercury was prepared, the air was reduced one-sixth in volume and it asphyxiated animals and extinguished flames. Heating the red calx produced an "eminently respirable" air which restored to the residual air the good qualities of common air when the two were mixed.

In discussing candle flames in another memoir Lavoisier explored the reason why candles were extinguished in a short time when they burned in an enclosed space. The phlogistonists argued that air absorbed phlogiston but became saturated after a time and hence could absorb no more. Lavoisier placed a candle in oxygen and observed that it continued to burn when two-thirds of the gas had been converted to fixed air and absorbed. He concluded that the burning continued because "eminently respirable" air was still present to support combustion.

Lavoisier carried out further respiration experiments from 1782 to 1784 with Laplace. They developed an ice calorimeter which enabled them to measure the heat evolved during chemical changes. In one experiment a guinea pig was confined in the calorimeter for ten hours, during which time 13 oz. of ice melted. They also measured the carbon dioxide the pig exhaled in ten hours. They then burned a weighed amount of carbon to determine the amount of carbon dioxide formed. With the ice calorimeter they measured the amount of ice melted when a weighed amount of carbon was burned. These figures enabled them to calculate the quantity of ice that would melt by burning enough carbon to produce the amount of carbon dioxide exhaled by the guinea pig in ten hours. Their calculations indicated 10.5 oz. Although the agreement between the two figures was by no means exact, they still felt justified in concluding that respiration was a form of combustion. It was not a mechanical process for cooling the lungs, but a chemical process which provided the body with heat and removed carbon dioxide from the blood.

The term *oxygine* was used for the first time in a memoir dated September 5, 1777, which dealt with the nature of acids. The term was derived from the Greek and meant *acid former* (to form or beget acid). This memoir developed the theory that acid character resulted from the presence of oxygen. In other words, all acids contained oxygen.

The burning of hydrogen created problems. According to Lavoisier's concept of oxygen as an acid former, the combustion of hydrogen should produce an acid. Various investigators had burned hydrogen, but no one seems to have noticed the formation of water until about 1777, when Macquer noted it on porcelain held over a hydrogen flame. His observation apparently went unnoticed, for the formation of "dew" in vessels in which hydrogen was burned was reported again in 1781 by Priestley, acting on Volta's suggestion that an electric spark be used to ignite inflammable air.[24] His friend John Warltire, lecturer in natural history at Birmingham,

[24] J. Priestley, *op. cit.*, 1881, vol. 5, p. 395.

continued these experiments and believed there was a weight loss after the explosion. He concluded that heat was a ponderable substance; i.e., heat was liberated by the explosion, a weight loss accompanied the evolution of heat, therefore heat had weight.

Priestley communicated these results to certain friends like Watt, Wedgewood, and Cavendish. Watt drew some conclusions regarding the formation of water from "inflammable" and dephlogisticated air, but public announcement was delayed on account of certain anomalous results which Priestley had been attempting to rectify. In the meantime Cavendish, on the basis of extensive experimentation, discovered the compound nature of water.

Cavendish, a believer in the imponderability of heat, began experimenting when Warltire reported evidence for weight loss due to heat loss. Warltire used copper vessels in most of his experiments in order to avoid danger from the shattering of vessels during the explosion. Cavendish substituted heavy glass. Though he could detect no significant weight loss, he too, like Priestley and Warltire, observed dew inside the vessel. He prepared large quantities of this dew and found that it had the properties of water. He failed to publish his results at once, because he was disturbed by the acid character of the liquid produced when dephlogisticated and "inflammable" airs were exploded. His observations were finally discussed by Priestley in a paper in 1783.[25] In it Priestley called attention to experiments of his own that were intended to verify those which Cavendish reported to him; according to the latter, water was produced by the explosion of "inflammable" and dephlogisticated airs and was equal in weight to the quantity of the two airs used. Priestley purportedly verified this, though his observations are open to grave doubt because he prepared his "inflammable air" from charcoal which produced only a modicum of hydrogen admixed with a variety of other combustible gases. His method of weighing the resulting liquid could hardly yield reliable figures and he failed to indicate how he measured his gases, if he even did so.

Cavendish finally reported his results after ascertaining the source of acidity in the liquid.[26] He found that the acid was nitrous (nitric) acid and that it was not formed if the volume of "inflammable air" present was more than twice that of dephlogisticated air. Addition of mephitic air (N_2) to the mixture of gases before explosion was responsible for increased acidity when the ratio of dephlogisticated air was increased. But if there was an excess of "inflammable air," even the addition of mephitic air did not produce an acid liquid. Cavendish expanded his study of the formation

[25] J. Priestley, *Phil. Trans.*, **73**, 426 (1783).

[26] H. Cavendish, *Phil. Trans.*, **74**, 119 (1784). The paper was read before the Royal Society on January 15, 1784. This paper and the one on the composition of nitric acid are available as Alembic Club Reprint No. 3.

of nitric acid the following year.[27] He showed that when air enriched with oxygen was sparked over mercury and exposed to potash solution, the air decreased markedly in volume. The original air almost disappeared when the excess oxygen was taken up by exposure to liver of sulfur (K_2S).[28] Niter could be isolated from the potash solution, thus demonstrating that nitric acid was formed from oxygen and nitrogen (dephlogisticated and mephitic airs).

The 1784 paper is important because it showed the quantitative combination of oxygen and hydrogen. Cavendish reported that when two measures of "inflammable air" and five measures of common air were exploded, no "inflammable air" remained, and the decrease in volume of the common air amounted to one-fifth; a colorless liquid remained in the vessel. He burned a large measured amount of hydrogen in $2\frac{1}{2}$ times its volume of common air in such a way as to collect 135 grains of the liquid, which proved to be water. Results were similar with dephlogisticated and "inflammable" air.

Though the evidence clearly pointed toward hydrogen and oxygen combining to form water, Cavendish insisted on explaining his results on the basis of phlogiston. Dephlogisticated air, he suggested, was water deprived of its phlogiston. Cavendish preferred the latter explanation because it seemed unlikely that high temperatures would be necessary to cause dephlogisticated air to accept phlogiston in the form of "inflammable air" when it so readily accepted phlogiston from iron (as in rusting), liver of sulfur, nitrous air, etc. The formation of water by burning "inflammable air" might be explained thus:

$$(\text{Water} + \phi) \quad + \quad (\text{Water} - \phi) \quad \rightarrow \quad \text{Water}$$
Inflammable air   Dephlogisticated air

Lavoisier learned about these experiments when Charles Blagden, Cavendish's assistant who soon became secretary of the Royal Society, visited Paris in 1783. Lavoisier, who had been uncertain about the role of oxygen in the combustion of hydrogen, immediately fitted the information into the theory he was developing. He verified the composition of water by synthesis and then took the next step, namely, using analysis to prove its compound nature. He passed steam through a heated gun barrel and observed that the steam was decomposed; calx of iron was formed inside the gun barrel and "inflammable air" was the exhaust gas.

In November, 1783, Lavoisier reported to the Académie on the experiments made in association with Pierre Simon de Laplace (1749–1827) in a paper entitled "On the Nature of Water and on Experiments that Appear

---

[27] H. Cavendish, *Phil. Trans.*, **75**, 372 (1785); Alembic Club Reprint No. 3, p. 39.
[28] Cavendish reported that not more than 1/120 of the original air remained. Clearly he had a sample of a rare gas, but recognition of these elements was not possible for another century.

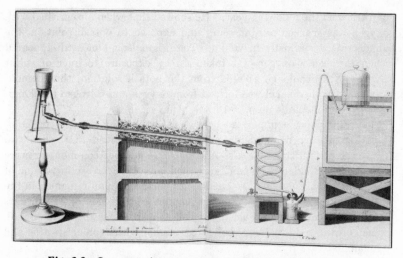

Fig. 3.3. LAVOISIER'S EXPERIMENT ON THE DECOMPOSITION
OF WATER.
(From *Œuvres de Lavoisier*, J. B. Dumas, Ed. [1862], vol. 1,
Plate I.)

to Prove that this Substance is not Properly Speaking an Element, but
can be Decomposed and Recombined."[29] Again Lavoisier showed his
superior theoretical mind by fitting the new facts into his own philosophical
system at a time when other chemists were attempting to explain the
formation of water by reference to the phlogiston theory.

About the same time Gaspard Monge (1746–1818), the inventor of
descriptive geometry, carried out experiments in which he burned measured
volumes of the two gases and obtained water in an amount approximately
equal to the weight of the gases.[30]

These various investigations of water created considerable controversy,
the remnants of which still flare up from time to time. When he read
Cavendish's paper, Watt felt that Cavendish had plagiarized his ideas.
His letters expressed his indignation and resentment; in one of them he
mentions the piracy of rich men, apparently a reference to Cavendish and
Lavoisier. Recently discovered letters of Priestley's resolve the question of
priority in Watt's favor.[31] However, Watt's ideas were confused with
phlogistic concepts, as were Cavendish's; but, unlike the latter's, they were

---

[29] A. L. Lavoisier, *Observations sur la physique*, **23**, 452 (1783); it appears in more
extended form in *Mémoires de l'Académie*, 1781, p. 269 (published in 1784).

[30] G. Monge, *Mémoires de l'Académie*, 1783, p. 78 (published in 1786).

[31] S. Edelstein, *Chymia*, **1**, 123 (1948). More recently Robert Schofield discovered
a letter from Priestley to De Luc which shows clearly that Watt anticipated Cavendish in explaining Priestley's experiments (private communication from Schofield,
July 29th, 1961).

based only on descriptions of Priestley's observations. Cavendish based his ideas on careful experimental measurements; his work also derived from Priestley's.

Lavoisier assumed credit for discovering the compound nature of water and made only slight references to the work of his contemporaries. There is no question that he was unfair in this respect, because it was Blagden's remarks that set him on the trail. Lavoisier immediately recognized the real nature of the phenomenon, verified his conclusions by hasty experiments, and achieved publication in an unusually short time. Cavendish clearly deserves credit for showing that water is formed when "inflammable air" is burned in dephlogisticated air and for determining the quantitative relationships; Lavoisier, for correctly interpreting the reaction.[32]

Burning hydrogen as a source of water suggested to Lavoisier the reason for the discrepancy between the amount of ice melted by burning carbon (10.5 oz.) and by a respiring guinea pig (13 oz.). His earlier respiration experiments did not take into account the moisture in the guinea pig's exhaled breath. Food materials—e.g., fatty oils—were found to contain hydrogen as well as carbon and consequently produced more heat than would be obtained by burning an amount of carbon equivalent to the carbon in the compound.

Lavoisier embarked on a new series of experiments, using his young assistant, Armand Séguin (1765–1835), as a human guinea pig. Séguin was weighed before and after the experiments; during them he was enclosed in a varnished silk bag that was open only at the mouth. A brass face mask enabled him to inhale only oxygen from a reservoir; his exhaled breath was collected and analyzed for oxygen, carbon dioxide, and moisture. Two sketches by Madame Lavoisier describe the experiments; one shows Séguin at rest, the other shows him at work operating a treadle with his foot. Because of Lavoisier's many other obligations and the outbreak of the French Revolution, the experiments were never finished. A preliminary report was presented to the Académie in 1790, and after Lavoisier's death Séguin attempted to collect all the existing notes on the results of these experiments.

RISE OF THE NEW CHEMISTRY

Lavoisier was not one to be deterred by unfinished experiments or lack of precise data. Since his reading of the Easter Memoir to the Académie

[32] The whole controversy, including Lavoisier's claims, has been treated very thoroughly by George Wilson, *Life of the Honourable Henry Cavendish*, Cavendish Society, London, 1851; and by J. R. Partington, *Composition of Water*, Bell, London, 1928. See also James P. Muirhead, *Correspondence of the Late James Watt on His Discovery of the Composition of Water*, Murray, London, 1846; and S. Edelstein, *op. cit.*

in 1775 he had been convinced that air combined with combustible
substances and with metals during calcination. Before long he realized
that the air was a mixture of an active ingredient, oxygen, and an inactive
component which he named *azote*. He found that combustion and calcination could be explained as the combination of the reacting substance with
oxygen, thus dispensing with the mysterious phlogiston. However, he had
little success at first in winning converts to his beliefs; mathematicians
like Laplace were convinced, but not chemists.

Gradually, as Lavoisier came to understand oxygen in relation to the
composition of air and water, to acids and calces, and to combustion and
respiration, he became bold enough to attack the phlogiston theory
openly. He submitted a memoir entitled *Reflections on Phlogiston*[33] to the
Académie in 1783. In it he presented a logical analysis of the problem,
showing how each phenomenon could be explained in terms of oxygen and
at the same time showing that the explanation in terms of phlogiston was
more complicated or absurd, or both. He credited Stahl with discovering
that calcination of metals was a form of combustion and with observing
that combustibility could be transmitted from one body to another; i.e.,
calx of lead when heated with charcoal changed to lead which could then
be calcined, the carbon being consumed in the process and hence losing its
combustibility to the lead. Although phlogiston could explain the reduction of calces, a problem arose in explaining the carbon dioxide formed
during the reduction. Oxygen could account for every change during the
process, whereas phlogiston became a liability in explaining all the reaction
products.

By 1785 Berthollet had been won over to the new ideas, and within the
next two years Guyton de Morveau and Fourcroy publicly indicated their
adherence to the new theory. Monge, Chaptal, and Meusnier soon followed.
Overjoyed when Black wrote of his conversion to the new system,
Lavoisier answered with a flowery letter of appreciation. In general,
however, acceptance was slow and there was a good deal of open antagonism. Macquer (1718–1784), professor at the Jardin du Roi, author of
popular textbooks and the *Dictionnaire de chimie* (1766; 2nd ed., 1778),
and director of state dyeing industries and of porcelain manufacture at
Sèvres, complained about this young upstart who would upset all of
chemical science. Bergman and Scheele died before the new chemistry
made any real impact, but both remained phlogistonists to the end.
Richard Kirwan, one of the leading chemical figures in London, was
vigorous in his defense of phlogiston. Cavendish remarked that Lavoisier's
ideas explained his experiments about as well as phlogiston did, then
devoted the remainder of his life to physical studies. Priestley remained

---

[33] A. L. Lavoisier, *Mémoires de l'Académie*, 1783, p. 505 (published in 1786).

violently opposed; two of his last publications presented a vigorous defense of phlogiston.[34]

Three of the earliest adherents of the new chemistry—Berthollet, Guyton de Morveau, and Fourcroy—were closely associated with Lavoisier at this time. Claude Louis Berthollet (1748–1822) was born of French

Fig. 3.4. BERTHOLLET AND LAVOISIER.
(Courtesy of the Edgar Fahs Smith
Collection.)

parents in Savoy, Italy, and studied medicine in Turin. Upon becoming physician to Madame de Montesson in 1772 he began chemical studies in Paris. Chemistry quickly became his major interest and he pursued it

[34] J. Priestley, *Experiments and Observations relating to the Analysis of Atmospherical Air—including considerations on the Doctrine of Phlogiston,* and *The Doctrine of Phlogiston Established and that of the Composition of Water Refuted.*

The efforts to save the phlogiston theory have been competently examined by J. R. Partington and D. McKie, *Ann. Sci.,* **2,** 361–404 (1937); **3,** 1–58, 337–371 (1938); **4,** 113–149 (1939). Also see J. H. White, *The Phlogiston Theory,* Arnold, London, 1932; and Joshua C. Gregory, *Combustion from Heracleitos to Lavoisier,* Arnold, London, 1934. Robert Siegfried has made a study of the efforts to save the theory in the United States; see *Isis,* **46,** 327 (1955).

ably during much of his life. He was active in the promotion of scientific education and the application of science to industrial pursuits. His early studies of chlorine led to the discovery of its bleaching power and its ultimate introduction in the textile industry. He prepared hypochlorites by passing the gas into potassium hydroxide solution, the resulting product being marketed as *eau de Javelle*; he also discovered potassium chlorate. Lavoisier became interested in the latter compound as a possible substitute

Fig. 3.5. ANTOINE FRANÇOIS DE FOURCROY.
(Courtesy of the Edgar Fahs Smith Collection.)

for saltpeter in gunpowder and did research on it, but he abandoned this study after a violent explosion in which two lives were lost.

Antoine François de Fourcroy (1755–1809) was the youngest of the group around Lavoisier. After a medical education he became professor of chemistry at the Jardin du Roi in 1784, following the death of Macquer. Berthollet succeeded to Macquer's post as superintendent of the French dyeing industry at the same time.

Louis Bernard Guyton de Morveau (1737–1816) was the author of *Digressions académiques* (1772) which set forth his ideas on crystallization and on phlogiston, besides demonstrating by means of carefully performed experiments that metal calces are heavier than the metals from which the calces are formed. This point had been in dispute for more than two centuries because investigators had not always used pure metals and had frequently been careless regarding the cleanliness of their apparatus. Despite his fine experimental work, Morveau missed the conclusions that logically resulted from his study by arguing that phlogiston had a natural levity; hence calces were heavier than the original metal because phlogiston had the effect of buoying up metals in which it was present. Later Lavoisier convinced him of the superiority of the oxygen theory.

Around 1780, Morveau made an issue of the unsatisfactory state of chemical nomenclature. Many names were traditional and meaningless (powder of Algaroth, turbith mineral, colcothar); some substances were named after a person (Glauber's salt) or a place (Epsom salt); still other names were superficially associated with the appearance (oil of vitriol, butter of arsenic, liver of sulfur) or the effect of such substances on the sense organs (sugar of lead). The names of substances used in the kitchen were particularly objectionable because many of these products were corrosive or toxic. Lavoisier agreed with Morveau regarding the need for reform in this field.

Morveau had already attempted to devise a nomenclature that would give information regarding the chemical constituents of a substance. He was joined in this by Lavoisier and by Berthollet and Fourcroy. The four men set out to formulate a nomenclature within the framework of Lavoisier's theories of chemistry. Their efforts culminated in the publication of the *Méthode de nomenclature chimique* in 1787. The adoption of Latin and Greek words made the terms internationally useful, and a systematic pattern of word endings made it possible to distinguish a great many compounds very easily. Thus salts formed from *acide sulfurique* were called *sulfates*. The *-ique* (Engl. *-ic*) ending on a name indicated acids saturated with oxygen; the *-ite* ending referred to salts of *-eux* (*-ous*) acids. The *-ure* (*-ide*) ending indicated compounds not in an acid state, i.e., *sulfure* (*sulfide*) used for sulfur-containing compounds like liver of sulfur (K_2S). The *-ide* ending indicated oxides of the metals.

The new nomenclature did not meet with enthusiasm in all quarters. It had a noncommittal reception when presented before the Académie; the reviewing committee was composed of Baumé (who completed a book based on the phlogiston theory in 1787), Sage (another phlogistonist), D'Arcet, and Cadet. Publication was approved, but there was no recommendation that it be adopted by the Académie. Both Black and Kirwan in Britain expressed disapproval. As late as 1802, Thomas Thomson's textbook was highly critical. However, in quarters that adopted the new chemistry, the nomenclature was usually adopted with it.

Lavoisier's last important step in securing acceptance for his concepts was the publication of his *Traité élémentaire de chimie* in 1789. In this text he treated the subject of chemistry in terms of the oxygen theory. His experimental work was described to bring out the uselessness of phlogiston and the importance of oxygen. Chemistry was thus systematized, and the new nomenclature was used. In the preface Lavoisier indicated his indebtedness to the philosopher de Condillac (1715–1780) who had stressed in his *Logique* (1780) the importance of language to clear thinking. Lavoisier's book on the whole had a profound effect in winning converts to the new chemistry. In addition to being reprinted in French, it was translated into German by S. F. Hermbstadt (1760–1833) in 1792 and into English by Robert Kerr (1755–1813) in 1790. Dutch, Italian, and Spanish translations were also available.

The first part of the book dealt with heat and gases. The term "gas" derived from van Helmont and had been recently introduced into French terminology by Macquer. Lavoisier proposed its adoption for the elastic fluids which his contemporaries had been calling "airs." The term has been in general use since that time. Heat and light, which are associated with so many chemical changes, were a problem in Lavoisier's system. He gradually developed his doctrine of *caloric*, which treated heat, and possibly light, as an imponderable substance that caused other substances to expand when it was added to them—i.e., expansion of liquids and metals when heated. Gases he supposed to be a "base of gas" containing caloric. Thus during combustion of carbon the "base of oxygen" combined with carbon to form carbon dioxide; its weight was equivalent to that of the combined carbon and oxygen, and the caloric was liberated as heat and light. He believed that the "base of oxygen" had a greater affinity for hot metals than for caloric, and therefore calces—now called oxides—were formed.

In discussing the reactions of metals with acids he reasoned that the metal took oxygen from the water and liberated hydrogen, whereupon the acid reacted with the metal oxide to form salt. Thus, oxygen had a direct role in salt formation.

Lavoisier considered that oxygen was responsible for the acidic properties of acids; in fact, for many years the term "acid" was applied to the oxides of carbon, sulfur, and phosphorus, the water being considered merely an absorbent. He also regarded vegetable acids as oxides, a radical consisting of carbon and hydrogen being attached to the acidifying principle, oxygen.

In his section on chemical substances Lavoisier treated the element as an indecomposable substance, that is, anything that chemical analysis had failed to break down into simpler entities. He understood the shortcomings of this working definition but felt it still worth while. In the *Traité* he lists thirty-three substances which he proposed to consider elemental, at least until they were further broken down. The majority of

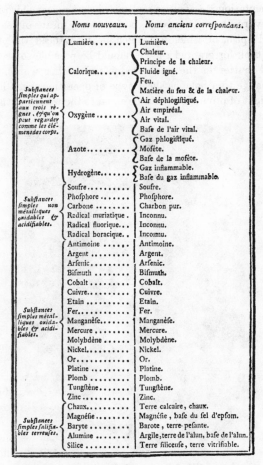

	Noms nouveaux.	Noms anciens correspondans.
Substances simples qui appartiennent aux trois règnes, & qu'on peut regarder comme les élémens des corps.	Lumière	Lumière.
	Calorique........	Chaleur. Principe de la chaleur. Fluide igné. Feu. Matière du feu & de la chaleur.
	Oxygène	Air déphlogistiqué. Air empiréal. Air vital. Base de l'air vital.
	Azote..........	Gaz phlogistiqué. Mofète. Base de la mofète.
	Hydrogène.......	Gaz inflammable. Base du gaz inflammable.
Substances simples non métalliques oxidables & acidifiables.	Soufre..........	Soufre.
	Phosphore	Phosphore.
	Carbone	Charbon pur.
	Radical muriatique .	Inconnu.
	Radical fluorique...	Inconnu.
	Radical boracique..	Inconnu.
Substances simples métalliques oxidables & acidifiables.	Antimoine	Antimoine.
	Argent	Argent.
	Arsenic.........	Arsenic.
	Bismuth	Bismuth.
	Cobalt	Cobalt.
	Cuivre..........	Cuivre.
	Etain	Etain.
	Fer.............	Fer.
	Manganèse.......	Manganèse.
	Mercure	Mercure.
	Molybdène	Molybdène.
	Nickel..........	Nickel.
	Or.............	Or.
	Platine	Platine.
	Plomb	Plomb.
	Tungstène.......	Tungstène.
	Zinc	Zinc.
Substances simples salifiables terreuses.	Chaux..........	Terre calcaire, chaux.
	Magnésie........	Magnésie, base du sel d'epsom.
	Baryte	Barote, terre pesante.
	Alumine	Argile, terre de l'alun, base de l'alun.
	Silice	Terre siliceuse, terre vitrifiable.

Fig. 3.6. LAVOISIER'S TABLE OF THE ELEMENTS.
(From Lavoisier, *Traité élémentaire de chimie* [1789].)

these appear in our present tables of elements. Even at that time Lavoisier was of the opinion that the earths, such as magnesia, baryta, and alumina, were oxides of unknown metals. His *radical muriatique* (for chlorine) created confusion because it showed acid properties with water and hence must be the oxide of an unknown nonmetal. The *radical fluorique* caused a similar problem. Though the element fluorine was not isolated for another century, fluorides and hydrofluoric acid pointed to its existence. The element in the *radical boracique* was discovered early in the next century.

The concept of conservation of mass was suggested in the section on vinous fermentation.[35] Lavoisier recognized that sugar, which is composed of carbon, hydrogen, and oxygen, is converted by means of yeast to carbon dioxide and spirit of wine, for which he introduced the Arabic term *alcohol*. He took for granted that it should be possible to account for all the original matter in the final products. In other words, he reduced chemical change to a balance-sheet type of operation when he suggested that the weight of products should be equal to the weight of the reactants,

Fig. 3.7. M. V. LOMONOSOV.
(Courtesy of the Edgar Fahs Smith Collection.)

an idea utilized but not expressed earlier by Black and Cavendish. Even earlier, the Russian scientist Mikhail Vasilevich Lomonosov (1711–1765) had tacitly assumed that when one body gained weight some other body lost an equivalent amount of weight. Lomonosov was unusually talented as a philologist, poet, educator, and scientist. He applied physical and mathematical principles to chemistry in his studies of solutions. In studying combustion and calcination in the 1740's, he noted the weight gain of the products and attributed it to some substance taken from the air.

[35] A. L. Lavoisier, *Oeuvres*, vol. 1. p. 101; *Elements of Chemistry*, transl. by Kerr, p. 130.

He was a believer in phlogiston. His rudimentary ideas about particles in motion anticipated the kinetic-molecular theory of a century later. However, his position at the University of Moscow kept him away from scientific centers and his ideas failed to stimulate western minds.

As has been said, Lavoisier's *Traité* had a marked influence in establishing the new chemistry. It was well organized and well written. Its factual content had little that was new, for it was based on the experiments and concepts its author had developed from the experiments and concepts of other chemists. However, it represented a synthesis, in that it gathered together all the minor experiments and made them into an organized whole. At the same time much of the earlier chemistry was discarded. The comparison between this book and the one by Baumé, published about the same time, is striking. Baumé's book is like many others published in the preceding century—ponderous, credulous, inconsistent, almost alchemical. Lavoisier's, on the other hand, is modern; it can be read with understanding even today. Consequently, it carried conviction among its readers, particularly the younger ones whose minds were not closed to new concepts. Older chemists did not always read it with sympathy. Much was written in an effort to reconcile the phlogiston and oxygen concepts, but the results were unconvincing. Phlogiston was unnecessary, so why make complicated explanations in order to save it? The phlogiston concept gradually disappeared as its elder adherents were removed from the scene by death.

Among the important converts to the new chemistry were Kirwan and Klaproth, chemists who were highly respected among scientists in their respective countries. Richard Kirwan (1733–1812) had risen to prominence among London scientists and those of his native Ireland as the result of his studies of chemical affinity and his *Elements of Mineralogy* (1784). As *An Essay on Phlogiston* (1787), another of his books, reveals, he was at first bitterly opposed to the new chemistry. This book was translated into French by Madame Lavoisier and published; each argument was carefully answered by Lavoisier and his colleagues. This in turn was translated into English by William Nicholson in 1789 with added remarks and replies by Kirwan. However, Kirwan gradually changed his ideas, and in 1792 wrote Berthollet, "At last I lay down my arms and abandon Phlogiston." The second edition of his *Elements of Mineralogy* (1794–1795) agreed with Lavoisier's concepts.

William Higgins (1766–1825) was the first English chemist to publish a book approving the new concepts. His *Comparative View of the Phlogistic and Antiphlogistic Theories* (1789) was a refutation of Kirwan's *Essay on Phlogiston*. However, since Higgins was a young man who had no significant influence in chemical circles, his book had little impact.

The new doctrines might be expected to have the poorest reception in Germany because Stahl had strong followers there. This was particularly true in Halle. Although F. A. Carl Gren (1760–1798) vigorously defended Stahl's position there, he was forced to alter his ideas as his position became steadily more untenable. Gren's admiration of things German led him to defend Stahl in the university where the latter had taught. (Gren is otherwise remembered as the founder of the *Journal der Physik*. This was continued after his death by L. W. Gilbert as the *Annalen der Physik* and is still published today.)

Martin Heinrich Klaproth (1743–1817) was largely responsible for the early acceptance of Lavoisier's doctrines among German chemists. Although trained as a pharmacist, Klaproth developed a profound interest in analytical chemistry which he pursued during the remainder of his life; he did much to place quantitative analysis on a sound footing. As an analyst he was able to appreciate Lavoisier's reasoning. Thorough testing of combustion and calcination reactions convinced him of the soundness of the new concepts and made him a firm adherent of Lavoisier's views.

SCIENCE AND THE FRENCH REVOLUTION

During the last five years of his life Lavoisier had little time for chemistry. The economic instability of the French government, coupled with the unwillingness of the nobles and clergy to face their responsibility in maintaining financial solvency, had brought France to the brink of revolution by 1789. In the turbulent years that followed, Lavoisier fell victim to the guillotine, along with his father-in-law and other members of the Ferme Générale.

Lavoisier's Public Services

Lavoisier's many services to France were forgotten during the Terror, and his membership in the hated tax-collecting firm was held against him when trials were a mockery of justice. Had his contributions to chemistry been the only thing to offset his tax-collecting activities, the revolutionary tribunal's harsh treatment would have been understandable; but his whole adult life had been given to public service.

As a member of the Ferme Générale he instituted reforms in collecting the salt tax and brought smuggling under control by having a wall built around Paris. As a member of the Académie des Sciences he assisted in drafting report after report, many of them covering investigations made at the request of the government.[36] Of particular importance were those dealing with prison reform, hospital reform, ballooning, and mesmerism.

---

[36] D. McKie, *Antoine Lavoisier, Scientist, Economist, Social Reformer* (Schuman, New York, 1952, pp. 75–76), devotes almost two solid pages to merely listing the reports in which Lavoisier had a hand.

In the last-named he served with Jean Bailly, astronomer and mayor of Paris (1789); Joseph Guillotin, inventor of the machine for humane execution; and Benjamin Franklin.

As an agriculturist, Lavoisier operated his estate near Fréchines as an experimental farm. Here he searched for improved methods which would bring greater success to French agriculture, applying the same balance-sheet methods as he used as businessman and chemist. As a follower of the physiocrats he believed that the land was the ultimate source of national wealth. However, running his farm convinced him that interest rates, because of the French debt, were too high to encourage farmers to invest their profits in agricultural improvement. He was a founder of the Society of Agriculture and was appointed by Callonne to the Committee of Agriculture when it was formed in 1785.

The Metric System

France's system of weights and measures had always varied from province to province and hence had always been chaotic. As early as 1790 the National Assembly passed a decree, proposed by Talleyrand, to simplify and unify weights and measures. The necessary study was delegated to the Académie des Sciences. A committee was set up consisting of the philosopher Condorcet and the mathematicians Laplace, Lagrange, Borda, and Monge. Lavoisier later succeeded Laplace on the committee, immediately becoming secretary and treasurer. Their report issued in 1791, stated their decision to use a new unit of length—one ten-millionth of the earth's quadrant between the pole and equator—and to base subdivisions and multiples on the decimal system. A survey between Dunkirk and Barcelona was started the next year, but was slowed up by financial and military difficulties. At the same time a subcommittee composed of Lavoisier and the Abbé René Just Haüy was determining the weight of a volume of distilled water in vacuum at the temperature of maximum density.

The work of the committee was gravely handicapped by the troubles besetting the Académie des Sciences. There was much opposition at this time to learned societies because they were associated with the aristocracy. Jacques Louis David (1748–1825), founder of the French classical school of painting, headed this opposition. Although he had been court painter to Louis XVI and had painted the beautiful portrait of Lavoisier and his wife, he sympathized with the revolutionists and attained great influence as an associate of Robespierre. Highly antagonistic toward the Académie des Arts, he placed other academies in the same classification and worked for the abolition of them all. Fourcroy, who had been rebuffed in his efforts to purge the membership list of the names of suspected counter-revolutionaries, also bore a grudge against the Académie des Sciences. A bill debated before the National Convention on August 8, 1793, called for certain academies to be abolished but excluded the scientific body because of the important work it had in progress for the government. However, David's oratory swept the members to vote to abolish all of them. When it was realized that this would halt the work on weights and measures, provision for reestablishing the Académie des Sciences was made, but Fourcroy was successful in preventing it from becoming effective. In the end, a new commission which included many of the members of the earlier committee was created to carry on the work. Fourcroy was placed in charge.

The commission was greatly hampered in its work by political conditions; but despite this, the meter was adopted as the provisional standard of length in 1795. The commission finally finished its work in 1798. A platinum bar was prepared and marked to serve as the standard meter; two replicas were also made. At the same time the standard kilogram was established; this corresponds to the mass of a cubic decimeter of distilled water at the temperature of maximum density. The new standards were legalized in 1799.

French resistance to the new standards was widespread and various compromises were made. In 1812 Napoleon permitted the use of a parallel system involving the old terms, but decreed that the metric system must be taught in schools and used in official transactions. Only in 1840 did using other than metric weights and measures become a penal offence in France.

When the metric system was devised the hope was that it would gain international acceptance. As early as 1790 the Royal Society had been invited to participate in the establishment of standards, but the British showed a singular lack of interest. Although several other countries sent representatives to assist in the final preparation of standards, the system did not receive general acceptance outside France for many years.[37] An international commission meeting in 1872 advocated the construction of metric models to be held by an international bureau near Paris, with provision for replicas to be available to the cooperating nations. A metric convention was established three years later; England, however, did not join until 1884.

The Last Days of Lavoisier

There has been much speculation as to why Lavoisier's former associates, some of whom held important positions in the revolutionary courts, exerted no influence to save his life. The only conclusion possible is that

[37] The metric system has been officially adopted by the following countries: Switzerland, 1822; Spain, 1860; Italy, 1861; Germany and Portugal, 1872; Austria, 1876; Norway, 1882; Romania, 1884; Bulgaria; 1892; Denmark, 1912; U.S.S.R. and Greece, 1922.

in this period of hostility and suspicion none of them felt strong enough to speak up in Lavoisier's behalf for fear of the consequences to himself. Much of the popular resentment against Lavoisier had been whipped up even before the Reign of Terror. Jean Paul Marat (1743–1793) had made Lavoisier his particular target in *L'ami du peuple*, the pamphlet he issued in 1791. Marat had from the beginning worked actively to incite the populace to violence and his influence increased greatly after Danton and Robespierre led the Jacobin cause to victory over the more moderate Girondists.

Fig. 3.8. JEAN PAUL MARAT.
(Courtesy of the Edgar Fahs Smith Collection.)

Life had not been good to Marat. Being of humble birth, he obtained a medical education only with great difficulty. He was short of stature and misshapen in body; his pocked face was constantly distorted by a nervous tic. His ambition to achieve scientific recognition led to the publication in 1780 of his *Recherches physiques sur le feu*, in which he claimed to have observed the element of fire. The book was not well received; Lavoisier himself commented disparagingly about it. Marat's studies of electricity failed to impress either Franklin or French scientists. Thwarted in his desire to be elected to the Académie des Sciences, he vainly sought the favor of Frederick the Great. Finally, however, he succeeded in winning Spanish favor and an appointment to the directorship of the Madrid Academy. His hatred and suspicion of others, coupled with his venomous attacks in his pamphlets, made him an influential force in the Revolution until his assassination.

In attacking Lavoisier, Marat called emphatic attention to the wall around Paris, which, he charged, ruined the air for the inhabitants. He further accused Lavoisier of adulterating tobacco with water. The Ferme Générale controlled tobacco processing as well as tobacco taxes, and Lavoisier had introduced the practice of moistening the leaf to a certain standard content to make the tobacco less dry and brittle. Since it was sold on the basis of dry weight there seems to have been no intent to defraud the public. Another charge made by Marat concerned the transfer of powder from the Arsenal at a time that endangered public safety, a charge that again was based upon partial rather than complete facts.[38]

In the uncertain days of the Reign of Terror, such former associates of Lavoisier as Hassenfratz and Monge were apparently too concerned with their own safety and ambitions to want to make an issue of the trial of the Fermiers-Généraux. Guyton de Morveau and Fourcroy had risen to positions of eminence in the revolutionary government and conceivably might have raised their voice in Lavoisier's behalf without fear of repercussions; but either they were afraid or they felt resentment toward him. Fourcroy, an ambitious and vindictive man, apparently went out of his way to call attention to Lavoisier's efforts to save the Académie des Sciences and its work on the metric system. Guyton de Morveau also seems to have harassed his former associate about the committee on weights and measures.

It was Lavoisier's less important friends who attempted to help him. Cadet and Baumé argued that Lavoisier never favored the watering of tobacco. Pluvenet, an obscure druggist who sold chemicals to Lavoisier, utilized a tenuous connection to extract an agreement from Dupin, the prosecutor, whereby Lavoisier would be transferred to another prison before the trial, provided Madame Lavoisier personally made the request. Appearing not as a cringing suppliant but as a righteous wife, she argued that the charges against all the members of the Ferme Générale were false, and thereby effectively closed this avenue of appeal. Two of Lavoisier's old colleagues on the powder commission reminded the Committee of Public Safety of a still unwritten report which required his attention. In behalf of the Bureau of Arts and Crafts, Jean Noël Hallé, a former *abbé*, presented at the trial a statement calling attention to Lavoisier's many services to France. According to a legend, apparently incorrect,[39] this statement provoked Coffinhal, the presiding judge, to remark, "The Republic has no need for scientists. Let justice take its course." Accordingly Lavoisier went to the guillotine on May 8, 1794.

[38] McKie examines the charges and facts carefully in his *Antoine Lavoisier*, pp. 318–321.
[39] James Guillaume, *Etudes révolutionnaires*, 1st series, Bibliothèque Historique, Paris, 1908.

Prestige of Science in an Era of Crisis

Despite its treatment of Lavoisier, the revolutionary government, and later Napoleon, found scientists useful persons to have around, particularly where the country's defense was concerned. The National Convention in 1794–1795 revealed its reliance on science in numerous ways. The Académie des Sciences was soon reestablished as a branch of the Institut de France, the other branches being literature and social sciences. The Jardin du Roi was reorganized as a museum of natural history and was called the Jardin des Plantes. The professorships in the Jardin were continued. Military, medical, and trade schools were founded and given a scientific orientation. The École Polytechnique was established to train the best French students in scientific and engineering subjects. Its distinguished staff included Laplace, Lagrange, Monge, and Berthollet; and such later chemists as Gay-Lussac, Thenard, Vauquelin, Dulong, and Petit attended it.

Leading scientists also participated actively in national affairs. Monge investigated the casting and boring of cannon upon becoming Minister of the Navy. The mathematician, Lazare Carnot (1753–1823), served as Minister of War; he was spoken of as the "organizer of victory." Fourcroy continued Lavoisier's studies of saltpeter and gunpowder and became Napoleon's Minister of Public Instruction. Berthollet and Guyton de Morveau also carried on studies in connection with gunpowder manufacture.

French science during this period was both highly experimental and practical. Industrial applications were stressed; prizes were frequently offered for useful discoveries. Though neither scientist nor mathematician, Napoleon enjoyed the company of French scientists and was on intimate terms with some of them. When he set out on his Egyptian expeditions in 1798, Monge, Berthollet, and the mathematician Jean Baptiste Joseph Fourier accompanied him.

CONCLUSION

The period from 1770 to 1800 may truly be looked upon as a chemical revolution. Although van Helmont and Boyle had expressed discontent with the state of chemistry in the seventeenth century and both had moved in the direction of reform, they were unable to sweep aside the accumulated errors of two millennia, particularly when they themselves rather firmly believed in the possibility of transmuting metals.

A century after Boyle, Lavoisier's clear skeptical mind was adequate to the task of bringing about the much needed revolution. The advances in pneumatic chemistry called attention to the key element, oxygen, and gave the death blow to the concept of air and water as elements. Guyton de Morveau's conclusive demonstration that metals gain weight on calcination, together with Priestley's preparation of oxygen from the calx of mercury, led Lavoisier to perform the experiments described in the Easter Memoir. Lavoisier's further work on respiration and on the compound nature of water led to the conclusion that the phlogiston theory was unnecessary in explaining major chemical phenomena. Despite his unfortunate doctrine of caloric, his explanation of combustion, calcination, and respiration in terms of oxygen set chemistry on a fruitful path.

The success of Lavoisier's revolution was enhanced by his reform of the nomenclature of chemistry in collaboration with Berthollet, Fourcroy, and Guyton de Morveau. The principle that a name must reflect the elemental composition of a compound established a pattern which gave chemistry a clarity frequently lacking in other sciences.

Finally, Lavoisier insured a complete hearing for his concepts when he published his *Traité élémentaire de chimie*, which revealed the consistency and clarity of the new point of view. Although much was still not understood, Lavoisier cleared the road for such progress when he eliminated the four elements of antiquity and the intangible phlogiston, and introduced the concept of elements as the fundamental substances of chemistry which could undergo no further simplification. This foundation enabled chemistry to move ahead dramatically during the next two centuries.

CHAPTER 4

CHEMICAL COMBINATION
AND THE ATOMIC
THEORY

RISE OF ANALYTICAL CHEMISTRY

The period following the death of Lavoisier was marked by steady and important advances in quantitative analysis, the work of Klaproth in Germany, Vauquelin in France, and Wollaston in England being particularly important.[1] Of the three, Klaproth was perhaps the best analyst and his teachings were of great importance in laying down the principles of sound analytical procedure. Vauquelin was a competent analyst, though he never achieved Klaproth's skill. His importance is due primarily to the leading chemists he trained—Thenard, Chevreul, Orfila, Humboldt, Leopold Gmelin, and Stromeyer. Wollaston still reflected the tradition of the wealthy amateur. Although he established neither fundamental principles nor a school of followers, his versatile interests led to important developments in various fields. These three men found chemistry a qualitative science. They left it a quantitative science.

Martin Heinrich Klaproth (1743–1817) was born the same year as Lavoisier, whose new chemical concepts he introduced into Germany. Trained as an apothecary, he became interested in chemistry and avidly educated himself in it. In 1771 he entered the pharmaceutical laboratory of Valentin Rose (1736–1771), a former student of Marggraf and the first of a famous family of pharmacist-chemists. Rose died within a short

[1] Elsie G. Ferguson, *J. Chem. Educ.*, **17**, 554 (1940); **18**, 3 (1941).

time, leaving Klaproth as director of the laboratory, and with the responsibility for the upbringing of Rose's son, Valentin the Younger (1762–1807), who later became Klaproth's assistant. The younger Rose's sons, Gustav and Heinrich, contributed to mineralogy and analytical chemistry during the next generation.

Klaproth achieved prominence in German chemistry both as a teacher and as a member of the Berlin Academy. The field of quantitative analysis was reformed as a result of his work. Whereas earlier chemists had been in the habit of recalculating their results so they would total 100 per cent, or of arbitrarily readjusting them to coincide with some preconceived notion—even Bergman and Lavoisier were guilty of this—Klaproth instituted the accurate reporting of results. He published the weights of samples and precipitates so that any errors present could be detected. He gave much attention to the sources and elimination of errors. The purity of reagents was emphasized and procedures were devised for purifying them. Apparatus was carefully chosen so that the constituent being analyzed would not be introduced into the sample by the mortar and pestle, crucible, beaker, or flask. He was the first to use agate, flint, and diamond mortars, and he corrected for weight losses in the mortar and crucible. He introduced the practice of drying precipitates to constant weight. When ignition produced a more stable residue without decomposition, this procedure was used. Klaproth studied the composition of pure compounds extensively so that this knowledge might be available in analysis. Bergman before him had originated the idea of precipitating silver as the chloride, calcium as the oxalate or sulfate, and lead as the sulfide or sulfate. Klaproth extended this idea in a number of new directions.

In the course of his mineral analyses he frequently encountered samples for which the analytical results failed to total 100 per cent. Since this suggested the presence of additional elements, Klaproth followed up the leads and either discovered, or verified the discovery of, zirconium, uranium, tellurium, and titanium. In analyzing zircon from Ceylon in 1789 he found that 70 per cent consisted of an earth (oxide) which Bergman had previously confused with alumina, lime, and iron oxide. Klaproth suggested the name *Zirconerde* for the earth. Zirconium metal was not isolated until 1824 when Berzelius heated potassium zirconium fluoride with potassium.

Studies of pitchblende revealed the presence of a new earth which Klaproth named *uranium*, after the new planet Uranus discovered by William Herschel in 1781. He succeeded in isolating a black powder which he believed was the element but which was really an oxide. The metal was not isolated until 1841 when Eugène Melchior Peligot (1811–1890) successfully reduced uranous chloride (UCl_4) with potassium.

The oxide of tellurium was first reported by Franz Joseph Müller von

Fig. 4.1. MARTIN HEINRICH
KLAPROTH.
(Courtesy of the Edgar Fahs Smith
Collection.)

Fig. 4.2 LOUIS NICOLAS
VAUQUELIN.
(Courtesy of the Edgar Fahs Smith
Collection.)

Reichstein of Austria in 1782. Klaproth verified the discovery by isolating
the metal; he named it after *tellus*, the Latin word for earth. He also
verified William Gregor's discovery of titanous earth; the name he chose
was based on *Titans*, the mythological sons of the earth goddess.

Louis Nicolas Vauquelin (1763–1829), the son of a farmer, pursued his
pharmaceutical studies in Paris. There he came to the attention of
Fourcroy, and quickly became almost a member of the family. Before
long he began to assist Fourcroy with his teaching. In 1795 he was made
inspector of mines and gave a course in assaying at the École des Mines.
Later he became professor of chemistry at the Collège de France, at the
Jardin des Plantes, and, after Fourcroy's death, at the École de Médecine.
His interest in chemistry was broad and included organic as well as
inorganic analysis. With Fourcroy he studied tartaric and pyromucic
acids and their occurrence in plants. He investigated respiration in insects
and worms, the phosphate in bones, the coloring matter of blood, and the
brains of fish and men.

The Abbé René Just Haüy (1743–1822) at the École des Mines supplied
him with samples of minerals for analysis. These analyses were generally
inferior to those done by Klaproth because Vauquelin paid less attention
to impurities in reagents and to the quality of his samples. He did, however,
discover chromium and he recognized the oxide of beryllium. He isolated
chromium from crocoite, a red ore of lead which had been observed in a
lead-mining region in Siberia. Vauquelin became interested in this after
widely varying analytical results were reported by others. When he fused

the red ore with potassium carbonate he obtained lead carbonate and a yellow solution. The solution gave a red precipitate with mercuric salt and a yellow one with lead salt; it became green when treated with stannous salt. Vauquelin converted the chromate to the green oxide and in 1798 successfully reduced this to the metal by heating it with carbon. Because of the variety of colored compounds formed by the element, Fourcroy and Haüy suggested the name *chromium*.

Vauquelin encountered beryllium oxide as the result of crystallographic studies made by the Abbé Haüy. This famous mineralogist, who established the foundations of crystal science on a geometric basis, observed that emerald and beryl appeared to be similar if not identical. Vauquelin showed that the two minerals were chemically alike except for a trace of chromium in emerald. During his analysis he discovered the presence of a new "earth" which he himself, Bergman, and others in their earlier analyses of beryl, and Klaproth in his analysis of emerald, had reported to be alumina. Vauquelin found that, unlike alumina, this earth was not soluble in excess sodium hydroxide. It differed further in forming no alum and in forming salts that had a sweet taste. This latter was the reason for the new earth being named *glucinia*, meaning sweet. Klaproth called the earth *beryllia* because yttria also forms sweet salts. The names *glucinium* and *beryllium* both continued to be used for the element for a long time, the latter gradually being favored. The element itself was not prepared until 1828, when Friedrich Wöhler and Antoine-Alexandre Brutus Bussy independently succeeded in reducing beryllium chloride with potassium.

William Hyde Wollaston (1766–1828), the son of an English clergyman, studied languages at Cambridge before turning to medicine. After establishing a prosperous medical practice, he suddenly abandoned it when he was thirty-four in order to devote his life to scientific studies. He was an enthusiastic and talented investigator in various fields of chemistry and physics who was not averse to prospering from his discoveries. He invented the reflecting goniometer in 1809 for measuring the angles between crystal faces, and the *camera lucida* in 1812. He observed the more prominent dark lines of the solar spectrum in 1802, but interpreted them erroneously as being boundaries between colors. Independently of Johann Wilhelm Ritter he discovered the ultraviolet spectrum when he noted that the darkening effect of the spectrum on silver chloride extended into the dark region beyond the violet.

In chemistry Wollaston's most important work was done with the platinum metals. Ever since platinum became known to Europeans around 1750 it had received a great deal of attention from chemists. Its chemical inertness, its high temperature of fusion, and its erratic physical and chemical behavior aroused continuous interest that led, however, to little progress in understanding its real properties. Pierre-François Chabaneau (1754–1842), a French chemist at the Spanish court of Charles II, was induced to study the metal extensively, but the erratic results he obtained discouraged him. He succeeded in preparing malleable platinum in 1783 and supplied it for jewelry, medals, chemical vessels, and similar small items. Because it cost less than gold, the Spanish government actually ordered that the platinum separated from Colombian gold by amalgamation be thrown into the rivers to prevent it being used to debase the coinage.

Wollaston discovered how the platinum sponge prepared by the thermal decomposition of ammonium chloroplatinate could easily be made malle-

Fig. 4.3. WILLIAM HYDE WOLLASTON
(Courtesy of the University of Wisconsin.)

able by compression and hot hammering. The process, which was kept secret until his death, netted him a tidy income. More important for chemistry, his studies of platinum resulted in the discovery of palladium and rhodium in 1803. These metals, along with osmium and iridium, frequently caused anomalous behavior in platinum. Palladium was named after Pallas, a recently discovered asteroid. Rhodium, its name deriving from the rose color of its salt, was discovered in the residues of a crude platinum solution from which the platinum and palladium had been removed.

Two more elements in the platinum group were discovered by a close friend of Wollaston, Smithson Tennant (1761–1815), the son of a Yorkshire clergyman and a former student of Joseph Black. In 1803 Tennant noted that a black residue remained after crude platinum had been dissolved in aqua regia. Others who had made the same observation concluded that the residue was graphite. Tennant believed it to be metallic and in 1804 demonstrated that it consisted of two new elements, osmium and iridium. *Osmium* was so named because of its odor, *iridium* because of the colors of its salts.

Ruthenium, the sixth of the platinum metals, was not discovered until 1844 when it was encountered by Karl Karlovich Klaus (1796–1864), professor of pharmacy and chemistry at the University of Kazan. By that time platinum metals were being found in the Ural Mountains and Klaus became the foremost authority on their chemistry. The name ruthenium derives from the Ruthenian region of Russia.

AFFINITY CONCEPTS

The nature of the forces which hold bodies together in complex substances had received attention as early as the thirteenth century when Albertus Magnus spoke of an *affinitas* between bodies which combined. In the seventeenth century Glauber, Boyle, and Mayow had all sought to deal with replacement reactions from the viewpoint of relative affinities. In 1718 the Parisian apothecary, Étienne François Geoffroy, published a *Table des rapports* in which he attempted to compare affinities of various acids and bases. A half century later Torbern Bergman sought to refine the Geoffroy table by taking reaction conditions into account.

Bergman studied extensively the relative ability of bases and acids to replace one another in salts. For example, when *caustic terra ponderosa* ($Ba(OH)_2$) was added to *vitriolated tartar* (K_2SO_4), *ponderous spar* ($BaSO_4$) was formed, leaving a liquor that contained *caustic vegetable alkali* (KOH); hence he concluded that the vitriolic acid (H_2SO_4) had a greater affinity for barytes than for potash.[2] In similar fashion he worked out relative affinities of many other acid-base combinations; however, not all combinations were as clear cut as the above example.

Bergman and to a lesser degree Macquer, Kirwan, and Guyton de Morveau, among others, were obsessed with the idea that substances could be ranked in a relative order of affinity for a test substance, the substances with the greatest affinity being capable of displacing all those with lesser affinity from compounds in which they were present. It was conceded that affinity relations differed under wet and dry conditions. Bergman was convinced that a substance with lesser affinity could not expel one with greater affinity from a salt.

It was this latter conviction which was attacked by Berthollet in 1799 in Cairo when he read a paper entitled "Recherches sur les lois de l'affinité." Berthollet expanded these ideas in his book, *Essai de statique chimique*

---

[2] "Potash" is used here with reference to the "basis of potash" rather than in its more conventional sense of potassium carbonate, the salt prepared by leaching wood ashes and evaporating the solution in an iron pot. The "basis of potash" was suspected before the discovery of potassium in 1807, as is evident in such terms as caustic potash (KOH), vitriol of potash (K_2SO_4), and muriate of potash (KCl). "Soda" was similarly used to refer to sodium carbonate and to the "basis of soda" in various sodium compounds.

(1803). Berthollet argued that when two substances were competing for a third, the ratio of the division depended not alone on the ratio of the affinities but upon the quantities of the reacting substances. Using as example the reactions of alkalies and alkaline earths with acids, he sought to establish the importance of the quantity of the reacting substances. He also wanted to investigate the properties which influenced the direction of a reaction. He found that an insoluble salt like barium sulfate could be made soluble by boiling with fresh potash (K_2CO_3) solution, decanting the soluble portion (containing traces of SO_4^{-2}), and repeating. This contradicted Bergman's belief that reactions could go in only one direction.

Berthollet obtained further evidence for the reversibility of reactions when he evaporated potash solution in contact with solid calcium phosphate and obtained evidence for the formation of potassium phosphate and lime. When solutions of sodium hydroxide and potassium sulfate were evaporated to dryness and the alkaline hydroxides were removed with alcohol, the residue contained both sodium and potassium sulfates. Berthollet interpreted these findings as showing that the products of a reaction are the result not only of affinities but of the relative quantities of the reacting substances. He cited numerous instances in which solubility had a significant effect on the products of a chemical reaction. When potash or soda competed with barium for sulfuric acid, he pointed out that the insolubility of barium sulfate caused the active mass of that compound to remain low, thus handicapping the potash or soda in the competition. He recognized that since solubility of compounds varies with temperature, the temperature has an important part in reactions between the same compounds. For example, when soda and potash were mixed with hydrochloric and nitric acids the product which crystallized from a cold solution was potassium nitrate, whereas sodium chloride crystallized from a hot solution.

From such observations Berthollet concluded that affinity alone could not determine the direction of a chemical reaction, but that the active masses of the reacting substances were of primary importance because of the relation between solubility and concentration. Mass effects, he felt, made it impossible to measure the relative affinities of two substances toward a third. In drawing these conclusions Berthollet was on sound ground; actually he was working toward the law of mass action, which was not formulated until after 1864.

EQUIVALENCES AND COMBINING PROPORTIONS

Berthollet's rejection of affinities was tied up with another problem being discussed at that time, namely, the proportion in which substances combined, for his skepticism regarding affinities carried over into the latter area. As we shall see, the unsoundness of his position here led chemists to overlook the soundness of his reasoning with respect to affinities; as a result the latter field of chemistry was given little attention for many years.

Some chemists—among them Kunckel, Lemery, Stahl, Homberg, and Wenzel—had been studying the weights of metals displaced by other metals, and of acids neutralized by bases, in order to arrive at figures for equivalent ratios. In 1767 Cavendish spoke of a certain amount of fixed alkali (K_2CO_3) as being equivalent to a certain amount of lime. Bergman reported that when one metal replaced another from its neutral salt, neutrality was maintained; he interpreted this as indicating that phlogiston was merely transferred from one metal to the other.

Iron   +   Copper sulfate →   Copper   +   Iron sufate
(Iron calx + φ)   (Acid + copper calx)   (Copper calx + φ)   (Acid + iron calx)

Carl Friedrich Wenzel (1740–1793), director of the Freiberg foundries, observed, on the basis of numerous analyses, that acids and bases reacted in constant proportions. Although his studies of the composition of salts were not as accurate as those by Klaproth, they pointed the way toward the principles of quantitative analysis.

The problem was attacked most effectively by Jeremias Benjamin Richter (1762–1807) in two books[3] which laid the foundations of stoichiometry. Richter was an engineer in the department of mines in Silesia who was later connected with the porcelain works in Berlin. In the belief that chemistry might be reduced to a mathematical system, he set out to do this in his *Anfangsgründe der Stöchiometrie*; it is crude and ponderous, and resorts to phlogistic concepts. Observing that when two neutral salts are brought together in solution the products are neutral, he concluded that the total quantities must remain unchanged. He thereupon started to determine the amounts of different bases required to neutralize 1000 parts of several acids. (See Table 4.1.[4])

According to the table, the ratios of potash to soda are:

$$\frac{1606}{1218} = \frac{2239}{1699} = \frac{1143}{867} = \frac{1.318}{1}$$

This suggested that the neutralizing powers of the two alkalies are in the proportion of 1.318 parts potash to 1 part soda. Richter was aware of the equivalence inherent in chemical reactions, but because of his obsession

---

[3] J. B. Richter, *Anfangsgründe der Stöchiometrie*, Hirschberg, Breslau, 3 vols., 1792–1794; *Über die neueren Gegenstände der Chemie*, Hirschberg and Lissa, Breslau, 11 vols., 1791–1802.

[4] From Ida Freund, *The Study of Chemical Composition*, Cambridge Univ. Press, Cambridge, 1904, p. 175.

Table 4.1. CORRELATION OF RICHTER'S EQUIVALENTS[a]

Alkali	Sulfuric Acid	Hydrochloric Acid	Nitric Acid
Potash	1606	2239	1143
Soda	1218	1699	867
Volatile alkali	638	889	453
Baryta	2224	3099	1581
Lime	796	1107	565
Magnesia	616	858	438
Alumina	526	734	374

[a]The figures represent the parts of base required to neutralize 1000 parts of acid.

Table 4.2. FISCHER'S CORRELATION OF RICHTER'S EQUIVALENTS

Base		Acid	
Alumina	525	Hydrofluoric	427
Magnesia	615	Carbonic	577
Ammonia	672	Sebacic	706
Lime	793	Muriatic	712
Soda	859	Oxalic	755
Strontia	1329	Phosphoric	979
Potash	1605	Formic	988
Baryta	2222	Sulfuric	1000
		Succinic	1209
		Nitric	1405
		Acetic	1480
		Citric	1583
		Tartaric	1694

with mathematical relationships which did not exist he frequently doctored data to fit his preconceived notions. He believed, for example, that the combining weights of bases followed an arithmetical progression, those of acids a geometric one. Such ideas, together with his difficult style of writing, brought Richter's two books little attention until Ernst Gottfried Fischer (1754–1831) summarized them in his German translation of Berthollet's *Recherches sur les lois de l'affinité* (1802). Fischer combined the Richter tables so as to show the equivalence of various acids and bases in terms of 1000 parts of sulfuric acid. (See Table 4.2.[5]) Fischer's table and note were

[5] *Ibid.*, p. 176.

included in Berthollet's *Essai de statique chimique* in 1803 and thereafter became generally known. Richter published a similar table for 18 acids and 30 bases the same year.

Definite Proportions

From the published literature of Lavoisier and other leaders of late eighteenth-century chemistry it is evident that there was general acceptance of both the concept of the conservation of matter and the concept of definite proportions in chemical compounds. The analytical methods of Klaproth and others were clearly based upon these concepts. However, Berthollet's work on chemical combination made the entire concept of definite proportions questionable because from the correct conclusion regarding the effect of active masses on the distribution of reaction products he drew an erroneous conclusion regarding the effect of these masses on the distribution of constituents within a compound. He believed that the composition of a compound might vary over a very considerable range.

Berthollet had to admit that there were numerous compounds whose composition was proved by analysis to be fixed, but he argued that these were just cases in which solubility conditions were such that the least soluble fraction was separated before variation could occur. The solution still contained compounds of the same elements, but their composition differed. Since it was not possible to analyze the compounds still in solution, there was no way to prove or disprove this assertion. He maintained that the compounds which separated out were merely those in which the greatest condensation occurred; i.e., they were the most insoluble. Since, in neutralizing an acid, the maximum amount of heat was liberated at the point of neutrality, as revealed by an indicator, the neutral salt obtained on evaporation represented the salt at its most stable point, but did not exclude other compositions of that salt.

As far as gases were concerned, he pointed out that the volume of ammonia is significantly less than that of the elemental gases of which it is composed—in fact, one-half; hence the composition is fixed because the cohesive force is at a maximum. In the case of nitric oxide there is little or no contraction; therefore the gas can combine with additional oxygen. Berthollet thought that this combination took place gradually but continuously.

For positive evidence to support his theory of varying composition Berthollet mentioned solutions, alloys, glasses, metallic oxides, and what we now know to be basic salts. The numerous incorrect analyses reported at the time provided him with examples in the case of oxides and salts. Furthermore, the fact that some of the metals formed several oxides led him to believe that the change in composition was continuous rather than intermittent. "So lead forms an oxide which to begin with is gray, then passes through various shades of yellow, and then finishes by being red. . . . Iron also passes, as the oxidation proceeds, through different shades of color, and assumes different colors."[6] Basic salts obtained by the action of alkali on the salts of copper, bismuth, and mercury provided him with examples to support his contention, though here again his analyses were incomplete and incorrect.

Because of his prominence, Berthollet's views attracted considerable attention. Despite the fact that prominent analysts like Klaproth and Vauquelin based their methods on the principle of constant composition, they showed no tendency to fight for that principle. It was Joseph Louis Proust (1754–1826) who waged an eight-year controversy with Berthollet which ended with the law of constant composition being firmly upheld. Unfortunately, in discrediting the concept of variable composition, Proust swept aside Berthollet's sound observations regarding the effects of active masses on the direction of chemical reactions.

Proust, the son of a pharmacist in Angiers, was educated in Paris. He operated a hospital pharmacy there before going to Spain where he held several academic positions during the reign of Charles IV. His laboratory at Madrid was splendidly fitted out for analytical work, with an abundance of platinum vessels and other equipment. His experiments were done carefully and interpreted clearly; hence there was no question as to the soundness of his ideas. His work in Madrid was terminated and his laboratory destroyed when French troops occupied Spain in 1808. Although reduced to poverty, he refused Napoleon's offer to become supervisor of manufacture of grape sugar, a substance he had discovered in grape juice. When Louis XVIII came to power he was granted a pension and made a member of the French Academy.

It was in 1799 that Proust demonstrated that artificially prepared copper carbonate was identical in composition with that found in nature. He was actually working with the basic carbonate, $CuCO_3 \cdot Cu(OH)_2$, from which he obtained water on mild heating and carbon dioxide on vigorous heating. He correctly analyzed the black oxide which remained in order to ascertain its composition. This led him to set up a criterion for chemical compounds, as distinct from mixtures and solutions. Thus, when Berthollet's views were published, Proust was in an excellent position to defend the opposing point of view. His numerous papers in the *Journal de Physique* between 1802 and 1808 attacked and demolished Berthollet's position with sound experimental evidence. The controversy was carried on with the utmost courtesy by both men, but in the end—Dalton's atomic theory was just beginning to be discussed—there was no question of the soundness of the law of definite proportions.

In dealing with Berthollet's contention that the composition of metallic oxides varied continuously between limits, Proust proved that there were

[6] *Ibid.*, p. 134.

Fig. 4.4. Joseph Louis Proust.
(Courtesy of the Edgar Fahs Smith Collection.)

two oxides of tin and that the composition of each one was fixed. If any analyses showed an intermediate percentage of oxygen, this was due to the presence of mixtures. Again, in the case of the sulfide of iron, Proust demonstrated the existence of two distinct compounds. He proved that minerals have the same composition regardless of their place of origin. His conclusions are well summarized in this statement:

> . . . The properties of true compounds are invariable as is the ratio of their constituents. Between pole and pole, they are found identical in these two respects; their appearance may vary owing to the manner of aggregation, but their properties never. No differences have yet been observed between the oxides of iron from the South and those from the North. The cinnabar [mercuric sulfide] of Japan is constituted according to the same ratio as that of Almaden. Silver is not differently oxidized or muriated in the muriate [chloride] of Peru than in that of Siberia. In all the known parts of the world you will not find two muriates of soda [rock salt], two muriates of ammonia, two saltpetres, two sulphates of lime, of potash, of soda, of magnesia, of baryta, etc. differing from each other. . . . The native oxides

follow the same relations of composition as the artificial. This is a fact which analysis confirms at every step. We find in the bosom of the earth copper oxide containing 25 per cent of oxygen, arsenic with 33, lead with 9, antimony with 30, iron with 28 and 48, and others still.[7]

Proust showed that some of Berthollet's oxides were actually hydrated and therefore could not have the suggested composition. He concluded that hydrates had the characteristics of genuine compounds. His investigation of the oxides and sulfides of antimony, cobalt, copper, and nickel served further to strengthen the case for fixed composition.[8]

His analyses were, in general, carefully done and good enough to win support, but they were not entirely free of shortcomings. At times he made assumptions which were not justified. He was convinced of the correctness of his position and in a systematic fashion broke down Berthollet's arguments. Proust performed a service to the chemistry of his time, for such a law was especially needed when Dalton was struggling to establish his theory of chemical atoms. The law of definite proportions was a major bulwark for Dalton's theory, because if compounds had been proved to have variable proportions the fact would have been a serious detriment to any theory of atomic combinations.

Despite the tenacity and brilliance of Proust's attack on the concept of variable proportions, he did have a certain amount of good fortune in establishing the validity of his position. Even though his analyses did not attain the same degree of excellence as those done by Berzelius and Stas in the next generation, these later workers assumed the validity of Proust's law, and their results were of such a character as to uphold it.

There was naturally a general failure to recognize and study compounds in which ions of similar size and valence can interchange within a crystal lattice and produce a variation in composition. The latter are frequently referred to as *berthollides*, in contrast to *daltonides*, the compounds whose composition is fixed. Perhaps it was fortunate that no one encountered berthollides at this time, for although they would have saved Berthollet's prestige, they would have introduced an uncertainty which would have greatly handicapped the acceptance of the atomic theory.

HIGGINS, DALTON, AND THE ATOMIC THEORY

The development of a chemical atomic theory was a natural outgrowth of investigations and speculations regarding chemical combination. At the

---

[7] From Freund's translation in *ibid.*, p. 137.
[8] J. L. Proust, *J. physique*, **53**, 89 (1801); **54**, 89 (1802); **55**, 324 (1802); **59**, 260, 265, 321, 343, 350, 403 (1804); **63**, 364, 421 (1806); *Ann. chim.*, **32**, 26 (1799).

end of the eighteenth century a number of chemists utilized some concept of an ultimate particle in their explanations of chemical phenomena. This was a heritage of the seventeenth century, when corpuscular ideas attained considerable popularity. However, the particles were primarily physical, and the chemists had not given much real thought to them in explaining the nature of chemical compounds.

The most serious attempt to develop a chemical atomism before the nineteenth century was made by William Higgins in 1789. Higgins, it will be remembered, was the Irish chemist who, while still an Oxford student, published a book comparing the views for and against phlogiston and found greater virtues in the latter. While speculating on chemical combination in this book, he used a system of ultimate particles. By means of diagrams of the oxides of nitrogen he attempted to explain the forces that hold particles together; in other words, he was groping toward primitive concepts that underlay valence, bonding energies, and the law of multiple proportions. He seems to have paid no attention to relative weights of particles. His ideas were based partly on experimental information but largely on speculation derived from such information. However, the time was not yet ripe for chemical atomism because analytical chemistry was still in a rudimentary state. Higgins' concepts, although well thought out, were premature and attracted little attention. Furthermore, he failed to pursue his ideas beyond the point of their value in supporting the anti-phlogiston theory.

It was John Dalton (1766–1844) who established a continuing tradition of chemical atomism. The son of a poor Quaker weaver of Eaglesfield in the Cumberland area, he acquired most of his education by his own efforts; he had little formal schooling. A wealthy Quaker relative provided instruction in science and mathematics, subjects the boy excelled in, and later he came in contact with John Gough, a blind philosopher, who interested him in studying Newtonian science. Dalton began to teach in a local school at the age of twelve. In 1793 he became professor of mathematics and natural philosophy in New College, a Dissenting academy in Manchester. Six years later the academy moved to York, but Dalton remained in Manchester for the rest of his life, supporting himself in a modest way by private tutoring.

Soon after coming to Manchester he became a member of the Literary and Philosophical Society, a learned society which had been formally organized in 1781; it stemmed from a more informal group that had been meeting for some years under the guidance of Thomas Percival, a former student of Priestley and a graduate of the medical school in Edinburgh. Its discussion concerned literature, natural philosophy, general politics, commerce, and the arts, but excluded religion and British politics. Publication of the more important papers presented before the society began in 1785 in a series of *Memoirs*. Dalton read his first paper in 1794; its subject was color blindness, an affliction from which he himself suffered and which is still frequently referred to as Daltonism.

In 1787 he made his first entry in a notebook entitled "Observations on the Weather." He diligently continued this practice for fifty-seven years; the last entry was made the evening before his death. His *Meteorological Observations and Essays*, published in 1793, was chiefly descriptive

Fig. 4.5. JOHN DALTON.
(Courtesy of the Edgar Fahs Smith Collection.)

of instruments used in the study of weather; it included thermometers and barometers, and an instrument for determining the dew point. It was his interest in meteorology that was responsible for his greatest contribution to chemistry, the atomic theory.

Dalton did no important experimental work until nearly the end of the century, when he became interested in the nature of the atmosphere. A series of papers he read before the Manchester Society between 1799 and 1801 discussed water vapor in relation to vapor pressure. He found that the amount of water vapor in the air increased with the temperature and he proved that this was true when other gases were substituted for air. In other words, the quantity of water vapor in a gas was independent of the type of gas but dependent on the temperature. Dalton formulated the concept of the vapor pressure of water and soon extended his studies to include the phenomenon in other liquids. In these same studies he apparently recognized the expansion of volume shown by different gases when heated, an observation made some years before by Charles but not generally followed up until Gay-Lussac's work early in the nineteenth century. Dalton noted that the expansion of various gases was the same for a given temperature rise, but he pursued the matter no further.

His studies of water vapor in the air convinced him that, contrary to the teachings of Lavoisier and others, this vapor was not chemically combined with the nitrogen and oxygen in the atmosphere. He very early convinced himself that the theory of the air being a chemical compound was untenable. His own analyses of air from various localities showed that it was composed of 21 per cent of oxygen and 79 per cent of nitrogen, and that the water vapor varied up to a maximum, the maximum increasing with temperature. He demonstrated that the degree of saturation of the air with water vapor could be determined by observing the number of degrees by which the air must be cooled to reach the dew point; this has been the basis of hygrometry ever since.

It was during these studies—1801—that Dalton formulated the law of partial pressures. In his work on vapor pressure he noticed that adding water vapor to dry air increased the total pressure by the amount of the vapor pressure. Hence, the pressure of the water vapor in the mixture was identical with the pressure it would exert if no other gases were present in the same total volume. He therefore concluded that the pressure exerted by each gas in a mixture was independent of the pressure exerted by the other gas in the mixture, and that the total pressure was the sum of the pressures exerted by each gas.

It was probably this sort of reasoning that led him at this time to his hypothesis regarding the diffusion of mixed gases. According to this concept, particles of unlike gases exerted neither attraction nor repulsion for one another. He developed his hypothesis to explain why gas mixtures did not form layers according to their differences in density. For example, in a mixture like air the oxygen is the densest, the nitrogen slightly less dense, and the water vapor definitely less dense. Should they not form layers with oxygen next to the earth, nitrogen above the oxygen, and water vapor above that? Yet analysis showed that the air was a uniform blend of the three gases. Some chemists, particularly the French school, explained this by suggesting that the gases of the atmosphere were present in some form of chemical combination. In rejecting this hypothesis Dalton reasoned that the number of particles of water vapor must be far less than those of oxygen, which in turn were far outnumbered by nitrogen particles. Any compound formed of the components of the atmosphere would have to be unduly rich in nitrogen. This, together with his work on vapor pressure, caused him to reject combination as an explanation for the homogeneity of air.[9]

Seventeenth-century physics postulated that the pressure of gases was due to a repulsion which the particles of a gas exerted toward one another. Dalton modified this concept, restricting the repulsion to particles that have the same chemical identity. Unlike particles could then diffuse freely through other such particles and produce a homogeneous mixture. Incidentally, we should mention that for Dalton, particles were tiny solids surrounded by an envelope of caloric. It was this heat layer which created the repulsion in the particles. This idea was a natural consequence of the caloric theory, for it was readily observed that heat flowed from a hot substance to a cooler one; hence, heat substance was mutually repulsive. Dalton wished to modify the traditional theory of gaseous particles so that it would account for unlike particles being inert toward one another.

Still another observation which was being discussed at this time was Henry's law dealing with the solubility of gases in liquids. William Henry (c. 1774–1836), the son of a Manchester apothecary, spent a great deal of his time in chemical pursuits. His textbook, *Elements of Experimental Chemistry*, was first published in 1799 and went through eleven editions in thirty years. He was perhaps Dalton's closest friend; his son, William Charles Henry, wrote a biography of Dalton. William Henry first formulated the law which bears his name in 1801. He found that the volume of a nonreactive gas which dissolves in water is independent of pressure; that is, if a certain amount of water dissolves 10 ml. of gas at a pressure of 1 atm., it will still dissolve 10 ml. of gas at a pressure of 2 atm. Of course, the weight of gas dissolved at 2 atm. will be doubled; therefore the law is frequently stated so as to indicate that the mass of gas dissolved by a liquid is directly proportional to the pressure on the gas. Henry further observed that each gas in a mixture behaves as if the others were absent.

According to Dalton, these observations indicated that the dissolving process was purely physical. He himself investigated the solubility of mixed gases and found that different gases varied considerably in this respect. A paper he read before the Manchester Literary and Philosophical Society in October, 1803, revealed his concern at this variation in solubility: ". . . I am nearly persuaded that the circumstance depends upon the weight and number of the ultimate particles of the several gases: Those whose particles are lightest and single being least absorbable, and the others more according as they increase in weight and complexity. An enquiry into the relative weights of the ultimate particles of bodies is a

---

[9] Dalton here shows the influence of Newton's *Principia*, 23rd Prop., Book 2.

subject, as far as I know, entirely new: I have lately been prosecuting the enquiry with remarkable success."[10] (See Table 4.3.[11])

This is the first public intimation on Dalton's part of his grasp of the idea of particle weights, the basis of chemical atomism. An entry in his notebooks dated September 6, 1803, describes his earliest effort to draw up a table of comparative atomic weights. His calculations are based upon

Table 4.3. DALTON'S 1803 TABLE OF THE RELATIVE WEIGHTS OF THE ULTIMATE PARTICLES OF GASEOUS AND OTHER BODIES

Hydrogen	1.0	Nitrous oxide	13.7
Azot	4.2	Sulphur	14.4
Carbone	4.3	Nitric acid	15.2
Ammonia	5.2	Sulphuretted hydrogen	15.4
Oxygen	5.5	Carbonic acid	15.3
Water	6.5	Alcohol	15.3
Phosphorus	7.2	Sulphureous acid	19.9
Phosphuretted hydrogen	8.2	Sulphuric acid	25.4
Nitrous gas	9.3	Carburetted hydrogen	6.3
Ether	9.6	Olefiant gas	5.3
Gaseous oxide of carbone	9.8		

the analyses of Lavoisier for water and carbonic acid, of Austin for ammonia, and of Chenevix for sulfuric acid; the respective formulas are assumed to be HO, CO_2, NH, and SO_2. Dalton up to this time had shown little interest in chemical matters; but now his gas studies had apparently led him to his particulate philosophy of gases from which stemmed his concept of atomic weights, and from this in turn, his expanded chemical atomism.[12]

Not long after this, Dalton formulated the law of multiple proportions. In working with the particle weights of carburetted hydrogen (methane, CH_4) and olefiant gas (ethylene, C_2H_4) he observed that twice as much hydrogen combined with a given weight of carbon in carburetted hydrogen as in olefiant gas. Since he had no way of knowing the formulas of these gases, he assigned CH to olefiant gas and CH_2 to carburetted hydrogen.

[10] John Dalton, *Mem. Lit. and Phil. Soc. Manchester*, [2], **1**, 286 (1805).

[11] *Ibid.*, p 287.

[12] Here I follow L. K. Nash, *Isis*, 101 (1956), in believing that Daltons' studies on gases, particularly those on mixed gases, led him to the chemical atomic theory. But there are several other hypotheses. Dalton himself gave three different versions, none of them mutually consistent. On the various aspects of the question see, in addition to Nash, T. Thomson, *History of Chemistry*, Colburn & Bentley, London, 1830–1831; William C. Henry, *Memoir of the Life and Scientific Researches of John Dalton*, Cavendish Soc., London, 1854; A. N. Meldrum, *Mem. Lit. and Phil. Soc. Manchester*, **55**, nos. 3, 4, 5, 6, 19, 22 (1910–1911); J. R. Partington, *Ann. Sci.*, **4**, 245 (1939); H. E. Roscoe and A. Harden, *A New View of the Origin of Dalton's Atomic Theory*, Macmillan, London, 1896. The last-named is of particular value because it contains extensive quotations from Dalton's notebooks. These notebooks were destroyed in the bomber raids on Manchester during World War II.

Multiple proportions naturally became apparent in compounds containing the same elements, once the concept of atomic weights began to develop. As time went on, Dalton observed this in other sets of compounds. The subject is discussed casually in his book, particularly in the case of the oxides of carbon, sulfur, and nitrogen.

Both Thomson and Wollaston observed multiple proportions in 1808, Thomson working with the oxalates of potassium and strontium and Wollaston with the oxalates and sulfates of potassium. These men noted that two salts could be obtained, one containing twice the amount of the acid radical as the other did (acid and normal salts). Multiple proportions should have been evident in many of Proust's analyses, but he failed to detect the relationship. This failure has been attributed to the way he expressed his results, for many times he used percentages; this would hide any multiplicity of proportions. However, he frequently used "parts of B combined with 100 parts of A," which is exactly the form of expression which should bring out multiple proportions. Meldrum[13] is probably correct in thinking that Proust missed this relationship because he was not thinking in terms of an atomic framework that would suggest combination in multiple proportions.

Dalton's ideas about atomism appear to have developed rapidly after the autumn of 1803, the initial emphasis turning from the usefulness of the concept in explaining physical behavior of gases to its value in explaining chemical combination. Dalton once again revised his theory of mixed gases on the basis of his idea that the particles of different elements had different weights, for he concluded that the sizes of such particles must also differ. Hence, in discussing their repulsive force toward one another he abandoned the idea that unlike particles had no effect on one another in favor of a new concept—that an atom of a gas was the solid core of the element itself, plus the envelope of heat surrounding it. He says in a paper he wrote in 1805 that when he originally formulated his theory of mixed gases it had not occurred to him that the particle sizes of different gases would differ. Therefore he thought that particles of dissimilar gases would neither attract nor repel one another, though he realized that this would involve a different force for every kind of gaseous atom. Specialized forces (i.e., magnetism) were not unknown, but this system was far more complicated than one in which the repulsion between particles was due solely to heat. Accordingly Dalton revised his theory of mixed gases so it would take into account the difference in size of particles. He reasoned that as long as all the particles were the same size, the heat forces radiating out

---

[13] A. N. Meldrum, *op. cit.*, **55**, no. 6, p. 12 (1910), is of the opinion that Dalton's experiments with the reaction of air with nitric oxide under various conditions suggested the law of multiple proportions and the atomic theory. However, the reaction would show a 1 : 2 ratio so randomly that exact results would hardly be obtained by a poor experimenter like Dalton. See L. K. Nash, *op cit.*, p. 104; J. R. Partington, *A Short History of Chemistry*, Harper, New York 1960, p. 171.

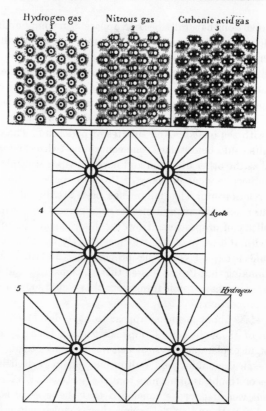

Fig. 4.6. Dalton's representation of atoms with their caloric envelopes. Note that the force lines radiating out from the nitrogen (azote) atoms meet uniformly at level 4. Thus, the nitrogen atoms will show a uniform repulsion for one another but this will not prevent layering. The caloric force lines of nitrogen and hydrogen do not meet at level 5 and this unbalance of forces will create an "intestine motion" which will serve to keep the gases mixed.
(From Dalton, *New System of Chemical Philosophy* [1808].)

from the solid core would be the same and the particles would form uniform layers like a pile of shot. But if they were unlike in size, there could never be a stability of forces and an "intestine motion" would be set up which would keep the gases mixed.

Dalton's ideas on the physical behavior of atoms never attained great influence, but the chemical aspects of his atomism permanently changed the nature of chemistry. In August, 1804, Dalton informed Thomson about his ideas on atoms. Thomas Thomson (1773–1852), a former student of Black, was then giving lectures and laboratory instruction in chemistry at Edinburgh. From 1817 on, he was professor of chemistry at the University of Glasgow. While still a student, Thomson wrote scientific articles for the *Encyclopædia Britannica* and in 1802 he published the first edition of his famous textbook, *System of Chemistry*. The third edition, published in 1807, contained an account of Dalton's theory of atoms. Thomson was highly enthusiastic about the theory and did a great deal in subsequent years to promote Dalton's ideas. Dalton himself called general public attention to the theory in his *New System of Chemical Philosophy*; the first volume was published in two parts in 1808 and 1810; the second volume appeared in 1827.

Application of the Atomic Theory

The major assumptions of Dalton's theory are well known. It held that all elements are made up of small indivisible particles called atoms. The atoms of a given element are all alike with respect to mass and properties but differ from the atoms of every other element. Chemical combination occurs when the atoms of two or more elements form a firm union. The particles resulting from such combinations were also called "atoms," thus creating confusion in chemical terminology.

A little-known assumption in Dalton's theory is referred to as the "rule of greatest simplicity." In estimating atomic weights, Dalton was confronted with certain grave difficulties. Since it is obviously impossible to weigh single atoms, any system of atomic weights must be formulated on a comparative basis. The atom of some element must be arbitrarily selected as a reference weight. Dalton chose the hydrogen atom and assigned *one* as its weight. The atomic weight of oxygen could then be found by either (1) comparing the weights of equal numbers of oxygen and hydrogen atoms or (2) finding by analysis the combining weights of oxygen and hydrogen in water. Dalton considered the first approach but rejected it. Since to him the atoms in a gas were analogous to a pile of shot, and since he believed that atoms of different gases varied in diameter, therefore equal volumes of different gases could not contain equal numbers of atoms. The second approach he considered valid. However, reflection will show that it is valid only when the ratio of atomic combination is known. (Chemical formulas were of course unknown at this time, because they are a natural outgrowth of atomic theory.)

Dalton overcame this difficulty to his own satisfaction by assuming that combination would always be of the simplest type. For example, when two elements formed only one compound, he considered that in most cases the compound was binary, that is, one atom of A to one atom of B. If two elements formed two compounds, he considered that the more common compound was binary and the second one ternary—composed of three atoms, A_2B or AB_2. If more than three compounds were involved, as with

Fig. 4.7. COMPARISON OF SIZES OF ATOMS AND THEIR CALORIC ENVELOPES. The sixteen figures represent the atoms of different elastic fluids, drawn in the centers of squares of different magnitude, so as to be proportionate to the diameters of the atoms as they have been determined. *Fig. 1* is the largest, and they gradually decrease to *Fig. 16*, namely, as under:—

1. Superfluate of silex	1.15		9. Oxymuriatic acid	0.981
2. Muriatic acid	1.12		10. Nitrous gas	0.980
3. Carbonic oxide	1.04		11. Sulphurous acid	0.95
4. Carbonic acid	1.00		12. Nitrous oxide	0.947
5. Sulphuretted hydrogen	1.00		13. Ammonia	0.909
6. Phosphuretted hydrogen	1.00		14. Olefiant gas	0.81
7. Hydrogen	1.00		15. Oxygen	0.794
8. Carburetted hydrogen	1.00		16. Azote	0.747

(From Dalton, *New System of Chemical Philosophy* [1808].)

the oxides of nitrogen, it was necessary to consider quaternary and higher combinations.

Since water was then the only known compound of hydrogen and oxygen, Dalton considered it binary (HO). Using Lavoisier's analysis of 85 per cent oxygen and 15 per cent hydrogen, Dalton calculated that the atomic weight of oxygen was 5.66. When more accurate analyses of water were made, he assigned oxygen an atomic weight of 7.

In working with the oxides of nitrogen, Dalton concluded that "nitrous gas," the lightest, was binary (NO). On the basis of density and analysis, he assigned "nitrous oxide" or laughing gas the ternary formula N_2O,

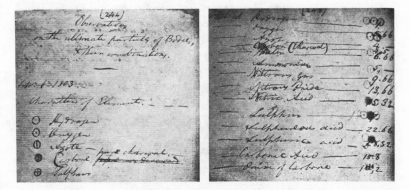

Fig. 4.8. SYMBOLS AND PARTICLE WEIGHTS FROM DALTON'S
NOTEBOOK.
(From H. E. Roscoe and A. Harden, *A New View of the Origin of
Dalton's Atomic Theory* [1896].)

and the brown oxide, termed "nitric acid," the ternary formula NO_2.
Using analytical values and the atomic weight for oxygen, he could now
determine the atomic weight of nitrogen. Similar reasoning in regard to
the oxides and hydrides of commonly known elements enabled him to
determine other atomic weight values.

BERZELIUS AND ATOMIC SYMBOLS

Early in his studies Dalton introduced the use of characteristic symbols
for the elements and compounds. The use of symbols was not unique with
him; on the contrary, they had been used empirically by the alchemists
and to some extent by such chemists as Bergman and Lavoisier. Those
proposed by Hassenfratz and Adet were appended to the book on nomenclature written by Lavoisier and his associates. However, until there was a
sound basis for representing a characteristic quantity of an element, these
symbols could be no more than a form of shorthand. The rise of the
atomic concept enabled the symbol to stand for more than the name of
the element.

Dalton's symbols were part of his atomic theory, along with his calculations of atomic weights. Quite in keeping with his idea of atoms, his
symbols were circles which he varied so they would be distinguishable, one
from another. An open circle stood for oxygen, a black circle for carbon;
circumscribed letters were frequently used for metals. "Atoms" of compounds were represented by the appropriate number of elemental atoms
shown in contact with one another.

Although Dalton and a few of his close followers continued to use them, the circular symbols never proved popular with chemists generally. They were cumbersome to write, especially when compounds were involved, and they created special problems in printing. Even so, Dalton's idea of having a symbol represent an atom was retained by Berzelius in the more practical system of symbols he devised, starting in 1811.

Fig. 4.9. FORMULAS OF SOME COMPOUNDS FROM DALTON'S NOTEBOOK.
(From H. E. Roscoe and A. Harden, *A New View of the Origin of Dalton's Atomic Theory* [1896].)

Jöns Jakob Berzelius (1779–1848), despite the misgivings of his teachers who found that his interest in science failed to carry over to other subjects in the medical curriculum, was destined to become one of the major figures in chemistry during the first half of the nineteenth century. On graduating from Uppsala he took a hospital job in Stockholm but devoted all his spare time to work in chemistry. He soon attracted the attention of the medical faculty at the university and in time became a professor. His earliest researches of importance dealt with electrochemical phenomena, but his major interest in analytical chemistry was apparent in 1810, when he began publishing his studies on combining proportions. He was an early adherent of the atomic philosophy and his most important work was done in connection with determining accurate atomic weights. Careful and painstaking in his experimental work, he often repeated an analysis many times before he was satisfied with his results.

In 1817 Berzelius discovered selenium; its name comes from *selene*, the Greek for moon. The element was isolated from sulfuric acid made in a

Fig. 4.10.   Jöns Jakob Berzelius.
(Courtesy of the University of Wisconsin.)

Swedish plant of which Berzelius and Gahn were part owners. It was first confused with tellurium but later was definitely identified as a new element. Berzelius reported that it had the properties of a metal, in addition to those of sulfur.

In 1822 Berzelius began publishing the famous *Jahres-Bericht*, an annual review which reported not only the results of his own researches, but those which came to his attention through scientific journals and correspondence. He was shrewd in judging scientific work and had rare insight into the significance of the observations made by others. Around 1820, as the result of his critical evaluation of new scientific work, he became the veritable lawgiver of chemistry. He was the author of a popular and comprehensive textbook published in 1808, which went through five editions and was translated into German and French. His manual on blowpipe analysis was standard for many years. His private laboratory was occasionally opened to selected students who sought instruction from him. Friedrich Wöhler and Eilhardt Mitscherlich were his most famous private students; Nils Gabriel Sefström, Johan August Arfwedson, Carl Gustav Mosander, Christian Gottlob Gmelin, Heinrich and Gustav Rose, and Gustav Magnus also studied under him. All except Arfwedson attained prominence as teachers of chemistry in Sweden and Germany. Arfwedson, the discoverer of lithium, became a Swedish industrialist.

Berzelius made several long trips to England, France, Switzerland, and Germany and became acquainted with the important scientists in those countries. His wide correspondence kept him in touch with scientific developments everywhere.

In 1813 Berzelius submitted to Thomson's *Annals of Philosophy* a long "Essay on the Cause of Chemical Proportions, and Some Circumstances Relating to Them, Together with a Short and Easy Method of Expressing Them." The essay was published in English in several issues.[14] The third part dealt with chemical symbols and their use in expressing chemical proportions. Berzelius argued that letters should be used for chemical symbols because they could be written more easily than other signs and did not "disfigure" the printed book. He suggested using the initial letter of the Latin name of each element because Latin was used more widely than any other languages for scientific terms. If two elements had names starting with the same letter, Berzelius suggested:

1. In the class which I call *metalloids* [non-metals], I shall employ the initial letter only, even when the letter is common to the metalloid and to some metal. 2. In the class of metals I shall distinguish those that have the same initial with another metal or a metalloid, by writing the first two letters of the word. 3. If the first two letters be common to two metals, I shall, in that case, add to the initial letter the first consonant which they have not in common: for example, S = sulphur, Si = silicium, St = stibium (antimony), Sn = Stannum (tin), C = carbonicum, Co = cobaltum (cobalt), Cu = cuprum (copper), O = oxygen, Os = osmium, &c.

The chemical sign expresses always one volume [sic] of the substance. When it is necessary to indicate several volumes, it is done by adding the number of volumes; for example, the *oxidum cuprosum* (protoxide of copper) is composed of a volume of oxygen and a volume of metal, therefore its sign is Cu + O. The *oxidum cupricum* (peroxide of copper) is composed of 1 volume of metal and 2 volumes of oxygen; therefore its sign is Cu + 2O. In like manner the sign for sulphuric acid is S + 3O, for carbonic acid, C + 2O, for water 2H + O, &c.

[14] J. J. Berzelius, *Ann. Phil.*, **3**, 51 (1814).

When we express a compound volume of the first order, we throw away the $+$, and place the number of volumes above the letter: for example, $CuO + \overset{3}{SO}$ = sulphate of copper, $\overset{2}{CuO} + 2\overset{3}{SO}$ persulphate of copper. These formulas have this advantage, that if we take away the oxygen we see at once the ratio between the combustible radicles. As to the volumes of the second order, it is but rarely of any advantage to express them by formulas as one volume; but if we wish to express them in that way, we may do it by using the parenthesis, as is done in algebraic formulas: for example, alum is composed of 3 volumes of sulphate of alumina and 1 volume of sulphate of potash. Its symbol is $3(\overset{2}{AlO} + 2\overset{3}{SO}) + (\overset{2}{Po} + 2\overset{3}{SO})$. As to the organic volumes, it is at present very uncertain how far figures can be successfully employed to express their composition. We shall have occasion only in the following pages to express the volume of ammonia. It is $6H + N + O$, or $\overset{6}{H}NO$.[15]

Nearly all the symbols suggested by Berzelius are in use today. A few have been changed, i.e., Ch to Cr, Tn to W, Cl (columbium) to Nb, Pl to Pt, Pa to Pd, Ma to Mn, Gl (glucinium) to Be, Ms to Mg, So to Na, Po to K, and M (muriatic radical) to Cl.

At first symbols were not very popular among chemists. As late as 1837, Dalton complained that "Berzelius' symbols are horrifying; a young student in chemistry might as well learn Hebrew as make himself acquainted with them. They appear like a chaos of atoms . . . to equally perplex the adept of Science, to discourage the learner as well as to cloud the beauty of the Atomic Theory." Failure to use symbols reflected the uncertainty regarding the theory upon which they were based.

Berzelius himself, after suggesting an excellent system of symbols, later clouded the issue by introducing arbitrary variants in it. In line with his idea that elements frequently occur in compounds in "double volumes," i.e., H_2O, he introduced a barred symbol to represent this. Thus, water became $\bar{H} + O$. Then he began using dots above the symbol to represent oxygen in the compound; water became $\dot{H}$, sulfuric acid became $\overset{\cdots}{S}$. Extending this, he used a comma for sulfur ($\overset{,}{\bar{H}}$ for H_2S), a $+$ for selenium ($\overset{+}{\bar{H}}$ for H_2Se), and a $-$ for tellurium ($\bar{H}$ for H_2Te).

A more practical development in the use of symbols was the introduction of superscripts to indicate the number of atoms, i.e., H^2O, Cu^2O, SO^3. This presented no typographical problem because they had long been used in mathematical publications and hence were already available in type. Confusion developed, however, when chemists began to place the superscript before the symbol to represent hydrogen atoms and after it

[15] *Ibid.*, pp. 51–52.

to represent oxygen atoms. Thus ammonia was symbolized ^{3}N and sulfuric acid S^3. Subscripts were also used to some extent.

By 1830 there was some use of symbols, but most chemists followed a more or less arbitrary system of their own devising. The result was chaos. Furthermore, there was a tendency to discredit the use of symbols, whereupon the newly founded British Association for the Advancement of Science decided to study the problem. Their report,[16] made in 1834, recommended that Berzelius' symbols for the elements be accepted and used consistently. This recommendation was slow to win acceptance, and consequently formulas were not widely used before the middle of the century. Mineralogists clung tenaciously to Berzelius' use of dots and commas for oxides and sulfides, although Berzelius himself abandoned this idea in his later years. The use of subscripts to indicate the number of atoms gradually became standard practice except in France, where superscripts have continued to be used for this purpose almost to the present time.

GAY-LUSSAC AND COMBINING VOLUMES

Strong support for Dalton's chemical atomism appeared in 1808 with the publication of Gay-Lussac's paper on volume relations in reactions of gases. Gay-Lussac made no attempt to interpret his results, but this was easily done in terms of combinations of atoms in fixed ratios. Dalton apparently never realized the significance of Gay-Lussac's law and always questioned its validity.

Joseph Louis Gay-Lussac (1778–1850) was educated at the École Polytechnique where he studied under Berthollet, Guyton de Morveau, and Fourcroy. Berthollet took him into his private laboratory at Arcueil where he spent a number of years on chemical and physical studies. In 1802 Gay-Lussac investigated the expansion of gases with increasing temperature. His determination of the coefficient of expansion (1.3750 vol./100° C. temp. change) was distinctly more accurate than Dalton's (1.3912), but still above the correct value, 1.3663. Both Gay-Lussac and Dalton believed that the coefficient of expansion was identical for all gases, a conclusion that Henri Victor Regnault (1810–1878) showed was wrong in the 1840's. Gay-Lussac worked on improving thermometers and barometers, and studied vapor pressure, hydrometry, and capillary effects during these years.

The great center of French science at this time was Arcueil, just south of Paris, where Berthollet had settled following his return from Napoleon's abortive Egyptian campaign. Laplace had a neighboring estate. These two

[16] Committee on Chemical Notation, Brit. Assoc. Adv. Sci., *Reports*, **5**, 207 (1836).

Fig. 4.11. JOSEPH LOUIS GAY-LUSSAC.
(Courtesy of the Edgar Fahs Smith Collection.)

great figures of French science attracted some of the most promising scientific figures of the day, who gathered regularly at Berthollet's home for scientific discussions and good fellowship, just as the scientific great of an earlier day had met at Lavoisier's laboratory at the Arsenal. In 1805 Napoleon approved the formal organization of the Société d'Arcueil. Besides Berthollet and Laplace, its members included Arago, Bérard, Biot, Amédée Berthollet (Claude's son), Chaptal, Collet-Descostils, de Candolle, Dulong, Gay-Lussac, Humboldt, Malus, Poisson, and Thenard. The Société was something of a miniature Académie des Sciences and was the center of French science while France clearly was the world leader in the field. Following the fall of Napoleon the Société was abandoned, but not until publication of the third volume of the *Mémoires de la Société d'Arcueil,* which included its members' important contributions to chemistry and physics.

The second volume of the *Mémoires,* published in 1808, contained Gay-Lussac's paper on combining volumes of gases. He had been led into studying this subject by some faulty analyses of the oxygen content of air published by the German naturalist Alexander von Humboldt. Gay-Lussac's criticism of these analyses caused Humboldt to seek him out and together they began efforts to devise a reliable test for gaseous oxygen, based upon the shrinkage in volume when air was exploded in an excess of hydrogen. By passing a spark through the mixed gases in a eudiometer they found that a volume shrinkage occurred which was equivalent to three-fifths of the volume of air in the mixture. Cavendish and Priestley had earlier performed similar experiments which apparently indicated that a volume of hydrogen combined with half a volume of oxygen. In following up their work, Gay-Lussac found that the combining ratio of hydrogen and oxygen was extremely close to 2 volumes to 1.

In his studies of the expansion of gases with temperature, Gay-Lussac found that all gases appeared to expand equally; hence he wondered if there was a whole-number relationship in the combining volumes of all gases. Experiments revealed that 2 volumes of carbon monoxide combined with 1 volume of oxygen and formed 2 volumes of carbon dioxide. He also discovered that single volumes of ammonia combined with single volumes of muriatic acid (HCl), fluoboric acid (BF_3), and carbonic acid (CO_2) to form salts. He found further that single volumes of the last two combined with 2 volumes of ammonia, forming a second series of salts. Although how these gaseous reactions were carried out is not clearly described, Gay-Lussac remarked in one place that it (the salts) can be obtained only through the intervention of water.[17]

Calculations from the work of others revealed that 1 volume of nitrogen and 1 volume of oxygen were present in 2 volumes of nitric oxide. Experiments done by Amédée Berthollet demonstrated that ammonia contained nitrogen and hydrogen in a ratio of 1:3 by volume. Using Davy's analyses of the oxides of nitrogen, Gay-Lussac showed that the volume ratios of nitrogen and oxygen varied little from 2:1, 1:1, and 1:2. Working with other experimental results, Gay-Lussac consistently found that volume ratios in gaseous reactions are whole numbers.

Near the end of his paper, Gay-Lussac referred to Dalton's theory and remarked that his own studies supported it. However, he was closely enough associated with Berthollet to add that the idea about variable proportions had many facts in its favor, and that the two might be reconciled. In concluding, he reiterated his observations that volumes of gaseous substances reacted with one another in very simple ratios such as 1 volume with 1, 2, or 3 volumes. Regular ratios such as these were not observed in liquids and solids, nor was there such a relation with respect to weights.

The law of combining volumes was well received in some quarters but not in others. Berzelius used it in determining such formulas as H_2O for water and NH_3 for ammonia. Dalton was so vigorously opposed to it that he never admitted its validity. He recalculated some of Gay-Lussac's ratios and accused the French chemist of rounding off figures when it suited his purpose. It is true that Gay-Lussac did so on occasion, but he made no effort to hide the fact. His paper mentioned cases in which the volume ratios are close to whole numbers and suggested that better analyses and density values would definitely show that they are whole numbers.

[17] J. L. Gay-Lussac, *Mém. Soc. Arcueil*, **2**, 207 (1808), or Alembic Club Reprint No. 4, p. 11.

AVOGADRO'S HYPOTHESIS

Gay-Lussac's law of combining volumes suggested rather clearly that equal volumes of different gases, under similar conditions of temperature and pressure, contain the same number of reactive particles. Although Gay-Lussac, doubtless out of respect for Berthollet, failed to emphasize this idea, it could not be disregarded. If 1 volume of ammonia gas combines with exactly 1 volume of muriatic acid gas to form a salt, it is natural to conclude that each volume must contain an equal number of ammonia and of muriatic acid particles.

Dalton had at one time examined the idea that equal volumes of different gases contained equal numbers of particles, but he now questioned it. For example, he knew that the density of carbonic oxide (CO) was less than that of oxygen. But if the concept were correct, a gas whose particles consisted of atoms of carbon and oxygen should be denser than a gas whose particles consisted solely of oxygen atoms. Dalton also knew that the density of steam was less than that of oxygen, and that the density of ammonia was less than that of nitrogen. Hence, he rejected Gay-Lussac's law, feeling quite naturally that it must be invalid. He was also of the opinion that gases resembled piles of shot, the caloric envelopes of the particles of gas being in physical contact. This led him to reason that particles of a compound gas would occupy a greater volume than particles of an elemental gas; hence the concept of equal volumes containing equal numbers of particles could not be valid.

Another cause of trouble in accepting this concept lay in Gay-Lussac's observations themselves. For example, 2 volumes of carbonic oxide and 1 volume of oxygen combine to form 2 volumes of carbonic acid (CO_2).

Carbonic oxide + Oxygen → Carbonic acid

2 vol.   1 vol.   2 vol.

Now if each volume of gas contains n particles of carbonic oxide, oxygen, or carbonic acid, we would expect that 1 volume of carbonic oxide would unite with 1 volume of oxygen to form 1 volume of carbonic acid. If we reflect for a moment, we will see there is a discrepancy, in many gaseous reactions, between the observed and the expected ratio of volumes if equal volumes contain equal numbers of particles. Thus,

2 vol. of hydrogen + 1 vol. of oxygen → 2 vol. of steam (observed)

2 vol. of hydrogen + 1 vol. of oxygen → 1 vol. of steam (expected)

and

3 vol. of hydrogen + 1 vol. of nitrogen → 2 vol. of ammonia (observed)

3 vol. of hydrogen + 1 vol. of nitrogen → 1 vol. of ammonia (expected)

Obviously, 1 volume of oxygen or nitrogen, containing n particles, can produce no more than n particles of H_2O or NH_3 and these n particles will occupy a single volume. But 2 volumes are seen to be produced; hence the validity of the concept is open to suspicion.

It was the Italian physicist Amedeo Avogadro (1776–1856) who made clear the desirability of adopting the hypothesis that equal volumes of gases contain the same number of particles (molecules). He further proposed an assumption regarding the molecules of elemental gases which would solve the problem Dalton raised regarding the comparative densities of gases, and also the problem discussed in the preceding paragraph.

Fig. 4.12. AMEDEO AVOGADRO.
(Courtesy of the Edgar Fahs Smith Collection.)

Avogadro's paper of 1811,[18] based upon Gay-Lussac's law and Dalton's atomic theory, reconciled the two problems. Starting with the assumption that equal volumes of all gases contain equal numbers of molecules under similar conditions, Avogadro proceeded to analyze the facts of gaseous combination. Using the examples discussed by Gay-Lussac, he showed that the ambiguities disappeared if he assumed that the molecules involved in typical reactions might split into "half-molecules"; that is, he supposed the existence of molecules of elemental gases which contained more than a single atom. He did not use the term "atom", but always used the term "half-molecule" as its equivalent. Avogadro also mentioned several types of molecules. The *molécule intégrante* referred to molecules in general, but most often to molecules of compounds; the *molécule constituante* referred to molecules of elementary gases such as oxygen, hydrogen, etc.; the *molécule élémentaire* was used for atoms of elements.

[18] A. Avogadro, *J. physique*, **73**, 58 (1811), or Alembic Club Reprint No. 4.

Avogadro showed, basing his argument on combining ratios and gas densities, how the density of water vapor could be less than that of oxygen. If the molecules in 1 volume of oxygen split into "half-molecules" (atoms) during combination with 2 volumes of hydrogen, one would expect 2 water molecules to be formed for every molecule in the original oxygen. If equal numbers of molecules occupied equal volumes, the molecules of water vapor must occupy twice the volume of the oxygen gas. Hence the density of water vapor was naturally less than that of oxygen.

Avogadro similarly explained the combination of the elements in ammonia, in the oxides of nitrogen, and in hydrochloric acid, and the reactions of the oxides of carbon and sulfur. He saw how the theory of combining volumes should be applied in determining formulas; and he pointed out that the weight of oxygen atoms must be 15, rather than 7.5 as suggested by Dalton, because each "half-molecule" of oxygen was combined with two "half-molecules" of hydrogen. Similar reasoning about the composition of nitrous gas (NO) and of ammonia led him to reject Dalton's value for the atomic weight of nitrogen and to arrive at 14.156 instead. So also he computed atomic weights for sulfur and chlorine which are of a correct order of magnitude. Since his calculations were based upon the values for densities and compositions then current, his results are only as accurate as these values; but even so, they indicated an approach to the problems then facing chemists.

Despite the soundness of Avogadro's reasoning, his hypothesis was generally rejected or ignored. Dalton could never bring himself to appreciate its significance because he refused to accept the validity of Gay-Lussac's law. Although he readily accepted and used Gay-Lussac's law, Berzelius objected to the concept of diatomic molecules of elements because of his dualistic theory. He accepted the assumption that equal volumes of elemental gases contain equal numbers of molecules (single atoms), but refused to extend it to compound gases. In this way he avoided having to explain why 2 volumes of compound gas were so frequently obtained from 1 volume of elemental gas.

Other chemists apparently took little notice of Avogadro's paper. It has frequently been charged that the paper was poorly written and filled with ambiguous terminology and complex mathematics, but examination of it makes this hard to accept. The paper showed clear reasoning, and its author's use of mathematics in dealing with the subject was simple and convincing. The terminology was new, but there was no ambiguity for any one who was willing to read with discernment. It seems more likely that the paper was ignored because the chemical world was not ready for this kind of thought. Even Dalton's concept of atoms was rejected by many chemists; for example, prominent chemists like Davy used equivalent weights without accepting atoms.

Avogadro's hypothesis was revived in 1814 by André Marie Ampère

(1775–1836), the famous student of electrical phenomena,[19] but he was no more successful than Avogadro in winning converts to it. After this, the hypothesis received almost no attention. In the meantime the future of the atomic theory also became uncertain. The determination of atomic weights, although worked on diligently by Berzelius and others, presented rather formidable problems of interpretation which were not solved for many years. Thus, it is not surprising that many chemists were content to use equivalents as a working tool but did not accept the atomic theory.

There is a common misconception that Avogadro's hypothesis resulted from the physical properties of gases, that is, the tendency of all gases to follow Boyle's and Charles' laws. Although it is true that the hypothesis follows logically from the gas laws, there is no evidence that Avogadro was influenced by the pressure and temperature behavior of gases. But reading his paper shows clearly that he based his thinking entirely upon Gay-Lussac's work on the volume relations of gases during chemical combination, and that Avogadro was searching for an explanation of the volume ratios Gay-Lussac observed.

CONCLUSION

During the period when Lavoisier's views were gaining acceptance a major interest in quantitative analysis began to be evident. Klaproth's fundamental contributions to analytical operations were particularly important; and his work, together with that of Wollaston and Vauquelin, led to the discovery of a number of new metallic elements. Proust's controversy with Berthollet results in the law of definite proportions becoming a working concept in chemistry, but at the same time Berthollet's ideas on mass action failed to achieve their proper place.

The atomic theory was a natural development of the studies on gases and the analytical work being carried on at the turn of the century. The theory was of unique value in explaining weight and volume relations in chemical combinations but it created problems of its own in connection with atomic weight determinations because formulas were highly uncertain. Such relationships as comparative densities and Gay-Lussac's law of combining volumes could have been helpful but they were frequently misunderstood and were therefore rejected.

Avogadro and Ampère showed how volume relations in gaseous reactions could be interpreted and correlated with Dalton's atomic theory, but their papers failed to carry conviction among chemical leaders like Berzelius, Dalton, and Davy.

[19] A. M. Ampère, *Ann. chim. phys.*, **90**, 45 (1814).

Berzelius recognized the importance of the atomic theory more than anyone else did. His analytical studies of combining weights of the elements were unusually sound and he was reasonably successful in deriving usable atomic weights from them. His soundly conceived system of symbols is the basis of modern symbolism in chemistry. Although he was becoming the great authority on chemical matters, at the same time he was frequently confused and confusing. Not for several decades was the atomic theory sufficiently clarified to become the unifying concept in chemistry.

CHAPTER 5

ELECTROCHEMISTRY
AND THE DUALISTIC
THEORY

THE DISCOVERY OF CHEMICAL ELECTRICITY

Following the publication of Gilbert's *De magnete* in 1600, electricity received a good deal of attention from scientific investigators. Frequently the interest lay in a naive and childish demonstration of the spectacular, but there was also a slow but steady series of significant discoveries. Gilbert had already developed a primitive electroscope for detecting whether a body carried a charge. Around the middle of that century, von Guericke devised a form of static machine upon a rotating sphere of sulphur; and this machine, after a series of improvements, stimulated electrical researches during the next century. Stephen Gray succeeded in gaining an understanding of conductivity. The Leyden jar was invented around 1745 independently by a Pomeranian clergyman, E. G. von Kleist, and by Pieter van Musschenbrock, a Dutch mathematician and physicist. In the meantime Charles François du Fay had distinguished two kinds of electricity, which he called vitreous and resinous, terms for which Benjamin Franklin substituted positive and negative. Franklin made extensive studies with the Leyden jar and formulated the one-fluid theory; it held that positive and negative charges were due to a surplus or deficiency of electrical fluid in the charged body. In 1784 Charles Coulomb carried out experiments with the torsion balance which enabled the effect of charged bodies on one another to be understood.

All the studies up to this time were on static electricity. Although successive improvements were made in devices for generating it, and for storing and using it, experimenters were limited by the characteristics of static charges. Current electricity was discovered at the end of the eighteenth century as the outgrowth of researches by two Italian investigators, the physiologist Galvani and the physicist Volta.

Luigi Galvani (1737–1798) noticed about 1780 that a freshly dissected frog's leg, grounded through a piece of metal, started to twitch whenever an electrical discharge occurred in the vicinity. It was well known that the muscles of dead animals twitched when they received electrical discharges from a static machine, a Leyden jar, or a torpedo (electric eel), but Galvani was surprised that the frog's leg twitched without being in contact with the source of electricity. He undertook a long series of experiments to investigate the phenomenon, and published the results in 1791.[1]

During these experiments it became apparent that the twitching might be caused by contact of the nerves and muscles with a circuit composed of dissimilar metals. Galvani concluded that the phenomenon was evidence of electricity in the animal organism and looked upon the animal as a storage vessel for electricity, resembling a Leyden jar. His ideas aroused a great deal of interest and were widely accepted, but were attacked by Volta, who held that animal electricity was not involved in any way.

Alessandro Volta (1745–1827) was professor of physics at Pavia. He had first been interested in gases, but by the time Galvani's work was published he had made important studies of the electrophorus and condenser, and had devised a straw electrometer for measuring small quantities of electricity. Following publication of Galvani's experiments Volta began to extend his studies. At first he accepted the idea of animal electricity, but soon found it necessary to modify it in view of his work with dissimilar metals. He found that if two dissimilar metals touched the tongue and were brought into contact, a bitter taste resulted; when they touched the eye, contact between them created a sensation of light. Such findings soon convinced him that the metals not only served as conductors but generated electricity when brought into contact.

Volta's next experiments abandoned the physiological system in favor of sensitive condensers and electrometers. By 1794 he reported in a letter to Joseph Banks of the Royal Society that the electrical effects became more vigorous the farther apart the conducting substances were in the series: zinc, tin, lead, iron, copper, platinum, gold, silver, graphite, charcoal. He showed that momentary contact between two metals left

[1] L. Galvani, *De Bononiensi Scientiarum et Artium Instituto atque Academia Commentarii*, **7**, 363 (1791). A German translation appears in Ostwald's *Klassiker der exakten Naturwissenschaften*, no. 52. There is a partial translation into English in William F. Magie, *A Source Book of Physics*, McGraw-Hill, New York, 1935.

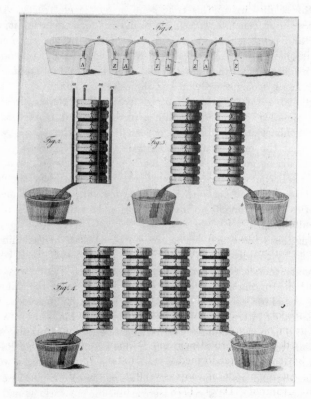

Fig. 5.1. DIAGRAMS OF VOLTA'S APPARATUS FOR THE
PRODUCTION OF ELECTRIC CURRENT.
(From *Phil. Trans. Royal Soc.* [1800].)

them oppositely charged; for example, disks of zinc and copper, after
contact, had respectively positive and negative charges.

Further studies demonstrated that an electrical force was generated
when a metal was in contact with a fluid. Volta found that a current
was produced when two dissimilar metal disks such as silver and zinc
were separated by a moist conductor—blotting paper soaked in brine,
for example—and brought in contact by a wire. He also found that when
he made a "pile" of these units—first dissimilar metal disks in contact
with each other, then a piece of felt or paper soaked in brine, then disks,
then felt or paper, and so on—the current was intensified so that when a
person touched the top disk with one hand and placed his other hand in
a dish of brine that was connected to the bottom metal disk by a strip
of metal, he experienced a feeble and continuous shock. Volta reported

that the pile had the properties of the Leyden jar to a very weak degree but that it offered an advantage because it did not need to be charged from the outside.[2]

Another arrangement for producing electricity utilized the *couronne de tasses* (crown of glasses). Here dissimilar metals were soldered together and placed in conducting solutions of salt or acid in glass vessels. Volta observed that the strength of the current varied with the concentration of the salt solution, but he apparently failed to notice any chemical processes associated with the flow of current.

INVESTIGATIONS OF CHEMISTRY AND ELECTRICITY

The voltaic pile attracted great interest among scientists, especially in England and France. Anthony Carlisle (1768–1840), surgeon and professor of anatomy in London, immediately constructed a pile from half crowns, zinc disks, and pasteboard soaked in salt water. During experiments with it, he and William Nicholson (1753–1815) observed that bubbles were released in water that was in contact with one of the metal disks. Subsequent experiments resulted in the electrolysis of water, with hydrogen and oxygen observed as decomposition products.[3] Water had been decomposed by van Troostwyk and Deiman in 1789, using static electricity, but the gas thus formed was a mixture.

The most important of the early electrochemical investigations were done by Humphry Davy (1778–1829) during the first decade of the nineteenth century; his extensive experiments had great practical and theoretical significance. Davy was the son of a farmer in Penzance, Cornwall. On his father's death in 1794 he became assistant to a local physician but was discharged because of his liking for explosive experiments. However, his interest in experimentation brought him to the attention of Thomas Beddoes, a Bristol physician who founded the Pneumatic Institution for the purpose of investigating the physiological effects of the numerous gases which Priestley and others had discovered. Davy was placed in charge of the gas studies in 1798.

He set about preparing and inhaling gases, sometimes to the detriment of his own health. He soon discovered that nitrous oxide (laughing gas) induced a giddy intoxication. Although this discovery was of no practical use at the time—chemical anesthesia was not developed until the 1840's—inhaling laughing gas became a popular fad. Davy and others presided

[2] A. Volta, *Phil. Trans. Royal Soc.*, **90**, 430 (1800). A facsimile reprint is available in *Isis*, **15**, 129 (1931).

[3] W. Nicholson and A. Carlisle, in Nicholson's *J. Natl. Phil. Chem. Arts*, **4**, 179 (1800).

over public meetings in which volunteers inhaled the gas for the amusement of the audience. These meetings were sufficiently common to attract the attention of such caricaturists as Cruikshank and Gillray. Davy proved to be popular as a lecturer and quickly attracted a large following among the public; this may have been a factor in his being selected for a professorship at the Royal Institution in 1801.

Fig. 5.2. PNEUMATIC EXPERIMENTS AT THE ROYAL INSTITUTION (a caricature by James Gillray). The lecturer is supposed to be Thomas Garrett, first professor of chemistry. The assistant holding the bellows is Davy. Count Rumford is the man standing nearest the door at the right.
(Courtesy of the British Museum.)

The Royal Institution was founded in 1800 by Count Rumford, who persuaded influential London friends to help him establish an institution where lectures could be given on the latest developments in science and the technological arts in order to assist workingmen to improve their lot. Benjamin Thompson (1753–1814), who became Count Rumford when knighted by the Elector of Bavaria in 1784, was born in Massachusetts.

Because his sympathies lay with the Royalists during the Revolutionary War, life in his native New England was sufficiently unfriendly so that he left the colonies; he never returned. He spent a great deal of time in Bavaria, where he was director of munitions manufacture, but he also had important connections in London and Paris. He was familiar with the Conservatoire Nationale des Arts et Métiers which the French government had founded in 1794 as a mechanic's institute or Sorbonne

Fig. 5.3. HUMPHRY DAVY.
(Courtesy of the Edgar Fahs Smith Collection.)

industrielle. Rumford was frequently active in schemes for curbing drunkenness and promoting industry among the poor.

King George III chartered the Royal Institution in 1800, but unlike the Paris Conservatoire it received no government aid. Rumford soon quarreled with his associates and left England, whereupon the Institution developed along entirely different lines than had been intended. Davy as professor there planned his lectures to appeal to wealthy patrons, and his researches to attract groups interested in practical objectives; the Institution soon had sufficient income to make it financially solvent.

Among Davy's first researches were studies of new materials suitable for tanning, though he had no real interest in that trade, and of problems that related chemistry to agriculture. These latter, conducted between 1802 and 1812, were undertaken at the request of Arthur Young, Secretary of the Board of Agriculture, and resulted in the publication of Davy's *Elements of Agricultural Chemistry* in 1813. In 1816 he developed the safety lamp at the request of the Society for the Study and Prevention of Mine Explosions. The safety lamp represented the application of the principles learned in his study of flames—that heat is dissipated rapidly by wire gauze, and that explosions of firedamp (methane) in mines can be prevented by enclosing the flame of an oil lamp inside a wire gauze shield.

Along with these more practical projects Davy found time to carry on extensive studies in electrochemistry. These investigations, begun soon after the introduction of the voltaic cell, had far-reaching theoretical and practical implications. In the Bakerian Lecture delivered before the Royal Society in 1806, Davy summarized the results of his researches. He pointed out that hydrogen, metals, metallic oxides, or alkalies could be observed around the negative pole during electrolysis, and oxygen or acids around the positive. In careful experiments on the electrolysis of distilled water he demonstrated that the alkali in the solution came from the glass vessels, the nitric acid from the nitrogen of the air. When distilled water was electrolyzed in gypsum cups, lime formed around the negative pole, sulfuric acid around the positive. He showed that an electric current decomposes many minerals into an earthy or alkaline base and an acid.

It appeared to Davy that since the electric current overcame the normal chemical affinity which held elements together in compounds, this affinity must be electrical in nature. If this reasoning were valid, it should be possible to decompose such substances as the alkalies, which were believed to be compounds but had resisted every effort to break them down into simpler substances.

In 1807 Davy succeeded in decomposing both potassium hydroxide and sodium hydroxide and obtaining the corresponding metals. He had batteries available that were made up of many large zinc and copper plates submerged in solutions of alum and nitrous acid. When a current was passed through alkali solutions, only water decomposed; but when the current was applied to the dry fused hydroxides, he obtained different results. When he laid a piece of potassium hydroxide on a platinum disk connected to the negative pole of a 100-plate battery, and connected the positive pole and the hydroxide with a platinum wire, he obtained metallic globules of potassium. Dry, solid potassium hydroxide failed to conduct; but once the surface became slightly moist, fusion occurred at the points of contact with the platinum. Violet effervescence began to appear at the positive wire; at the negative plate "small globules having a high metallic luster, and being precisely similar in visible characters to quicksilver, appeared, some of which burnt with explosion and bright flames, as soon as they were formed, and others remained and were merely tarnished, and finally covered by a white film which formed on their surfaces."[4]

Sodium metal was discovered in a similar manner a few days after potassium was isolated. The properties of both metals were studied extensively in the days that followed and the Royal Society was informed of the results in the Bakerian Lecture for 1807, delivered on November 19. By this time Davy had established that gaseous oxygen was combined with the metal in alkalies and that the metals could be preserved in naphtha.

The next year Davy continued his electrical researches in an effort to isolate metals from the alkaline earths. At first he was unsuccessful, but he finally developed a technique, based on work done by Berzelius and Pontin, in which the alkaline earth element was obtained as an amalgam that could be distilled to recover the metal. This procedure enabled him to isolate magnesium, calcium, strontium, and barium. Although his efforts to reduce alumina, beryllia, and zirconia failed, he did produce an amalgam of ammonia. Here again he followed up a lead supplied by Berzelius and Pontin, who had observed that when mercury in contact with a solution of ammonia was negatively charged it expanded and solidified. Davy concluded that a metallic substance, which he named "ammonium," was associated with the mercury in the amalgam; however, he never succeeded in isolating it. This created considerable confusion in his thinking and was in part responsible for his belief that hydrogen was a component of numerous substances which were considered elemental.

Electrochemistry was studied enthusiastically in other laboratories, particularly in that of Berzelius in Sweden. Working on electrolysis of salts with his friend Wilhelm Hisinger (1766–1852) in 1803, Berzelius observed that acids formed around the positive pole and bases around the negative pole. This led him to form ideas similar to Davy's and resulted ultimately in his electrochemical or dualistic theory which will be discussed in the next section.

The mechanism of electrolysis attracted attention as soon as the dissimilar nature of the electrode reactions became evident. Why did hydrogen and alkali form at the negative pole, and oxygen and acid at the positive? Even when the poles were separated from each other the migration of acid and base was undeniable.

Christian J. D. von Grotthuss (1785–1822) evolved a rather popular explanation in 1805 which was based on the alternate decomposition and

[4] H. Davy, *Phil. Trans. Royal Soc.*, **98**, 333 (1808); or *Alembic Club Reprint* No. 6, p. 8.

recombination of particles in the electrolyte.[5] In the electrolysis of water, for example, the negative pole attracted a particle of hydrogen from an adjacent water particle. The isolated oxygen particle then robbed an adjacent water particle of its hydrogen, and the free oxygen in turn took hydrogen from the next water particle. This continued until the last oxygen particle on the other side of the cell was liberated at the positive pole. This theory was accepted for a long period of time.

BERZELIUS AND THE DUALISTIC THEORY

The dualistic theory was formulated by Berzelius in an elaborate attempt to systematize chemical behavior around relationships revealed in electrochemical studies. The germ of the theory lies in Davy's early work as well as in his own, but it was Berzelius who painstakingly developed it as the basis for understanding chemical combination. The theory was advanced in tentative form in 1812,[6] though Berzelius' thinking had been moving in this direction for several years. He presented it with extensive conclusions in his *Versuch über die Theorie der chemischem Proportionen und über die chemischen Wirkunge der Eliktrizitat.*[7] Here he clearly developed his theory from experimental evidence and extended it to the whole field of chemistry.

In his dualistic theory Berzelius followed Davy in assuming that chemical and electrical attraction were essentially identical. Atoms were held to be electrically charged because compounds were decomposed by an electric current and the released elements formed at one or the other pole. Atoms were further believed to show polarity, with positive or negative polarity predominating in the different elements. Thus, the atoms of certain elements were electropositive, others were electronegative. Oxygen was rated the most electronegative of all the elements. Metals were generally electropositive.

Chemical combination resulted from mutual neutralization of opposite charges. However, the compound thus formed was not necessarily neutral because the opposite charges of the combining atoms were not necessarily equal in magnitude. Furthermore, since atoms themselves exhibited polarity they might be negative toward one element and positive toward another. For example, sulfur was considered negative toward hydrogen and metals, and positive toward oxygen.

[5] C. J. D. von Grotthuss, *Ann. chim. phys.*, **58**, 54 (1806).
[6] J. J. Berzelius, in Schweigger's *J. Chem. Phys.*, **6**, 119 (1812).
[7] J. J. Berzelius, *Versuch über die Theorie der chemischen Proportionen und über die chemischen Wirkungen der Eliktrizitat*, Arnold, Dresden, 1820. A Swedish edition had been published in 1814, a French translation in 1819.

Some examples of combinations are the following:

$$\overset{+}{Cu} + \overset{-}{O} \rightarrow CuO \text{ (slightly positive)}$$

$$\overset{+}{S} + 3\overset{-}{O} \rightarrow SO_3 \text{ (slightly negative)}$$

Since the oxides formed are not neutral, they can combine with each other:

$$\overset{+}{CuO} + \overset{-}{SO_3} \rightarrow CuO \cdot SO_3 \text{ (modern, } CuSO_4)$$

Even a salt like the above was not necessarily neutral and could be expected to combine further.

$$\overset{+}{CuO \cdot SO_3} + 5\overset{-}{H_2O} \rightarrow CuO \cdot SO_3 \cdot 5H_2O$$

In the case of alum there was first a combination of potassium, aluminum, sulfur, and hydrogen to form oxides. The oxides of potassium and sulfur then formed potassium sulfate; those of aluminum and sulfur formed aluminum sulfate. These two sulfates, together with water, united, forming crystalline alum. In Berzelius notation:[8]

$$\overset{\cdot}{K}\overset{\cdots}{S} + \overset{\cdots}{Al}3\overset{\cdots}{S} + 24\overset{\cdot}{H}$$

The dualistic theory had an important vogue in chemistry. For a time it apparently explained composition successfully, but it soon outlived its usefulness. With the advent of organic chemistry in the 1830's and 1840's the dualistic theory actually became a handicap in understanding organic compounds. Some scientists felt that mineral chemistry was unique as far as dualistic precepts were concerned and that it should be excluded from the obviously confusing chemistry of organic compounds. Dualistic thinking continued to be popular among mineralogists and analysts, and even today vestiges of it and of Berzelius' dualistic notation remain. For example, in quantitative analysis, elements are frequently calculated as the oxide rather than as the element. This is particularly true in mineral analysis and in commercial practices. American farmers buy fertilizer on the basis of its K_2O and P_2O_5 content, rather than on the basis of its potassium and phosphorus content.

FARADAY AND THE ELECTROCHEMICAL LAWS

It has frequently been said that the greatest of all of Davy's discoveries was Faraday. Michael Faraday (1791–1867) was born of humble parents in Newington Butts, Surrey. At the age of thirteen he was apprenticed to a bookbinder in London; apparently he became a reader of books as

[8] J. J. Berzelius, *Jahres-Bericht*, 7, 78 (1928; for the year 1826).

Fig. 5.4. MICHAEL
FARADAY.
(Courtesy of the J. H. Walton
Collection.)

Fig. 5.5. FARADAY AT WORK IN THE ROYAL
INSTITUTION LABORATORY.
(Courtesy of the Edgar Fahs Smith
Collection.)

well as a binder. Among the works brought to the shop for binding were
Marcet's *Conversations in Chemistry* and the sections on electricity in
the *Encyclopædia Britannica*. Jane Marcet (1769–1858), the wife of a
London physician, had prepared a small book on chemistry for the instruction of a niece. The conversational style and elementary approach to the
subject proved so popular that the work was used as a textbook; a number
of editions were printed not only in England, but in America as well.

Young Faraday, his appetite for science thus whetted, was soon attending weekly lectures of the City Philosophical Society and carrying out
such experiments in chemistry and electricity as his meager income
permitted. In the spring of 1812 one of the patrons of the shop where he
worked took him to hear several of Davy's lectures at the Royal Instition.
As a result of his great interest, he wrote up and illustrated the lectures
and sent them to Davy with a plea to be employed as his assistant. The
request was finally granted and Faraday's association with Davy began
in March, 1813.

That autumn Sir Humphry and Lady Davy began an extended tour
of France and Italy; Faraday accompanied them as secretary and scientific
assistant. Although Lady Davy's inclination to treat him as a personal

servant made the trip unpleasant, it was not totally without reward, for Sir Humphry was then at the height of his scientific career. A storehouse of scientific information, he was welcomed in Napoleonic France when it was at war with England. He was likewise welcomed in the foremost laboratories where he frequently stayed to perform an experiment. Thus, Faraday not only drew upon his master's scientific knowledge but came in contact with such French and Italian scientists as Ampère, Chevreul, Gay-Lussac, and Volta.

On returning to London, Faraday was reemployed at the Royal Institution. Davy had given up lecturing in 1812 because of poor health and was made an honorary professor. William T. Brande became professor of chemistry. Faraday's responsibilities increased as his scientific acumen became manifest. He was made director of the laboratory in 1825 and was later appointed to the Fullerian Professorship of Chemistry, a special research chair created for him in 1833.

Faraday's researches included both chemistry and physics. From the first he investigated the mysteries of electricity, returning to this subject repeatedly after work in other fields. His most productive years were those between 1820 and 1835. During this period he studied alloys of steel with J. Stodard, a maker of scientific instruments, but the results were not definitive.[9]   As a member, with John Herschel and John Dollond, of a Royal Society committee on optical glass, Faraday prepared and studied very dense glass. Although these studies led to no improvements in telescopes, he found them useful twenty years later in his work on magneto-optical rotation.

In 1823 he succeeded in liquefying chlorine by a combination of pressure and cooling. Davy liquefied hydrogen chloride gas by an improved technique which Faraday used successfully in liquefying sulfur dioxide, hydrogen sulfide, carbon dioxide, nitrous oxide, euchlorine ($Cl_2 + ClO_2$), cyanogen, and ammonia. In 1845 he returned to these experiments and liquefied arsine, hydrogen iodide, and hydrogen bromide. In the meantime Thilorier substituted metal cylinders for glass tubes in the apparatus and liquefied carbon dioxide in quantity; he obtained it in the solid state.[10] Faraday failed to liquefy the six permanent gases—O_2, N_2, H_2, CH_4, CO, NO—even though he used solid carbon dioxide and ether to obtain low temperatures.[11]

---

[9] Faraday's samples of alloys were found in a box many years later. Sir Robert A. Hadfield, a leading British metallurgist at the turn of the century, had the samples analyzed and reported that Faraday had discovered stainless steel. This was not rediscovered as a commercial alloy until the end of the nineteenth century. See Robert A. Hadfield, *Faraday and His Metallurgical Researches*, Chapman and Hall, London, 1931.

[10] C. S. A. Thilorier, *Compt. rend.*, **1**, 194 (1835); *Ann. chim. phys.*, **60**, 432 (1835).

[11] M. Faraday, *Phil. Trans. Royal Soc.*, **135**, 155 (1845), or his *Experimental Researches on Chemistry and Physics*, Taylor & Francis, London, 1859, p. 96.

The experiments on electromagnetic induction were resumed from time to time. Following Oersted's discovery that a moving current set up a magnetic field, Faraday succeeded on Christmas Day, 1821, in producing mechanical motion by means of a permanent magnet and an electric current—an ancestor, so to speak, of the electric motor. In his experiments Faraday was driven by his belief in a uniformity in nature which would permit the creation not only of magnetism by electricity, but of electricity by magnetism. He did not succeed in this for another decade. His work was dominated by his attempt to create electric currents with magnets, but he ignored the place of relative motion in the scheme of things.

Meanwhile, François Arago discovered that the oscillations of a magnet could be damped by a piece of a nonmagnetic metal like copper, and that the needle of a magnet followed a rotating copper disk. Babbage and Herschel soon found that a disk would follow a rotating magnet. For some reason, these other experiments, together with his own and Oersted's, failed to impress upon Faraday the importance of motion in connection with his problem. Not until 1831 did he observe the induction of a current in a wire coil when the current in an adjacent coil started or stopped. In following up this observation, Faraday made the discoveries in electrical and magnetic induction on which the early development of electrical ˙generators and transformers was based. He was anticipated by Joseph Henry in America, but it was Faraday's work that attracted more attention.

Faraday was content to uncover the secrets of nature, leaving it to others to develop the major theoretical ideas and to reap the practical benefits. Thus the practical development of the generator and transformer occurred later, and was done by other men. He was little interested in mathematics or theory; for example, when his ideas on magnetic fields were extensively developed later by James Clerk Maxwell (1831–1879), Faraday was little concerned with the results. His own scientific career was characterized by simple ideas and simple experiments.

The terminology of electrochemistry was developed in 1833 by Faraday in collaboration with William Whewell (1794–1866). In addition to being an ordained minister, Whewell was a classical scholar, philosopher, mathematician, educator, and historian of science, and hence admirably equipped to work with Faraday on his problem. Together they devised the terms *electrode, anode, ion, cathode, anion, cation (cathion), electrolyte,* and *electrolysis.*[12] Faraday used the terms anion and cation for the portions of the electrolyte which were discharged at the anode and cathode repectively; today, however, they have a different meaning. Faraday was never convinced that the electrodes attracted any portion of the electrolyte. He substituted the term *electrode* for *pole* in order to avoid a term which implied polarity or attraction.

[12] M. Faraday, *Phil. Trans. Royal Soc.*, **124**, 77 (1834).

The discovery of the electrochemical laws was an outgrowth of Faraday's studies of the conductivity of solutions in 1833. During a period of cold weather he observed that the flow of electricity stopped when a film of ice formed on the electrode and he therefore concluded that ice is a nonconductor. Extending his observations, he found that such substances as lead chloride, though nonconductors in the solid state, became conductors when melted. Davy had maintained that water was necessary for electrolytic decomposition. Faraday proved that this was wrong, that actually pure water was a very poor conductor, and that its conductivity immediately increased when soluble salts were added. He then demonstrated that water is not necessary in voltaic cells. Using copper and platinum, or iron and platinum electrodes, he produced a current when the electrodes were immersed in molten salts such as potassium nitrate, sodium sulfate, lead chloride, or bismuth chloride.

His early researches on electrolysis led Faraday to suspect that there was a quantitative relationship between the amount of substance decomposed and the quantity of current that passed through the solution. The discovery of this relationship was not easy, because secondary reactions frequently complicated the results. However, his work proceeded smoothly after he developed the *volta-electrometer* as a means of measuring the quantity of current used. This device—its name was soon changed to *voltmeter* and since 1902 it has commonly been known as a *coulometer*—was simply an electrolytic cell so designed that the gases evolved in the decomposition of water could be collected individually or in a mixture. In this way the quantity of electricity which caused the liberation of 1 g. of hydrogen could be studied in relation to its ability to cause decomposition in cells placed in series with the voltmeter. Faraday proved that the size and number of plates in the batteries made no difference, nor did the dimensions of the electrodes or the distance between them. The only factor of importance was the quantity of electricity. He was able to demonstrate that the quantity of electricity which liberated 1 g. of hydrogen liberated other substances in an amount equal to the chemical equivalents of those substances. In his own words: "I have proposed to call the numbers representing the proportions in which they are evolved *electro-chemical equivalents*. Thus, hydrogen, oxygen, chlorine, iodine, lead, tin are ions; the four former are anions, the two metals are cations, and 1, 8, 36, 125, 104, 58 are their electro-chemical equivalents nearly."[13]

Faraday's electrochemical equivalents should have assisted the chemists of that day in solving their big problem—atomic weight values. However, chemists were far from agreement as to the reality of atoms and, as we shall see in the next chapter, there was a great deal of confusion regarding atoms, equivalents, and molecules. Davy and Faraday had never found it necessary to believe in atoms, though Davy used equivalent weights.

[13] *Ibid*, p. 111.

Faraday had never had to deal with chemical quantities until he encountered electrochemical equivalents, and had no interest in extending his work in the direction of atomic weight estimations. Berzelius might have done this had he not refused to accept the validity of Faraday's electrochemical laws.[14]

The discovery that a given quantity of electricity liberates elements from their compounds in amounts that are proportional to their equivalent weights was made independently by Carlo Matteucci (1811–1868), a young Italian investigator.

Faraday's health broke down during his great researches in the 1830's and the rest of his life was marked by impaired vigor and faulty memory. He continued his researches and lectures at a reduced pace, interrupted by periods of travel and convalescence. He was still able to study the magneto-optical properties of substances around 1845. He found that many substances acquired the ability to rotate plane-polarized light when placed in a strong magnetic field, and that substances were either attracted toward a magnetic field (paramagnetic) or repelled (diamagnetic). These studies indicated a relationship between light, magnetism, and electricity, and further suggested that all substances have a magnetic character within themselves. However both these phenomena were developed by others.

CONCLUSION

The discovery of current electricity had a more immediate impact on chemistry than on physics. The ability of an electric current to decompose various compounds led to the discovery of new elements and gave rise to new thinking about chemical bonding.

The introduction of a useful new tool has always stimulated the development of science. The pneumatic trough led to the discovery and understanding of new gases. The electrolytic cell was responsible for the discovery of the alkali metals and the alkaline earth metals. A half century later, another instrument, the spectroscope, had a similar impact.

At the same time the electrochemical studies led to new ideas on chemical affinity. Berzelius' dualistic theory was useful in dealing with many of the common compounds. Unfortunately, however, it proved too attractive and was used to interpret types of compounds in which it only introduced confusion.

Faraday's discovery of the relationship between quantity of electricity and electrochemical equivalents was of fundamental importance but was misunderstood; consequently it did not have the proper impact until a half century later. Because of his lack of confidence in the concept of

[14] Rosemary G. Ehl and Aaron J. Ihde, *J. Chem. Educ.*, **31**, 226 (1954).

atoms, Faraday did not extend his equivalents into the region of atomic weights; and Berzelius' confusing quantity of electricity with potential, led him to discount the significance of Faraday's electrochemical laws.

CHAPTER 6

THE PERIOD OF
PROBLEMS

THE ATOMIC WEIGHT PROBLEM

Concurrently with the developments in electrochemistry described in Chapter 5 there was concern over the estimation of atomic weights. Whereas chemists like Davy and Faraday were content to work without accepting the atomic concept, there were others like Dalton and Berzelius who recognized the value of the concept and struggled to make it more useful. Even those who questioned the validity of atoms found that equivalent weights were useful in many ways.

Atomic weights created a problem from the beginning because their estimation involved calculating them from combining weights and the correct formulas of compounds. Combining weights could be obtained more or less accurately in the analytical laboratory. The formulas, however, raised a more serious problem. Today the beginner in chemistry quickly learns to calculate a formula from analytical values and atomic weights, but Berzelius and his contemporaries had to make their own atomic weight tables, and formulas were important in doing this. The dilemma is evident—formulas are needed to calculate atomic weights, and atomic weights are needed to determine formulas.

We have already described Dalton's efforts toward establishing atomic weights. He solved the formula problem by arbitrarily adopting the rule of greatest simplicity. This was useful only until it became evident that the rule frequently did not hold for compounds; thereafter it should have been abandoned for all time.

Using comparative densities of gases proved valuable to Dalton in the case of the oxides of nitrogen, but in later years he argued against the validity of the concept that equal volumes of gases contain equal numbers of particles. Other chemists used comparative densities from time to time, but apparent inconsistencies made their use impossible in determining atomic weights. Berzelius used this approach to some extent; he held that in elemental gases like hydrogen, nitrogen, and oxygen, equal volumes under similar conditions contained the same number of atoms. In other words, there was an equivalence between gaseous atoms and volume. Accordingly, the densities of these gases would be in the same proportion as their atomic weights. However, most elements are not gaseous, and therefore such comparisons could not be widely applied. Berzelius was unwilling to extend the "equal volumes—equal numbers" concept to molecules of compounds.

BERZELIUS' EARLY EFFORTS

There were no other approaches to the problem when Berzelius began working on atomic weight values. Despite this fact, he undertook an extensive analytical program aimed toward formulating an atomic weight table. He reported some values as early as 1814 and in 1818 he published his first table, which was by no means complete. He continued this work for many years and in 1826 published an extensively revised table.

Berzelius' work as an analyst was outstanding. He utilized carefully purified reagents and devised well-planned gravimetric procedures to attain his objectives. He patiently repeated his analyses until his values for combining weights of the elements could be refined no further. In general, the analytical values he established stand up well when compared with those determined by modern analysts.

One problem in atomic weight determination involved the selection of an element as a standard for purposes of comparison. As was said earlier, in his crude estimation Dalton set the weight of the hydrogen atom equal to 1. This is logical because all other atomic weights will then be greater than unity. But hydrogen fails to form hydrides with many of the elements; hence direct determination of the combining weights of such elements with hydrogen is impossible. Since such elements can be compared with hydrogen only in terms of an intermediate element, a degree of uncertainty is introduced whose magnitude depends on the accuracy with which the combining weights of that element and hydrogen can be determined. Furthermore, it was known that hydrides seldom lend themselves to precise analysis. This is true of water, for example; for some years it was reported to contain 13.27 per cent of hydrogen. In 1819, Dulong determined the correct figure as 11.1 per cent, when Berzelius, visiting Paris, stated that his analyses of oxides of lead and

Table 6.1. BERZELIUS' ATOMIC WEIGHTS

Although Berzelius calculated his atomic weights using oxygen = 100 as a base, this value is shown only in the 1826 list. His values for each of the three tables have been recalculated to O = 16 and are shown in the appropriate columns. The present value appears in the last column at the right. The oxide formulas are those he considered appropriate in 1814 and 1826. No formulas are given for 1818 because most were still the same as those for 1814.

The data for 1814 are taken from *Annals of Philosophy*, **3**, 362 (1814); those for 1818, from *Essai sur le théorie des proportions chimiques*, Paris, 1819, pp. 191 ff.; and those for 1826, from *Jahres-Bericht*, **7**, 73 (1828).

The elements in Table 6.1 are listed alphabetically, except for a few of the very common nonmetals and metals at the beginning of the table. All the elements reported by Berzelius are included.

Element	1814 Oxides	1814 Atomic Weight (recalc.)[a]	1818 Atomic Weight (recalc.)	1818 Oxides	1826 Atomic Weight (O = 100)	1826 Atomic Weight (recalc.)	1960 Modern Atomic Weight
Oxygen		16.00	16.00		100.000	16.00	16.000
Hydrogen	$2H + O$	1.062	0.995	H^2O	6.2398	0.998	1.008
Carbon	$C + O, C + 2O$	12.02	12.05	CO, CO^2	76.437	12.25	12.01
Nitrogen	$N + O, N + 2O, N + 3O$	12.73	12.36	N^2O, NO, etc.	88.518	14.16	14.008
Sulfur	$S + 2O, S + 3O$	32.16	32.19	SO^2, SO^3	201.165	32.19	32.066
Calcium	$Ca + 2O$	81.63	81.93	CaO	256.019	40.96	40.08
Iron	$Fe + 2O, Fe + 3O$	110.98	108.55	FeO, Fe^2O^3	339.213	54.27	55.85
Potassium	$Po + 2O$	156.48	156.77	KO	489.916	78.39	39.10
Sodium	$So + 2O$	92.69	93.09	NaO	290.897	46.54	22.99
Silver	$Ag + 2O$	430.11	432.51	AgO	1351.607	216.26	107.88
Aluminum	$Al + 3O$	54.72	54.72	Al^2O^3	171.167	27.39	26.98
Antimony	$Sb + 3O$	258.07	258.06	Sb^2O^3, Sb^2O^5	806.452	129.03	121.76
Arsenic	$As + 3O, As + 6O$	134.38	150.52	As^2O^3, As^2O^5	470.042	75.21	74.91
Barium	$Ba + 2O$	273.46	274.22	BaO	856.88	137.10	137.36
Beryllium		(BeO^3)	106.01	Be^2O^3	331.479	53.04	9.01
Bismuth	$Bi + 2O$	283.84	283.80	Bi^2O^3	1330.376	212.86	209.00
Boron	$B + 2O$	11.72	11.15	B^2O^6	135.983	21.75	10.82

Element						
Chromium	Ch + 3O, Ch + 6O	113.35	351.819	Cr^2O^3, CrO^3	56.29	52.01
Cobalt	Co + 2O	117.22	368.991	CoO, Co^2O^3	59.04	58.94
Copper	Cu + O, Cu + 2O	129.04	395.695	Cu^2O, CuO	63.31	63.54
Fluorine		9.6	116.900		18.70	19.00
Gold	Au + O, Au + 3O	397.41	1243.013	Au^2O^3	198.88	197.00
Iodine	(IO^3)	202.67	768.781	I^2O^5	123.00	126.91
Lead	Pb + 2O, Pb + 3O	415.58	1294.498	PbO, Pb^2O^3, PbO^2	207.12	207.21
Lithium	(LO^2)	40.90	127.757	LO	20.44	6.94
Magnesium	Ms + 2O	50.47	158.353	MgO	25.34	24.32
Manganese	Ma + O, Ma + 2O, Ma + 3O	113.85	355.787	MnO, Mn^2O^3, MnO^2	56.93	54.94
Mercury	Hg + O, Hg + 2O	405.06	1265.822	Hg^2O, HgO	202.53	200.61
Molybdenum	Mo + O	96.25	598.525	MoO^3	95.76	95.95
Nickel	Ni + 2O	117.41	369.675	NiO	59.15	58.71
Niobium	Cl					
Palladium	Pa + 2O	225.21	714.618	PdO	114.34	106.4
Phosphorus	2P + 3O, 2P + 5O	26.80	196.155	P^2O^3, P^2O^5	31.38	30.975
Platinum	Pl + O	193.07	1215.220	PtO^2	194.44	195.09
Rhodium	R + 3O	238.45	750.680	Rh^2O^3	120.11	102.91
Selenium			494.582	SeO^2	79.13	78.96
Silicon	Si + 2O	34.66	277.478	SiO^3	44.40	28.09
Strontium	Sr + 2O	226.90	547.255	SrO	87.56	87.63
Tantalum	(TaO^2)	291.70	1153.715	TaO^3	184.59	180.95
Tellurium	Te + 2O	129.04	806.452	TeO^2	129.03	127.61
Tin	Sn + 2O, Sn + 4O	235.29	735.294	SnO, SnO^2	117.65	118.70
Titanium	Ti + O, Ti + 2O	288.16	389.092	TiO^2	62.25	47.90
Tungsten	W + 4O, W + 6O	387.88	1183.200	WO^3	189.31	183.86
Uranium	(UO^2, UO^3)	503.50	2711.360	UO, U^2O^3	433.82	238.07
(Yttrium)	Y + 2O	141.07	401.840	YO	64.29	88.92
Zinc	Zn + 2O	129.03	403.226	ZnO	64.52	65.38
Zirconium			420.238	Zr^2O^3	67.24	91.22

[a]Formulas in parentheses in this column are the bases of Berzelius' 1818 atomic weights.

copper produced less water than he had expected. Dulong's careful analysis, in which hydrogen was converted to water over copper oxide, revealed that the discrepancy reported by Berzelius was due to the percentage of hydrogen in water being incorrect.

As a result, there was an attempt to find a more suitable element for use as a standard. Since most elements form stable and well-defined oxides and oxides can usually be accurately analyzed, the element oxygen was adopted as the standard by most chemists interested in atomic weight determinations. The weight value assigned to it varied widely; Thomson gave its weight as 1, Wollaston as 10, and Berzelius as 100. The modern value, 16, did not come into consistent use until Stas used it as the standard in his excellent studies on atomic weights around the middle of the nineteenth century.

Although men like Thomas Thomson and Wollaston were content to stop with equivalent weights, Berzelius went on to grapple with the tougher problem involving combining ratios and atomic weight. He used Gay-Lussac's law from the beginning—water was H_2O, ammonia NH_3, hydrogen chloride HCl. Frequently reasoning by analogy, he accepted H_2S for hydrogen sulfide on account of the similarity between oxides and sulfides. His work with selenium and tellurium indicated the relation of these elements to sulfur and he assumed their hydrides to be H_2Se and H_2Te.

It was impossible to determine the formulas of more than a few compounds on the basis of Gay-Lussac's law or of analogies. However, Berzelius cleverly devised methods for establishing certain relationships. For example, he oxidized lead sulfide, PbS, to lead sulfate, $PbSO_4$, with nitric acid and proved that neither lead nor sulfur was lost in the acid treatment. Hence, the ratio of lead to sulfur must be identical in the sulfide and sulfate. He then established the ratio of oxygen atoms to sulfur atoms in the acid portion of the salt to be $3:1$ (SO_3), and the ratio in the basic part of the compound to be the same as in the common oxide litharge (PbO).[1]

Because such demonstrations were somewhat limited in number, he found it necessary to set up arbitrary working rules to cover inorganic compounds. He held that when 1 atom of element A is present in a compound, it must combine with either 1, 2, 3, or 4 atoms of B. At first he believed that 4 was the upper limit for atoms of B, but in 1828 he dropped this idea. When there are 2 atoms of A in the compound, there must be either 3 or 5 atoms of B. He maintained that this same rule held for combinations of simple compounds such as a metal oxide with an acidic oxide; for example, CaO combined with CO_2 in a $1:1$ ratio.

In his earliest tables of atomic weights (1814, 1818) Berzelius considered the metal oxides as primarily dioxides (AgO_2, FeO_2), or, when more than

[1] For a more detailed discussion of Berzelius' reasoning see Ida Freund, *The Study of Chemical Composition*, Cambridge Univ. Press, Cambridge, 1904, pp. 180–182.

one oxide is formed, as trioxides (FeO_3). He held to the idea that metals tended to form oxides that had the same formulas. In his 1826 table he discarded his notion about dioxides in favor of monoxides, but he treated such metals as potassium, sodium, and silver just as he did calcium, magnesium, and zinc. He now correctly interpreted the formulas of such compounds as Fe_2O_3, Al_2O_3, and Cr_2O_3. His values for many of the elements, when recalculated on the basis $O = 16$, are strikingly close to those of the present day. His atomic weights for the monovalent metals were of course double what they should have been. Two major developments, both of them in 1819, were responsible for his change in thought with respect to the atomic weight of metals. One was the law of Petit and Dulong; the other, the law of isomorphism.

THE LAW OF PETIT AND DULONG

The relationship between specific heat and atomic weight, known as the law of Petit and Dulong, was reported by its discoverers to the French Academy in the spring of 1819 and was published soon thereafter.[2] Pierre Louis Dulong (1785–1838) was trained as a physician but quickly turned to the study of chemistry and physics. He worked as an assistant to Berthollet and served as chemistry instructor in the normal school at Alfont before becoming professor of physics at the École Polytechnique in 1820. He was made director of the École in 1830. His researches were done primarily on heat phenomena after early experiments which led to the discovery of nitrogen trichloride—and the loss of an eye and two fingers when the compound exploded violently. Alexis Thérèse Petit (1791–1820) was a student at the École Polytechnique and later became a professor of physics there. With Dulong he developed methods for determining thermal expansion and specific heat of solids.

These two investigators noticed that in the case of a number of solid elements the product of specific heat and atomic weight was a constant. Working on the assumption that individual atoms might have the same specific heat, they demonstrated that when the specific heat of an element was multipled by its atomic weight an approximate constant, called the "atomic heat," was obtained. The atomic weights available in 1819 were extremely uncertain, and Petit and Dulong adjusted some of them to fit their law. For example, Berzelius' value for lead was 2589 on the basis $O = 100$. Petit and Dulong reported this value as 12.95 on the basis $O = 1$, obviously half of Berzelius' value. A similar halving is evident in the case of gold, tin, zinc, tellurium, copper, nickel, and iron. The value for silver is quartered; that for bismuth has no obvious relation to Berzelius'

[2] A. T. Petit and P. L. Dulong, *Ann. chim. phys.*, [2], **10**, 395 (1819).

value. The values for platinum and sulfur were not changed, and that for cobalt is one-third.

The law actually proved to be a fair approximation despite the uncertainty regarding atomic weights and the rather arbitrary assumptions made by the authors. Its promulgation represents an act of faith rather than being a sound scientific generalization, but this is not the only time this has occurred in science.

Table 6.2. THE RELATIONSHIP DISCOVERED BY PETIT AND DULONG

Element	Specific Heat	Atomic Weight O = 1	Product of Atomic Weight and Specific Heat
Bismuth	0.0288	13.30	0.3830
Lead	0.0293	12.95	0.3794
Gold	0.0298	12.43	0.3704
Platinum	0.0314	11.16	0.3740
Tin	0.0514	7.35	0.3779
Silver	0.0557	6.75	0.3759
Zinc	0.0927	4.03	0.3736
Tellurium	0.0912	4.03	0.3675
Copper	0.0949	3.957	0.3755
Nickel	0.1035	3.69	0.3819
Iron	0.1100	3.392	0.3731
Cobalt	0.1498	2.46	0.3685
Sulfur	0.1880	2.11	0.3780

Petit and Dulong suggested that this relation might be used to determine atomic weight values. Actually, however, the constant is only approximate, and the authors' high hopes for calculating atomic heats were never attained. Therefore, the law could give no more than an approximation in fixing atomic weights, but this approximation would be in the proper range of magnitude and would therefore suggest the figure by which the equivalent weight should be multiplied to obtain the accurate atomic weight.

Berzelius gave the law his attention when he published his third major table of atomic weights in 1826. He abandoned the MO_2 formula for metal oxides in favor of MO. But at the same time he failed to make full use of the clues offered by the law when he retained the formulas AgO, NaO, KO and ended up with atomic weights for these metals that were twice as large as they should have been. Obviously, he was willing to accept the law to only a limited extent, and there is evidence that he mistrusted it. Later chemists, like Regnault, Kopp, and Weber gave it considerable attention.

The atomic weights of some elements, however, such as those of carbon, silicon, and boron, deviate markedly from those predicted by the law. Later workers learned that the specific heat of elements varies with temperature, and that the above nonmetals follow the law more closely when the specific heat is determined at high temperatures.

In 1831, Franz Neumann[3] proved that a similar relation exists between specific heat and molecular weight of compounds. In working with carbonates and sulfates he found a striking agreement. Regnault later extended these investigations to show that compounds containing the same number of similar atoms have the same heat capacity. Somewhat later, Kopp showed that heat capacity was additive in nature; actually, however, this was of distinctly limited importance.

THE LAW OF ISOMORPHISM

The studies on crystallography being made at this time commonly held that identity of crystal form implied identity of composition. This principle had been enunciated by Haüy, the founder of crystallography, and it had many adherents, in spite of certain exceptions which had been observed. Klaproth in 1788 established the chemical identity of calcite, which occurs in the rhombohedral form, and aragonite, which occurs in the rhombic form (both are $CaCO_3$). In 1816 Gay-Lussac observed that crystals of potassium alum grew normally in a solution of ammonium alum, and about the same time Gehlen succeeded in preparing typical alum crystals from a sodium-containing solution. In 1817 J. N. von Fuchs called attention to the similarity of crystal forms in aragonite, strontianite ($SrCO_3$), and cerussite ($PbCO_3$).

It was Eilhardt Mitscherlich (1794–1863) who clearly established the relationship between chemical composition and crystalline forms. This was formulated in his law of isomorphism which was published after work done in 1819.[4] In order to determine if crystalline form depends upon the chemical nature of the elements, he studied salts of the phosphates and arsenates. (Berzelius had already established the close relationship of these two sets of salts in properties.) Mitscherlich found that the phosphates and arsenates each formed three distinct series of salts, and he reported that "every arsenate has its corresponding phosphate, composed entirely in the same proportions."[5] He observed similar analogies in the various sulfates and carbonates.

The law of isomorphism, which states that compounds which crystallize in the same form are similar in chemical composition, was well received.

[3] Franz Neumann, in Poggendorff's *Ann. phys.*, **23**, 1 (1831).
[4] E. Mitscherlich, *Ann. chim. phys.*, [2], **14**, 172 (1820); **19**, 350 (1821).
[5] *Ibid.*, **19**, 410 (1821).

Mitscherlich recognized that similarity in chemical composition resulted in approximate rather than absolute isomorphism. Supporting data accumulated rapidly.

Mitscherlich's career in chemistry was delayed by his early interest in Oriental philology. While he was a student at Göttingen he came under the influence of Friedrich Stromeyer (1776–1835), noted for his discovery of cadmium in 1817, and for his early support of laboratory instruction in chemistry. Mitscherlich went to Berlin in 1818 for further studies in

Fig. 6.1. EILHARDT MITSCHERLICH.
(From Mitscherlich's *Gesammelte Schriften*
[1896].)

chemistry in the laboratory of H. F. Link, the botanist. It was here that he began his studies on arsenates and phosphates of potassium and confirmed Berzelius' reports regarding properties and composition. In starting his work on crystals he received instruction from Heinrich Rose's brother Gustav, mineralogist at the University of Berlin. The promising nature of Mitscherlisch's work led to his being invited to work in Berzelius' laboratory in Stockholm.

Soon after his return from Sweden Mitscherlich was appointed to the chair of chemistry, once held by Klaproth, at the University of Berlin. Here he continued his researches and wrote his *Lehrbuch der Chemie* (1829–1830). In his crystallographic studies he observed the variation in thermal expansion along dissimilar axes of a crystal, and the different crystal forms of sulfur. The property whereby a given substance could have more than one crystal form he called *dimorphism*. In 1827 he prepared selenic acid and showed that the selenates are isomorphous with the sulfates. Later he demonstrated isomorphism in the manganates, chromates, and sulfates, and the permanganates and perchlorates. The isomorphism of the two latter salts revealed the composition of perchloric and permanganic acids.

Mitscherlich utilized the isomorphism of the sulfates and selenates to determine the atomic weight of selenium; he reasoned that sulfate and potassium selenate have identical atomic ratios, except for the atom of sulfur in one, and of selenium in the other. As Table 6.3 shows, the weights of sulfur and selenium combined with equal weights of potassium and oxygen thus are 18.39 and 45.40 respectively. Since the atomic weight of sulfur is 32, that of selenium must proportionately be 79.

Table 6.3. USE OF THE LAW OF ISOMORPHISM TO CALCULATE ATOMIC WEIGHTS

Potassium Sulfate		Potassium Selenate		Recalculated to 127.01 Parts
K	44.83%	K	35.29%	44.83
O	36.78	O	28.96	36.78
S	18.39	Se	35.75	45.40

Berzelius meanwhile used the law of isomorphism in connection with his atomic weight estimations. The isomorphism of sulfates and chromates led him to revise the formula of green chromic oxide from CrO_3 to Cr_2O_3; he assigned the formula CrO_3 to chromic anhydride. The isomorphism of chromic oxide with the oxides of aluminum, iron, and manganese enabled him to assign correct formulas to the oxides of these metals. This led to his halving the atomic weights of these metals that he published in 1818. He subsequently halved the atomic weights of the other metals as well, thus bringing the results into accord with the law of Dulong and Petit. However, there were four exceptions. In the case of silver and cobalt, he doubted the validity of the law; and in the case of sodium and potassium, no specific heats were available.

DUMAS AND THE DETERMINATION OF VAPOR DENSITIES

In 1826 the young French chemist Dumas developed a procedure for determining vapor densities of substances which are liquid or even solid at ordinary temperatures. Up to then, comparison of the densities of equal volumes of gases provided a basis for comparing atomic weights; but the only gaseous elements known at the time were hydrogen, nitrogen, oxygen, and chlorine. Dumas' procedure significantly extended the number of comparisons that could be made.

Jean Baptiste André Dumas (1800–1884) was the son of a municipal clerk in Arles, a city on the lower Rhone. At fifteen he was apprenticed to a local apothecary but, finding the work dull, he set out for Geneva where he had relatives. There he attended lectures by de Candolle in botany, Pictet in physics, and Gaspard de La Rive in chemistry. He became part of the scientific circle of the city and, in association with local physicians, carried on research which revealed iodine in calcined sponge and investigated the composition of the blood. His papers attracted the attention of Humboldt, who persuaded him to go to Paris to continue his studies. He soon became *répétiteur* in chemistry for Thenard at the École Polytechnique and, before long, professor of chemistry at the Athenæum. During the next thirty years he held important chairs in several colleges in Paris and became the most prominent figure in French chemistry. His authority was widely respected and his influence regarding the advancement of younger men in the field was important. After 1848 Dumas became active in the Ministry of Education and ceased to be a major figure in chemical research and theorization.

In his researches done in 1826, Dumas revived the generally ignored hypothesis of Avogadro and Ampère, hoping to extend the comparison

Fig. 6.2.   JEAN BAPTISTE ANDRE DUMAS.
(Courtesy of J. H. Mathews.)

of gas densities for atomic weight determinations to include elements which are not gases under usual conditions of measurement.[6] In the procedure he devised, the solid or liquid whose vapor density was to be measured was placed in a glass bulb and heated in a bath of water,

[6] J. B. A. Dumas, *Ann. chim. phys.*, [2], **33**, 337 (1826), **50**, 337 (1832).

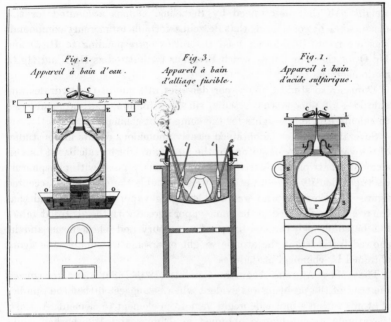

Fig. 2.
Appareil à bain d'eau.

Fig. 3.
Appareil à bain
d'alliage fusible.

Fig. 1.
Appareil à bain
d'acide sulfurique.

Fig. 6.3. DIAGRAMS OF APPARATUS USED BY DUMAS FOR
MEASURING VAPOR DENSITY.
(From *Ann. chim. phys.* [1826].)

sulfuric acid, or metal alloy to a sufficiently high temperature to be
completely vaporized. At this point the neck of the bulb was sealed off
with a torch. The temperature and pressure of the vapor in the bulb
were known from the temperature of the bath and the barometer reading.
The bulb was then cooled and weighed to obtain the weight of the vapor
which had condensed. The volume of the bulb was obtained by determining the weight of water it could hold.

In his paper Dumas discussed the possibilities of extending the range
of atomic weight determinations by using vapor density measurements.
Maintaining that Ampère's concept of equal volumes containing equal
numbers of particles led to a natural correlation of relative densities with
atomic weights, he set out to establish atomic weights for a number of
elements that either could be vaporized directly or formed gaseous or
easily vaporizable compounds. He found that the density of iodine vapor,
calculated in terms of modern values, corresponded to a molecular weight
of 254. He preferred to assume that the molecules were diatomic, which
makes the atomic weight 127; this is consistent with the value derived
from chemical analyses.

In the case of mercury, Dumas' calculations suggested an atomic weight half that determined by Berzelius. Dumas accounted for this discrepancy by considering that Berzelius' formula for mercury compounds was incorrect. He himself used the ratios corresponding to Hg_2O and Hg_4O for the compounds which Berzelius formulated as HgO and Hg_2O, as do chemists today.

Dumas also studied the vapor densities of some of the hydrides and chlorides of phosphorus, arsenic, silicon, titanium, and tin, and tried to calculate atomic weights for the component elements.

Several years later he obtained clearly anomalous results in his studies of the vapor density of sulfur, for the atomic weight he calculated for this element was triple that arrived at by Berzelius. So also with phosphorus; its vapor density gave an atomic weight double the commonly accepted figure. Mitscherlich, who was also studying vapor densities found that the results he obtained for bromine vapor agreed with the accepted value; but he confirmed Dumas' figures for mercury and phosphorus and he showed further that the atomic weight of arsenic was double the figure obtained by chemical procedures.[7]

That Marc Antoine Augustin Gaudin[8] (1804–1880) alone recognized the crux of the problem was evident when he suggested that the number of atoms within a molecule might vary from element to element. At least, the atomic weight figures obtained by Dumas and Mitscherlich could be reconciled with those of Berzelius if mercury vapor were considered monatomic and sulfur vapor hexatomic. Gaudin set up an acceptable table of atomic weights and showed by means of diagrams how atomic and molecular relations in gaseous reactions could be explained. But the uncertainty regarding the whole theory of atoms created an unfavorable climate for his concepts, and he won few followers. He also failed to develop his concepts further in subsequent years because his activities in chemistry touched only the fringe of the subject. He was for many years a mathematician with the Bureau of Longitudes in Paris, and he was interested in crystal forms and photography.

The anomalies in the atomic weights obtained by the vapor density method resulted in a general lack of interest in the procedure. There was a widespread feeling that Avogadro's hypothesis was open to grave doubt and was of no help in clarifying questions concerning atomic weights. Berzelius reiterated his earlier belief that the hypothesis was valid only for elements that are naturally gaseous at ordinary temperatures. Others, however, considered the whole theory of atoms itself open to question, and, to the extent that they dealt with quantitative relations, used the equivalent weights advocated by Leopold Gmelin.

The heart of the problem, of course, was the fact that no one made a clear-cut distinction between atoms and molecules of an element. To the

---

[7] E. Mitscherlich, in Poggendorff's *Ann. phys.*, **105**, 193 (1833).

[8] M. A. A. Gaudin, *Ann. chim. phys.*, [2], **52**, 113 (1833).

extent that these terms were used, they were used loosely—physical atoms (our molecules) and chemical atoms (our atoms)—or were contradictory as in the case of Dumas' atoms and "half-atoms." Furthermore, there was a tendency to believe that molecules of the different elements contained the same number of atoms.

EQUIVALENTS

Richter had been aware of the concept of equivalence even before Dalton developed his atomic theory. In his *System of Chemistry*,[9] Thomson listed the weights of acids and bases which neutralized each other. On the basis of published analytical work, Wollaston in 1814 extended the concept to include twelve elements and forty-five compounds.[10]

Wollaston believed that equivalent weights were less theoretical than atomic weights, and of greater practical importance. He argued that Dalton's atomic weights were based on arbitrary assumptions, wherefore such weights were completely hypothetical. Equivalent weights, however, were based only on analytical values and therefore were universally reliable. In order to facilitate their use he invented a "logarithmic scale of chemical equivalents adopted for experimental and manufacturing chemists," a sort of slide rule for calculating amounts produced in chemical reactions.

Equivalents were favorably received by a number of working chemists. This was particularly true in the fourth decade of the century, following Leopold Gmelin's adoption of the concept and Faraday's discovery of electrochemical equivalence. Gmelin devised a system in which compounds were assigned formulas on the basis of the equivalents present. The equivalent weights of some common elements were as follows: $H = 1$, $O = 8$, $S = 16$, $C = 6$; HO was the formula for water, HS for hydrogen sulfide, SO_2 for sulfur dioxide, and CH_2 for methane. Gmelin's primary concern was simplicity, for he considered the kind of speculation necessary in determining atomic weights was not only confusing but dangerous because of possible errors it involved.

Leopold Gmelin (1788–1853) became professor of chemistry and medicine at Heidelberg in 1813. He was the son of Johann Friedrich Gmelin, pharmacist and chemical historian, and the brother of Christian Gottlob Gmelin (1792–1860), professor of chemistry and pharmacy at Tübingen. Leopold discovered potassium ferricyanide in 1822, and Christian developed a process for the production of ultramarine. The first edition of Leopold's well-known *Handbuch der Chemie* was published in 1817–1819.

[9] T. Thomson, *System of Chemistry*, Bell & Bradfute, Edinburgh, 4th ed., 1810, p. 630.
[10] W. Wollaston, *Phil. Trans.*, **1814**, part 1, p. 1.

Fig. 6.4. LEOPOLD GMELIN.
(Courtesy of the Edgar Fahs Smith Collection.)

Equivalents provided a useful stoichiometric tool but were useless as a universal concept in chemical formulation. And not only that, it introduced as much confusion as it eliminated because numerous elements have varying valences and therefore more than one equivalent weight. These considerations were either ignored or settled arbitrarily.

PROUT'S HYPOTHESIS

In 1815 there appeared in Thomson's *Annals of Philosophy* an anonymous paper which pointed out the fact that gas densities were proving to be exact multiples of the density of hydrogen. William Prout, the author of this and of a subsequent paper, suggested in these articles that all atomic weights are exact multiples of that of hydrogen and that hydrogen is the *protyle* or "first matter" of which all other elements are composed.[11]

This hypothesis was a somewhat logical outgrowth of the knowledge available at that time. The concept of the element was operational, and there was still a great deal of uncertainty as to which substances might safely be considered elemental and which were compounds. Some of Davy's work had suggested the presence of hydrogen in sulfur,

[11] W. Prout, *Ann. Phil.*, **6**, 321 (1815); **7**, 111 (1816).

phosphorus, and some of the other "elements." The results of other investigations also suggested that hydrogen might be recovered from certain
elements. Furthermore, the phlogiston theory still survived in memory;
hence the association of hydrogen with phlogiston retained a certain appeal.
Thus, Prout's deduction had material antecedents; moreover, it provided
an appealing and simple hypothesis at a time when atomic weight determinations were few in number and highly suspect as to validity.[12]

Thomas Thomson found the hypothesis extremely tenable and struggled
for several years to fit atomic weights into the concept. His figures,
however, were very unsatisfactory because they were based on an attempt
to fit experimental results into preconceived conclusions. His book,
An Attempt to Establish the First Principles of Chemistry by Experiment
(2 vols., 1825), was vigorously attacked by Berzelius, who said: "The
experiments described in it appear to have been performed at the writing
desk rather than in the laboratory; and the greatest consideration which
contemporaries can show to the author is to treat his book as if it had
never appeared."[13]

Berzelius' atomic weights in his 1826 table differed markedly from those
calculated by Thomson, and from those based on Prout's hypothesis.
Continental chemists who utilized atomic weights accepted those of
Berzelius; but British chemists generally used Thomson's until 1833,
when Edward Turner, in a report to the British Association for the
Advancement of Science,[14] emphasized their inaccuracy and the lack of
support for Prout's hypothesis.

Despite this, the hypothesis continued to have a certain appeal and was
given serious attention from time to time. Dumas turned to it around
1840 when he and Jean Servais Stas (1813–1891) carefully redetermined
and corrected the atomic weight of carbon. Dumas believed that the
atomic weights of the common elements were in accord with Prout's
hypothesis. However, subsequent work done by Stas clearly showed that
it failed to hold rigorously.

In 1859 Dumas returned to the hypothesis in a memoir[15] in which he
suggested that three classes of elements were apparent: a large class in
which the atomic weights are whole multiples of hydrogen atoms taken
as unity, a second class in which the hydrogen is taken as 0.5, and a third
in which the hydrogen is taken as 0.25. Stas' careful atomic weight
determinations, however, revealed cases in which atomic weights were
definitely not multiples of 0.25. Marignac also supported the hypothesis
until 1860, but was forced to admit its failure in view of Stas' work.

[12] Robert Siegfried, *J. Chem. Educ.*, **33**, 263 (1956); O. T. Benfey, *ibid.*, **29**, 78 (1952).

[13] J. J. Berzelius, *Jahres-Bericht*, **6**, 77 (1827—for the year 1825). Quoted from the translation in Ida Freund, *op. cit.*, p. 597.

[14] E. Turner, *Rept. Brit. Assoc. Adv. Sci.*, **3**, 399 (1834—for the meeting in 1833).

[15] J. B. A. Dumas, *Ann. chim. phys.*, [3], **55**, 129 (1859).

CONCEPTS ABOUT ELEMENTS AND COMPOUNDS

The Hydrogen Theory of Elements

Early in the nineteenth century there was considerable uncertainty regarding the chemical nature of certain substances that are today clearly recognized as elements but were then considered to be compounds. These included the alkali metals as well as sulfur and phosphorus.

Davy's procedure for producing sodium and potassium from the fused hydroxide was quickly followed by a technique developed by Gay-Lussac and Thenard in France. They used hot iron to reduce the alkalies, and obtained the elements in quantity and in a fairly pure state. Studies of their properties carried on by Davy and the two French chemists led to general confirmation of the experimental facts but to frequent conflicts regarding theory.

Davy observed that the alkali metals, which appeared at the negative pole of his electrolytic apparatus, could reduce metallic oxides. He believed that the original alkalies were produced when the sodium and potassium were burned in oxygen. The low density of the two metals created a suspicion as to their metallic character. Following studies of ammonium amalgam, which revealed that its properties resembled those of metal amalgams, Davy suggested that hydrogen was a component of sodium and potassium. This would explain their vigorous combustibility, and was in accord with Cavendish's identification of phlogiston with hydrogen. Davy extended his theory that hydrogen was a constituent of combustible substances to include sulfur and phosphorus following experiments which he thought demonstrated its presence.[16]

Gay-Lussac and Thenard came to a similar conclusion with respect to sodium and potassium as the result of their experiments with ammonia.[17] They found that a green substance (potassium amide, KNH_2) and hydrogen gas were formed when ammonia reacted with potassium. The amount of hydrogen evolved was equivalent to that evolved when the same amount of potassium reacted with water. When the green substance reacted with water, potassium hydroxide and ammonia were produced. They interpreted these experiments as showing that potassium contains hydrogen which can be liberated by either ammonia or water.

Davy soon realized that the hydrogen might be formed by the decomposition of ammonia (or water) by the potassium, and he maintained

[16] Robert Siegfried, *op. cit.*

[17] According to a commonly held viewpoint, ammonia was an oxygen compound. However, Amédée Berthollet showed this to be in error, for his analysis revealed that ammonia is composed of nitrogen and hydrogen. See A. Berthollet, *Mém. Soc. Arcueil*, **2**, 268 (1809).

Fig. 6.5. Louis Jacques Thenard.
(Courtesy of the Edgar Fahs Smith Collection.)

that the green substance was a compound made up of potassium and the residue of ammonia. Gay-Lussac and Thenard, however, continued to advocate the hydrogen concept, in which they were supported by Berthollet and D'Arcet, who regarded water as a constituent of fused potassium hydroxide. The latter two adopted Davy's viewpoint in 1811 when they found that the substance formed when potassium was burned differed from caustic potash (KOH) in being richer in oxygen. They reasoned that the oxide had to contain water if the potassium contained hydrogen, because no water was liberated during combustion. They also observed that the potassium oxide reacted with carbon dioxide to form the carbonate with the liberation of oxygen gas whereas the reaction of caustic potash with carbon dioxide gave the same carbonate and water.

These experiments removed all doubt as to the elemental nature of sodium and potassium. Davy's idea that sulfur and phosphorus contained hydrogen, never very firmly established, was obliterated by the experimental work done by Gay-Lussac and Thenard.

Rejection of the Oxygen Theory of Acids

As was said earlier Lavoisier believed that oxygen was a universal constituent of acids and that the oxide (anhydride) was the true acid. The hydrogen compound (our present-day acid) he considered as merely an adventitious form of the acid that contained water similar to water of hydration. Thus, CO_2 was carbonic acid, SO_3 was sulfuric acid, P_2O_5

was phosphoric acid. This idea prevailed in chemical circles for years despite the opposition of Claude Berthollet, whose analysis of hydrocyanic acid in 1787 showed that it contained only hydrogen, carbon, and nitrogen. Scheele had demonstrated even earlier that sulfuretted hydrogen (H_2S) contained only hydrogen and sulfur.

Nearly all chemists, however, held to the oxygen theory. They believed that hydrochloric acid, known then as muriatic acid, consisted of oxygen combined with an undiscovered element, *muriaticum*. Even Berthollet was in agreement on this. Chlorine gas was considered to be oxygenated muriatic acid. Berthollet's experiments on the bleaching action of chlorine, and his observation that oxygen gas was liberated when chlorine water stood in the sunlight, seemed to confirm this view. When William Henry passed electric sparks through hydrogen chloride gas confined over mercury, hydrogen was produced and the mercury underwent a change. He believed that water was a constituent of hydrogen chloride, hence the liberation of hydrogen and the accompanying attack of mercury by oxygen.

Davy in 1808 decomposed hydrogen chloride with potassium, obtaining hydrogen and potassium chloride. He also produced the salt by direct combination of potassium and chlorine. In subsequent experiments he sought to extract oxygen from chlorides and chlorine, but failed completely; for when he treated chlorine or chlorides with powerful reducing agents such as phosphorus, no oxygen could be extracted from muriatic acid or the muriates. These repeated failures caused him to conclude that no oxygen was present, and that chlorine was an element. Davy called the element *chlorine* because of the greenish color of the gas.

Davy's failure to extract oxygen from chlorine actually was no proof of the elemental nature of chlorine, but only a strong suggestion that it was not an oxygen compound. Gay-Lussac and Thenard were unwilling to accept the elemental hypothesis, for it would wreak havoc on the Lavoisierian concept of acids. In 1811 they compared the two theories regarding chlorine; but despite the neutral character of the gas, Davy's failure to extract oxygen, and the greater simplicity of Davy's concept, they clung to the muriaticum theory. Two years later, however, their own experiments with iodine and cyanogen led them to agree with Davy's position.

Bernard Courtois (1777–1838), the discoverer of iodine, was the son of a saltpeter manufacturer in Dizon. The elder Courtois assisted Guyton de Morveau when the latter lectured on chemistry at the Dijon Academy. After a pharmaceutical apprenticeship, Bernard was given an opportunity to study under Fourcroy at the École Polytechnique of which Morveau was director. For a time young Courtois was active in pharmaceutical circles, but he joined his father when the saltpeter business was faced with financial difficulties.

At that time, the ashes of seaweed collected along the coasts of Normandy and Brittany served as a source of sodium and potassium salts. One day in 1811 young Courtois observed clouds of purple vapor rising from mother liquor which had been acidified with sulfuric acid. The vapors, which had an irritating chlorine-like odor, condensed on cold objects in the form of dark crystals with a metallic luster. A study of the properties of these led Courtois to suspect that he had discovered a new element. However, the press of business activities and the inadequacy of his laboratory facilities caused him to turn over his chemicals to Charles-Bernard Desormes and Nicolas Clement, two chemist friends. These men reported the new substance in 1813. Courtois became active in the manufacture of iodine, but others investigated its chemistry. Davy and Gay-Lussac independently established it as an element. Its relationship to chlorine was immediately apparent, and the oxygen-free nature of hydrogen iodide was generally accepted.

Berthollet's early experiments on hydrocyanic acid were repeated and extended by Gay-Lussac; his work showed that the compound contains hydrogen, nitrogen and carbon, but no oxygen. During his studies of cyanide salts Gay-Lussac discovered cyanogen gas (C_2N_2), which is formed when mercuric cyanide is heated. The resemblance of cyanogen to chlorine and iodine was striking.

The existence of acids without oxygen could no longer be denied. Hence, there was a new search for the acidfying principle which oxygen failed to satisfy. Davy suggested that hydrogen might be it, since hydrogen, rather than oxygen, apparently was present in the acids whose composition was then known. Of course, there were numerous hydrogen compounds which were not acids, so the acidifying role of hydrogen could not be accepted *in toto*, but it could be useful.

Many were still reluctant to put aside Lavoisier's oxygen concept and there was a tendency to refer to *oxyacids* as distinct from *hydracids* which contained no oxygen. Berzelius, in particular, was disinclined to abandon the concept that chlorine is an oxide and that acids could exist without oxygen, but even he adopted the new views by 1820.

Bromine and Fluorine

The other halogens, bromine and fluorine, were studied during this period. Bromine was isolated in the mid-1820's by several chemists, including Carl Löwig (1803–1890), Justus von Liebig (1803–1873), and Antoine-Jérôme Balard (1802–1876). Balard first recognized the reddish brown liquid as an element related to chlorine and iodine. He isolated it from a deposit of salts left by mother liquors from which sodium chloride had been separated, and observed that it was a trace component of many salt deposits and of sea water.

The existence of the fourth halogen, fluorine, was suspected during these years and many laboratories sought to isolate it. Scheele prepared hydrogen fluoride by the action of sulfuric acid on fluorspar (CaF_2). The corrosive action of this acid on glass and other silicates was recognized immediately and the acid soon came into use as an etching agent.

Davy established the absence of oxygen in hydrofluoric acid, and Ampère suggested that the acid was a compound of hydrogen and an undiscovered element. Its properties were predicted; they suggested that the element would be an intensely active substance related to chlorine. Repeated attempts were made to isolate it, but without success until 1886, when Ferdinand-Frédéric Henri Moissan (1852–1907) succeeded in electrolyzing potassium hydrogen fluoride dissolved in anhydrous liquid hydrogen fluoride.

CONCLUSION

During the first quarter of the nineteenth century a large body of chemical information accumulated and certain concepts were clarified, but the period came to an end with more confusion than enlightenment. Although Lavoisier's antiphlogistic hypothesis and the new nomenclature were generally accepted, the recently introduced atomic theory—which could have been a new unifying concept in chemistry—created its own formidable problems. Berzelius made considerable progress in establishing a workable set of atomic weights, but there were enough uncertainties to cause many investigators to content themselves with the use of equivalents. All methods for atomic weight determination contained limitations which led to ambiguities. These shortcomings might have been resolved by a combination of methods, but no one had the breadth of vision to do this.

Although the period saw the clarification of the nature of acids and a clearer distinction between elements and compounds, new discoveries kept matters confused. This was particularly true with respect to the studies being made on carbon compounds in biological materials. The abundance and variety of such compounds created new problems of classification and understanding which could not really be solved until problems involving atomic weights and formulas were solved. Knowledge about these organic compounds ultimately aided in the solution of such problems, but this occurred only after a long period during which factual knowledge grew rapidly but theoretical concepts suffered one failure after another.

CHAPTER 7

ORGANIC CHEMISTRY I.
RISE OF ORGANIC
CHEMISTRY

KNOWLEDGE OF ORGANIC COMPOUNDS
IN 1800

An examination of contemporary textbooks on chemistry and medicine reveals that organic chemistry was in a very rudimentary state when the nineteenth century began. Mineral chemistry was beginning to receive a certain amount of systematic treatment as the result of the application of fundamental chemical principles and the findings made possible by improved analytical procedures. At least, there was some interest in composition and relationships. Organic chemistry, however, was frequently ignored; at best, it was treated cursorily under such headings as "animal chemistry" and "vegetable chemistry." The approach was medical and the treatment was largely descriptive. Materials were discussed from the standpoint not of composition but of biological source and medical use.

Primary attention was given to blood, bile, saliva, urine, skin, hair, horn, gelatin, fibrin, etc., from animals, and to sugar, gum, camphor, etc., from plants.

There was some commercial production of organic materials, mostly at home or in small shops. A few operations—sugar refining, soapmaking, dyeing, distillation—were carried out on a larger scale.

A small number of compounds were known in a comparatively pure state. A few of them dated back several centuries. The production of alcohol in about the eleventh century was mentioned in an earlier chapter. Ether, produced by the action of sulfuric acid on alcohol, was described by Valerius Cordus in the sixteenth century. Two other "ethers" (esters) formed by the action of nitric acid on alcohol (nitric ether = ethyl nitrate) and of hydrochloric acid on alcohol (salt ether = ethyl chloride) date from a slightly later time. Only four organic acids were known before Scheele's brilliant studies: acetic and formic, produced by the distillation of vinegar and ants respectively; and benzoic and succinic, sublimed respectively from gum benzoin and amber. Salts of oxalic and tartaric acids were known, but the acids themselves remained unknown until Scheele's discovery of them around 1780.

Scheele demonstrated that the calcium salt of benzoic acid is appreciably soluble in water, whereas the free acid is not. The acid could be prepared by boiling gum benzoin with calcium hydroxide, then acidifying to separate the benzoic acid. A variant of this approach was used by Scheele to separate such water-soluble acids as oxalic, malic, tartaric, citric, lactic, uric, gallic, pyrogallic, mucic, and pyromucic acids from their natural juices or degradation products, taking advantage of the fact that their calcium salts are insoluble. Thus, he precipitated the acids by adding calcium hydroxide, then decomposed the salts with sulfuric acid to liberate the organic acid. Oxalic acid was also produced by the oxidation of cane sugar with nitric acid; and mucic acid, by the similar oxidation of milk sugar. Glycerol, which was formed by the saponification of oils and fats, Scheele considered a sweet principle (*Ölsüss*) related to the sugars, since its oxidation with nitric acid gave oxalic acid.

At about the same time H. M. Rouelle discovered urea in human urine, and hippuric acid in the urine of cows and camels. He believed that the hippuric acid was benzoic acid, an error corrected by Liebig in 1829.

Lavoisier, who was deeply interested in organic chemistry, observed that in the mineral kingdom all the oxidizable and acidifiable "radicals" (elements) are simple, whereas in the plant and animal kingdoms they always contain carbon and hydrogen and frequently also nitrogen and phosphorus. Noting that elements often form more than one oxide, he concluded that this should also be true of organic radicals. He considered that sugar was a neutral oxide, and oxalic acid its higher oxide.

VITALISM

Lavoisier clearly thought of organic chemistry as an integral part of chemistry, not as something uniquely related to living organisms merely because the organic compounds were produced by these organisms. He classified all the acids together, but subdivided them as mineral, vegetable, and animal acids as Lemery had done a century before. Some but by no means all chemists followed him in doing this. In his *Gundriss der Chemie* (1797) Gren treated all the organic compounds in a separate chapter; he considered them as proximate principles present in animal and vegetable organisms and incapable of being synthesized artificially.

This uncertainty prevailed for several decades. Some chemists felt that the organic origin of this type of compounds set them apart from mineral compounds. The latter could frequently be obtained in a very pure state and followed the law of definite proportions. But it was often difficult to obtain the organic compounds in any state of purity and consequently their analysis gave highly variant results. Coupled with this was the fact that organic analysis was in a very unsatisfactory state at the time. In 1811 Berzelius was of the opinion that organic compounds did not obey the law of definite proportions. Although later improvement of analytical methods led him to change his position, it is evident that at various times during his long career he was uncertain as to the place of these compounds in the science of chemistry.

Leopold Gmelin maintained that inorganic and organic compounds must be kept in separate categories, the difference between them being more easily perceived than defined. According to him, simple inorganic compounds were composed of two elements, whereas even the simplest organic compounds contained three elements. In keeping with this notion, he treated methane, ethylene, cyanogen, and similar compounds in the section of his book that dealt with inorganic chemistry. He believed that inorganic compounds could be synthesized directly from the elements, whereas organic compounds required a plant or animal for their production, the chemist being capable of making only minor modifications.

A widespread belief, called vitalism, held that organic compounds were produced through the agency of a vital force that was present only in living plants and animals. These compounds could be converted into other compounds in the laboratory or in a brewer's or a vinter's establishment, but they could not be synthesized from the elements. However, vitalism received a serious blow at the hands of Friedrich Wöhler in 1828.

The son of a schoolmaster in Escherheim, near Frankfurt-am-Main, Friedrich Wöhler (1800–1882) studied medicine at Marburg and Heidelberg.

Fig. 7.1. FRIEDRICH WÖHLER.
(Courtesy of the Edgar Fahs Smith Collection.)

His work at Heidelberg brought him to the attention of Leopold Gmelin, who interested him in making chemistry his career. After obtaining his medical degree, Wöhler was accepted as a student in Berzelius' laboratory in Sweden. On his return to Germany, he taught in a technical school in Berlin and, in 1831, moved to Cassel where he continued to teach. After Stromeyer's death in 1835, Wöhler became professor of chemistry at Göttingen in 1836, a post he held until his death.

Although his contributions to the field of organic chemistry were of great significance, the major portion of his work was done in the field of inorganic chemistry. It has been said that there were hardly any known elements he did not study. He isolated impure aluminum in 1827, just two years after Oersted did. He did extensive work on boron, titanium, and silicon compounds, and discovered silicon hydrides. He just missed discovering vanadium, an element isolated by Nils Gabriel Sefström (1787–1845), a Swedish physician and chemist.

One of Wöhler's greatest achievements in the field of organic chemistry was the synthesis of urea, which started the decline of the idea of vitalism. This occurred in 1828, when Wöhler, who had been studying cyanates for several years, attempted to synthesize ammonium cyanate and obtained urea instead. He treated silver cyanate with ammonium chloride solution and obtained a white crystalline material which showed none of the properties of the cyanates. By treating lead cyanate with ammonium hydroxide, he obtained the white crystals in an uncontaminated state after separation from the lead oxide. The crystals had organic properties, and Wöhler suspected the compound to be an alkaloid. It failed to give the tests typical of alkaloids, but behaved like the urea described by Proust and Prout, among others. Comparison with urea obtained from urine showed that the compounds were identical. Wöhler triumphantly wrote to Berzelius, "I must tell you that I can make urea without the use of kidneys, either man or dog. Ammonium cyanate is urea."[1]

Actually, Wöhler had obtained urea as early as 1824 in experiments in which he succeeded in converting cyanogen into oxalic acid by treatment with aqueous ammonia. He clearly described oxalic acid and a white crystalline substance which was urea, although he failed to recognize it as such until four years later. Urea had been prepared in 1811 by Humphry Davy's brother John, but not identified. Davy obtained it by the action of ammonia on phosgene, the latter prepared by exposing carbon monoxide and chlorine gases to sunlight.

Wöhler regarded the synthesis of urea from cyanate and ammonia as significant, but admitted that some might argue that it was not synthesized from inorganic materials. Both cyanic acid and ammonia were ultimately derived from organic substances. The *Naturphilosoph* might still claim that the organic portion had not disappeared from the animal charcoal or from the cyanogen compounds made from it. Wöhler evidently overlooked the fact that in 1783 Scheele had prepared potassium cyanide by heating ammonium chloride and potassium carbonate not only with animal charcoal, but with graphite.

Wöhler's synthesis of urea was recognized by his contemporaries although it had by no means a dramatic impact. Berzelius, Liebig, and Dumas, among others, commented favorably, and the synthesis was usually included in any listing of Wöhler's accomplishments, but neither Wöhler nor his associates put forth any claims that this marked the death of vitalism. The fact that vitalism could have no serious place in organic chemistry was realized gradually as knowledge of organic compounds and their synthesis accumulated.

The real demonstration of the synthesis of organic compounds from completely inorganic materials was achieved by Kolbe in 1844 when he synthesized acetic acid. Berthelot clinched this in the 1850's with his synthesis of a number of organic compounds—i.e., methane and acetylene —reactions carried on at high temperatures.

INVESTIGATION OF NEW COMPOUNDS

Following Scheele's death, there was little further progress in isolating new organic compounds until 1800. Johann Tobias Lowitz (1757–1804),

[1] F. Wöhler to J. J. Berzelius, Feb. 22, 1828. In O. Wallach (ed.), *Briefwechsel zwischen J. Berzelius und F. Wöhler*, Engelmann, Leipzig, 1901, vol. 1, p. 206.

Court Apothecary in St. Petersburg, prepared glacial acetic acid in 1789 and trichloracetic acid four years later. He observed the effectiveness of charcoal in removing color from organic solutions and introduced its use in the clarification of acids, alcoholic beverages, and sugar syrups. He succeeded in isolating glucose and fructose from honey, and showed that they differed from sucrose. He prepared absolute alcohol in 1796.

Although hydrocarbons were encountered as products of dry distillation, they were seldom pure and were generally not understood. Marsh gas (methane, CH_4) was burned in the eudiometer by Volta in 1776; fixed air was shown to be a combustion product. A decade later Berthollet demonstrated that methane is a compound of carbon and hydrogen, i.e., carburetted hydrogen. Olefiant gas (ethylene, C_2H_4) was prepared in 1794 by the Dutch chemists J. R. Deiman, A. Paets van Troostwyk, N. Bondt, and A. Lauwerenburgh by the action of concentrated sulfuric acid on ethyl alcohol. On the basis of Berthollet's analysis, the compound was believed to contain oxygen, but Henry disproved this in 1805. Chlorination resulted in an oily liquid ($C_2H_4Cl_2$) which became known as the "oil of the Dutch chemists."

The term "ether" was applied to volatile liquids formed by the action of acids on ethyl alcohol. That the acid was believed to be present in the ethers is shown by the names sulfuric ether, nitric ether, and muriatic ether. In 1807 Thenard demonstrated that alcohol and acid were regenerated by boiling nitric and muriatic ethers with alkalies, but not on heating sulfuric ether with alkali. Fourcroy and Vauquelin proved the sulfuric ether contains no sulfur.

The first two decades of the nineteenth century saw striking progress in isolating new compounds. Fourcroy and Vauquelin were active in investigating substances that were of medical interest. They prepared urea in a highly pure state, proved that the pyroligneous acid in the aqueous liquor from wood distillate was identical with the acid in vinegar, and isolated camphoric acid. Vauquelin isolated allantoin from the allantoic fluid of cows (working with Buniva in 1800), asparagine (with Robiquet in 1805), quinic acid (with Valentin Rose in 1805), and cyanic acid in 1818. In 1811 he discovered uric acid in the excrement of birds. William Prout, who was also active in physiological research, found this acid in the excrement of snakes. In addition to formulating his law of definite proportions, Joseph Louis Proust made an extended study of the saccharine juices of plants from 1799 to 1808 and identified three sugars they contained—glucose, fructose, sucrose. He proved that the glucose from grape juice was identical with that discovered in honey by Lowitz. Proust isolated mannitol from manna, and leucine from the decomposition products of casein. He studied the ripening of cheese, the composition of urine, and the production of gunpowder.

In 1811 Gottlieb Sigismund Kirchhoff (1764–1833), an apothecary in St. Petersburg, observed that when starch was heated with sulfuric acid glucose could be isolated from the syrup thus formed. This discovery created wide interest and within a year Döbereiner had interested the local prince in establishing a glucose factory near Weimar. In 1819 Henri Braconnot (1781–1855), professor and director of the botanical gardens at Nancy, found that when old linen rags were digested with sulfuric acid, glucose was produced. During the next year he obtained sweet crystals after glue was similarly digested. This product, named glycocoll (glycine), was later found to contain nitrogen; hence it did not belong among the sugars. The amino acid leucine, which Braconnot isolated from beef about this time, had been isolated from cheese slightly earlier by Proust.

Analyses by Gay-Lussac and Thenard revealed that sugars, starch, and cellulose all contained hydrogen and oxygen in the same proportions as in water. These compounds were placed in the saccharine class by Prout in 1827; the other classes of natural substances were the oily and the albuminous. The term carbohydrate was first used for saccharine substances in 1844 by Karl Schmidt.

The class of alkaloids was definitely recognized by 1820. Crystalline morphine was isolated from opium by Friedrich Wilhelm Sertürner (1783–1841) in 1805. In a lengthy publication in 1816, he reported that it was alkaline in character, and therefore a vegetable alkali. Sertürner, who was a pharmacist in Einbeck, and later in Hameln, also isolated meconic acid. The presence of this acid in opium became the basis for the ferric chloride color test in cases of suspected opium poisoning. Actually, morphine alkaloids had been isolated in impure form as early as 1803 by Charles Louis Derosne (1780–1846), a Parisian pharmacist, and by Armand Séguin; but both investigators worked with mixtures of narcotine and morphine and neither made as careful a study as Sertürner. The latter's work became a model for alkaloid research and a great deal of progress was made during the next few years, especially by Pelletier and Caventou.

Two professors at the École de Pharmacie in Paris, Pierre Joseph Pelletier (1788–1842) and Joseph Caventou (1795–1877), between 1818 and 1820 isolated strychnine and brucine from St. Ignatius' beans (botanically related to nux vomica, the common source of strychnine), and quinine and cinchonine from cinchona bark. During this same period piperine was isolated from pepper by Oersted, emetine from ipecacuanha root by Pelletier and Magendie, caffeine from coffee beans by Runge, veratrine from sabidilla seeds by Meissner, solanine from the berries of *Solanum nigrum* by Desfosses. By 1835 some thirty aklaloids had been isolated and studied.

Perhaps the most significant studies of the period were those made by Chevreul on fats. Michel Eugène Chevreul (1786–1889) studied in Vauquelin's laboratory and served as an assistant at the Jardin des Plantes before becoming professor at the Lycée Charlemagne in 1813. He succeeded Vauquelin as professor in the natural history museum of the Jardin in 1830, a position he held for more than thirty years. His most important publication was his *Recherche chimique sur les corps gras d'origine animale* (1823), which summarized studies covering more than a decade.

Fig. 7.2.   MICHEL EUGÈNE CHEVREUL.
(Courtesy of the Edgar Fahs Smith Collection.)

When Chevreul began his work on fats, common belief held that soap was a combination of alkali and fat. The significance of Scheele's discovery of glycerol in 1783 was completely ignored, as was Geoffroy's observation that the insoluble fatty material obtained when a soap was acidified differed from the fat used in making the soap. Chevreul was able to demonstrate that the alkali converted the fat into soap and glycerol during the soapmaking process. The soap, in turn, was converted into an insoluble acid material on the addition of a mineral acid. Therefore the fat must be a compound of the acid material and glycerol.

In his early analyses of a potassium soap prepared from lard he obtained an acidic crystalline substance resembling mother-of-pearl which he named margaric acid. From the mother liquor he obtained an oily acid; he called it oleic acid. That the two acids were not present as such in the fat, but were combined with glycerol, was shown by the fact that the two acids and the glycerol weighed more than the fat from which they were made. Glycerol, the *Ölsuss* of Scheele, was named *glycérine*. Chevreul isolated stearin and olein as glycerides; from stearin he obtained stearic acid, which melted at a higher temperature than margaric acid. He obtained butyric acid from butterfat, caproic and capric acids from goat fat, and delphinic or phocenic acid from the fat of dolphins and porpoises.

After investigating many fatty substances of animal origin, Chevreul found that saponification almost always resulted in the formation of soaps of fatty acids—usually stearic, margaric, and oleic acids in proportions that varied with the source of the fat—and glycerol. He failed to find glycerol in spermaceti and biliary calculi, but instead obtained ethal (cetyl alcohol) and cholesterol respectively. Saponification was thus a chemical action in which an inorganic base replaced an organic base combined with a fatty acid. The fats might therefore be considered true salts. This idea of the fatty molecule was confirmed several decades later by synthesis when Berthelot prepared glycerides from fatty acids and glycerol. Chevreul's conclusions stood up exceedingly well in their broader aspects and frequently in their detailed aspects as well. Margaric acid (C_{17}) was shown later to consist of equimolecular amounts of stearic (C_{18}) and palmitic (C_{16}) acids. For the most part, later research on fats involved extensions of his work, rather than corrections.

Chevreul's studies also were important industrially. For example, soapmaking became a practical rather than an empirical operation, for knowledge of the composition of various fats made possible greater uniformity of product. So also with candlemaking. At that time candles were made of tallow; they were soft, dripped badly, and gave a smoky malodorous flame. In 1819 Braconnot sought to improve the tallow used for candles by removing the more fluid portion of the fat (the olein). In 1825 Chevreul and Gay-Lussac patented a process in which stearic acid was used in candles, but efforts to commercialize it were unsuccessful. Several years later two young physicians named M. Motard and Adolphe de Milly produced an improved candle which found a good market. De Milly improved the wick by impregnating it with ammonium borate, cheapened the production of stearic acid by substituting lime for the more expensive alkalies, and introduced hot pressing for eliminating liquid acids more efficiently.

Chevreul's researches on fats achieved their great success as a result of his insight into the principles of proximate analysis. In order to understand the nature of natural substances it is necessary to isolate their constituents in unchanged form, or at least in an identifiable form. Chevreul introduced the use of inert solvents for making separations without destroying or greatly altering the original constituents of a mixture. He also was the first to use melting points as a criterion of purity; thus a fatty acid was considered pure only when, after successive recrystallizations from alcohol, its melting point did not rise. His principles of proximate analysis were clearly set forth in his *Considérations générales sur l'analyse organique et sur ses applications*, published in 1824. His methods clearly show the application of the rules formulated two centuries earlier by Descartes.[2]

In 1824 Chevreul became director of dyeing at the tapestry works at Gobelin. He did no further work on fats except for minor studies on the drying of oils and on wool fat during the 1850's. His research on the drying of oils showed that drying represented not a loss of moisture but a combination of the oil with oxygen. The rest of his studies were concerned with dyeing and with color relationships. Even in his earlier work with Vauquelin he had been interested in dyes and had separated hematoxylin from logwood, brazilin from brazilwood, quercitrin from oak bark, morin from fustic, and luteolin from dyer's weed (*Reseda luteola*). He obtained the colorless form of indigo by reducing the blue form in 1812.

At the Gobelin tapestry works Chevreul set up a laboratory in order to study dyeing processes in a systematic manner. His innovations met with hostility on the part of workmen accustomed to personal rule-of-thumb methods, but he was able to demonstrate the superiority of his systematic approach. In 1828–1830 he published a two-volume work, *Leçons de chemie appliquée à la teinture*. He then turned to the problem of color contrasts and pioneered in developing laws of color harmonies. His *De la loi contraste simultané des couleurs* (1839) was translated into German and English. Not only did his studies influence the dyeing industry, but they provided a stimulus for the impressionistic school of painting.

ISOMERISM

The rapid discovery of new compounds brought a new problem to plague investigators. There had hitherto been a tendency to believe that chemical properties were dependent upon the composition and temperature of a substance. It was well known that a given substance might be a solid at one temperature, a liquid at another, and a gas at still another. However, if analysis showed a certain composition, the identity of the substance was accepted as established. But during the 1820's some substances were found to have the same composition but different properties.

[2] P. Lemay and R. E. Oesper, *J. Chem. Educ.*, **25**, 68 (1948).

One of Wöhler's earliest researches was made on the cyanates, the same group of compounds which led to the synthesis of urea. The silver salt of cyanic acid was carefully analyzed and reported in the literature— 77.23 per cent silver oxide, 22.77 per cent cyanic acid. At about the same time Justus von Liebig, working in Gay-Lussac's laboratory, completed analyzing silver fulminate—77.53 per cent silver oxide, 22.47 per cent cyanic acid. The analyses indicated an identity, but the compounds clearly differed in properties. Liebig, certain that Wöhler's analysis was wrong, apparently confirmed this by preparing and analyzing silver cyanate; it contained 71 per cent of silver oxide. A recheck by Wöhler verified his own original results; Liebig's analysis was shown to be in error because of impure cyanate. The two chemists were faced with the dilemma of having two compounds identical (within experimental error) in composition, but drastically different in properties. A study on which Liebig and Wöhler collaborated revealed an even worse situation, for cyanic acid (HOCN) was shown to be identical in composition with fulminic acid (HONC) and also with cyanuric acid ($H_3O_3C_3N_3$), the white solid which when heated forms cyanic acid and which in turn precipitates from cyanic acid on standing.

Berzelius took notice of the fact that the composition of cyanates and fulminates was apparently identical, but was unable to arrive at an explanation. Gay-Lussac, convinced of the correctness of the analyses, attributed the difference in properties to variation in the way the elements were combined.

Illuminating gas, produced by the destructive distillation of coal had come into extensive use in London and other large cities early in the century. Faraday in 1825 examined an oil which separated out in the cylinders of the gas produced by the Portable Gas Company, a company which was marketing, for small-scale use, a gas prepared by heating whale oil. When the gas was compressed in the cylinders, an oily substance separated. From this Faraday isolated two new "carburets of hydrogen," the first of which was benzene (C_6H_6). The other came from the more volatile portion of the oil. It was liquid at the pressure in the cylinders, but at atmospheric pressure it vaporized at temperatures above −14° C. On analyzing the gas (butylene, C_4H_8), Faraday found that its composition was the same as that of the well-known olefiant gas (ethylene, C_2H_4), but that its density was double the latter's. Both gases combined volume for volume with chlorine, but the chloride of the new gas contained twice as much carbon and hydrogen as the chloride of the olefiant gas. Faraday, familar with Liebig's and Wöhler's findings on fulminates and cyanates, believed that this was another case in which the same elements combined in the same proportions but in different ways.

Faraday isolated naphthalene in 1825 from Persian naphtha—naphthalene had been discovered in 1819 in coal tar by Alexander Garden, an English industrialist—and studied its composition. He observed that when it is treated with sulfuric acid, two sulfonated derivatives of identical composition are obtained. However, he turned his attention to other matters and never followed up the problem.

Baffling problems involving inorganic substances were also being encountered. Berzelius knew that there were two different oxides of tin with the same composition. In 1825, he and Engelhart noted differences in two forms of phosphoric acid whose composition they believed was identical. Three years later Wöhler obtained urea from reagents that should have produced ammonium cyanate. When the latter was finally prepared, it proved to have the composition but not the properties of urea. In 1823 Mitscherlich observed that the sulfur crystals which formed when liquid sulfur cooled slowly had a different form (monoclinic) from those produced by the evaporation of a carbon disulfide solution (rhombic). Also surprising was the fact that aragonite crystals are composed of calcium carbonate, just as calcite crystals are. Both iron pyrites and marcasite are iron disulfide; rutile and anastase are different forms of titanium dioxide. It had been known for some time that diamond and graphite are carbon.

Finally, in 1830, Berzelius sought to devise a system that would cover such cases after he convinced himself of the identical composition of tartaric and racemic acids. Racemic acid was first obtained by Kestner, operator of a chemical works in Thann, as a by-product in the production of tartrates. It was much less soluble than tartaric acid and was sold as oxalic acid until 1819, when Johann Friedrich John, of Berlin, proved that it was neither oxalic nor tartaric acid. When Gay-Lussac visited Thann in 1826 he obtained a supply of the acid from Kestner. After careful study, he concluded that the neutralization capacity of both racemic and tartaric acid was identical, but he found that racemic acid was less soluble and that it failed to form Rochelle salt, a double salt, when crystallized from a mixture of sodium and potassium salts in solution.

To account for such cases, Berzelius proposed the concept of *isomerism* and attempted to distinguish different variations. *Polymerism* he applied to compounds which contained the same elements in the same proportions, but whose molecular weights had a multiple relation, i.e., ethylene and Faraday's butylene. *Metamerism* he applied to compounds that had the same composition, but different constituent parts. Although his best example was hypothetical because one of the members was unknown, he believed that there might be two metamers of tin and the acids of sulfur, i.e., $\mathrm{Sn} + \overset{..}{\mathrm{S}}$ and $\overset{..}{\mathrm{Sn}} + \mathrm{S}$; or in more familar notation, $\mathrm{SnO} + \mathrm{SO_3}$ and $\mathrm{SnO_2} + \mathrm{SO_2}$. From the viewpoint of his dualistic concept of salts, this is a good example. If he could have prepared stannic sulfite, however, analysis would have shown that it was not metameric with stannous sulfate. Berzelius also considered that cyanic and cyanuric acids were metamers. He believed that cyanic acid was the oxide of cyanogen combined with water $(C_2N_2 \cdot O + H_2O)$ and that cyanuric acid was an oxide of the radical $C_2H_2N_2$. Since cyanic acid is produced by heating cyanuric acid, he thought that heat was responsible for the change from one form to the other. Actually, cyanuric acid is a trimer of cyanic acid. The term metamer could be meaningful only as long as there was confusion regarding composition, although Gay-Lussac indicated that methyl acetate and ethyl formate were authentic metamers of each other. However, as knowledge of structure was clarified, the term lost its usefulness. *Isomer* (equal parts) gradually acquired its present-day meaning— compounds with the same composition and molecular weight; *polymer* (several parts) was applied to compounds that have the same composition but whose molecular weights are multiples of one another.

The term *allotrope* was proposed in 1841 by Berzelius for the different forms of an element. The term *polymorphism* had come into use for compounds that occurred in more than one crystalline form. As polymorphism became generally known, it threw doubt upon the value of the law of isomorphism as a basis for atomic weight determinations.

ELEMENTARY ORGANIC ANALYSIS

Organic analysis was still highly alchemical at the beginning of the nineteenth century. An idea of its nature can best be gained from Fourcroy's tabulation of the general methods that were important in 1801, when his book was published.[3]

1. Natural mechanical analysis (separation by nature).
   Exudates of plants—saps, gums, manna, resins, rubber.
2. Artificial mechanical analysis (separation by presses, mortars).
   Juices and oils. The product is unaltered.
3. Distillation.
   Forms products which may not have been present as such in the plants.
4. Combustion analysis.
   Produces quantity of carbon and ash.

[3] A. F. Fourcroy, *Système des connaissances chimiques*, 10 vols., Baudouin, Paris, 1801; vol. 7, pp. 44 ff.

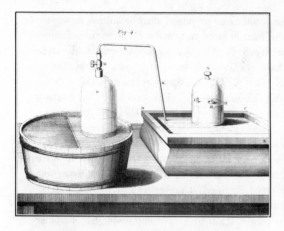

Fig. 7.3. LAVOISIER'S APPARATUS
FOR ANALYZING ALCOHOL AND
OTHER ORGANIC SAMPLES.
(From *Œuvres de Lavoisier,*
J. B. Dumas, Ed. [1862], vol. 2.)

Fig. 7.4. LAMP APPARA-
TUS USED BY LAVOISIER
FOR THE ANALYSIS OF
ALCOHOL.
(From *Œuvres de Lavoisier,*
J. B. Dumas, Ed. [1862],
vol. 2.)

5. Analysis by water.
    a. Soaking after maceration.
    b. Soaking with agitation.
    c. Infusion (boiling water poured over macerated tissues).
    d. Digestion (tissues in cold water are heated slowly until
       boiling point is reached).
    e. Decoction (tissues are boiled with water for several hours).
   The various forms of analysis by water result in progressively
   greater alteration of the tissue components.
6. Analysis by acids and alkalies.
   Treatment may be similar to analysis by water, but alteration
   of principles is generally greater.
7. Analysis by alcohol, ether, or oils.
   Results in a selective dissolving action; i.e., alcohol dissolves
   essential oils but no fixed (fatty) oils.
8. Analysis by fermentation.

As is clearly evident, the above analytical procedures are, at best,
capable only of separating mixtures of related substances (proximate
principles). Frequently the separation is achieved only after significant
chemical alteration. The analyses could have only superficial value in

leading to an understanding of organic materials; in many instances they were downright misleading. The time was becoming ripe for a more sophisticated approach, one which demanded pure, unaltered compounds which could be analyzed for their component elements, and studied for their characteristic properties.

The first serious attempt at the quantitative analysis of organic materials was made by Lavoisier. In order to determine the volume of oxygen consumed and of carbon dioxide formed, he burned weighed amounts of charcoal over mercury in a bell jar containing oxygen, and measured the volume of the gas formed both before and after absorbing the carbon dioxide on alkali. He also attempted to estimate the carbon content of alcohol, fats, and waxes by burning them in a spirit lamp so as to measure the oxygen consumed and the carbon dioxide produced; from these figures he calculated the composition. He already knew that these

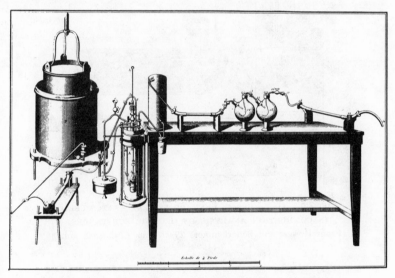

Fig. 7.5. Combustion apparatus used by Lavoisier in analyzing oils.
(From *Œuvres de Lavoisier*, J. B. Dumas, Ed. [1862], vol. 2.)

substances gave only carbon dioxide and water on combustion; this indicated that carbon, hydrogen, and oxygen were the sole components of his samples.

Lavoisier also analyzed difficultly combustible substances like sugars and resins, using mercuric oxide or red lead as his source of oxygen. Because of inadequate data on the composition of carbon dioxide and water[4] all his results were inaccurate, as the following examples show:

$$CO_2 \quad 28\% \text{ C, } 72\% \text{ O; correct values, } 27.2\% \text{ C, } 72.8\% \text{ O}$$

$$H_2O \quad 13.1\% \text{ H, } 86\% \text{ O; correct values, } 11.1\% \text{ H, } 88.9\% \text{ O}$$

They were, however, based on a carefully conceived method and were sufficiently accurate to show that hydrogen and oxygen were present in sugar in the same ratio as in water. He found that the ratio of hydrogen to oxygen was much higher in fats and lower in vegetable acids.

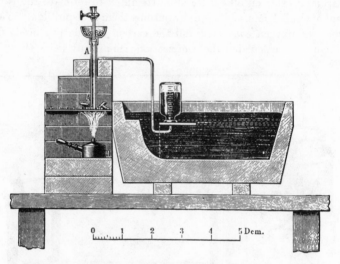

Fig. 7.6. GAY-LUSSAC-THENARD TYPE APPARATUS USED
IN COMBUSTION ANALYSIS OF ORGANIC SUBSTANCES.
(From Roscoe and Schorlemmer, *Treatise on Chemistry*, vol. 3
[1890].)

Dalton in 1803 developed a method for analyzing methane and ethylene which involved exploding the gases in oxygen in a eudiometer and analyzing the products. Similar experiments were made by de Saussure and Thenard in 1807. Although none of these procedures were highly successful, the results enabled Dalton to use methane and ethylene to illustrate the law of multiple proportions.

In 1810 Berthollet passed samples of sugar, oxalic acid, and vegetable gum through a heated tube, thus decomposing them into carbon, water,

---

[4] A. L. Lavoisier, *Elements of Chemistry*, transl. by Robert Kerr, Creech, Edinburgh, 1790, pp. 101–102.

carbon dioxide, and hydrocarbons. He then detonated these products with oxygen and measured the carbon dioxide and water.

Nicolas de Saussure analyzed alcohol and ether by three different methods between 1807 and 1814. He noted that alcohol corresponded to olefiant gas plus water, and ether to two units of olefiant gas plus water.

Gay-Lussac and Thenard introduced the use of potassium chlorate as an oxidizing agent in 1811.[5] The powdered sample and the chlorate were compressed into a tablet which was introduced into a vertical tube through a stopcock with a cavity large enough to hold the tablet. The lower end of the tube was heated in a charcoal fire, so that combustion was vigorous when the tablet was dropped into the tube. Decreasing the size of the samples made it possible to keep the shattering of the combustion tube to a minimum. The carbon dioxide formed from the sample was absorbed by potassium hydroxide. The results of many analyses confirmed Lavoisier's conclusions regarding the ratio of hydrogen to oxygen in carbohydrates, acids, and oils.[6]

Gay-Lussac substituted cupric oxide for potassium chlorate in 1815, because he found it less dangerous and more reliable, particularly for compounds containing nitrogen, such as cyanides and uric acid. Such compounds always caused trouble in analyses, because the potassium chlorate oxidized much of the organic nitrogen to the oxides. This was avoided by using copper turnings in the upper part of the combustion tube.

Berzelius, recognizing the importance of accurate analyses in determining whether organic compounds obeyed the laws of chemical combination, modified the combustion procedure used by Gay-Lussac and Thenard by mixing the sample with potassium chlorate and sodium chloride in order to temper the reaction. The sample was placed in a slanting but almost horizontal tube, which was then heated. The water was absorbed for weighing by passing the combustion products through a tube of calcium chloride. The oxygen and carbon dioxide were collected in a bell jar over mercury, the carbon being absorbed by alkali for weighing.

Berzelius chose to begin his analytical studies with what he considered the simplest organic compounds, the acids. These he analyzed as lead salts because these salts were readily prepared and were considered pure. Furthermore, he quickly established that whether water or lead oxide was combined with the acid (anhydride), the same quantity of oxygen was present. Based on this approach, later and more accurate analyses led to such formulas as C_2O_3 for oxalic acid and $C_4H_6O_3$ for acetic acid. If we add H_2O to these, the formulas are satisfactory, except for the doubling of the formulas of monobasic acids such as acetic acid. With other acids, such as citric and succinic, the formulas were much more uncertain.

[5] J. L. Gay-Lussac and L. J. Thenard, *Recherches physico-chimiques*, Paris, 1811, vol. 2, p. 265.
[6] *Ibid.*, pp. 321 ff.

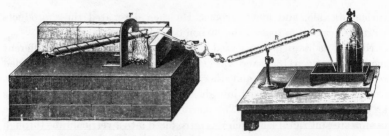

Fig. 7.7. BERZELIUS' APPARATUS FOR COMBUSTION
ANALYSIS OF ORGANIC SUBSTANCES.
(From Roscoe and Schorlemmer, *Treatise on Chemistry*, vol. 3
[1890].)

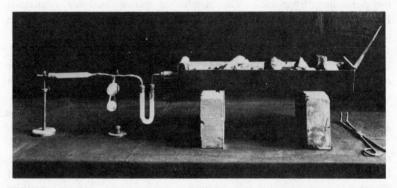

Fig. 7.8. LIEBIG'S ORIGINAL APPARATUS FOR THE ELEMENTARY ANALYSIS
OF ORGANIC COMPOUNDS; now in the Deutsches Museum, Munich. The combustion tube passes through the charcoal furnace at the right and leads
successively into the U-tube for absorbing water vapor, the *Kaliapparat*,
and a small horizontal absorption tube to retain moisture lost from the *Kaliapparat*.
(Courtesy of the Edgar Fahs Smith Collection.)

The slanting tube used by Berzelius and Gay-Lussac's copper oxide were
quickly adopted by such leading analysts as Chevreul, Döbereiner, Bussy,
Liebig, and Wöhler.

The final major modification in combustion analysis for carbon and
hydrogen was introduced by Liebig when he developed the *Kaliapparat*
and thereby eliminated the use of the bell jar and mercury for collecting
gases. The *Kaliapparat* was a glass tube that was blown into a series of
five bulbs containing potassium hydroxide solution which absorbed the
carbon dioxide produced during the combustion. In front of the *Kaliapparat* was a calcium chloride tube to absorb the water vapor formed from

the hydrogen in the sample; in the back of it was another calcium chloride tube to retain any water lost to the gases as they passed through the potassium hydroxide solution. The second tube was weighed with the *Kaliapparat* and in time became part of the device. The combustion was carried out in a long glass tube heated by charcoal. At the end of the analysis the closed tip of the combustion tube was broken so that air could be drawn through the apparatus and any residual water vapor and carbon dioxide drawn into the absorption tubes.

Liebig's apparatus was so well conceived that reliable analyses became commonplace. The method developed by Liebig, Gay-Lussac, and Berzelius has not been supplanted to the present day, although details have been modified. For example, Brunner and von Hess in 1838 used an open-end combustion tube instead of the closed-tipped tube, thus permitting purified air or oxygen to be swept through the apparatus during combustion. Charcoal heating gave way to gas heating after the Bunsen burner was invented and to electrical heating in the twentieth century. The *Kaliapparat* has given way to simple tubes using solid alkaline absorbents for carbon dioxide. The major change has been the conversion of the apparatus to microanalysis, largely through Pregl's work early in the present century.

Justus von Liebig

Justus von Liebig (1803–1873) was the son of a dealer in drugs, dyes, oils, and chemicals in Darmstadt. As in Davy's case, early experiments probably led Liebig toward a chemical career. Watching an itinerant showman taught him how to prepare silver fulminate by dissolving silver in nitric acid and adding alcohol. After being apprenticed to a pharmacist, Liebig continued his experiments with the fulminates and successfully removed the shop's attic window—whereupon he found himself removed from his apprenticeship. His father then sent him to the newly founded university at Bonn where he studied chemistry under Karl Kastner (1783–1857); he went to Erlangen when Kastner took a position there. There his activities in a student political organization led to his arrest and ultimately caused him to flee from Bavaria and return home to Hesse. Friends interceded in his behalf and persuaded the Grand Duke Ludwig to grant him funds for study in Paris. Soon thereafter, Kastner persuaded the faculty at Erlangen to award Liebig a degree of doctor of philosophy *in absentia*.

In Paris Liebig finally found the kind of chemical activities he had long been seeking. While he was still in school, his teachers had found him a mediocre pupil, more interested in experimenting with the chemicals in his father's shop than in studying languages and the other subjects taught in the Darmstadt schools. In Erlangen he had access to Kastner's laboratory in his capacity of student assistant, and he frequently initiated

Fig. 7.9. JUSTUS VON LIEBIG. The *Kaliapparat* on the
cushion at his right is a side view of that which is shown in
a front view in Fig. 7.8.
(Courtesy of the Edgar Fahs Smith Collection.)

chemical experiments of his own. Although Kastner was reputedly the
leading professor of chemistry in Germany and the author of several
widely used books, he was not active in original research. In Paris,
however, things were different, for not only were there chemists of the
stature of Vauquelin, Gay-Lussac, Thenard, Dulong, Proust, and Chevreul,
but an enthusiastic group of men of lesser stature was interested in promoting chemistry in both its theoretical and applied fields. Paris had a
rich heritage—in chemistry, stemming from Lavoisier and Berthollet;
and in the physical sciences and mathematics, stemming from Laplace,
Lagrange, Monge, Fourier, Biot, and Coulomb, among others. Under
both the Revolution and Napoleon's rule, scientific activity had been
encouraged so that the city became the world leader in this realm.

Liebig soon won admission to Gay-Lussac's laboratory where he continued his studies on the fulminates. He came to the attention of Humboldt,
who was instrumental in his being appointed extraordinary professor of
chemistry in the small Hessian University of Giessen at age twenty-one.
A year later, after the death of the senior professor, Wilhelm Ludwig
Zimmerman (1780–1825), Liebig was made ordinary professor and placed
in complete charge.

During the following quarter century Liebig exhausted himself by making Giessen the foremost institution for chemical instruction in the world. In spite of a meager salary and inadequate appropriations, he brought practical instruction in chemistry to such a level that the university attracted students not merely from Germany, but from France, England, Italy, the United States, and Mexico. He and his students made many experimental contributions, particularly in the areas of analytical, organic, and agricultural chemistry. His great energy and enthusiasm made him an inspiring teacher and a prolific writer. At the same time he was a vigorous propagandist and an impulsive dogmatist. For a time he attained the authoritative position once held by the now aging Berzelius. His reputation in the field of agricultural chemistry was so great that when his book on the subject was published in England and Germany, pirated editions were printed in Boston and Philadelphia soon after copies of it reached the shores of America.

Despite his standing, he was not infallible. He frequently made hasty, dogmatic statements and faulty generalizations. His books contained many mistakes and erroneous conclusions, and he was guilty of taking credit for work done by others. He was quick to attack other men's ideas and during much of his life was a bitter participant in scientific polemics. Probably the only prominent chemist he did not repeatedly attack was Wöhler, who was too kind and too wise to spend his time arguing. As a matter of fact, Wöhler frequently brought Liebig back to his senses by pointing out that he should be working in the laboratory instead of wasting his energy in a meaningless battle with Mitscherlich, Mulder, or Pasteur. One letter from Wöhler to Liebig reads:

Göttingen, March 9th, 1843

To make war against Marchand, or indeed against anyone else, brings no contentment with it and is of little use to science. You merely consume yourself, get angry, and ruin your liver and your nerves—finally with Morrison's pills. Imagine yourself in the year 1900, when we are both dissolved into carbonic acid, water, and ammonia, and our ashes, it may be, are part of the bones of some dog that has despoiled our graves. Who cares then whether we have lived in peace or anger; who thinks then of thy polemics, of the sacrifice of thy health and peace of mind for science? Nobody. But thy good ideas, the new facts which thou hast discovered—these, sifted from all that is immaterial, will be known and remembered to all time. But how comes it that I should advise the lion to eat sugar?[7]

[7] F. Wöhler to J. von Liebig, March 9, 1843. In A. W. Hofmann (ed.), *Aus Justus Liebig's und Friedrich Wöhler's Briefwechsel*, Vieweg, Braunschweig, 2 vols., 1888, vol. 1, p. 224. The translation follows W. A. Shenstone, *Justus von Liebig, His Life and Work*, Macmillan, New York, 1895, p. 37; and F. R. Moulton, *Liebig and After Liebig*, Am. Assoc. Adv. Sci., Washington, 1942, p. 3.

Liebig's vigorous activity soon led to a decline of his health which plagued him throughout the remainder of his life. By 1852 his health was such that when he received an attractive offer from the University of Munich, he accepted on condition that he be permitted to abandon laboratory instruction. During his last twenty years he continued to publish actively and engaged in some of his bitterest controversies, but his influence in chemical circles was clearly declining.

Determination of Nitrogen

Most chemists after Lavoisier sought to determine nitrogen along with carbon and hydrogen in combustion analyses, but without real success. Liebig's long study of combustion analysis led him to conclude that the nitrogen analysis must be done separately from that for carbon and hydrogen. Several chemists worked on nitrogen analysis, but it was Dumas who first devised a reliable procedure.

Dumas made two important innovations. (1) He substituted carbon dioxide for air in the combustion tube, and (2) he used concentrated potassium hydroxide instead of mercury as the liquid over which to collect the nitrogen. Using carbon dioxide in the combustion tube enabled him to measure directly the nitrogen liberated by the sample, instead of having to correct for the air in the tube. Furthermore, the carbon dioxide gas dissolved in the potassium hydroxide solution as the nitrogen was collected. Dumas produced the carbon dioxide by heating lead carbonate in an extension of the combustion tube; but in 1838 O. L. Erdmann and R. F. Marchand devised a gas generator which produced carbon dioxide by the action of acid on a suitable carbonate. In 1868 Hugo Schiff introduced an improved azotometer that replaced the graduated cylinder Dumas used for collecting the nitrogen gas.

An alternative procedure for the determination of nitrogen was introduced in 1841 by two of Liebig's former students, Varrentrapp and Will, who converted the nitrogen to ammonia by heating the sample with soda lime. Their method was satisfactory for the analysis of many industrial and physiological materials such as proteins, but it lacked the reliability and universal applicability of the Dumas method and was superseded by the Kjeldahl method in 1883. The Dumas method is in use today; it has been modified only with respect to heating methods and scale of operation.

Determination of Sulfur and the Halogens

Besides his contributions to carbon and hydrogen analysis, Liebig is also responsible for devising methods for analyzing elemental sulfur and the halogens. He oxidized organic materials with a nitrate in an alkaline solution. The sulfate formed could then be determined by precipatation as barium sulfate. This method is not suitable for all organic sulfur compounds because many of them volatilize before the oxidation to the sufate is completed; but it works well for proteins and is still used. The same method of oxidation was applied in analyzing the organic halogens, the halide being precipitated and weighed as the silver salt. This method also is still used, but to a somewhat limited extent because of the ease with which many organic halogen compounds are volatilized, even under alkaline conditions.

A more universal method for analyzing elemental sulfur, chlorine, bromine, and iodine was developed in 1864 by George Ludwig Carius (1829–1875), a professor at Heidelberg in 1861 and at Marburg in 1865. He studied the thermal decomposition of nitric acid, the identity of phenaconic acid and fumaric acid, and the reactions of organic halogen compounds. In his method for sulfur and halogen analysis, the sample is oxidized with concentrated nitric acid in a sealed glass tube. The sulfuric acid formed is precipitated and weighed as barium sulfate, the hydrohalogen acid as the silver salt. The method is widely applicable, but there is the risk of explosion that is always present in handling hot, sealed glass tubes.

ATOMIC WEIGHT OF CARBON

There remained one difficulty in determining empirical formulas from analysis—the inaccuracy of the atomic weight of carbon. The then current value, 76.4 ($O = 100$, or 12.23 for $O = 16$), had been used for years. It was based upon Dulong's comparison of the relative densities of carbon dioxide and oxygen. The value 138.218:100 suggested that 38.218 g. of carbon combined with 100 g. of oxygen. Gay-Lussac and Dumas assumed that the two elements were present in equal atomic ratios. On this basis the atomic weight should be 38.218 (6.12 for $O = 16$). In order to use CO as the formula of the lower oxide, Berzelius assigned the formula CO_2 to carbon dioxide, making the atomic weight of carbon 76.436 (12.23).

This value caused difficulty from time to time, particularly with naphthalene which was discovered in coal tar by Garden in 1819. Faraday, who studied the sulfonic acid derivatives in 1826, derived a formula of $C_{20}H_8$ ($C = 6$) for naphthalene. Oppermann, one of Liebig's students, obtained a higher value in his carbon analysis and assigned the formula C_3H_2 ($C = 12$). A determination of the vapor density made by Dumas in 1832 suggested the validity of Faraday's formula, but taken as $C_{10}H_8$. Since there was no way of reconciling the $C_{10}H_8$ formula with C_3H_2, the whole question remained open. A careful analysis made by Mitscherlich in 1836 gave widely varying results, but he assumed the discrepancy to be due to an experimental error. Dumas was concerned about the discrepancy, particularly because it appeared consistently when the formula of naphthalene was calculated from analytical results; this led him to question the accuracy of the accepted atomic weight of carbon. Dumas, working in association with Jean Servais Stas (1813–1891), carefully determined the weight of carbon dioxide produced by the combustion of weighed samples (five diamonds and nine pieces of graphite) of pure carbon.[8] The atomic weight proved to be exactly 75 (O = 100, or 12 when O = 16). They reported that previous values had been erroneous because of incomplete combustion of the carbon.

GROWTH OF THE RADICAL THEORY

When Lavoisier was studying the nature of acids he expressed the idea that they were binary compounds of a radical with oxygen. In mineral acids the radical was a single element like sulfur, phosphorus, carbon, the radical "boracique," the radical "fluorique," or the radical "muriatique."

Table 7.1.   FORMULAS OF SOME ORGANIC COMPOUNDS

Compound	Berzelius' Formula	Modern Formula	Anhydride
Citric acid	CHO	$C_6H_8O_7$	$C_6H_6O_6$ or $(CHO)_6$
Tartaric acid	$C_4H_5O_5$	$C_4H_6O_6$	$C_4H_4O_5$
Oxalic acid	$C_{12}HO_{18}$	$C_2H_2O_4$	C_2O_3
Succinic acid	$C_4H_4O_3$	$C_4H_6O_4$	$C_4H_4O_3$
Acetic acid	$C_4H_6O_3$	$C_2H_4O_2$	$C_4H_6O_3$
Gallic acid	$C_6H_6O_3$	$C_6H_6O_3$	
Mucic acid	$C_6H_{10}O_8$	$C_6H_{10}O_8$	
Benzoic acid	C_5H_3O	$C_7H_6O_2$	
Tannin	$C_6H_6O_4$		
Sugar	$C_{12}H_{21}O_{10}$	$C_{12}H_{22}O_{11}$	
Milk sugar	CH_2O	$C_{12}H_{22}O_{11}$	
Gum	$C_{13}H_{24}O_{12}$		
Starch	$C_7H_{13}O_6$	$C_6H_{10}O_5$	

Acetic acid, however, contained a radical made up of carbon and hydrogen which was brought to the acid state by oxygen. Since such acids as malic and tartaric contained the same elements, the proportion of carbon and hydrogen in the radical might be different.

Berzelius, in 1815, sought to calculate formulas from his analytical values; the resulting formulas are shown in Table 7.1.[9] Since he based his

[8] J. B. A. Dumas and J. S. Stas, *Compt. rend.*, **11**, 991 (1850); *Ann. chim. phys.*, [3], **1**, 5 (1841).

[9] J. J. Berzelius, *Ann. chim. phys.*, **94**, 1, 170, 296, 323 (1815); **95**, 51 (1815). The table is from T. M. Lowry, *Historical Introduction to Chemistry*, Macmillan, London, 1915, p. 393.

analyses on the lead salts wherever possible, his formulas really represent the anhydrides. Modern formulas of the acids and their anhydrides are shown for comparison.

In following years Berzelius found the dualistic theory very successful for formulating inorganic compounds, but very confusing in the case of organic counterparts. About the best that could be done was to treat acids and their metallic salts in dualistic terms. It was difficult to formulate other compounds in a dualistic manner.

When Gay-Lussac carried out his researches with hydrogen cyanide in 1815 and discovered cyanogen, $(CN)_2$, he noted that here was a compound that behaved like an element. It showed the characteristics to be expected in an organic radical and came to be generally symbolized as Cy rather than CN. Berzelius listed cyanogen compounds in his 1839 *Jahres-Bericht* and compared them with corresponding chlorine compounds; his list is shown in Table 7.2.[10]

Table 7.2. RELATION OF CYANOGEN AND CHLORINE COMPOUNDS

Cyanogen gas	C_2N_2 or Cy_2	Chlorine gas	Cl_2
Cyanogen chloride	ClCN or ClCy		
Prussic acid	HCN or HCy	Muriatic acid	HCl
Potassium cyanide	KCN or KCy	Potassium chloride	KCl
Silver cyanide	AgCN or AgCy	Silver chloride	AgCl
Mercuric cyanide	HgC_2N_2 or $HgCy_2$	Mercuric chloride	$HgCl_2$

Gay-Lussac determined the vapor density of alcohol and ether and found that the density could be expressed in terms of olefiant gas (C_2H_4) and water vapor, thus:

1 vol. alcohol vapor = 1 vol. olefiant gas + 1 vol. water vapor
1 vol. ether vapor = 2 vol. olefiant gas + 1 vol. water vapor

He also observed that during fermentation 51.34 parts of alcohol and 48.66 parts of carbon dioxide were obtained. This suggested that the formula of sugar is $C_6H_{12}O_4$.

No particular emphasis was placed upon the concept of radicals until 1828 when Dumas and Pierre F. G. Boullay (1777–1869), a French pharmacist, advanced the etherin theory. This suggested that the compounds related to alcohol might be understood if they were considered to be addition products of ethylene[11] (C_2H_4) just as ammonium compounds were considered addition products of ammonia. Such a comparison is

---

[10] J. J. Berzelius, *Jahres-Bericht*, **18**, 120 (1839).

[11] J. B. A. Dumas and P. F. G. Boullay, *Ann. chim. phys.*, **37**, 15 (1828). Table 7.3 is from T. M. Lowry, *op. cit.*, p. 402.

presented in Table 7.3. Dumas and Boullay concluded that the ethylene played the role of a strong alkali and that it would show the alkaline reactions of ammonia if it were soluble in water.

Berzelius was not enthusiastic about the proposal despite the fact that it incorporated a certain element of dualism. The absence of oxygen in

Table 7.3. COMPARISON OF ETHERIN AND AMMONIUM COMPOUNDS

Etherin (ethylene)	$C_2H_4{}^a$	Ammonia	NH_3
Alcohol	C_2H_4, H_2O	Ammonium hydroxide	NH_3, H_2O
Sulfuric ether	$2C_2H_4, H_2O$	Ammonium oxide	$2NH_3, H_2O$
Hydrochloric ether	C_2H_4, HCl	Ammonium chloride	NH_3, HCl
Hydroiodic ether	C_2H_4, HI	Ammonium iodide	NH_3, HI
Nitric ether	C_2H_4, HNO_2	Ammonium nitrite	NH_3, HNO_2
Acetic ether	$C_2H_4, C_2H_4O_2$	Ammonium acetate	$NH_3, C_2H_4O_2$
Sulfovinic acid	C_2H_4, H_2SO_4	Ammonium bisulfate	NH_3, H_2SO_4

^{a}Dumas and Boullay used the atomic weight, C = 6, and consequently their formula for ethylene was C_4H_4.

the ethylene portion of the formula bothered him. According to him, Dumas and Boullay's suggestion was an interesting way of representing the relationships between these compounds but he doubted that the compounds were actually formed in this way.

Table 7.4. LIEBIG AND WÖHLER'S BENZOYL DERIVATIVES

Compound	Formula	Modern formula
Benzoyl hydride	$C_{14}H_{10}O_2 \cdot H_2$	$C_7H_5O \cdot H$
Benzoic acid	$C_{14}H_{10}O_2 \cdot (OH)_2$	$C_7H_5O \cdot OH$
Benzoyl chloride	$C_{14}H_{10}O_2 \cdot Cl_2$	$C_7H_5O \cdot Cl$
Benzoyl cyanide	$C_{14}H_{10}O_2 \cdot C_2N_2$	$C_7H_5O \cdot CN$
Benzamide	$C_{14}H_{10}O_2 \cdot N_2H_4$	$C_7H_5O \cdot NH_2$
Ethyl benzoate^a		$C_7H_5O \cdot OC_2H_5$
Benzoin^a	_____	$C_{14}H_{12}O_2$

^{a} These compounds were prepared by Liebig and Wöhler but the formulas were not completed.

In 1832 Liebig and Wöhler published their famous paper on oil of bitter almonds.[12] In their researches they encountered evidence for the existence of the benzoyl radical, $C_{14}H_{10}O_2$. The seeds of bitter almonds contain a glycoside, amygdalin, which on hydrolysis yields glucose, hydrogen cyanide, and benzaldehyde, C_6H_5CHO, the last-named compound

[12] J. von Liebig and F. Wöhler, *Ann.*, **3**, 249 (1832).

being referred to as oil of bitter almonds. Liebig and Wöhler learned that benzaldehyde can be oxidized to benzoic acid, chlorinated to benzoyl chloride, and converted to benzoin. Benzoyl chloride can be converted to the bromide, iodide, cyanide, amide, and ethyl ester by double decomposition. All these compounds indicated the presence of the benzoyl radical, $C_{14}H_{10}O_2$ (really C_7H_5O). Thus benzoyl compounds could be formulated as in Table 7.4.

Berzelius was delighted with this discovery. The benzoyl radical, unlike the previously studied radicals such as ammonium, cyanogen, and ethylene, contained three elements, including oxygen, and behaved as if it were a single element. In a letter to Liebig and Wöhler that was published with their paper, Berzelius wrote:

> The facts that you have ascertained suggest so many considerations, that they may well be regarded as the beginning of a new day in vegetable chemistry. I would therefore propose, to call the first example of a compound radical containing more than two substances *Proin* (from πρωί, the beginning of the day . . .) or *Orthrin* (from ὀρθρός, dawn of the morning).[13]

He proposed that the benzoyl radical be represented by the symbol Bz, as he was already doing in the case of ammonium, Am. He also proposed the name amide for compounds that contain NH_2 and are formed by removing an atom of hydrogen from ammonia:

$$Bz \cdot NH_2 \quad \text{Benzamide}$$
$$K \cdot NH_2 \quad \text{Potassamide}$$
$$Na \cdot NH_2 \quad \text{Sodamide}$$

Berzelius was now ready to accept ethylene as a radical and proposed that it have the symbol E. After further consideration, however, his enthusiasm cooled and he sought to abandon both the benzoyl and etherin radicals in favor of a revision which brought back oxygen more nearly to the position of importance it had held under Lavoisier. In 1833 he reiterated the importance of binary structures for organic as well as inorganic compounds, and renounced the idea of oxygen-containing radicals. Benzoyl must be the oxide of $C_{14}H_{10}$; anhydrous benzoic acid, the peroxide. Ether, $C_4H_{10}O$, was the suboxide of the ethyl radical, C_2H_5; this made ether the equivalent of metal oxides. Ethers (esters) then became, when combined with acid, the equivalent of salts, i.e., $CaO \cdot SO_3$, $CaO \cdot C_4H_6O_3$, $C_4H_{10} \cdot C_4H_6O_3$. Alcohol, according to Berzelius' concepts, was considered to be the oxide of a different radical, C_2H_6. Thus, the relationship between alcohol and ether was obscured, since

---

[13] J. J. Berzelius, *Ann.*, **3**, 282 (1832).

the vapor densities of alcohol (46) and ether (74) suggested no relationship between the two compounds. Thus he arrived at the correct molecular composition for ether while failing to show its relationship with alcohol.

Liebig, quickly aware of this difficulty, suggested that ether and alcohol were derivatives of the same radical—ethyl, C_4H_{10}. Ether was called the oxide, $C_4H_{10}O$, and alcohol was called the hydrate of ether, $C_4H_{10}O \cdot H_2O$. Here he erred in making the formula for alcohol double its real value. He realized the discrepancy, but considered it inconsequential because of his lack of faith in vapor density values.

The advantages of the ethyl theory formulated by Berzelius and Liebig were readily apparent, despite the discrepancies, for such compounds as ethyl chloride and ethyl iodide fitted into the system as addition products of the halogen to the ethyl radical. This was true also of mercaptan, recently discovered by Zeise; ethyl mercaptan, C_2H_5SH, was formulated as the sulfide of the ethyl radical combined with sulfuretted hydrogen, $C_4H_{10}S + H_2S$. Furthermore, this theory agreed with the observation that two of the hydrogen atoms played a different role than the others did. Benzoic ether (ethyl benzoate) was formulated as $C_4H_{10}O \cdot C_7H_{10}O_3$. Glacial acetic acid was regarded as being $C_4H_6O_3$, so ethyl acetate became ethyl oxide plus acetic acid, $C_4H_6O_3 \cdot C_4H_{10}O$, just as calcium acetate was $C_4H_6O_3 \cdot CaO$. Following the same line of thought, Berzelius believed that the ammonium radical was N_2H_8, with N_2H_8O as its hypothetical oxide. Then ammonium sulfate would be $N_2H_8O \cdot SO_3$, and ammonium acetate would be $N_2H_8O \cdot C_4H_6O_3$.

For the time being, Liebig and Berzelius were pleased with the way a number of common compounds could be formulated as derived from the ethyl radical. Liebig was able to secure Dumas' acceptance of the theory when he visited Paris in 1837. These two leaders of European organic chemistry announced their intention to work together rather than in opposition and to publish jointly a natural classification of organic compounds based upon a detailed study of component radicals.[14] They agreed to prepare a joint report on organic formulation for the Liverpool meeting of the British Association in 1838, and Dumas and Graham were listed as collaborators in the publication of Liebig's *Annalen der Pharmacie und Chemie* between 1838 and 1841. However, the era of good feeling was short-lived, for by the time the intention to collaborate was publicly announced, Dumas was already muddying the waters with his experiments on substitution. The Liverpool paper failed to materialize and Liebig soon discarded the ethyl radical in favor of the acetyl radical.

Liebig's abandonment of the ethyl radical stemmed from newly discovered chlorides which contained too little hydrogen to be included in the ethyl series, but were clearly derived from ethyl alcohol. One of his students, Henri Victor Regnault (1810–1878), had shown that when the

[14] J. von Liebig and J. B. A. Dumas, *Compt. rend.*, **5**, 567 (1837).

oil of the Dutch chemists ($C_2H_4Cl_2$) was treated with potassium hydroxide a compound was formed; analysis indicated that its formula was $C_4H_6Cl_2$ (really C_2H_3Cl, chloroethylene). Regnault found further that ethylene, when treated with bromine or iodine, gave the analogue of the "oil of the Dutch chemists," whereupon treatment with alkali produced bromoethylene and iodoethylene, formulated as $C_4H_6Br_2$ and $C_4H_6I_2$.

Since the formula of acetic acid was $C_4H_6O_3$, Liebig named the C_4H_6 radical the acetyl radical; acetic acid was considered to be its oxide. Etherin (ethylene) could be thought of as a compound of acetyl and hydrogen, $C_4H_6 + H_2$, and ethyl could be considered either as $C_4H_6 + 2H_2$ or as etherin $+ H_2$. Thus, it was possible to classify a larger number of compounds as derived from the acetyl radical than from the ethyl radical. At the same time, however, Liebig failed to realize that at each step he was setting up an artificial radical solely for the purpose of including more compounds.

Despite the confusion in deciding on suitable radicals, the radical theory continued to gain support during the 1830's. New series of compounds suggested additional radicals. Dumas and Peligot in 1834 discovered the methyl radical in their experiments on "wood spirit." Boyle had observed that the distillate of wood could be separated into an acid and a neutral portion, but it was not until this decade that the neutral portion was shown to be an alcohol. The French chemists named it methyl alcohol, meaning "wine of wood," and prepared gaseous dimethyl ether and several esters. These compounds were considered derivatives of the radical, C_2H_3 (CH_3), which Berzelius called methyl.

These same two investigators examined Chevreul's ethal in 1836. They showed that it was an alcohol, which they named cetyl alcohol ($C_{16}H_{33}OH$), and that it gave characteristic reactions of an alcohol and of a cetyl radical. An Italian student of Dumas, Raphaele Piria (1815–1865), was responsible for the discovery of cinnamic acid ($C_6H_5 \cdot CH{=}CH \cdot COOH$) and of a cinnamyl series of compounds.

Three years later Bunsen reported his investigations on cacodyl, the heavy, fuming, stinking, spontaneously flammable liquid that the pharmacist Cadet had obtained in 1760 on distilling potassium acetate with arsenious oxide. From this liquid Bunsen obtained an oxide which he assigned the formula $C_4H_{12}As_2O$. On treatment with acids, this oxide yielded the chloride, cyanide, iodide, and fluoride. When the chloride was heated with zinc in an atmosphere of carbon dioxide, it gave the free radical, $C_4H_{12}As_2$, $[(CH_3)_2As{-}As(CH_3)_2]$. This series of experiments resulted in a new radical and further strengthened the radical theory.

Robert Wilhelm Bunsen

The son of a philology professor at Göttingen, Robert Wilhelm Bunsen (1811–1899) received his doctorate there in 1830, presenting a thesis on various kinds of hygrometers. Following a period of travel, he became a docent at Göttingen. He succeeded Wöhler at Cassel in 1836, became professor at Marburg in 1839, and taught for a year at Breslau before accepting a professorship at Heidelberg in 1852, from which he retired in 1889. His research on cacodyl marked the end of his studies in organic chemistry. Cacodyl compounds are not only spontaneously flammable, but explosive and highly toxic. In 1843 an explosion of cacodyl cyanide resulted in the loss of his right eye and the fumes left him ill for weeks.

His subsequent career was in the field of inorganic analysis. He was highly inventive and had a hand in developing not only the Bunsen burner (1853) but the spectroscope (1859), the Bunsen electrochemical cell (1840), the grease-spot photometer (1844), the absorptiometer (1855), the actinometer (with Roscoe in 1856), the effusion apparatus (1857), the filter pump (1868), and an ice calorimeter (1870). He used simple methods with great effectiveness in his experiments and abhorred speculative theorization. His study of English blast-furnace operations, made with Lyon Playfair in 1845, was important because it showed that only 15 per cent of the reducing power of the fuel was utilized, the rest being lost as carbon monoxide. The study also revealed the possibility of recovering ammonium and cyanogen compounds.

His studies on blast-furnace gases led Bunsen to examine gas analysis as a whole, and his book, *Gasometrische Methoden*, published in 1857, became a classic on the subject. A trip to Iceland in 1846 led to significant studies on the nature of geysers. During the 1850's, working with his English student Henry Enfield Roscoe (1833–1915), he made extensive studies on the effect of light on chemical reactions.

Confusion in Formulas

The end of the 1830's saw the radical theory rather generally accepted as a means of formulating organic compounds. However, there was no real agreement among chemists and, as we have seen, even prominent chemists frequently changed their ideas. The whole problem was further confused by the lack of consistency in formulas resulting from the lack of agreement on atomic weights. Berzelius tended to use atomic weights which, calculated on the modern basis, approximately agree with present-day values. But he was prone to use double-volume formulas, that is, formulas that agreed with the vapor density of two volumes of gas or with the equivalent obtainable from one volume of hydrogen gas. Thus, he symbolized common gases as $\overline{HCl}$, $\overline{NH}_3$, and $\overline{HI}$. Methane became $\overline{CH}_4$ or C_2H_8. Liebig's inclination was to use equivalent weights for carbon and oxygen which had the effect of doubling his formulas. Dumas used C = 6 and O = 16 for a time. As a result of these different approaches, formulas were meaningless unless the system used by a particular chemist and his followers was known.

Nevertheless, the radical theory, including as it did the methyl, cetyl, cinnamyl, benzoyl, and cacodyl radicals, had the effect of bringing a group of compounds into one system. Since the greatest number of well-known compounds were drived from ethyl alcohol, this series naturally received the greatest attention—and was subject to the greatest confusion with

Atomic Weights in Common Use

	H	C	O
Berzelius	1	12	16
Liebig	1	6	8
Dumas	1	6	16

the advent of the etherin, ethyl, and acetyl theories. The relationships between these radicals can perhaps best be shown in tabular form, as in Table 7.5. Modern formula equivalents are used wherever possible, but when they become meaningless Liebig's double formula, in brackets, is listed.

Table 7.5. COMPARISON OF FORMULAS OF DERIVATIVES OF ETHYL ALCOHOL

	Modern	Etherin (Dumas)	Ethyl (Berzelius and Liebig)	Acetyl (Liebig)
Radical		C_2H_4	C_2H_5	C_2H_3
Oxide			C_2H_5O	C_2H_3O
Alcohol	C_2H_5OH	$C_2H_4 \cdot H_2O$	$C_2H_5O \cdot H$	$C_2H_3O \cdot 3H$
Ether	$(C_2H_5)_2O$	$2C_2H_4 \cdot H_2O$	$[C_4H_{10}O \cdot H_2O]$	$[C_4H_6O \cdot 6H]$
Ethyl chloride	C_2H_5Cl	$C_2H_4 \cdot HCl$	$C_2H_5 \cdot Cl$	$C_2H_3Cl \cdot 2H$
Ethyl iodide	C_2H_5I	$C_2H_4 \cdot HI$	$C_2H_5 \cdot I$	$C_2H_3I \cdot 2H$
Ethyl cyanide	C_2H_5CN	$C_2H_4 \cdot HCN$	$C_2H_5 \cdot CN$	$C_2H_3CN \cdot 2H$

SUBSTITUTION AND THE TYPE THEORY

During all the debate about radicals there was general agreement that they were fixed in composition and that there was no place in them for highly electronegative elements like oxygen and chlorine. At the same time, there had been evidence that chlorine could be substituted for hydrogen ever since Gay-Lussac prepared cyanogen chloride by the action of chlorine gas on hydrogen cyanide. Using direct substitution, Faraday had prepared hexachloroethane from ethylene chloride and Liebig and Wöhler had prepared benzoyl chloride from benzaldehyde. The phenomenon of

Fig. 7.10. AUGUSTE LAURENT (after a
painting by his daughter).
(Courtesy of the Edgar Fahs Smith Collection)

substitution received no particular notice, however, until the mid-1830's
when Dumas and Laurent brought it into prominence.

Auguste Laurent (1807–1853) was born in the small village of La Folie
in northeastern France. The son of a wine merchant, it was expected that
he would enter his father's business, but he showed no interest in it;
instead, his teachers urged his father to let him prepare for college.
He attended a small college near his home before entering the École des
Mines in 1826. On receiving his engineering diploma he became assistant
to Dumas. After studying analytical procedures, Laurent began work on
naphthalene, reporting on its preparation from coal tar, confirming the
formula suggested by Faraday, and describing its reactions with bromine
and nitric acid.

In 1833 Alexandre Brongniart (1770–1847) appointed Laurent to direct
the testing laboratory of the porcelain works at Sèvres. He remained here
for two years during which he developed a procedure for analyzing
silicates by means of hydrofluoric acid. He then opened a school where he
gave instruction in laboratory techniques. When he found that this left
him no time for research, and when his funds were exhausted, he went to
work for a perfume manufacturer. After he accumulated sufficient funds
he embarked upon an industrial venture which failed. In 1838 he was
employed at a porcelain works in Luxembourg, but left before long and
became a professor at the University of Bordeaux. Later he returned to

Paris where he offered private instruction in chemistry while continuing his researches.

His papers on naphthalene indicate that Laurent was beginning to grasp the nature of substitution. His understanding of it improved as he extended his studies to include the chlorination of ethylene chloride, esters, and the products of coal tar, particularly phenol.

In the meantime, Dumas was carrying on chlorination studies with various substances. His interest in substitution seems to have originated with the "incident of the smoky candles." At a ball held in the Tuileries in Paris, fumes given off by the candles made the guests start coughing. The problem of finding the cause was referred to Brongniart, whom the king frequently consulted. Brongniart passed it on to his son-in-law, Dumas, who soon found that the irritant gas was hydrogen chloride, the candles having been made from wax that had been bleached with chlorine. In his studies of the chlorination of wax, oil of turpentine, and similar substances, Dumas observed, as Gay-Lussac had earlier in connection with the bleaching of wax, that for every volume of hydrogen eliminated, a volume of chlorine was absorbed. In 1834 Dumas studied the mechanism for converting alcohol into chloral and chloroform, compounds that had been discovered by Liebig in 1831 when he treated alcohol with chlorine. (Chloroform was independently discovered by Soubeiran in France and Guthrie in the United States when they treated alcohol with bleaching powder.)

Dumas stated his law of substitution as follows:

1. When a substance containing hydrogen is exposed to the dehydrogenising action of chlorine, bromine, or iodine, for every volume of hydrogen that it loses, it takes up an equal volume of chlorine, bromine, etc.
2. When the substance contains water, it loses the hydrogen corresponding to this water without replacement.[15]

The second rule was proposed in order to explain the formation of chloral and at the same time uphold his etherin radical which had been under fire by Berzelius and Liebig. Dumas believed that in alcohol, $C_8H_8 + 2H_2O$ ($C = 6$), the eight hydrogen atoms that were combined with carbon behaved differently from the four that were combined with oxygen, thus:

$$(C_8H_8 + 2H_2O) + 4Cl \rightarrow C_8H_8O_2 + 4HCl$$
Aldehyde

$$C_8H_8O_2 + 12Cl \rightarrow C_8H_2Cl_6O_2 + 6HCl$$
Chloral

[15] J. B. A. Dumas, *Ann. chim phys.*, [2], **56**, 113 (1834). Quoted from A. Ladenburg, *Lectures on the History of the Development of Chemistry*, Alembic Club, Edinburgh, 1900, p. 141.

The chlorine thus removed the oxygenated hydrogen atoms without substitution, but substituted, atom for atom, for the hydrogen atoms that were combined with carbon.

He went further in regarding oxidation as substitution. In the conversion of alcohol into acetic acid he believed that every hydrogen atom eliminated was replaced by a half atom of oxygen.

$$(C_8H_8 + H_4O_2) + O_4 \rightarrow (C_8H_4O_2 + H_4O_2) + H_4O_2$$
$$\text{Alcohol} \qquad\qquad \text{Acetic acid}$$

$$C_{28}H_{10}O_2 \cdot H_2 + O_2 \rightarrow C_{28}H_{10}O_2 \cdot O + H_2O$$
$$\text{Oil of bitter almonds} \qquad \text{Benzoic acid}$$

Thus Dumas supposed that hydrogen, chlorine, bromine, and iodine were equivalent, and that oxygen had double their substituting value. He introduced the name metalepsy for substitution reactions. The subject was followed up energetically in the following years by Dumas himself and by Laurent, Peligot, Regnault, and Malaguti.

Laurent in 1835 studied the relationship between the original and the substituted compound. Observing that the properties were not much altered, he concluded that chlorine took on the role of hydrogen. A year later he proposed what came to be known as the nucleus theory; it held that compounds contained nuclei or radicals within which substitution might take place. According to him, there were original nuclei (*radicaux fundamentaux*) and derived nuclei (*radicaux dérivés*). The latter were produced from the former by the substition of hydrogen atoms by chlorine, bromine, or oxygen atoms, or even by compound radicals such as amide or nitro groups. The nucleus theory was an abandonment of the radical theory since radicals were no longer considered to have a permanent composition. The *radicaux fundamentaux* were thought to be hydrocarbon nuclei in which the hydrogen could be replaced without destroying the character of the nucleus.

In his doctoral thesis, which he presented in December, 1837, and defended before Dumas and Dulong, and two mineralogists named Beudant and Despretz, Laurent developed his nucleus theory. The fundamental radicals were presented as being like geometric figures. He used a four-sided prism as an example of a fundamental radical, C^8H^{12}. The eight corners were occupied by carbon atoms. This fundamental nucleus might serve as an acceptor of suitable atoms or groups of atoms. For example, the two narrow sides of the prism might accept atoms of hydrogen, chlorine, or oxygen.

$$C^8H^{12} + H^2 \qquad \text{Hyperhydride}$$
$$C^8H^{12} + Cl^2 \qquad \text{Hyperchloride}$$
$$C^8H^{12} + O \qquad \text{Aldehyde}$$
$$C^8H^{12} + O^2 \qquad \text{Acid}$$

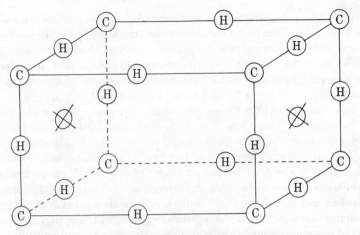

Fig. 7.11. LAURENT'S NUCLEUS THEORY. The rectangular prism represents a nucleus consisting of 8 carbon atoms joined with 12 hydrogen atoms to form a fundamental radical, C^8H^{12}. The two positions marked with a circle with crossed lines may be occupied by hydrogen, chlorine, or some other element. Substitution at these positions is possible without destroying the fundamental radical, or nucleus.

In certain reactions the addition products might be removed, leaving the fundamental nucleus alone.

Laurent supposed further that if an atom of hydrogen were removed from the fundamental nucleus the effect would be to destroy the edge of the prism unless the hydrogen were replaced by a suitable atom like chlorine or a half atom of oxygen. This prism $C^8(H^{11}Cl)$ would then be a derived nucleus. This derived nucleus might then accept hydrogen, chlorine, or oxygen just as the fundamental nucleus did, i.e.

$$C^8(H^{11}Cl) + H^2$$
$$C^8(H^{11}Cl) + Cl^2$$

Or more of the hydrogen atoms might be substituted in the derived nucleus, i.e. $C^8(H^{10}Cl^2)$, $C^8(H^9Cl^3)$, $C^8(H^8Cl^4)$, etc. Thus the concept of the radical would be utilized in a way that permitted alteration of the radical. But this was directly opposed to the traditional concept of the radical. Also in effect it abandoned the dualism which Berzelius considered so important.

Laurent's nucleus theory was received very unfavourably. Liebig charged that it was unscientific and pernicious. Berzelius, rising to the attack, erroneously linked Dumas with the theory, and dismissed the whole matter with the remark that detailed criticism was superfluous. Dumas was quick to disclaim any responsibility for the theory, pointing out that his sole contribution was the law of substitution. Berzelius refused to believe that a highly electronegative element like chlorine could replace electropositive hydrogen without drastically altering the character of the substance.

In the summer of 1838 Dumas prepared trichloracetic acid and studied its reactions and derivatives, noting its striking similarity to acetic acid. During the next two years he developed the type theory. Comparison of the properties of trichloracetic acid and acetic acid, chloral and acetaldehyde, and methane and chloroform led him to accept the idea that chlorine might take on the role of hydrogen; this contradicted his earlier disclaimer when Berzelius had charged him with advocating ideas actually developed by Laurent. In a memoir on the type theory presented before the Académie in 1840, Dumas attributed the idea of mechanical types to Regnault and referred to Laurent in passing as a man who had maintained that the role of chlorine was identical with that of hydrogen in substituted compounds before there was evidence to substantiate it. Dumas ignored Laurent's development of the mechanical type, and discounted the significance of the latter's fine experimental work on the substitution of halogens and nitric acid in benzene, naphthalene, and phenol; at the same time he suggested that his own work on trichloracetic acid provided the important evidence against the dualistic theory and in favor of the type theory.

Laurent justifiably claimed priority for the type theory, or derived radicals. Even earlier he had accused Dumas of duplicity in connection with his ideas.

> I have not been able to dismiss an emotion of indignation in seeing certain chemists first call my theory absurd, then much later when they have seen that the facts are in agreement with my theory better than they are with all the others, pretend that I have taken some ideas of M. Dumas. If it fails, I shall be the author, if it succeeds another will have proposed it. M. Dumas has done much for the science; his part is sufficiently great that one should not snatch from me the fruit of my labors and present the offering to him.[16]

There followed a lengthy series of papers in which Laurent and Dumas each sought to establish priority for the type theory.

Laurent continued to carry on experimental studies in the primitive laboratory he set up at Bordeaux with his own meager funds. He continued his studies of coal tar distillates, isolating phenol in 1841 and studying its nitration to the already known picric acid. In his work on naphthalene derivatives he ascertained the correct formula for phthalic acid and

[16] A. Laurent, *Ann. chim. phys.*, [2], **66**, 326 (1837). The translation is from Clara de Milt, *Chymia*, **4**, 95 (1953).

studied the anhydride, imide, and the acid formed on nitration. From pine tar he isolated and identified pimaric acid; he isolated isatin from indigo. As early as 1837 he had begun to study fatty acids in order to establish relationships between them. By treating oleic acid with concentrated nitric acid he obtained the already known suberic acid and the new pimelic, adipic, and azelaic acids. Besides his research on organic compounds he continued his work on the atomic weight of chlorine and on borates and silicates.

Resistance to Substitution and the Type Theory

Despite the accumulating mass of experimental evidence, the work on substitution and types was not favorably received in some of the important chemical centers. In part, this was due to personalities. Dumas had a reputation for speculation, in some cases on the basis of meager experimental evidence. Laurent, a young man who was already showing an unusual capacity for achieving significant experimental results, was even then setting his brilliant mind to the problem of formulating a systematic concept of organic chemistry. His intuitive capacity for theoretical synthesis was leading him toward a unitary view of chemical formulas which naturally rankled Berzelius, who was finding it increasingly difficult to maintain the authority of his dualistic theory.

A contribution in French, which appeared in the *Annalen der Pharmacie und Chemie* in 1840,[17] gave evidence of the reluctance to accept Dumas' enthusiastic effusions on substitution and types. The author, one S. C. H. Windler, reported the results of experiments on the substitution of chlorine in manganese acetate, $MnO + C_4H_6O_3$. Windler first found that the hydrogen in the acetate was replaced by chlorine; but as the treatment with chlorine was continued, the oxygen and manganese in the base were replaced, then the oxygen in the acid, and finally even the carbon. Thus, a compound was produced that analyzed 100 per cent chlorine, but had all the properties of manganese acetate!

The hoax was a satire on Dumas' unsubstantiated speculations on substitution, which finally went so far as to suggest that carbon atoms were susceptible to substitution. The idea of the hoax originated with Wöhler, who always loved a joke; he wrote the piece and sent it to Berzelius. A copy of it reached Liebig who, in his capacity as editor of the *Annalen*, printed it in the journal after adding a footnote reporting that cotton cloth produced in England by chlorine substitution, and containing 100 per cent chlorine, was in great demand in fashionable circles in London.

On the more serious side, Berzelius was engaged in a last effort to save his dualistic concept in the face of the evidence for substitution. Dumas vigorously denounced dualism, Liebig and Berzelius as vigorously upheld it. Liebig was willing to admit the fact of substitution but regretted the

[17] S. C. H. Windler (F. Wöhler), *Ann*, **33**, 308 (1840).

wide extension of the principle. For five years after 1838, Berzelius protested the type theory and maintained that electronegative elements could not enter into the composition of radicals.

As a means of explaining the existence of compounds like trichloracetic acid Berzelius, following an 1839 idea of Gerhardt, developed the copula theory in 1841. He suggested that the radicals in the substituted compound differed in composition from those of the original compound. For example:

Acetic acid: $\qquad$ $C_4H_6 + O_3 + H_2O$

Trichloracetic acid: $\qquad$ $C_2Cl_6 + C_2O_3 + H_2O$

The substitution caused a drastic realignment in which chlorinated carbon was copulated (united) with oxalic acid, whereas in acetic acid the compound was the oxide of the acetate radical.

In 1842 Melsens showed that trichloracetic acid could be converted into acetic acid by treatment with potassium amalgam. Since this clearly established the similarity between the two acids, Berzelius was forced to change his position further. He now held that substitution might be possible in a limited part of the compound and proposed that both acids be considered as being derived from a radical combined with oxalic acid. Thus:

Acetic acid $\qquad$ $C_2H_6 + C_2O_3 + H_2O$

Trichloracetic acid: $\qquad$ $C_2Cl_6 + C_2O_3 + H_2O$

Such reasoning, however, could not stop the onslaught of unitary ideas and Berzelius was a lonely figure during his last years. Even Liebig, in view of the experiments on chlorine and bromine derivatives of aniline, carried out in the Giessen laboratory by his own student, Hofmann, could no longer go along with Berzelius' attempts to save his theory by desperate rearrangements of radicals. Liebig's attitude is seen in the statement:

> During the last years [of his life] Berzelius ceased to take an experi-
> mental share in the solution of the problems of the time, and turned
> the whole force of his mind to theoretical speculations; but these not
> being the result of his own observations or supported by them, found
> no echo or approval in the science.[18]

Furthermore, Liebig's own work on acids had the effect of turning him away from dualism.

STUDIES OF POLYBASIC ACIDS

After Davy and Dulong formulated their concept that hydrogen some-how was of significance in acids but that oxygen was not essential, there

[18] J. von Liebig, *Ann.*, **50**, 297 (1844). Translation from E. Meyer, *History of Chemistry*, Macmillan, London, 1891, p. 268.

was little further advancement in the understanding of acids until the studies of polybasic acids made by Graham and Liebig in the 1830's.

Thomas Graham (1805–1869) studied at Glasgow under Thomas Thomson and at Edinburgh under Thomas Hope before becoming professor of chemistry in the Andersonian University (now the Royal Technical College) in Glasgow. He then went to University College, London, where he filled the chair of Edward Turner. He finally left teaching to become Master of the Mint in 1854. His *Elements of Chemistry* (1841) became a popular textbook that was translated into German. His researches on the diffusion of gases and liquids and on phosphoric acids are of fundamental importance.

In 1833 Graham published the results of his studies on phosphoric acids. It was known at this time that there were apparently two forms of phosphoric acid which produced a variety of salts. Common sodium phosphate (Na_2HPO_4) gave a yellow precipitate with silver nitrate and left the solution acidic. But when the phosphate was heated above 350° C., the salt that was formed gave a white precipitate with silver nitrate and the solution remained neutral.

The dualistic formulation in use at that time was instrumental in covering up the real nature of the phosphoric acids. Thus the normal sulfate of potassium was formulated as $KO \cdot SO_3$. The acid sulfate was formulated as $KO \cdot 2SO_3$ and was known as potassium bisulfate. The hydrogen in the latter was considered to be water somewhat similar to water of crystallization. Bicarbonates were formulated as $KO \cdot 2CO_2$, bichromates as $KO \cdot 2CrO_3$.

In the case of the phosphates of sodium, both the pyrophosphate ($Na_4P_2O_7$) and the neutral phosphate (Na_2HPO_4) appeared to have the formula $2NaO \cdot P_2O_5$. The existence of two such salts was considered a form of isomerism. Graham found that when crystals of the neutral phosphate of soda (formulated $PO_5 \cdot 2NaO \cdot 25HO$) were heated at the temperature of boiling water, 24/25 of the water was lost. The last unit of water was not lost until the temperature was much higher; the salt thus formed was the pyrophosphate which gave the white precipitate with silver salts. The difference between the neutral phosphate and the pyrophosphate was due to the molecule of water that was lost on heating. This must be other than water of crystallization because solutions of the two salts were not identical. Hence Graham concluded that water might play the role of a base (metal oxide) in a salt, and even in the acid itself.

He went on to show that a third type of phosphate (NaH_2PO_4) was derived from the acid phosphate of sodium; its solution was acid to indicators. On heating this salt, he obtained the metaphosphate ($NaPO_3$). Therefore he arrived at the conclusion that there were three phosphoric acids which differed from one another in the amount of water that was combined with the oxide of phosphorus. He further concluded that one or more units of water were replaced by units of base (basic oxide) in the formation of salts, and that the water acted as a base in the acid. This is illustrated in Table 7.6, which shows formulas used by Graham, the formulas corrected for the proper valence of sodium, phosphorus, and hydrogen, and the modern formulas for the various compounds.

Table 7.6. GRAHAM'S CLASSIFICATION OF THE PHOSPHORIC ACIDS

Acid or Salt	Ratio Acid/Base	Graham's Formula	Corrected Graham Formula	Modern Formula
(Ortho)phosphoric acid	1:3	$PO_5 \cdot 3HO$	$P_2O_5 \cdot 3H_2O$	H_3PO_4
Pyrophosphoric acid	1:2	$PO_5 \cdot 2HO$	$P_2O_5 \cdot 2H_2O$	$H_4P_2O_7$
Metaphosphoric acid	1:1	$PO_5 \cdot HO$	$P_2O_5 \cdot H_2O$	HPO_3
Neutral sodium phosphate	1:3	$PO_5 \cdot 2NaO \cdot HO$	$P_2O_5 \cdot 2Na_2O \cdot H_2O$	Na_2HPO_4
Basic or pyro- phosphate	1:2	$PO_5 \cdot 2NaO$ or $PO_5 \cdot NaO \cdot HO$	$P_2O_5 \cdot 2Na_2O$ or $P_2O_5 \cdot Na_2O \cdot H_2O$	$Na_4P_2O_7$ $Na_2H_2P_2O_7$
Sodium meta- phosphate	1:1	$PO_5 \cdot NaO$	$P_2O_5 \cdot Na_2O$	$NaPO_3$

Graham reported that the same relationships were apparent in the case of the arsenic acids and the arsenates.

When Liebig and Dumas agreed to stop arguing and pool their ideas in 1837, they published one paper before they again went their individual ways. The significant part of this paper deals with organic acids which apparently show the multiplicity of character that Graham found in the phosphoric and arsenic acids. The work is clearly Liebig's and he published an even more important paper on the subject the following year.[19]

Liebig's researches dealt with such organic acids as fulminic, cyanic, meconic, comenic, tartaric, malic, and citric. He at one time thought that cyanates, fulminates, and cyanurates represented a series like the meta-, pyro-, and ortho-phosphates. However, his studies soon revealed the futility of treating organic groups the same way as inorganic oxides were treated. Tartaric acid, formulated $C_4H_4O_5$ (really $C_4H_6O_6$, i.e., $C_4H_4O_5 \cdot H_2O$), was considered to be capable of combining with one unit of base until Liebig, reasoning from the existence of tartar emetic, Rochelle salt, and potassium and ammonium tartrate that the acid must be able to combine with two units of base, doubled the formula to $C_8H_8O_{10}$.

He now began to think of acids as monobasic, dibasic, and tribasic.

[19] J. von Liebig, *Ann.*, **26**, 113 (1838).

A dibasic acid was one in which two units of water could be replaced by two units of base i.e., $C_8H_8O_{10} + 2H_2O$. In such an acid the replacement of one unit of water would produce an acid salt. The two units of water in such an acid could also be replaced by two different bases, as in the case of Rochelle salt (sodium and potassium are both present in the tartrate) and tartar emetic (potassium and the antimonyl group are present).

In his search for a comprehensive understanding of acids, Liebig soon recognized the inadequacy of the role supposedly played by water. Each of Graham's phosphoric acids was definitely stable even in solution. If the difference between them was due simply to the presence of one, two, or three units of water combined with phosphorus oxide, why did not the meta- and pyro-phosphates immediately revert to the ortho-phosphates in solution? The hydroacids also presented a problem for they could not be explained on the basis of water because no oxygen was present. Liebig resurrected the ideas of Davy and Dulong on the role of hydrogen in acids, thus abandoning the attempt to retain Berzelius' dualism in favor of a unitary concept. His new views were expressed as follows:

Davy's theory arose from the behavior of potassium chlorate . . . whose decomposition at an elevated temperature into oxygen and potassium chloride without change in neutrality made it necessary . . . to conclude that the potassium was not present as the oxide. . . . Davy concluded as follows: hydrochloric acid is a compound of chlorine and hydrogen, $Cl_2 + H_2$.

The radical of hydrochloric acid may take up one or several atoms of oxygen without changing its saturation capacity for . . . this facility is dependent only upon the hydrogen of the acid which is located outside the radical.

Hydrochloric acid	$Cl_2 \quad + H_2$
Hypochlorous acid	$Cl_2O_2 + H_2$
Chlorous acid	$Cl_2O_4 + H_2$
Chloric acid	$Cl_2O_6 + H_2$
Perchloric acid	$Cl_2O_8 + H_2$

Acids are, according to this view, hydrogen compounds in which the hydrogen may be replaced by metals.

The salts of phosphoric acid receive the following form:

$P_2O_8 + H_6$	Phosphoric acid
$P_2O_8 + \left.\begin{array}{c} H_2 \\ 2K \end{array}\right\}$	So-called neutral salt
$P_2O_8 + \left.\begin{array}{c} H_4 \\ K \end{array}\right\}$	Acid salt
$P_2O_8 + 3K$	So-called basic salt . . .

If phosphoric acid be exposed to a higher temperature a part of the hydrogen outside the radical combines with an equivalent of oxygen of the latter, water is evolved and two new acids, pyro- and metaphosphoric acid, are formed.

$$P_2O_7 + H_4 \text{ Pyrophosphoric acid}$$
$$P_2O_6 + H_2 \text{ Metaphosphoric acid}$$

All the properties of cyanuric, meconic and citric acids indicate that they contain no water in the dry state. . . . The salts of these acids are composed in a manner analogous to the phosphorus acids.

Durch die Nacht führt unser Weg zum Lichte.[20]

CONCLUSION

It was inevitable that organic chemistry should go through a chaotic period in the first half of the nineteenth century. As long as the other fields of chemistry were in a state of uncertainty, they could be of little help in clarifying organic chemistry. Furthermore, the very nature of organic chemistry made it unlikely that its secrets would be revealed easily.

Up to this point, progress had been of an empirical sort, stemming from the discovery of new compounds and new reactions. The perfection of methods for elemental analysis led in the direction of a new level of understanding, but progress in this direction was retarded by the lack of agreement on atomic weights, which in turn was reflected in formulas. Another two decades were necessary to achieve clarification of the field of organic chemistry.

---

[20] *Ibid.*, pp. 181–189. The translation is from H. Leicester and H. S. Klickstein, *A Source Book of Chemistry*, McGraw-Hill, New York, 1952, pp. 318–320.

CHAPTER 8

ORGANIC CHEMISTRY II.
ORGANIZATION

THE CHAOS OF THE 1840's

As the fifth decade of the nineteenth century began, the dualistic view of organic compounds had begun to give way to the unitary view that a compound must be considered as an entity rather than as a composite made up of two units which were themselves complete compounds. However, many chemists were still uncertain as to where they stood. Despite the facts of substitution, the radical was still in favor in many laboratories and some workers were certain that radicals could exist in the free state. There was uncertainty regarding their composition and, as we saw in Liebig's use of them, they were frequently manipulated to fit preconceived notions or experimental facts.

There was equal uncertainty about nuclei and types among supporters of these concepts. Dumas' enthusiasm about substitution led to ridicule and disbelief. The insight needed to bring about a consolidation of radicals and types was not available in this period when there was no clear understanding of the distinction between equivalents, atoms, and molecules. In many cases, the equivalent weights advocated by Wollaston and Gmelin were preferred to Berzelius' atomic weights. Berzelius and his followers still clung to double-volume formulas for simple gases like hydrogen chloride, (H_2Cl_2). To Berzelius, the size of an acid molecule was the amount that combined with a molecule of metal oxide; the latter was written as AgO and KO. This naturally showed no correlation with the volume of gases and concealed the true formulas of polybasic acids.

Formula notation was still primitive, frequently meaningless, and always confusing. The formula of water was written variously as HO, H_2O, ~~HO~~, and ~~HO~~. The ~~H~~O formula could be taken to mean two atoms of hydrogen combined with one atom of oxygen, or two volumes of hydrogen combined with one volume of oxygen, or one equivalent of oxygen united with two equivalents of hydrogen. There was such chaos in a world of atoms, equivalents, radicals, and types that it is scarcely surprising that many chemists were contemptuous of all manner of speculation. That Faraday did not accept the atom and that Wöhler regarded organic chemistry as a primeval forest so densely matted that one hesitated to enter, should not strike us as surprising. To these chemists the search for facts offered satisfaction, but the idea of giving the whole body of facts a logical organization offered only distress. Perhaps it is not unfair to say that chemistry a century ago was in the state of confusion that character-izes nuclear physics today.

THE ORGANIZING EFFORTS OF LAURENT AND GERHARDT

The two figures who stand out in the 1840's were the young French chemists, Laurent and Gerhardt. Laurent has already been mentioned in connection with substitution and the nucleus theory. Gerhardt attained his major stature somewhat later, only after he came in contact with Laurent early in that decade. Although the individual contributions of the two men are not easily dissociated, it appears today that Laurent was responsible for the original ideas, and Gerhardt for their development.

Charles Frédéric Gerhardt (1816–1856) was born in Strasbourg; his father was a manufacturer of white lead. He studied with Erdmann at Leipzig where he did noteworthy research on the formulas and classifica-tion of natural silicates. After an unhappy period of work in his father's factory and of service in the army, he studied with Liebig and worked on the formula of picric acid. A second period of work in his father's manu-facturing plant resulted in a permanent break with his father, whereupon he completed his doctorate in Paris in 1841. He took a teaching position at Montpellier but gave it up in 1848 in order to go to Paris, then the center of interest in chemistry. However, Dumas' antagonism made it impossible for him to secure a connection with any college there. He subsisted only on what he made from tutoring private students and from his books; the Scottish relatives of his wife also provided financial assistance. He took a position at the University of Strasbourg in 1855, just a year before his very sudden death. His brilliance as a systematizer of organic chemistry was recognized early, but his personality created many enemies. Dumas definitely stood in the way of his advancement; furthermore, few chemists were in a position to promote his ideas during his lifetime.

Fig. 8.1. CHARLES FRÉDÉRIC GERHARDT.
(From *Charles Gerhardt, sa vie, son œuvre, sa correspondence,* ed. by E. Grimaux and Ch. Gerhardt [1900].)

By 1839 Gerhardt had proposed his theory of residues—also known as the second radical theory because the residues often had the same formulas as the dualistic radicals. Gerhardt refused to admit this identity, for his residues had no electrical characteristics and he never claimed that they were present as such in the molecule. He believed that even though simple substances like water, ammonia, and hydrogen chloride were reaction products, this was no proof that they were present as water, ammonia, and hydrogen chloride in the compounds from which they were obtained; moreover, these reaction products could not be recompounded directly into the original organic compounds from which they were formed. His concepts indicated that he was groping toward structural ideas, but he refused to believe that structures could be determined. He even wrote variant formulas for the same compound in order best to reveal its role in chemical reactions. Thus, barium sulfate he formulated variously as BaO, SO_3; BaS, O_4; and BaO_2, SO_2.

According to the theory of residues, certain inorganic compounds are so stable that they are formed from organic compounds with great ease. This leaves organic residues which combine and form new compounds. For example, hydrogen readily separates from one compound, oxygen from another, to produce water, a stable molecule; the residues then combine to form a stable molecule of a new organic compound. Mitscherlich's nitrobenzene was formed by such a reaction between benzene and nitric acid. In modern terms:

$$C_6H_5\cdot H + HONO_2 \rightarrow H_2O + C_6H_5\cdot NO_2$$

Gerhardt, of course, did not regard the residues as being preexistent in the reactants or intact in the product. Ethyl nitrate was formed by the reaction of ethyl alcohol and nitric acid, as follows (again in modern terms):

$$C_2H_6O + HNO_3 \rightarrow C_2H_5NO_3 + H_2O$$

In Gerhardt's notation:

$$C_4H_6O + HNO_5 \rightarrow C_4H_5, NO_5 + H_2O$$

Ammonia and benzoyl chloride reacted as follows:

$$Cl(C_7H_5O) + NH_3 \rightarrow NH_2(C_7H_5O) + HCl$$

whereas potassium hydroxide reacted with the benzamide to give potassium benzoate:

$$NH_2(C_7H_5O) + KOH \rightarrow KO(C_7H_5O) + NH_3$$

Chemical reactions thus became decompositions which were followed by exchanges of residues, and these in turn by combination.

Essentially Gerhardt's residues were radicals that were capable of rearrangement. He made it plain that radicals must be looked upon only as residues that could combine with other radicals (residues) to form new compounds, not as radicals that could be isolated. This concept of radicals was accepted to a certain extent because it did not demand the rigidity of composition embodied in the older theory of radicals. Further, the residues were satisfactory in explaining substitution reactions; thus the two older theories began to be brought together. In substitution, the element eliminated is replaced by an equivalent of another element or the residue of the reacting compound. By 1843 Gerhardt regarded all compounds formed by the elimination of water, such as those involving the reaction of nitric or sulfuric acid with alcohols or hydrocarbons, as products that illustrated his concept of the combination of residues.

Gerhardt attacked another major problem in 1842, following up ideas first brought to his attention by Laurent. This involved atomic and molecular weights.[1] Liebig and Gerhardt had adopted the use of four-volume formulas, as against Berzelius' two-volume formulas. The four-volume formula represented the volume of vapor that occupied the same space as four volumes of hydrogen, i.e., alcohol, $C_4H_{12}O_2 = H_4$. Gerhardt now undertook to collate all the organic formulas believed to be reason-

[1] C. F. Gerhardt, *J. prakt. Chem.*, **27**, 439 (1842).

ably valid. He observed that these formulas could be divided by 2; this resulted in formula weights compatible with Ampère's hypothesis. (Gerhardt always referrred to Avogadro's hypothesis as Ampère's, never as Avogadro's.) Gerhardt realized that, according to the older formulas, water, ammonia, carbon dioxide, and hydrogen chloride must always be eliminated as double molecules, H_4O_2, N_2H_6, C_2O_4, and H_2Cl_2. Halving the organic formulas meant that these products were eliminated in accordance with Berzelius' formulas. This idea of halving was further supported by work Regnault was doing on specific heats of compounds, which indicated an approximate relationship between specific heat and molecular weight akin to that postulated by the law of Petit and Dulong for elements. Gerhardt confused the whole issue, however, by concluding that "atoms, volumes, and equivalents are synonymous terms."

The net effect of Gerhardt's innovation was the replacement of equivalents like $C = 6$ and $O = 8$ in organic formulas by their atomic weights, i.e., $C = 12$ and $O = 16$. In effect, Gerhardt was using without acknowledgment the atomic weights of Berzelius, thus bringing organic formulas into line with inorganic formulas. He then adopted atomic weights for metals which halved those calculated by Berzelius. He assumed that the oxides of metals followed the pattern of water, Ag_2O, K_2O, Na_2O, etc., and thus corrected the weights arrived at by Berzelius, who had assigned the general formula, MO, to these metal oxides. Unfortunately Gerhardt extended the reasoning to various other metals and consequently the correct values for such elements as zinc, calcium, and copper were halved. He used the laws of isomorphism and of Petit and Dulong when they suited his purpose, but did not apply them consistently. His atomic weights were an improvement in certain cases, a worsening in other cases. They were not generally adopted even though, in the case of organic compounds, they had the effect of making the formula agree with the weight of "two volumes"; that is, the weight of a volume of vapor was equal to the weight of 2 g. (two "volumes") of hydrogen. Gerhardt was attempting to apply Avogadro's hypothesis without really understanding it. He used the old formulas in the first three volumes of his *Traité de chimie organique*, published in four volumes between 1853 and 1856, because he believed this was necessary to attract purchasers. Only in the last volume, published after his death, were the new formulas used.

Laurent finally clarified the distinction between *atom* and *equivalent* in 1846 when he adopted Gerhardt's atomic weights. The atomic weight of an element, Laurent maintained, must be taken as the smallest weight of it present in a molecular weight of its compounds. According to him, the molecule was the smallest quantity that could be used in forming a compound, and the value of the equivalent must vary with the nature of the reaction. Since he held that molecular weights must be related to vapor densities, he adopted the practice of using as the molecular weight, except in a few cases, the weight of vapor whose volume corresponded to the volume occupied by 2 g. of hydrogen.

Laurent maintained that hydrogen, nitrogen, oxygen, and chlorine molecules occurred as double atoms, which he called *dyadides*. Gerhardt welcomed this idea, for it enabled him to treat substitution by chlorine as a double decomposition, one atom of the chlorine serving as a residue for combination with an organic residue.

In his two-volume *Précis de chimie organique*, published in 1844–1845, Gerhardt advocated using empirical formulas exclusively because he saw no way of establishing the real structure of organic compounds. Any special arrangement assigned to atoms in a formula was merely for convenience in describing a particular reaction or concept, and hence had no actual meaning. Laurent deplored this arbitrary attitude, for he believed that certain values could be derived from special arrangements.

Gerhardt, with his penchant for classification, followed the suggestions made by Laurent and other chemists and devised three series of organic classifications. He extended and systematized the *homologous series*, introduced earlier by J. Schiel.[2] Schiel had pointed out the uniform difference in composition in the seven known alcohols (methyl, ethyl, etc.). Dumas had done this for fatty acids (formic, acetic, propionic, butyric, etc.).[3] Hermann Kopp's researches had revealed that the similarities between such compounds were physical as well as chemical.[4] Gerhardt now gathered together the compounds which made up distinct series which were alike in general properties and differed only in having one or more CH_2 groups. He did not mention Dumas' papers on fatty acids, and this was the beginning of the bitter feeling between the two chemists, for Dumas claimed priority for the idea of homology. Dumas himself had failed to recognize Schiel's earlier work and had ignored Laurent's even earlier ideas on the relation of acids and esters.[5] That Laurent also helped to clarify Gerhardt's ideas on classification by his criticisms of the first volume of the *Précis* is evident from the fact that the second volume reflects a clarification that was based on these criticisms.

In the *isologous series* Gerhardt included compounds whose chemical behavior was related, but which showed no evidence of homology, i.e., ethyl alcohol and phenol, acetic acid and benzoic acid. His *heterologous series* included compounds whose chemical behavior differed but which showed a genetic relationship—ethyl alcohol and acetic acid, and amyl alcohol and valeric acid, for example. Compounds in this series usually have the same number of carbon atoms.

[2] J. Schiel, *Ann.*, **43**, 107 (1842).

[3] J. B. A. Dumas, *Compt. rend.*, **15**, 935 (1842); J. A. B. Dumas, J. B. Boussingault, and A. Payen, *Ann. chim. phys.*, [3], **8**, 63, (1843).

[4] H. Kopp, *Ann.*, **41**, 79, 169 (1842).

[5] A. Laurent, *Ann. chim. phys.*, [2], **65**, 294 (1837).

THE NEW TYPE THEORY

The collecting of compounds into related series made inevitable the development of a new kind of type. Both Laurent and Gerhardt were moving in this direction at the time that Wurtz and Hofmann's experiments on amines led to the establishment of the ammonia type and Williamson's work on ethers led to the development of the water type.

The Ammonia Type

Although the alkaloids had attracted a great deal of interest, the simple amines were not studied seriously until the 1840's. In 1845 Hofmann investigated the chlorination and bromination of aniline. This organic base, first prepared in 1826 by Unverdorben by the decomposition of indigo, Berzelius regarded as a copulated compound of ammonia and hydrocarbon, $NH^3 \cdot C^{12}H^8$. Hofmann agreed with this concept for a time.

August Wilhelm von Hofmann (1818–1892) was born in Giessen. He attended the local university where he studied philology and law until he became interested in Liebig's work. He remained at Giessen as Liebig's assistant for several years after taking his doctorate; then, on Liebig's recommendation, he became director of the newly founded Royal College of Chemistry in London.

While still with Liebig, Hofmann embarked upon an extensive study of coal tar and its derivatives, among which was aniline. His interest stemmed from his desire to clear up the confusion resulting from the similarity between the *Kyanol* that Runge found in coal tar in 1834, the *Anilin* that Fritzsche prepared from indigo in 1840, and a base that Zinin prepared by reducing nitrobenzene with alcoholic ammonium sulfide. Hofmann isolated Runge's *Kyanol* from coal tar and proved its identity with the compounds prepared by Fritzsche and Zinin.

Nikolai Nikolaevich Zinin (1812–1880) began his career as a mathematician, after studying with Lobachevski at the University of Kazan. He taught at the university, turning from physics to chemistry, which became his lifetime interest. A traveling fellowship in 1837 permitted him to study under Mitscherlich and Rose in Berlin, and to spend a year at Giessen where he worked on benzoyl compounds. On his return to Kazan he was made professor of technology, but continued his research on organic compounds. In his first studies he discovered that aromatic nitro compounds can be reduced by sulfide. He submitted a manuscript on this to the *Annalen*, which was read by Carl Julius von Fritzsche (1808–1871), who recognized Zinin's product as aniline. After further studies by Zinin revealed the general nature of the reaction, he prepared other amines such as phenylenediamine and aminobenzoin by reducing the appropriate nitro analogue. Zinin was an important figure in Russian chemistry around mid-century, first at Kazan and later at the Medico-Surgical Academy in St. Petersburg. He was an early adherent of the ideas of Laurent and Gerhardt and had great influence on such students as Butlerov and Borodin.

Nitrobenzene was prepared and studied by Mitscherlich in the years following 1832. Hofmann, in 1845, improved on the preparation of aniline by using zinc and hydrochloric acid in reducing the nitrobenzene.

The next important step in the study of amines was made by Wurtz in 1849 when he obtained methylamine and ethylamine by treating the cyanic esters with potassium hydroxide. Wurtz recognized these compounds as bases comparable to ammonia but in which an equivalent of hydrogen had been replaced by a methyl or ethyl radical. Liebig had predicted such compounds a decade earlier. Wurtz's synthesis produced concrete evidence for the existence of an ammonia type of compound.

Charles Adolphe Wurtz (1817–1884) was born in Strasbourg, Gerhardt was a schoolmate of his. Wurtz studied under Liebig and Balard, then became an assistant to Dumas at the École de Médecine; he succeeded him there in 1853. He became professor at the Sorbonne in 1875. Besides his synthesis of amines, Wurtz synthesized ethylene glycol and glycolic acid and devised the well-known Wurtz synthesis for preparing aliphatic hydrocarbons by the action of sodium metal on alkyl iodides. Despite his friendship for Dumas, Wurtz was an admirer of the ideas of Laurent and Gerhardt and was influential in giving them proper prominence.

It was Hofmann who brought the ammonia type to its logical conclusion. He synthesized ethylamine by the reaction of ammonia on ethyl iodide, then went on to prepare the secondary and tertiary analogues of ethylamine. According to this work, the ammonia type of organic compound could represent the replacement of one, two, or three hydrogen atoms by organic radicals in the ammonia molecule.

$$
\left.\begin{array}{l} H \\ H \\ H \end{array}\right\}N \qquad
\left.\begin{array}{l} C_2H_5 \\ H \\ H \end{array}\right\}N \qquad
\left.\begin{array}{l} C_2H_5 \\ C_2H_5 \\ H \end{array}\right\}N \qquad
\left.\begin{array}{l} C_2H_5 \\ C_2H_5 \\ C_2H_5 \end{array}\right\}N
$$

  Ammonia   Ethylamine   Diethylamine   Triethylamine

There was no limit to the number of combinations in the substituted ammonias, as was shown by experiments in which aniline was allowed to react successively with ethyl and amyl bromides.

$$
\left.\begin{array}{l} C_6H_5 \\ H \\ H \end{array}\right\}N \qquad
\left.\begin{array}{l} C_6H_5 \\ C_2H_5 \\ H \end{array}\right\}N \qquad
\left.\begin{array}{l} C_6H_5 \\ C_2H_5 \\ C_5H_{11} \end{array}\right\}N
$$

   Aniline   Phenylethylamine   Phenylethylpentylamine

All these compounds retained a basic character.

Hofmann further observed that when triethylamine was treated with ethyl iodide a white crystalline addition product, tetraethylammonium iodide, was formed. He correctly interpreted the compound as one in which an ammonium radical in which ethyl groups were substituted for hydrogen played the same role as sodium or potassium did in a metal iodide. The action of silver oxide on the iodide produced a strong base. The analogy with ammonia was complete. However, even further evidence of the analogy was at hand in Hofmann's observation that the amine

Fig. 8.2. CHARLES ADOLPHE WURTZ.
(Courtesy of the Edgar Fahs Smith Collection.)

Fig. 8.3. ALEXANDER WILLIAM WILLIAMSON.
(Courtesy of the Edgar Fahs Smith Collection.)

hydrochloride behaved like ammonium chloride in forming insoluble salts with platinum chloride.

Organic bases now began to be found frequently. Methylamine was prepared from caffeine by von Rochleder. Trimethylamine was isolated from herring brine by von Wertheim in 1851. Although he believed that it was propylamine, Hofmann revealed the true composition of the compound which is liberated when fish are being pickled, and also when they are spoiling. Thomas Anderson (1819–1874), one of Liebig's former students and now professor of chemistry at Glasgow, obtained from tar a base that had the same formula as aniline, but different properties; he called it picoline (methyl pyridine). The dry distillation of bones yielded additional bases—pyridine, lutidine (dimethyl pyridine), and collidine (trimethyl pyridine). Anderson realized that the four compounds formed a new homologous series of bases, starting with pyridine, and that as yet there were no other examples. When they reacted with ethyl iodide they behaved like Hofmann's quaternary ammonium compounds. He held that the radical C_5H_5 took the place of three atoms of hydrogen in ammonia. Anderson also isolated pyrrole from bone oil and correctly analyzed it; Runge had previously isolated it from tar. Pyridine was isolated from coal tar in 1855 by C. G. Williams.

Studies by Hofmann and Auguste Cahours (1813–1891) in 1855 showed that ammonia type had phosphinium analogues. Cahours, who had been trained under Chevreul and Dumas, taught for many years at the École Centrale.

The Water Type

The water type of compound was explained by Alexander William Williamson (1824–1904) in connection with his method of preparing ethers by the action of potassium alcoholates on alkyl iodides. Born in London of Scottish parents Williamson studied chemistry under Graham, Leopold Gmelin, and Liebig, and mathematics under Auguste Comte (1798–1857) at the École Polytechnique. In 1849 he succeeded Fownes as professor of analytical chemistry at University College, London, and five years later became professor of general chemistry when Graham gave up his professorship.

Williamson began work in 1850 with the intention of preparing substituted alcohols in the way Hofmann had prepared substituted ammonias. When he treated ethyl iodide with potassium ethylate he was surprised when ethyl ether was produced. He recognized the relationship between ether and alcohol at once and succeeded in clearing up the confusion regarding them. Most chemists still regarded ethyl alcohol and ether on the basis of one of the older radical hypotheses. Representative formulas were:

	Dumas (C = 6)	Berzelius	Liebig
Alcohol	C^8H^8, H^4O^2	C^2H^6O	$C_4H_{10}O, H_2O$
Ether	C^8H^8, H^2O	$C^4H^{10}O$	$C_4H_{10}O$

The concept of vapor densities had led Laurent in 1846 to conclude that the compounds were analogous to water.

$$\left.\begin{array}{l}H\\H\end{array}\right\}O \qquad \left.\begin{array}{l}C_2H_5\\H\end{array}\right\}O \qquad \left.\begin{array}{l}C_2H_5\\K\end{array}\right\}O \qquad \left.\begin{array}{l}C_2H_5\\C_2H_5\end{array}\right\}O$$

Water   Alcohol   Alcoholate   Ether

Williamson's synthesis suggested that the following might occur. (He did not use brackets in his formulas; Gerhardt was the first to use them.)

$$\begin{array}{l}C_2H_5\\K\end{array}O + C_2H_5I \rightarrow KI + \begin{array}{l}C_2H_5\\C_2H_5\end{array}O$$

Of course, some still argued that the potassium ethylate decomposed first to potassium oxide and ether, and that the oxide then reacted with the iodide to form more ether. Using the Liebig formulas with $C = 6$, the reaction was:

$$C_4H_{10}O, KO \rightarrow KO + C_4H_{10}O$$
$$C_4H_{10}I + KO \rightarrow KI + C_4H_{10}O$$

Williamson eliminated this possibility by repeating the synthesis, using methyl iodide and potassium ethylate. According to the last mechanism, this should have given a mixture of dimethyl and diethyl ethers; according to his own method, methylethyl ether should be formed. The product was entirely methylethyl ether.

$$\left.{C_2H_5 \atop K}\right\} O + CH_3I \rightarrow KI + \left.{C_2H_5 \atop CH_3}\right\} O$$

Laurent followed up these experiments with others, all of which pointed to the fact that alcohols and ethers agreed with the concept of the water type. The significance Laurent attributed to the difference in vapor densities of alcohols and ethers was justified.

Williamson showed that methylethyl ether could also be produced from ethyl iodide and potassium methylate. He used amyl iodide to prepare mixed ethers containing the amyl radical with ethyl and with methyl radicals. Ether could also be synthesized by the action of sulfuric acid on alcohol; Williamson demonstrated the role of sulfovinic acid (ethyl sulfate) as an intermediate:

$$\left.{H \atop H}\right\} SO_4 + \left.{C_2H_5 \atop H}\right\} O \rightarrow \left.{C_2H_5 \atop H}\right\} SO_4 + \left.{H \atop H}\right\} O$$
$$\left.{H \atop C_2H_5}\right\} SO_4 + \left.{C_2H_5 \atop H}\right\} O \rightarrow \left.{C_2H_5 \atop C_2H_5}\right\} O + \left.{H \atop H}\right\} SO_4$$

In 1851 Williamson set out to show that most salts can be formulated as a water type. Monobasic acids caused no problem.

$$\left.{NO_2 \atop H}\right\} O \qquad\qquad \left.{NO_2 \atop K}\right\} O$$

Nitric acid   Potassium nitrate

However, dibasic acids like sulfuric acid were more difficult. At first he treated the sulfates as a water-like type.

$$\left.{H \atop H}\right\} SO_4 \qquad \left.{H \atop K}\right\} SO_4 \qquad \left.{K \atop K}\right\} SO_4 \qquad \left.{H \atop C_2H_5}\right\} SO_4$$

Then he hit upon the multiple water type, a concept which treats the compounds as if they belonged to a type represented by doubled water molecules.

$$\left.\begin{array}{l}H_2\\H_2\end{array}\right\}O_2 \qquad \left.\begin{array}{l}SO_2\\H_2\end{array}\right\}O_2 \qquad \left.\begin{array}{l}SO_2\\HK\end{array}\right\}O_2 \qquad \left.\begin{array}{l}SO_2\\K_2\end{array}\right\}O_2 \qquad \left.\begin{array}{l}CO\\K_2\end{array}\right\}O_2$$

$$\text{Water} \qquad \text{Sulfuric} \qquad \text{Potassium} \qquad \text{Potassium} \qquad \text{Potassium}$$
$$\text{acid} \qquad \text{acid sulfate} \qquad \text{sulfate} \qquad \text{carbonate}$$

William Odling in 1855 developed the multiple type further when he treated phosphates and bismuth compounds as triple water types.

$$\left.\begin{array}{l}3H\\3H\end{array}\right\}3O \qquad \left.\begin{array}{l}PO\\3H\end{array}\right\}3O \qquad \left.\begin{array}{l}PO\\2HK\end{array}\right\}3O \qquad \left.\begin{array}{l}3H\\Bi\end{array}\right\}3O \qquad \left.\begin{array}{l}3NO_2\\Bi\end{array}\right\}3O$$

$$\text{Water} \qquad \text{Phosphoric} \qquad \text{Potassium} \qquad \text{Bismuth} \qquad \text{Bismuth}$$
$$\text{acid} \qquad \text{acid phosphate} \qquad \text{hydroxide} \qquad \text{nitrate}$$

Acetic acid was included in the water type by Williamson as:

$$\left.\begin{array}{l}C_2H_3O\\H\end{array}\right\}O$$

By heating sodium valerate with acetic acid he prepared methyl butyl ketone. This confirmed his suspicion that acetone, derived from acetates, was dimethylketone.

$$\left.\begin{array}{l}CH_3\\CH_3\end{array}\right\}CO \qquad\qquad \left.\begin{array}{l}CH_3\\C_4H_9\end{array}\right\}CO$$

$$\text{Acetone} \qquad\qquad \text{Methyl butyl ketone}$$

The proposal that acids should be included in the water type led Gerhardt to reason that acid anhydrides should be possible, and that they should be produced by the type of reaction used by Williamson. If an acid chloride reacted with the sodium salt of the acid, sodium chloride should separate, leaving the acid residues in combination. Using the appropriate benzoyl derivatives, Gerhardt successfully produced the anhydride.

$$\left.\begin{array}{l}C_7H_5O\\Na\end{array}\right\}O + C_7H_5OCl \rightarrow NaCl + \left.\begin{array}{l}C_7H_5O\\C_7H_5O\end{array}\right\}O$$

$$\text{Sodium} \qquad \text{Benzoyl} \qquad\qquad \text{Benzoic}$$
$$\text{benzoate} \qquad \text{chloride} \qquad\qquad \text{anhydride}$$

Benzoyl chloride and potassium acetate gave the mixed anhydride. For several years benzoyl chloride was the only acid chloride known, but the studies made by Cahours on the chlorides of phosphorus and their action on organic compounds made it possible to prepare acetyl chloride. Gerhardt used this chloride for the preparation of acetic anhydride. Vapor density determinations confirmed the doubling phenomenon, just as in the case of alcohol and ether. The formulas corresponding with two volumes of hydrogen (2 g.) are:

$C^2H^4O^2$	Acetic acid	C^2H^6O	Alcohol
$C^4H^6O^3$	Acetic anhydride	$C^4H^{10}O$	Ether

Other Types

During his last years, Gerhardt brough the new type theory to a high level of development, partly through his own experimental work, but largely through his ability to see and exploit relationships. The germ of the idea was clearly Laurent's; the stimuli from experimental work came from Wurtz, Hofmann, and Williamson, but the systematization was largely the work of Gerhardt. According to his scheme, organic compounds could be classified in four types: water, hydrogen, hydrogen chloride, and ammonia.

The water type included, on the one hand, the neutral compounds like the alcohols and ethers, and on the other, the acidic compounds like acids and acid anhydrides. Esters occupied an intermediate position in the water type. The relationships may be shown as follows, the water type being shown first.

$$\left.\begin{array}{l}H \\ H\end{array}\right\}O \qquad \left.\begin{array}{l}C_2H_5 \\ H\end{array}\right\}O \quad \text{Alcohol} \qquad\qquad \left.\begin{array}{l}C_2H_3O \\ H\end{array}\right\}O \quad \text{Acid}$$

$$\left.\begin{array}{l}C_2H_5 \\ C_2H_5\end{array}\right\}O \quad \text{Ether} \qquad\qquad \left.\begin{array}{l}C_2H_3O \\ C_2H_3O\end{array}\right\}O \quad \begin{array}{l}\text{Acid} \\ \text{anhydride}\end{array}$$

$$\left.\begin{array}{l}C_2H_5 \\ CH_3\end{array}\right\}O \quad \begin{array}{l}\text{Mixed} \\ \text{ether}\end{array} \quad \left.\begin{array}{l}C_2H_5 \\ C_2H_3O\end{array}\right\}O \quad \text{Ester} \quad \left.\begin{array}{l}C_2H_3O \\ C^7H_5O\end{array}\right\}O \quad \begin{array}{l}\text{Mixed} \\ \text{anhydride}\end{array}$$

The hydrogen type included hydrocarbons, ketones, and aldehydes. (The diketone class of compound remained hypothetical until diacetyl, the first representative of the class, was discovered in 1887.)

$$\left.\begin{array}{l}H \\ H\end{array}\right\} \qquad \left.\begin{array}{l}C_2H_5 \\ H\end{array}\right\} \qquad\qquad\qquad \left.\begin{array}{l}C_2H_3O \\ H\end{array}\right\} \quad \text{Aldehyde}$$

$$\left.\begin{array}{l}C_2H_5 \\ C_2H_5\end{array}\right\} \qquad \left.\begin{array}{l}CH_3 \\ C_2H_3O\end{array}\right\} \begin{array}{l}\text{Ketone} \\ \text{(acetone)}\end{array} \quad \left.\begin{array}{l}C_2H_3O \\ C_2H_3O\end{array}\right\} \quad \text{Diketone}$$

The hydrogen chloride type included alkyl chlorides and acid chlorides. Organic bromides and iodides were variants of this type.

$$\left.\begin{array}{l}H \\ Cl\end{array}\right\} \qquad \left.\begin{array}{l}C_2H_5 \\ Cl\end{array}\right\} \quad \text{Alkyl chloride} \qquad \left.\begin{array}{l}C_2H_3O \\ Cl\end{array}\right\} \quad \text{Acid chloride}$$

The ammonia type included amines and amides. Phosphorus compounds could be included in the type as a variant.

$$\left.\begin{array}{l} H \\ H \\ H \end{array}\right\}N \qquad \left.\begin{array}{l} C_2H_5 \\ H \\ H \end{array}\right\}N \qquad \left.\begin{array}{l} C_2H_3O \\ H \\ H \end{array}\right\}N \quad \text{Acid amide}$$

$$\left.\begin{array}{l} C_2H_5 \\ C_2H_5 \\ H \end{array}\right\}N$$

$$\left.\begin{array}{l} C_2H_5 \\ C_2H_5 \\ C_2H_5 \end{array}\right\}N$$

Some of Gerhardt's formulations are farfetched, for obviously they were made to fit into a system. Gerhardt had no faith in any formula which sought to reveal the arrangement of atoms; he placed complete reliance only on empirical formulas. He did not envisage his type formulas as structures. To him they constituted a system of classification which combined the usefulness of the type idea and the usefulness of the radical as a reasonably stable entity, and thus enabled him to treat a great variety of reactions as double decomposition reactions.

THE CONCEPT OF VALENCE

Gerhardt's theory of types was not received kindly by Kolbe and Frankland, both of whom were still disciples of Berzelius' dualism. Although these two men were young enough to realize the strength of the unitary concept of chemical compounds, they still preferred to regard radicals as stable atomic groupings rather than as arbitrary arrangements made solely for classification purposes. Although their work during this period confused the issue further, ultimately it brought about further clarification.

Adolf Wilhelm Hermann Kolbe (1818–1884) studied under Wöhler and assisted Bunsen in Marburg and Playfair in London before taking up editorial duties in connection with chemical publications in 1847. Four years later he was appointed to Bunsen's chair at Marburg. In 1865 he moved to Leipzig where he remained until he died. He became editor of the *Journal für praktische Chemie* upon Erdmann's death in 1869. Kolbe was known as a brilliant thinker and a caustic critic. His outspoken and sometimes ill-natured remarks embroiled him in many arguments and lost him the support of chemists who might otherwise have followed him.

Fig. 8.4. ADOLF WILHELM HERMANN KOLBE.
(Courtesy of the Edgar Fahs Smith Collection.)

Fig. 8.5. EDWARD FRANKLAND.
(Courtesy of the Edgar Fahs Smith Collection.)

Edward Frankland (1825–1899) studied under Playfair at the time Kolbe was his assistant. He also studied for short periods under Bunsen and Liebig. In 1851 he began teaching in Owens College in Manchester. Six years later he moved to St. Bartholomew's Hospital in London. He taught a short time at the Royal Institution before succeeding Hofmann at the School of Mines in 1865.

Realizing that the radical theory could not ignore the fact of substitution, Kolbe proceeded to formulate a new body of concepts which retained as much of Berzelius' dualism and of the radical theory as was possible without disregarding the facts of experimental chemistry. He supported Berzelius' copula theory, but maintained that substitution within part of a compound could influence the character of the rest of the compound. A great deal of Kolbe's work and thought was concerned with the nature of the organic acids and the compounds related to them.

In the studies he made between 1843 and 1847 on chlorination of organic compounds he achieved a total synthesis of acetic acid from inorganic substances. He recognized the significance of synthesis in clarifying relationships between compounds and used this approach frequently. In this case he discovered that carbon tetrachloride was formed on the chlorination of carbon disulfide. The tetrachloride had first been prepared by Regnault in 1839 by the chlorination of chloroform. Kolbe's preparation of it from carbon disulfide meant that it could be prepared from nonorganic substances, since carbon disulfide was prepared by heating iron pyrites with carbon. The latter reaction was discovered in 1796 by Wilhelm August Lampadius (1772–1842), professor of chemistry in the Freiberg School of Mines. In 1844 Kolbe discovered that tetrachloroethylene, a liquid first prepared by Faraday, could be converted to trichloracetic acid by the action of chlorine and water in sunlight. Melsens had shown in 1842 that trichloracetic acid could be converted to acetic acid. Kolbe accomplished this by using electrolytic hydrogen. The connecting link in the total synthesis was found when Kolbe discovered that carbon tetrachloride could be converted to tetrachloroethylene by passing the vapor through a red-hot tube.

Synthesis of Acetic Acid from Non-organic Materials

$$FeS_2 + C \rightarrow CS_2 \qquad\qquad \text{Lampadius, 1796}$$

$$CS_2 + 2Cl_2 \rightarrow CCl_4 + 2S \qquad\qquad \text{Kolbe, 1843}$$

$$2CCl_4 \xrightarrow{\Delta} C_2Cl_4 + 2Cl_2 \qquad\qquad \text{Kolbe, 1844}$$

$$C_2Cl_4 + 2H_2O + Cl_2 \xrightarrow{\text{sunlight}} CCl_3COOH + 3HCl \quad \text{Kolbe, 1844}$$

$$CCl_3COOH + 3H_2 \rightarrow CH_3COOH + 3HCl \qquad\qquad \text{Melsens, 1842}$$

In 1847 when Kolbe and Frankland were acting as assistants to Playfair they discovered that acetic acid could be produced by the action of alkali on methyl cyanide.[6] Dumas, Malaguti, and Felix Leblanc prepared propionic acid from ethyl cyanide, a method that became generally used in the synthesis of fatty acids.

Since cyanogen could be converted to oxalic acid, Kolbe regarded the convertibility of alkyl cyanides to the corresponding fatty acids as evidence of the presence of the oxalic acid copula in these acids. He supported Berzelius' concept of acetic acid as methyl copulated with oxalic acid, $C_2H_3 \cdot C_2O_3$, a view that was strengthened by his results on the electrolysis of potassium salts of fatty acids in 1849. In the case of potassium acetate, for example, hydrogen was liberated at the cathode, and a mixture of carbon dioxide and "free methyl" at the anode. The "free methyl" was actually ethane; but since the analysis for carbon and hydrogen agreed with the calculated content of these two elements in the methyl radical, this misinterpretation was plausible. Kolbe believed that the electric current caused a transfer of oxygen from water to oxalic acid to form the carbonic acid, with the consequent liberation of hydrogen and methyl.

$$C_2H_3 \cdot C_2O_3 + H_2O \rightarrow C_2H_3 + 2CO_2 + H$$

Acetic acid   Water   Methyl   Carbonic   Hydrogen
acid

About the same time, Frankland was carrying on experiments which

[6] W. H. Kolbe and E. Frankland, *Ann.*, **65**, 269 (1850).

suggested the formation of the free ethyl radical. When ethyl iodide was treated with zinc, zinc iodide and "free ethyl" (actually butane) were produced. According to Frankland:

$$C_4H_5I \quad + \quad Zn \quad \rightarrow \quad C_4H_5 + ZnI$$

Ethyl iodide $\qquad\qquad$ Ethyl

Or in modern terms:

$$2C_2H_5I + Zn \rightarrow C_4H_{10} + ZnI_2$$

Butane

As a by-product of the reaction some zinc ethyl was produced.

$$C_4H_5I + 2Zn \rightarrow C_4H_5Zn + ZnI$$

The modern equation is:

$$2C_2H_5I + 2Zn \rightarrow (C_2H_5)_2Zn + ZnI_2$$

This completely unexpected product was the forerunner of the whole class of organometallics.

It was the combination of organic fragments with metals that led Frankland to the concept of combining capacities or *valences*. The genesis of the concept is apparent in his paper, "On a New Series of Organic Compounds Containing Metals."[7] Actually there was a great deal of accumulated evidence which pointed toward a valence concept. The laws of definite and multiple proportions certainly carried the implication that the capacity of atoms to combine was numerically exact and limited. It was well known that a unit of phosphorus combined with three and with five units of chlorine, but there was as yet no knowledge of saturation limits. The number of compounds known was sufficiently large to include a variety of anomalous compounds, thus concealing otherwise clear relationships. Furthermore, the confusion of atom and equivalent prevented recognition of relationships between compounds, and uncertainty regarding atomic weights and formulas resulted in such confusion that no effort to determine these relationships was made. Strangely enough, the valence concept first became apparent in the field of organic chemistry, rather than in the field of inorganic chemistry, with its fairly simple formulas.

Substitution phenomena should have suggested valence relationships. Dumas in 1834 noticed that one atom of chlorine replaced one atom of hydrogen, but only a half atom of oxygen. Liebig's studies on polybasic acids revealed differences in the replacement capacity of metals. Thus he observed that one atom of antimony was equivalent to three atoms of hydrogen, whereas one atom of potassium was equivalent to only one atom of hydrogen.

[7] Frankland, *Phil. Trans. Royal Soc.*, **142**, 417 (1852); *Ann.*, **85**, 329 (1853).

In thinking about Bunsen's cacodyl radical and his own zinc and stanno alkyls, Frankland began to consider the organometallics as derivatives of inorganic compounds in which oxygen or a similar element is replaced by an equivalent amount of hydrocarbon radical. Stanno ethyl might then be explained as SnO in which the oxygen is replaced by ethyl, SnC_4H_5 (C = 6). Cacodyl oxide would be arsenius acid, AsO_3, in which methyl groups have replaced two of the oxygen atoms:

$$
\begin{array}{c}
C_2H_3 \\
\diagup \\
As-C_2H_3 \\
\diagdown \\
O
\end{array}
$$

Similarly, zinc ethyl would be zinc oxide in which ethyl has been substituted for the oxygen.

That Frankland proceeded to extend his ideas of saturation capacity into the inorganic field is clear from the following:

> When the formulae of inorganic chemical compounds are considered, even a superficial observer is impressed with the general symmetry of their construction. The compounds of nitrogen, phosphorus, antimony, and arsenic, especially, exhibit the tendency of these elements to form compounds containing 3 or 5 atoms of other elements; and it is in these proportions that their affinities are best satisfied: thus in the ternal group we have NO_3, NH_3, NI_3, NS_3, PO_3, PH_3, PCl_3, SbO_3, SbH_3, $SbCl_3$, AsO_3, AsH_3, $AsCl_3$, etc.; and in the five-atom group, NO_5, NH_4O, NH_4I, PO_5, PH_4I, etc. Without offering any hypothesis regarding the cause of this symmetrical grouping of atoms, it is sufficiently evident from the examples just given, that such a tendency or law prevails, and that, no matter what the character of the uniting atoms may be, the combining power of the attracting element, if I may be allowed the term, is always satisfied by the same number of these atoms.[8]

Because formulas were commonly inadequate, Frankland's concept had certain unconvincing aspects. Nevertheless, it was a fruitful beginning which led gradually to the idea of treating each atom in a formula as a unit with a characteristic capacity for combination. Other chemists did not adopt this concept enthusiastically, for they were still imbued with radicals and types, and confused by uncertain formulas based upon uncertain atomic weights. The fact that many elements had more than one valence did not help to make the concept more acceptable. However, Frankland brought the idea of valence out into the open where it could begin to exert influence.

---

[8] *Ibid.*, p. 440.

Williamson was groping toward the idea of equivalent combining capacities in 1851 when he represented the relation between potassium carbonate and potassium hydroxide as members of the water type:

$$\left.\begin{array}{l}H_2 \\ H_2\end{array}\right\}O_2 \qquad \left.\begin{array}{l}K_2 \\ (H_2)\end{array}\right\}O_2 \qquad \left.\begin{array}{l}K_2 \\ (CO)\end{array}\right\}O_2$$

Water   Potassium hydroxide   Potassium carbonate

William Odling (1829–1921), professor of chemistry at Oxford, extended Williamson's concept of multiple types in 1855 when he used one or more ticks at the upper right of a symbol to represent substitution values within formulas; this reflected close adherence to the type theory. Thus, Bi‴ denoted that one atom of bismuth had the value of three hydrogen atoms, H′. He represented water and alum as:

$$\left.\begin{array}{l}H' \\ H'\end{array}\right\}O'' \qquad \left.\begin{array}{l}2SO_2'' \\ K'Al_2'''\end{array}\right\}4O''$$

Both Odling and Frankland continued to use the Gmelin equivalents. Hence, Odling considered that two atoms of aluminum were tervalent; this was also his conception of ferric iron, Fe_2'''. Thus, using the double and triple water type enabled him to devise a formula for compounds like glycol and glycerol.

Kolbe's Formulations

Between 1855 and 1859 Kolbe developed a system of organic formulas in which the copulae became integral parts of the compound rather than devices used to explain substitution, a system which treated organic compounds as derivatives of simple inorganic compounds, particularly carbonic acid. Thus, acetic acid was methyl carbonic acid—that is, carbonic acid, C_2O_4—in which an equivalent of oxygen was replaced by C_2H_3, forming $C_2H_3 \cdot C_2O_3$. Organic sulfonates could be similarly derived from SO_3.

Kolbe's work on formulas, although frequently ambiguous and without permanent impact, became valuable as a means of detecting relationships between compounds. His study of alcohols led him to predict the probable existence of secondary and tertiary alcohols, a prediction that was proved true before many years had passed, when isopropyl alcohol was synthesized by Friedel and tertiary butyl alcohol by Butlerov.

Kolbe's experimental researches were of further importance in clarifying relationships. He demonstrated that lactic acid is hydroxypropionic acid, and alanine is aminopropionic acid. He discovered the relationship between glycollic acid, H_2C—OH, and glycine, H_2C—NH_2. He showed that

         | |

      COOH   COOH

salicylic acid is hydroxybenzoic acid, and that benzamic acid is the corresponding aminobenzoic acid. Following his suggestions, Smitt in 1860 converted both malic and tartaric acids into succinic acid. Kolbe revealed the nature of asparagine and of aspartic acid and converted cyanoacetic acid into malonic acid, thereby proving the structure of the latter.

THE CONTRIBUTIONS OF KEKULÉ AND COUPER

Two developments during the 1850's brought order out of the chaos in which chemists found themselves. One of them, the work of Kekulé and Couper, represented a final clarification of the ideas which had been slowly germinating among chemists for the last twenty years. The other, proposed by Cannizzaro, embodied a reevaluation of an old relationship; it is discussed in the next section.

Friedrich August Kekulé (1829–1896) was a student of architecture at Giessen when he came under Liebig's influence and turned to chemistry. He also studied under Dumas in Paris and had contacts with Gerhardt. In 1853 he served as assistant to Stenhouse in London where he knew Williamson, Frankland, and Odling. In 1856 he became a docent at Heidelberg, where Bunsen was professor. Two years later Kekulé accepted a professorship at Ghent where he remained for nine years. He spent the rest of his life in Bonn. He was an independent thinker of great imaginative powers, but made no major contributions as an experimenter.

Archibald Scott Couper (1831–1892) was born near Glasgow. He studied philosophy at Glasgow and Berlin, then suddenly became interested in chemistry and went to Paris where he studied with Wurtz. He worked on coal tar derivatives, particularly the reactions of salicylic acid. He became assistant to Playfair at Edinburgh in 1859, but soon suffered a breakdown in health which permanently ended what had started out as an apparently brilliant scientific career.

Kekulé seems to have grasped the real significance of the valence concept while he was in London. While studying the reaction of acetic acid with phosphorus pentasulfide he obtained thioacetic acid, a reaction which he compared with the formation of acetyl chloride from acetic acid and phosphorus pentachloride.

$$5 \left.\begin{matrix} C_2H_3O \\ H \end{matrix}\right\} O + P_2S_5 \rightarrow 5 \left.\begin{matrix} C_2H_3O \\ H \end{matrix}\right\} S + P_2O_5$$

$$5 \left.\begin{matrix} C_2H_3O \\ H \end{matrix}\right\} O + 2PCl_5 \rightarrow \frac{5C_2H_3OCl}{5HCl} + P_2O_5$$

Fig. 8.6. ARCHIBALD SCOTT COUPER.
(Courtesy of the Edgar Fahs Smith Collection.)

That in drawing the analogy he was aware of the difference in combining capacity is evident from his observation that with the chloride the product was broken up, whereas with the sulfide this did not happen since "the quantity of sulfur which is equivalent to two atoms of chlorine is not divisible." Using Gerhardt's atomic weights, he reasoned that "sulfur, like oxygen, is dibasic (*zweibasisch*) so that one atom is equivalent to two atoms of chlorine."

In 1857 Kekulé introduced the term, the *marsh gas type*, but applied it to only a limited number of compounds. Now in Heidelberg, he used the double carbon atom which was still popular in Germany and obtained formulas such as the following:

C_2HHHH	Marsh gas	$C_2HClClCl$	Chloroform
C_2HHHCl	Methyl chloride	$C_2(NO_4)ClClCl$	Chloropicrin

During the next year he went back to using Gerhardt's atomic weights, representing them at first with barred symbols, Ꮯ, Ꮎ, Ħ, etc.

Odling also suggested the marsh gas type in a lecture at the Royal Institution, but since he failed to publish his ideas they did not come to the attention of organic chemists.

Kekulé and Odling also developed the concept of mixed types, thereby improving on Gerhardt's four types by including, in a single formula, the component types present in a particular compound. Gerhardt had had to resort to two or more type formulas when he wished to show more than a single characteristic of a compound. Odling found the mixed type useful for representing such salts as sodium thiosulfate. The mixed type as he used it was actually an extension of the multiple water type.

$$\left.\begin{array}{l}2H'\\2H'\end{array}\right\}2O''$$

Double
water

$$\left.\begin{array}{l}SO_2{}''\\2Na'\end{array}\right\}2O''$$

Sodium
sulfate

$$\left.\begin{array}{l}2H'\\2H'\end{array}\right\}O''+S''$$

Water and
hydrogen sulfide

$$\left.\begin{array}{l}SO_2{}''\\2Na'\end{array}\right\}O''+S''$$

Sodium
thiosulfate

$$\left.\begin{array}{l}Na'\\SO_2{}''\\Na'\end{array}\right\}\begin{array}{l}O''\\S''\end{array}$$

Sodium thiosulfate,
modified to show
multiple type

Kekulé realized the great variety of formulas made possible through use of the mixed types. A few examples are:

$$\left.\begin{array}{l}H\\H\\H\\H\end{array}\right\}O$$

H_2, H_2O
type

$$\left.\begin{array}{l}H\\SO_2\\H\end{array}\right\}O$$

Sulfurous
acid

$$\left.\begin{array}{l}Cl\\H\\H\\H\end{array}\right\}O$$

HCl, H_2O
type

$$\left.\begin{array}{l}Cl\\SO_2\\H\end{array}\right\}O$$

Chlorosul-
fonic acid
(SO_3 + HCl)

$$\left.\begin{array}{l}H\\H\\H\end{array}\right\}N \quad \left.\begin{array}{l}H\\H\end{array}\right\}O$$

NH_3, H_2O
type

$$\left.\begin{array}{l}H\\H\\CO\\H\end{array}\right\}N\quad \left.\right\}O$$

Carbamic acid
(parent
substance
of ammonium
carbonate,
$2NH_3$ + CO_2)

Kekulé's greatest contributions were made in 1858. They were actually twofold—the tetravalence of the carbon atom and the ability of carbon atoms to link with one another.[9] Couper came to the same conclusions at exactly the same time. Early in 1858 he submitted to Wurtz a manuscript which set forth these ideas; he asked that Wurtz arrange for its presentation before the Académie. However, Wurtz was not a member of the Académie and hence had to obtain the cooperation of a member in submitting the paper. Meanwhile, Kekulé's paper appeared in the May number of the *Annalen*. Since Couper's paper had not yet been presented before the Académie, Kekulé had clear priority. When Couper protested about the delay to Wurtz, the latter expelled him from his laboratory.[10] Dumas had the paper read before the Académie on June 14 and it was subsequently published.[11] Kekulé regarded Couper's paper as meritorious, but claimed priority for himself in view of the relative publication dates.

[9] F. A. Kekulé, *Ann.*, **106**, 153 (1858).

[10] Leonard Dobbin, *J. Chem. Educ.*, **11**, 335 (1934) sheds light on the incident in a letter, from A. Ladenburg to R. Anschütz dated May 12, 1906, which he includes in this article.

[11] A. S. Couper, *Compt. rend.*, **46**, 1157 (1858); *Phil. Mag.*, [4], **16**, 104 (1958).

Couper made no further effort to uphold his position, and dropped out of scientific activities because of poor health. His obscurity was so complete that Richard Anschütz, Kekulé's biographer, had great difficulty obtaining information about him early in the present century.[12]

In his earlier paper on the methane type, Kekulé stated that carbon is *vierbasisch* or *vieratomig*; i.e., the carbon atom is attached to four hydrogen atoms, any or all of which may be displaced by other radicals. The carbon binds four atoms of monatomic elements or two of diatomic elements. Kekulé developed this concept further in his 1858 paper both to account for compounds with one carbon atom like methane, methyl chloride, chloroform, carbon tetrachloride, phosgene, carbon dioxide, carbon disulfide, and hydrogen cyanide, and to show clearly that this type of combination is present in ethane, ethyl chloride, ethylene chloride, methyl cyanide, cyanogen, acetaldehyde, acetyl chloride, etc. He pointed out that in these cases the apparent valence of the two carbon atoms is not 8 but 6, because one of the valences of each of the two carbon atoms is satisfied by the combination of carbon to carbon and hence these two valences are not available to hydrogen or other atoms. Thus, in reactions involving such compounds as ethyl alcohol, acetaldehyde, acetic acid, ethyl chloride, glycollic acid, oxalic acid, etc., the two carbon atoms remain fixed as a carbon skeleton; only the attached atoms can change. Couper's paper handled the subject in a similar fashion. The effect of the two papers was to show that an understanding of organic formulas could be gained from a study of the component elements and their manner of combination, rather than from the study of radicals.

Kekulé observed that double decomposition, as Gerhardt used it, could not explain all types of chemical reactions. Accordingly he set forth his own ideas of reaction types:

1. Direct addition:

$$CO + Cl_2 \rightarrow COCl_2$$

2. *Umlagerung* (combination and rearrangement):

$$H_2O + SO_3 \rightarrow \left.\begin{matrix} SO_2 \\ H_2 \end{matrix}\right\} O_2$$

3. Double decomposition:

a	b	a	b	a	b
a_1	b_1	a_1	b_1	a_1	b_1
Before		During		After	

---

[12] Leonard Dobbin, *op. cit.*, pp. 331–338.

CANNIZZARO AND THE RETURN TO AVOGADRO'S HYPOTHESIS

One problem still remained to be solved before chemical formulation could be done with confidence. This involved clarifying the nature of the molecule, particularly with respect to its weight. Isomerism made it evident that different compounds with the same empirical formula were not only possible but abundant.

Although structural analysis could, of course, account for numerous cases of isomerism, it was hardly adequate to deal with polymers— compounds with the same empirical formula but different molecular weights. An understanding of them required molecular weights which could be accepted with confidence.

Fig. 8.7. STANISLAO CANNIZZARO.
(Courtesy of the Edgar Fahs Smith Collection.)

The calculation of molecular weights was still a matter of uncertainty at mid-century. Guesswork and arbitrary rules were commonplace. The disagreement regarding the atomic weights of even carbon and oxygen has been mentioned frequently. Gerhardt, as was said earlier, corrected Berzelius' values for silver and sodium by halving them, and then proceeded to halve the correct values for zinc and calcium, thus making them incorrect.

One chemist succeeded in bringing order out of this chaos, for he saw that Avogadro's hypothesis could provide the thread that could tie so many troublesome facts together. This chemist was Stanislao Cannizzaro (1826–1910), who was born in Palermo, on the island of Sicily. He began the study of medicine in the university in Palermo, then went to Pisa to study chemistry under Piria. However, the Sicilian revolt soon broke out and Cannizzaro took part in the capture of Messina. The failure of the revolt forced him to flee to France where he resumed his studies in Chevreul's laboratory. In 1851 he returned to Italy and took a teaching position in Alessandria. It was here that he completed his study of the action of alkali on benzaldehyde, now known as the Cannizzaro reaction for the preparation of benzyl alcohol and benzoic acid. He became professor of chemistry in Genoa in 1855, but resigned five years later to join Garibaldi's campaign in Sicily. With the consolidation of Italy proceeding smoothly, Cannizzaro returned to Palermo and a professorship there. In 1871 he was installed as professor of chemistry in the new University of Rome, where he remained until the year before his death.

In the application of Avogadro's hypothesis Cannizzaro found a way to simplify the teaching of chemistry. Furthermore, chemical facts began to fit together and make sense. In 1858 Cannizzaro published "Sunto di un Corso di Filosofia Chimica" in *Il Nuovo Cimento*, a scientific journal founded four years earlier by Piria and Matteucci. This paper summarized his application of Avogadro's hypothesis in the teaching of chemistry.

Cannizzaro clearly recognized the value of the hypothesis in settling the major problems confronting chemists at mid-century, but at the same time he recognized that historical developments in chemistry between 1811 and 1858 had cast doubt on the validity of Avogadro's two assumptions. In his paper he set himself the task of showing how such doubts could be eliminated and a system of chemistry which was internally consistent could be devised if complete faith were placed in the two assumptions.

By comparing the density of gases and vapors with the density of hydrogen it was possible accurately to determine the molecular weights of elements and compounds. Cannizzaro reasoned that, since the hydrogen molecule contained two atoms, the vapor densities in terms of hydrogen must be doubled, since the molecular comparisons were being made against a diatomic hydrogen molecule. His wisdom was shown in his acceptance of comparative densities determined with the balance, and his consequent avoidance of preconceived notions regarding the composition of the molecule—in particular, the pitfalls Dumas encountered with respect to sulfur, mercury, phosphorus, and arsenic molecules. According to the vapor densities of these four elements, their molecules contained six, one, four, and four atoms respectively; and Cannizzaro was willing to accept this, even though the molecules of hydrogen, oxygen, chlorine, bromine, and nitrogen were diatomic. Bineau had shown that the vapor density of sulfur at temperatures above 1000° C. was 32 times that of the same volume of hydrogen. On the basis of Bineau's results Cannizzaro reasoned that sulfur molecules, which he considered to be hexatomic at temperatures just above the vaporization point, dissociated into diatomic molecules at 1000° C. (It is now known that the apparent hexatomic molecules are really a mixture of S_8 and S_2.)

One of the objections to using vapor densities in the immediately preceding period was the anomalous vapor densities of such substances as ammonium chloride, ammonium carbonate, phosphonium bromide, and phosphorus pentachloride. Cannizzaro held that these compounds formed vapors which dissociated at the temperatures used for measurement, and hence the results were not in accord with the accepted formula. Kopp and Kekulé came to this same conclusion independently. Wurtz demonstrated the validity of this explanation for phosphorus pentachloride in 1869.

Cannizzaro proved·that the correct molecular formula of a compound could be calculated from the vapor density and the analytical values of the constituent elements. He showed further that Avogadro's hypothesis was of value in determining the atomic weight of elements which were not volatile, but whose compounds could be vaporized.

The Karlsruhe Congress

Despite its comprehensiveness and its cold logic, Cannizzaro's paper attracted no particular attention when it was published. Two years later, however, about 140 very prominent chemists met at Karlsruhe in an attempt to resolve their differences with respect to atoms, molecules, equivalents, and rational nomenclature. Kekulé supplied the spark that initiated the international meeting, but the arrangements and correspondence were carried on largely by Professor Carl Weltzien of the Technische Hochschule in Karlsruhe. A circular bearing the names of 45 prominent chemists—among them Boussingault, Bunsen, Cannizzaro, Deville, Dumas, Frankland, Hofmann, Kekulé, Kopp, Liebig, Mitscherlich, Pasteur, Stas, Williamson, Wöhler, Wurtz, and Zinin—was sent out in July, 1860; and on the third of September the chemists assembled in the pleasant Rhineland city to begin their deliberations. The list of those in attendance is notable for the absence of Liebig, Wöhler, Hofmann, Frankland, Mitscherlich, and Pasteur,[13] but perhaps no great significance should be attached to this fact, for several of these men were well on in years and several were too far away from Karlsruhe to make the trip conveniently. Those who attended included many of the younger men whose names became important in the decades which followed—men like Baeyer, Beilstein, Erlenmeyer, Friedel, Mendeleev, Lothar Meyer, Roscoe, Schiff, and Wislicenus. Some, like Kekulé, were already firm adherents of the atomic weights proposed by Gerhardt, but others preferred Berzelius' weights or Gmelin's equivalents.

[13] A list of 127 names recorded by C. A. Wurtz, the secretary is given in Clara de Milt, *Chymia*, 1, 162 (1948).

A steering committee composed of Kekulé, Cannizzaro, Fresenius, Erdmann, Béchamp, Schischkoff, Strecker, and Wurtz and headed by Kopp drafted a series of questions about atoms, molecules, radicals, and equivalents to be presented on the second day. There was an extended discussion on the floor, but no agreement. Cannizzaro described the ideas he had been using in his teaching, but failed to sway the audience in his favor. On the third day he spoke at length on the significance of the work done by Gerhardt, who had based his molecular weights on the hypothesis of Avogadro and of Ampère. He emphasized Dumas' studies on vapor densities and Gaudin's reasoning concerning polyatomic molecules of elemental gases.

Cannizzaro then explained the application of Avogadro's hypothesis in calculating atomic and molecular weights, and pleaded for the adoption of weights based upon this principle. There followed a discussion in which Strecker expressed his intention of accepting the proposed atomic weights and Kekulé agreed to do likewise, with certain reservations. Kopp and Erdmann argued that no vote could be taken on scientific questions, for every scientist must be entirely free to make up his own mind.

At the close of the meeting, Angelo Pavesi, professor of chemistry at the University of Pavia and a friend and follower of Cannizzaro, distributed copies of the latter's paper which had appeared two years earlier in *Il Nuovo Cimento*. Although most of the reprints may have gone the way of most handouts, that at least one man read the reprint with understanding is evident from the statement of Julius Lothar Meyer (1830–1895) regarding its influence on him.

I also received a copy which I put in my pocket to read on the way home. Once arrived there I read it again repeatedly and was astonished at the clearness with which the little book illuminated the most important points of controversy. The scales seemed to fall from my eyes. Doubts disappeared and a feeling of quiet certainty took their place. If some years later I was myself able to contribute something toward clearing the situation and calming heated spirits no small part of the credit is due to this pamphlet of Cannizzaro. Like me it must have affected many others who attended the convention. The big waves of controversy began to subside, and more and more the old atomic weights of Berzelius came to their own. As soon as the apparent discrepancies between Avogadro's rule and that of Dulong and Petit had been removed by Cannizzaro both were found capable of practically universal application, and so the foundation was laid for determining the valence of the elements, without which the theory of atomic linking could certainly never have been developed.[14]

[14] L. Meyer, in the German translation of Cannizzaro's paper, published in Ostwald's *Klassiker der exakten Naturwissenschaften*, No. 30. An English translation is available as *Alembic Club Reprint*, No. 18, Alembic Club, Edinburgh, re-issue edn., 1947.

Meyer, who was then a docent at Breslau, spent the next years writing his book, *Die modernen Theorien der Chemie* (1864), which developed theoretical chemistry on the basis of Avogadro's hypothesis. This book was significant in influencing chemists to apply the hypothesis and in directing attention toward the physical aspects of chemistry.

CONCLUSION

Although the period between 1840 and 1850 continued to reflect the chaos of the previous two decades of organic chemistry, there were certain advances which would ultimately lead to clarification of the major problems. At first these advances, particularly the contributions of Laurent and Gerhardt, failed to have an impact because these men were often partially in error and their ideas were seldom understood.

The increasing knowledge of experimental organic chemistry, reflected in the work of Wurtz and Hofmann on amines, of Williamson on ethers, of Kolbe on the products of the electrolysis of acids, and of Frankland on metal alkyls, led to recognition of classes or types of compounds. The concept of valence also grew out of the work in experimental chemistry, culminating in the ideas of Kekulé and Couper on the tetravalence of carbon and the capacity of carbon atoms to combine with carbon atoms.

The final necessary tool—a means for ascertaining molecular size, and thereby accurate formulas—was shown by Cannizzaro to have been available for a half century in the form of Avogadro's hypothesis. Once this was forced on the attention of the chemical world and sufficient time had elapsed for its usefulness to become apparent, chemistry was ready for a steady series of experimental and conceptual successes.

CHAPTER 9

CLASSIFICATION OF THE ELEMENTS

DISCOVERY OF NEW ELEMENTS

With the stabilization of atomic weights which followed the application of Avogadro's hypothesis, and with the introduction of the valence concept, chemistry was ready for the next great step—organizing the elements into an orderly system which would bring out relationships between them. Attempts had been made at classification before 1860, but they could not be completely successful because of the chaotic situation regarding atomic weights. That such efforts had any degree of success was due to recognition of empirical relations between elements and to correct atomic weight values arrived at fortuitously.

For a sound system of classification to be devised, two essentials were needed—reliable atomic weights and enough known elements so that relationships and variations would be apparent. These conditions were met by the 1860's. The previous century had been extremely fruitful in bringing new elements to the attention of chemists, with the result that at least something was known about more than 60 elements.

Lavoisier's rule that a substance was to be considered elemental only if it failed to break down into simpler substances when treated chemically constituted an approach that served chemists well for fully a century. Although his list of 33 elements included a few forms of energy and a few compounds in addition to a few postulated elements, he directed chemical thought toward a group of well-known substances that were significant as elements. His errors were not sufficiently serious to handicap chemists and his correct hunches provided a foundation on which others could build.

It was natural that the formulation of an operational definition of an element should draw attention to elements and provide a climate in which chemists would search more carefully for new ones. Thus minerals were analyzed with greater care than had been usual, and there was marked interest in new minerals which had not hitherto been studied seriously. The analytical studies made by Klaproth, Wollaston, and Vauquelin set the pattern that Berzelius, Gay-Lussac, Thenard, and Wöhler followed so well. In the period between 1790 and 1844, 31 new elements were discovered. (See Appendix I.)

The discovery of cesium in 1860 marked the beginning of a new era in increasing the number of known elements. Only five new elements had been discovered during the preceding three decades. Such techniques as ordinary analysis, electrochemical decomposition, and replacement by potassium had been exploited to the limit of their effectiveness. But from 1860 on, new elements were discovered largely as the result of deliberate searches for trace amounts, frequently with the aid of new tools, one of which was the spectroscope.

The colored spectrum had been known since ancient times—Seneca had discussed it. The refraction of light had been studied by the Arabs and by Roger Bacon, and in more recent times by Kepler and Descartes. Newton first demonstrated systematically that a prism separated white light into its component colors and that a second prism could reconstitute these colors into white light. The spectrum was extended into the infrared in 1800 when the astronomer, William Herschel, discovered the heating effect of the dark region beyond the red. Its extension into the ultraviolet occurred the following year when J. W. Ritter and W. H. Wollaston independently discovered the blackening effect on silver chloride of the dark region beyond the violet.

Wollaston's paper on the ultraviolet also reported the presence of dark lines in the spectrum of sunlight when viewed through a prism, but he interpreted these lines incorrectly as representing boundaries between the colored bands. These dark lines, plus many additional lines, were rediscovered in 1814 by Joseph Fraunhofer (1787–1826), a Bavarian lens manufacturer. While studying the refractive index of samples of glass for various colors, he saw a pair of bright yellow lines in the spectrum of the flame he was using. These same lines appeared in the flame of burning alcohol, sulfur, oil, and tallow and were useful in determining refractive indices. When he passed sunlight through a prism and examined the spectrum with a small telescope, there were no yellow lines, but he observed a great number of dark lines against the colored background. After mapping and studying these lines, he came to the conclusion that they were characteristic of sunlight rather than being effects of the instrument or optical illusions. His letter designations of A to H for the more prominent lines are used today. He observed similarly placed lines in the spectrum from the moon and the planets. Light from the stars also contained lines, but not always in the same place as the solar lines. His early death prevented him from carrying these fruitful studies further.

Studies of the spectrum continued during the following years.[1] The Scottish physicist David Brewster (1781–1868) discovered in 1834 that lines were produced when sunlight was passed through "nitrous acid gas." After learning that sulfur vapor absorbed light from the violet end of the spectrum and iodine vapor absorbed it from the middle part, he suggested "the discovery of a general principle of chemical analysis in which simple and compound bodies might be characterized by their action on definite parts of the spectrum."[2] However, this hope did not materialize for several decades. Meanwhile, Brewster found that the spectrum of brown oxide of nitrogen was crossed by hundreds of lines, and he showed that the lines became more intense as the density of the gas was increased and when the gas was heated. The lines did not coincide with those in the solar spectrum. Brewster's studies were extended to halogen vapors by John Frederic Daniell and William Hallowes Miller, and to a variety of other gases by William Allen Miller.

Thomas Melvill (1726–1753) observed the bright lines in the spectra of metallic salts as early as 1752. Andreas Marggraf was using flame colors to distinguish sodium and potassium salts by 1758. John Herschel showed that when the flame colors of boric acid and of the chlorides of barium, calcium, strontium, and copper were passed through a prism they were resolved into characteristic lines which could be used for identification purposes. Henry Fox Talbot (1800–1877), the pioneer in photography, made similar observations during the mid-1830's.

In his analysis of the salts in mineral water, Bunsen used flame colors for identification purposes. He invented the burner which is named after him for the express purpose of securing a colorless gas flame that would not create confusion in the results obtained in his chemical experiments. When Bunsen became professor at Heidelberg in 1852 he was promised a new laboratory to replace the old monastery laboratory Gmelin had used. Since gas had just been introduced for street lighting in Heidelberg, Bunsen immediately decided to equip the new laboratory with gas, thus abandoning the charcoal and coal furnaces and the alcohol and oil lamps hitherto used for heating. However, the gas burners then available were designed to produce light rather than heat. Even laboratory burners, such as the Argand type which Bunsen's student, Roscoe, brought from London, were difficult to regulate; their flame was large and unsteady, and they gave off very little heat. To solve this problem, Bunsen thought of changing the construction of the burners so that air would be mixed with gas in the burner tube before combustion took place; and in 1855,

[1] T. H. Pearson and A. J. Ihde, *J. Chem. Educ.*, **28**, 267 (1951).
[2] D. Brewster, *Trans. Royal Soc. Edinburgh*, **12**, 519 (1834).

Fig. 9.1. Robert Wilhelm Bunsen. (Courtesy of the Edgar Fahs Smith Collection.)

Fig. 9.2. The spectroscope developed by Bunsen and Kirchhoff. (From *Annalen der Physik* [1860].)

the university mechanic, Peter Desaga, succeeded in producing such a burner. Neither Bunsen nor Desaga patented the burner; hence it was quickly imitated by other instrument-makers in Germany.[3]

Gustav Robert Kirchhoff (1824–1887), a native of Königsberg, had worked with Bunsen in Breslau in 1850. They became colleagues once more when Kirchhoff accepted the physics chair at Heidelberg in 1854, a post which he held until he joined Helmholtz in Berlin twenty years later. Besides his studies on light he made important contributions in the field

[3] G. Lockemann, *J. Chem. Educ.*, **33**, 20 (1956).

of electrical conductivity. Kirchhoff suggested that Bunsen might get better results when working with salt mixtures if, instead of using colored solutions and glasses to filter out interfering colors, the light were passed through a glass prism and viewed as a spectrum. Together the two men developed the spectroscope for this purpose. The instrument embodied no new principles but brought together the necessary collimating and viewing tubes on a single stand for convenient operation.

In contrast to Bunsen, Kirchhoff was a skilled mathematician and had a speculative mind; hence he immediately grasped the relationships between dark-line and bright-line spectra. He demonstrated in experiments that the D line not only appeared as an emission line in light from a sodium flame, but was present as an absorption line when white light from an incandescent source was passed through a sodium flame before it was passed through the spectroscope. A paper he published in 1859 set forth the Kirchhoff laws of spectroscopy:

1. An incandescent body gives off a continuous spectrum.
2. An excited body gives off a bright-line spectrum.
3. White light passed through a vapor has dark lines where the vapor ordinarily emits light.

He realized that hot gases absorb the same kind of light as they emit. The phenomenon of the Fraunhofer lines was finally explained. In doing this, Kirchhoff provided astronomers with a powerful tool for gaining an understanding of the chemical composition of the sun and the other stars. At the same time, chemists acquired a unique and sensitive tool for analytical work.

The effects were quickly apparent. Bunsen and Kirchhoff announced the discovery of cesium in 1860, of rubidium the following year. Cesium was first detected in the salts in mineral waters from Dürkheim, because of its blue spectral lines. The name is derived from *caesius,* the word by which the Romans designated the blue of the firmament. The dark red lines of rubidium were first observed in alkaline compounds separated from a mineral called lepidolite.

At the end of all their early papers on spectroscopy Bunsen and Kirchhoff announced their intention to continue research in this area and reserved this field exclusively for themselves. Naturally this reservation was not respected and numerous scientists began reporting the results of their own spectroscopic studies. The spectroscope figured prominently in the discovery of thallium by William Crookes in 1861 (it was isolated independently by C. A. Lamy), and of indium in 1863 by Ferdinand Reich and H. T. Richter of the Freiberg School of Mines. Spectroscopy subsequently played an important role in the discovery of gallium, the rare earths, and the rare gases.

EARLY ATTEMPTS AT CLASSIFICATION

Even at a time when atomic weights were highly uncertain and the number of known elements was limited, chemists were tempted to search for relationships between elements. Prout's hypothesis sought to relate all the elements to hydrogen as a building block. Other chemists were not content with such an oversimplified concept of the chemical elements but sought in other ways for a mode of classification. All such schemes utilized atomic weights as the principal device around which to design groupings of elements.

Döbereiner's Triads

The earliest actual attempt at a systematic arrangement of the elements into groups was apparently made by Döbereiner. As early as 1816, following his work on celestite, he observed that the atomic weight of strontium appeared to be 50, the mean of the then accepted values for calcium (27.5) and barium (72.5). Finding a similar mean in the specific gravity of the sulfates of these three elements led him to question the independent existence of strontium.[4]

His idea of a triad relationship did not reach full fruition until 1829. By then he had available the atomic weights Berzelius published in 1826, as well as knowledge of several newly discovered elements. Döbereiner, who claimed to have prophesied in his lectures at Jena that the atomic weight of the newly discovered bromine would be the mean of those of chlorine and iodine, was overjoyed when this was confimed by Berzelius. His triads included elements with similar properties, the atomic weight of the central member being the mean of the other two atomic weights. Some of his well-defined triads are:

1. Lithium	Calcium	Chlorine	Sulfur	Manganese
2. Sodium	Strontium	Bromine	Selenium	Chromium
3. Potassium	Barium	Iodine	Tellurium	Iron

The middle member of certain triads he believed was not known:

1. Boron	Beryllium	Yttrium
2.		
3. Silicon	Aluminum	Cesium

Magnesium he considered as an isolated element, not part of a triad. Fluorine, as yet undiscovered but its existence clearly known, he did not include in the halogen triad. Although the atomic weight of nitrogen fell exactly between that of carbon and oxygen, Döbereiner regarded the three elements as isolated non-metals rather than as members of a triad. He also treated hydrogen as an isolated element.

[4] J. W. Döbereiner, in Gilbert's *Ann. phys.*, **56**, 332 (1816); **57**, 436 (1817).

Gmelin adopted the Döbereiner classification in his *Handbuch* but it obtained no major notice elsewhere. Berzelius commented that apparent number relationships could frequently be found between elements, but that subsequent revision of atomic weights would alter these relationships and therefore it was not safe to make speculative assumptions.[5]

Other Numerical Classification Systems

The interest in number relations flared up again in certain quarters around mid-century. P. Kremers observed that the atomic weights of certain non-metals in which he saw a similarity differed by 8; i.e., O = 8, S = 16, Ti = 24.12, P = 32, Se = 39.62. At the same time, the atomic weights of certain metals fell between those of successive non-metals; i.e., Mg = 12, Ca = 20, Fe = 28. When divided by 4, the atomic weights of the non-metals were an even number, those of the metals an odd number. Although Kremers developed this theory extensively, it was of no great significance because his atomic weights were not well stabilized and his list of non-metals included several elements that are really metals.

Max von Pettenkofer had pointed out even earlier that the combining weights of related elements often differed by increments of 8 or multiples thereof:

Li (7) 7	Mg (12) 12
Na (23) 7 + 16	Ca (20) 12 + 8
K (39) 23 + 16	Sr (44) 20 + 24
	Ba (68) 44 + 24

J. H. Gladstone suggested three types of relationships between elements in 1853: (1) Elements whose atomic weights were nearly identical, i.e., chromium, manganese, iron, cobalt and nickel (around 28); palladium, rhodium, and ruthenium (around 52). (2) Elements in which there was a multiple proportion between atomic weights, i.e., the palladium group (52), the platinum group (99), and gold (197). (3) Elements in which the atomic weights of successive members differed by a common increment, i.e., Li, Na, K. The first type he compared with allotropy; the second, with polymerism in organic chemistry; the third, with homologous series. Gladstone also tried to arrange the elements in order of increasing atomic weight, but discerned no pattern other than the concentration of several elements around atomic weights 28, 53, and 99. His atomic weights came from Liebig's *Jahresberichte* for 1851 and represent equivalents; furthermore, he included a number of elements whose atomic weight was uncertain.

About this same time Josiah Parsons Cooke (1827–1894) of Harvard developed a classification based upon six series of elements, each series

[5] J. J. Berzelius, *Lehrbuch der Chemie*, **3**, 1178 (1845).

derived from the atomic weight by a mathematical formula. For example, the sixth series, composed of hydrogen and the alkali metals, was based on the formula, $1 + (n \times 3)$. Thus:

$$H\,(1) = 1 + (0 \times 3) \qquad Na\,(23) = 1 + (7 \times 3)$$
$$Li\,(7) = 1 + (2 \times 3) \qquad K\,(39) = 1 + (13 \times 3)$$

Cooke was of the opinion that the series brought out relationships as homology did in organic compounds. He did not confine himself solely to atomic weights but included such factors as electronegative properties and crystalline forms. In his classification scheme a given element could appear in more than one series.

A British scientist, David Low, suggested in 1859 that hydrogen and carbon were the real constituents of many of the elements. He assigned an atomic weight of 6 to carbon, which made it possible to explain nitrogen as C_2H_2 and oxygen as CH_2 ($O = 8$). He referred to Davy's experiments which suggested that hydrogen was a constituent of sulfur and phosphorus. He held that inability to break down a substance was not sufficient basis for regarding it as an element.

William Odling published a thirteen-group classification, essentially a modified triad arrangement, in 1857. Seven years later he devised a classification based on increasing atomic weights. The fact that similar elements were separated by intervals of 16, 40, 44, or 48 led him to conclude that 4 might be of some significance in atomic weights.

Dumas' Speculations

From time to time Dumas pondered the relationships between elements. At the meeting of the British Association at Ipswich in 1851 he participated in a discussion in which he advanced some highly speculative ideas. They were not reported in full at the time, and he did not prepare and publish a paper on the subject until 1859.[6]

Dumas held that the existence of triads suggested a close connection between related elements. An incremental unit, CH_2, differentiated successive members in homologous series of organic compounds. An arithmetical progression was apparent, for example:

$$H = 1 + (0 \times 14)$$
$$CH_3 = 1 + (1 \times 14)$$
$$C_2H_5 = 1 + (2 \times 14)$$
$$C_3H_7 = 1 + (3 \times 14)$$
$$C_4H_9 = 1 + (4 \times 14)$$

General formula, molecular weight $= 1 + (n \times 14)$

[6] J. B. A. Dumas, *Ann. chim. phys.*, [3], **55**, 129 (1859).

He professed to see a similar relationship in certain series of elements.

$$F = 19 \qquad\qquad\qquad\qquad = 19$$
$$Cl = 19 + \qquad 16.5 \qquad\qquad = 35.5$$
$$Br = 19 + (2 \times 16.5) + \qquad 28 \quad = 80$$
$$I = 19 + (2 \times 16.5) + (2 \times 28) + 19 = 127$$

$$O = 8 \qquad\qquad = 8$$
$$S = 8 + \qquad 8 = 16$$
$$Se = 8 + (4 \times 8) = 40$$
$$Te = 8 + (7 \times 8) = 64$$

$$N = 14 \qquad\qquad\qquad = 14$$
$$P = 14 + 17 \qquad\qquad = 31$$
$$As = 14 + 17 + \qquad 44 = 75$$
$$Sb = 14 + 17 + (2 \times 44) = 119$$
$$Bi = 14 + 17 + (4 \times 44) = 207$$

$$Mg = 12 \qquad\qquad\qquad = 12$$
$$Ca = 12 + \qquad 8 = 20$$
$$Sr = 12 + \quad (4 \times 8) = 44$$
$$Ba = 12 + \quad (7 \times 8) = 68$$
$$Pb = 24 + (10 \times 8) = 104$$

Dumas speculated that since such regular changes appeared in a series of elements, transmutation of elements might be possible after all. Might the central member of a triad not represent a half-completed transmutation? To him, the existence of allotropic forms of elements and the ability of the electric arc to change diamond into graphite were significant. He was impressed by the fact that similar elements such as cobalt and nickel were found associated in ores. Might the alchemists' failure to achieve transmutations be due to their persistence in trying to transmute unrelated elements like lead into gold, and mercury into silver? His speculations renewed the interest in triads and elemental relationships. Faraday even commented that perhaps elements were not such unchanging entities as had been supposed.

Dumas' ideas were attacked by César Mansuéte Despretz, professor of physics at the Sorbonne, who conducted a series of experiments to determine whether common elements such as copper, lead, zinc, and other metals revealed any tendency toward condensation or decomposition. In these experiments he used electrolysis, fractional precipitation, and distillation, but failed to observe any change in the metals. In defending his position, Dumas maintained that the methods Despretz used were not adequate to deal with the problem.

IMMEDIATE PRECURSORS OF THE PERIODIC TABLE

Cannizzaro's work, as we have seen, solved the problem of inaccurate atomic weights, and efforts were made in several quarters to effect an arrangement of the elements based upon increasing atomic weights. The most important of these plans, prior to the work of Mendeleev and Lothar Meyer, were the systems formulated by de Chancourtois and Newlands.

De Chancourtois' Telluric Helix

A. E. Béguyer de Chancourtois (1820–1886), professor of geology at the École des Mines, conceived the idea of plotting the elements according to atomic weights on the surface of a cylinder. The circumference of the cylinder was divided into sixteen sections since the atomic weight of oxygen was 16. The elements were plotted on a line or helix that descended at an angle of 45° with the top of the cylinder. There was a striking resemblance in elements that were on the same vertical line. De Chancourtois also claimed that secondary helices on the cylinder frequently brought out other relationships.

As a result of his work with the telluric helix, as he called this device, de Chancourtois asserted that the properties of elements were the properties of numbers. He observed that atomic weights followed the formula $n + 16n'$, the value of n frequently being 7 or 16. He expressed atomic weights as whole numbers, in accordance with Prout's hypothesis. Gaps in the helix were considered as indicating not unknown elements but different varieties of known elements. He believed, for example, that there was a form of carbon whose atomic weight was 44.

The telluric helix was a step in the right direction, but lacked precision because it failed to bring similar elements together, owing to certain flaws that are readily apparent to the present-day critic. De Chancourtois first reported this idea to the French Academy in 1862, but his reports were obscure and failed to include a diagram of the helix. The idea gained no supporters.

Newlands' Law of Octaves

The other major precursor of Mendeleev and Lothar Meyer was John Alexander Reina Newlands (1837–1898), a student of Hofmann. He volunteered with the Garibaldi forces in 1860, and became assistant to

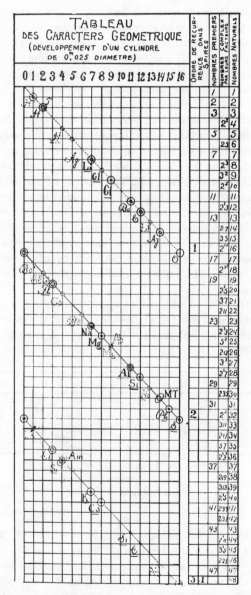

Fig. 9.3. THE TELLURIC HELIX OF DE CHANCOURTOIS.
(From F. P. Venable, *The Development of the Periodic Law* [1896].)

the chemist of the Royal Agricultural Society before setting up his own analytical laboratory in 1864. Four years later he was appointed chief analyst for J. Duncan's sugar refinery, where he made improvements in processing methods and, with his brother, wrote a treatise on sugar refining. In 1886 he and his brother established their own consulting firm.

Newlands' first publication dealing with the classification of the elements appeared in 1863. It was essentially a reworking of Dumas' ideas, for Newlands was looking for numerical relationships between elements. A year later he adopted the atomic weights recommended by Cannizzaro and published a table of 37 elements subdivided into ten families. A crude repetitive character was evident, and blank spaces were left for undiscovered elements. In a table in a subsequent paper he assigned numbers to the elements, which were listed in order of increasing atomic weights. He used numbers for the elements in all his work thereafter. In August, 1865, he published an eight-column table listing 62 elements in order of increasing atomic weights; they were subdivided into seven horizontal families. It was in this table that he saw an analogy to the octave in music. The eighth element resembled the first element, the fifteenth resembled the first and the eighth; in other words, an interval of seven elements separated similar elements. Soon afterwards he suggested that the numerical relationships observed by earlier chemists were due to this law of octaves.

In March, 1866, Newlands read a paper, "The Law of Octaves and the Causes of Numerical Relations Between Atomic Weights," before the members of the Chemical Society. It presented a revised table which,

No.	No.	No.	No.	No.	No.	No.	No.
H 1	F 8	Cl 15	Co & Ni 22	Br 29	Pd 36	I 42	Pt & Ir 50
Li 2	Na 9	K 16	Cu 23	Rb 30	Ag 37	Cs 44	Os 51
G 3	Mg 10	Ca 17	Zn 24	Sr 31	Cd 38	Ba & V 45	Hg 52
Bo 4	Al 11	Cr 19	Y 25	Ce & La 33	U 40	Ta 46	Tl 53
C 5	Si 12	Ti 18	In 26	Zr 32	Sn 39	W 47	Pb 54
N 6	P 13	Mn 20	As 27	Di & Mo 34	Sb 41	Nb 48	Bi 55
O 7	S 14	Fe 21	Se 28	Ro & Ru 35	Te 43	Au 49	Th 56

Fig. 9.4. NEWLANDS' OCTAVE ARRANGEMENT.
(From *Chemical News* [1866].)

unlike the 1864 table, left no blank spaces for new elements. This omission was criticized, as was the location of some of the heavier elements. Gladstone saw a marked resemblance between individual elements in the last column and elements in the same horizontal row. Another scientist criticized the separation of manganese from chromium, and of iron from cobalt and nickel. One Carey Foster, who holds no other claim to fame, rose to ask facetiously if Newlands had ever sought to classify the elements in alphabetical order. The unsympathetic hearing was reported in *Chemical News*, but the paper itself was returned by the editor of the *Journal of the Chemical Society* as "not adapted for publication in the Society's Journal."

Apparently Newlands took up other activities following this rebuff, though his interest in classification continued. Only after the work of Mendeleev and Meyer aroused new interest in the subject did he again seek to rekindle attention for his own contributions. The Royal Society belatedly honored him with its Davy Medal in 1887, five years after it had similarly honored Mendeleev and Meyer.

Newlands must be credited for taking a pioneering step toward the discovery of the periodic law, for he detected the repetition of properties when elements are arranged according to increasing atomic weights. He noted that this relation was evident only if Cannizzaro's atomic weights were used. He used blank spaces for unknown elements but failed to do this consistently, holding that perhaps the interval between repetitions was eight or nine and that this could be dealt with as new elements were discovered.[7]

THE PERIODIC LAW OF MENDELEEV AND MEYER

The systematic classification of the elements was carried further by Mendeleev in Russia and Lothar Meyer in Germany, both men working independently. Their thoughts on the subject germinated for several years before their first papers appeared in 1869, and the subject was completely developed within the next two years. The periodic law did not have a startling impact at first, but it began to receive serious notice following the discovery of gallium which had been predicted by Mendeleev.

Mendeleev's Periodic Tables

Dmitriĭ Ivanovitch Mendeleev (1834–1907) was the youngest of seventeen children. His father was a school teacher in Tobolsk, Siberia. His mother operated a glass factory after his father became blind. Following her husband's death and the burning of the factory, she took young Dmitriĭ, who had already shown a talent for the sciences, to St. Petersburg where he continued his education. He graduated in 1855 after preparing a dissertation on isomorphism. Because of his poor health he was excused from fulfilling the teaching obligations that graduation ordinarily entailed. He continued his studies at the University of St. Petersburg; here he wrote a dissertation on specific volumes. His work won him a traveling fellowship which took him to Heidelberg. There he studied the expansion

[7] W. H. Taylor, *J. Chem. Educ.*, **26**, 491 (1949).

Fig. 9.5. Dmitriĭ Ivanovich Mendeleev.
(Courtesy of the Edgar Fahs Smith Collection.)

of gases and liquids, which led him to the concept of critical temperatures, usually attributed to Andrews. Before returning to Russia he had the opportunity to attend the Karlsruhe Congress.

Back in St. Petersburg he taught briefly in a technical school, then became professor of chemistry at the university. He remained there until

	Typische Elemente						
			K = 39	Rb = 85	Cs = 133	—	—
			Ca = 40	Sr = 87	Ba = 137	—	—
			—	?Yt = 88?	?Di = 138?	Er = 178?	—
			Ti = 48?	Zr = 90	Ce = 140?	?La = 180?	Th = 231
			V = 51	Nb = 94	—	Ta = 182	—
			Cr = 52	Mo = 96	—	W = 184	U = 240
			Mn = 55	—	—	—	—
			Fe = 56	Ru = 104	—	Os = 195?	—
			Co = 59	Rh = 104	—	Ir = 197	—
			Ni = 59	Pd = 106	—	Pt = 198?	—
H = 1	Li = 7	Na = 23	Cu = 63	Ag = 108	—	Au = 199?	—
	Be = 9,4	Mg = 24	Zn = 65	Cd = 112	—	Hg = 200	—
	B = 11	Al = 27,3	—	In = 113	—	Tl = 204	—
	C = 12	Si = 28	—	Sn = 118	—	Pb = 207	—
	N = 14	P = 31	As = 75	Sb = 122	—	Bi = 208	—
	O = 16	S = 32	Se = 78	Te = 125?	—	—	—
	F = 19	Cl = 35,5	Br = 80	J = 127	—	—	—

Fig. 9.6. Mendeleev's vertical table of 1869.
(From *Annalen der Chemie*, supplemental vol. 8 [1872].)

1890, when he resigned because of a controversy with the Ministry of Education regarding student political activities. His liberal views made him somewhat suspect by government officials and were a factor in his never being elected to the Russian Academy of Sciences. He became director of the Bureau of Weights and Measures in 1893, a post which he held until his death. During his life he was active in many areas of chemistry and was frequently consulted regarding industrial problems, especially those connected with petroleum.

In the period immediately preceding 1869 Mendeleev was apparently preoccupied with examining the properties of the known elements. He prepared a set of cards on which the properties of each element were tabulated on a separate card. Arranging and rearranging these cards

Reihen	Gruppe I. — R²O	Gruppe II. — RO	Gruppe III. — R²O³	Gruppe IV. RH⁴ RO²	Gruppe V. RH³ R²O⁵	Gruppe VI. RH² RO³	Gruppe VII. RH R²O⁷	Gruppe VIII. — RO⁴
1	H=1							
2	Li=7	Be=9,4	B=11	C=12	N=14	O=16	F=19	
3	Na=23	Mg=24	Al=27,3	Si=28	P=31	S=32	Cl=35,5	
4	K=39	Ca=40	—=44	Ti=48	V=51	Cr=52	Mn=55	Fe=56, Co=59, Ni=59, Cu=63.
5	(Cu=63)	Zn=65	—=68	—=72	As=75	Se=78	Br=80	
6	Rb=85	Sr=87	?Yt=88	Zr=90	Nb=94	Mo=96	—=100	Ru=104, Rh=104, Pd=106, Ag=108.
7	(Ag=108)	Cd=112	In=113	Sn=118	Sb=122	Te=125	J=127	
8	Cs=133	Ba=137	?Di=138	?Ce=140	—	—	—	— — — —
9	(—)	—	—	—	—	—	—	
10	—	—	?Er=178	?La=180	Ta=182	W=184	—	Os=195, Ir=197, Pt=198, Au=199.
11	(Au=199)	Hg=200	Tl=204	Pb=207	Bi=208	—	—	
12	—	—	—	Th=231	—	U=240	—	— — — —

Fig. 9.7. MENDELEEV'S HORIZONTAL TABLE OF 1871.
(From *Annalen der Chemie*, supplemental vol. 8 [1872].)

ultimately led him to realize that the properties recurred periodically when the cards were arranged on the basis of increasing atomic weights. Accordingly he prepared a table in which the elements were listed in vertical columns; it was published in 1869.[8]

Although the table had some obvious imperfections, the periodicity of properties and the tendency to bring similar elements together are evident. It also incorporated several principles which contributed to the ultimate acceptance of the periodic law—the listing according to increasing atomic weight, the separation of hydrogen from the immediately following elements, the blank spaces for unknown elements, the uncertainty regarding the placing of the heavier elements.

In his paper Mendeleev pointed out certain conclusions. There was of

[8] D. Mendeleev, *J. Russ. Phys. Chem. Soc.*, **1**, 60 (1869). The table was reprinted as a note in *Z. Chem.*, **12**, 405 (1869). Also see Ostwald's *Klassiker der exakten Naturwissenschaften*, No. 68.

course the repetition of properties at periodic intervals. Elements with similar properties were grouped together in certain cases where the atomic weights were nearly identical (e.g., Fe, Co, Ni); in other cases, where the atomic weights increased by uniform amounts (e.g., K, Rb, Cs), the arrangement into groups brought elements with similar valences together. The blanks in the table indicated the existence of undiscovered elements. Mendeleev also believed that he could detect inaccuracies in accepted atomic weights. For example, tellurium, whose atomic weight was taken as 128, showed properties that logically placed it in Group VI, ahead of iodine, whose atomic weight was 127. He assumed that the atomic weight of tellurium must fall between 123 and 126, thus avoiding an inversion of atomic weight order. Time proved him wrong in this deduction, although it was entirely logical in his day when many atomic weights were known to be inaccurate. There are several atomic weight inversions in the modern periodic table.

Mendeleev published a much improved table two years later.[9] The elements were listed in 12 horizontal rows, called series, so that related elements were in vertical columns, or groups, numbered I to VIII. Group VIII was unique in that it remained empty until iron, cobalt, and nickel were placed in the same series in the column. Copper was placed here provisionally, but Mendeleev also listed it in Group I. The clustering of elements occurred again with the two series of platinum metals. The rare earth elements created a problem that Mendeleev was incapable of understanding because not enough of these elements were known for him to recognize their unique similarity to one another. Consequently the known rare earth elements were mixed up with the heavier elements. His arrangement also put chromium, molybdenum, tungsten, and uranium in the same group with oxygen, sulfur, selenium, and tellurium. Such anomalies occurred in every group—a shortcoming that Mendeleev recognized, but could handle only by treating the anomalous elements as members of subgroups. He also used an even series and an odd series. This causes a certain amount of nonconformity in Series 2 and 3 but after that it works quite well.

Mendeleev used the recently discovered indium to demonstrate the ability of the periodic classification to correct erroneous chemical judgments. The element's equivalent weight was found to be 38.3. Since the element was considered to be divalent, its atomic weight was 76.6, which would have placed it in the table between arsenic and selenium. But no space was available because these two elements logically belonged in Groups V and VI. Tripling the equivalent weight made the atomic weight high enough so that indium fell in Group III in a blank space between cadmium and tin. Mendeleev maintained that indium belonged here by

    [9] D. Mendeleev, *J. Russ. Phys. Chem. Soc.*, **3**, 25 (1871). A German version of the paper appeared in *Ann.*, Supplement 8, 133 (1872).

demonstrating the relationship between the properties of indium and its compounds and the properties of the analogues of cadmium and tin, and of aluminum and thallium. Thus he definitely established indium's position in the periodic table. He also determined the specific heat, 0.055; this also, in accordance with the law of Dulong and Petit, indicated an atomic weight close to 115.

The periodic classification was useful in other cases where uncertainty regarding valences created uncertainty about atomic weights. Beryllium, for example, formed compounds resembling those of aluminum, and it dissolved in alkali with the liberation of hydrogen; hence it was assumed to be trivalent. This would make the atomic weight 13.65 (3 × 4.55), which would put it in Group IV or V in the periodic table, in a position properly occupied by carbon or nitrogen. Mendeleev was able to justify placing beryllium in Group II by pointing out the steady gradation of properties in the series, Li—Be—B. He also showed that the resemblance between beryllium and aluminum was paralleled by similar resemblances among neighboring elements.

$$\begin{matrix} Li & Be & B & C \\ & \searrow & \searrow & \searrow \\ Na & Mg & Al & Si \end{matrix}$$

The resemblance between lithium and magnesium had been detected by Newlands, and that between boron and silicon was also clear, so beryllium was not unique in its resemblance to aluminum, which was situated in the next series diagonally from beryllium. That 9.1, the smaller figure, was the correct atomic weight of beryllium, was finally established by Nilson and Pettersson when they measured the vapor density of beryllium chloride. The molecular weight of 81.7 indicated a divalent metal with an atomic weight around 9.[10]

Uranium had an equivalent weight of 59.6, and caused similar problems. Newlands, assuming it to be divalent, placed it between cadmium and tin. But Mendeleev saw analogies to chromium, molybdenum, and tungsten which indicated an atomic weight of 240 (4 × 60) and a position at the end of Group VI.

Some elements were seen to be out of place when the elements were arranged strictly in order of increasing atomic weights. This suggested to Mendeleev that errors in atomic weights, based in turn on slightly incorrect equivalent weights, were responsible, wherefore he ceased to adhere strictly to increasing atomic weights. Thus he placed gold (196.2) after osmium, iridium, and platinum (198.6, 196.7, 196.7). Later atomic weight determinations demonstrated that this was correct.

Os	Ir	Pt	Au
190.2	193.1	195.2	197.2

[10] L. F. Nilson and O. Pettersson, *Compt. rend.*, **98**, 988 (1884).

Other inversions—cobalt and nickel, tellurium and iodine—were not corrected by means of improved methods of determining atomic weights but were proved to be inversions by Moseley's work on atomic numbers in 1913.

Mendeleev's real insight was revealed in his 1871 paper in connection with the vacant spaces in the periodic table. For example, he reasoned that the unknown element for the space following calcium should be closely related to boron. He gave it the provisional name *eka boron*, the word *eka* being Sanskrit for "first." From its position in the table he deduced the properties of the element and its compounds. In similar fashion he predicted the properties of *eka aluminum* and *eka silicon*, the missing elements between zinc and arsenic in Series 5 of the table. These predictions were verified in striking fashion during the next two decades by the discovery of the three elements.

The first to be discovered was *eka aluminum*, in 1875, by Paul Émile Lecoq de Boisbaudran (1838–1912). In these years, when France's defeat in the Franco-Prussian War rankled in the breast of every Frenchman, it was natural that the new element should be named *gallium* in honor of Gaul. The correspondence of Mendeleev's predictions and the observed properties of the new element is shown in the accompanying tabulation.

Property	Eka Aluminum	Gallium
Atomic weight	Approx. 68	69.9
Density	5.9	5.93
Melting point	Low	30.1°C.
Formula of oxide	Ea_2O_3	Ga_2O_3

Mendeleev correctly predicted that the metal would not be volatile, and would be attacked only slowly by air, water, acids, and alkalies; the oxide would be more basic than alumina but less basic than magnesia; the gelatinous hydroxide would be soluble in acids and bases; the sulfide would be precipitated by hydrogen sulfide and would not be soluble in ammonium sulfide; the metal would form a trichloride, a sulfide, and an alum, all of them soluble; the element would probably be discovered by spectroscopic analysis.

Mendeleev's *eka boron* was discovered by Lars Fredrik Nilson (1840–1899) of Sweden in 1879 in the mineral euxenite, while he was studying the rare earth elements. He named it *scandium* because it had been observed only in ores found in Scandinavia. The accompanying tabulation shows how Mendeleev's predictions agreed with the observed properties of scandium. Mendeleev was correct in predicting that the oxide would

Property	Eka Boron	Scandium
Atomic weight	44	44.1 (1956, 44.96)
Oxide	Eb_2O_3	Sc_2O_3
Density of oxide	3.5	3.8

be more basic than alumina, less basic than yttria and magnesia; that the salts would be colorless and would form gelatinous precipitates with potassium hydroxide and with sodium carbonate; that the alkali double sulfates would not be alums; that the anhydrous chloride would be less volatile than aluminum chloride; and that the element would not be discovered spectroscopically.

Eka silicon was discovered in 1886 by Clemens Winkler (1838–1904), of the Freiberg School of Mines, in a new silver ore, argyrodite. In typically nationalistic fashion he named it *germanium*. The agreement of its properties with Mendeleev's predictions is shown in the accompanying tabulation. Mendeleev was wrong in one prediction—that the element could be

Property	Eka Silicon	Germanium
Atomic weight	72	72.32 (1956, 72.6)
Specific gravity	5.5	5.47
Atomic volume	13	13.22
Valence	4	4
Specific heat	0.073	0.076
Specific gravity of GeO_2	4.7	4.703
Boiling point of $GeCl_4$	Under 100° C.	86° C.

melted and vaporized only with difficulty. Lothar Meyer had predicted the opposite.

Mendeleev did not fare so well with some of his other predictions. He noted that the long period following cesium was almost completely empty, but he believed that this was hardly accidental but was due to the nature of these elements. He also made some speculations regarding *eka manganese* (at. wt. = 100), *dwi manganese*, and *tri manganese*, but could not realize the difficulties involved in filling the spaces represented by element number 43 and related elements.

Meyer's Periodic Arrangement of the Elements

Lothar Meyer is entitled to credit as an independent discoverer of the periodic law. As early as 1864, in his *Die modernen Theorien der Chemie*, he included a table of elements arranged horizontally so that analogous elements came under one another. The change in valence with atomic weight was easily apparent. The difference in atomic weight from one series to the next was also clear. Some elements were not included in the list; furthermore, inaccuracies reduced the effectiveness of the table. The idea of natural families of elements, which was becoming common at this time, seems to have been the basis for Meyer's arrangement. The numerical increase in atomic weights was incidental. Like Newlands, he was groping toward the periodic law but was not yet aware of it.

Fig. 9.8. JULIUS LOTHAR MEYER.
(Courtesy of the Louis Kahlenberg Collection.)

	4 val.	3 val.	2 val.	1 val.	1 val.	
	....	....	....	....	Li 7.03	(
Diff.	....			....	16.02	
	C 12.0	N 14.4	O 16.00	F 19.0	Na 23.5	N
Diff.	16.5	16.96	16.07	16.46	16.08	
	Si 28.5	P 31.0	S 32.0	Cl 35.46	K 39.13	(
Diff. 8·9/1	44.45	44.0	46.7	44.51	46.3	
	....	As 75.0	Se 78.8	Br 79.97	Rb 85.4	
Diff. 8·9/2	44.55	45.6	49.5	46.8	47.6	
	Sn 117.6	Sb 120.6	Te 128.3	I 126.8	Cs 133.0	
Diff. 8·9·4/2	44.7	8·7·4/2 43.7	....	....	35.5	
	Pb 207.0	Bi 208.0	....	....	(Tl 204.0?)	B

	4 val.	4 val.	4 val.	2 val.	1 val.
{ Mn 55.1 { Fe 56.0		Ni 58.7	Co 58.7	Zn 65.0	Cu 63.5
Diff { 49.2 { 48.3		45.6	47.3	46.9	44.4
Ru 104.3		Rh 104.3	Pd 106.0	Cd 111.9	Ag 107.94
Diff. 2·2·2/2 46.0		2·2·2/2 46.4	9·3/2 46.5	2·2·2/2 44.5	2·2·2/2 44.4
Pt 197.1		Ir 197.1	Os 199.0	Hg 200.2	Au 196.7

Fig. 9.9. MEYER'S EARLIEST TABLE, 1864.
(From F. P. Venable, *The Development of the Periodic Law* [1896].)

SUGGESTION FOR A SYSTEM OF ELEMENTS BY LOTHAR MEYER. SUMMER 1868.

1	2	3	4	5	6	7	8
Cr=52.6	Mn=55.1 49.2 Ru=104.3 92.8=2.46.4 Pt=197.1	Al=27.3 2·8·7=14.8 Fe=56.0 48.9 Rh=103.4 92.8=2.46.4 Ir=197.1	Al.=27.3 Co=58.7 47.8 Pd=106.0 93=2.465 Os=199.	Ni=58.7	Cu=63.5 44.4 Ag=107.9 88.8=2.44.4 Au=196.7	Zn=65.0 46.9 Cd=111.9 88.3=2.44.5 Hg=200.2	C=12.00 16.5 Si=28.5 8·9·1=44.5 8·8·1=44.5 Sn=117.6 89.4=2.41.7 Pb=207.0

9	10	11	12	13	14	15
N=14.4 16.96 P=31.0 44.0 As=75.0 45.6 Sb=120.6 87.4=2.43.7 Bi=208.0	O=16.00 16.07 S=32.07 46.7 Se=78.8 49.5 Te=128.3	F=19.0 16.46 Cl=35.46 44.5 Br=79.9 46.8 I=126.8	Li=7.03 16.02 Na=23.05 16.08 K=39.13 46.3 Rb=85.4 47.6 Cs=133.0 71=2·35.5 Te=204.0	Be=9.3 14.7 Mg=24.0 16.0 Ca=40.0 47.6 Sr=87.6 49.5 Ba=137.1	Ti=48 42.0 Zr=90.0 47.6 Ta=137.6	Mo.=92.0 45.0 Vd=137.0 47.0 W=184.0

Fig. 9.10. MEYER'S UNPUBLISHED TABLE OF 1868.
(From F. P. Venable, *The Development of the Periodic Law* [1896].)

In 1868, Lothar Meyer wrote out for Professor A. Remele, his successor at Neustadt-Eberswald, a suggested system of the elements. This somewhat more advanced table did not come to public attention until 1895, when Karl Seubert, one of Meyer's former pupils, reported the incident.[11]

[11] K. Seubert, *Z. anorg. Chem.*, **9**, 334 (1895).

The table included 52 elements arranged in 15 columns, but numerous imperfections are readily apparent.

Meyer first wrote on the subject in December, 1869, the paper being published the following March.[12] In it the elements were arranged vertically on the basis of increasing atomic weights. The horizontal arrangement of families which resulted brought out periodic relations more clearly than Mendeleev's first table. This is particularly true with respect to double periodicity, for elements with analogous properties recur in alternate columns rather than in adjacent ones. Mendeleev recognized this relationship, but his first table failed to reveal it clearly.

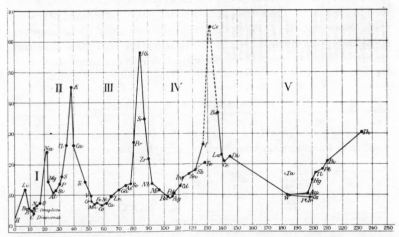

Fig. 9.11. CURVE SHOWING ATOMIC VOLUME PLOTTED AGAINST ATOMIC WEIGHT as redrawn by T. Bayley, for publication in *Philosophical Magazine*, 1882. Lothar Meyer's curve, on which this is based, appeared in *Annalen der Chemie*, supplement 7, in 1870. Meyer's curve was published on graph paper, which does not reproduce clearly; hence the substitution.
(From *Philosophical Magazine* [1882].)

This paper also included the curve resulting when Meyer plotted atomic weights against atomic volumes; he observed that similar elements appeared at similar places on the curve. Alkali metals are at all of the peaks on the curve; non-metals are on the ascending sides, and metals on the descending sides and in the valleys. This mode of representing atomic weights led him to examine other physical properties, such as hardness, compressibility, boiling points of analogous compounds, etc., and to plot them against atomic weights.

[12] L. Meyer, *Ann.*, Supplement 7, 354 (1870).

SUBSEQUENT DEVELOPMENTS

Recognition of the usefulness of the periodic table also made chemists aware of certain deficiencies. Although the discovery of a new element posed no problem—it was placed in the vacant place left for it in the table —a more serious problem faced chemists when the inert gases were discovered in the 1890's, for there were no spaces in the table for a whole new family of elements. The problem was solved by adding a column headed zero, at the left-hand side of the table for all these inert gases with a valence of zero.

Because little was known about the rare earth elements when Mendeleev designed his table, it was difficult to determine the best method of treating them. Moreover, the problem deepened when supposed elements proved to be mixtures. Chemists began slowly to realize that these rare earth elements failed to show the differences which would cause them to be

Periodisches System der Elemente (volle Gestalt).

Reihe	Gruppe 0	Gruppe I	Gruppe II	Gruppe III	Gruppe IV	Gruppe V	Gruppe VI	Gruppe VII	Gruppe VIII			
	---	—	—	—	RH_4	RH_3	RH_2	RH	—			
	R	R_2O	RO	R_2O_3	RO_2	R_2O_5	RO_3	R_2O_7	RO_4			
1		1 H										
2	He 4	Li 7	Be 9	B 11	C 12	N 14	O 16	F 19				
3	20 Ne	23 Na	24 Mg	27 Al	28 Si	31 P	32 S	35.5 Cl				
4	A 40	K 39	Ca 40	Sc 44	Ti 48	V 51	Cr 52	Mn 55	Fe 56	Co 59	Ni 59	Cu 63
5		63 Cu	65 Zn	70 Ga	72 Ge	75 As	79 Se	80 Br				
6	Kr 82	Rb 85	Sr 87	Y 89	Zr 90	Nb 94	Mo 96	−100	Ru 102	Rh 103	Pd 106	Ag 108
7		108 Ag	112 Cd	114 In	119 Sn	120 Sb	128 Te	127 J				
8	Xe 128	Cs 133	Ba 137	La 139	Ce 140 Pr 141 Nd 144 −145							
					−147 Sm 148 Eu 151 −152							
					−155 Gd 156 −159 −160							
					Tb 163 Ho 165 Er 166 −167							
					Tm 171 Yb 173 −176							
					−178	Ta 182	W 184	−190	Os 191	Ir 193	Pt 195	Au 197
9		197 Au	200 Hg	204 Tl	207 Pb	209 Bi	212−	214−				
10	−218	−220	Rd 225?	−230	Th 233	−235	U 239					

Fig. 9.12. BRAUNER'S TABLE OF 1902 (short form).
(From *Z. anorg. Chem.* [1902].)

placed in series across the table. Because they were unique in their similarities, they had to be lumped together. One solution of the problem was made by the Czech chemist Bohuslav Brauner in 1902 when he brought them together into a sort of miniature table within the standard table,[13] but even this was not completely satisfactory. It was gradually realized that all of these elements were best located with lanthanum in

[13] B. Brauner, *Z. anorg. Chem.*, **33**, 1 (1902).

Group III. Brauner's table was one of the earliest to include the rare earth family.

The elements which did not fit properly into their assigned groups—the elements now referred to as the transition elements—remained a source of trouble which led to various efforts toward modifying the periodic table so that they could be logically treated. Numerous variants of the table have been proposed, particularly in the present century, but all stem from nineteenth-century antecedents. In view of the many proposed modifications, it might be thought that no inorganic chemist considers he has achieved prominence until he has devised a new form of table. Fundamentally, all modifications attempt to show the elements classified in one of four basic types of table: the short form, long form, spiral, or three-dimensional.

The first short-form table was the one Mendeleev drew up in 1871. It provided for the inclusion of newly discovered elements, and was capable of receiving the inert gas elements and the rare earth elements without

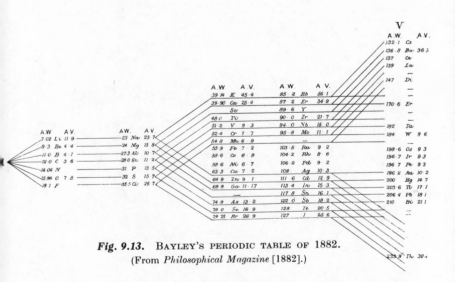

Fig. 9.13. BAYLEY'S PERIODIC TABLE OF 1882.
(From *Philosophical Magazine* [1882].)

fundamental alteration. It was very popular during the first three decades of the present century and is still reproduced frequently today. Its major shortcoming is that subgroup and main-group elements are listed in the same columns, but this is not too serious for anyone who recognizes the distinction between them.

The long form of table was the result of efforts to segregate the transition elements from the main-group elements. Mendeleev's 1869 table was

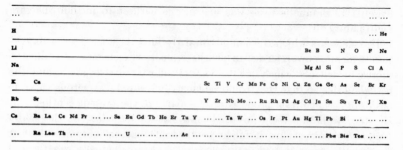

Fig. 9.14. WERNER'S LONG-FORM PERIODIC TABLE OF 1905.
(From *Berichte* [1905].)

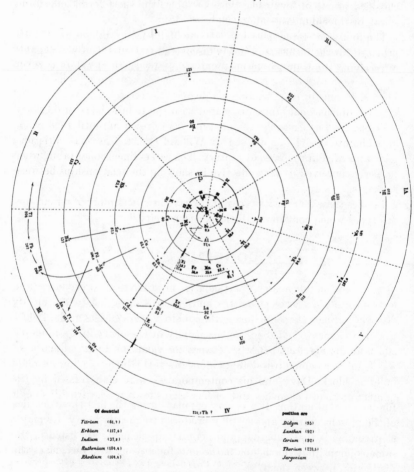

Fig. 9.15. BAUMHAUER'S SPIRAL PERIODIC TABLE OF 1870.
(From F. P. Venable, *The Development of the Periodic Law*
[1896].)

actually a not-too-unsatisfactory attempt at such a table. Other chemists' efforts along these lines slowly led to more satisfactory results. There are both vertical and horizontal versions of the long form. T. Bayley in 1882 proposed a vertical table in which hydrogen was connected to a period of seven elements. These elements in turn were connected to another period of seven elements, from which double lines led main-group and subgroup elements into two long and one very long period.[14] Bayley's table is the direct antecedent of the well-known Bohr form so frequently used today. One of the earliest long-form horizontal tables was devised by Werner in 1905.[15] It dealt with the transition elements satisfactorily but was unnecessarily long because the rare earth elements were arranged side by side. In subsequent tables these elements were assigned to one place, thus making for a more condensed table.

A spiral table was proposed as early as 1870 by H. Baumhauer. Hydrogen was at the center of the table and the rest of the elements were arranged in continuous spiral. He then divided the spiral into seven wedges, each containing related elements. Since then, variants have been proposed, some of them attaining a degree of success.

Three-dimensional tables originated with the telluric helix of de Chancourtois. No further tables of this type were drawn up until 1898, when an elaborate model was devised by William Crookes. Several variations have appeared in the present century. The three-dimensional table, when properly conceived, is very effective in showing the relationships between elements.

Besides these four major types, there have been a number of miscellaneous types which contributed little except novelty.[16]

CONCLUSION

Advances in chemistry, particularly in analytical procedures and instrumentation (electrochemistry and spectroscopy), led to the steady discovery of new elements during the hundred years ending in the 1860's. The new elements added to the complexity of chemistry and stimulated efforts to discover relationships between the elements. Prout's hypothesis was a futile attempt to relate all elements to hydrogen as the ultimate building block. Davy and his contemporaries were embarrassed by the number of known elements and, as has been shown by Siegfried,[17] sought

[14] T. Bayley, *Phil. Mag.*, [5], **13**, 26 (1882).

[15] A. Werner, *Ber.*, **38**, 914 (1905).

[16] The development of the various types of periodic tables has been carefully examined by G. N. Quam and Mary B. Quam, *J. Chem. Educ.*, **11**, 27, 217. 288 (1934). An early study which is still valuable is F. P. Venable's *The Development of the Periodic Law*, Chemical Publishing Co., Easton, Pa., 1896.

[17] R. Siegfried, in E. Farber, ed., *Great Chemists*, Interscience, New York, 1961, p. 379; *Isis*, **54**, 247 (1963); *Chymia*, **9**, 117 (1964).

to rationalize some of them as very stable compounds. However, the elemental nature of these substances could not be rationalized away and a method had to be found to fit them into a system of chemistry.

Early attempts to classify elements were based on atomic weight relationships, but none of the schemes were sufficiently useful to gain general acceptance. In part, this was attributable to the uncertainties in many of the atomic weights, in part to the fact that a significant number of elements were still unknown. Only after 1860, when Cannizzaro directed chemists toward correct atomic weights, was it possible to establish a system of classification which was sound.

Mendeleev and Lothar Meyer, using different approaches, pointed the way to a useful method of classification. Mendeleev's scheme was sufficiently sound to lead him to predict the properties of several unknown elements, predictions which were shown to be remarkably accurate when several of the elements were discovered during his lifetime. The system was also sufficiently versatile to permit the addition of a whole family of elements when the inert gases were discovered.

The periodic system of classification, which was developed on empirical grounds, suggested a systematic relationship between the various elements but clarification of this matter would not occur for nearly another half century.

III

THE GROWTH OF
SPECIALIZATION

THE NINETEENTH CENTURY MUST
be looked upon as the century of the atom. During the first half of the
century the atom was becoming established as the fundamental unit in
chemical thought. Certain leading chemists moved doggedly ahead toward
solving the problem of atomic weight, despite doubts regarding its value.
Once this problem was mastered, they were free to devote themselves
to the relation of atoms and molecules; in other words, structural problems
could be attacked. The last half of the nineteenth century saw rapid
progress in the organic field, for the early structural concepts proved
adaptable to the new problems. But in inorganic chemistry there was a
long stalemate caused by too close adherence to organic structural concepts. Not until the end of the century, as the result of Werner's work,
did inorganic structural chemistry take a profitable direction.

Thoughts about structure even extended to solutions, where colloid
chemistry and ionic theory proved challenging. Nevertheless, there were
physical chemists of importance who failed to see the significance of atoms,
and who advocated solving chemical problems with the aid of thermodynamic principles which involved no atomic concepts.

Besides bringing the atom into general acceptance as the fundamental
unit of chemistry, the decade of the 1850's marked the end of an era in
the development of the science. Its foundations were laid; hence it was

now possible to proceed with studying the visible structure. Chemistry took on a new character that differed from what it had been in the preceding century.

The day of the general chemist was passing. Up to 1850 most of the important chemists could do competent work in any area of the science. They turned from minerals to organic substances and back to minerals with ease. They not only were capable analysts, but could adapt existing methods to their special needs and create new methods when there were none. They were bold in building new theoretical constructs and in tearing apart the theories of others.

After 1860 a new type of chemist—the specialist—emerged. No longer could the whole province of chemistry be grasped by individuals. The foundations having been laid, the chemist no longer had to be equally at home with inorganic and organic compounds, with analytical and theoretical aspects. He could now enjoy the luxury of studying a limited area in depth. The science was broken down into specific areas; the major figures became identified with organic, or inorganic, or analytical chemistry. The field of physical chemistry was opened up with great enthusiasm in the last decade of the century; agricultural and physiological chemistry developed as applied fields; and by the end of the century some universities were offering courses in chemical engineering.

The chemist began to have a place in industry as a valuable specialist with unique contributions to make. This was particularly true in Germany where a vigorous chemical industry developed during the last three decades of the century.

CHAPTER 10

THE DIFFUSION OF
CHEMICAL KNOWLEDGE

CHEMICAL EDUCATION TO 1825

In discussing the growth of chemistry as a science, it is desirable to pause long enough to note the changes which took place around the mid-nineteenth century with respect to the extension of chemical knowledge. As we have seen, up to the beginning of that century scientific development was very much in the hands of the gifted amateur, frequently either wealthy himself or able to secure a wealthy patron. Scientific progress often centered around a talented enthusiast who could pass on his enthusiasm to his friends and encourage younger men. From the seventeenth century on, the scientific society was a center where interested persons could exchange ideas and promote new investigations. Government assistance was irregular at best, and often nonexistent.

There was hardly any systematic scientific education before 1800. To the extent that it was recognized at all in educational institutions, chemistry was generally an adjunct to medicine, occasionally to mathematics. The best place to learn it was not in a university but in a pharmacist's shop.

As was said earlier, the situation began to change during the French Revolution and the change continued under Napoleon. Whether it was due to governmental enlightenment is open to question. Nevertheless, French science rose to great heights during the last half of the eighteenth century. This was particularly true in the field of chemistry, where the names of Lavoisier, Berthollet, Fourcroy, Chaptal, Guyton de Morveau, and Monge stand out.

Unquestionably the primary purpose of the founding of the École Polytechnique in 1794 was the training of military engineers, men competent in mathematics, mechanics, and drawing. But the other sciences were not neglected.[1] Berthollet, the first professor of chemistry at the École, lacked the ability to reach the students; because of his host of other activities he was soon replaced by Guyton de Morveau. Chemistry instruction took on new importance in other Parisian schools as well. In view of the group of outstanding chemists who were working in Paris early in the nineteenth century, no school found it difficult to secure a distinguished faculty, even though the salaries were frequently so low that professors taught at more than one institution or held commercial positions in order to obtain an adequate income.

Laboratory work was not usually a part of the instruction in these schools. Instead, the professors maintained private laboratories to which chosen students were admitted. Fourcroy provided such an opportunity for Vauquelin, who served as his associate at the Jardin des Plantes for some years. Vauquelin provided a similar opportunity for Thenard, who continued the tradition in Paris, and for Stromeyer, who operated a teaching laboratory for his students at Göttingen. Gay-Lussac, an early student of Berthollet, was important in teaching laboratory work to private pupils—among them, Dumas and Liebig.

None of these teachers was influential in establishing a "school" in the sense of a personally trained group of disciples. In the first quarter century, young men interested in chemistry naturally gravitated to Paris where they heard lectures by Vauquelin, Gay-Lussac, Thenard, Chevreul, Dulong, Biot, and others; the fortunate ones might spend some time in the laboratory of one of these men. Great teachers provided inspiration, but not close guidance. Youths with great personal initiative flourished in this environment, but the less able youths must have drifted rather aimlessly.

One other place, apart from Paris, attracting chemistry enthusiasts was Berzelius' laboratory in Stockholm. Here again, laboratory instruction was not provided by the school, but Berzelius was willing to accept one well-recommended student in his private laboratory each year. Heinrich Rose, Mitscherlich, and Wöhler, among others, studied with him.

In Germany there were no important centers for instruction in chemistry. We saw earlier that Liebig studied with Kastner, who was considered the foremost chemistry professor in the country. His enthusiasm soon chilled, however, and he went to Paris, where he made disparaging remarks about the quality of Kastner's chemistry, despite the fact that Kastner had befriended him on more than one occasion.

English scientists had always been an individualistic lot; hence there was nothing resembling a school of chemistry in that country. In Scotland

[1] W. A. Smeaton, *Ann. Sci.*, **10**, 224 (1954).

in 1817, Thomas Thomson was offering systematic instruction in chemistry, including laboratory work. However, he lacked the stature of his French contemporaries, both as an experimentalist and as a theorist.

DEVELOPMENT OF LABORATORIES; LIEBIG AND THE GIESSEN LABORATORY

When Liebig became professor of chemistry at Giessen he immediately took steps to offer laboratory instruction in the science. A deserted barracks was converted into a laboratory where students were put to work learning chemistry experimentally. They were first trained in qualitative and quantitative analysis, they then prepared organic compounds, and finally carried out a special investigation on a problem suggested by Liebig.

The laboratory at Giessen received a great deal of attention and attracted students from many parts of the world. It was frequently said to be the first laboratory for student instruction, a statement fostered by Liebig himself, as when he wrote: "At that time, chemical laboratories in which instruction was given in analysis did not exist anywhere; what people called such, were rather kitchens, filled with all sorts of furnaces and utensils for carrying out metallurgical or pharmaceutical processes. Nobody understood how to teach analysis." This is a characteristic Liebig effusion, for Stromeyer had been offering laboratory work to students at Göttingen since 1806.[2]

Friedrich Stromeyer (1776–1835) was the son of a professor of medicine at Göttingen. He studied pharmacy there under Johann Friedrich Gmelin and continued his studies at the École Polytechnique under Vauquelin until 1800, when he took his medical degree. His interests were now centered distinctly in chemistry. He habilitated as *Privatdocent* until he succeeded Gmelin at Göttingen in 1805. The next year he offered laboratory work in analytical chemistry. The laboratory continually expanded. It was there that Mitscherlich's interest was diverted from Oriental languages to chemistry, for he was at the laboratory about the time that Stromeyer, an ardent student of mineralogy and an expert analyst who served as inspector of apothecaries for the kingdom of Hanover, discovered cadmium in a sample of impure zinc oxide.

Student laboratories were also operated by J. N. von Fuchs at Landshut after 1807, by Döbereiner at Jena after 1811, and by N. W. Fischer at Breslau after 1820.

Outside Germany there were laboratories offering instruction in chemistry that antedated Liebig's. Thomas Thomson's laboratory in Scotland has already been mentioned. Frederic Accum (1769–1839), who was

[2] G. Lockemann and R. E. Oesper, *J. Chem. Educ.*, **30**, 202 (1953).

prominent in English industrial chemistry during the first two decades of the nineteenth century, opened a private laboratory in London for students about 1800. He opened a shop that sold scientific apparatus, and the school taught prospective customers of his shop. In the United States, Amos Eaton established a teaching laboratory at the newly founded Rensselaer Polytechnic Institute in Troy, New York, in 1824. Even earlier than any of these was the laboratory that Lomonosov opened in St. Petersburg in 1748.[3]

The earliest known university laboratories date back to the early part of the seventeenth century. On his appointment as professor of chemistry at the University of Marburg in 1609, Johannes Hartmann (1563–1631) opened a laboratory for students. Its purpose was instruction not in analysis—still in a very rudimentary state—but in pharmaceutical preparations; Hartmann's manual, *Praxis chymiatrica*, was used. A similar laboratory was opened by Johann Moritz Hofmann (1621–1698) at the University of Altdorf in 1683.

Possibly, as Liebig said, the student laboratories then in existence may have been kitchens, but they did exist and some of them did teach analytical chemistry. However, Liebig's laboratory was unique in the results it produced. Perhaps analysis was emphasized more there than in earlier laboratories, and certainly the methods were sounder. But the most important factor was the enthusiasm and inspiration of Liebig himself. An expert analyst, he made numerous contributions to analytical chemistry, as we have already seen. He had a talent for devising unique methods and new apparatus for accomplishing a particular purpose. Although not all the apparatus which bears his name was original with him—i.e., the Liebig condenser—he was certainly important in extending their use. The large number of students of his who rose to prominence is testimony to his greatness as a teacher. Among these students were A. W. von Hofmann, Fresenius, Pettenkofer, Kopp, Fehling, Volhard, Will, Varrentrapp, Erlenmeyer, Strecker, and Kekulé from Germany; Henneberg from Denmark; Wurtz, Regnault, and Gerhardt from France; Williamson, Playfair, and Muspratt from England; and Horsford, O. Wolcott Gibbs, and J. Lawrence Smith from the United States.

There developed in the laboratory an *esprit de corps* which was a factor in spreading its fame. Liebig lived in the building and the students spent their entire day there; Aubel, the caretaker, complained about not being able to get them to leave. Liebig, a highly energetic man, had numerous projects under way at the same time. He gave the younger students little actual instruction in the laboratory, relying instead on his older students to act as his assistants in guiding the beginners in their work. The older students worked on original problems, turning in a report each morning

---

[3] B. N. Menschutkin, *J. Chem. Educ.*, **4**, 1075 (1927); T. L. Davis, in *ibid.*, **15**, 204 (1938), and in *J. Am. Chem. Soc.*, **34**, 109 (1912).

Fig. 10.1. LIEBIG'S LABORATORY IN GIESSEN, 1842 (after a lithograph by Trautschold). Shown in the print are the following men: *1* Ortigosa, a Mexican. *2* and *3* Unidentified men. *4* The janitor. *5* Wilhelm Keller, later a pharmacist in Philadelphia. *6* Heinrich Will, Liebig's successor at Giessen. *7* Aubel, Liebig's laboratory servant. *8* A. F. C. Strecker, later at Tübingen and Würzburg. *9* Wydler, who came from Aarau. *10* Franz Varrentrapp, later warden of the mint in Braunschweig. *11* Johann Josef Scherer, later professor of medicine in Würzburg. *12* An unidentified man. *13* Emil Boeckmann, later director of an ultramarine factory in Heidelberg. *14* A. W. von Hofmann.

(Courtesy of the American Chemical Society Center for History of Chemistry, University of Pennsylvania.)

on their progress the day before. Liebig discussed these reports with the various students in planning their future work. Thus there was a great deal of activity of different kinds, and the students educated one another. The research work done in the Giessen laboratory covered a wide range of subjects.

During the earlier years Liebig's interests leaned strongly toward organic chemistry and he was instrumental in the isolation and analysis of various new compounds. Gradually, however, he turned toward applied fields, particularly agricultural chemistry. About this time too Liebig began his "Familiar Letters on Chemistry," which appeared first as letters in the *Augsburger Allgemeine Zeitung* and were then gathered together for publication in book form. In them he attempted to bring his chemical researches and ideas before the general public, for he felt that people should benefit by the application of chemistry to practical problems.

EXTENSION OF EXPERIMENTAL CHEMISTRY

The pattern of chemical education established at Giessen was quickly extended to other German universities. Liebig's friend Wöhler was the first to adopt the pattern. On his return from Stockholm, Wöhler took a teaching position in a technical school in Berlin. He had no opportunity there to create a laboratory like Liebig's, but he carried on his own researches. In 1831 he went to teach at the newly founded technical school in Cassel. Not until 1836, by which time his fame was well established, did he receive a university appointment—at the George Augustus University in Göttingen, where he was selected to fill the chair left vacant at Stromeyer's death. A laboratory was already in operation at the university, but in Wöhler's hands it was operated more nearly along the lines of the one at Giessen.

The laboratory attracted numerous students, particularly during Wöhler's last thirty years. His career was, in many respects, the antithesis of Liebig's, just as his personality was. Wöhler was friendly, easygoing, inclined to see the humor in a situation. He accepted poorly prepared students in his laboratory, and with great patience helped them develop. On one occasion, when chided for wasting so much time on American students, he answered, "I know, I know, but they are all such nice boys."[4] His patience was rewarded, for many of his students exerted a major influence in the extension of chemistry in both the industrial and educational fields. Among these students of his were Kolbe, Fittig, and Hübner in Germany; and Ira Remsen, Edgar Fahs Smith, Stephen Moulton Babcock, J. C. Booth, Henry Nason, G. C. Caldwell, and C. F. Chandler from the United States.[5]

Other centers of activity began to develop. Bunsen's laboratory at Heidelberg also had wide appeal after mid-century; Bunsen was responsible for launching many young men on a career in chemistry. German universities in every principality began to model their laboratories after the pattern set by Liebig, and graduates of the Giessen laboratory were eagerly sought. Eventually very elaborate laboratories were built as a means of attracting leading chemists. When Hofmann was brought to

---

[4] Told to the author by Louis Kahlenberg in 1940. Kahlenberg had the story from Stephen M. Babcock, who received his doctorate from Göttingen.
[5] H. S. Van Klooster, *J. Chem. Educ.*, **21**, 158 (1944).

Bonn in 1864 he was provided with a palatial new home and laboratory, but even this inducement was insufficient to hold him in Bonn when the University of Berlin offered to build him even more elaborate quarters.

The Liebig pattern was not copied in France, for academic positions there were largely under the control of a centralized authority such as the national government or the Académie des Sciences. A few prominent chemists like Dumas, and later Marcellin Berthelot, exerted a great deal of influence over appointments, particularly the desirable ones in Parisian

Fig. 10.2. GROUP OF BUNSEN'S STUDENTS AT HEIDELBERG IN 1855. Back row, left to right, Gaupillet, Frapolli, A. Wagner, Henry Roscoe, Lothar Meyer, Paresi, Beilstein. Seated, Bahr, Landolt, Carius, Kekulé, von Pebal.
(Courtesy of the J. H. Walton Collection.)

schools. As a result, France failed to maintain its renown as a center of chemistry, and Germany became the attraction for students.

In England there was a desire in certain quarters to follow Liebig, but it was far from universal. Queen Victoria's prince consort, Albert, was prince of Saxe-Coburg-Gotha, a region near Liebig's part of Germany. When Victoria and Albert were married in 1840, Liebig was already a popular figure in Britain, having traveled and lectured in England and Scotland in 1837. He had concerned himself particularly with British agriculture and food processing, and on every occasion had emphasized how science might be a boon to agriculture and industry.

Prince Albert, who was active in promoting scientific education in England, was able to interest a number of British landowners and industrialists in founding a college of chemistry. When Liebig visited London in 1842 he met Sir Robert Peel, the prime minister, and plans were drawn up, the collection of funds being put in the hands of James Clark, the queen's physician. In 1845 the Royal College of Chemistry was opened, with Liebig's former student, A. W. von Hofmann, as its first director. Hofmann entered into his duties with enthusiasm and the college began to attract English youths interested in the teaching and application of chemistry. The laboratory was run in the same fashion as its prototype at Giessen. Along with his teaching, Hofmann carried on a research program, for while still with Liebig he had become interested in compounds derived from coal tar. This research he naturally continued in England, for coal tar was plentiful there, and virtually without use.

The scientific climate in England, unfortunately, was not conducive to success for the Royal College. Hofmann's reputation drew pupils for a time, but there was discontent among the institution's financial supporters who disagreed with Hofmann's emphasis on basic science. English industrialists wanted scientists who would immediately pay their way in established industries by introducing more efficient processes, developing new products, and finding profitable uses for by-products. There was little interest in supporting long-term research projects which might or might not be profitable, and there was no interest in starting new industries based on scientific foundations. As time went on, the Royal College enrollment dropped and it was finally consolidated with the Royal School of Mines. Hofmann resigned in 1864 to take a position at Bonn. A year later he was appointed to the professorship at the University of Berlin left vacant by the death of Mitscherlich.

In Germany, Hofmann was free to pursue the fundamental researches on organic nitrogen compounds which he had carried on under duress in London. British industrialists had had little patience for such research since they were interested in immediate applications and quick profits. Hofmann held that it was first necessary to understand fundamental relationships; after these were understood, the practical arts could truly benefit from the application of science.

The older English universities paid little attention to chemistry. Oxford and Cambridge continued to emphasize the classical curriculum; no chemist of prominence was on their faculty. However, London's University College brought in Thomas Graham in 1837.

In Scotland the situation was somewhat better, for Glasgow and Edinburgh never completely lost their strong position that dated from the time of Cullen and Black. Medical aspects continued to receive major emphasis, but practical applications of science were made in Scottish industry and agriculture.

Black succeeded Cullen at Glasgow University in 1756, and was in turn succeeded in 1766 (when Black went to Edinburgh) by John Robison, who stayed until 1769. From then on, into the present century, the professorship was held by William Irvine (to 1787), Thomas Charles Hope (to 1791), Robert Cleghorn (to 1817), Thomas Thomson (to 1852), Thomas Anderson (to 1874), and John Ferguson, who died in 1915.

Black remained at the University of Edinburgh until his death in 1799. His chair at Edinburgh was filled by Thomas Charles Hope (1766–1844), one of the early adherents of Lavoisierian chemistry in Great Britain. Although competent, Hope never attained Black's stature. His most important scientific work was done on strontium compounds and the maximum density of water.

All these men were educated at Edinburgh or Glasgow, usually in medicine. The older ones, of course, came under Black's immediate influence.

CHEMICAL EDUCATION IN AMERICA

In the United States, chemistry had been given sympathetic treatment in the universities from the end of the eighteenth century when Priestley came to this country. Science as a whole had few full-time devotees in the New World in this period, but it attracted the enthusiastic support of such leaders as Benjamin Franklin and Thomas Jefferson. Franklin, desirous of creating an institution whose curriculum would not be fettered by the deadening hand of tradition, was active in founding the College of Philadelphia (ultimately part of the University of Pennsylvania). The Philadelphia Medical College, founded in 1765, was planned with the same ideas in mind; basic sciences were given proper emphasis. John Morgan, one of Cullen's students at Edinburgh, was named professor of chemistry and materia medica at the new college; and in 1769 a chair devoted only to chemistry was established. Benjamin Rush, the first man to hold this professorship, had received his doctorate in medicine at Edinburgh in 1768, where he had studied under Black.

When the Philadelphia colleges were consolidated into the University of Pennsylvania, Rush gave up his professorship of chemistry to become professor of medicine. The chair of chemistry was held for several years by Casper Wistar, but he also transferred to medicine. In 1794 the vacancy was offered to Priestley who had just arrived in this country, but he declined it, whereupon the appointment went to James Woodhouse (1770–1809), who only a few years before had been a student of Rush.

In 1767, Dr. James Smith, a graduate of Leyden, began to teach in the medical school at King's College (its name was changed to Columbia College during the American Revolution) and within two years became professor of chemistry there.

Even before 1767, chemistry had acquired some status as part of a course known as natural philosophy. It was included in such courses by James Madison at William and Mary in 1774, William Smith at the College of Philadelphia in 1756, and Charles Morton at Harvard in 1787. All three of these men had been educated as clergymen. The practice of having clergymen serve as science instructors continued to be common throughout the nineteenth century in American liberal arts colleges.[6]

At the beginning of the nineteenth century six eastern colleges offered serious instruction in chemistry. Besides the University of Pennsylvania and the College of William and Mary, with Woodhouse and Madison respectively, these included the Harvard Medical School, with Aaron Dexter (began teaching in 1783); the Dartmouth Medical School, with Lyman Spaulding (1898); Columbia, with Samuel Latham Mitchill (1792); and the College of New Jersey (now Princeton), with John Maclean (1795). All these teachers, except Madison, had been trained in medicine, Mitchill and Maclean having studied with Black. Spaulding was a leader in preparing the first United States Pharmacopoeia in 1820. Interestingly enough, all six men were followers of Lavoisier's system of chemistry. Spaulding and Mitchill were active in introducing the new nomenclature into America. Maclean, of Scottish birth, had become acquainted with Lavoisier's theories before coming to America. He and Woodhouse vigorously opposed Priestley when the latter sought to uphold the phlogiston theory. Mitchill, whose journal, the *Medical Repository*, carried the polemical series between Priestley and Maclean, tried unsuccessfully to effect a compromise between the two systems as late as 1810.[7]

The most important figure in American science and scientific education in the first half of the nineteenth century was the elder Benjamin Silliman (1779–1864). Although trained for law, he was invited in 1802 by the president of Yale to become the first professor of chemistry there. In order to prepare himself, Silliman spent a year at Philadelphia where he heard Woodhouse lecture on chemistry, and Rush, Wistar, and Benjamin Barton on medicine. On his way back to Yale, Silliman stopped at Princeton where Maclean gave him further assistance in preparing for his work. At the end of a year of teaching, Silliman was voted a liberal grant for laboratory apparatus. He spent the next year in Europe buying equipment, much of it from Accum, and expanding his scientific knowledge. He met such men as Davy, Dalton, Fulton, Watt, Banks, Wollaston, Cavendish,

---

[6] R. Siegfried and A. J. Ihde, *Trans. Wisconsin Acad. Sci.*, **42**, 25 (1953).
[7] R. Siegfried, *Isis*, **47**, 327 (1955).

Thomson, and Hope; received scientific instruction from Accum; and attended lectures in London and Edinburgh.

Silliman's position as a leading figure in American science was due not so much to his original contributions to knowledge as to his enthusiasm in making his countrymen aware of science. His texts on geology and chemistry were extremely popular. He was responsible for the training of a large number of young men who later made their mark in education, medicine, and industry—men like Charles V. Shepard (South Carolina Medical College, and Yale), James Dwight Dana (Yale), Denison Olmsted (College of North Carolina, and Yale), Oliver P. Hubbard (Dartmouth), Benjamin Silliman, Jr. (Louisville Medical College, and Yale), Robert Hare (Pennsylvania Medical School), and Amos Eaton (Rensselaer). In his fifties Silliman achieved widespread fame as a popular lecturer, his tours taking him as far away from New England as St. Louis and New Orleans.

By mid-century American science underwent a new phase in its development, the founding of scientific schools. The encroachment of science into the college curriculum had been quite generally resisted by the defenders of classical studies, for colleges had as their primary purpose the training of young men for the ministry or law. Hence scientific studies were considered important only to the extent that they provided knowledge that enabled one to defend the Bible against heretical ideas. The student interested in science was almost forced to pursue this work at one of the few existing medical schools. Even Silliman, despite his prestige and enthusiasm, could not have a really scientific program incorporated in the regular curriculum at Yale. Interested students generally returned after graduation to work more closely with Silliman as assistants or special students. It was the presence of these advanced students that led to the foundation of a scientific school at Yale.

In 1846 Silliman proposed to the governing board that Yale provide for a chair of agricultural chemistry and vegetable and animal physiology. As a result, a school was established with funds obtained by subscription. Its principal benefactor in its formative years was Joseph E. Sheffield, the school finally becoming known as the Sheffield Scientific School of Yale College. Its first faculty consisted of Benjamin Silliman, Jr. (1816–1885), in practical chemistry and John Pitkin Norton in agricultural chemistry. Both men were former assistants of the elder Silliman. Norton had been greatly influenced by Liebig's writings and had spent some time studying agricultural chemistry with Johnston at Edinburgh and Mulder at Utrecht. On Norton's untimely death in 1852 his professorship went to one of his students, Samuel William Johnson (1830–1909), who had had two years of study under O. L. Erdmann at Leipzig and under Liebig at Munich.

At Harvard, Abbott Lawrence provided funds for the founding of the

Lawrence Scientific School in 1847. Its first director was Eben Norton Horsford (1818–1893), just returned from work in Liebig's laboratory. He was made Rumford Professor of Science as Applied to the Arts, and instruction was patterned on the Giessen model. Horsford's studies concerned the chemistry of foods, and he obtained a patent on a new baking powder based on calcium acid phosphate. He resigned in 1863 in order to pursue his business interests more actively at the Rumford Chemical Works, which he had founded in Providence, Rhode Island, in 1856 to manufacture baking powder. His position at Harvard was filled by O. Wolcott Gibbs (1822–1908), another of Liebig's students.

The popularity of these schools of science gradually wore down the resistance toward work in this field, and opportunities broadened during the last half of the century. It became increasingly popular to go to Germany for doctoral studies, and this brought a consequent decrease in the influence of medical schools in training scientists. Graduate schools gradually appeared in America in imitation of the German university. Yale granted its first Ph.D. in science to Josiah Willard Gibbs in 1863, and Harvard granted its first doctorate in chemistry to Frank Austin Gooch in 1877. Although at the end of the century Germany was still favored for graduate study, respectable graduate schools were developing at Johns Hopkins, Pennsylvania, Harvard, Columbia, Michigan, Chicago, and Wisconsin.

CHEMICAL PUBLICATIONS

Until well into the seventeenth century, scientific communication was carried on primarily by word of mouth, private letters, manuscripts, and printed books. As scientific societies were established, the publication of printed transactions or memoirs constituted a significant part of their work. These general periodicals that covered all fields of science were adequate as long as scientific activity was slow; but when it accelerated during the eighteenth century, such publications proved inadequate. The volume of material that was submitted made it impossible for the secretaries of the various societies to evaluate its worth. Furthermore, the financial resources of the societies did not permit the printing of the material that was accepted; hence publication frequently lagged by several years.

By the end of the eighteenth century, privately sponsored journals had been established—some of them of a general nature, some restricted to a particular field of science. The earliest journals which restricted their pages to chemistry were published by Lorentz von Crell (1744–1816), who was professor of chemistry and counselor of mines in Helmstadt. The titles

Fig. 10.3. LORENTZ VON CRELL.
(Courtesy of the Edgar Fahs Smith Collection.)

and emphasis changed frequently.[8] These journals represent the state of chemical activity in Germany during this period; a great deal of the material pertained to the phlogiston problem. Summaries of work done in other countries were also included.

The *Allgemeines Journal der Chemie* was founded in 1798 by Alexander Nicolaus Scherer (1771–1824); the name was changed to *Neues allgemeines Journal der Chemie* in 1804 when Gehlen became editor. Gehlen's interests caused a further change in its name to *Journal für die Chemie, Physik, und Mineralogie* in 1807. Another change in name occurred in 1811 when Johann S. C. Schweigger (1779–1857) became editor and dropped mineralogy from the title.[9] In 1834 it was combined with the *Journal für*

[8] The Crell journals were published under the following titles:
Chemisches Journal für die Freunde der Naturlehre, Artzneygelahrtheit, Haushaltungskunst und Manufacturen, Lemgo, 6 vols., 1778–1781, continues as:
Die neusten Entdeckungen in der Chemie, Leipzig, 13 vols., 1781–1786.
Chemische Annalen für die Freunde der Naturlehre, Helmstadt and Leipzig, 40 vols., 1784–1803.
Beytrag zu den chemische Annalen, Helmstadt and Leipzig, 6 vols., 1786–1799.
Chemisches Archiv, Leipzig, 2 vols., 1783, continued as:
Neues chemisches Archiv, Leipzig, 8 vols., 1784–1791, continued as:
Neustes chemisches Archiv, Weimar, 1 vol., 1798.
[9] The first 12 volumes of the *Journal für die Chemie und Physik* were alternatively known as *Beitrage zur Chemie und Physik.* Volumes 31 to 60 were also called *Jahrbuch der Chemie und Physik.* Volumes 61 to 69 had the alternative title, *Neues Jahrbuch der Chemie und Physik.*

technische und ökonomische Chemie as the *Journal für praktische Chemie* under the editorship of Otto Linne Erdmann (1804–1869).

The most important of the German publications was the *Annalen der Pharmacie*, established by Liebig in 1832 through the merger of two pharmaceutical journals, one of which Scherer had started in 1817. Liebig's *Annalen* carried reports of the major researches of the time, particularly those on organic chemistry. The name was changed to *Annalen der Chemie*

Fig. 10.4. LIEBIG IN HIS STUDY IN MUNICH.
(Courtesy of the Edgar Fahs Smith Collection.)

und Pharmacie in 1839 as the emphasis on chemistry became more apparent. Dumas and Graham were made co-editors during the "era of good feeling" in 1838, but Liebig controlled the policy. Wöhler became co-editor in 1840, and other men served in this capacity from time to time. In 1873, following Liebig's death, the name of the journal was changed to *Justus Liebig's Annalen der Chemie*, its present name. This journal has always been a leader in the field of organic chemistry.

Rozier's journals[10] had an important part in the early period of French chemistry, for papers that were read before the Académie frequently were published in these journals several years before they or revised versions appeared in the official memoirs. In 1789 the group around Lavoisier founded the *Annales de chimie* as a medium for their studies; Lavoisier was one of the original editors. However, many changes occurred during and after the Revolution. Thus the title was changed to *Annales de chimie et de physique* in 1815, and a century later—in 1914—the journal was split up into the *Annales de chimie* and the *Annales de physique*.

In Britain, Nicholson founded the *Chemical Journal* in 1798.[11] Immediate competitors were the *Philosophical Journal* and the *Philosophical Magazine*. The last-named proved to be the hardiest of the three and absorbed the other two in 1814. The year before this, Thomas Thomson founded the *Annals of Philosophy*,[12] which played a vigorous role in scientific publication during the next decade. However, it became apparent that British science could not support a number of private journals, and Thomson's *Annals* was incorporated into the *Philosophical Magazine* in 1827.

In the United States, scientific activity had not yet made any great progress, and reports of investigations were published in various periodicals. Medical journals like Mitchill's *Medical Repository* (New York, 1797–1824) were a natural outlet because investigations were frequently conducted by physicians. Priestley used this journal for his final papers defending the phlogiston theory, and John Maclean used the same journal to answer him. Other publications included the *Transactions of the American Philosophical Society*, dating from 1771; the *Memoirs of the American Academy of Arts and Sciences* of Boston; and equivalent periodicals of state and local academies. The more prominent investigators were able to have their work published in European journals like the *Philosophical Magazine* and the *Annals of Philosophy*, and even, in a few cases, in journals published in France and Germany.

In 1818 the *American Journal of Science* was founded by Benjamin Silliman to take the place of the short-lived *American Mineralogical Journal* which Dr. Archibald Bruce had founded in 1810. Silliman's *Journal* held an important place in American science during the nineteenth century and today is a privately published journal.

---

[10] The titles of the Rozier journals varied as follows:
  Introduction aux observations sur la physique, vol. 1, 1773.
  Observations et mémoires sur la physique, sur l'histoire naturelle et sur les arts et métiers, vols. 2–43, 1773–93.
  Journal de physique, de chimie, de histoire naturelle et des arts, vols. 44–96, 1794–1823.

[11] The title was soon changed to *Nicholson's Journal of Natural Philosophy, Chemistry and the Arts*. It was commonly known as *Nicholson's Journal*.

[12] A short-lived *Annals of Philosophy* was founded by Thomas Garnett in 1800, but only three volumes were published.

THE RISE OF CHEMICAL SOCIETIES

With the development of chemistry as a science there was a need for societies devoted exclusively to the subject. Generalized societies like the Royal Society and the Académie des Sciences were no longer capable of filling the needs of the strong and rapidly growing discipline. Furthermore, the decline they suffered in the nineteenth century prevented them from realizing the direction in which they would have to expand if they were to serve the specialized sciences. The Royal Society, in particular, was vigorously attacked by Charles Babbage, a mathematician. Babbage maintained that the Society was declining because many of its members not only made no scientific contributions but had no real interest in science; their sole interest was the honor of being a member.

Babbage's attacks ultimately brought about a reform in the Society's membership policy, but not until after the British Association for the Advancement of Science had been formed in 1831. The new Association immediately set to work on accumulated problems; thus one committee on chemical formulas sought to set up a standard governing the use of symbols and numerals in formulas. An American counterpart of the British Association was formed in 1848, led by the Association of American Geologists.

France had no equivalent organization, but the Académie des Sciences was vigorous enough to meet the new demands—among them, an organ for the rapid publication of scientific investigations. This need was met by founding a weekly, *Comptes rendus hebdomadaires des séances de l'académie des sciences*, in 1835. This journal has been important in disseminating French science by providing a medium in which short papers can be published without delay. Although all the sciences are included, many papers have dealt with chemistry. The somewhat equivalent weekly journals, *Nature* in England and *Science* in the United States, were not started until 1869 and 1883 respectively. *Science* became the official journal of the American Association for the Advancement of Science in 1895.

The first society to be limited exclusively to chemistry was founded in London in 1841,[13] at the time when the enthusiasm for chemistry that Liebig and his disciples engendered in Britain was at its height. The

---

[13] This is not actually the first chemical society, but it is the first one to be composed of mature chemists and to have had a continuous existence. There were student societies of record at the University of Edinburgh and of Glasgow (J. Kendall, *Endeavour*, **1**, 106 (1942)), and at Philadelphia. The Columbian Chemical Society of Philadelphia was made up of adult members of the locality, but it was short-lived, passing out of existence around 1811; a volume of *Memoirs* was published. See Edgar F. Smith, *Chemistry in America*, Appleton, New York, 1914, p. 208; W. Miles, *Chymia*, **5**, 145 (1959).

Chemical Society began publication of the *Quarterly Journal* in 1847; this was renamed the *Journal of the Chemical Society* when it became a monthly in 1861. The founding of a journal published by a society perhaps occurred earlier in England than in France and Germany, where interest in chemistry was more vigorous, because there was no English equivalent of the *Annales de chimie et de physique* and the *Annalen der Chemie und Pharmacie.*

A chemical society was organized in Paris in 1855 and began publication of the *Bulletin de la société chimique de France* in 1858.

There was no German society until Hofmann returned from England in 1864. The Deutsche chemische Gesellschaft was founded in Berlin in 1866, and publication of its *Berichte* began in 1868.

A Russian physical chemical society was founded in the same decade; the first issue of the *Journal of the Russian Physical Chemical Society* appeared in 1869.

The centennial of the discovery of oxygen was the stimulus for the founding of a chemical society in the United States. The occasion was commemorated by a number of American chemists who gathered at Priestley's grave in Northumberland, Pennsylvania. They were conscious of the need for a permanent organization, and accordingly the American Chemical Society was founded in New York City in 1876. For nearly two decades the Society was dominated by members who lived in New York, but it became truly national after a shake-up in the 1890's. Publication of the *Journal of the American Chemical Society* began in 1879, the same year that Ira Remsen of Johns Hopkins launched the *American Chemical Journal.* The latter was for some years the more important of the two; but as the Society's *Journal* improved in quality, the Remsen *Journal* lost ground and was finally merged with the Society's *Journal* in 1913. It had already absorbed a competitor, the *Journal of Analytical Chemistry,* founded by Edward Hart in 1887. Another chemical publication that preceded the Society's *Journal* was the *American Chemist,* published by C. F. Chandler of Columbia from 1870 to 1877. The *American Chemist* stemmed from Chandler's connection with the *Chemical News,* a paper that was published in London. Reprints were published in New York, together with an American supplement prepared by Chandler.

CONCLUSION

The professionalization of chemistry took place during the nineteenth century when students could attend universities where specialization in a scientific discipline was possible. The École Polytechnique, founded just before the century began, was the first such school of importance, but the major growth took place in Germany under the stimulus of Liebig,

Wöhler, and Bunsen. By the end of the century there was a well-established pattern of chemical education in the German universities and, to a lesser extent, in those of other countries. Graduates found employment in educational institutions, in industry, and in governmental laboratories.

Coincident with the growth of chemical education there was a development of chemical societies and journals devoted to chemistry. These several factors intermeshed to give chemistry a vigor which had been completely unanticipated at the beginning of the century.

CHAPTER 11

ANALYTICAL
CHEMISTRY I.
SYSTEMATIZATION

THE PLACE OF ANALYTICAL CHEMISTRY

It is not incorrect to ascribe to analytical chemistry a position of
primary importance since only through chemical analysis can matter in
its variety of forms be dealt with intelligently. The stimulus given to
chemistry by new analytical approaches, either qualitative or quantitative,
has been repeatedly observed. In general, however, analytical chemistry
has never achieved recognition in keeping with its importance because
the application of new techniques has resulted in new descriptive or
theoretical knowledge that completely overshadows the technique which
made the knowledge possible. Furthermore, in the unspecialized eras thus
far discussed, the analyst not only invented his procedures, but drew
sweeping conclusions from their results. Although their superiority as
analysts made certain chemists stand out, nearly every chemist had a
certain competence in the field.

Throughout the development of analytical chemistry, three objectives
—speed, selectivity, and sensitivity—have been uppermost in the minds
of chemists in designing new methods and improving old ones. Ideally,
an analytical method is most useful when it combines all three of these
objectives; but in practice, one or even two may be sacrificed when the
third must be given unusual emphasis. For example, in industrial analysis
the need to obtain results without delay may dictate the use of methods

not noted for their sensitivity. In toxicological analysis the ability to detect a poison in quantities of less than 1 part per million may be of primary importance. The purpose of the analysis frequently determines the choice of method. However, the objective of the analytical chemist is to improve the tools which the chemist uses; and he accomplishes this most satisfactorily when he can work for all three objectives simultaneously.

The basic patterns of analytical chemistry were established by 1850. Gravimetric methods were preferred to volumetric procedures, except where the latter offered clear advantages, as in multiple analyses. The blowpipe was still extremely popular and was used very skillfully by certain chemists. Organic analysis was soundly based upon the procedures developed by Liebig and Dumas. Gas analysis was taking a new direction as a result of the interest aroused in knowing more about the composition of industrial gases like illuminating gas and blast-furnace gas.

Analytical chemistry remained important for chemists who specialized in other areas. Some of these men gave so much attention to analytical methods that they are properly regarded as analytical specialists. Particularly important in this connection is the name Fresenius, for it represents not a single chemist but a dynasty. The analytical laboratory founded by the first chemist of that name celebrated its centennial in 1948, with the grandson of the founder as its manager.[1]

THE FRESENIUS LABORATORY

Karl Remegius Fresenius (1818–1897) was born in Frankfurt-am-Main, and eventually became an apprentice apothecary there. His interest in chemistry took him to Bonn in 1840. There was no student laboratory in Bonn, but Carl Marquart, the professor of pharmacy, allowed him to work in his private laboratory. Here Fresenius did analytical work and compiled a manual for qualitative analysis that so impressed Marquart that he urged its young author to have it published. The manual, *Anleitung zur qualitativen chemischen Analyse*, appeared in print in 1841 and received prompt and enthusiastic acceptance.

Fresenius now went to Giessen to continue his education under Liebig. The latter, recognizing his ability, quickly made him an assistant. Fresenius took his doctorate in 1842, with the second edition of his manual on qualitative analysis serving as his thesis. He became a *Privatdocent* in 1843, but left Giessen two years later to become professor of chemistry, physics, and technology at the Ducal Nassau Agricultural Institute near Wiesbaden. His *Anleitung zur quantitativen chemischen Analyse*, published in 1846, was received with the same enthusiasm as his earlier work. The

[1] R. Fresenius, transl. by R. E. Oesper, *Register of Phi Lambda Upsilon*, **33**, 69 (1948).

two books were revised repeatedly to keep them up-to-date. In the early editions the uncertain notation and nomenclature of the period were used, but Cannizzaro's reforms were incorporated in the later editions. For the remainder of the century the books were a veritable bible for analytical chemists, not only in Germany but in other countries as well. An American edition, translated by Samuel W. Johnson, was brought out in 1869.

The fact that there was no laboratory at the Wiesbaden institute was a source of dissatisfaction to Fresenius. When the authorities repeatedly refused to provide funds for one, he turned to his father, a prosperous

Fig. 11.1. KARL REMEGIUS FRESENIUS.
(Courtesy of Beta Chapter, Phi Lambda Upsilon.)

attorney, and with the money received from him purchased a building which was converted into a laboratory dwelling; it was opened to students in the spring of 1848. The laboratory assumed a dual role; analytical chemistry was taught there and chemical analyses were done there. Government agencies, police departments, and chemical industries began to consult it about their problems. As purchasers became convinced of the wisdom of buying materials on the basis of analytical specifications, the laboratory was commissioned to develop methods for specialized testing. In addition, it analyzed mineral water, wines, foods, and physiological specimens. Its reputation eventually was so widespread that it was frequently called upon for assistance in legal problems and for decisions regarding analyses that were questioned.

A great many famous analysts received their training in the laboratory and some maintained a long association with it. Emil Erlenmeyer became an assistant when the laboratory was opened. Carl Neubauer, who worked in the laboratory in its early period, won a considerable reputation for his food and physiological analyses. He and F. Luck had an important part in developing methods for analyzing artificial fertilizers. Leo Grünhut, Ernst Hintz, W. Dick, and W. Hartmann were active in the work on mineral waters. Fresenius' son-in-law, Ernst Hintz (1854–1934), specialized in the mineral water studies and was active in developing the internationally accepted Lange-Hintz-Weber method for determining the presence of sulfur in pyrites.

Both of Fresenius' sons, R. Heinrich (1847–1920) and Theodor Wilhelm (1856–1936), were competent analysts who participated actively in the affairs of the laboratory from the time they reached maturity and took over its management after their father's death. Their own sons, Ludwig (1866–1936) and Remigius (b. 1878), in turn entered the laboratory and assumed management responsibilities in the 1920's. The two World Wars seriously affected laboratory activities; the physical plant was badly damaged in a bombing raid in 1945. Operations were resumed soon after the end of the war under the management of Remigius and of Ludwig's only son Wilhelm (b. 1913), who represented the fourth generation of the family.

Soon after he founded the laboratory, Karl Remigius Fresenius realized the need for a journal devoted exclusively to analytical chemistry, and in 1862 he founded the *Zeitschrift für analytische Chemie*. This journal, which has had a leading role in the development of the field, has throughout its history been published by the laboratory and the Fresenius family in association.

QUALITATIVE ANALYSIS

Although qualitative tests had been used since antiquity, they were highly empirical and amounted to unorganized spot tests. As the result of his interest in reagents and reactions, Boyle drew attention to numerous specific or partially specific tests that had qualitative importance. Although Bergman used hydrogen sulfide as a reagent for the separation of metals, and Klaproth and Heinrich Rose improved the procedure, qualitative identification of the metals and the non-metal radicals did not become systematized until the time of Fresenius. Even his contributions did not involve a systematic sequence of group separations and tests so much as a collating of the reactions of the various metals and radicals. His book showed in great detail how these reactions could be used for identification purposes, but the choice of procedure was not highly systematized.

The use of hydrogen sulfide as a reagent for the separation of groups of metals was extended by Fresenius, for he recognized the suitability of the gas for precipitating a large number of metals under suitable conditions of acidity and alkalinity. The development of a generator about 1862 by P. J. Kipp (1806–1864), an apothecary in Delft who also dealt in physical and chemical apparatus, facilitated the use of hydrogen sulfide. The Kipp generator was widely used in analytical and preparative experiments that required a convenient source of the gas. In addition, it was a very convenient source of carbon dioxide in the Dumas method for nitrogen analysis.

The actual development of a systematic basis for group separations was accomplished gradually by numerous analysts. It was not until the present century that procedures became highly standardized, and this occurred only as the result of advances in physical chemistry that led to an understanding of equilibrium processes and oxidation-reduction reactions.

The development of systematic methods of precipitation did not immediately render the traditional qualitative methods obsolete. The blowpipe, and flame and spot tests continued to be very popular in certain quarters where they were used as adjuncts to systematic analysis, or in place of it. Mineralogists in particular considered blowpipe analysis their principal tool and adopted other procedures only reluctantly. Of course, since their analyses were frequently made in the field, procedures calling for hydrogen sulfide were hardly practical.

GRAVIMETRIC METHODS

The gravimetric methods which had been so effectively used by Bergman, Klaproth, Proust, Vauquelin, and Berzelius continued to be popular in the last half of the nineteenth century. The contributions during this period did not consist so much of new and unique methods as of refinements of procedures and apparatus. This trend had been established earlier by Berzelius, who introduced the use of small samples, the decomposition of silicates with hydrofluoric acid, improved methods of igniting precipitates, and, in the incineration of filter paper, correction for the ash from the paper itself.

Gravimetric procedures were looked upon with favor because the presence of a weighable precipitate gave the analysis an air of certainty. Actually, the analysis was no more accurate than the state of chemical knowledge permitted. There was, first of all, the completeness of precipitation. This varied with the nature of the precipitates and reactants. Fortunately for early gravimetric analysis, the precipitates selected—$BaSO_4$, $AgCl$, and $MgNH_4PO_4$, for example—were sufficiently insoluble so as to create no serious difficulties. Another problem involved identifying the precipitate. As interest turned to new minerals consisting of new

elements and compounds, the accuracy with which the composition of precipitates could be determined frequently left much to be desired. Some of
the newer elements showed marked tendencies to form insoluble compounds that varied in composition. Furthermore, the uncertain atomic
or equivalent weights which were used until after 1860 gave rise to
additional difficulties.

The quality of the materials used for filtering improved greatly during
the nineteenth century. The term "filter" is derived from *filtrum*, the
medieval Latin word for felt. This stems from the fact that felt bags were
used by the alchemists for filtration; cotton or linen cloths mounted on
wooden frames were also used. Lavoisier, however, concluded that unsized
paper was cleaner than cloth. Berzelius introduced the use of very pure
absorbent paper. J. H. Munktells, a Swedish papermaker, began to
manufacture this paper around 1810. For many years filter paper was sold
only in sheets, the chemist having to cut out circular pieces of the proper
size; when used for quantitative purposes he also had to determine the
weight of the ash.

Analytical papers, double-washed with hydrochloric and hydrofluoric
acids to remove mineral matter, were first produced commercially by Carl
Schleicher and Schüll, papermakers of Duren, Germany, in 1883. The
Munktells Company marketed an acid-washed paper five years later.
Other types of chemical paper were developed about this same time.
Hardened filter papers for organic work were placed on the market in
1890, ether-treated extraction thimbles in 1894.

The introduction of the Gooch crucible in 1878 marked an important
development in handling precipitates which are unstable in the conditions
prevailing during the incineration of filter papers. Actually, however, it
was nearly a quarter of a century before the crucible was generally
accepted because the problem of preparing an analytical grade of asbestos
was not solved until then. Frank Austin Gooch (1852–1929) was a student
of O. Wolcott Gibbs and Josiah Parsons Cooke at Harvard where he
received his doctorate in 1877. After studying abroad, he held several
positions as an analyst for the United States government before becoming
professor of chemistry at Yale in 1885. Here he had an important part in
the standardization of analytical methods, the development of iodometric
methods, and the study of electrolytic analyses.

A significant contribution in the field of silicate analysis was made by
J. Lawrence Smith (1818–1883), who studied with Dumas and Liebig
after graduating from the Medical College of South Carolina in 1840. This
was the J. Lawrence Smith fusion for the decomposition of silicates,
which was introduced in 1871. By means of this process, the sodium and
potassium in rocks was determined by fusing them with calcium carbonate
and ammonium chloride; this converted the two alkali metals to their
soluble chlorides, leaving the alumina and silica in insoluble form.

ANALYTICAL BALANCES

The double-pan beam balance dates back to ancient Egypt. The papyrus of Hunnafer (c. 1300 B.C.) shows such a balance being used to weigh souls against a feather on the other pan. These balances apparently continued to be used in commercial transactions. According to illustrations in books on Renaissance mining, they were used as assay balances during that period. They were also a standard item in apothecary shops.

The importance of quantitative methods in chemistry is evident in the work of Black, Cavendish, and Lavoisier. Black used an ordinary apothecary's balance, but both Cavendish and Lavoisier had specially constructed precision balances which were enclosed to protect them from air currents. From that time forward, the manufacture of chemical balances was regarded as a fine art in instrument-making.

At first, chemical balances were custom-made by artisans who specialized in scientific instruments. Jesse Ramsden made a balance that was sensitive to 0.5 mg. for the Royal Society in 1789; it had a 24-inch beam consisting of two hollow cones united at the base. The commercial manufacture of balances began about 1823 when a London instrument-maker named Robinson introduced a balance that had a perforated triangular beam.

The graduated beam and rider was introduced about mid-century. Although several others claimed credit for its invention, L. Oertling, a Berlin-born instrument-maker who worked in London, received an award at the International Exposition of 1851 for this type of balance. Paul Bunge of Hamburg, a bridge engineer, was the first to challenge the belief that a long beam provided greater sensitivity; in 1866 he built a short beam balance that functioned successfully. F. Sartorius, who had studied at Göttingen under Wöhler and learned to build apparatus in Frederic Apel's shop—Apel was the university mechanic—opened his own shop in 1870 and introduced the aluminum beam with compensating hangers suspended from three points.

Christopher Becker, who emigrated from Holland to New York, designed and began to manufacture analytical balances in the early 1850's. His balances had a beam arrest consisting of two arms that pivoted through the same arc as the supporting points of the beam. He and his sons Ernst and Christian were associated in business under the firm name of Becker and Sons. During the Civil War the firm went back to Amsterdam where his other sons joined the company. Becker and Sons returned to New York after the war, and the other sons established a firm in Rotterdam known as Becker's Sons. One of these brothers soon left the Rotterdam firm and founded H. L. Becker Fils in Brussels. Ernst and Christian left their father's firm in 1884 and went into business as Becker Brothers.

Becker and Sons closed when Christopher Becker died in 1888. Becker Brothers became Christian Becker on Ernst's death in 1892; it was incorporated with the Torsion Balance Co. in 1914 as Christian Becker, Inc. Christian's son, Christopher A. Becker, developed the chainomatic balance and patented it in 1915. The chainomatic principle was anticipated as early as 1890 by several European patents, but its successful application did not occur until a quarter century later.

Other successful manufacturers of balances in this country made their start during this period, among them William A. Ainsworth and Son, G. P. Keller Mfg. Co., and Voland and Sons. The elder Ainsworth was of English birth; Keller and Voland were German. Both Ainsworth and Keller established their firms in the Rocky Mountain area in response to the need for assay balances in connection with mining activities there.

The Torsion Balance

The torsion balance was of American origin. In 1882 F. A. Roeder of the University of Cincinnati and Alfred Springer developed a balance based on the torsion principle; in it all knife edges and bearings with their attendant friction were eliminated. Various types of torsion balances were extensively used in rough analytical work as well as in commercial fields like pharmacy; they were manufactured by the Springer Torsion Balance Company.

Specific-Gravity Balances

Crude specific-gravity balances had been in use from fairly early times. A hydrostatic balance constructed by F. Hawksbee of London in 1710 is the oldest such balance known to have been used in scientific work; it is on exhibit in the Ashmolean Museum in Oxford.

Mohr developed a specific-gravity balance for liquids and solids which indicated the specific gravity by the position of a rider on the beam. George Westphal of Hanover improved this, but his was for liquids only. He used a thermometer-containing plummet that was immersed in the liquid, the specific gravity being read directly from the position of a rider on the beam.

ATOMIC WEIGHT DETERMINATIONS

The determination of accurate atomic weights was an important goal of analytical chemists during the half century following the Karlsruhe Congress. Although the work of such pioneers as Berzelius, Dumas, and Stas had some merit, the acceptance of Avogadro's hypothesis provided a consistent basis for deciding questions which no one previously had been able to settle with finality. Furthermore, there were theoretical reasons

Fig. 11.2. JEAN SERVAIS STAS. (From Henry M. Smith, *Torchbearers of Chemistry* [1949] by permission of Academic Press, Inc.)

Fig. 11.3. EDWARD WILLIAMS MORLEY. (Courtesy of Frank Hovorka, Western Reserve University.)

for having more precise values available, and there was need for fundamental ratios that would be as accurate as possible. These problems were attacked not solely by chemists whose chief interest was the field of analytical chemistry, but also by a group of chemists desirous of obtaining extremely accurate results and with the patience needed for the tedious experimental work required for such results.

The work done by Stas was of such high quality that the decades immediately following 1860 were characterized by the consolidation of existing information rather than a search for greater precision. Cannizzaro's approach established the order of magnitude of atomic weights of elements that could be vaporized or that formed compounds which could be vaporized. The introduction of the periodic table revealed the need of correcting some atomic weights; furthermore, the atomic weights of newly discovered elements had to be determined. There was continuous but not concerted effort along these lines.

The atomic weight of indium had been only roughly evaluated by Reich and Richter; but after Mendeleev corrected it, thus bringing it into the proper magnitude, it was carefully determined by Clemens Winkler and by Bunsen. The suspected inaccuracies in the atomic weights of the platinum metals were confirmed and corrected by Karl Seubert (1851–

1942), a professor of chemistry in Tübingen. Vanadium, encountered by del Rio in 1801 and Sefström in 1828, was finally isolated in 1867 by Roscoe, who determined its atomic weight. The atomic weights of Mendeleev's three *eka* elements were determined by their respective discoverers. Nilson also established the atomic weight of thorium. That of aluminum was accurately determined in 1880 by John William Mallet (1832–1912). As a result of analyses made by this Dublin-born Wöhler-trained professor at the University of Virginia, the figures for the atomic weight of gold had to be changed.

Despite the importance of the elements hydrogen and oxygen, their combining ratios had not been accurately determined. As was said earlier, Berzelius and Dulong attacked the problem in 1819 and made some changes in the earlier values. In 1842, Dumas passed hydrogen over hot copper oxide and collected the water that was formed. Measuring the weight loss of the oxide and the weight of the water formed, he reported 7.98 to 1.0 as the ratio of oxygen to hydrogen in water. In spite of extensive precautions to eliminate possible sources of error, he realized that some errors were present. His ratio was generally accepted throughout the next half century, though later experimenters found that his figure for oxygen seemed too high.

Several laboratories worked on the problem in the 1880's. In the method devised by E. H. Keiser, the hydrogen was adsorbed on palladium and then weighed. The hydrogen was then driven off the palladium by the application of strong heat. The most significant experiments were done by Edward Williams Morley (1836–1923), a professor of chemistry at Western Reserve University in Cleveland, Ohio. He exploded weighed amounts of carefully purified hydrogen and oxygen and weighed the water that was formed. As the mean of twelve experiments in which 400 g. of water were formed, he obtained an oxygen-hydrogen ratio of 7.9396 : 1. He also measured the density of hydrogen and oxygen and their combining volumes; the oxygen-hydrogen ratio here was 7.9395 : 1. His book on the composition of water was published in 1895. Morley collaborated with Albert Michelson (1852–1931) in the famous ether drift experiments at the Case School of Applied Science in 1886–1889. These experiments demonstrated that the stationary ether the physicists claimed was required for the transmission of light waves did not exist, and paved the way for Einstein's theory of relativity.

Further studies on the composition of water were made by William Albert Noyes (1857–1941) at the U.S. Bureau of Standards. Improving a technique he had used in 1890 while at Rose Polytechnic Institute, he burned hydrogen from palladium in pure oxygen; the water that was formed condensed in a cooled extension of the tube. He also reduced copper oxide with hydrogen. The most satisfactory results gave an oxygen-hydrogen ratio of 8 : 1.00787 (7.9375 : 1).

Official Atomic Weights

During these years chemists disagreed as to whether $H = 1$ or $O = 16$ should be used as the basis for atomic weight comparisons. The most generally adopted system of atomic weights used $H = 1$, but many chemists preferred to use $O = 16$ because this made the atomic weights of a number of elements whole numbers, even though that of hydrogen was raised slightly above 1.

No table of atomic weights had any official status until December, 1893, when Frank W. Clarke prepared such a table at the request of the American Chemical Society. The Deutsche chemische Gesellschaft published a table of approved values in 1898. The next year the chemical societies sponsored a movement to set up an International Committee on Atomic Weights. This was finally organized in 1903; a working subcommittee consisted of Clarke of the United States, Seubert of Germany, and Thorpe of England. Moissan of France was appointed to it in 1904. International tables were issued annually from 1903 until World War I brought an end to such cooperation.

The Work of T. W. Richards

The twelve atomic weights Stas had established in the 1860's were considered the ultimate in accuracy for the next four decades. But in 1894, Theodore William Richards (1868–1928) noticed that the atomic weight of strontium was higher by 0.033 unit when referred to Stas' figure for chlorine, than when referred to Stas' value for bromine. Richards suspected that the chlorine figure might be wrong but failed to report his suspicion in his writings.

Richards studied at Haverford College under Lyman B. Hall, a student of Wöhler; he took his doctorate at Harvard in 1888. The next year he spent in Germany on a traveling fellowship, where he studied under Hempel, Jannasch, and Victor Meyer. He returned to Harvard to teach analytical chemistry and do research on atomic weights and problems of physical chemistry. In 1894 a year's leave of absence enabled him to study physical chemistry with Ostwald at Leipzig and with Nernst at Göttingen. In 1901 he declined a proffered chair at Göttingen to remain at Harvard. His laboratory became a mecca for graduate students in the first quarter of the twentieth century.

Richards had worked with Josiah Parsons Cooke on the atomic weight of oxygen as a graduate student; he had also worked on the atomic weight of copper. He found that when copper oxide is prepared by heating copper nitrate, the gases that are occluded cause discrepancies in the results obtained. He next found that chlorides and bromides of metals are more suitable than the oxides for work on atomic weights because they can be fused in hydrogen or hydrogen chloride gas without decomposing. Furthermore, these salts were excellent starting materials from which

Fig. 11.4. THEODORE WILLIAM RICHARDS.
(Courtesy of Mrs. James B. Conant.)

silver chloride could be precipitated and weighed. On his return from his first trip to Germany he worked extensively with the chlorides of barium, strontium, zinc, and magnesium. After 1895 he extended his studies to the chlorides of nickel, cobalt, iron, calcium, uranium, and cesium. None of these elements were among the twelve Stas had studied. While working with sodium bromide in 1904, Richards observed that Stas' figure for sodium was incorrect. His knowledge of physical chemistry enabled him to point out the source of Stas' errors and make the proper corrections. The atomic weights of a total of 28 elements were studied closely by Richards and his students. His painstaking attention to detail made him the foremost authority in this field. He was awarded the Nobel prize in 1914, the first American chemist to be so honored. Two of his students —Gregory Paul Baxter (1876–1953) of Harvard and Otto Hönigschmidt (1878–1945) of the University of Munich—did important work on atomic weights.

VOLUMETRIC ANALYSIS

According to Francis Home's book on bleaching, the "value" of pearl ashes was measured in 1756 by noting the number of teaspoonfuls of dilute nitric acid which had to be added before effervescence ceased. This was apparently the first clear instance of using a volumetric approach to chemical analysis, although there had been even earlier attempts to measure the "strength" of chemical substances by adding measurable amounts of a reactant. The most noteworthy was C. L. Geoffroy's comparison of the strength of vinegar samples in 1729. He added solid potassium carbonate to the sample until effervescence ceased, the amount of carbonate used being determined by weighing it. C. Neumann in 1732 and H. T. Scheffer in 1751 used a similar method for measuring the "strength" of samples of nitric acid.[2]

During the next half century titrimetric methods were used occasionally for examining chemical materials, particularly for evaluating acids and bases intended for technical use and determining the hardness of water. The results were comparative and estimation of end points was generally based on the cessation of effervescence as observed by eye or ear. Guyton de Morveau utilized litmus, curcuma, and brazilin as indicators in acid-base titrations.

The added solution was measured by counting drops or teaspoonfuls, or by weighing before and after the titration. Guyton de Morveau first used a graduated cylinder in 1782 and Achard devised a crude transfer pipette four years later. F. A. H. Descroizilles used calibrated glass tubes in 1791 for the titration of hypochlorite solutions with indigo solutions, and he continued the development of titrimetric methods during the following years, particularly for measuring the strength of acidic and basic substances. In 1806 he introduced the burette as a device for measuring the volume of solution added to a sample. His pioneering efforts were extended by Vauquelin, but the methods in use at this time were slow and empirical, and their applicability was generally restricted.

The real foundations of volumetric analysis were laid by Gay-Lussac between 1824 and 1832. First he measured the efficiency of bleaching powder by observing its decolorizing action on an indigo solution. He then extended quantitative titrations to the neutralization of acids and bases. In 1832 he published a paper describing a method for determining the amount of silver in bullion; the sample was dissolved in nitric acid, then titrated with sodium chloride solution.

Volumetric methods had not acquired much popularity by mid-century. The recognition of equivalence points was extremely uncertain, there was as yet no agreement on expressing the concentration of standard solutions, and the volumetric apparatus itself was generally unsatisfactory. The pharmacist Karl Friedrich Mohr (1806–1879) did much to overcome these difficulties.

The son of a pharmacist in Coblenz, Mohr studied at Bonn, Heidelberg, and Berlin, where he came under the influence of Leopold Gmelin and Heinrich Rose. Although primarily a manufacturer of pharmaceuticals,

---

[2] E. Rancke Madsen, *The Development of Titrimetric Analysis Till 1806*, Gad, Copenhagen, 1958, p. 30.

throughout his life he was interested in chemical research, particularly that involving analytical problems. He had an unusual talent for improving apparatus and procedures; when only twenty-six years old he made improvements in the analytical balance. He designed the specific-gravity balance. His reflux condenser, which consisted of concentric tubes through which water could be poured, was the precursor of the Liebig condenser. He introduced the sets of cork-borers which are standard equipment in most chemistry laboratories today. The stopcock on the burette then in use leaked and stuck. Mohr redesigned it with a pinch clamp and tip, and he designed volumetric glassware such as flasks and pipettes. He secured the acceptance of the liter as the volume occupied by 1000 g. of water at 17.5° C.; this, however, was revised later to the volume occupied by 1000 g. of water at 4° C.

Fig. 11.5.　Karl Friedrich Mohr.
(Courtesy of the Edgar Fahs Smith Collection.)

The analysis of chlorides by titration with silver nitrate solution was made practical by Mohr's innovation of using potassium chromate as an internal indicator. At this time, precipitation titrimetry was a tedious form of analysis, for it was customary to estimate the end point by noting when the addition of more standard solution caused no further precipitate to form. This meant that the analyst had to undertitrate, let the precipitate settle, and then add the solution drop by drop at widely separated intervals so he could see whether more precipitate formed when another drop was added. Titration of a set of samples was frequently a daylong operation.

Mohr also introduced oxalic acid as a primary standard for alkalimetry, and ferrous ammonium sulfate (Mohr's salt) as a primary standard for oxidizing agents. His textbook, *Lehrbuch der chemisch-analytischen Titriermethode* (1855), was very influential in popularizing volumetric methods; in it he introduced the idea of back-titration and advocated the use of normal solutions.

In 1846, Margueritte proposed the use of permanganate solution for the quantitative determination of iron, and in 1850 Penny suggested the use of dichromate for this purpose.

Gay-Lussac had devised a procedure for estimating the available chlorine in bleaching powder that was based on the reducing power of arsenious oxide; Penot improved this procedure in 1852.

Bunsen introduced iodometric methods in 1853. In a careful study of the reaction between sulfurous acid and iodine, he showed that the oxidation-reduction reaction is quantitative only when the concentration of sulfur dioxide is below 0.04 per cent by weight. Bunsen also perfected the indirect quantitative determination by a number of oxidizing agents. The iodine liberated by their action on an acid solution of potassium iodide could be titrated with standard sulfurous acid solution. Soon thereafter, Schwartz used sodium thiosulfate in this titration. As a result of its greater stability, this reducing agent quickly replaced sulfurous acid.

GAS ANALYSIS

Analysis of gases, particularly air, received considerable attention in the time of Priestley, Cavendish, and Volta. Priestley's test for the "goodness" of air and Cavendish's experiments to determine the composition of water and of air with respect to nitrogen and oxygen have already been mentioned. Volta devised a eudiometer for use in determining the composition of combustible gases by exploding them with oxygen, a technique used by Dalton in his studies on methane and ethylene. Gay-Lussac's test for determining the purity of air was instrumental in his formulating the law of combining volumes.

Although gas analysis was frequently used in the early part of the nineteenth century, the techniques for it were usually quite restricted in their application and, furthermore, not noted for their accuracy. The systematic development of procedures for gas analysis began when Bunsen became interested in analyzing blast-furnace gases in England. He used solid absorbents which removed a particular gas from a mixture, the concentration being measured by the diminution of volume. His methods were described in *Gasometrische Methoden* (1857), a standard work which was made immediately available in an English translation by Bunsen's student, Henry Enfield Roscoe.

Bunsen's methods were accurate, but tedious. Hence attempts were soon made to use absorbent solutions instead of his solid absorbents; but this technique sacrificed accuracy for convenience. No successful improvements were made until the 1870's, when Walther Hempel (1851–1916) designed gas pipettes which solved the problem of using liquid absorbents with gas samples. The technological developments in the production and utilization of gaseous fuels then being made created interest in simple and accurate methods of analyzing gases. Clemens Winkler, who published an important book on gas analysis in 1876,[3] was important not only for his work on quantitative methods but also for his systematization of qualitative gas analysis. By using absorbing agents systematically, he was able to divide gases into analytical groups similar to those used in metal analysis. Besides his work and that of Hempel, important contributions were made by Frankland, Pettersson, Orsat, Coquillion, and Bunte.

INSTRUMENTAL METHODS OF ANALYSIS

The last quarter of the nineteenth century saw the introduction of instrumental techniques into analytical chemistry as the correlation between physical properties and chemical composition became apparent. Particularly adaptable to analytical needs were such optical and electrical instruments as the spectroscope, the polarimeter, the refractometer, and the electrolytic cell.

The Spectroscope

The spectroscope—its development has already been discussed—became a powerful tool in qualitative analysis, for its great sensitivity enabled the detection of trace elements that were frequently overlooked in more traditional methods of analysis. Improved spectroscopic techniques soon made possible the detection of elements that did not emit spectral lines in the ordinary burner flame. The spectroscope achieved its greatest importance in the analysis of astronomical bodies. Kirchhoff's interpretation of absorption spectra made it possible to associate elements with particular lines in the spectrum of the sun and the stars. What had once been considered unanswerable now became rather simple, for astronomers could now identify the chemical elements present in the stars.

Electrochemical Analysis

Electrolysis began to acquire importance in analytical chemistry in the last quarter of the nineteenth century, though some studies of it had been made in the 1860's. O. Wolcott Gibbs (1822–1908) published the

---

[3] C. Winkler, *Anleitung zur chemischen Untersuchung die Industriegase*, 1876–1877. This was followed in 1885 by his *Lehrbuch der technischen Gasanalyse*. W. Hempel's *Gasanalytische Methoden* (1890) quickly became the standard text on the subject.

results of his investigation of the electrolytic determination of copper and nickel in 1864; the same report described methods for analyzing silver and bismuth as metals, and lead and manganese as the dioxides. The following year C. Luckow stated that he had used an electric current for the determination of copper twenty years earlier, but he did not publish his results until Gibbs' report appeared.

Numerous publications in the field of electrolytic analysis followed those by Gibbs and Luckow as the technique was applied to a variety of analytical problems. The importance of differences in potential in electrodeposition was first recognized by Kiliani. Edgar Fahs Smith of the University of Pennsylvania was responsible for developing rotating electrodes and metal gauze electrodes. The laboratory of Alexander Classen (1843–1934) in Aachen was a center for research on electrolytic analysis. Classen invented useful apparatus and wrote the *Handbuch des chemischen Analyse durch Electrolyse.*

Optical Instruments

Optical instruments began to have a place in instrumental analysis with the improvements in optical glass which were made during the century. The imperfections inherent in glass had long seriously handicapped the builders of telescopes and microscopes; this is why the reflecting telescope was the favored type of instrument during the eighteenth century. Microscopes were of almost no significance because aberrations made them unreliable. Chromatic aberration was lessened in 1758 by the English optician, John Dollond (1706–1761), who combined suitably convergent and divergent lenses made of crown and flint glass; this led to the improvement of the refracting telescope. These principles, along with improvements made in optical glass to satisfy scientific demands, were sufficient to make the microscope a practical instrument by the end of the 1820's.

The Italian astronomer and physicist, Giovanni Battista Amici (1786–1863), was particularly adept at improving optical apparatus. In 1827 he devised an achromatic lens system for the microscope which was quickly adopted by instrument builders. This paved the way for spectacular progress in the field of biology—Schleiden and Schwann's cell theory, Virchow's cellular pathology, Pasteur's rejection of spontaneous generation, and the progress in the field of bacteriology. Amici also pioneered in using liquid-immersion objectives.

Further improvements in optical instruments resulted from the work of Abbe and Schott. Ernst Abbe (1840–1905) was educated at Göttingen and Jena. He became a member of the physics faculty at Jena in 1863, and in 1878 director of the astronomical and meteorological observatories. In 1866 he began work with the optical firm established by Carl Zeiss (1816–1888) in 1840. Abbe was made responsible for the development and improvement of the apparatus it produced. He invented the apochromatic objective with compensating ocular in 1886, and the refractometer which bears his name. His oil-immersion objectives contained at least eight compensating lenses and were so well designed that magnification up to 3000 diameters was possible. Even earlier (1872) he had devised a condenser to be used under the microscope stage which made it possible to concentrate light so highly that tiny high-power objective lenses no larger than a pinhead could be used. At Zeiss' death in 1888, Abbe—he had been

Fig. 11.6.   ERNST ABBE.
(Courtesy of the J. H. Walton Collection.)

made a partner in 1875—became sole owner of the firm. He established the Zeiss Foundation under which the officials, the workmen, and the University of Jena shared cooperatively in the firm's profits.

Otto Schott (1852–1935), whose father operated a plate-glass factory in Westphalia, became associated with Zeiss' firm in connection with a research project aimed toward improving optical glass. He and Abbe demonstrated that using baryta, borates, phosphates, and zinc made it possible to produce glass whose properties were very different from those of flint and crown glass. The Jena Glass Works was founded in 1884 for the production of this new glass. It also manufactured glassware for chemical purposes. The resistance of this glass to chemicals and to thermal and mechanical shock was so much better than that of ordinary glass that the Jena Glass Works held a virtual monopoly until World War I.

Microscopes were, of course, far more important in biology than in chemistry, but the progress made in improving them benefited chemistry in several ways. The improvements in lens systems were apparent in improved refractometers, spectroscopes, and polarimeters. The microscope itself became important in the detection of adulterated food (e.g., the polarizing microscope is used to distinguish between butter and oleomargarine), and in crystallography and metallurgy.

Henry Clifton Sorby (1826–1908) of Sheffield pioneered in the latter field in establishing the relationship between the properties of metals and their crystalline structure. He devised techniques whereby the crystalline structure of alloys could be revealed microscopically by treating their polished surface with suitable etching agents.

Polarimetry

Polarized light responds to optically active compounds; hence the use of the polarimeter, developed by Biot in 1840, in analytical work was obvious. Since sugars are in such great demand as foods, the instrument came rapidly into use for routine analyses of syrups.

The phenomenon of polarization was first observed by Huygens, who described the unusual behavior of light when it was passed through a crystal of Iceland spar. In 1813 Jean Baptiste Biot (1774–1862), French physicist and crystallographer, reported that when light was passed through a piece of quartz that was cut perpendicular to the axis, the plane of polarization was rotated. The amount of rotation was proportional to the thickness of the quartz. Some pieces rotated the light to the right, others to the left. Biot also discovered that polarized light was rotated similarly when passed through certain liquids such as turpentine, laurel oil, lemon oil, or an alcoholic solution of camphor.

Improvements in the polarimeter came rapidly. The prism devised by William Nicol (1768–1851) in 1828 for separating polarized light into two beams was applied to the polarimeter by Ventzke in 1842. Mitscherlich introduced the use of monochromatic light. Soleil, the French physicist, made the use of white light practical by devising a quartz wedge compensator. The half-shadow principle was formulated by Jellett of Dublin.

Commercial development of the polarimeter occurred in both Germany and France. Its value in sugar analysis, already mentioned, took on immediate practical importance when the Prussian government levied a tax on refined sugar in 1860. Such taxes or duties were rapidly imposed by other countries, making sugar analysis an important port activity. Eventually, a specialized polarimeter was made; this was known as a saccharimeter because it was devised as a direct-reading instrument for sugar analysis.

ORGANIC ANALYSIS

The foundations of elementary organic analysis were so carefully laid by Liebig and Dumas around 1830 that the methods underwent essentially no changes during the rest of the century except for minor modifications involving refinements in design and heating methods. This was particularly true in the case of carbon and hydrogen.

But in the case of nitrogen, an element not present in all organic compounds but important in proteins and certain other biochemically significant substances, there was a recognized need for a less intricate and time-consuming procedure than Dumas' method. In 1840 this method still required refinement with respect to the source of carbon dioxide, the degradation of nitrogen oxides, and the collection of nitrogen. Although these shortcomings were eliminated during the next two decades, the procedure was still complex and tedious. It was universally respected for reliability with all types of organic samples; but whenever other methods could be relied upon to give accurate results, they were preferred. Therefore, there were continuous attempts to develop alternative methods.

Most promising was the procedure introduced in 1841 by Heinrich Will (1812–1890) and Franz Varrentrapp (1815–1877), both associated with Liebig's laboratory and both active later in life in developing improved analytical methods. Will, who became professor of chemistry at Giessen in 1845, had been an assistant to P. L. Geiger and Leopold Gmelin in Heidelberg, where he developed a quantitative procedure for sodium and potassium. Varrentrapp, as professor of chemistry at Braunschweig (where he was later director of the mint), made various contributions to the analysis of fatty acids.

The Will–Varrentrapp procedure was a refinement of an earlier method for organic nitrogen developed by the French industrial chemist, Anselme Payen (1795–1871), in which the sample was exposed to red heat and the gaseous decomposition products were collected in dilute sulfuric acid. However, the nitrogen was not always converted completely to ammonia, so the results obtained with Payen's method were unreliable. Will and Varrentrapp found that quantitative conversion of the nitrogen to ammonia was more certain if the sample was heated with soda lime in a glass tube so designed that the ammonia produced was collected in a receiver that contained an acid solution. At first, the ammonia was determined by precipitation as the chloroplatinate; later it was absorbed in standard acid. This technique was satisfactory for proteins and amines, but not for nitro compounds.

Despite its limitations and its tediousness, the Will–Varrentrapp procedure was extensively used for about a half century after its introduction, but was rapidly superseded by the Kjeldahl procedure following its publication in 1883.[4] Johan G. C. T. Kjeldahl (1849–1900) was educated in Copenhagen and was then employed by Carl Jacobsen, a brewer who wanted to introduce scientific methods into brewing. In 1876 the Jacobsen family founded the Carlsberg Laboratory for broad-scale fermentation research, the results of which were to be published for everyone's use. Kjeldahl and bacteriologist Emil Christian Hansen (1842–1909) were

Fig. 11.7. JOHAN G. C. T. KJELDAHL.
(Courtesy of J. H. Mathews.)

transferred to the laboratory, where Hansen was made responsible for introducing the use of pure yeast cultures into the brewing process, and Kjeldahl studied the sugar and protein content of sprouting and fermenting grain. In connection with his work with proteins, Kjeldahl realized the need for a method of determining nitrogen which could be used in rapid and large-scale analyses.

In earlier attempts to develop such procedures, the sample was pretreated with an appropriate reagent to make the soda-lime treatment in the Will–Varrentrapp procedure more rapid and complete. James Alfred Wanklyn (1834–1906) in Edinburgh developed a procedure for oxidizing the samples with alkaline permanganate. At least two other investigators,

[4] J. G. C. T. Kjeldahl, *Z. anal. Chem.*, **22**, 366 (1883). H. A. Schuette and F. C. Oppen review the determination of organic nitrogen in *Trans. Wisconsin Acad. Sci.*, **29**, 355 (1935).

Grete and Dreyfus, subjected wool, horn, leather, and fertilizer to preliminary digestion with sulfuric acid. Kjeldahl found that satisfactory results might be obtained if the samples were digested with concentrated sulfuric acid supplemented by permanganate at the end. The digested samples were then made alkaline and the ammonia distilled into standard acid. The excess acid was determined by iodometric titration.

His method was tested in various agricultural experiment stations in Germany, and the results were favorable. Not only did it provide an accurate method for determining protein nitrogen in biological materials, but it did not require detailed attention from skilled chemists; furthermore, numerous samples could be digested and distilled simultaneously. The procedure was ideal for research on protein in grains and similar materials. Improvements which were rapidly introduced made it even more satisfactory.

Functional Groups

The determination of functional groups, such as carboxyl, ester, unsaturated bonds, hydroxyl, and methoxyl, took on importance in connection with two different types of activities—the proximate analysis of foods and agricultural materials, and the establishment of the constitution of new organic compounds. Proximate analysis is discussed below; here, however, it must be remarked that methods useful in constitution studies are also useful in proximate analysis, and vice versa.

Perhaps the most commonplace functional group is the carboxyl or organic acid group. When this group was present, titration with standard bases was the obvious method, and this procedure was commonly used. When a single acid of known identity was present, it permitted direct analysis; and it aided identification of pure acids whose identity was unknown, since the neutralization equivalent provided a guide to the molecular weight. The major drawback in the titration of organic acids was the fact that most such acids are weak and can be titrated correctly only with the proper indicator. Hence until there was better understanding of indicators, uncertainty was inherent in this method.

A similar approach could be used with esters, where a carboxyl group is formed upon hydrolysis. In one procedure, the sample of ester was saponified in a standard alkali; this was followed by back-titration with a standard acid. Although it is equally applicable to the study of pure esters, the technique was developed by Koettstorfer in connection with the analysis of fats—actually in response to the need for a test which would distinguish between oleomargarine and butter. First proposed in 1879, it was soon found to be valuable in the study of fatty oils, waxes, and other esters; it gave rise to a standardized value that is known as the saponification number.

Unsaturation was commonly detected by the decolorizing action of a compound on permanganate or bromine solutions. The ability of the double bond to absorb halogen was made the basis of a quantitative procedure, the iodine number, by Arthur Freiherr von Hübl, a student of Rudolf Benedikt at the Technische Hochschule in Vienna. The iodine number was determined by treating the sample of oil with an alcohol solution of mercuric chloride and iodine. Iodine monochloride was added quantitatively to the unsaturated bonds; the unused halogen was titrated with standard thiosulfate solution. The method proved satisfactory for fatty oils, waxes, ethereal oils, resins, and many pure unsaturated organic compounds, and thus became widely used. However, Hübl's solution presented some difficulties in preparation and was in time supplanted by the methods suggested by J. A. Wijs and Josef Hanuš, using solutions of I·Cl and I·Br respectively.

The presence of hydroxyl groups was best determined by acetylation, followed by saponification of the ester, separation of the acetic acid by steam distillation, and titration. The acetyl number was first developed by Benedikt and Ulzer in connection with the analysis of fats like castor oil, which contain esters of ricinoleic acid (12-hydroxystearic acid). Subsequent studies by Julius Lewkowitsch (1857–1913) improved the procedure. Lewkowitsch, who became one of the leading authorities on the chemistry of fatty oils, was of Russian birth but did most of his scientific work in England.

A method for determining methoxyl and ethoxyl groups was introduced in 1885 by Simon Zeisel (1854–1933) while he was an assistant to Adolf Lieben (1836–1914), professor of pharmaceutical chemistry at the University of Vienna. In working with the alkaloid, colchicine, that came from the seeds of the autumn crocus, Zeisel obtained methanol as a hydrolysis product. He converted this to methyl iodide by treating it with hydriodic acid, but the compound was too volatile to be weighed accurately. However, if the vapors were bubbled through alcoholic silver nitrate, silver iodide was precipitated and could be collected and weighed. A quantitative procedure applicable to methoxyl and ethoxyl compounds was published. Zeisel's method was extended to the determination of methyl and ethyl groups attached to nitrogen toward the end of the century by Josef Herzig and Hans Meyer.

Proximate Analysis

Proximate analysis attained great importance in connection with the research on agricultural and food chemistry which was pursued so vigorously in the last part of the nineteenth century. In the case of proteins, fats, carbohydrates, essential oils, and minerals it was usually impossible to isolate a particular compound quantitatively. Furthermore, the procedure would have been of little value because the isolated compound would only represent other very similar ones in the sample. Knowledge regarding the quantity of a class of compounds present was much more important than knowledge of the quantity of a particular representative of the class. Hence it became usual to report the presence of proximate principles, such as proteins, carbohydrates, crude fiber, and fats, in biological materials.

Proteins were found to have a nitrogen content of approximately 16 per cent. After Kjeldahl's convenient method for nitrogen determination became available, protein content could be reported merely by multiplying the percentage of nitrogen by 6.25. However, this gives the true protein content only if all the nitrogen present is proteinaceous, as is essentially the case with many biological materials.

The marked differences in character between individual compounds made carbohydrates a special problem, because it was not particularly informative to report them simply as carbohydrates. In connection with animal feed, the practice of reporting "crude fiber" and "nitrogen-free extract" became popular after 1863. These terms were introduced by William Henneberg (1825–1890), one of Liebig's pupils, and a leading authority on animal nutrition; he became director of the Agricultural Institute at Weende, near Göttingen. Crude fiber referred to the inert residues left after successive treatment with sulfuric acid and sodium hydroxide; it was the portion of the feed—cellulose and similar compounds with high molecular weights—which was assumed to be indigestible. Nitrogen-free extract was the remainder after the sum of the percentages of fat, crude fiber, protein, and ash, on a moisture-free basis, was subtracted from 100 per cent. Supposedly it represented digestible carbohydrates such as the sugars and starch. However, since it was estimated by subtraction, its value was subject to the cumulative errors in the determination of the other food substances.

Frequently a more exact knowledge of carbohydrate content was desired. The soluble sugars could be determined optically or by reduction methods. The polarimeter was well adapted to estimating single sugars after the sample had been properly clarified to remove opaque material from it. Clerget developed a procedure for the optical determination of sucrose even though contaminating sugars were present; it was based on the change in the optical rotation of sucrose after hydrolysis or inversion. Reduction methods for sugar analysis all depended upon the reducing effect of certain sugars on cupric salts. This reaction, observed by Hermann Fehling (1812–1885) of Stuttgart in 1849, became the basis of both qualitative and quantitative procedures for sugar analysis. Gravimetric, volumetric, and electrometric methods are now used for measuring the cuprous oxide formed during the reduction.

Fats were determined in biological material by extracting them with a suitable solvent, usually ethyl ether. The procedure developed by Franz von Soxhlet (1848–1926), professor of agricultural chemistry at the

Fig. 11.8. STEPHEN MOULTON BABCOCK.
(Courtesy of the University of Wisconsin News
Service.)

technical school in Munich, was used widely because it made extraction
of multiple samples possible with a minimum of attention. Certain substances—for example, milk—required special procedures. In the Adams
paper-coil method the milk sample was absorbed on a coil of filter paper
which was then dried and extracted. But this was slow and tedious in the
face of a growing dairy industry which made clear the need for a simple,
rapid, and accurate test.

The challenge was met by Stephen Moulton Babcock (1843–1931),
professor of agricultural chemistry at the University of Wisconsin.
Babcock was educated at Tufts College and in Göttingen, receiving his
doctorate from Göttingen in 1879. Soon after coming to Wisconsin he
devised a procedure for liberating the fat in milk; the sample was digested
with concentrated sulfuric acid and the fat was then collected in the neck
of a calibrated bottle by centrifuging and dilution. The amount of fat in

the sample was determined directly by measuring the height of the column of fat. The whole procedure is so simple that any creamery operator can learn it in a short time. It has been said that the Babcock test did more to promote honesty among dairymen than reading the Bible ever did![5] The procedure was so well conceived that it is still used in dairies all over the North American continent. Appropriate variants have been devised for cream, skim milk, and other dairy products. Variants of it are used in Europe; among them is the Gerber technique in which smaller samples, less acid, amyl alcohol for fat separation, and different glassware are used.

Moisture and Ash

The moisture content of organic materials was usually estimated in terms of the weight lost when the sample was heated at or slightly above the boiling point of water. Vacuum ovens were introduced for heating samples—e.g., syrups—that are unstable at 100° C.

Mineral matter was generally determined by burning off the organic matter in a muffle furnace and weighing the ash. The ash might be further analyzed for individual elements by traditional inorganic methods. It was of course realized that the minerals in the sample were in a form that differed from the oxides, carbonates, sulfates, phosphates, and chlorides which made up the ash. Progress in studying mineral-containing compounds in organic materials naturally was very slow.

CONCLUSION

The nineteenth century was a period of consolidation in the field of chemical analysis. As a result of Klaproth's trail-blazing at the beginning of the century, the discipline began to have a sound empirical basis. This was important because it led to an understanding of the weight relationships in chemical combination, and hence to the atomic theory. The latter in turn forced improvements to be made in analytical chemistry to solve the problem of atomic weights and formulas. The stabilization which resulted in this area, following Cannizzaro's suggestions, was merely one of degree; the need for continuous refinement of atomic weights soon became apparent.

Gravimetric methods underwent continuous refinement and extension during the century, but progress here was largely empirical. The progress in physical chemistry came too late to affect gravimetric analysis significantly.

[5] The statement was made by William D. Hoard, dairyman, publisher, and onetime governor of Wisconsin. See H. L. Russell, *Science*, **74**, 86 (1931); and Paul de Kruif, *Hunger Fighters*, Harcourt, New York, 1928, p. 278.

Volumetric methods gained favor during the century. There was no established tradition of their use except for a few crude industrial assays when the century began; but by mid-century several procedures had been favorably accepted. In the last half of the century the volumetric approach was extended in a variety of directions and provided sound and accurate methods of analysis. The availability of sensitive internal indicators was instrumental in bringing about acceptance of titrimetric methods even though little was known about the theory of indicators.

Instrumental methods became important in analytical chemistry as the century drew to a close. The spectroscope was received enthusiastically as a qualitative tool almost from the moment of its discovery. The polarimeter and refractometer had an important place, not only in purely scientific laboratories, but in government laboratories and industrial operations, especially in connection with food and drug control. The microscope also had an important role. The speed and sensitivity of instrumental analysis clearly justified the cost of all these instruments, especially when repeated analyses had to be made.

CHAPTER 12

ORGANIC CHEMISTRY III.
CONSOLIDATION

With the settling of the problem of molecular composition as the result of the contributions of Kekulé and Cannizzaro and their contemporaries, organic chemistry was in a position to make rapid progress. The adoption of a uniform notation, the realization of the importance of molecular formulas based on a consistent system of atomic weights, and a backlog of empirical knowledge about compounds, reactions, and techniques, made it possible to produce a multitude of new compounds, many of them unknown in nature. Organic chemistry immediately became a specialty. Contributions from men of broad interests like Berzelius, Gay-Lussac, Dumas, Wöhler, Bunsen, and Laurent were now rare; instead they came from the laboratories of men who had little interest in such fields as inorganic and physical chemistry. As a result, there began a period of fission during which chemists were either organic or nonorganic in their interests, and frequently knew little and cared less about the other field. Each field was of sufficient magnitude to present plenty of problems for competent men without being concerned about integration between the two.

THE PROBLEM OF MOLECULAR STRUCTURE

Berzelius' recognition of the possibility of isomerism brought gradual realization that structural differences between compounds with identical molecular formulas must account for differences in properties. As long as there was no agreement regarding molecular formulas the problem was not

```
                        C| O...OH
                         | H²
                         ⋮
C⎰O...OH    C⎰...OH      C...H²        C⎰ O...O ⎰C
 ⎱H³         ⎱...H²      ⋮              ⎱ H²  H² ⎱
            ⋮            C...H³        ⋮
            C ...H³                    C...H³  H³ C

   a            b           c              d              Acide salicylique.
                                                          C⎰C...H²
                                                           ⎱C...H
                                                          ⋮
  ⎰O...OH    C⎰O...OH   C⎰O...OH   C⎰O...OH   C⎰C...H
C⎱O²          ⎱O²        ⎱H²        ⎱O²        ⎱C...O...OH
  ⎱H         ⋮           ⋮          ⋮          ⋮
            C...H³      C⎰H²        C⎰O²       C⎰O²
                         ⎱O...OH     ⎱O...OH    ⎱O...OH

   e            f           g           h           i
```

Fig. 12.1. COUPER'S STRUCTURAL FORMULAS. *a* methyl
alcohol. *b* ethyl alcohol. *c* propyl alcohol. *d* ethyl ether.
e formic acid. *f* acetic acid. *g* ethylene glycol. *h* oxalic acid.
i salicylic acid.
(From *Compt. rend.*, 46: 1158 ff. [1858].)

given concerted attention; but once the way was paved for consistency
in formulas the structural problem became of primary importance.

We have already noted the suggestions toward its solution before 1860.
The type theories contained an implied concept of structure. The valence
concept proposed by Frankland and Kekulé was a necessary step. The
final steps occurred rapidly around 1860, although Kolbe must be given
credit for some cumbersome efforts along these lines in the preceding
decade.

Kekulé introduced the concept of the chemical bond but did not take
the obvious next step of representing it symbolically by a dot or line.

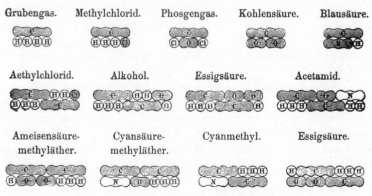

Grubengas.   Methylchlorid.   Phosgengas.   Kohlensäure.   Blausäure.

Aethylchlorid.   Alkohol.   Essigsäure.   Acetamid.

Ameisensäure-   Cyansäure-   Cyanmethyl.   Essigsäure.
methyläther.   methyläther.

Fig. 12.2. KEKULÉ'S "SAUSAGE" FORMULAS.
(From Kekulé, *Lehrbuch der organischen Chemie*, vol. i [1861].)

Couper[1] did this in 1858, when he presented structural formulas in a form readily understandable to the modern reader. Since he accepted O = 8, his oxygen atoms are always paired.

Kekulé introduced a different type of graphic formula in 1859 and used them to some extent in the first volume of his *Lehrbuch der organischen Chemie* (1861).[2] In them he attempted—successfully—to show both valence and bonding relationships. But like Dalton's circles (see Chapter 4), they were tedious to draw and required special printing techniques; hence they

Fig. 12.3. ALEXANDER MIKHAILOVICH BUTLEROV.
(Courtesy of the Edgar Fahs Smith Collection.)

never attained popularity. Jokesters, particularly Kolbe, referred to them as "Kekulé's sausages." Kekulé himself abandoned them, in the second volume of his *Lehrbuch* (1865), in favor of using lines for bonds and letters for symbols.

About the time that Couper introduced his structural formulas, Alexander Mikhailovich Butlerov (1828–1886) began to discuss the "structure of a chemical compound"; Erlenmeyer used the word "constitution" in a similar sense. Butlerov was educated at Kazan, where he studied under Klaus and came under Zinin's influence. He taught chemistry there after Zinin went to St. Petersburg and Klaus to Dorpat. After receiving his

[1] A. S. Couper, *Compt. rend.,* **46**, 1157 (1858).
[2] F. A. Kekulé, *Lehrbuch der organischen Chemie,* Enk, Erlangen, **1861**, vol. i, pp. 162 ff. Issued in fascicles from June 1859.

doctorate in 1854, Butlerov visited Zinin and was converted to the unitary ideas of Laurent and Gerhardt. He returned to Kazan for three years more, then traveled in western Europe, spending some time with Mitscherlich, Kekulé, and Wurtz, and visiting several laboratories. Imbued with the concepts set forth by Kekulé and Couper, he again returned to Kazan. Although he used Gerhardt's types, he could not agree that the structure of organic types would remain forever unknown. Experimental studies, he felt, would provide knowledge of atomic groupings that would lead to an understanding of molecular structure.

His theoretical ideas developed with his experimental work. At a scientific meeting at Speyer in 1861, he read a paper, "The Chemical Structure of Compounds." Soon thereafter he prepared tertiary butyl alcohol, a compound which Kolbe had predicted. Butlerov continued his work on tertiary alcohols, improving the methods of preparation and studying derivatives. His book on organic chemistry (Russian ed., 1864; German ed., 1867) presented his ideas on structure very fully; it was well received in the West. In certain respects it departed from Gerhardt's types more than Kekulé's book did, and it utilized the concept of tetravalence of carbon to the fullest extent in dealing with structure. Butlerov joined the faculty of the University of St. Petersburg in 1869 where he continued his researches on alcohols and unsaturated hydrocarbons. In 1876 he presented a paper on diisobutylenes in which he hinted at what later became the theory of tautomerism.

Hofmann quickly grasped the importance of structural ideas and demonstrated a well-advanced understanding of the concept in a lecture delivered before the Royal Institution in 1865. He presented several

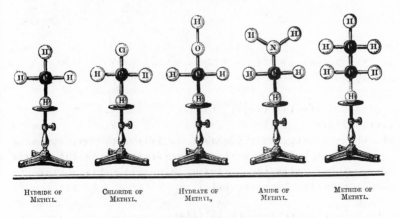

HYDRIDE OF METHYL.   CHLORIDE OF METHYL.   HYDRATE OF METHYL.   AMIDE OF METHYL.   METHIDE OF METHYL.

Fig. 12.4. HOFMANN'S CROQUET-BALL MODELS.
(From Hofmann, *Proc. Royal Inst.*, 4: 416 ff. [1865].)

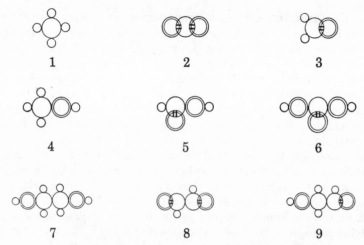

Fig. 12.5. LOSCHMIDT'S FORMULAS. *1* methane. *2* carbon dioxide. *3* formaldehyde. *4* methyl alcohol. *5* formic acid. *6* carbonic acid. *7* ethylene glycol. *8* glyoxal. *9* glycolic aldehyde.

(From Loschmidt, *Chemische Studien I* [1861].)

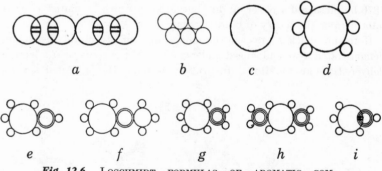

Fig. 12.6. LOSCHMIDT FORMULAS OF AROMATIC COMPOUNDS. *a* benzene nucleus (trial). *b* benzene nucleus (trial). *c* benzene nucleus. *d* benzene. *e* phenol. *f* methoxybenzene. *g* aniline. *h* diaminobenzene. *i* a hypothetical imide.

examples of homologous series in which the difference between successive compounds was a uniform increment such as O, NH, or CH_2. In his lectures Hofmann used models constructed of sticks and croquet balls to represent some of the compounds.[3]

An unheralded attempt to establish the theoretical foundations of structural chemistry was made in 1861 by a comparatively unknown

[3] A. W. Hofmann, *Proc. Royal Inst.*, **4**, 421 (1865).

school teacher in Vienna named Josef Loschmidt (1821–1895), in his book, *Chemische Studien I*. The book was privately printed and because of its limited circulation was never widely known.[4] In the first part of it Loschmidt discussed "Constitutional Formulas of Organic Chemistry in Graphic Representation." He represented carbon atoms as circles ○ , hydrogen atoms as much smaller circles ○ , oxygen atoms as double circles ◎ , and nitrogen atoms as triple circles ◉. With these symbols, reminiscent of Dalton's, he built up the structural formulas of several hundred compounds. He devised a sort of double and triple bond to represent the combination of atoms in compounds like carbon dioxide, ethane, and acetylene. He was puzzled regarding how to represent the "phenyl series" consisting of such compounds as benzene, phenol, aniline, benzoic acid, and benzaldehyde. He considered representing the nucleus by the forms shown in Fig. 12.6a and b, but abandoned them in favor of a large circle (c). Thus he formulated benzene as a large circle with six tiny circles attached (d). This formula clearly suggested a ring of carbon atoms but ignored the problem of disposing of the fourth valence of carbon.

In the same year, 1861, the Scottish chemist Alexander Crum Brown[5] (1838–1922) began using formulas with bonds like those used by Couper, but the symbols of the elements were encircled as in Fig. 12.7a. Three years later Crum Brown abandoned the barred symbols he had originally

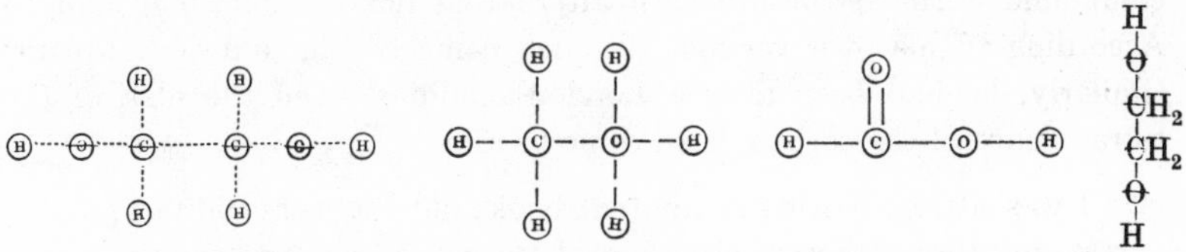

a   *b*   *c*   *d*

Fig. 12.7. FORMULAS OF CRUM BROWN AND WURTZ. *a* Crum Brown's formula for ethylene glycol (1861). *b* Crum Brown's formula for ethane (1864). *c* Crum Brown's formula for formic acid (1868). *d* Wurtz' formula for ethylene glycol. (From J. Walker, *J. Chem. Soc.*, 123: 942 [1923], with permission of the publisher.)

used for carbon and oxygen, showing that he had replaced the atomic weights 6 and 8, with the values 12 and 16. His 1868 formula for formic acid is shown in *c*. Wurtz' 1864 formula for ethylene glycol is shown in *d*.

Frankland also was groping with structural concepts at this time, but

---

[4] M. Kohn, *J. Chem. Educ.*, **22**, 381 (1945); R. Anschütz, *Ber.*, **45**, 539 (1912). Loschmidt's *Chemische Studien I*, Privately Printed, Vienna, 1861, is very rare. A reprint edited by Anschütz was published by Ostwald's Klassiker (No. 190) in 1913.

[5] James Walker, *J. Chem. Soc.*, **123**, 942 (1923).

he conservatively continued to use the brackets associated with the type theory.[6]

Unsaturation

The concept of unsaturation was developing even before Hofmann produced his croquet ball model for ethylene. We have seen how Loschmidt handled the problem in 1861. Crum Brown used a double bond in his formula for ethylene in 1864. Lothar Meyer, in his book published the same year, applied the term "unsaturated" to this kind of compound. Erlenmeyer, however, had used a triple bond in his acetylene formula two years earlier. J. Wilbrand's study of the combination of carbon with carbon reveals unsaturation.[7]

Aromatic Structures

The formulation of structures for aliphatic compounds posed few major problems once the general approach was revealed. Benzene and related compounds offered a more difficult problem on account of the low ratio of hydrogen to carbon. This ratio suggested a high degree of unsaturation, but benzene did not show the additive behavior of typical unsaturated compounds.

Kekulé, one of the most imaginative chemists in the organic field, solved the benzene structure successfully in 1865 when he assigned the compound a ring structure with alternating double and single bonds. According to his own version, the idea came to him during a dream; similarly, he had been atop a London omnibus when the idea of the tetravalence of the carbon atom came to him.

I was sitting, writing at my text-book; but the work did not progress; my thoughts were elsewhere. I turned my chair to the fire and dozed. Again the atoms were gambolling before my eyes. This time the smaller groups kept modestly in the background. My mental eye, rendered more acute by repeated visions of the kind, could now distinguish larger structures, of manifold conformation: long rows, sometimes more closely fitted together; all twining and twisting in snake-like motion. But look! What was that? One of the snakes had seized hold of its own tail, and the form whirled mockingly before my eyes. As if by a flash of lightning I awoke; and this time also I spent the rest of the night in working out the consequences of the hypothesis.

"Let us learn to dream, gentlemen," adds Kekulé, "then perhaps we shall find the truth . . . but let us beware of publishing our dreams before they have been put to the proof by the waking understanding."[8]

[6] E. Frankland, *J. Chem. Soc.*, **19**, 372 (1866).
[7] J. Wilbrand, *Z. Chemie*, **8**, 685 (1865).
[8] Quoted from R. Japp, "Kekulé Memorial Lecture," *J. Chem. Soc.*, **73**, 100 (1898).

In developing the ring structure of benzene, Kekulé was insistent on retaining the tetravalence of carbon just as he had in his other formulas. He observed that benzene and its derivatives never contained less than six carbon atoms. He realized that six such atoms might be linked together by alternate double and single bonds.

If one assumes that six carbon atoms are attached to one another according to this law of symmetry, one obtains a group which, regarded as an *open chain*, contains *eight* unsaturated units of affinity. By making the further assumption that the two carbon atoms at the ends of the chain are linked together by one unit of affinity each, a *closed chain* (a symmetrical ring) is obtained which still contains six unsaturated units of affinity. [See Fig. 12.8a, b.]

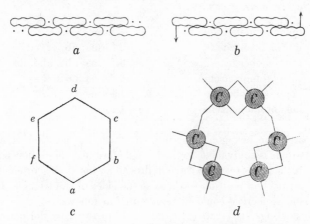

a   b

d

e   c

f   b

a

c   d

Fig. 12.8. KEKULÉ'S FORMULAS FOR BENZENE. *a* open chain, C_6H_8. *b* closed chain, C_6H_6. *c* hexagon formula. *d* ring formula showing double bonds.
(From Kekulé, *Lehrbuch der organischen Chemie*, vol. ii [1866] and *Ann.*, 137: 158 [1866].)

From this *closed chain* all the substances usually designated as "aromatic compounds" are derived. . . .

In all aromatic substances a common nucleus may be assumed: it is the closed chain C_6A_6 (in which A denotes an unsaturated affinity).

The six affinities of this nucleus may be satisfied by six monatomic elements. They may also, wholly or at least in part, be satisfied by one affinity of polyatomic elements, the latter necessarily bringing with them other atoms into the compound, thus producing one or more *side chains*, which in their turn may be lengthened by the addition of other atoms.

The satisfying of two affinities of the nucleus by one atom of a diatomic element, or of three affinities by one atom of a triatomic element, is, according to the theory, impossible. Compounds of the molecular formulae, C_6H_4O, C_6H_4S, C_6H_3N, are, therefore, unthinkable.[9]

In his early publications on aromatic compounds Kekulé generally preferred the sausage formulas satirized by Kolbe, though he used the simple hexagon in 1865 (Fig. 12.8c) and the hexagon with double bonds (d) a year later.

He clearly recognized the possibilities of isomerism in substitution products of benzene containing two or more substituents. Using the lettered hexagon in Fig. 12.8c, he discussed the possibilities of isomerism as follows:[10]

1. monobromobenzene   1 form
2. dibromobenzene   3 forms: ab, ac, ad
3. tribromobenzene   3 forms: abc, abd, ace
4. tetrabromobenzene   3 forms: as in 2
5. pentabromobenzene   1 form
6. hexabromobenzene   1 form

He compared the hexagon structure, in which the position of all the hydrogen atoms is equivalent, to the triangle formula where their position is not equivalent, because the three hydrogen atoms at the corners would not be equivalent in position to those at the sides of the triangle. In this case, possible isomerism could be represented as follows:

1. monobromobenzene   2 forms: a and b
2. dibromobenzene   4 forms: ab, ac, bd, ad
3. tribromobenzene   6 forms: abc, bcd, abd, abe, ace, bdf
4. tetrabromobenzene   4 forms: abcd, abde, bedf, abce
5. pentabromobenzene   2 forms: abcde, bcdef

In deciding about the alternate structures, Kekulé drew on the work of Paul Schützenberger (1827–1897) on the iodine-substituted benzenes, as well as on studies of bromination and nitration products of benzene, toluene, phenol, and similar aromatic compounds. All this work indicated the acceptability of the hexagon formula.

He solved the problem of the fourth valence of carbon by supposing an alternation of double and single bonds in the benzene ring, but other chemists were not willing to accept this concept. "Aromatic" compounds,

[9] *Ibid.*, p. 133.
[10] F. A. Kekulé, *Ann.*, **137**, 158 (1866).

as Kekulé called benzene and its relatives (the term "aliphatic" for open-chain compounds was introduced by Hofmann), failed to show the addition reactions characteristic of typical unsaturated compounds. Hence, the presence of double bonds was open to question.

Adolf Claus (1840–1900), in particular, was concerned about the absence of typical addition reactions in benzene. He suggested that each carbon

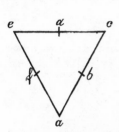

Fig. 12.9. KEKULÉ'S TRI-ANGULAR FORMULA. (From Kekulé, *Ann.*, 137, 160 [1866].)

Fig. 12.10. FRIEDRICH AUGUST KEKULÉ VON STRADONITZ.
(Courtesy of the Edgar Fahs Smith Collection.)

atom in the C_6 nucleus must have three valences satisfied by three other carbon atoms:

Claus I   Claus II   Dewar

He personally preferred the first formula and used it in his later publications. Claus had studied with Wöhler and Kolbe, and had taught at the university in Freiburg-im-Breisgau.

Another alternative was suggested in 1867 by James Dewar (1842–1923). A student of Kekulé, he became professor of natural experimental philosophy at Cambridge in 1875 and, two years later, Fullerian Professor of Chemistry at the Royal Institution. He proposed that the bonds in the benzene ring be placed as shown above.

Albert Ladenburg (1842–1911), *Privatdocent* at Heidelberg, pointed out that the Kekulé formula required the existence of two *ortho* isomers, depending on whether there was a single or a double bond between the carbon atoms that carried the two substituents.

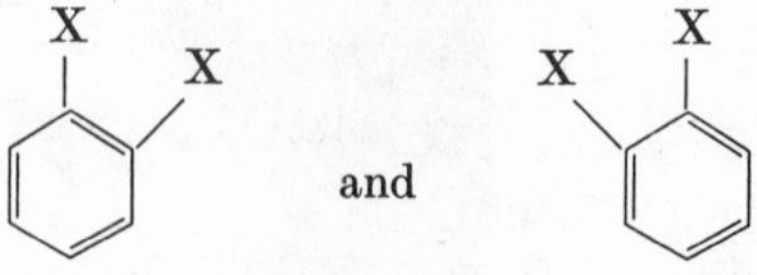

and

Since no such isomers were known, Ladenburg reasoned that each carbon atom must be bonded to three others and suggested three alternative structures.

I   II   III

Number I is identical with the Claus II, but the prism and star formulas were original with Ladenburg. In a subsequent paper he stated that the prism formula was preferable. It could explain the number of di- and tri-substituted isomers as long as no importance was attached to mirror images, but it could not explain the double bonds in dihydro- and tetrahydrobenzene.

Kekulé attempted to deal with the problem posed by Ladenburg in 1872, in a paper on an aldehyde condensation product.[11] After reviewing the various benzene structures that had been proposed, Kekulé assembled arguments that supported the ring structure he had devised, and then proposed an oscillatory mechanism which made all ring positions equivalent and made two *ortho*-disubstituted isomers unnecessary. He suggested that the atoms in a molecule oscillate around an equilibrium position, colliding regularly with neighboring atoms, and that these collisions per unit time are related to the valence of the atom and are part of the forces holding the molecule together. While a hydrogen atom is making one oscillation that results in one collision, a carbon atom makes one oscillation that results in four collisions. For example, C^1 in the formulas in

[11] *Ibid.*, **162**, 77 (1872).

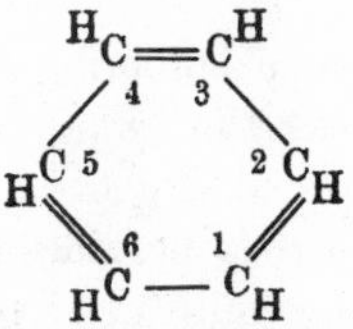

$$\begin{array}{c} H_C = C^H \\ 4 \quad 3 \\ H C\,5 \qquad 2\,C H \\ 6 \quad 1 \\ H C - C H \end{array}$$

so kann man die Stöfse, welche es in der ersten Zeiteinheit erfährt, ausdrücken durch :

1.   2, 6, h, 2,

worin h den Wasserstoff bedeutet. In der zweiten Zeiteinheit wendet sich dasselbe Kohlenstoffatom, welches gerade von **2** kommt, zunächst zu dem Kohlenstoff 6. Seine Stöfse während der zweiten Zeiteinheit sind :

2.   6, 2, h, 6.

Während die Stöfse der ersten Zeiteinheit durch die oben geschriebene Formel ausgedrückt werden, finden die der zweiten ihren Ausdruck in der folgenden Formel :

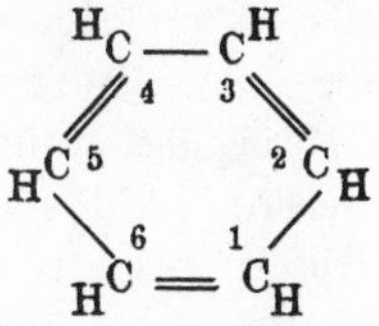

$$\begin{array}{c} H_C - C^H \\ 4 \quad 3 \\ H C\,5 \qquad 2\,C H \\ 6 \quad 1 \\ H C = C H \end{array}$$

Fig. 12.11. PAGE FROM THE *Annalen* SHOWING KEKULÉ'S
TREATMENT OF THE FOURTH VALENCE IN BENZENE.
(From *Ann.*, 162: 88 [1872].)

Fig. 12.11 may, in one oscillation, collide with C^2, C^6, H, and C^2; in a reverse oscillation the sequence is C^6, C^2, H, and C^6. Thus every fourth collision will be with the hydrogen atom, in accordance with the period of the hydrogen atom. The sequence of collisions of C^1 with neighboring atoms during successive intervals is:

$$|C^2C^6HC^2|C^6C^2HC^6|C^2C^6HC^2|C^6C^2HC^6| \dots$$

During the first interval the structure might be represented as at the top of Fig. 12.11; during the next as at the bottom, the single and double lines representing the number of collisions per interval.

Later, many chemists interpreted Kekulé's formulas as indicating a rapid alternation of double and single bonds, but his concept is actually somewhat more subtle than this. The arrows in the above collision sequence indicate that delaying the timing interval by one collision reverses the number of collisions with the neighboring carbon atoms in the next interval. Since the start of the timing is arbitrary and since the two formulas in the figure have no real structural significance, it follows that Kekulé's collision hypothesis merely provides for an equivalence of all six carbon atoms at all times, without specifying single or double bonds between particular carbon atoms at a particular moment.[12]

In the second edition of his *Modernen Theorien der Chemie* (1872) Lothar Meyer also discussed the fourth valence of carbon in the ring. He introduced the idea that each carbon atom in the ring had a free affinity. Thus, the structure of benzene might best be represented as in the accompanying

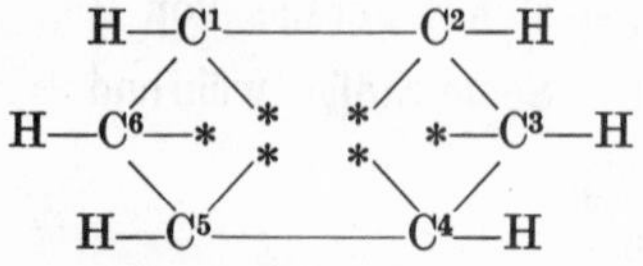

Meyer 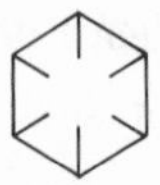   Baeyer 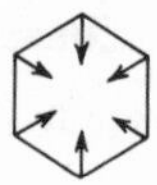   Armstrong

diagram. He agreed that the existence of not more than three disubstitution products might be explained equally well by his formula or Kekulé's. His formula, commonly designated as the "centric" formula, was brought forth again in 1887 by both Baeyer and Armstrong.

Later, John Norman Collie (1859–1943) attempted to reconcile the centric and Kekulé formulas. He suggested that the benzene molecule was a vibrating system in which the two Kekulé structures should be considered as the oscillatory extremes, the centric structure being an intermediate form.

All the research on aromatic compounds following the introduction of the ring formula proved to be consistent with the concept of a planar rather than a three-dimensional molecule. Both the chemical and physical evidence called for abandoning the prism formula. Although the disposition of the fourth valence of carbon in the ring was still unsettled, most organic chemists found the Kekulé formula a satisfactory working concept; hence the chemistry of aromatic compounds advanced rapidly during the final three decades of the last century.

The fact that aromatic compounds did not show typical addition reactions was investigated by Johannes Thiele (1865–1918) at the end of the century. He developed the theory of partial valences in attempting to explain the addition reactions of compounds with conjugated double bonds. In a compound like butadiene, addition takes place at the 1,4 positions with the transposition of a double bond.

$$H_2\overset{1}{C}\!\!=\!\!CH\!-\!CH\!\!=\!\!\overset{4}{C}H_2 + Br_2 \rightarrow H_2C\!-\!CH\!\!=\!\!CH\!-\!CH_2$$
$$\underset{Br}{|} \qquad\qquad \underset{Br}{|}$$

[12] A. Gero, *J. Chem. Educ.*, **31**, 201 (1954).

In order to explain this, Thiele suggested that some double bonds have an unsatisfied valence, or a potentially available bond. When double bonds are on adjacent carbon atoms the central partial valences become ineffective, and the outermost positions are more reactive.

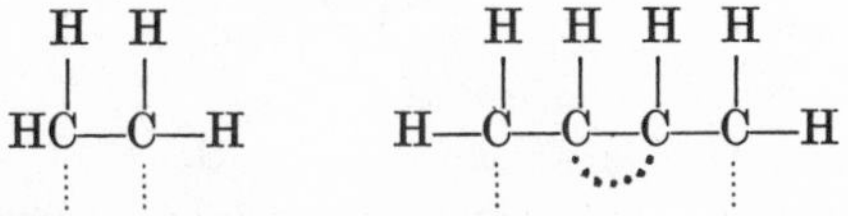

All the partial valences are adjacent in the benzene ring; hence they would be ineffective and the molecule would participate in addition reactions only under vigorous conditions.

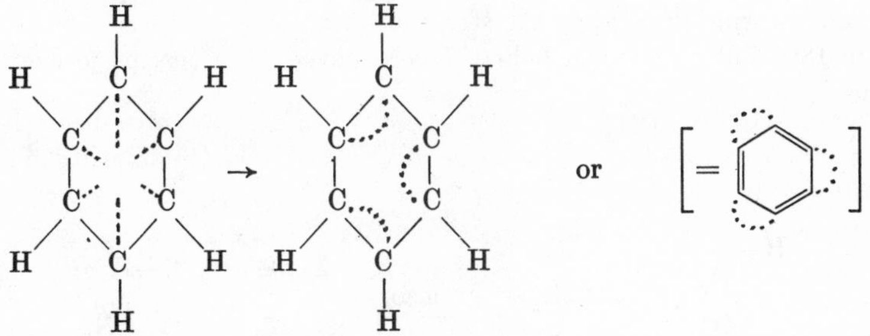

Ring Positions

A problem which immediately arose when the benzene ring was introduced involved assigning proper positions to the polysubstituted derivatives. Kekulé realized at once that there would be three isomeric compounds of formula $C_6H_4X_2$ or C_6H_4XY:

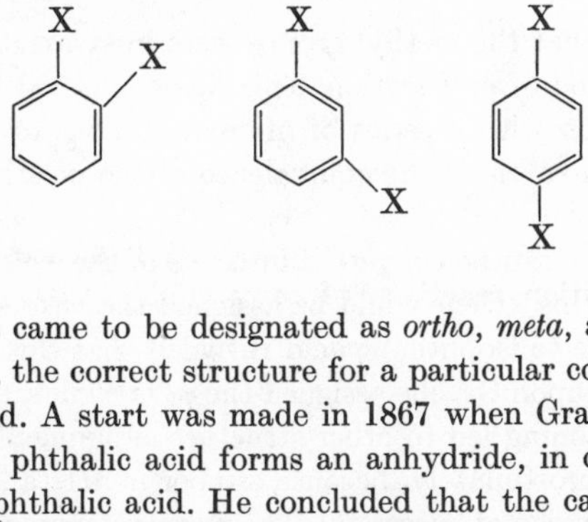

These structures came to be designated as *ortho*, *meta*, and *para*, respectively. However, the correct structure for a particular compound was not easily determined. A start was made in 1867 when Graebe observed the ease with which phthalic acid forms an anhydride, in contrast with the stability of terephthalic acid. He concluded that the carboxyl groups in phthalic acid must be on adjacent carbon atoms. (The term "carboxyl" was used for the COOH group in 1865.) Kekulé had prepared terephthalic

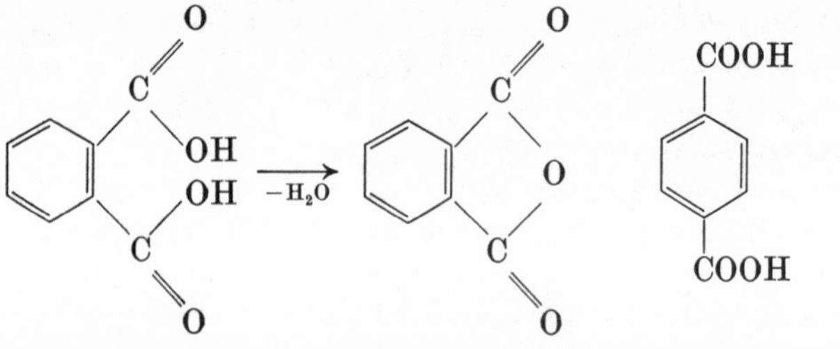

Phthalic acid   Phthalic anhydride   Terephthalic acid

acid from dimethyl benzene in 1865, but the structure of the dimethyl benzene and its derivative acid was still unknown.

In the meantime, Erlenmeyer had determined the structure of naphthalene. Baeyer's oxidation of naphthalene to phthalic acid proved that Graebe's surmise regarding the structure of phthalic acid was correct.

In 1866 Fittig confirmed Robert Kane's conversion of acetone to mesitylene.

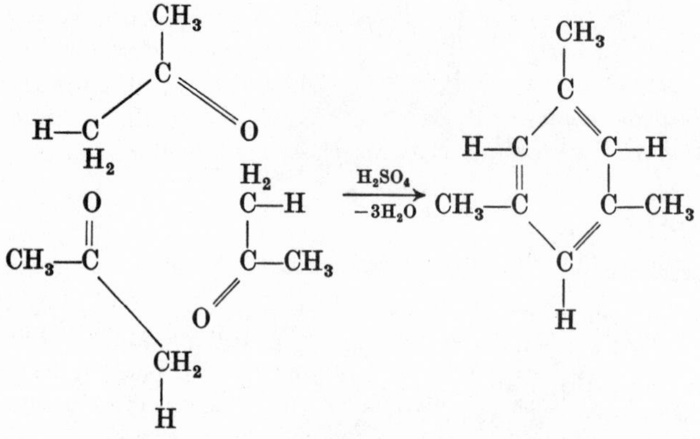

3 Acetone   Mesitylene

Baeyer deduced that the methyl groups were in symmetrical positions in view of the method of preparation. This was confirmed by Ladenburg in 1874 in connection with a series of interconversions of nitro and amino derivatives which revealed the equivalence of the positions not occupied by methyl groups.

Since mesitylene can be converted into one of the xylenes by removing a methyl group, this xylene could be assigned the *meta* structure. Oxidation of this xylene to isophthalic acid revealed that this acid is the *meta* compound, whereupon Graebe assigned the *para* structure to terephthalic acid. Similar reasoning led to other structure assignments, some of them successful; but unfortunately, because of poor understanding of aromatic reactions, many were not successful. The proper structure for a particular compound was not determined easily. William Körner (1839–1925) demonstrated in 1874 that the problem could be analyzed numerically if the disubstituted benzene were converted to a trisubstituted product. In the case of the three dibromobenzenes he showed that nitration to nitro-dibromobenzene could distinguish the compounds by the number of products formed; the *ortho* compound yielded two isomers, the *meta* compound three, the *para* one. Similar results were obtained by brominating the three dibromobenzenes.

The method was equally applicable in assigning formulas to the tri-substituted benzenes. The nitration of the three possible tribromobenzenes gives nitro derivatives as follows:

1,2,3-tribromobenzene	2 isomers
1,2,4-tribromobenzene	3 isomers
1,3,5-tribromobenzene	1 isomer

Körner's procedure provided a fine method for structure analysis; its success depended only on the ability of chemists to prepare all the possible isomers required for a correct interpretation. Actually, various alternatives were possible. As early as 1872, Peter Griess, using reverse logic, obtained information about structures when he showed that, of the six possible diaminobenzoic acids, three were decarboxylated to the same diamino-benzene (*meta*), two to another (*ortho*), and one to a third (*para*).

Fused Rings

The structure of many other aromatic compounds was successfully studied during this period. The ring concept was extended to naphthalene when Erlenmeyer proposed a structure of two fused rings in 1866. This was proved correct two years later by Graebe on the basis of degradation reactions of chlorinated quinones. The double-ring structure assumed two monosubstitution products for naphthalene. Many years earlier, Faraday had prepared two naphthalene monosulfonic acids and other isomeric pairs had been prepared since that time.

The structure of anthracene was formulated by Graebe and Liebermann as tribenzene with alternative formulas.

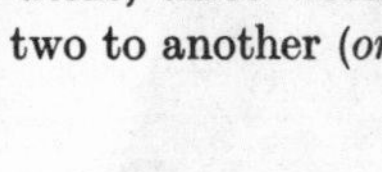

I 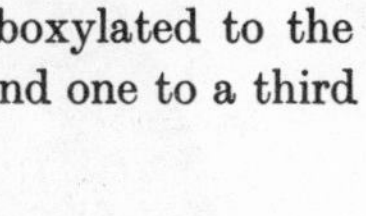   II

Graebe and Liebermann preferred the second formula, but with the discovery of phenanthrene they realized that the first formula was correct. Phenanthrene, which is isomeric with anthracene, was investigated simultaneously by Graebe and Glaser, and by Fittig and Ostermeyer. Other fused aromatic ring systems were also studied during this period.

Heterocyclic Compounds

The similarity between the formulas of benzene, C_6H_6, and naphthalene, $C_{10}H_8$, and those of pyridine, C_5H_5N, and quinoline, C_9H_7N, suggested to Körner and to Dewar, independently, ring structures in which a CH group is replaced by N.

Pyridine   Quinoline

Pyridine, along with other homologous bases, had been discovered by Thomas Anderson (1819–1874) in bone oil. Substituted pyridines posed a new type of problem because the nitrogen in the ring destroys the equivalence of the carbon atoms. The presence of pyridine and quinoline nuclei in various alkaloids made structural problems of practical importance, the

Fig. 12.12. (Johann Friedrich Wilhelm) Adolf von Baeyer. (Courtesy of the Edgar Fahs Smith Collection.)

major work in solving them being done in the laboratories of Wilhelm Königs in Munich, Adolf Baeyer in Berlin, and Zdenko Skraup in Vienna.

Other heterocyclic rings became known with Baeyer's discovery of indole as a degradation product of indigo in 1866. Pyrrole, discovered by Runge in 1834, was shown to be a five-numbered ring that contains a nitrogen atom. Furan, the oxygen analogue, was prepared by Limprich and Rohde in 1870. Victor Meyer discovered thiophene, the sulfur analogue, in 1882.

STEREOCHEMISTRY

Biot's discovery, mentioned in Chapter 11, that turpentine, sugar, and certain other organic compounds rotate the plane of polarized light not only in pure form but in solution as well, led to the realization that their optical activity is not due to their crystal form, as is the case with quartz, but is a property inherent in the compounds themselves and not related to their crystalline form.

Pasteur's Studies

The crystallographic study of tartrates was undertaken by Louis Pasteur (1822–1895), the son of a tanner of Dôle in the Jura, when he

Fig. 12.13. LOUIS PASTEUR.
(Courtesy of the Edgar Fahs Smith
Collection.)

completed his studies at the École Normale. While studying for the licentiate in science he made such a good impression on Balard, his chemistry professor, that the latter interceded with the authorities, who intended to assign him to a *lycée* far from Paris. As a result, Pasteur carried on his work in association with Balard. In 1846 Balard admitted

Auguste Laurent, who had returned to Paris, to his laboratory. This was the beginning of a fruitful association between Laurent and Pasteur. Laurent, himself a competent crystallographer, encouraged Pasteur in studying the tartrates and brought to his attention some apparently pure substances, like sodium tungstate, which had more than one crystal form.

Tartar, or crude potassium acid tartrate, had long been known to vintners as a solid which separates as a sludge from wine during fermentation; it is poorly soluble in alcohol. Tartaric acid, a normal constituent of grapes, was first isolated and studied by Scheele and subsequently was produced commercially. This was the *dextro*-rotatory form of the acid. Around 1820 Charles Kestner, a manufacturer of chemicals in Thann in the Haut-Rhin, encountered a form of tartaric acid which behaved differently than the usual product; he was unable to produce more of it. The unique acid was studied by Johann Friedrich John of Berlin, later by Gay-Lussac, who named it racemic acid (L. *racemus*, grape), and still later by Berzelius, who called it paratartaric acid. Biot showed that racemic acid and its salts do not influence polarized light.

In a careful study of the sodium ammonium salts of tartaric and racemic acids, Mitscherlich in 1844 reported that the salts have the same crystalline form, their only difference being that tartaric acid is *dextro*-rotatory and racemic acid is inactive.

Pasteur suspected that Mitscherlich and other crystallographers might have overlooked a dissymmetry in the crystals. His own painstaking investigation of the crystals of these salts showed that the tartrate crystals were truly hemihedral. The racemic crystals, which he expected to be symmetrical, he found were also hemihedral. Closer examination revealed that in the tartrate crystals the hemihedral faces were all oriented in the same way, but that in the racemic crystals the faces of some were oriented toward the right whereas the faces of others were oriented toward the left.

Pasteur laboriously separated the right-handed and left-handed crystals and dissolved each kind in water. He noted that one solution rotated polarized light toward the right and that the other solution rotated the light to the left. When equal weights of the two kinds of crystals were dissolved, the resulting solution had no effect on polarized light. By precipitating and acidifying the lead salts, Pasteur in 1848 obtained the free acids, one of them the well-known *dextro*-rotatory form, the other the hitherto unknown *levo*-rotatory acid. When the elderly Biot learned about these results, he remained unconvinced until Pasteur repeated the experiments in his presence, using chemicals supplied by him. After the crystals of sodium ammonium tartrate had been separated, Biot himself made up the solutions and placed them in the polarimeter. He saw that the results were exactly what Pasteur had reported and thereupon became an ardent supporter of the young scientist.

No one realized until later that Pasteur had been exceedingly fortunate in preparing his crystals, in the choice both of compound and of working conditions. When sodium ammonium tartrate crystallizes from a hot concentrated solution, the crystals are fully symmetrical and the monohydrate contains equal proportions of the *dextro-* and *levo*-rotatory molecules. But there is a transition point at 28° C. When the tartrate crystallizes below this temperature it forms the tetrahydrate and half the crystals are pure *dextro-* and the other half pure *levo*-rotatory molecules. Under these conditions the hemihedral faces are apparent, and the crystals are mirror images. Since then, hemihedral crystals large enough to be hand-sorted under a lens have been reported in only nine cases.

Pasteur continued his studies of tartrates for another six years before becoming involved in fermentation research and restricting his subsequent work largely to biological problems. During this time he developed the two other methods used for resolving optically active isomers. He found in 1858–1860 that when the mold *Penicillium glaucum* was grown on a nutrient solution that contained racemic acid, the solution gradually became *levo*-rotatory. The mold consumed the naturally occurring *dextro* acid, but showed a distinctly lessened ability to metabolize the unnatural *levo* acid. Although the method is useful for certain separations of racemics, the natural isomer is destroyed.

The other procedure developed by Pasteur was the result of his methodical preparation and study of all sorts of tartrates. He not only prepared a variety of metallic tartrates, but included salts formed by the combination of tartaric acids with organic bases. He observed not only that the solubility and other physical properties of *d-* and *l*-tartrates of sodium, ammonium, and aniline were identical, but that this was no longer true when optically active bases such as asparagine, quinine, quinidine, brucine, cinchonidine, and strychnine were used. The solubilities and other properties of the salts derived from *d-* and from *l*-tartrates differed so markedly that solubility differences served as a basis for separation. The relationship can perhaps best be shown as follows:

$$d\text{-Tartaric acid} + l\text{-Cinchonidine} \rightarrow dl \text{ salt}$$
$$l\text{-Tartaric acid} + l\text{-Cinchonidine} \rightarrow ll \text{ salt}$$

The two acids, called *enantiomorphs*, are identical except in their effect on polarized light. The two salts, called *diasterioisomers*, are no longer identical in properties, hence can be separated by crystallization.

Pasteur, puzzled about the occurrence of racemic acid, visited a number of plants in Germany and Austria that produced tartaric acid before he found the answer. Several producers had encountered racemic acid in the initial purification of tartar, but discarded it without understanding its real nature. Research showed that crude tartar frequently contained both *d-* and *l*-acids, the former being predominant. During the initial purification, racemic acid separated and was discarded as a useless by-product, after which a pure d-acid could be prepared.

Pasteur found that *dextro*-tartaric acid could be converted to the racemic form by prolonged heating of the cinchonate. He also produced the inactive or *meso*-tartaric acid in this procedure. Thus, there were four forms of the acid: the *dextro*, the *levo*, the racemic—optically inactive because of equal proportions of the *dextro* and *levo* forms—and the *meso*, a truly inactive form of the acid.

The experiments on the tartrates clearly suggested a relationship between molecular configuration and optical activity. Organic chemistry, however, had not progressed to the point where structural considerations had become meaningful. Pasteur gave much thought thereafter to the factors which might be responsible for turning light toward either the right or the left. The concept of asymmetry could not be avoided in view of the mirror-image relationship between the sodium ammonium tartrate crystals. In 1860 Pasteur speculated on the possibility of the molecules being right-handed and left-handed helices. He referred to the asymmetry of living organisms in terms of right- and left-handedness and expressed the belief that life must be an asymmetrical process. Whereas optically active compounds were frequently isolated from biological material, the same compounds produced by chemical synthesis were optically inactive. He believed that the synthetic form was inherently inactive. Only after 1860, when Perkin and Duppa prepared racemic acid from succinic acid, was he forced to realize that synthesis could give equal quantities of the optical antipodes.

Pasteur, fascinated by asymmetry, projected the asymmetry of biological products to asymmetry of forces acting in the universe; he suggested that such forces influenced the formation of asymmetrical products. He pointed out that although the earth is round, its image is not superimposable upon itself; it turns on its axis and its mirror image turns in the opposite direction. Thus, in the whole universe there are motion and consequent asymmetry. In order to test his ideas regarding the effect of asymmetrical forces on chemical synthesis he experimented at Strasbourg with powerful magnets to determine whether crystallization was affected. At Lille, where he became dean of sciences in 1854, he devised a mechanical procedure for reversing the direction of the sun's rays to see whether plants, illuminated in reverse from dawn to dusk, would produce the opposite optical form of compound. Although lack of time forced him to abandon these weird experiments, he frequently reverted to the problem in his scientific discussions.

Van't Hoff and Le Bel

Further light on optical activity was not forthcoming until 1874 when van't Hoff and Le Bel independently suggested the concept of an asymmetrical carbon atom. Born in Rotterdam, Jacobus Henricus van't Hoff (1852–1911) studied at Delft, Leyden, Bonn, and Paris before taking his doctorate at Utrecht. In school he showed a flair for mathematics and Byronic poetry. At Bonn he began study under Kekulé with great enthusiasm, but the two men failed to impress each other; van't Hoff found Wurtz' laboratory far more congenial. In 1876 he took a teaching position at the Veterinary College in Utrecht and two years later became a professor in Amsterdam. After eighteen years there, while establishing a solid reputation as a leader in physical chemistry, he was called to Berlin. His earliest work was done in organic chemistry.

Fig. 12.14. JOSEPH ACHILLE LE BEL.
(From Henry M. Smith, *Torchbearers of Chemistry* [1949] with permission of Academic Press, Inc.)

Joseph Achille Le Bel (1847–1930), a nephew of the agricultural chemist Boussingault, was educated in Paris where he served as an assistant to Balard and later to Wurtz. He and van't Hoff were acquainted when they were working with Wurtz, but both apparently arrived at their ideas on stereochemistry independently. Van't Hoff's paper first appeared in a Dutch journal in September, 1874; it was published in expanded form as a pamphlet in French early the next year.[13] Le Bel's paper appeared in

13 J. H. van't Hoff, *Archiv. néerland. sci. exact. nat.*, **9**, 445 (1874). The expanded version is the pamphlet *La chimie dans l'espace*, published in 1875. G. M. Richardson's *The Foundations of Stereochemistry*, American Book Co., New York, 1901 includes an English translation of this paper, as well as those by Pasteur, Le Bel, and Wislicenus.

the November, 1874, issue of the *Bulletin* of the French Chemical Society.[14]
Van't Hoff developed his arguments along geometric lines, following up
Wislicenus' observations on lactic acid a few years earlier. Le Bel's paper
showed the influence of Pasteur's ideas regarding asymmetrical molecules.

Lactic acid, which Scheele had isolated from sour milk in 1770, was
subsequently found in association with various fermentation products.
In 1807 Berzelius discovered a similar acid in extracts of muscle tissues.
Liebig demonstrated that this acid had the same composition as fermentation lactic acid and it was named sarcolactic (muscle) acid in 1832. It was
difficult to go further in comparing the two acids because they are very
soluble and, when isolated, highly hygroscopic. In 1849 Engelhardt
reported studies of various salts which showed that fermentation and

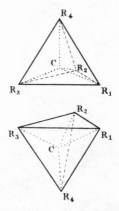

Fig. 12.15. VAN'T HOFF'S TETRAHEDRAL REPRESENTATION
OF THE MIRROR-IMAGE RELATIONSHIP.
(From van't Hoff, *La chimie dans l'espace* [1875].)

muscle lactic acids are chemically distinct substances even though they
have the same composition. Muscle lactic acid was also found to be *dextro-*
rotatory, whereas fermentation lactic acid is optically inactive.

Johannes Wislicenus (1835–1902) approached the structure of the lactic
acids from the standpoint of both synthesis and degradation. He showed
that the structure of β-hydroxypropionic acid cannot be that of either
form of lactic acid, wherefore both must be the α-hydroxy form. Wislicenus
concluded: "If molecules can be structurally identical and yet possess
dissimilar properties, this can be explained only on the ground that the
difference is due to a different arrangement of the atoms in space."[15]

Van't Hoff and Le Bel observed that when a carbon atom is attached to

[14] A. Le Bel, *Bull. soc. chim.*, **22**, 337 (1874).
[15] J. Wislicenus, *Ann.*, **166**, 47 (1873).

four different atoms or atomic groups, the four substituents can be arranged in two different ways, and the resulting molecules will be mirror images of each other. Van't Hoff showed that when the carbon atom is treated as a tetrahedron, with the attached atoms at the four corners, two geometric arrangements are evident. Every compound which in solution rotates the plane of polarized light must have such an asymmetrical carbon atom. Le Bel said nothing about the geometry of the atom; he simply argued for a system which permitted two arrangements of different substituents around an asymmetrical carbon atom.

The isomerism of the tartaric acids was explained by two asymmetrical carbon atoms which resulted in a *dextro* and *levo* form, plus a molecule which internal compensation made optically inactive.

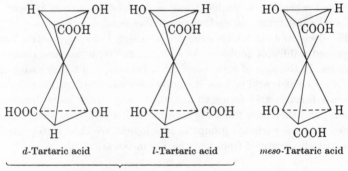

| *d*-Tartaric acid | *l*-Tartaric acid | *meso*-Tartaric acid |

Racemic tartaric acid

The concept of the asymmetrical carbon atom was accepted in certain quarters, frowned upon in others. In typically cantankerous fashion, Kolbe wrote about the "fanciful nonsense" and "supernatural explanations" of the two "unknown" young chemists. That he was particularly vitriolic toward van't Hoff is seen in the following:

A Dr. J. H. van't Hoff, of the veterinary school at Utrecht, finds as it seems, no taste for exact chemical investigation. He has thought it more convenient to mount Pegasus (obviously loaned by the veterinary school) and to proclaim in his "La chimie dans l'espace" how during his bold flight to the top of the chemical Parnassus, the atoms appeared to him to have grouped themselves throughout universal space.[16]

Fittig, Claus, Lossen, and Hinrickson criticized van't Hoff's proposals on various grounds, particularly their lack of agreement with current ideas on physical forces. However, great progress was made during the next two decades by such organic chemists as Wislicenus, Baeyer, Wallach,

[16] H. Kolbe, *J. prakt. Chem.*, [n.s.] **15**, 473 (1877).

and Emil Fischer. The term *stereoisomers* was applied to these compounds by Victor Meyer in 1888.

In his 1874 paper, van't Hoff also dealt in a theoretical fashion with another kind of structural isomerism, commonly known as *geometric isomerism*. This occurred, he pointed out, in compounds in which double bonds prevent free rotation within the molecule and in which at least two dissimilar substituents are present on the double-bonded carbon atoms. For example, maleic and fumaric acids had been known for some time to be isomeric, but the cause of the isomerism was not explained until van't Hoff studied the double-bond relationship.

The isomers are not mirror images and neither member is optically active. The isomerism results from the fixed character of the double bond which prevents free rotation around the bond. This isomerism may be expected whenever a double bond in a molecule carries at least two different substituents on each of the double-bonded carbons.

The decision as to which structure to assign to two geometric isomers presented a difficult problem. As a result, the terms *cis* and *trans* came into use. In the case of maleic and fumaric acids, van't Hoff assigned the *cis* form to maleic acid because it readily loses water to form the anhydride, whereas fumaric acid forms an anhydride only under more drastic conditions, and then only following transformation to maleic acid. This suggested that the carboxyl groups in maleic acid are close to one another, and spatially removed from one another in fumaric acid.

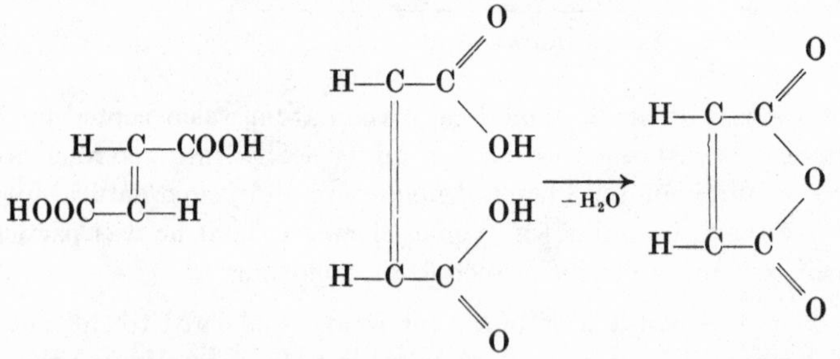

Fumaric acid   Maleic acid   Maleic anhydride

Other pairs which showed a relationship that conformed to van't Hoff's interpretation were brommaleic and isobrommaleic acids, citraconic and mesaconic acids, and the liquid and solid crotonic acids. Wislicenus was foremost in interpreting structural problems, both those involving optical and those involving geometric isomerism; his studies on the interrelationships between the fumaric-maleic acid pair, the tartaric acids, and the bromosuccinic acids are particularly masterful.[17] Wislicenus taught at Würtzburg and, after 1885, at Leipzig.

[17] J. Wislicenus, *Ber.*, **20**, 1008 (1887); *Abhandl. königl. sachs Gesellsch. Wissensch.*, *Math-Phys. Klasse*, **14**, 1 (1887).

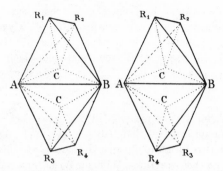

Fig. 12.16. Van't Hoff's representation of
geometric isomers.
(From van't Hoff, *La chimie dans l'espace* [1875].)

By 1890 geometric isomerism was recognized in such nitrogen compounds as the oximes and dioximes. Arthur Hantzsch and Alfred Werner suggested that if the three valences of nitrogen were not in the same plane, geometric isomers might be expected. Hantzsch (1857–1935) was active at Leipzig for many years in clarifying the structure of nitrogen compounds.

Tautomerism

Another structural problem concerned the phenomenon of tautomerism. In 1863 the compound now known as acetoacetic ester was independently discovered by Geuther and by Frankland and Duppa. Geuther named the compound ethyl diacetic ester; Frankland and Duppa named it acetone carboxylic ester. The confusion arose from the fact that Geuther studied the salt-forming characteristics of the ester, whereas the other two men studied its ketonic character. Numerous investigations of the compound gave uncertain results because it is highly reactive. It gradually became evident that the substance was present in two forms in a state of equilibrium. The name *tautomerism* (Gr. *tauto*, the same) was proposed by Conrad Laar when he discussed the problem in 1885. Paul Jacobson (1859–1923), a Berlin professor, proposed the name *desmotropy* (Gr., *desmos*, ligament; *tripein*, to change), but this gradually fell into disuse. Brühl suggested the terms *keto* and *enol* for the two forms.

$$CH_3-\overset{\overset{\displaystyle O}{\|}}{C}-CH_2-\overset{\overset{\displaystyle O}{\|}}{C}-OC_2H_5 \rightleftarrows CH_3-\overset{\overset{\displaystyle OH}{|}}{C}=CH-\overset{\overset{\displaystyle O}{\|}}{C}-OC_2H_5$$

Keto form   *Enol* form

Acetoacetic ester

In various other cases, the properties of a compound were described better by two formulas than by a single formula. As early as 1877 Butlerov came across such a compound in his study of the diisobutylenes. Baeyer and Erlenmeyer also encountered such compounds. Laar called attention to several such cases a decade later. He postulated a change in the position of a hydrogen atom that resulted in two structures being present in equilibrium.

In 1896 Wilhelm Wislicenus (1861–1922) succeeded in isolating two ethyl formyl phenyl acetates that had *keto* and *enol* properties. Ludwig Claisen isolated two forms of dibenzoyl acetone at this time that also suggested *keto* and *enol* structures. Wilhelm Wislicenus and Ludwig Knorr recognized the reversibility of the change from one form to the other, and Wislicenus proved that equilibrium could be obtained by starting with either the *enol* or the *keto* form.

It was not until 1911 that each form of the acetoacetic ester was isolated free of contamination by the other form. Using temperatures around −78° C., Knorr and K. H. Meyer, working independently, succeeded in obtaining pure *keto* and *enol* isomers.[18]

DEVELOPMENT OF METHODS OF SYNTHESIS

Although a fairly large number of organic compounds had been synthesized by 1860, the decades which followed were characterized by the enthusiasm of organic chemists for achievements in synthesis. Several incentives were responsible for this activity: (1) the challenge of producing, in the laboratory, both compounds of natural origin and compounds not known in nature; (2) the need for special compounds to test current theories; (3) the elaboration of structural problems; and (4) the demand of the growing chemical industry for compounds that were of value as dyes and drugs. This was the period during which many of the name reactions in organic chemistry were developed and extended.[19]

Since it is obviously impossible to mention most of the compounds which were successfully synthesized during this period,[20] the following

---

[18] For more details regarding geometric isomerism and tautomerism, see A. J. Ihde, *J. Chem. Educ.*, **36**, 330 (1959).

[19] For an excellent work on name reactions see Alexander R. Surrey, *Name Reactions in Organic Chemistry*, Academic Press, New York, 1954, 2nd ed., 1961. See also Kurt G. Wagner, *Autoren-Namen als chemische Begriffe*, Verlag Chemie, Weinheim, 1951.

[20] Those interested are referred to A. Ladenburg, *Lectures on the History of the Development of Chemistry*, Alembic Club, Edinburgh, 1900, Lecture 14; and to T. P. Hilditch, *A Concise History of Chemistry*, Methuen, London, 2nd ed., 1922, chaps. 7, 8, and Appendix B.

discussion is restricted to major methods, particularly those of value in the preparation of important series of compounds.

Berthelot's Studies of Synthesis

Marcellin Berthelot (1827–1907) studied medicine in the Collège de France. Here he came under the influence of Pelouze, Dumas, Regnault, and Balard as his interests turned toward chemistry. As Balard's assistant he had the time and opportunity to pursue his own researches. In 1859 he became a professor at the École Supérieure de Pharmacie and a year

Fig. 12.17. MARCELLIN BERTHELOT.
(Courtesy of the Louis Kahlenberg Collection.)

later the Collège de France created a special research professorship for him. He held both chairs simultaneously until 1876, when he resigned from the faculty of the École and became inspector of higher education. He had an influential part in the academies of medicine and science, as well as in political affairs. He was made a senator for life in 1881, and later served as minister of public instruction and foreign minister. During this latter period his researches were primarily agricultural and historical.

In his earlier researches Berthelot studied the constitution of polyhydric alcohols, particularly glycerol. He succeeded in preparing fats by heating stearic and oleic acids with glycerol, thus confirming Chevreul's idea that fats are compound ethers (esters). During this work he found that three units of acid combine with one unit of glycerol; this established the trihydroxy nature of glycerol. Berthelot's work on glycerol enabled Wurtz to synthesize ethylene glycol, the dihydroxy intermediate between simple alcohols and glycerol. These two men clarified the whole problem of polyhydroxy alcohols, just as Liebig had done for the polybasic organic acids. They demonstrated the analogy between alcohols and inorganic bases, an instance of which Odling pointed out when he indicated the similarity between glycerol and bismuth hydroxide. The problem was also related to Frankland's valence concept. Berthelot carried on various studies of alcohols and introduced acetylation as a technique for recognizing alcoholic character. He also studied the reducing action of hydrogen iodide on glycerol, obtaining a mixture of isopropyl iodide and allyl iodide; he converted the latter into mustard oil (allyl isothiocyanate).

This work was followed by Berthelot's synthesis of ethyl alcohol from ethylene, based on the addition reaction of sulfuric acid with ethylene. He then hydrolyzed ethyl sulfuric acid (its synthesis had been carried out by Henry Hennell, a London apothecary, in the 1820's). Propyl alcohol Berthelot prepared similarly from propylene obtained by the degradation of propyl iodide. He showed that the higher olefins could be converted first to the chlorides, by adding hydrogen chloride, and then to the corresponding alcohol; thus he introduced a general method for the preparation of alcohols.

Olefins, in turn, could be prepared from alcohol by the removal of water by concentrated sulfuric acid. Berthelot now realized that this was not a one-way process, since water could be added to the olefin to obtain alcohol. This suggested that it might be possible to reverse other dehydrations, such as the one in which carbon monoxide is produced from formic acid by the action of sulfuric acid. He heated carbon monoxide over potassium hydroxide for three days and found that the gas was completely absorbed. After acidification and distillation, he obtained formic acid.

The following year, 1856, Berthelot prepared methane by heating barium formate to a high temperature; he collected ethylene, propylene, and acetylene as by-products. Using still another approach, he passed carbon disulfide and hydrogen sulfide over hot copper; he reasoned that the copper would combine with the sulfur in the two compounds, liberating carbon and hydrogen in a highly reactive state. He identified methane, ethylene, and naphthalene as the reaction products. The results were similar when he substituted iron for copper; he also obtained methane when phosphine or steam was substituted for hydrogen sulfide. The ethylene that was present was treated with hydrogen chloride and hydrolyzed to ethyl alcohol. Berthelot was convinced he had demonstrated that important organic compounds could be produced from non-organic sources. He introduced the term *synthesis* for the process of building up compounds in this manner.

In 1860 Berthelot published his *Chimie organique fondée sur la synthèse* in which he set forth the general principles and methods. It was his contention that organic chemists have an obligation to synthesize compounds out of clearly inorganic materials in order to demonstrate that plants and animals are not unique in their ability to synthesize. This approach constituted a major attack on the concept of vitalism, which had not been completely overthrown by Wöhler's synthesis of urea. Kolbe's synthesis of acetic acid had pointed the way. Now Berthelot sought to settle the issue by synthesizing all sorts of natural and unnatural compounds without living intermediaries.

In 1862 he demonstrated the production of acetylene (a compound discovered by Edmund Davy in 1836) from the elements when hydrogen gas was passed through an electric arc, and four years later he produced benzene by passing acetylene through a heated tube. The barrier between inorganic and organic chemistry was broken. His investigation of the acetylides of copper and silver led him to suggest that the mineral oils in the earth might have originated from acetylides of the alkali metals.

In studying the behavior of hydrocarbons at red heat he noted that they rarely decomposed directly into the elements; instead, they either polymerized or yielded hydrogen and such compounds as ethylene or acetylene. He postulated that when elemental carbon was obtained by high-temperature reactions, it was a complex of many atoms, corresponding to the complex hydrocarbons that underwent decomposition. However, later studies did not verify this concept; Edward Thorpe, for example, demonstrated that aliphatic hydrocarbons decompose by forming an unsaturated and a saturated shorter chain molecule.

Classical Methods of Synthesis

Studies made around the 1850's by Kolbe, Frankland, Laurent, and Wurtz, among others, resulted in the production of various aliphatic hydrocarbons, halides, alcohols, acids, and amines at a time when the chemical nature of these compounds was not understood. This problem was clarified in the 1860's; hence the earlier methods were found to be highly useful in synthetic processes. Alcohols were converted to acids by means of the halide, cyanide sequence by Frankland and Kolbe in 1848. Frankland's preparation of zinc alkyl iodides and zinc alkyls provided good intermediates for hydrocarbon synthesis, as did Kolbe's electrolytic procedure in which potassium salts of organic acids were used. Williamson's synthesis provided a classical method for preparing ethers. In 1855 Wurtz first used sodium in condensing alkyl halides to produce hydrocarbons, a technique Fittig extended to aromatic compounds in 1864 when he produced toluene by condensing chlorobenzene and methyl iodide in the presence of sodium. The same technique was used by Fittig and Tollens for synthesizing xylene and ethylbenzene.

The organic halogen compounds were exceedingly valuable intermediates in synthetic processes because they could be readily replaced by hydroxyl groups in preparing alcohols and acids. The direct substitution of chlorine in organic compounds, discovered by Dumas and Laurent, was seldom used with aliphatic compounds because the reaction could not be restricted so that the desired products would be obtained. However, the alcohols and acids were easily converted to alkyl halides and acyl halides by reaction with phosphorus halides. Cahours introduced the use of phosphorus pentachloride for preparing acid chlorides from acids. Gerhardt used phosphorus oxychloride, and Béchamp used phosphorus trichloride, as a reagent for replacing hydroxyl groups with chlorine. The acyl chlorides, first discovered by Liebig and Wöhler as benzoyl chloride, facilitated the synthesis of amides, esters, and cyanides. Gerhardt produced the acid anhydrides from the acid chlorides. Bromides and iodides were equally valuable as intermediates in synthetic work; they were readily produced from the phosphorus halides. Aromatic chlorides and bromides were prepared easily by direct halogenation of the hydrocarbons.

The Friedel-Crafts Reaction

The use of anhydrous aluminum chloride as a catalyst in the condensation of aromatic hydrocarbons and alkyl halides was introduced by Charles Friedel and James Mason Crafts in 1877. Through the formation of intermediary compounds with the catalyst, the hydrogen on the aromatic hydrocarbon nucleus is removed and replaced by alkyl groups. Besides being valuable in the synthesis of hydrocarbons, the Friedel-Crafts reaction is extremely versatile and has been successfully adapted to the synthesis of aromatic ketones, aldehydes, and amides, as well as other types of compounds.

Condensation Reactions

Oxygen compounds received a great deal of attention. There was naturally great interest in acids, and aldehydes and ketones proved very useful in synthesizing reactions because of the activating influence of the carbonyl group on nearby hydrogen atoms (alpha hydrogen). The potential reactivity of the hydrogen in malonic acid was suggested by van't Hoff in 1874. The use of the malonic ester in synthesis dates from the experiments of Max Conrad and Carl Adam Bischoff in 1880; the ester was extremely useful because of the ease with which one carboxyl group is destroyed following hydrolysis.

The activation of alpha hydrogen by a carbonyl group had been encountered empirically before van't Hoff's speculation. Use of the aldol condensation resulted from Wurtz's studies of aldehydes; the isolation of aldol was reported in 1872. The dehydration of the hydroxy aldehyde on warming produced crotonaldehyde. The aldol condensation proved to be a general reaction of aldehydes with some ketones, provided hydrogen is present on the alpha carbon atom. In Kiel, Ludwig Claisen extended its use in several directions; he condensed aldehydes with aldehydes, ketones, and esters under alkaline conditions and obtained valuable compounds, among them acetoacetic ester. This latter, discovered by Guether in 1863, was very useful in synthetic work.

Perkin developed his procedure for preparing α,β-unsaturated aromatic acids in 1868. He used benzaldehyde and acetic anhydride in the presence of potassium acetate to obtain cinnamic acid. Subsequent studies revealed that the reaction worked well with substituted benzaldehydes. With *ortho*-hydroxybenzaldehyde Perkin obtained coumarin, which has the odor of new-mown hay. It was used extensively in the perfume industry and for food flavors; however, when it was found to have harmful physiological effects around 1950, it was withdrawn from use in foods.

The Cannizzaro reaction, in which benzaldehyde is converted to benzoic acid and benzyl alcohol in the presence of concentrated alkali, had been known from 1853. Study of the reaction showed clearly that it could be used as a general action for aldehydes having no hydrogen on the carbon alpha to the carbonyl group.

Karl Ludwig Reimer (1856–1921) produced *ortho*-phenolic aldehydes in Hofmann's laboratory in Berlin in 1876, by the reaction of chloroform with phenol in an alkaline medium. Reimer studied the reaction, working with Ferdinand Tiemann (1848–1899), and observed that guaiacol was converted to vanillin. Reactions of this sort are commonly known as Reimer-Tiemann reactions. Both men were active later in the German chemical industry.

In the Schotten-Baumann reaction, a phenol and benzoyl chloride react to form an ester. This reaction was discovered by Eugen Baumann (1846–1896) and Carl Schotten (1853–1910), the latter a professor in Berlin. Baumann studied in Tübingen, where he became an assistant to Hoppe-Seyler; he later taught and carried on research in physiological chemistry in Berlin and Freiburg.

The Grignard Reaction

Metal organic compounds had been valuable in synthesis since the time Frankland discovered the zinc alkyls. Zinc compounds were used in a number of ways thereafter. One use involved the reaction discovered in 1887 by Sergius Reformatsky (1860–1934), a professor at Kiev. In this reaction, zinc is added to an α-bromo ester and allowed to react with a ketone. A β-hydroxy ester is formed on hydrolysis.

The major metal organic compounds of interest in synthesis are the Grignard reagents, discovered in 1899 by Philippe Antoine Barbier (1848–1922), a professor at the University of Lyons. Barbier, a student of terpenes, sought a metal which might be substituted for zinc in reactions in which a methyl group was introduced into organic compounds. Although zinc increased the reactivity of methyl iodide, it was difficult to work with the resulting zinc compound because of its flammability on contact with air. Barbier was successful in substituting magnesium for zinc. Magnesium, in combination with organic iodides in anhydrous ether, formed a magnesium-organic iodide-ether complex which reacts with various reagents and forms a wide variety of organic compounds—hydrocarbons, alcohols, acids, ketones and amides.

Barbier introduced François Auguste Victor Grignard (1871–1935), a pupil of his at the University of Lyons, to the study of magnesium organic halides. Asked to study the reaction further, Grignard made it the subject of his doctoral dissertation in 1901. Immediately recognizing the usefulness of the reaction, Grignard continued his work on it and made it widely applicable. The reaction was at one time known as the Barbier-Grignard reaction, but Barbier insisted that credit for its development should go to Grignard, although he himself had first prepared methyl magnesium iodide.

Nitrogen Compounds

The synthesis and reactions of nitrogen compounds received a great deal of attention, particularly in view of their importance in the dye industry and in connection with natural products. The work of Wurtz and Hofmann on amines has already been mentioned. Nitrobenzene was first prepared by Mitcherlich after he converted benzoic acid to benzene in 1832. He prepared various benzene derivatives, including chlorobenzene and bromobenzene, benzene sulfonate, and azobenzene. Zinin discovered the process whereby nitrobenzene is converted to aniline by reduction. Hofmann began to work on the reactions of aniline in 1843, in connection with his studies of coal tar derivatives. Charles Mansfield (1819–1855), a student of his at the Royal College, did research on the fractional distillation of coal tar fractions and perfected a process for separating aromatic compounds from each other. Hofmann conclusively demonstrated the presence of benzene in coal tar in 1845, though Leigh's work had indicated this fact three years earlier. Mansfield developed the distillation process for large-scale separations but died as the result of burns suffered when, in preparing benzene for the Paris Exposition, liquid in the still pot boiled over and caught fire.

Although azo compounds were discovered by Mitscherlich, and related compounds were prepared by Zinin and Hofmann, the potentialities of the class were not realized until Peter Griess (1829–1888) began his work. Griess studied in Jena and Marburg, serving as Kolbe's assistant in Marburg and then as Hofmann's assistant at the Royal College. He became a chemist in the Allsopp and Sons brewery in Burton-on-Trent in 1862. He began to study the azo compounds while still with Kolbe and continued his work on them after he left Kolbe. He studied the process whereby diazonium salts were formed by the reaction of nitrous acid with aniline in the cold, and then established that these salts could be used in coupling reactions with phenol, aniline, benzoic acid, and other agents. In 1864 he demonstrated that this procedure could be used in deamination, i.e., aniline to benzene. The first azo compound of the aliphatic series was obtained in 1883 when Theodor Curtius (1857–1928) prepared diazoacetic acid ester; later he prepared other compounds of this type. He was professor of organic chemistry at Heidelberg for many years.

No aliphatic nitro compounds were known until 1872, when Kolbe prepared nitroethane and Victor Meyer prepared nitromethane. In 1882 Meyer introduced the preparation of oximes by the reaction of hydroxylamine with aldehydes and ketones.

Emil Fischer discovered phenylhydrazine while studying for his doctorate under Baeyer in Strasburg. The compound proved extremely useful in identifying aldehydes and ketones because the derivatives that are formed are frequently solids with definite melting points. The compound was found to form osazones with simple sugars like glucose and fructose. His work with these compounds led Fischer to embark on his lengthy study of the structure of the simple sugars which will be discussed in the next chapter.

The Schotten-Baumann reaction, mentioned earlier, was found to be applicable to the condensation of amines with acid chlorides to form amides.

The Hofmann degradation, in which acid amides are converted to amines by hypobromite or hypochlorite, was reported in 1882.

The reaction reported by Curtius in 1894 for the preparation of amines from acids involved the preparation and decomposition of an acid azide. The azide could also be prepared by the reaction of hydrazine on an acid chloride, followed by treatment with nitrous acid.

A procedure for preparing primary amines was introduced in 1887 by Siegmund Gabriel (1851–1924), then working in Hofmann's laboratory, but later a professor in Berlin. In this procedure, which proved to be widely applicable, potassium phthalimide reacts with an alkyl halide, and hydrolysis then splits out the amide. The Gabriel reaction is specific for primary amines because the other valences of the nitrogen are blocked.

In 1886, Ernst Otto Beckmann (1853–1923) discovered that ketoximes treated with acids, acid chlorides, or phosphorus pentachloride rearranged to the amide. This Beckmann rearrangement, as it was called, attracted a great deal of interest from the standpoint of both preparation and theory.

The reaction discovered in 1887 by Conrad Heinrich Willgerodt (1841–1930) of Freiburg involves the conversion, by means of yellow ammonium sulfide (sulfur in ammonium sulfide), of partially aromatic ketones to the corresponding arylacetic amide with some reduction of the ketone.

Traugott Sandmeyer (1854–1922) of Zurich discovered in 1884 that aromatic amines can be converted to the corresponding halide with diazonium salts in the presence of a cuprous chloride catalyst. Production of the corresponding cyanide is possible with a cuprous cyanide catalyst. The reaction was later modified when Ludwig Gattermann (1860–1920) found that it will take place if copper powder is used as a catalyst. Gattermann received his doctorate at Göttingen in 1885, presenting a dissertation on a problem suggested by Hübner who died in 1884. Victor Meyer and his lecture assistant, Sandmeyer, came to Göttingen in 1885 but Sandmeyer soon returned to Zurich and Gattermann succeeded to his assistantship. He moved to Heidelberg with Meyer where he remained until 1900 when he became head of the chemistry department at Freiburg. He was adept as an analyst and a synthetic organic chemist. His *Die Praxis des organischen Chemie* was widely used as a laboratory manual.

The Walden Inversion

An interesting phenomenon regarding the reactions of optically active compounds was discovered in 1893 by Paul Walden (1863–1958). He found that when a group attached directly to an asymmetrical carbon atom was replaced, an inversion to the opposite configuration frequently took place. This reaction, known as the Walden inversion, made it possible to produce a missing enantiomorph from an avaliable one.

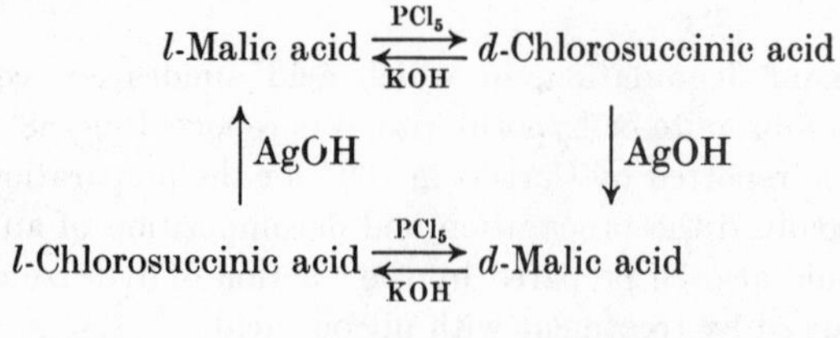

$$l\text{-Malic acid} \underset{\text{KOH}}{\overset{\text{PCl}_5}{\rightleftarrows}} d\text{-Chlorosuccinic acid}$$

$$\uparrow \text{AgOH} \qquad\qquad \downarrow \text{AgOH}$$

$$l\text{-Chlorosuccinic acid} \underset{\text{KOH}}{\overset{\text{PCl}_5}{\rightleftarrows}} d\text{-Malic acid}$$

The Walden inversion, in ideal cases, permits quantitative production of the desired enantiomorph, whereas ordinary racemization leaves a 50-50 mixture which must still be separated. Of course the reaction is not always ideal; at best it depends upon the selection of suitable reagents and conditions. The reagent was found to be important in determining whether an inversion will take place; thus in the above reactions potassium hydroxide causes an inversion, silver hydroxide does not. The nature of the optically active compound, the solvent, and the temperature were all found to have an influence on whether inversion takes place.

COMMUNICATION

Nomenclature

The problem of communication grew increasingly complex for organic chemists as the number of compounds snowballed. The introduction of structural formulas made for improved understanding among chemists, but names were needed. Commonplace names could be of value only as long as there were few compounds; even so, these terms did not uphold the dictum of Lavoisier and his colleagues that the name of a compound should give information about its chemical nature.

In the early days of organic chemistry the name of a new chemical was usually related either to the compound it was derived from, or to one it was clearly related to. Members of clearly defined series were given the same class names—i.e., acids, alcohols (or carbinols, following Kolbe), ethers (used for both the true ethers and the esters), amines. In substituted compounds, the letters of the Greek alphabet were used to designate position, i.e., α-hydroxypropionic acid for lactic acid. Körner introduced the terms *ortho*, *meta*, and *para* to indicate ring positions. Number systems also came into use to represent positions of substituent groups.

By the late 1880's organic nomenclature was becoming sufficiently troublesome to receive official recognition. Although textbook writers like Gerhardt, Kekulé, Kolbe, and Meyer influenced nomenclature, there was frequent lack of agreement and resultant confusion. Accordingly, a study commission was established at the International Congress of Chemists held in Paris in 1889, to prepare a report for the meeting scheduled three years later.

The next meeting of the International Congress was held in Geneva in 1892, with Friedel as the presiding officer. The study commission, of which Alphonse Combes (1858–1896) was the most active member, advanced nomenclatural propositions based on its work over the three-year period. These were approved by some forty chemists in attendance, and an official nomenclature was established for organic chemistry.

The basic nomenclatural principles were based primarily on suggestions originally made by Laurent. Hydrocarbons were made the base of compound names, any substituents being treated as substitution products. The names of branched-chain hydrocarbons were based on the longest hydrocarbon chain, the branches being treated as methyl, ethyl, or other appropriate substituent radicals. A numbering system was introduced to indicate the position of substituents. The names of all saturated hydrocarbons ended in *-ane*. Names of hydrocarbons with one double bond ended in *-ene*; with two double bonds, in *-diene*; triple bonds were indicated by *-ine* or *-yne*. These endings had been used to some extent since 1865, when Hofmann suggested them. The presence of functional groups was indicated by appropriate endings: *-ol* for alcohols, *-al* for aldehydes, *-one* for ketones, *-oic acid* for acids.

The Geneva nomenclature has received official acceptance among international chemical groups. However, certain commonplace names are less cumbersome than their Geneva equivalents and have naturally remained in use—lactic acid rather than α-hydroxypropanoic acid, glycine rather than 2-aminoethanoic acid. The official nomenclature has met great resistance in industry, archaic names being retained even when they are misleading; i.e., benzol, toluol, and xylol for benzene, toluene, and the xylenes respectively, muriatic acid for hydrochloric acid, and oil of vitriol for sulfuric acid.

Compendia

The rapid increase in the number of known organic compounds during the nineteenth century made the problem of keeping up with knowledge about them more and more formidable. Leopold Gmelin, who dealt with it during the first half of the nineteenth century in successive editions of his *Handbuch der Chemie*[21] was particularly troubled by the problem of nomenclature and classification. The organic part of the fourth edition was rapidly outmoded as the result of the new concepts introduced by Kekulé and his contemporaries. The organic part was completely abandoned by Karl Kraut when he began editing the sixth edition and the *Handbuch* became a leading compendium of inorganic chemistry. An entirely new treatment of organic chemistry was needed as soon as the ideas on structural chemistry came into general use. This need was filled by Beilstein's famous *Handbuch der organischen Chemie*.

Friedrich Konrad Beilstein (1838–1906) was born in St. Petersburg, of German parents. His early education in the German schools of that city was followed by his *Wanderjahre* in which he studied with Bunsen at Heidelberg and with Liebig at Munich; he took his doctorate under Wöhler at Göttingen. After further work with Wurtz at the École de Médecine he became a division head in Wöhler's laboratory, where he was closely associated with Fittig and Hübner in editing the *Zeitschrift für Chemie*. In 1866 he returned to St. Petersburg and succeeded Mendeleev as professor at the Imperial Technological Institute.

Although Beilstein is best known for his work as an editor, he carried on extensive experimental research. His studies of the action of phosphorus pentachloride as a chlorinating agent were of considerable importance in clarifying the structure of chlorinated hydrocarbons, and, together with his work on substituted benzoic acids, were important in establishing the validity of Kekulé's ring structure for aromatic compounds. In the course of these studies he developed his test for halogens in organic compounds. In this test the sample is dropped on hot copper oxide powder; a Bunsen burner flame becomes greenish-blue when halogen is present, because of

---

[21] The first edition was published in 1817–1819 in 3 volumes as *Handbuch der theoretischen Chemie*. A greatly enlarged fourth edition, titled *Handbuch der Chemie*, was published in 10 volumes in 1843–1870. The sixth edition, edited by Karl Kraut, appeared in 3 volumes in 5 parts in 1871–1886 as *Handbuch der anorganische Chemie*. The eighth edition, edited first by R. J. Meyer and then Erich Pietsch, began to appear in 1924 and is only now being completed.

the great volatility of copper halides. Following his return to Russia he began his studies of Caucasian petroleum and demonstrated that, contrary to popular belief in Russia, these petroleum oils had greater illuminating power than imported American oils. Eventually this was found to be due to the presence of ring hydrocarbons. This led to Markownikov's studies on the naphthenes, begun in 1883.

It is not known just when Beilstein conceived the idea for his *Handbuch*. While in Göttingen, he compiled extensive notes on organic compounds

Fig. 12.18. FRIEDRICH KONRAD
BEILSTEIN.
(Courtesy of the Edgar Fahs Smith Collection.)

and his activity along this line increased on his return to Russia. At any rate, he completed the manuscript for the first edition late in the 1870's. The two-volume first edition, consisting of 2201 pages, was brought out in 1880–1882 by a Hamburg publisher. The demand was so great that the edition was out of print within a few months. The publisher wanted to reprint it, but Beilstein insisted on making corrections and additions.

A second edition of the *Handbuch*, consisting of three volumes (4080 pages), was published between 1886 and 1889. All the work on the manuscript was done by Beilstein with the help of one assistant. He worked on it almost constantly, even vacations outside Russia being planned so he would have access to good libraries. That he refused professorships in German universities in order to retain the freedom to write that he enjoyed in Russia is clear from a letter to a German friend written in 1895:

To be sure I could write my "Handbuch" only in Russia and therefore I have declined invitations to come back to Germany. At a
Russian technical school the professors need not undertake research
for here it carries no weight with the students, but in Germany this
would be regarded askance.[22]

Beilstein began work on a third edition as soon as the second was published. By this time the number of organic compounds was increasing to
such an extent that one author could scarcely deal adequately with them
any longer. Beilstein's original plan of classification was breaking down,
whereupon he decided to make only the most necessary revision in
classifying existing compounds and to incorporate new ones in later
supplements. He asked Paul Jacobson of Heidelberg to prepare the
supplement to the third edition. Realizing the magnitude of the task,
Jacobson declined but suggested that the Deutschen chemischen Gesellschaft prepare the supplement to the *Handbuch*. After the necessary
negotiations, the Gesellschaft voted to accept the proposal. Paul Jacobson
was appointed editor of the supplement and began work on it in 1897. At
the same time, the Gesellschaft consolidated certain other operations by
discontinuing the publication of abstracts in the *Berichte* and taking over
the old journal of abstracts, the *Chemisches Centralblatt*, which became the
Chemisches Zentralblatt. The major portion of the third edition of the
Handbuch was published between 1892 and 1899 from the manuscript
prepared by Beilstein. The supplement was published in 1906, just before
his death.

After Beilstein died, the fate of the *Handbuch* immediately became the
subject of thought. A proposal to publish a second supplement that would
include new compounds was rejected by Jacobson because the need for a
revised classification was becoming even more apparent. The number of
compounds known when the first edition was published was approximately
15,000, but had increased to perhaps 150,000 by 1910.

Jacobson recommended that a completely revised fourth edition be
prepared; it would require sixteen years to complete and would run to
16,000 pages. The Deutschen chemischen Gesellschaft agreed to undertake
it, and appointed Jacobson and Bernard Prager the editors. In 1907 the
two men developed a new system of classification, and spent the next five
years reclassifying old material in the new system. It was decided to
bring the survey of the literature down to 1910. Editorial work continued
even after the start of World War I and printing was begun on the first
volume in 1916. Twenty-seven volumes were necessary for the basic
edition; the last one did not come off the press until 1937. Preparation
of the first supplement, dealing with the literature between 1910–1919,
was carried on simultaneously with the publication of the basic edition.

[22] F. Richter, *J. Chem. Educ.*, **15**, 310 (1938).

This supplement, which also ran to 27 volumes, was completed in 1938. Work on the second supplement, covering the literature between 1920–1929, was halted by World War II.

CONCLUSION

After the solution of the problem of molecular formulas, organic chemistry entered upon a very fruitful period. Structural concepts, as developed out of the work of Butlerov and Kekulé and augmented by the ideas of LeBel and van't Hoff, proved to be capable of revealing atomic relationships of considerable subtlety before the century ended. This kind of progress was to contribute significantly to the success of the synthetic dye industry and to the understanding of the chemistry of such natural products as the carbohydrates, terpenes, purines, and proteins.

The period from 1860 to 1900 was also one in which great progress was made in designing methods for the synthesis of a wide variety of organic compounds. Consequently, it was possible to prepare molecules of considerable complexity from simple and readily available starting materials. Such success was particularly evident in the work with aromatic compounds originating from coal tar. The resultant increase in the number of known compounds placed a heavy burden on those who were involved with the literature of chemistry since unambiguous nomenclature and indexing were of crucial importance.

CHAPTER 13

ORGANIC CHEMISTRY IV.
NATURAL PRODUCTS

Besides being interested in synthetic and structural problems for their own sake, organic chemists were motivated in their researches by the economic aspects of the subject and by the interest always manifest in substances of natural origin. The direct economic aspects, involving in particular the development of dyes and drugs, will be treated in a subsequent chapter. The interest in such compounds as the sugars, terpenes, purines, and proteins also had an economic aspect, but was less important than in the case of dyes and drugs. These compounds aroused curiosity primarily because they played an important role in life itself.

CARBOHYDRATES

The carbohydrates received major attention during the last three decades of the century. At the beginning of this period the simple sugars —glucose, fructose, galactose, and sorbose—were known. Sucrose was a valuable commercial sugar that was known to hydrolyze to fructose and glucose. Lactose was recognized to be milk sugar, and galactose and glucose were known as products of hydrolysis. Since 1811, when Kirchhoff of St. Petersburg hydrolyzed starch with sulfuric acid, starch had been known to be a complex of simple sugars, the ultimate unit of which was glucose.

Analytical investigations of sugars had been carried on with considerable success before 1870, with both optical and chemical methods being used. Studies by Fehling and by Bernhard Tollens (1841–1918) had shown the value of copper and silver complexes in detecting "reducing sugars."

Sucrose was peculiar in not undergoing reducing reactions. Since the reducing reactions were known to result from aldehyde or ketone groups, it was evident that these groups were lost when glucose and fructose united.

It was also known that hydroxyl groups were present in the sugars. In 1871 Rudolf Fittig (1835–1910) proposed a formula for sucrose; it revealed the presence of two C_6 chains, the proper number of hydroxyl groups, and indefinitely defined oxygen atoms. Tollens later attempted to devise a structure which would tie the two sugar fragments together by a carbon-to-carbon linkage. As early as 1870 A. Colley had used the oxygen ring to explain the fact that the aldehyde reactions of glucose are more sluggish than those of regular aldehydes. The linkage suggested by Tollens, however, was untenable because sucrose is easily hydrolyzed to its simple sugars.

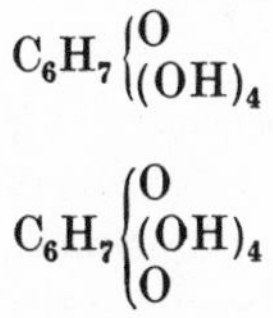

Fittig (1871) 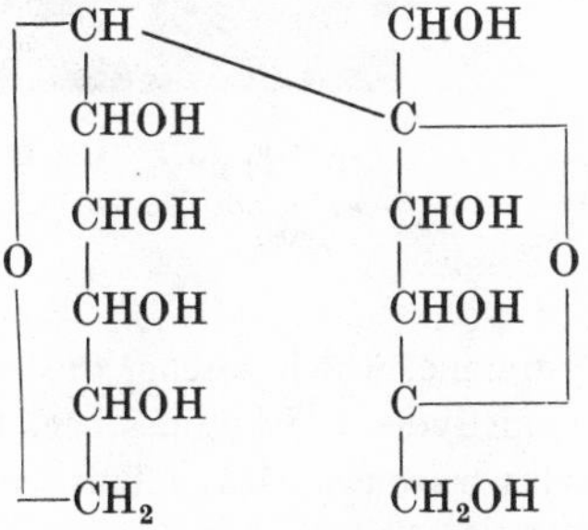   Tollens (1883)

Although the functional groups in the sugar molecules were known, there were still many unanswered questions regarding structure. All four of the simple sugars had the same empirical formula, $C_6H_{12}O_6$; all four had an aldehyde or ketone group, and all four were polyalcohols. The difference between them was due to the location of the carbonyl group and the configuration of the hydrogen and hydroxyl groups around the central carbon atoms. Not until asymmetrical carbon atoms were discovered by Le Bel and van't Hoff could the major problem be attacked. The structures were worked out after a long period of investigation by Emil Fischer and his students, by means of very effective use of all the tools, both theoretical and experimental, available at the time.[1]

Emil Fischer (1852–1919) studied briefly under Kekulé in Bonn, then with Baeyer in Strassburg and Munich. He remained with Baeyer until 1882, then taught successively at Erlangen, Wurtzburg (1885), and Berlin (1892). Despondent over deteriorating health (cancer) and the loss of two sons during World War I, he committed suicide. The eldest son, Hermann O. L. Fischer (1888–1959) became an important organic chemist.

[1] C. S. Hudson, *J. Chem. Educ.*, **18**, 353 (1941).

Fig. 13.1 EMIL FISCHER.
(Courtesy of the Edgar Fahs Smith Collection.)

Phenylhydrazine, which Fischer discovered in 1875, was valuable as a reagent for aldehydes and ketones because with many of them it formed solid phenylhydrazones with definite melting points. Fischer reported between 1884 and 1887 that with sugar phenylhydrazine formed not only phenylhydrazones but attacked the adjacent hydroxyl group as well; the osazones that resulted were poorly soluble solids with definite melting points and were useful in identifying sugars.[2] However, glucose and fructose formed identical osazones. Mannose, a new sugar obtained by Fischer in 1887, also formed this osazone. Hence he concluded that the configuration of these three sugars must be identical below the second carbon atom.

Consideration of steric relations made it clear to Fischer that a sugar like glucose must have four asymmetrical carbon atoms; therefore, according to van't Hoff's rule, the number of possible isomers must be 2^4 or 16, consisting of 8 pairs of enantiomorphs. Fructose, with three asymmetrical carbon atoms, would have 2^3 or 8 isomers. In Fig. 13.2, showing the configurations of the sixteen isomers of glucose, the vertical line stands for the chain of carbon atoms and the horizontal lines represent the side where an OH group is found, as follows:

$$\left|- \;=\; H-\overset{|}{\underset{|}{C}}-OH\right.$$

[2] Emil Fischer, *Ber.*, **17**, 579 (1884); **20**, 821 (1887). See also *Untersuchungen über Kohlenhydrate und Fermente*, Springer, Berlin, 2 vols., 1909, vol. 1, *passim.*

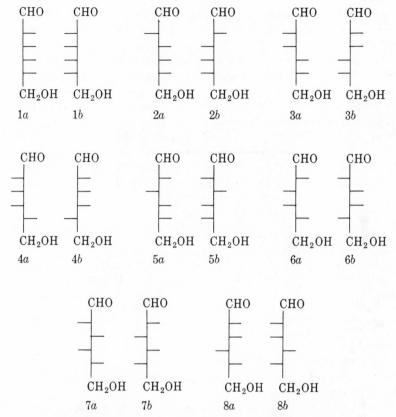

Fig. 13.2. POSSIBLE STRUCTURES OF ALDEHYDE SUGARS WITH THE FORMULA, $C_6H_{12}O_6$. Any superimpositions must be made only by rotating the formula within the plane of the paper.

The formula of glucose had to be one of these, that of mannose another, and that of galactose a third.

In attacking the problem, Fischer utilized the available information and the reactions developed by his predecessors, plus devices of his own. Several valuable contributions came from Heinrich Kiliani (1855–1945). Born in Würzburg, Kiliani studied under the senior Erlenmeyer at the Technische Hochschule in Munich and took his doctorate at the University of Munich; his doctoral dissertation concerned the preparation of inulin. In working on inulin he observed that glucose is rapidly oxidized to gluconic acid by bromine, whereas fructose is only slowly oxidized over a period of weeks, glycolic acid being the oxidation product. This indicated rather clearly that glucose is an aldehyde, and fructose a ketone with the carbonyl oxygen on the second position in the chain. In another study,[3] Kiliani developed the addition reaction of hydrogen cyanide to sugars which Fischer used so effectively for lengthening carbon chains. When Kiliani applied this reaction to the recently discovered arabinose, the sugar proved to be a pentose rather than a hexose as had at first been believed. Kiliani hydrolyzed the cyanohydrin to the corresponding acid and, on reduction, obtained mannitol. Fischer, however, realized that the aldehyde sugar, mannose, must have been formed first by the reduction. He used the reaction in building pentoses to hexoses and synthesized sugars with up to nine carbon atoms. The Kiliani reaction may be represented as follows, the position of the H and OH around the carbon atoms having no significance:

$$
\begin{array}{ccccc}
& \text{CN} & \text{COOH} & \text{CHO} & \text{CH}_2\text{OH} \\
& | & | & | & | \\
\text{H—C}{=}\text{O} & \text{HCOH} & \text{HCOH} & \text{HCOH} & \text{HCOH} \\
| & | & | & | & | \\
\text{(HCOH)}_3 \xrightarrow{\text{HCN}} & \text{(HCOH)}_3 \xrightarrow{2\text{H}_2\text{O}} & \text{(HCOH)}_3 \xrightarrow{2\text{H}} & \text{(HCOH)}_3 \xrightarrow{2\text{H}} & \text{(HCOH)}_3 \\
| & | & | & | & | \\
\text{H}_2\text{COH} & \text{H}_2\text{COH} & \text{H}_2\text{COH} & \text{H}_2\text{COH} & \text{H}_2\text{COH} \\
\text{Arabinose} & \text{Arabinose} & \text{Mannonic} & \text{Mannose} & \text{Mannitol} \\
& \text{cyanohydrin} & \text{acid} & &
\end{array}
$$

In 1886 Kiliani used the cyanohydrin reaction to establish the position of the carbonyl group in glucose and fructose. When glucose was converted through the cyanohydrin to the C_7 acid and reduced with hydrogen iodide, n-heptylic acid was a product. Fructose, similarly treated, yielded isoheptylic acid.

Mild oxidation with bromine water readily converted the aldehyde group on an aldose sugar to a carboxyl group, thus forming the corresponding sugar acid, i.e., glucose to gluconic acid, mannose to mannonic acid. More vigorous oxidation with nitric acid converted the sugar to a dicarboxylic acid, i.e., glucose to saccharic acid, galactose to mucic acid. Reduction of a sugar with sodium amalgam converted the aldehyde group to alcohol, i.e., glucose to sorbitol, mannose to mannitol.

$$
\begin{array}{cccc}
\text{H}_2\text{COH} & \text{H—C}{=}\text{O} & \text{COOH} & \text{COOH} \\
| & | & | & | \\
\text{(HCOH)}_4 \xleftarrow[\text{NaHg}]{2\text{H}} & \text{(HCOH)}_4 \xrightarrow{\text{HBrO}} & \text{(HCOH)}_4 \xrightarrow{\text{HNO}_3} & \text{(HCOH)}_4 \\
| & | & | & | \\
\text{H}_2\text{COH} & \text{H}_2\text{COH} & \text{H}_2\text{COH} & \text{COOH} \\
\text{Alcohol} & \text{Sugar} & \text{Monocarboxylic} & \text{Dicarboxylic} \\
& & \text{acid} & \text{acid}
\end{array}
$$

---

[3] H. Kiliani, *Ber.*, **18**, 3060 (1885); **19**, 221 (1886).

With both the dicarboxylic acid and the alcohol the end groups are the same. Hence, any sugar with a symmetrical arrangement around the four asymmetrical carbon atoms should give an alcohol or dicarboxylic acid which is optically inactive because of internal compensation, as is the case with *meso*-tartaric acid. Such a situation exists in structures 1 and 6 (Fig. 13.2). There is a plane of symmetry; and once the end groups are made the same, the *a* and *b* forms become superimposable. Since mucic acid, produced by the nitric acid oxidation of galactose, is optically inactive, Fischer reasoned that structure 1 or 6 must apply. Saccharic acid and sorbitol, produced from glucose, must be unsymmetrical because they are both optically active.

After Kiliani had converted natural arabinose, a *levo*-rotatory sugar, into the corresponding cyanohydrin and hydrolyzed this to the acid, he isolated and identified the lactone of L-mannonic acid. Lactones had been recognized as dehydration products of γ-hydroxy organic acids, butyrolactone having been discovered in 1870 by Alexander Saytzev. The chemistry of this class of compound was clarified by Fittig and Bredt in 1880, when they showed that lactone formation is to be expected when a hydroxyl group occurs on a carbon atom gamma to a carboxyl group.

$$\overset{\gamma}{\text{C}}\text{H}_2 \cdot \overset{\beta}{\text{C}}\text{H}_2 \cdot \overset{\alpha}{\text{C}}\text{H}_2 \cdot \text{C}{=}\text{O} \qquad \xrightarrow{-\text{H}_2\text{O}} \qquad \text{CH}_2 \cdot \text{CH}_2 \cdot \text{CH}_2 \cdot \text{C}{=}\text{O}$$

γ-Hydroxybutyric acid   γ-Butyrolactone

Later, delta-hydroxy acids were also found to form lactones, but not as easily as when the hydroxyl group is on the gamma carbon atom.

Because of the presence of suitably placed hydroxyl groups, the sugar acids formed lactones with ease. Fischer observed that the lactone Kiliani had obtained from arabinose was identical with the one formed from mannose by oxidation and lactonization, except that Kiliani's was *levo*-rotatory and hence was the enantiomorph of the one from mannose. Fischer then isolated a second product from the lactone Kiliani obtained from arabinose—namely, L-gluconic acid—the enantiomorph of the acid formed by the mild oxidation of glucose.

The only possible conclusion was that the configuration around the asymmetrical carbon atoms in arabinose was exactly opposite that around the equivalent atoms in glucose. Further, extending the carbon chain to six atoms created a new asymmetrical center and hence two derivatives should result; this was proved experimentally. In subsequent studies Fischer converted glucose to heptose, and again obtained two epimeric sugars. (The term *epimer* was introduced by Emil Votoček in 1911 to designate sugars which are identical in configuration below carbon atom number 2, counting from the carbon atom at the carbonyl end of the chain.)

$$
\begin{array}{ccccc}
& & CN & CN & COOH & COOH \\
& & | & | & | & | \\
H-C=O & \xrightarrow{\text{HCN}} & H-C-OH & \text{and} \ HOCH & \xrightarrow{\text{H}_2\text{O}} \ HCOH & \text{and} \ HOCH \\
| & & | & | & | & | \\
(HCOH)_3 & & (HCOH)_3 & (HCOH)_3 & (HCOH)_3 & (HCOH)_3 \\
| & & | & | & | & | \\
H_2COH & & H_2COH & H_2COH & H_2COH & H_2COH \\
\text{L-Arabinose} & & \multicolumn{2}{c}{\text{2 Cyanohydrins}} & \multicolumn{2}{c}{\text{2 Hexose sugars}} \\
& & & & \multicolumn{2}{c}{\text{(epimers)}}
\end{array}
$$

In these same studies (1889), Fischer found that the sugar lactones could be reduced with sodium amalgam, and that if the reduction took place under only slightly acidic conditions it stopped at the aldehyde stage instead of proceeding to the sugar alcohol. (The position of H and OH in the following equation is of no significance.)

$$
\begin{array}{c}
\overset{\displaystyle O}{\underset{}{C}} \\
| \\
HCOH \\
| \\
HCOH \quad O \\
| \\
HC \\
| \\
HCOH \\
| \\
H_2COH
\end{array}
\xrightarrow[\text{Sl. acid}]{\text{Na-Hg}}
\begin{array}{c}
H \qquad O \\
\diagdown \quad \diagup \diagup \\
C \\
| \\
(HCOH)_4 \\
| \\
H_2COH
\end{array}
\xrightarrow[\text{acid}]{\text{Na-Hg}}
\begin{array}{c}
H_2COH \\
| \\
(HCOH)_4 \\
| \\
H_2COH
\end{array}
$$

In 1890 Fischer discovered that an aldonic acid can be converted to its epimeric acid by being heated with a mild alkali like quinoline. Thus, gluconic acid can be converted to mannonic acid, and vice versa—further evidence of the similarity of configuration below the second carbon atom.

At that time, Fischer was also working on an attempted sugar synthesis, based upon studies reported by Butlerov in 1861. Butlerov had produced a sweet, pale yellow syrup by treating with lime water the compound now known as trioxane. The syrup gave the common tests for glucose, but was optically inactive and was not fermented by yeast. Trioxane decomposes to formaldehyde. The latter, because of the similarity of its formula, CH_2O, to that of sugars, $C_6H_{12}O_6$, has been of particular interest to chemists as a possible precursor of sugars. Baeyer explained the reaction as being due to the condensation of six molecules of hydrate of

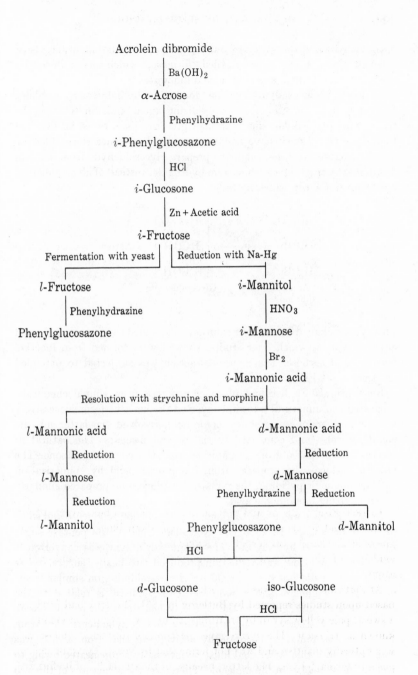

Fig. 13.3. FISCHER'S SYNTHESIS OF α-ACROSE AND PROOF
OF THE PRESENCE OF FRUCTOSE.

formaldehyde and the simultaneous loss of six molecules of water. Oscar Loew in 1885 obtained a syrup called "formose," which was produced by the reaction of formaldehyde with cold limewater.

In 1887, while studying osazone formation by polyvalent alcohols, Fischer treated glycerol with an oxidizing agent and, on treating the result with phenylhydrazine, obtained glycerosazone. Since he believed that the glycerol must have first been oxidized to either glyceraldehyde or dihydroxyacetone, he sought to prepare glyceraldehyde from acrolein dibromide by treating it with barium hydroxide. Instead of glyceraldehyde he obtained a syrup which he called acrose.

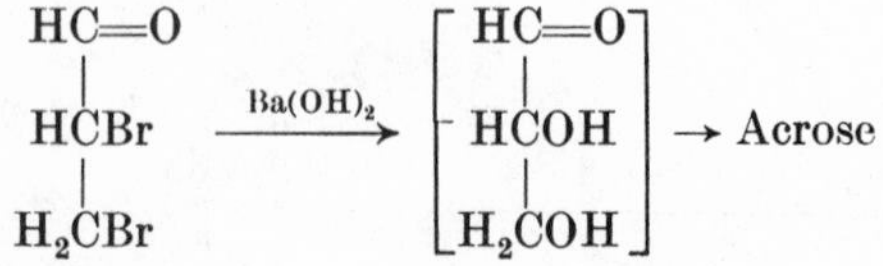

$$\begin{array}{c}
\text{HC}{=}\text{O} \\
| \\
\text{HCBr} \\
| \\
\text{H}_2\text{CBr}
\end{array}
\xrightarrow{\text{Ba(OH)}_2}
\left[\begin{array}{c}
\text{HC}{=}\text{O} \\
| \\
\text{HCOH} \\
| \\
\text{H}_2\text{COH}
\end{array}\right]
\rightarrow \text{Acrose}$$

Acrolein   Glyceraldehyde
dibromide

This syrup behaved like a sugar solution and yielded two osazones, termed alpha and beta, which later studies showed were formed from inactive fructose and sorbose. The α-acrose fraction was subjected to extended treatment; see Fig. 13.3.[4]

By synthesis Fischer had obtained fructose from simple chemicals, separated the unnatural L-form, converted the inactive form successively to mannitol, mannose, and mannonic acid, resolved the latter into its enantiomorphs, and converted to the D- and L-sugars. The natural D-fructose he prepared from D-mannose from the common osazone. The natural D-glucose he prepared from D-mannonic acid by epimerization with quinoline or pyridine; the unnatural L-glucose he prepared similarly by epimerizing L-mannonic acid.

In 1891 Fischer was ready to assign a configuration to glucose and some of the other sugars. In 1886 von Koch had isolated the pentose sugar xylose from wood hydrolysate. The cyanohydrin synthesis converted it into two sugar acids; and reduction produced two hexose sugars, *gulose* and *idose* that did not occur in nature. *Levo*-arabinose, on similar treatment, gives L-glucose and L-mannose. Thus, glucose is related to the arabinose structure, gulose to the xylose. When these two pentoses were converted to the corresponding trihydroxy dicarboxylic acids, that from arabinose was found to be optically active, and that from xylose was inactive. Hence this latter acid must be internally compensated owing to a plane of symmetry in the molecule. Therefore, the structures of the two acids and the parent sugars must be:

[4] Emil Fischer, *Ber.*, **23**, 2114 (1890).

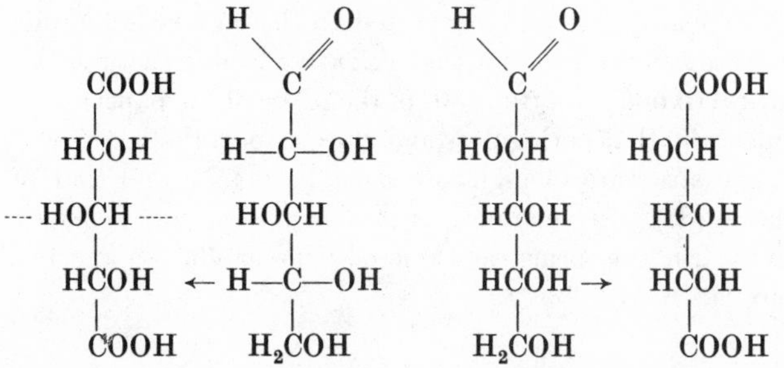

```
      COOH            H   O        H   O           COOH
       |               \ //        \ //             |
     HCOH              C           C              HOCH
       |               |           |                |
---- HOCH ----    H—C—OH         HOCH             HOCH
       |               |           |                |
     HCOH    ←       HOCH          HCOH     →      HCOH
       |               |           |                |
     COOH          H—C—OH         HCOH             HCOH
                       |           |                |
                    H₂COH        H₂COH            COOH
```

| Optically inactive | D-Xylose | D-Arabinose[a] | Optically active |

[a] The natural arabinose is L-arabinose. The D-form is shown here for consistency with D-xylose, the natural form.

The Kiliani synthesis, which converts xylose to gulose and idose, results in sugars with the following structures. (Note that 8a and 7a correspond to structures in Fig. 13.2.)

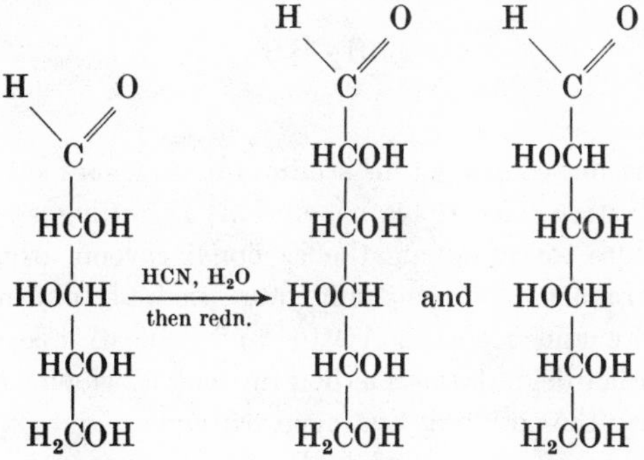

```
     H    O                  H    O         H    O
      \  //                   \  //          \  //
       C                       C              C
       |                       |              |
     HCOH       HCN, H₂O      HCOH           HOCH
       |       ─────────►      |              |
     HOCH       then redn.    HCOH           HCOH
       |                       |              |
     HCOH                     HOCH    and    HOCH
       |                       |              |
    H₂COH                     HCOH           HCOH
                               |              |
                            H₂COH          H₂COH
```

| D-Xylose | A (8a) | B (7a) |

The same treatment of arabinose, which gives glucose and mannose, must produce the following structures:

```
      H    O                  H    O         H    O
       \  //                   \  //          \  //
        C                       C              C
        |                       |              |
      HCOH       HCN, H₂O      HOCH           HCOH
        |       ─────────►      |              |
      HOCH       then redn.    HCOH    and    HCOH
        |                       |              |
      HOCH                     HOCH           HOCH
        |                       |              |
     H₂COH                     HOCH           HOCH
                                |              |
                             H₂COH          H₂COH
```

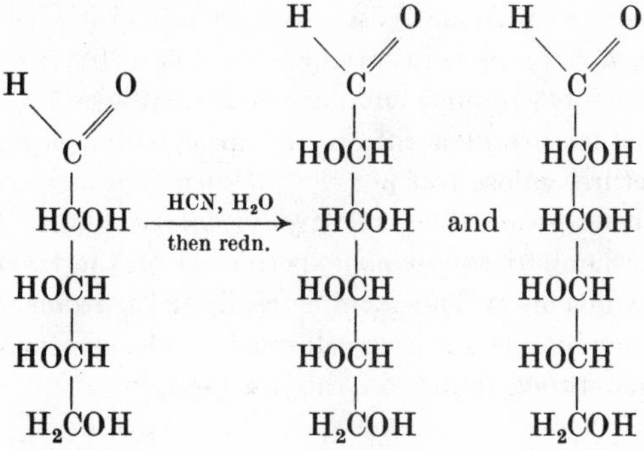

| L-Arabinose | C (5b) | D (3b) |

If these hexoses are now converted to the corresponding dicarboxylic acids it becomes evident that the derivative from structures A and C will be identical (if one is rotated 180° in the plane of the paper). It was known experimentally that both gulose and glucose form the same saccharic acid. Hence, the structure of L-glucose must be C (5b), and that of D-gulose must be A (8a). *Levo*-mannose is D (3b), and B corresponds to D-idose (7a). Since fructose forms the same osazone as glucose and mannose, its structure is:

$$
\begin{array}{c}
\text{H}_2\text{COH} \\
| \\
\text{C}=\text{O} \\
| \\
\text{HOCH} \\
| \\
\text{HCOH} \\
| \\
\text{HCOH} \\
| \\
\text{H}_2\text{COH}
\end{array}
$$

D-Fructose

Although Fischer continued his studies on sugars for several years, the results merely strengthened his conclusions regarding structure. Alfred Wohl's procedure for removing the carbonyl carbon atom from aldose sugars by degradation of the oxime was used to confirm the relation between glucose and arabinose. Otto Ruff, one of Fischer's students, developed another degradation method in 1898, in which the carbon atom was removed by the action of hydrogen peroxide.

Fischer established the practice of writing sugar structures as if all members of the D-series derive from D-glyceraldehyde, where the OH group on the carbon atom next to the primary alcohol group is written to the right when the carbonyl group is at the top. He used D to refer to this particular orientation; hence it lost its significance as indicating the direction of rotation. Frequently, a D-compound rotates polarized light to the left, a fact which is conventionally indicated by (−), as in D(−) fructose. Dextro rotations are similarly indicated by (+), as in D(+) glucose. The actual arrangement of the groups in space around the asymmetrical carbon atoms was not known with certainty until recently, but Fischer's concept of their relative orientation was correct. (See Chapter 23.) Fischer carried out his experiments on the assumption that the existing orientations around asymmetrical carbon atoms remained unchanged when new compounds were formed. Although discovery of the Walden inversion threw doubt on this assumption, there has been no evidence of this inversion in Fischer's sugar conversions.

Since he was interested in all aspects of sugar chemistry it was natural

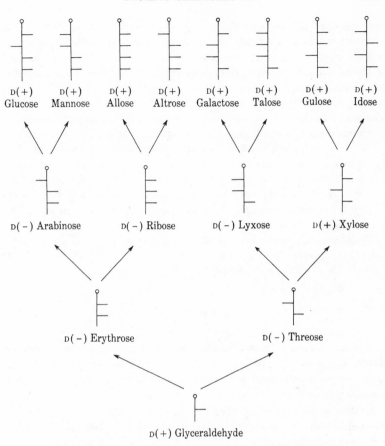

Fig. 13.4. FISCHER'S CORRELATION OF SUGAR STRUCTURES
TO AN ARBITRARY CONFIGURATION FOR D-GLYCERALDEHYDE.

for Fischer, despite a breakdown in health after being poisoned by phenylhydrazine, to turn to studying the type of combination which occurs in
disaccharides and glucosides. The glucosides are a class of natural substances like amygdalin, indican, and salicin which give glucose on being
hydrolyzed. In 1893 Fischer succeeded in condensing glucose and methyl
alcohol in the presence of anhydrous hydrogen chloride and obtained
methyl glucoside. A year later a second methyl glucoside was prepared by
W. Alberda van Ekenstein (1858–1937). This, together with such other
evidence as mutarotation and the feeble aldehydic properties of glucose,
suggested the tenability of some kind of ring structure such as Tollens had
postulated in 1883. Studies of the two methyl glucosides, named alpha
and beta, indicated the existence of an oxygen-containing ring in the

hexoses, thus creating asymmetry at a fifth carbon atom and explaining the existence of α- and β-glucosides, the possibility of mutarotation, and the apparent nonexistence of a reactive carbonyl group. Fischer and his students, Zach, Thierfelder, E. F. Armstrong, and B. Helferich now studied the stereospecificity of enzymes. Maltase, prepared from air-dried yeast, hydrolyzed α-methyl glucoside but was inert toward β-methyl glucoside. Emulsin, an enzyme in bitter almonds, showed the opposite behavior. Both enzymes were indifferent toward L-glucosides, D-galactosides, arabinosides, xylosides, rhamnosides, and gluco-heptosides. Lactose was hydrolyzed by emulsin but not by maltase; maltose was hydrolyzed by maltase but not by emulsin. These observations led Fischer to conclude that the two enzymes might be used to distinguish between α- and β-glucoside linkages. He looked upon the enzyme-substrate relationship as analogous to that between a lock and key.

Ring structure proved to be a formidable problem which was not satisfactorily solved until three decades later. Since this brings us well into the twentieth century it will be discussed later.

PURINES

Another group of natural substances whose structure was clarified during this period was the purines. Here again Emil Fischer was prominent; he first studied caffeine in 1881 and continued his investigations of these compounds until 1914.

Uric acid had been isolated from urinary calculi by Scheele in 1766, but its composition was not established until the 1830's. Liebig and Wöhler conducted extensive studies of uric acid and related compounds around 1837, and were responsible for the nomenclature associated with members of the purine group. Strecker and Baeyer each studied these compounds further. Ludwig Medicus (1847–1915), professor of chemistry and pharmacy at Würzburg, suggested a formula for uric acid in 1875.

Caffeine, which Fischer began to study six years later, had been isolated from coffee by Runge in 1820 and—as thein—from tea by Oudry in 1827. Jobst demonstrated the identity of the two substances in 1838. Fischer's early studies also included xanthine, theobromine, and guanine. Xanthine had been known since 1817, when Marcet reported it. Theobromine was discovered in cocoa by Woskressensky in 1842, guanine in guano by Unger in 1844. Kossel isolated adenine from pancreas and theophylline from tea not long after Fischer began his work. Xanthine and the closely related hypoxanthine, adenine and guanine were all associated with the blood, glandular tissues, and urine of animals. Caffeine, theobromine, and theophylline were obtained from plants and were stimulants and diuretics. Hence, both groups of compounds were of great interest.

The parent compound, purine, was derived from uric acid in 1897. From it Fischer established the structure of the class and systematized the nomenclature of the purine nucleus.

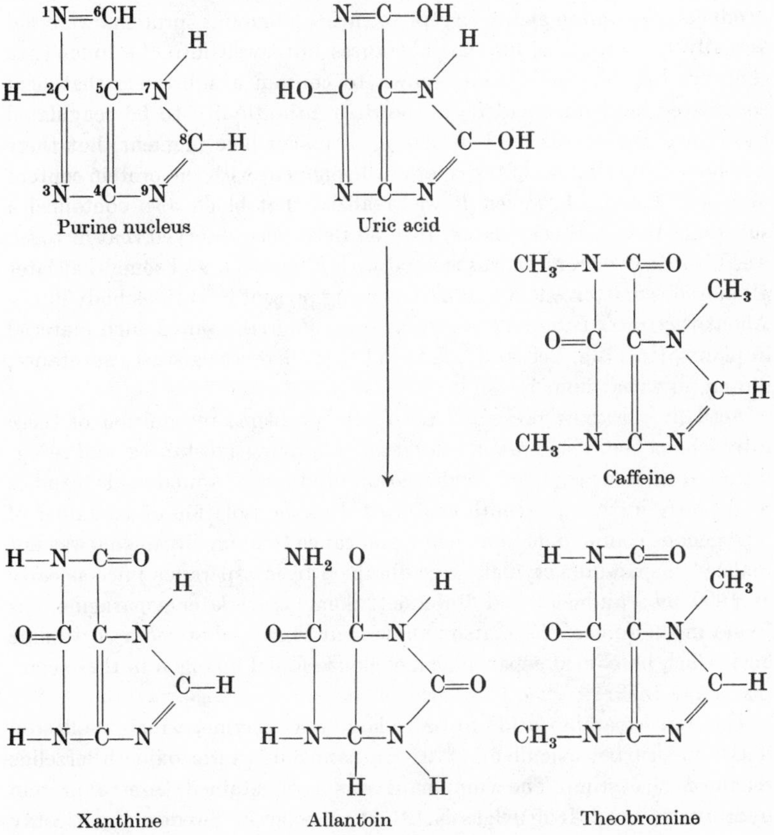

Purine nucleus

Uric acid

Caffeine

Xanthine

Allantoin

Theobromine

Thus the relationship between purines and their decomposition products, which Baeyer had partially clarified in the 1860's, was definitely established by Fischer and his collaborators before the new century began. They synthesized various purines and elaborated the chemistry of the group as a whole.

Uric acid proved to be the form in which purine nitrogen is eliminated in man and a few other animals. In most animals, however, it is decomposed further to allantoin before elimination. Uric acid was found to be the major end product of nitrogen metabolism in birds and reptiles; hence the occurrence of it and other purines in guano deposits. Snakes eliminate some uric acid with their skins when shedding. Urea is the form in which mammals eliminate the nitrogen derived from the breakdown of proteins.

PROTEINS

Closely allied to the purines are the proteins and their hydrolysis products, the amino acids. Except for a few albumins, proteins were not actually recognized as unique substances until well into the nineteenth century. Egg and milk were known to contain a substance that heat coagulated, and milk had been known from antiquity to be coagulated by rennet. Blood was used in making Prussian blue pigment, but there was no real association of the color of the pigment with the protein content of blood. Finally, however, it was realized that blood also contained a substance that heat coagulated. In 1780 Berthollet observed that nitrogen was liberated when meat was treated with nitric acid, and somewhat later albuminous matter was recognized as being present in various body fluids. About the turn of the nineteenth century, Fourcroy noted such material in plants. (In 1728, Beccaria had found that there was a sticky substance, gluten, in wheat flour.)

Actually, proteins presented a twofold problem: recognition of these nitrogenous bodies as unique constituents in living substances, and recognition of them as complex condensation products of amino acids. Studies made early in the nineteenth century led to the isolation of a number of nitrogenous compounds, but their significance to living organisms was not realized. Asparagine crystals were obtained from asparagus juice as early as 1806 by Vauquelin and Robiquet. Some years later asparagine was found in the root of the marshmallow, and Plisson obtained a crystalline acid which he named aspartic acid. Pelouze found nitrogen in these compounds in 1833.

The earliest amino acid[5] to be isolated was cystine, which Wollaston found in urinary calculi in 1810. He named it cystic oxide; Berzelius renamed it cystine. The compound was not obtained from a protein hydrolysate until 1899 when K. A. H. Morner in Sweden and Gustav Embden, working independently, found it in horn. No one had been able to do this before, because the compound was either discarded or reduced in the hydrolytic methods then in use.

Leucine was detected as a fermentation product in cheese by Proust in 1819, and a year later Henri Braconnot (1781–1855) isolated it from muscle fiber and wool. Braconnot also obtained glycine from gelatin at this time, although he was actually studying the acid hydrolysis of sugar-yielding substances. He had obtained sugar from wood, bark, straw, and hemp. He treated gelatin with sulfuric acid, neutralized it with calcium carbonate, filtered it, and evaporated it to a syrup. While it stood for a month, crystals formed which had a sweet taste; Braconnot named the

[5] H. B. Vickery and C. A. L. Schmidt give a good review of the history of the amino acids in *Chem. Reviews*, **9**, 169 (1931).

substance *sucre de gélatine*. However, although nitrogen was detected, the substance attracted interest only because of its sweetness.

Mulder isolated leucine and glycine from gelatin and meat by alkaline hydrolysis in 1838. The correct composition of glycine was determined in 1846 by Horsford, who was working in Liebig's laboratory, and by Laurent and by Mulder. Horsford observed that glycine combined with both acids and bases and proposed that it be named glycocoll to distinguish it from the true sugars; Berzelius suggested the name glycine. Its structure was established by Auguste Cahours, who synthesized it from chloroacetic acid and ammonia and, by treating it with nitrous acid, converted it into glycolic acid.

The structure of the amino acids created considerable difficulty; hence many were not established until the end of the century. Cahours in 1858 concluded that such amino acids as glycine, alanine, and leucine were related to the fatty acids—acetic, propionic, and caproic. However, the branched structure of leucine was not definitely known until 1891.

The proteins themselves remained a problem until late in the nineteenth century.[6] Early in the century only a few albuminous substances were known in a vague way. It was not until 1840, after Mulder proposed his protein hypothesis, that they interested Liebig.[7] Gerardus Johannes Mulder (1802–1880), then a professor of chemistry at Utrecht, suggested that albuminous substances are made up of a common radical, protein, which is combined with various amounts of hydrogen, sulfur, and phosphorus. At first Liebig was attracted by the hypothesis, but within a few years, work in his own laboratory led him to oppose it vigorously. (The word protein, meaning "in the first rank," which Mulder named this class of compounds after a Berzelius suggestion, has remained in use.)

Liebig and his contemporaries, underestimating the complexity of proteins, attempted to derive formulas from analyses of them. But the hydrolytic methods then available were usually severe; they consisted in heating the substance with molten caustic or concentrated sulfuric acid. Only the more stable amino acids could survive this treatment. Liebig isolated tyrosine from such an alkaline hydrolysate in 1846 and Carl Heinrich Ritthausen (1826–1912) obtained glutamic acid twenty years later. Ritthausen studied with Otto Linné Erdmann before becoming associated with the Agricultural Experiment Station at Mockern. Here he carried on experiments on plant products for many years. Eventually it became possible to use mild alkalies and acids in hydrolytic methods; Paul Schützenberger, who was particularly active in this work, succeeded in autoclaving proteins with barium hydroxide.

By 1875 enough was known about the reaction of proteins to heat,

---

[6] E. Farber, *J. Chem. Educ.*, **15**, 434 (1938).
[7] H. B. Vickery, *J. Chem. Educ.*, **18**, 73 (1942).

water, salt solutions, acids, and alkalies to suggest a system of classification to Felix Hoppe-Seyler (1825–1896). In 1856 Hoppe-Seyler served as assistant to Rudolf Virchow (1821–1902), the founder of cellular pathology, in Berlin. He became professor of physiological chemistry at Tübingen in 1864 and at Strassburg in 1872. A leader in the studies being made on blood and metabolism, he founded the *Zeitschrift für physiologische Chemie*.

The next three decades saw great progress in the purification and characterization of proteins. Additional amino acids were discovered and a number of useful color tests were devised. Curtius found that amino acid esters could be distilled at reduced pressure, a fact that Emil Fischer used extensively. Fischer also refined various precipitation methods for separating and purifying the acids.

Fischer then started research on the amino acid linkage in proteins, using synthetic methods as a point of departure. Curtius had already observed that glycine ester eliminates alcohol and forms a dimer, diketopiperazine. Boiling with concentrated hydrochloric acid opens the ring to give glycylglycine, $H_2N—CH_2 \cdot CO \cdot NH \cdot CH_2COOH$.

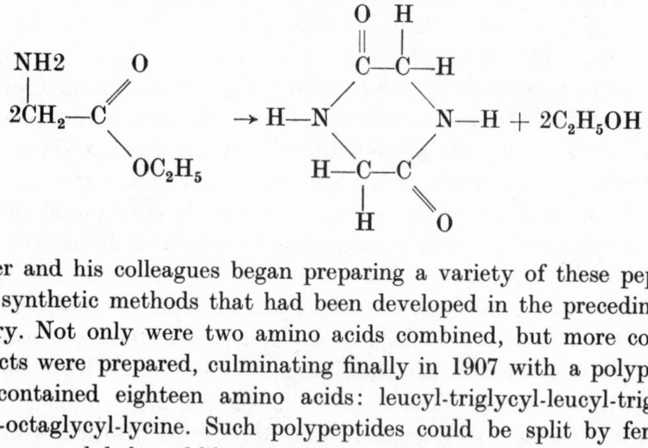

$$2CH_2—C \overset{\displaystyle O}{\underset{\displaystyle OC_2H_5}{<}} \quad \rightarrow \quad H—N \overset{\displaystyle C—C—H}{\underset{\displaystyle H—C—C}{} } N—H + 2C_2H_5OH$$

Fischer and his colleagues began preparing a variety of these peptides, using synthetic methods that had been developed in the preceding half century. Not only were two amino acids combined, but more complex products were prepared, culminating finally in 1907 with a polypeptide that contained eighteen amino acids: leucyl-triglycyl-leucyl-triglycyl-leucyl-octaglycyl-lycine. Such polypeptides could be split by ferments and, in general, behaved like some of the intermediate products of protein hydrolysis.

TERPENES

The fragrant oils that could be removed from plant materials by steam distillation had been known since the sixteenth century, but only a few of them had been subjected to careful chemical examination. The fact that several were used in perfumes and medicine naturally attracted the attention of organic chemists—among them, Baeyer, Tilden, Williams, Tiemann, Wallach, and William Perkin, Jr. C. Greville Williams (1829–

Fig. 13.5. OTTO WALLACH.
(Courtesy of the Richard Fischer Collection.)

1910) was an industrial chemist who for a time was associated with the elder Perkin. William Augustus Tilden (1842–1926) turned from pharmacy to chemistry as a result of attending Hofmann's lectures, and ultimately succeeded to the latter's chair at the Royal College.

Otto Wallach (1847–1931), who was clearly the outstanding figure in terpene chemistry,[8] studied at Berlin under Hofmann and at Göttingen under Wöhler. He became associated with Kekulé at Bonn, and spent a short time with the newly formed Aktien-Gesellschaft für Anilin-Fabrikation (AGFA) in Berlin. He returned to Bonn, where he took over pharmacy instruction after Mohr's death, and in 1889 succeeded Victor Meyer at Göttingen.

The essential oils known as the terpenes and their oxygenated derivatives, the camphors, taxed the ingenuity of organic chemists studying them. Most of these compounds are liquids, and many have approximately the same boiling points. They occur as mixtures. Stereoisomerism is usually possible. Unlike the amino acids, which in nature are all L-forms, the terpenes have no dominant form. For example, α-pinene, the major constituent of oil of turpentine, is *dextro*-rotatory in American turpentine but usually *levo*-rotatory in the European product. Furthermore, common reagents readily cause rearrangement and even ring opening in many of the terpenes.

[8] W. S. Partridge and E. R. Schierz, *J. Chem. Educ.*, **24**, 106 (1948).

As early as 1838 it became apparent that the terpenes were related to the aromatic compounds, for that year Dumas and Peligot converted camphor to p-cymene with phosphorus pentoxide. With the observation that terpenes could be converted into aromatic ring compounds, it naturally followed that many of the terpenes must be ring compounds. However, the aliphatic character of terpene reactions also indicated that the terpenes were more highly saturated than typical aromatic compounds.

The structure of camphor was not worked out until 1893 when Julius Bredt (1855–1937) of Aachen was successful in determining it. In 1903, Gustav Komppa (1867–1949) of Helsingfors succeeded in completely synthesizing the compound.

Wallach in 1887 propounded the isoprene rule which was valuable in showing interrelationships. According to this rule, the terpenes are derived from isoprene, C_5H_8, which is the fundamental unit in them. Isoprene had been isolated from rubber in 1860 by C. Greville Williams, by destructive distillation. In 1875 Bouchardat converted isoprene into dipentene, a constituent of many essential oils. Wallach showed that dipentene is the racemic form of limonene, whose structure he established in 1894.

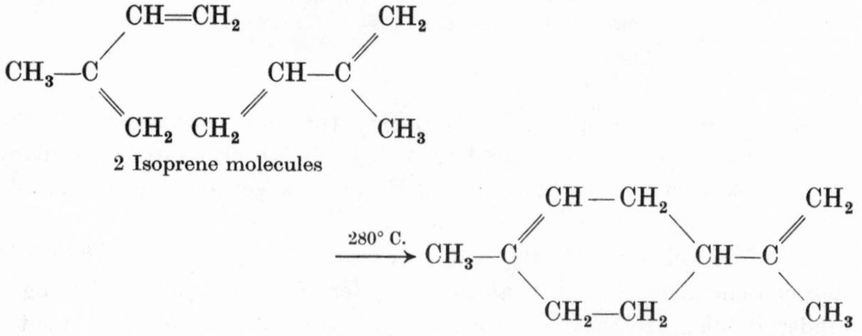

$$CH_3-C\begin{smallmatrix}CH=CH_2\\\\CH_2\end{smallmatrix} \quad CH-C\begin{smallmatrix}CH_2\\\\CH_3\end{smallmatrix}$$

2 Isoprene molecules

$$\xrightarrow{280^\circ\,C.} CH_3-C\begin{smallmatrix}CH-CH_2\\\\CH_2-CH_2\end{smallmatrix}CH-C\begin{smallmatrix}CH_2\\\\CH_3\end{smallmatrix}$$

DL-Limonene
(a dipentene)

CONCLUSION

The structure of various compounds present in biological products was elucidated during the last quarter of the nineteenth century, the work of Emil Fischer on sugars, purines, and proteins, and of Otto Wallach on terpenes being particularly noteworthy. Adolf Baeyer's work on alizarin and indigo might be placed in the same category but is more logically treated in Chapter 17.

Natural products offered very profound structural, analytical, and synthetic problems. The fact that chemists like Fischer and Wallach were able to make rapid progress not only reflects their genius, but also the fact that chemistry had progressed to the point where theory was capable of giving sound guidance.

CHAPTER 14

INORGANIC
CHEMISTRY I.
FUNDAMENTAL
DEVELOPMENTS

As specialized research developed toward the end of the nineteenth century, some investigators began to concentrate on the compounds of elements other than carbon. Actually, inorganic chemistry has never been as well defined as organic because most students of inorganic compounds have been concerned either with analytical problems or physical characteristics. Although this ambiguity has long existed, it is nevertheless appropriate to include in inorganic chemistry all studies that deal with the preparation and properties of compounds other than the hydrocarbons and their derivatives.

Many of the first studies in this field were made by general chemists, men like Berthollet, Fourcroy, Gay-Lussac, Thenard, Dulong, Berzelius, Davy, Faraday, Mitscherlich, Wöhler, and Bunsen. After 1860, however, there was clearly a channeling of effort into either the organic field, which received major emphasis, particularly in Germany, or the nonorganic, in which important but less spectacular work was carried out in many places.

The progress made on atomic weight determinations has already been discussed. The discovery of new elements also has been considered in connection with the development of the spectroscope and the influence of Mendeleev's predictions. The last half of the century was important too for additional discoveries in the rare earth field, the recognition of the rare gas family, and the final success of efforts to isolate fluorine. The discovery of the radioactive elements will be left for a later chapter.

Molecular structure, the great problem facing chemists during the last half of the nineteenth century, defied inorganic chemists, and consequently the discipline was in a chaotic state until 1900. Then Werner's concepts regarding coordination began to guide inorganic chemists to a fruitful course. However, the many years of drifting were damaging, and inorganic chemistry began to reflect a sense of direction only after the first quarter of the twentieth century.

GENERAL DEVELOPMENTS

Allotropy

In his discussion of isomerism in 1830 Berzelius asked whether a similar condition might not be found with respect to elements. He pointed to the existence of carbon in the form of diamond and graphite and to the difference between platinum prepared by the reduction of platinum salts with alcohol and by heating the chloride of platinum. Berzelius failed to pursue the question, despite his own work with silicon and his knowledge of Mitscherlich's preparation of monoclinic sulfur in 1823.

In 1831 Dumas speculated on the matter, noting that in the cases of platinum and iridium, silicon and boron, and molybdenum and tungsten the (then known) atomic weights were either the same or simple multiples. He suggested that the pairs might be different forms of the same element and that the carbon in organic compounds might be different in different parts of the compound, just as the nitrogen in ammonium nitrate might be considered electropositive in the ammonium portion and electronegative in the nitrate portion of the compound.

In the *Jahres-Bericht* for 1841 Berzelius introduced the term *allotropy* (another manner) for those cases where an element is found in different forms. He observed the transformation of white phosphorus to red but the preparation was worked out more systematically by Anton Schrötter in 1845. Metallic phosphorus was prepared by Hittorf in 1865.

A particularly interesting case of allotropy was revealed when a new form of oxygen was reported in 1839 by Christian Schönbein (1799–1868), professor of chemistry and physics at Basel. Using electrical equipment in a poorly ventilated laboratory he observed the characteristic odor of ozone (Gr., I smell), recognized it to be a definite chemical substance, and studied its properties. The odor had been reported as early as 1785 by the Dutch chemist van Marum, but neither he nor later experimenters

Fig. 14.1. CHRISTIAN FRIEDRICH SCHÖNBEIN.
(Courtesy of the Edgar Fahs Smith Collection.)

identified the odor with a unique form of oxygen. At first the odor led Schönbein to associate ozone with the halogens and to revive the earlier notion that the halogens were oxides of hypothetical elements—murium, bromium, etc. He also believed that ozone contained hydrogen, but found this position indefensible when Marignac and de la Rive in 1845 prepared ozone by passing pure dry oxygen gas through an electrical discharge. When it became evident that the gas was an allotrope of oxygen Schönbein suggested that there were three forms of oxygen: ozone, antozone, and ordinary oxygen, the latter a neutralization product of the first two. He diligently searched for antozone during the remainder of his life. The exact nature of ozone was not established until 1868 when J. Louis Soret established the formula O_3 by diffusion studies. This formula had been suggested by Odling in 1861 on the basis of the reaction of ozone with potassium iodide.

Much of Schönbein's research was concerned with oxidation reactions. He believed that oxidations were preceded by ozonization and associated ozone with hydrogen peroxide. Schönbein observed the accelerating effect of iron(II) and lead salts on the decomposition of hydrogen peroxide. He also noted that potatoes, blood, malt extract, and fungi had a similar effect. His observation that hydrogen cyanide inhibited these catalysts led him to suggest that compound as a preservative for meats—a suggestion fortunately not put into practice!

The subject of allotropy became a highly speculative one. Berzelius had suggested that in some cases isomerism might be attributable to allotropy, for example, the two forms of iron(II) sulfide which might contain rhombic sulfur in one case and monoclinic sulfur in the other. It was learned that allotropic changes are accompanied by thermal changes. Mitscherlich reported this in the case of sulfur. Favre and Silbermann determined by calorimetry the differences between heats of combustion of allotropes of the same element. These studies were extended by Berthelot, Troost and Hautefeuille, who pointed out the analogy with heats of fusion and vaporization, and Jungfleisch.

New Compounds

The first half of the century had been very productive in terms of the preparation and study of new compounds. Many leading chemists had contributed to the discovery of new hydrides, halides, oxyacids, and metallic salts as well as unusual compounds of sulfur, phosphorus, and nitrogen. Such investigations continued into the second half of the century, particularly those that sought to clarify the chemistry either of the newly discovered elements or of elements whose chemistry had been too complicated to make sense.

Berzelius' studies on molybdenum compounds were extended in several laboratories. Roscoe studied the chlorides of tungsten, while the complicated chemistry of the tungstic acids was attacked by Margueritte, Scheibler, Marignac, and von Knorre. This, along with the interpretation of the nature of the phospho-molybdic and phospho-tungstic acids, was not very satisfactorily pursued. The chemistry of uranium was pursued by Peligot, Roscoe, and Clemens Zimmerman, and vanadium was extensively studied by Roscoe, Gerland, and von Hauer. Niobium and tantalum chemistry was clarified through the work of Deville and Troost and of Krüss and Nilson.

DISCOVERY OF FLUORINE

Three of the halogens—chlorine, iodine, and bromine—were well known as elements by 1830. The existence of a fourth, fluorine, was also known at this time, but despite repeated efforts the element was not isolated for another half century.

The use of the mineral fluorspar (CaF_2) as a flux was mentioned by Agricola in 1529, and its thermoluminescence was recorded by Elscholtz in 1676. The etching of glass by acidified fluorspar was observed as early as 1670 by Schwanhard, a spectacle maker. A century later Marggraf showed that the mineral was chemically different from heavy spar ($BaSO_4$) and from selenitic spar ($CaSO_4$).

Scheele began an investigation of fluorspar in 1771 which resulted in the discovery of hydrofluoric acid. His retort was corroded when the spar was heated with sulfuric acid, and he showed that the acid of fluorspar could dissolve silica. Soon after this Priestley demonstrated the formation of a gaseous reaction product (SiF_4) from this reaction.

Gaseous hydrogen fluoride was obtained in pure form by Gay-Lussac and Thenard in 1809 when they distilled it in metallic vessels, and Davy published papers on fluorine compounds in 1813 and 1814. He first considered the acid to be similar to sulfuric acid (that is, consisting of an anhydrous nonmetal oxide plus water), but changed his views after he demonstrated the elemental nature of chlorine. Considering the nonmetal in hydrofluoric acid to be a genuine element, he named the substance *fluorine*, following Ampère's suggestion. Davy made determined efforts to prepare fluorine by electrolytic decomposition of its compounds, but succeeded only in ruining his health from their toxic effects.

The toxic effect of fluorine compounds deterred the isolation of the element. Gay-Lussac and Thenard also suffered from inhaling hydrogen fluoride fumes. George and Thomas Knox of the Irish Academy designed an apparatus of fluorspar in order to avoid the corrosive effects on metal containers, but they failed to isolate the element and were badly poisoned. The Belgian chemist Paulin Louyet died as the result of his persistent efforts, as did Jérôme Nicklès of Nancy.

Edmond Frémy (1814–1894) of the École Polytechnique obtained calcium at the cathode during the electrolysis of anhydrous calcium fluoride, while a gas, presumably fluorine, escaped at the anode. In these experiments, carried out in 1854, he was unable to isolate the gas. Fifteen years later a small amount of gas, which reacted explosively with hydrogen, was obtained by George Gore (1826–1908), an English electrochemist.

Success in preparing the element was finally achieved in 1886 by Moissan. A former student of Frémy, he benefited from his teacher's failures, but Moissan too found the problem to be treacherous. After a series of failures, which took their toll in health, he obtained the gas from potassium acid fluoride dissolved in anhydrous hydrofluoric acid. The reaction cell was a platinum V-tube, and the electrodes were made of a platinum-iridium alloy. The whole apparatus was cooled to $-23°$ C. by evaporating methyl chloride. The liberation of fluorine gas at the anode was demonstrated by its vigorous reaction with elemental silicon.

MOISSAN AND THE ELECTRIC FURNACE

Ferdinand Frédéric Henri Moissan (1852–1907), after a term as an apothecary apprentice, entered the Musée d'Histoire Naturelle to study chemistry under Frémy. After receiving his doctorate he taught and carried on research at the École de Pharmacie. His early studies on plant

Fig. 14.2.   HENRI MOISSAN.
(Courtesy of J. H. Mathews.)

chemistry were soon abandoned for the researches which made his laboratory the leading center of inorganic chemistry at the turn of the century. His work on the oxides of the iron group of metals and on pyrophoric metals was followed by an intensive study of fluorine compounds. By using silver fluoride and alkyl iodides as reagents, he prepared alkyl fluorides and greatly expanded the knowledge of fluorine and its compounds. His efforts to prepare diamonds by decomposing fluorides of carbon failed, but he did succeed in designing an electric furnace with which to pursue the problem.

Moissan's electric furnace was a lime block with two carbon electrodes admitted through holes in the top to provide an electric arc at a high temperature. In the diamond experiments iron and sugar charcoal were heated in a carbon crucible which then was plunged into cold water. According to theory, carbon dissolved in iron is subjected to great pressure as the mass solidifies and cools, and thereby particles of carbon coming out of solution should crystallize as diamonds. Moissan recovered flecks of transparent material after dissolving the iron in acids. These particles had a density between 3 and 3.5, scratched ruby, burned to carbon dioxide, and showed octahedral facets. At the time it was assumed that Moissan had successfully synthesized diamonds, a view now considered unsound.[1]

The electric furnace was used extensively for studying the preparation and properties of refractory oxides, metals, carbides, silicides, and borides. Moissan prepared crystalline lime, baryta, magnesia, and strontia. Through electric furnace reduction via the metal oxide and carbon, he isolated metallic chromium, manganese, molybdenum, tungsten, vanadium, uranium, zirconium, and titanium. Distilled copper, silver, gold, platinum, tin, iron, and uranium were prepared and carbon and silicon were volatilized.

Moissan's laboratory attracted many students, including some from other countries. He was awarded the Nobel Prize for chemistry in 1906.

THE RARE GASES

In 1890 no one anticipated the discovery of new elements in the atmosphere. Its composition had been carefully studied during the previous century and found to consist primarily of nitrogen and oxygen, with traces of carbon dioxide and water vapor and even more minute traces of compound gases such as ammonia. Bunsen established in 1846 that the oxygen content of air varies between 20.84 and 20.97 per cent. This variability was confirmed by Regnault, Angus Smith, A. R. Leeds, and Phillip von Jolly.

In 1882 Lord Rayleigh, stimulated by Prout's hypothesis, began measuring the relative densities of hydrogen and oxygen. Rayleigh, born John William Strutt (1842–1919), was Cavendish Professor of Physics at Cambridge at the time, but soon assumed the directorship of the Royal Institution. He was widely respected for his work on sound and optics when he began his studies on gas density measurements and the compressibilities of gases. In 1892 he published his final results which showed oxygen to be 15.882 times more dense than hydrogen. He used several methods to prepare oxygen—electrolysis of water, thermal

---

[1] F. P. Bundy, H. T. Hall, H. M. Strong, and R. H. Wentorf, *Nature,* **176,** 51 (1955).

Fig. 14.3. WILLIAM RAMSAY.
(Courtesy of the James H. Walton Collection.)

decomposition of chlorates, and thermal decomposition of potassium permanganate. In all cases there was no density variation beyond experimental error.

In similar studies on the density of nitrogen, however, Rayleigh found that nitrogen prepared from ammonia was lighter than nitrogen prepared from the air. Since the discrepancy was outside experimental error, he reported the results in *Nature*[2] and asked for comments. None was received. Rayleigh explored several ideas of his own, such as dissociation of N_2, contamination with residual hydrogen in the ammoniacal nitrogen, or the presence of N_3 (as an analogue to ozone), and ultimately found each hypothesis untenable.

William Ramsay discussed the subject with Rayleigh who encouraged him to examine the problem from a chemical standpoint. Ramsay ultilized information he had gained in earlier experiments where nitrogen and hydrogen were passed over heated metals in an effort to prepare ammonia.

[2] J. W. Strutt, *Nature*, **46,** 512 (1892).

Hot magnesium absorbed the nitrogen, forming the solid nitride. When atmospheric nitrogen (first freed of oxygen by passing air over hot copper) was passed over hot magnesium, the density of the "uncombined nitrogen" increased.

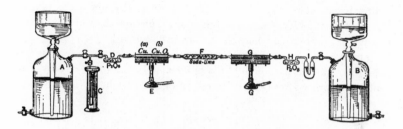

Fig. 14.4. RAMSAY'S APPARATUS FOR REMOVAL OF NITRO-
GEN AND OXYGEN FROM THE ATMOSPHERE.
(From W. Ramsay, *The Gases of the Atmosphere* [1896].)

In large-scale experiments atmospheric "nitrogen" was passed through a purification train of phosphorus pentoxide to remove moisture, heated copper and copper oxide to oxidize organic dust and contaminants, and soda-lime to remove carbon dioxide before being passed over hot magnesium to remove nitrogen. During the course of lengthy experiments the volume of uncombined gas was reduced to one eightieth of that of the original "nitrogen." The density of the gas was then 19.086 compared to hydrogen gas as 1. After their efforts to remove nitrogen, some of the gas was placed in a Plücker tube to examine the spectrum. In addition to the character-istic bands of the nitrogen spectrum there were clearly apparent lines corresponding to no known gas.

Rayleigh, meanwhile, prepared nitrogen from various chemical com-pounds and in all cases found it uniformly less dense than atmospheric "nitrogen." He then repeated Cavendish's work, sparking air enriched with oxygen over caustic solution to obtain the trace of residual gas Cavendish had observed. The amount of residual gas was too small for further studies, but during the sparking Rayleigh noted that the character of the spectrum was unusual. He also observed that the amount of residual gas remaining after the experiment was approximately proportional to the quantity of air used.

Both Rayleigh and Ramsay had assumed that they would isolate N_3 as an allotrope of nitrogen; now they independently concluded that they had discovered a new element in the atmosphere, which they reported in a joint paper submitted to the British Association at its Oxford meeting in August 1894. The gas was named *argon*, from the Greek, meaning "the lazy one." Their report was received with a good deal of incredulity.

They had not yet proved that the new gas was a normal constituent of the atmosphere, not something formed during chemical experiments. Two techniques were utilized in efforts to resolve this question. In one gaseous diffusion through clay tobacco pipes was used to separate the lighter from the heavier molecules. In the other, advantage was taken of the greater solubility of argon in water. Each method showed an increased density in the fraction where the argon was expected.

Ramsay now set out to prepare argon in large quantities, a tedious task. (The amount of argon in the air was ultimately shown to be only 1 per cent.) Two procedures were followed. In one, a variant of the Cavendish

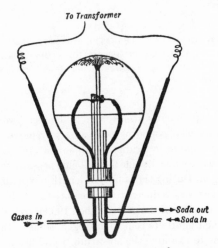

Fig. 14.5. RAMSAY'S SECOND APPARATUS FOR REMOVING
COMPONENTS FROM THE AIR.
(From W. Ramsay, *The Gases of the Atmosphere* [1896].)

experiment, oxygen-enriched air was sparked between platinum electrodes in a flask in which the oxides of nitrogen were continuously removed by a spray of sodium hydroxide solution. In the other, air was purified in a train in which oxygen was removed over hot copper and nitrogen over hot magnesium.

The properties of argon were quickly examined. Using a sample supplied by Ramsay, William Crookes plotted nearly 200 spectral lines. The gas was exposed to a variety of elements and compounds under mild and drastic conditions, with no evidence of chemical reaction. Workers in Ramsay's laboratory determined the specific heat at constant pressure

and constant volume and thereby established the monatomic character
of argon gas. A sample sent to Olszewski at Cracow was liquefied; a
constant boiling point at −186.9° C. was observed, and Olszewski concluded that the sample was that of a pure substance.

Soon after he had reported the discovery of argon, Ramsay learned
that in 1888 the American geochemist William Hillebrand had obtained
an inert gas, assumed to be nitrogen, upon heating uraninite with sulfuric
acid, Ramsay immediately heated cleveite, a related mineral, in similar
fashion and obtained an inert gas that showed not the expected argon
spectrum, but a bright yellow spectral line near, but not identical with,
the D lines of sodium. Samples of the gas were examined by Crookes
and by Lockyer who reported that the line coincided with the D$_3$ line
attributed to helium in the solar spectrum.

The D$_3$ line was first observed in the spectrum of the solar chromosphere
by the French astronomer Pierre Jules César Janssen during the solar
eclipse that occurred in India in 1868. He believed the line attributable
to hydrogen or to water vapor, but it could not be reproduced in the
laboratory. In England the same line was studied by the astronomer
Joseph Norman Lockyer and the chemist Frankland. Since the line could
not be attributed to any known element, they decided that it indicated
an element characteristic of the sun. Frankland proposed the name
helium, after the Greek *helios*, for sun.

The gas was prepared in quantity sufficient for study by Ramsay and
his co-workers Norman Collie and Morris William Travers who found it
to be a component of various minerals and mineral waters. Only later was
helium identified in the air by H. Kayser of Bonn and confirmed by
Siegfried Friedlander. It proved to be chemically inert and monatomic,
like argon. Cleve and Nils Abraham Langlet of Sweden were independent
discoverers of the gas in heated minerals.

When he announced the discovery of argon, Ramsay discussed its
position in the periodic table, suggesting that it might fall in a new column
between chlorine and potassium. Even though its atomic weight was
greater than potassium, it obviously did not belong between that element
and calcium, so a new group had to be created for it. In later speculations
Ramsay suggested that there might be other inert gases. In the first
edition of his *Gases of the Atmosphere* (1896) he included a periodic table
suggested by a former student Orme Masson of Melbourne. The pertinent
part of the table is shown on page 374.

Ramsay first sought the postulated elements in minerals, without
success. Next Ramsay tried systematic diffusion of argon gas and obtained
light and heavy portions. He decided to prepare a large quantity of argon
by fractionation of liquid air provided by William Hampson. In preliminary experiments Ramsay and Travers examined the last cubic centimeter
of liquid air remaining after a liter of the material had boiled away.

The spectrum of the resulting gas was rich in the lines of argon but also contained new yellow and green lines. The density of the gas was 22.47, higher than for any previous sample of argon; it was reported in 1898 as *krypton*, meaning "hidden."

Table 14.1. PARTIAL PERIODIC TABLE, RAMSAY, 1896

Hydrogen	1.01	Helium	4.2	Lithium	7.0
Fluorine	19.0	?		Sodium	23.0
Chlorine	35.5	Argon	39.2	Potassium	39.1
Bromine	79.0	?		Rubidium	85.5
Iodine	126.0	?		Cesium	132.0
?	169.0	?		?	170.0
?	219.0	?		?	225.0

From liquefied argon Ramsay and Travers boiled off a first fraction with a complicated spectrum. This gas was named *neon*, for "new." Finally, they isolated a heavier gas which was called *xenon*, for "stranger." The various gases were all prepared in pure form from liquid air and proved to be monatomic and chemically inert. A sixth rare gas was discovered later, as a component not of air but of radioactive materials; it will be considered in Chapter 18.

THE RARE EARTHS

Although investigation of the rare earth elements began in the eighteenth century, only at the end of the nineteenth century did order begin to emerge out of chaos. These elements are so similar in properties that their separation was accomplished only with tedious work. Mixtures were repeatedly reported as new elements, only to be fractionated further in later studies; and the elements themselves were obtained in reasonably pure form quite late in the nineteenth century. The metals were first recognized in the form of the oxides which, because they resemble the alkaline earth oxides, were spoken of as "earths." The term "rare" was used because for a time these elements were believed to occur in the earth's crust in trace amounts. Actually, some of the elements are moderately abundant.

In 1787 Carl Axel Arrhenius (1757–1824), a Swedish army lieutenant and amateur mineralogist, discovered a rock he named *ytterite* (now known as *gadolinite*) in a feldspar mine at Ytterby near Stockholm. This dense black mineral was analyzed in 1794 by Johan Gadolin (1760–1852), professor of chemistry at the University of Åbo in Finland, who found it to contain a new earth. Gadolin was the foremost Finnish chemist of his day. He had studied under Bergman and hence had been a phlogistonist until extensive study of Lavoisier's concepts converted him to the new chemistry. His textbook (1798) was the first in the Swedish language that dealt with chemistry from the new viewpoint.

The earth was investigated further by Ekeberg who confirmed Gadolin's work and introduced the name *yttria*, and Klaproth and Vauquelin also studied the earth. None of these investigators was able to separate a metal or even obtain the earth in a satisfactory state of purity.

In 1803, while investigating a heavy Swedish stone from Bastnas, Klaproth isolated a somewhat similar earth which he called *terre ochroite*. Berzelius and Hisinger discovered this earth at the same time and labelled it *cerium*, after the asteroid Ceres, discovered shortly before by Piazza. Klaproth and the Berzelius-Hisinger team had examined the mineral from Bastnas expecting to find yttria.

Vauquelin and Gahn's efforts to isolate metallic cerium were unsuccessful. Mosander later obtained a highly impure brown powder after passing potassium vapor over anhydrous cerous chloride. Wöhler also obtained an impure sample. In 1875 the Americans W. F. Hillebrand and Thomas H. Norton obtained a more satisfactory sample by electrolyzing molten cerous chloride. Finally, in 1911 an almost pure sample was prepared by Alcan Hirsch at the University of Wisconsin.

In 1839 Carl Gustav Mosander (1797–1858) established that ceria was not a pure oxide. He decomposed cerium nitrate by heat and found that only a portion of the oxide dissolved readily in dilute nitric acid. The soluble oxide was called *lanthana* (hidden); *ceria* continued to be used as the name for the insoluble oxide. Mosander was curator of minerals for the Swedish Academy of Sciences and professor of chemistry and mineralogy at the Caroline Institute. For many years he had been a student of and assistant to Berzelius, and when the latter retired from teaching, Mosander succeeded him at the Institute. Lanthana was independently discovered in 1839 by Axel Erdmann, one of Sefström's students, in *mosandrite*, a new Norwegian mineral.

Two years later, after difficult and discouraging studies, Mosander separated the lanthana fraction into two earths, *lanthana* and *didymia* (inseparable twin brother). The didymia gave rise to pink salts; the pinkness previously observed in cerium and lanthanum had resulted from this contamination.

Mosander turned next to yttria since he suspected that this too might be heterogeneous. His hunch proved correct. First he separated contaminant ceria, lanthana, and didymia. By 1843 he had separated the purified yttria into three earths: one of them colorless, for which he retained the name *yttria*; the second rose colored, named *terbia*; the third yellow, named *erbia* (all names after the Swedish town of Ytterby).

Mosander's work was confirmed in France by Delafontaine, Marignac,

Table 14.2.   DISCOVERY OF THE CERIUM EARTHS[a]

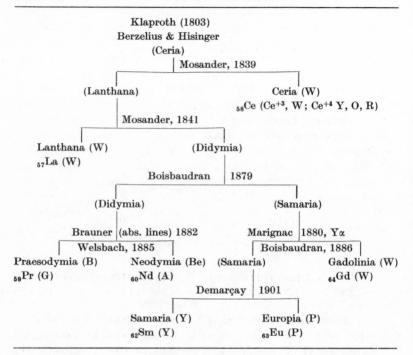

[a] Names of the earths (oxides) are used throughout the separation and discovery scheme rather than the names of the elements. The names of impure earths are enclosed in parentheses. The subscript preceding the symbol of the element is the atomic number. The letter in the parenthesis following the name of the earth indicates the color of the earth, and that following the symbol of the element gives the color of the salts, as follows: W = white or colorless, B = black, Y = yellow, Be = blue, P = pink, A = amethyst, R = rose, O = orange, and G = green.

and Boisbaudran, in America by J. Lawrence Smith, and in Sweden by Cleve. However, the names for erbia and terbia somehow became interchanged during the next forty years. Yttria, which yielded a crude metal to Wöhler in 1828, is not actually a member of the rare earth series, but the metal, because of its atomic diameter and atomic structure, behaves like the rare earth elements and is found associated with them.

By 1845 it was becoming evident that the chemistry of these elements was troublesome. The compounds closely resembled each other with respect to solubility and other properties; yet they showed recognizable differences. When Mendeleev and others attempted to classify all elements, the rare earths caused difficulties. Only several decades later was it recognized that these elements might best be handled by assigning all of them the same spot in the periodic table.

Table 14.3.   Discovery of the Yttrium Earths[a]

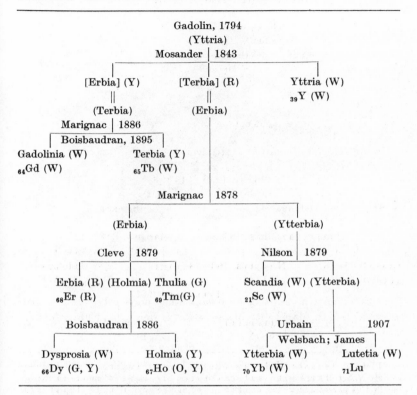

[a] See Table 14.2 regarding use of terms.

There was no significant progress in unraveling the chemistry of the rare earths from the forties, when Mosander was active, until the seventies, when Marignac concentrated on the subject. In 1878 he showed that erbia could be separated into two earths; the red earth he continued to call *erbia*, the colorless one he named *ytterbia*. A year later Nilson separated Marignac's *ytterbia* into *ytterbia* and *scandia*, the oxide containing Mendeleev's postulated *Eka-Boron*. It, like yttrium, is not strictly a member of the rare earths, but its properties are similar and it is found in admixture with these elements.

Marignac's erbia was, however, still not a pure oxide. Per Theodor Cleve (1840–1905), a professor at Uppsala, observed that atomic weight determinations of erbium were not constant and thereupon separated the earth into three fractions: *erbia, holmia* (named for Stockholm, Cleve's birthplace), and *thulia* (after Thule, the ancient name for Scandinavia).

Fig. 14.6. CHARLES MARIGNAC.
(Courtesy of J. H. Mathews.)

The spectral absorption bands for holmium had previously been detected in Switzerland by Marc Delafontaine and J. L. Soret who had postulated existence of an element X.

In 1879 Mosander's didymia was subjected to further investigation. As early as 1853 Marignac had suggested that didymia was an impure oxide. Now, independent spectral studies by Delafontaine and Boisbaudran showed that the spectral lines varied with the source of the oxide.

Boisbaudan separated another earth from didymia; the new earth was named *samaria*, from a samarskite, the mineral from which the impure didymia had been prepared.

In 1880 Marignac separated from samarskite still another earth which he designated "Yα." Yα was obtained in 1886 from samaria by Boisbaudran. Marignac assented to Boisbaudran's suggestion that the earth be named *gadolinia*, after gadolinite, the mineral from which it was obtained, and thereby after Johan Gadolin. Gadolinia was also separated from Mosander's yellow terbia in 1886. Samaria was further separated into samaria and europia by Eugene-Anatole Demarçay (1852–1904) at the École Polytechnique in 1901.

Marignac and Boisbaudran, and Cleve and the Bohemian chemist Bohuslav Brauner suspected that Boisbaudran's didymia was a mixture. Brauner noted in 1882 that didymium salts gave absorption spectra in the blue and the yellow. Carl Auer (1858–1929), who later became Baron von Welsbach, succeeded in 1885 in separating didymia into two new earths which he named *praesodymia* (green didymia after its leek-green salts) and *neodymia* (new didymia). Welsbach, who had studied under Bunsen, had a deep interest in the rare earths which eventually resulted in their being used in mantles for gas burners. The demand for ceria for his mantles made rare earth residues available to chemists in quantity.

Cleve's holmia was further fractionated in 1886 by Boisbaudran into *holmia* and *dysprosia* (for the Greek *dysprositos*, hard to get at).

Nilson's ytterbia also proved to be heterogeneous. In 1907 Georges Urbain (1872–1938) at the Sorbonne succeeded in making the separation into *neoytterbia* (which became merely *ytterbia*) and *lutecia* (now called *lutetia*, after an old name for Paris). These two earths were independently separated by Welsbach who named the elements *aldeberanium* and *cassiopeum*. Pure lutecia was also prepared independently in America by Charles James (1880–1928), a former student of Ramsay who had crossed the Atlantic to take a professorship at the University of New Hampshire where he made important contributions to rare earth research.

The rare earth picture was now complete except for element number 61. However, before the physical validity of atomic numbers was established and Bohr's concept of atomic structure was developed, there could be no clear inkling that one of the elements was missing. In fact, there was no assurance that there might not be several unknown rare earth elements. Clarification of this point could only come with the work of Moseley and Bohr in the next two decades.

WERNER AND THE COORDINATION COMPLEXES

Frankland's research on metallo-organic compounds permitted the formulation of the valency concept which soon was widely recognized as invaluable. In organic chemistry it permitted the rise of structural concepts, which resulted particularly from application of principles of atomaticity (valence) by Kekulé who held that the valence of an element was as constant as the atomic weight.

In the inorganic field the application of valence principles was less successful. Simple valence rules worked well with simple salts, but many salts were established as complex when subjected to careful analysis. Structural formulas for inorganic compounds were seldom completely satisfactory. Indeed, while structural principles opened the door to spectacular progress in organic chemistry, they actually retarded the theoretical growth of inorganic chemistry. Hydrates, ammonia complexes, cyanides, and double salts were handled only with great difficulty. To add to the confusion, many of the metals formed compounds where the valence of the metal appeared not to be constant.

In order to preserve constancy of valence, various devices were used, for example:

Compounds	Valence	Structure Proposed

$FeCl_3$
$FeCl_2$ $Fe = 3$

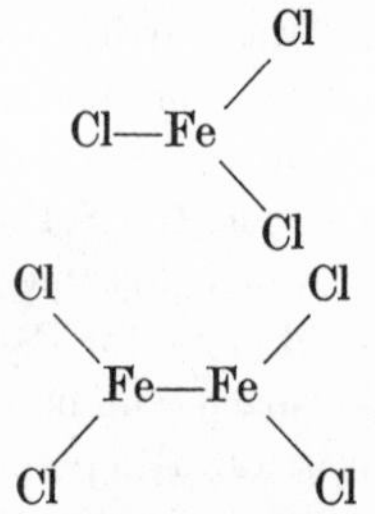

$$
\begin{array}{c}
Cl \\
Cl\!-\!Fe \\
\quad\diagdown \\
\qquad Cl \\
Cl \qquad\qquad Cl \\
\diagdown \qquad\quad \diagup \\
Fe\!-\!Fe \\
\diagup \qquad\quad \diagdown \\
Cl \qquad\qquad Cl
\end{array}
$$

$SnCl_4$
$SnCl_2$ $Sn = 4$

$$
\begin{array}{c}
Cl \qquad Cl \\
\diagdown \quad \diagup \\
Sn \\
\diagup \quad \diagdown \\
Cl \qquad\qquad Cl \\
Cl \qquad\qquad Cl \\
\diagdown \qquad\quad \diagup \\
Sn\!=\!Sn \\
\diagup \qquad\quad \diagdown \\
Cl \qquad\qquad Cl
\end{array}
$$

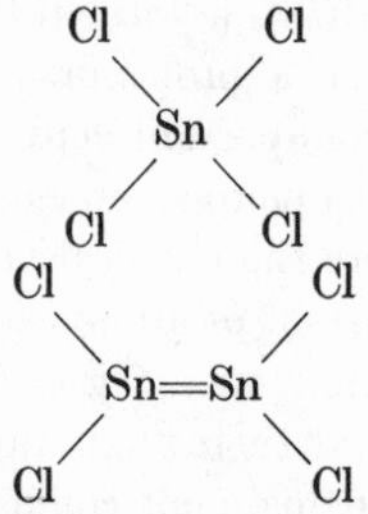

However, molecular weight determinations by the newer methods gave results inconsistent with certain postulated structures.

Formulas for acids had first been written on the basis of "chain" structures, but in certain quarters these gave way to "central" structures.

Empirical	Chain	Central

H_2SO_4 $H\!-\!O\!-\!O\!-\!S\!-\!O\!-\!O\!-\!H$

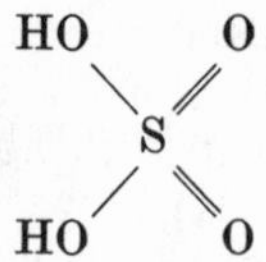

$$
\begin{array}{c}
HO \qquad\quad O \\
\diagdown \quad \diagup\!\!\diagup \\
S \\
\diagup \quad \diagdown\!\!\diagdown \\
HO \qquad\quad O
\end{array}
$$

$HClO_4$ $H\!-\!O\!-\!O\!-\!O\!-\!O\!-\!Cl$

$$
\begin{array}{c}
O \\
\diagup\!\!\diagup \\
HO\!-\!Cl\!=\!O \\
\diagdown\!\!\diagdown \\
O
\end{array}
$$

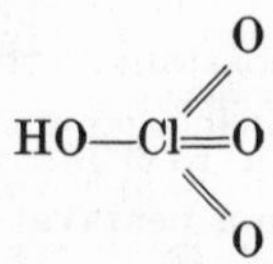

Although the tetrammine form of copper salt had been known to Libavius, compounds of this sort did not receive much attention until the nineteenth century. In 1798 Tassaert[3] observed that ammonia combined with cobalt salts. Real attention began to be given to such compounds after 1822,

---

[3] Tassaert, *Ann. chim.*, [1], **28**, 92 (1798).

when L. Gmelin prepared crystalline *luteo*-cobaltic oxalate, $(Co(NH_3)_6)_2$ $(C_2O_4)_3$, from an ammoniacal solution of a cobalt salt. Frémy in 1851 prepared a *purpureo*-cobaltic chloride, $(Co)(NH_3)_5Cl)Cl_2$, in which only a part of the chloride precipitated upon addition of silver nitrate solution. The remaining chloride could be precipitated only after prolonged boiling. During the next four decades there were extensive studies on the ammines of cobalt, chromium, and platinum, especially by O. Wolcott Gibbs (1822–1908) at Harvard, Blomstrand at Lund, and Sophus Mads Jørgensen (1837–1914) at Copenhagen.

Theoretical attempts to deal with these compounds date from 1819, when Berzelius[4] advanced his conjugate theory which suggested that a metal in conjugation with ammonia was still capable of combining with other substances, but his attempt to extend the theory to double salts and complex cyanides was unsuccessful. Two decades later Graham[5] advanced a theory that suggested that the ammoniated compounds were ammonium compounds in which one or more hydrogen atoms had been substituted by a metal atom. This theory was refined and extended by Rieset, Gerhardt, Wurtz, Hofmann, and Boedecker. Hofmann formulated $Co(NH_3)_6Cl_3$ as:

$$Co\left(\begin{array}{c} NH_2 \!-\! NH_4 \\ | \\ Cl \end{array}\right)_3 .$$

Jørgensen later showed the formulation to be unsound since it failed to account for the existence of similar complexes of tertiary amines and failed to show why the abstraction of one molecule of ammonia from the compound caused one of the chlorine atoms to become unreactive.

Klaus suggested in 1854 that ammonia combined with metals without influencing their capacity for further combination, the ammonia becoming passive in the process. The Klaus theory was violently attacked and received little further consideration, particularly after Blomstrand introduced the chain theory.

To deal with metal ammine salts Christian Wilhelm Blomstrand (1826–1897) in 1869 introduced his chain formulation in which the nitrogen took on a pentavalent state. At the time the only method for molecular weight determinations was vapor density measurement. Molecular weights of nonvolatile compounds were not determinable until the studies of Raoult and van't Hoff led to the introduction of methods based on the effect of solutes on the physical properties of solvents (after 1882, see

[4] J. J. Berzelius, *Essai sur la theorie des proportions chimique et sur l'influence chimique de l'electricitie*, Mequignon-Marvis, Paris, 1819.
[5] T. Graham, *Elements of Chemistry*, H. Baillière, London, 1837.

Chapter 15). Hence, since vapor density measurements indicated the formula Fe_2Cl_6 for iron(III) chloride, it was assumed that the formula of nonvolatile cobalt(III) chloride was Co_2Cl_6. Blomstrand therefore considered the formula of the *luteo* salt [hexammine cobalt(III) chloride] to be $Co_2Cl_6 \cdot 12NH_3$ and formulated this as:

$$Co_2\begin{cases} NH_3-NH_3-Cl \\ NH_3-NH_3-Cl \\ NH_3-NH_3-Cl \\ NH_3-NH_3-Cl \\ NH_3-NH_3-Cl \\ NH_3-NH_3-Cl \end{cases}$$

When *luteo* salt is heated, one-sixth of the ammonia is lost, one-third of the chloride no longer precipitates on addition of silver nitrate, and the *purpureo* salt [chloro pentammine cobalt(III) chloride] is formed. In 1884 Jørgensen, and also E. Petersen, obtained evidence for monomeric formulas for cobalt(III) chlorides on the basis of freezing point and conductivity measurements. Henceforth, the formula $CoCl_3 \cdot 6NH_3$ was accepted for the *luteo* salt. Jørgensen[6] formulated this as:

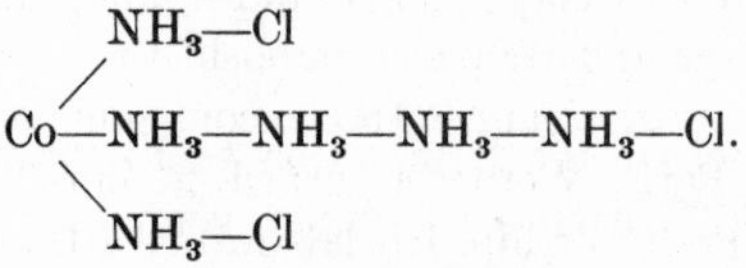

$$Co\begin{matrix} NH_3-Cl \\ NH_3-NH_3-NH_3-NH_3-Cl. \\ NH_3-Cl \end{matrix}$$

When Jørgensen observed that heating the *luteo* salt produced the *purpureo* salt, he postulated that the *purpureo* salt differed in having one chlorine atom bonded directly to the metal, making the chlorine unavailable for precipitation by silver nitrate.

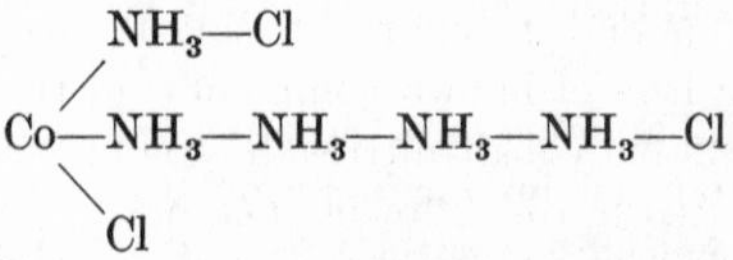

$$Co\begin{matrix} NH_3-Cl \\ NH_3-NH_3-NH_3-NH_3-Cl \\ Cl \end{matrix}$$

A fresh approach to the whole problem was introduced in 1892 by Alfred Werner (1866–1919), then a newly appointed lecturer at the University of Zurich. Werner was born in the Alsatian city of Mulhouse shortly before this region was taken by the Germans. He studied at the

[6] For a good account of Jørgensen's handling of the chain theory see J. B. Kaufmann, *J. Chem. Educ.*, **36**, 521 (1959).

Fig. 14.7. ALFRED WERNER.
(From Henry M. Smith, *Torchbearers of Chemistry* [1949], by courtesy of Academic Press, Inc.)

Polytechnicum in Zurich and then took a degree at the University where he was an assistant to George Lunge and studied under Hantzsch and Treadwell. In 1890 he and Hantzsch published the results of their study of the stereochemistry of organic nitrogen compounds. After further study with Berthelot in Paris, Werner returned to Zurich where he quickly advanced to a full professorship. His laboratory attracted many graduate students; Werner received the Nobel Prize in 1913.

Werner's first paper on the structure of inorganic compounds dealt with ammonia complexes. He continued to publish on this and similar subjects during the remainder of his life. His *Neuere Anschauungen auf dem Gebiete der anorganischen Chemie* (1905) systematically examined the whole subject. Werner, who devised his concepts by taking heed of the conductivity behavior and geometry of central metal atoms, began by carefully tabulating lists of known compounds of the ammonia complex type. In the ammoniated cobalt nitrites he found a series of compounds which might be treated in the following manner.

1. $[Co(NH_3)_6](NO_2)_3$ *Luteo*cobalt nitrite
2. $[Co(NH_3)_5NO_2](NO_2)_2$ *Xantho*cobalt nitrite
3. $[Co(NH_3)_4(NO_2)_2]NO_2$ *Croceo*cobalt nitrite
4. $[Co(NH_3)_3(NO_2)_3]$ Triammine cobalt nitrite
5. $[Co(NH_3)_2(NO_2)_4]K$ Erdmann's salt
6. $[Co(NH_3)(NO_2)_5]K_2$ Unknown
7. $[Co(NO_2)_6]K_3$ Potassium cobalt nitrite

The ammoniated platinum chlorides were similarly handled.

1. $[Pt(NH_3)_6]Cl_4$ Chloride of Drechsel's base
2. $[Pt(NH_3)_5Cl]Cl_3$ Unknown
3. $[Pt(NH_3)_4Cl_2]Cl_2$ Platinitetrammine chloride
4. $[Pt(NH_3)_3Cl_3]Cl$ Platinitriammine chloride
5. $[Pt(NH_3)_2Cl_4]$ Platinidiammine chloride
6. $[Pt(NH_3)Cl_5]K$ Cossa's second salt
7. $[PtCl_6]K_2$ Potassium platinum chloride

By using the metal as a central element, Werner surrounded it with a constant number (in these cases, 6) of molecules and ions. He suggested that attractions that held these molecules and ions to the central metal were secondary valences for which he introduced the term *coordination number*. For the above examples the coordination number of cobalt and platinum is 6. For other series the coordination number of the central metal was found to be 2, 4, 6, or 8. Odd-number coordination of 3, 5, and 7 have also been encountered.

As evidence for his interpretation Werner could draw upon conductivity behavior of these compounds. Figure 14.8 represents the molecular conductivities of the chlorides of the cobalt ammonia series; Fig. 14.9, that for the ammoniated platinum chlorides. The conductivity is highest for those compounds that dissociate into the largest number of ions (considering the part within the square brackets as forming only one ion). As soon as part of the negative ions enter the coordination sphere, the conductivity falls, reaching a zero value for $[Co(NH_3)_3(NO_2)_3]$ and $[Pt(NH_3)_2Cl_4]$, where dissociation is no longer possible.

Besides explaining the ammonia complexes in this manner, Werner was able to extend the same kind of reasoning to hydrates, cyanides,

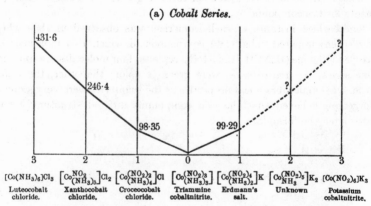

(a) *Cobalt Series.*

Fig. 14.8. CONDUCTIVITY OF COMPLEX COBALT COMPOUNDS.
(From A. Werner, *New Ideas on Inorganic Chemistry* [1911].)

(b) *Platinum Series.*

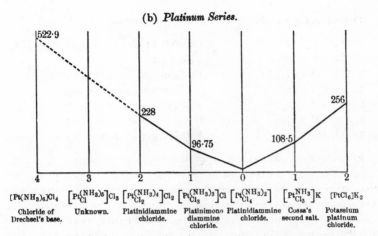

Fig. 14.9. CONDUCTIVITY OF COMPLEX PLATINUM COMPOUNDS.
(From A. Werner, *New Ideas on Inorganic Chemistry* [1911].)

thiocyanates, amine complexes, cyanates, carbonyls, and similar compounds. He also went on to examine the stereochemical implications. He reasoned that there must be isomerism due to the arrangement of ions in the inner, or coordination, sphere and the outer sphere where ordinary, or primary, valence relations held, e.g., $[Co(NH_3)_5Cl]Br_2$ and $[Co(NH_3)_5Br]ClBr$.

Of even greater import were the possibilities for geometric isomerism. Treating the 6-coordinated spheres in geometrical fashion, Werner demonstrated the possibilities for *cis* and *trans* isomerism. If the metal is considered to be at the center of a regular octahedron, the coordination valences will be directed to the six corners, giving two possible arrangements for the compound type MA_4B_2.

The earliest instance of such isomerism was observed in 1890 when Jørgensen prepared cobalt(III) compounds in which two molecules of ethylenediamine ($H_2NCH_2 \cdot CH_2 \cdot NH_2$) replaced four molecules of ammonia. One series, the *praseo* salts, were green in color; the other, the *violeo* salts, were violet. Chemical properties of the two series were very similar. Jørgensen believed that the two compounds differed structurally and postulated the formulas:

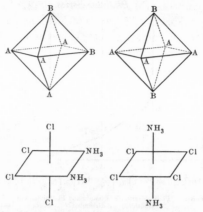

Fig. 14.10. SCHEMATIC REPRESENTATION OF
cis (left) AND *trans* (right) ISOMERS OF
COMPLEX METAL IONS.
(From A. Werner, *New Ideas on Inorganic Chemistry* [1911].)

Werner showed that the difference between the two compounds could be explained geometrically. While this form of isomerism was demonstrable in the case of ethylenediamine compounds (carrying what is now known

Cis; 1,2; or *violeo*   *Trans*; 1,6; or *praseo*

as a *bidentate*, or "double-toothed," group), it was much more difficult to obtain evidence for Ma₄b₂ isomers when a and b were represented by monodentate groups such as ammonia or chloride. Werner was finally successful in 1907 in preparing *violeo* [*cis*-dichlorotetramminecobalt(III)] salts isomeric with the *praseo* [*trans*-dichlorotetramminecobalt(III)] salts which had been known much earlier (see Fig. 14.10, bottom pair). It was ultimately possible to confirm this type of isomerism for various compounds of platinum, cobalt, and chromium. *Cis-trans* isomerism was also observed for compounds of the type Ma_2b_2, where a planar structure prevailed around the metal atom. (This isomerism, of course, cannot exist around a tetrahedral structure.)

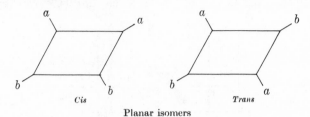

Cis   *Trans*

Planar isomers

Werner also recognized the possibility for optical isomerism in the case of the *cis* isomers carrying two or three bidentate groups. Difficulties connected with the preparation and separation of such compounds delayed demonstration of the phenomenon for many years. It was only in 1911 that Werner succeeded in preparing and resolving the optical antipodes of *cis*-dichlorobis (ethylenediamine) cobalt(III) chloride. He also

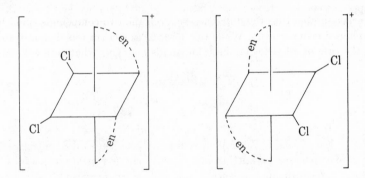

prepared and resolved triethylenediamine-cobalt(III) chloride. Such salts

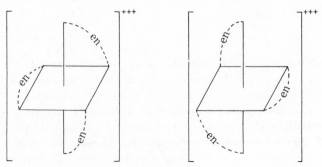

showed a strong optical rotation. The oxalate ion $C_2O_4^=$ also proved useful for preparing such optically active isomers. However, since both ethylene-diamine and the oxalate ion contain carbon atoms, it was argued that carbon was responsible for the optical activity even though neither compound contained asymmetric carbon atoms. In 1914 Werner prepared

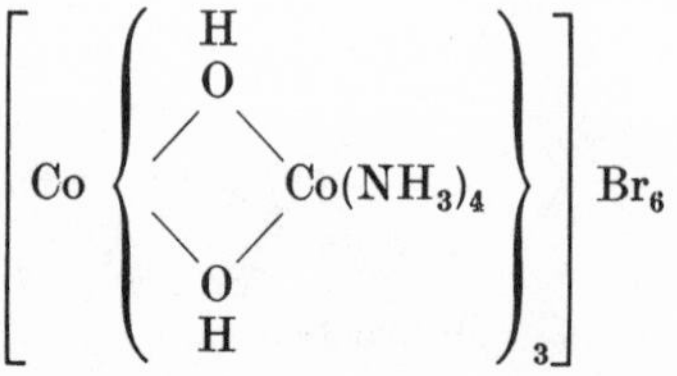

$$\left[Co \left\{ \begin{array}{c} H \\ O \\ / \ \backslash \\ \diamondsuit \quad Co(NH_3)_4 \\ \backslash \ / \\ O \\ H \end{array} \right\}_3 \right] Br_6$$

a compound, which contained no carbon atoms yet showed strong optical activity.

It was also possible for Werner to bring ammonium salts within the coordination concept. These compounds had brought about considerable confusion. While initially treated as addition compounds, i.e. $(NH_3)(HCl)$, there was a tendency, following van't Hoff, Blomstrand, and others, to suppose that the nitrogen went from trivalent to pentavalent form when salt formation took place. However, ammonium salts were found to be

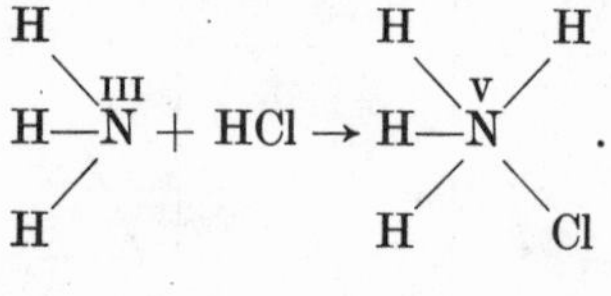

$$\begin{array}{ccc} H & H \quad H \\ \backslash III & \backslash V / \\ H\!-\!N \ +\ HCl \rightarrow H\!-\!N & . \\ / & / \ \backslash \\ H & H \quad Cl \end{array}$$

strong electrolytes. In 1902 Werner rejected the pentavalent nitrogen in favor of a structure permitting an NH_4 unit. He reasoned that in addition to the three principal valence bonds shown by nitrogen in ammonia there was a secondary valence which might take up a hydrogen atom. Thus, the formula of ammonium chloride might be represented as

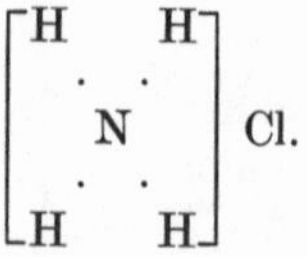

$$\left[\begin{array}{cc} H & H \\ \cdot \quad \cdot \\ N \\ \cdot \quad \cdot \\ H & H \end{array} \right] Cl.$$

Werner considered this to be more consistent with the reactions of ammonium compounds than the pentavalent nitrogen concept. Besides, it avoided the ludicrous results of extending the principle of increased valence, as, for example, in the case of the ferro- and ferricyanides.

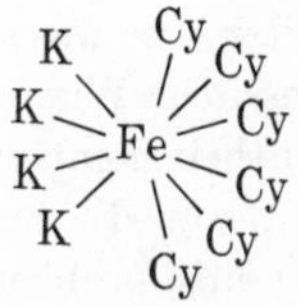

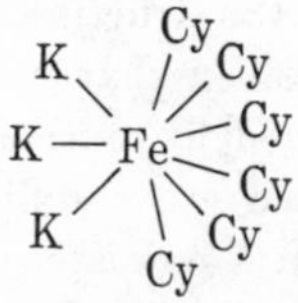

$$K_4[\overset{II}{Fe}(CN)_6] \quad \text{and} \quad K_3[\overset{III}{Fe}(CN)_6]$$

Werner realized that the iron in these compounds was best treated as divalent and trivalent, with 6-coordination by cyanide groups in each case.

$$K_4[\overset{II}{Fe}(CN)_6] \quad \text{and} \quad K_3[\overset{III}{Fe}(CN)_6]$$

The Werner concepts were developed as the result of the application of new principles which had recently come to the forefront in other fields of chemistry. The theory of ionization pointed the way toward an understanding of the make-up of the coordination sphere around metallic ions. Stereochemical studies of organic compounds had revealed the importance of giving attention to geometry in chemical bonding.

Jørgensen sought vainly to adapt Blomstrand's chain formulas to the needs of coordination chemistry. He and Werner each argued for his own point of view for more than a decade before the superiority of Werner's concepts became clearly evident. The facts of electrolytic conductivity and of isomerism brought about acceptance of Werner's position. Jørgensen dropped his opposition to the Werner concept of coordination chemistry in 1907.

Werner's concepts were used fruitfully in the succeeding years. The studies of Wolcott Gibbs on *iso-* and *hetero*poly acids of tungsten, molybdenum, arsenic, antimony, and phosphorus and those of Edgar Fahs Smith of Pennsylvania on the acids of tantalum and niobium were interpreted by A. Rosenheim in accordance with Werner's concepts. Of course, it must be realized that the coordination valence was no more explainable than the cause of primary valences. This was a problem remaining for solution later in the twentieth century.

CONCLUSION

The last four decades of the nineteenth century were hardly spectacular with respect to the field of inorganic chemistry. It was a field where only a few individuals stood out clearly, notably Moissan, Ramsay, and Werner —in sharp contrast to the many top rank organic chemists.

The reasons are not hard to find. Organic chemists dealt with the compounds of only a small number of elements, whereas inorganic chemists spanned an incomplete periodic table. While organic chemists dealt with an amazing number of compounds, including a large variety of classes, the structural theory which developed after 1860 was sufficiently well conceived that continuous progress was possible. When inorganic chemists sought to apply such structural ideas to their compounds, they encountered grave difficulties. These were only resolved when Werner introduced a new approach to structural and bonding problems toward the end of the decade. He recognized the usefulness of ideas coming out of the vigorously developing field of physical chemistry, and he used them effectively in seeking a solution to problems of inorganic structure, where ions play a much more profound role than they do in organic chemistry.

The period saw steady advances in empirical knowledge. The spectroscope was utilized effectively in uncovering and verifying the discovery of new elements. The tangle of rare earth elements was finally unraveled with the discovery of most of the members of the series. Progress in electrochemistry enabled Moissan to isolate fluorine, an element which had eluded chemists for almost a century. The electric furnace permitted the study of high-temperature reactions.

CHAPTER 15

PHYSICAL CHEMISTRY I.
ORIGINS

According to legend physical chemistry was established as a separate discipline in 1887, when Ostwald founded the *Zeitschrift für physikalische Chemie*. Actually, many developments prior to that year were physico-chemical in nature, but, since many chemists were physicists and many physicists were chemists, fundamental contributions were not classified as foundations of a physical chemistry. As the result of the work of such chemist-physicists as Dalton, Gay-Lussac, Avogadro, Berzelius, Dumas, Bunsen, Kopp, and Lothar Meyer, the development of physical chemistry was well underway by the time Ostwald announced his journal. Indeed, work based on the kinetic theory of gases and the relation of physical properties to chemical constitution was progressing rapidly when Ostwald was a student.

Kinetic theory was primarily the work of physicists, usually men with no real interest in or understanding of chemistry. As a result, chemistry was fragmented, especially during the last half of the nineteenth century. Descriptive chemists, particularly the organic group, were interested in theory only when it helped them directly with their problems. The physicists, on the other hand, believed that descriptive chemistry had little real significance.

PHYSICAL PROPERTIES AND
CHEMICAL CONSTITUTION

Although several researchers had investigated the relationship between physical properties and chemical constitution, for example, Mitscherlich's work which resulted in his law of isomorphism, no concerted search for relationships was made until Kopp's work, which dates from about 1840. His activities sparked widespread interest in the following decades.

Herman Kopp (1817–1892), who received his doctorate at Marburg for a thesis dealing with the density of oxygen, continued his studies of physical properties at Giessen and after 1863 at Heidelberg. Kopp sought to measure the physical properties of chemical substances. His approach, always straightforward, utilized simple but ingenious apparatus. In his work on specific heat he hoped to verify and extend Neumann's law which suggested that the product of molecular weight and specific heat is constant regardless of the composition of the compound. Kopp observed that to a certain degree the law was useful, but that the problem was actually very complicated and involved several factors.

At this time the boiling points of only a few compounds had been determined accurately. Kopp devised a reliable method and made numerous measurements. He also measured freezing points. His attempts to correlate boiling and freezing points with chemical composition were somewhat successful in the case of homologous organic acids, alcohols and esters.

In his work on atomic and molecular volumes Kopp was at first uncertain about the conditions under which comparisons might be valid. He finally assumed that comparisons should be made at the temperature at which the vapor pressure of the substance is equal to the atmospheric pressure, that is, the boiling point (approximately equal to two-thirds of what later became known as the *critical temperature*). Molecular volumes, calculated by dividing the molecular weight by the relative density at the boiling point, were found to be additive functions. Later workers showed that the additivity was only approximate and was influenced by types of combinations.

Kopp made many measurements of volume changes with temperature, including changes of volume with change of state. His studies on ice, phosphorus, sulfur, stearic acid, calcium chloride, sodium thiosulfate, and Rose's metal were so precise that the results are still sound.

When he investigated solutions, Kopp observed that nonreactive solutes have little influence on one another's solubility, except when a common radical is present. He noted the discontinuity in the solubility curve of sodium sulfate and interpreted it as indicative of a change in the degree of hydration. He also studied the partition of a solute between immiscible solvents and measured the cohesive forces in liquids.

Optical Characteristics of Chemical Substances

Optical properties of chemical substances also were investigated around 1850. Work on the rotation of polarized light was discussed in Chapter 12. Refractive index, magnetic rotation, and light absorption also were extensively studied.

In the seventeenth century Newton observed a proportionality between density and refractivity in his studies of camphor, olive oil, linseed oil, turpentine, amber, and diamond and deduced the absolute refractive power of a substance to be $(n^2 - 1)/d$ (where n = refractive index and d = density). This formulation was based upon his emission theory of light which assumed the refractivity to be independent of temperature, an assumption which was not entirely valid.

In 1858 John Hall Gladstone and Thomas Pelham Dale proposed a formula, $n - 1/d$, which although nearly independent of temperature was purely empirical in character. A more satisfactory formula $(n^2 - 1)/(n^2 + 2) \cdot 1/d$, was proposed in 1880 by Lorenz and Lorentz. This function was more independent of temperature than Gladstone and Dale's and had a definite theoretical basis. Hendrick Anton Lorentz (1853–1928), professor of theoretical physics at Leyden, based his formula on the electromagnetic theory of light proposed by Maxwell in 1865. Ludwig Valentin Lorenz (1829–1891), professor of physics at the Military High School in Copenhagen, independently arrived at the same equation from a consideration of the theory of dielectrics developed by O. F. Mosotti in 1850 and refined by Clausius.

Even before Lorenz and Lorentz published their formula, several workers had been investigating the correlation of refractivity and chemical composition. Their work suggested that molecular refraction is an additive property. If the specific refraction calculated by one of the above formulas is multiplied by the molecular weight, the product is known as the *molecular refraction*. In 1856 Berthelot reported that molecular refractivity increases with molecular weight in homologous series. Gladstone and Dale showed in 1863 that isomeric substances give the same refraction only if their constitutions are the same. Another pioneer in this field was Hans Heinrich Landolt, who studied fatty acids and esters and had some success in arriving at values for the refraction due to each element in a compound, thereby suggesting that molecular refraction is an additive property.

Gladstone showed in 1870 that molecular refraction for unsaturated compounds, particularly for benzene and the terpenes, was greater than that calculated from Landolt's values. Investigators now realized that refraction was not exclusively additive but was influenced by constitutive factors as well. The effect of the manner of combination was clarified in large degree by the studies of Julius Wilhelm Bruhl, professor of applied chemistry at the Technical College in Lemberg.

Magneto-optical Rotation

Magneto-optical rotation was discovered by Faraday in 1845 when he was searching for a fundamental relationship between light, magnetism, and electricity. This kind of rotation is independent of ordinary rotation such as that studied by Pasteur, van't Hoff, and Fischer. The latter is found only in compounds where an internal asymmetry exists or in crystals having right- and left-handed varieties. However, all transparent substances will rotate plane-polarized light when placed in a strong magnetic field. Faraday observed the phenomenon first in dense optical glass and subsequently in several other substances.

In 1845 Faraday distinguished between *paramagnetic* substances, which tend to move into a strong magnetic field, and *diamagnetic* substances, which move out of a magnetic field. His attempts to observe rotation in strong electrostatic fields were unsuccessful. This effect was observed in 1875 by John Kerr.

At first magneto-optical rotation attracted little interest among chemists and was explored only by physicists. In 1882 it was studied by William Perkin, who measured the rotatory power of a large number of organic compounds under carefully controlled conditions. Treating the data in terms of molecular magneto-optical rotatory power, he determined that the effect is both additive and constitutive. As in the case of refraction, the double bond was found to exhalt the optical effect. Of particular interest was Perkin's study of the molecular magnetorotation of hexatriene $\left(\begin{matrix} H \ \ H \ \ H \ \ H \\ H_2C=C—C=C—C=CH_2 \end{matrix} \right)$, which showed a value of 12.196. Benzene showed a value of 11.284 and hexane and cyclohexane values, of 6.6646 and 5.664, respectively, The difference in this latter pair was attributed to the removal of two hydrogen atoms together with ring closure. The similar difference between hexatriene and benzene was interpreted as physical verification of Kekulé's formula for benzene. Additional studies of the molecular refraction of these four compounds produced results consistent with Kekulé's formula.

Absorption Spectra

The study of absorption spectra of organic compounds was also pursued with enthusiasm around 1880. Soret and Rilliet[1] found that the ultraviolet absorption of alkyl nitrates shifts toward the red as methyl groups are added to the chain. Lieberman observed a connection between the fluorescence of organic compounds and their constitution.

Surface Tension and Viscosity

Other physical properties carefully examined were surface tension and viscosity. As with the properties just discussed, it was learned that these

[1] J. L. Soret and A. A. Rilliet, *Compt. rend.*, **89**, 747 (1879).

are partly additive and partly constitutive in character. These studies, which had been undertaken in the hope that physical measurements might solve problems of chemical composition, proved less promising than had been anticipated. The strong influence of consitution on certain physical properties largely ruled out the use of such measurements for the determination of composition. And the additive nature of the same properties ruled out their straightforward use for the determination of structure. At best, physical interpretations were useful in verifying structures already established from chemical evidence and for deciding between certain alternative structures. In the final analysis only optical rotation and absorption spectra proved to be of major significance in settling structural problems.

THE DEVELOPMENT OF KINETIC THEORY AND THERMODYNAMICS

Kinetic molecular theory, which with thermodynamics developed rapidly after 1850, proved of primary importance in chemical thought although the basic ideas were formulated entirely by physicists who for the most part ignored the chemical implications. Heat theory had wavered for two centuries between a material, or fluid, theory and a mechanical theory. Francis Bacon, Boyle, Hooke, and Newton had each toyed with the idea of heat being the motion of tiny corpuscles, the rate of motion increasing with temperature. During the eighteenth century, however, heat was considered *caloric*, a weightless fluid.

Not everyone was happy with caloric. Count Rumford, in particular, sought to overthrow the concept. While serving as superintendent of the munitions works in Munich, he observed the great quantities of heat given off during the boring of brass cannon. In planned experiments he demonstrated that the amount of heat produced was related to the amount of mechanical work expended during the boring process. When a blunt drill was used, very little metal was removed, but an enormous amount of heat was produced. Rumford's concept of heat as a form of mechanical motion was examined by Davy who showed that pieces of ice are melted by rubbing them together, even though the temperature remains at the freezing point.

Thomas Young, who was instrumental in reviving the wave theory of light, postulated in 1807 that heat, like light, might be a form of wave vibration. His suggestion was based on a study of the heating effect of the infrared portion of the spectrum and on the radiant heat given off by red-hot bodies.

Despite the work of these investigators, the caloric theory remained the popular one and most efforts to understand thermal phenomena were guided by the caloric framework. French engineers, who were trained in fundamental science at the École Polytechnique, concentrated on developing the caloric theory. In England there was far greater emphasis on practical applications, most notably the steam engine, which was the work of clever mechanics rather than of scientists. Watt was acquainted with British scientists such as Black and Priestley, but it is doubtful that even Black's studies on specific and latent heats contributed to his development of the steam engine. Watt worked out an indicator diagram which showed how steam pressure varies with the volume of the cylinder in a steam engine; he did not apply this information to engines he designed subsequently.

Among the French engineers who showed a flair for theoretical analysis in dealing with heat, Fourier, Carnot, and Clapeyron made especially significant contributions. Fourier examined the flow of heat in solids and developed a new method of mathematical analysis of the process. His studies, initiated in 1807, were published in 1822 as the *Théorie Analytique de Chaleur*. The subject was handled within the caloric framework and dealt solely with thermal conduction; mechanical effects of heat were not examined.

N. L. Sadi Carnot (1796–1832) reported an analysis of the factors involved in the production of mechanical energy from heat in his *Réflections sur la puissance motrice du feu* (Reflections on the motive power of fire, 1824). The eldest son of Lazare Nicolas Carnot (1753–1823), the Organizer of Victory during the Revolution, was familiar with his father's theoretical analysis of water-driven machines. Young Carnot believed heat a fluid analogous in some ways to water. Just as his father showed that water could be considered representative of liquids in general, Sadi showed that steam was representative of gases in general. He treated the behavior of the heat engine as an idealized cycle of heat transfers with accompanying expansions and compressions. Using the waterfall analogy, where height and quantity of water determine the power production, he argued that in the steam engine the temperature drop and quantity of caloric fulfilled a similar role. He fell into the error of supposing that no heat is lost or converted, but made the correct observation that power produced is dependent on change in temperature. Before his death during the cholera epidemic of 1832 he realized that some heat is converted to mechanical energy and is therefore lost, thus recognizing the failure of the water analogy. He abandoned the caloric viewpoint at this time but his views were not published until 1878, by which time they had been rediscovered by other investigators.

Carnot's ideas were extended by Benoit Paul Émile Clapeyron (1799–1864) in 1834. Clapeyron, who was a professor at the École des Ponts et Chaussees in Paris, translated Carnot's ideas into analytic form. He revived or rediscovered Watt's indicator diagram and showed that a measure of the work done in the cycle was afforded by the area under the pressure-volume curve.

Mechanical Equivalent of Heat

The contributions of Carnot and Clapeyron were not widely appreciated until the 1850's. At this time, when considerable attention was being given the mechanical theory of heat, the mechanical equivalent of heat was becoming established. Most of the fruitful clues to the nature of heat came from the biologists. During his metabolism studies Lavoisier had noted that the ratio of heat liberated to carbon dioxide produced in animals is roughly equivalent to the ratio of heat to carbon dioxide produced by burning candles.

Liebig, disagreeing with the vitalists who held that animal activity resulted from a "vital force," believed both heat and mechanical activity in animals to be derived ultimately from the combustion of food. Carl F. Mohr, in a paper rejected by Poggendorff's *Annalen der Physik* but published in Liebig's *Annalen der Pharmacie* in 1837, looked upon heat, light, electricity, magnetism, cohesion, and motion as mechanical forces. In 1839 Marc Séguin also published a work suggesting correlations of this sort. Other independent papers, which had no significance in the main development of the problem, came from the pens of Faraday (1840, pub. 1844), L. A. Colding (1843), Karl Holtzmann (1845), and G. A. Hirn 1854).[2]

Julius Robert Mayer (1814–1878), who had been educated in medicine at Tübingen, made a major contribution when he served as surgeon aboard a ship bound for Java. In the tropics he observed that the natives' venous blood had an intense red color. Recalling Lavoisier's theory that animal heat results from oxidation of the food, Mayer concluded that in the tropics not as much oxidation is necessary to maintain body temperatures, and hence venous hardly differs in color from arterial blood. He suggested that mechanical forces, chemical forces, and heat are interconvertible and, guided by what was known then about gaseous expansion, calculated a value for the mechanical equivalent of heat. Mayer conducted no experiments of his own; his calculations were based primarily upon the data others had accumulated on specific heats of air at constant pressure and constant volume. Since the quantity of heat required to raise the temperature of a given mass of gas is less at constant volume than at constant pressure (where the volume increases), Mayer assumed that the difference can be equated to the work done by the gas as it expands against the constant external pressure. Utilizing available data, Mayer

---

[2] A very fine study of the various contributions to the problem is Thomas S. Kuhn's "Energy conservation as an example of simultaneous discovery," in M. Clagett, ed., *Critical Problems in the History of Science*, Univ. of Wisconsin Press, Madison, 1959, p. 321.

Fig. 15.1. HERMANN VON HELMHOLTZ.
(Courtesy of the Edgar Fahs Smith Collection.)

calculated an equivalence of, in modern values, 3.6 joules/calorie. Because the data was unsatisfactory and he did not take into account work expended in overcoming cohesion within the gas, Mayer's falls somewhat below the accurate value, 4.185 joules/calorie.

Hermann Helmholtz (1821–1894) arrived at similar conclusions by way of biology while he was a military doctor at Potsdam. In his paper *Über der Erhaltung der Kraft* (1847) he clearly enunciated the principle of conservation of energy and applied mathematics to studies of the "force-equivalent of heat" and the "force-equivalent of electrical processes." In later years Helmholtz carried out many physiological investigations of sight, sound, and hearing, always from the physical point of view. A vigorous opponent of vitalistic doctrines, he argued that animals would be perpetual motion machines if they obtained energy not only from food but from some vital force. He held that there was a constant amount of energy in the universe and extended Leibnitz' principle of the conservation of mechanical energy to include heat, electricity, and other forms. *Correlation of Physical Forces*, another independent paper enunciating the conservation of energy, appeared in 1846. It was written by William Robert Grove (1811–1896), a lawyer whose hobby was physics.

The most significant work during the 1840's was that of James Prescott Joule (1818–1889), a Manchester brewer deeply interested in physical studies. In 1840 Joule, then only twenty-two, learned that the amount of heat produced by the flow of an electrical current is proportional to the resistance of the conductor and to the square of the current flowing (Joule's law). From these studies he concluded that the flow of electricity produces heat as a result of the resistance. He then measured the heat produced in water agitated by a paddle wheel activated by a measurable amount of mechanical energy. His best value for the mechanical equivalent of heat was 772 foot-pounds per B.T.U. (4.169 joules/calorie). The Royal Society rejected two of Joule's papers, but the importance of his work was pointed out by William Thomson at the meeting of the British Association in 1847.

William Thomson (1824–1907), better known as Lord Kelvin, was professor of natural philosophy at Glasgow from 1846 until 1899. In developing the consequences of Joule's doctrine of the interconvertibility of heat and mechanical work he arrived at the concept of absolute temperature in 1848. The observation that thermometers based on the expansion of various gases, liquids, or solids failed to agree led to the conclusion, based on a study of Carnot's theory, that perfect heat engines, operating between the same temperature limits, should be equally efficient regardless of the nature of the gas employed. He suggested that equal temperature increments on an absolute scale might be considered ranges over which a perfect heat engine would operate at equal efficiencies. In 1854, following the general abandonment of the caloric theory, Thomson proposed an absolute scale in which equal amounts of work are produced by equal temperature increments. This was shown to correspond reasonably well with the scale of the gas thermometer.

The term "energy" came into use in the mid-1800's. Some years earlier Thomas Young had pointed out the confusion resulting from use of the terms "force" and *vis viva* by various investigators. The acceptance of "energy" in a precise scientific sense was largely due to Thomson and William Rankine (1820–1872), a Scottish engineer and physicist.

Early Ideas on Molecular Motion

Coincident with the development of interest in heat as motion was a revival of the concept of gases as particles in motion. Seventeenth-century physicists, we have noted, advanced such hypotheses, but they had no experimental evidence with which to support their claims. They sought to explain why solids are cohesive whereas gases tend to expand to fill any containing vessel. Newton suggested that corpuscles must be looked upon as centers of force. According to him, when the corpuscles were close together there was an attraction; when at an intermediate distance the forces were neutralized and the substance was a liquid; and when far apart they repelled one another. The explanation failed to stand up under mathematical analysis. Boyle's law suggested that the repulsion of atoms, giving rise to pressure, would vary inversely with the first power of the distance between atoms, not with the square of distance as in gravitation. There was no reason why the repulsive force should be other than first power, but a new difficulty was encountered. The pressure of the gas not only should vary with the volume of the container, but also should be influenced by the shape of the container.

Even earlier, in 1649, Pierre Gassendi had sought to revive an atomic philosophy and Hooke, one of Newton's contemporaries, had suggested that the pressure exerted by a gas was attributable to the impact of moving particles of the gas against the walls of the container. Although Newton must have been aware of Gassendi and Hooke's ideas on moving molecules, he never pursued the matter.

Mathematical analysis of the problem was not handled with any degree of success until 1738 when Daniel Bernoulli correlated pressure with molecular motion to arrive at the relation, $p = \frac{1}{3}nmv^2/s^3$, where p is pressure, n the number of molecules, m the mass of a single molecule, v the mean velocity of the molecules, and s the length of a side of the cubic container. Since s^3 represents volume (V), the formula can be restated as $pV = \frac{1}{3}nmv^2$, a representation of Boyle's law $pV = $ constant.

Because the experimental study of gases had not reached a state of significance, it was not possible for Bernoulli or his contempories to extend the analysis further, and the subject remained dormant for more than a century. Bernoulli at least suspected that the mean velocity of molecules remains unchanged with changes in volume of the container only so long as the temperature remains constant, but he never asserted that temperature might be an expression of molecular velocities. In his day "energy" was at best primitively conceptualized. Leibnitz had groped toward an energy concept with his *vis mortua* and *vis viva*, terms related to what later became potential and kinetic energy, but his work provided no solid foundation for Bernoulli. Hence Bernoulli was unable because of conceptual and experimental limitations of the times, to carry his analysis through to a convincing level.

Molecules of gases figured again in early nineteenth-century scientific thought when Dalton, Gay-Lussac, Berzelius, Avogadro, and Ampère hypothesized about the characteristics of gases. As we have seen, leading scientists were at loggerheads each with the other and contributed little of value to an understanding of molecules. Finally, in midcentury, significant progress began to be made by investigators who were not concerned with the chemical behavior of molecules. The kinetic molecular theory was a model designed to provide an understanding of the physical behavior of gases. The theory received its initial impetus from Joule's work on the mechanical theory of heat; its applications were worked out between 1855 and 1890 under the leadership of Clausius, Maxwell, William Thomson, and Boltzmann. The Scottish physicist John James Waterston had developed a good theoretical treatment in 1845, but Waterston's paper lay unpublished in the Royal Society's archives until discovered by Lord Rayleigh around 1892.

Specific Heats of Gases

In 1848 Joule presented two papers dealing with the constitution of elastic fluids. He was convinced that heat represented a form of motion and had experimentally determined the mechanical equivalent in several ways. He also realized that a gas at 32° F. undergoes a volume change of one part in 491 for every degree F. of temperature change. When he calculated the specific heats of several gases at constant volume, he obtained values for hydrogen, water vapor, nitrogen, oxygen, and carbon dioxide which were all markedly lower than experimental values and sought to resolve the discrepancy by suggesting that the experimental values were in error. Regnault's subsequent reinvestigation of specific heats demonstrated the accuracy of the experimental values, and Joule abandoned the subject.

The next study of specific heats of gases was made by Rudolf Clausius (1822–1888), mathematical physicist at the University of Zurich. Clausius, who had already made significant contributions in thermodynamics, now set out to study the specific heat of gases at constant pressure. He reasoned that the specific heat at constant pressure should be $1\frac{2}{3}$ times the specific heat at constant volume. However the ratio of the two specific heats $(C_p/C_v = \gamma)$ as determined experimentally failed to approach 1.67 but fluctuated for various gases between 1.4 and 1.6. Not until 1876, when Kundt and Warburg measured γ for mercury vapor, was a gas found that approached the theoretical ratio.

On the basis of the observed discrepancy in γ Clausius concluded that not all of the heat absorbed by a gas has the effect of increasing the velocity of the molecules. If a gas molecule is composed of two atoms, Clausius recognized, some of the heat might have the effect of causing rotation of atoms around a common center or might cause them to vibrate to and fro. He termed the energy that brings about such effects "atomic energy" to distinguish it from "molecular energy," which causes an increase in the velocity of the molecule.

This reasoning suggested that γ may fall anywhere between 1.67 and

1.0, the former value holding for gases where all of the heat results in an increase in molecular velocity, the latter where all energy results in increased rotation and vibration of the atoms within a molecule, this latter value never being reached because an increase in temperature always results in some increase in translational energy. It became possible to measure γ with ease, since the velocity of sound in gases is related to the ratio of the specific heats at constant pressure and constant volume. Mercury vapor, being monatomic, was found to give the theoretical value. Later it was learned that the rare gases also give the theoretical value. The lower values commonly reported were clearly associated with diatomic and larger molecules in which rotational and vibrational changes might take place and thus lead to heat absorption in addition to an increase in translational motion.

Application of Kinetic Theory

Most of the consequences of kinetic theory were worked out during the following two decades. From the formula $pV = nmv^2/3$, it was possible to make correlations with Boyle's and Charles' laws, with Avogadro's hypothesis, and with Graham's law of gaseous diffusion (discovered experimentally in 1830). Calculations of molecular velocities at 0° C. gave hydrogen a value of 1,844 meters per second and oxygen a value of 461 m/s. The number of molecules in a cubic centimeter of gas at standard conditions was calculated by Loschmidt in 1865 to be 2.7×10^{19} (equivalent to 6.048×10^{23} per mole).

Using Gauss' law of error, Maxwell (1831–1879) and Ludwig Boltzmann (1844–1906) studied the distribution of molecular velocities. They showed that, while at a given temperature an average velocity might be assumed, individual molecules move both slower and faster than the average. This phenomenon might best be represented by a curve resembling an ordinary probability curve, suggesting that molecules must be treated as a statistical universe rather than as individuals.

Maxwell also developed a theory of gas viscosity based upon the study of viscous flow through fine capillaries made in 1844 by the French physiologist J. L. M. Poiseuille. Gas viscosity was related by Maxwell to the mean free path of molecules, making it possible to calculate this distance as well as the frequency of collision.

Entropy

The behavior of gas molecules was closely related to the second law of thermodynamics which underwent continuous study during the remainder of the century. It was recognized that while molecular collisions cause continually changing velocities, the natural trend in velocities, and therefore in temperature, is toward an intermediate value. Energy, researchers knew, can be utilized for work from a heat engine only if a temperature differential exists between cylinder and condenser, but in any system where such a differential exists there is a spontaneous flow of heat from the site of high temperature to that of low temperature. Hence, an energy redistribution causes energy to become unavailable when an average state, or a state of random distribution of velocities of molecules, is reached. Clausius introduced the term *entropy* to represent this degree of randomness which increases in spontaneous processes and announced that the "Entropy of the world tends to a maximum." This point of view had important philosophical implications since the universe might be looked upon as a closed system doomed to reach a steady state as energy differentials gradually are eliminated.

Maxwell focused attention on the concept of increasing disorder by suggesting that within a limited system, at least, one might conceive of changes in the direction of orderliness. He postulated a helpful demon who, without doing work, had the capacity to sort rapidly moving from slowly moving molecules. Through the intervention of Maxwell's demon the rule of increasing entropy would be invalidated and a substance might spontaneously grow hot at one end and cold at the other.

Van der Waals' Correction of Gas Laws

Nineteenth-century researchers learned that all gases deviate from the ideal gas laws at high pressures and low temperatures; that the deviation for compound gases is more pronounced; and that they deviate even at ordinary pressure and temperature. Van der Waals modified the familiar $pV = RT$ relationship in 1873 to achieve a closer reconciliation of theory and experimental results. His equation, $(p + a/v^2)(V - b) = RT$, takes into consideration the reduced pressure due to the attraction a/v^2 exerted by molecules for one another and the excluded volume b of the container due to the finite size of the molecules themselves.

Liquefaction of Gases

The concept of critical temperature was formulated in 1869 by Thomas Andrews at Belfast. In his studies of the continuity of the liquid and gaseous states he used carbon dioxide and found that there is a temperature above which a given gas will not liquefy, regardless of the amount of pressure applied. This clearly suggested that the *permanent* gases—hydrogen, oxygen, nitrogen, and carbon monoxide—were not unique but had never been cooled below their critical temperatures. In 1877 Raoul Pierre Pictet at Geneva and Louis Paul Cailletet, an ironmaster at Chatillon-sur-Seine, liquefied oxygen and nitrogen on a small scale. Large-scale liquefaction was achieved in 1883 by James Dewar in London and by Sigmund Florenty von Wroblevsky and Karol Stanislav Olszevski at the University of Cracow. Dewar, who was appointed Fullerian professor at the Royal Institution in 1877, introduced the thermos flask for handling cold liquids in 1893.

The first workers to liquefy permanent gases utilized the cooling produced when the gas is made to do external work during adiabatic expansion. Theoretically a gas expanding into a vacuum should not change in temperature since no external work is being done. Joule and William Thomson demonstrated that this is not true in 1852. Because of the internal work done to overcome attractions of molecules in a nonideal gas, the temperatures of all gases except hydrogen decrease with such expansion; the temperature of hydrogen tends to increase slightly. The Joule-Thomson effect was utilized for large-scale production of liquid air in 1895 by Karl von Linde (1842–1934) in Germany and W. Hampson in England. The cooling of the gas was made cumulative, and commercial-scale operations became possible. In 1898 Dewar established that even hydrogen shows a normal Joule-Thomson effect if first cooled to $-80°$ C. and succeeded in liquefying the gas. Helium, with a critical temperature of $-267°$ C., was liquefied finally in 1908 by Heike Kamerlingh Onnes (1853–1926) at the cryogenic laboratory at Leyden.

Chemical Thermodynamics

Thermodynamics were not quickly applied to chemistry even though there had long been an interest in the heat liberated during chemical reactions. Lavoisier and Laplace had studied heat output, both in combustion and respiration. Germain Henri Hess (1802–1850) had enunciated a limited form of the law of conservation of energy with his law of heat summation, in which he concluded that the heat liberated in a chemical process is independent of the path by which the process is carried out.

Beginning in 1852 more extensive measurements of heats of reaction were undertaken by Julius Thomsen (1826–1909) in Copenhagen and Marcelin Berthelot in Paris, who considerably refined their equipment and the techniques of thermochemical measurements during the next decade. The Berthelot bomb for measuring heats of combustion, developed in 1881, is essentially the one used today. For a time these studies were based on the assumption that chemical forces were proportional to the heat evolved during a chemical reaction. Gradually it was realized that this could be only partially true since reversible reactions would be impossible if compounds formed with maximum evolution of heat were of maximum stability.

The application of thermodynamics to chemistry was stimulated by the work of August Friedrich Horstmann (1842–1929) at Heidelberg from 1868. Studying the sublimation of ammonium chloride, he found that the change of vapor pressure with temperature corresponds to the equation formulated by Clapeyron for the vaporization of a liquid, namely

$$\frac{dp}{dT} = \frac{L}{T(V_v - V_l)},$$

where L stands for latent heat of vaporization, V_v the volume of the vapor, and V_l the volume of the liquid. Clausius recognized that the volume of the liquid is trivial compared to the volume of the vapor and hence that V_l can be ignored. Then, substituting RT/p for V gives (if molar amounts are involved)

$$\frac{dp}{dT} = \frac{pL}{RT^2},$$

or broadening it to include heat of sublimation by substituting Q for L,

$$\frac{dp}{dT} = \frac{pQ}{RT^2}.$$

Rearranged this becomes

$$\frac{1}{dT} \cdot \frac{dp}{p} = \frac{Q}{RT^2} \quad \text{or} \quad \frac{d\ln p}{dT} = \frac{Q}{RT^2}.$$

This is the Clausius-Clapeyron equation that found wide application. Horstmann, for example, used it in connection with heats of dissociation of hydrates and carbonates. Van't Hoff generalized the equation in 1889; replacing p by K, the equilibrium constant, he brought into its scope all equilibrium situations involving gases or dilute solutions.

Further progress in the application of thermodynamic principles to chemical equilibria was made by Josiah Willard Gibbs (1839–1903), professor of mathematical physics at Yale. His 321-page paper "On the Equilibrium of Heterogenous Substances," which appeared in the *Transactions of the Connecticut Academy of Science* in 1876–78, was only slowly recognized for its significance. Because of its abstract mathematical treatment and austere style, American physicists and chemists were unable to master its import. In Europe, except for Maxwell, who died suddenly in 1879, the paper failed to attract attention for almost a decade. Then its importance was recognized by van't Hoff, van der Waals, and Ostwald who made it a part of the rapidly developing physical chemistry.

The Phase Rule

Gibbs' paper contained the generalization now known as the *Phase Rule*. The equation $F = C - P + 2$ expresses the degrees of freedom (F) necessary in a system containing C components and P phases.[3]

---

[3] In the system where ice, water, and water vapor are in equilibrium, there is one component (H_2O) existing in three phases. Hence, $F = 1 - 3 + 2 = 0$, meaning that there can be no variability; temperature and pressure must be rigidly specified or one phase will cease to exist. On the other hand, liquid water in equilibrium with its vapor has one degree of freedom. Temperature and pressure can vary, but if one is specified, the state of the system becomes fixed.

Fig. 15.2. JOSIAH WILLARD GIBBS.
(Courtesy of the Yale University Press.)

H. W. B. Roozeboom (1854–1907) was the first to apply the *Phase Rule* to practical chemical problems. At the University of Leyden, where he was assistant to Jacob van Bemmelen, he studied the hydrates of sulfur dioxide. Van der Waals, the professor of physics, called his attention to Gibbs' work. Roozeboom found it useful in understanding the equilibrium of sulfur dioxide and water. After he moved to Amsterdam (1896) to occupy van't Hoff's chair when the latter went to Berlin, he applied the Phase Rule to the properties of alloys. His work in unraveling the phase diagram of the iron-carbon system was masterful.

Thermodynamics of Homogeneous Systems

Thermodynamics, applicable in the form of the Phase Rule to study of heterogeneous systems, was also fruitful in the study of homogeneous systems. Mixtures of gases or liquids in chemical equilibrium were examined by van't Hoff in his *Études de dynamique chimique* (1884) which also dealt with chemical kinetics. Here he stated his *principle of mobile equilibrium* which predicted that in an equilibrium system an elevation of temperature would favor the endothermic reaction.

At the same time Le Chatelier established the principle, commonly known by his name, that when a strain is placed upon a system in equilibrium, there will be a readjustment in the direction that will most effectively relieve the strain.[4] This was actually a theorem dealing with van't Hoff's principle of mobile equilibrium, but it was broad enough to deal with factors other than heat which might serve to displace a system at equilibrium. The principle was independently enunciated by F. Braun.

Henri Louis Le Chatelier (1850–1936) graduated from the École des Mines and after two years as a government engineer became professor of chemistry at the mining school. Research on the chemical nature of cements led him to the utilization of thermodynamic principles which suggested the theorem. While he later made important studies on alloys he never lost his interest in ceramic materials. His *Loi de stabilité de l'équilibre chimique* (1888) dealt primarily with the effect of changing pressures but was broad enough to cover other alterations of physical conditions. The theorem is, of course, based upon antecedent principles such as Maupertuis' concerning least action (expounded in the eighteenth century) and Fermat's principle of least time, developed in the seventeenth century in connection with studies of refraction of light. Le Chatelier was at the time unfamiliar with Gibbs' work and independently discovered parts of the phase rule. Later he figured prominently in the application of the phase rule to practical problems.

DYNAMICS OF REACTIONS

Although Berthollet's *Essai de statique chimique* contained the germ of the concept of mass action, his failure to establish the concept of indefinite proportions led to a decline of interest in his ideas, and consequently reaction rates and equilibria were ignored until the mid-1800's. In 1850, Ludwig Wilhelmy (1812–1864), a lecturer in physics at Heidelberg, studied the rate of hydrolysis of sucrose in the presence of acids. The change in rotation from *dextro* to *levo* as the sucrose was changed to invert sugar proved an ideal method of following the hydrolysis. Wilhelmy, who utilized a polarimeter, observed that the rate was proportional to the concentration of sucrose, which decreases with time according to $-(dC/dt) = kC$, where C represents concentration and t represents time. Integration gives $kt = 2.3 \log (C_0/C)$, where C_0 represents initial concentration. The reaction was shown to be of the first order with respect to sucrose.

About ten years later Berthelot and Péan de Saint-Gilles carried out rate studies of the formation and hydrolysis of esters. In 1866 A. Vernon Harcourt showed that the oxidation of hydrogen iodide with hydrogen peroxide is of the first order with respect to iodide and peroxide. The effect of varying concentration on the reduction of potassium permanganate

---

[4] For a critique of the Le Chatelier-Braun theorem see J. de Heer, *J. Chem. Educ.*, **34**, 375 (1957). This is based on the earlier criticism of P. Ehrenfest, *J. Russ. Phys. Soc.*, **41**, 347 (1909); *Z. physik. Chem.*, **77**, 227 (1911).

by oxalic acid was also studied. Harcourt developed a general treatment of reaction velocities which confirmed and extended Wilhelmy's work.

The whole problem of equilibrium reactions was carefully studied between 1864 and 1879 by the Norwegians Cato Maximilian Guldberg (1836–1902) and Peter Waage (1833–1900), professors of mathematics and chemistry, respectively, at the University of Christiania (Oslo). These brothers-in-law, working mostly with heterogeneous systems containing solids in contact with solutions, demonstrated experimentally that an equilibrium is reached in incomplete reactions. They also recognized that such an equilibrium can be approached from either direction. Treating such reactions mathematically, they expressed equilibrium conditions at a given temperature in terms of molecular concentration, which they termed "active mass." They stated early in their studies that "the driving force for a substitution is, under otherwise equal conditions, directly proportional to the product of the masses, each raised to some definite power."[5] Although their basic idea on the "law of mass action" was apparent in their *Études sur les affinités chimiques* (1867), the mathematical treatment of the law underwent a slow evolution. It was not until 1879 that Berthelot and Saint-Gilles' esterification equilibrium was expressed as:

$$\frac{(H \cdot C_2H_3O_2)(C_2H_5OH)}{(C_2H_5 \cdot C_2H_3O_2)(H_2O)} = \frac{1}{4}.$$

Van't Hoff had arrived at this expression two years earlier when he established, on the basis of thermodynamic principles, that equilibrium is reached when the velocities of opposing reactions become equal and that this dynamic state is related to the concentrations or active masses. Guldberg and Waage, most of whose publications were written in Norwegian, claimed priority on the basis of their 1867 paper. However, they had been groping toward the expression slowly, whereas van't Hoff recognized the relationship clearly once he put his mind to the problem. In his studies on various reactions he also showed that impurities and secondary reactions frequently beclouded his results.

Affinity

From 1800 until 1864 investigators devoted considerable attention to the forces that hold chemically dissimilar substances together. Like Newton, Geoffroy, Bergman, and Berthollet in earlier efforts, they made little progress. Various post-1864 studies, which failed to reveal the nature of affinities, did establish that relative affinities are amenable to mathematical treatment. Employing optical methods, Jellet studied the distribution of hydrogen chloride between two alkaloids when the amount of

[5] C. M. Guldberg and P. Waage as quoted by E. A. Guggenheim, *J. Chem. Educ.*, **33**, 545 (1956).

Fig. 15.3. J. H. VAN'T HOFF AND WILHELM OSTWALD.
(Courtesy of the J. H. Walton Collection.)

acid is insufficient to neutralize both. Utilizing calorimetric methods, Julius Thomsen studied the distribution of acids between bases in solution, arriving at a value for *avidities* of acids. In 1877 Ostwald studied the changes in volume occurring when acids are neutralized. Using carefully designed pyknometers, he was able to make very precise measurements. He also arrived at an order of acid strengths which corresponded to Thomsen's and made affinity a somewhat more precise concept. Ostwald later studied the hydrolysis of sucrose and esters in relation to the strength of acids.

Wilhelm Ostwald (1853–1932) was born in Riga where he completed the five-year course in the Realschule in seven years, not because of stupidity but because the breadth of his interests caused him frequently to become sidetracked. His doctorate was taken at Dorpat in 1871 under Carl Schmidt, a former student of Liebig. Ostwald taught at Dorpat and Riga and then at Leipzig, where he established a center for physical chemistry which attracted students from many quarters.

Temperature Effects on Equilibrium

The dissociation of hydrogen iodide was studied by Lemoine in 1877. His work proved that the same equilibrium is reached regardless of whether the reaction is started with acid or with a mixture of hydrogen and iodine and that temperature has a strong influence on the time required for the attainment of equilibrium; at 265° C., months are required, at 350° C., weeks, and at 440° C., only a few hours. Lemoine also learned that pressure has no effect on this reaction and that a constant is obtained if the product of the partial pressures of hydrogen and iodine are divided by the square of the partial pressure of hydrogen iodide. The results were confirmed two decades later by Bodenstein.

Temperature, it was discovered, affects rate of reaction and point of equilibrium in all cases. In 1884 van't Hoff clearly showed the influence on equilibrium while a few years later Svante Arrhenius (1859–1927) formulated the effect of temperature increases on reaction velocities. Arrhenius at that time introduced the concept of *activated molecules*, or molecules with much greater than average energy and therefore more susceptable to reaction. According to Arrhenius at higher temperatures a larger number of molecules have the necessary activation energies and the rate of reaction is greater.

Catalysis

The phenomenon of catalysis of reactions, a term suggested by Berzelius, was observed by Döbereiner and Davy early in the nineteenth century. Later in the century the influence of catalysts came under further investigation. In 1889 Ostwald made extensive studies of the velocities of acid-catalyzed reactions.

Enzyme-catalyzed reactions also received attention. O'Sullivan and Tomson studied the effect of yeast invertase in 1890; their time curve resembled Harcourt and Esson's where rate was related to sucrose concentration. They also found that every ten-degree temperature increase doubled the rate of inversion up to about 60° C., after which inactivation of the enzyme began to occur. These results were confirmed in 1908 by Hudson who showed the reaction to be of the first order.

In 1898 van't Hoff suggested that enzymes might be equally effective in bringing about synthesis. This was verified by Croft Hill who added yeast maltase to a concentrated glucose solution and was able to detect maltose (really isomaltose).

THEORY OF SOLUTIONS

Osmotic Pressure

Although many measurements had been made through the years, a theoretical understanding of solutions was not achieved until the 1880's. The breakthrough came from the work of van't Hoff who continued the successful application of thermodynamic principles to the problem of osmotic pressure, and Arrhenius who approached the problem from the direction of electrochemistry.

The phenomenon of osmosis was observed as early as 1748 when the Abbé Jean Antoine Nollet observed the rupture of a pig's bladder used to cover a bottle of alcoholic solution placed in water. Around 1830 the French physiologist René Joachim Henri Dutrochet made extensive studies of the passage of materials through membranes in contact with solutions. He introduced the term "osmosis" (from the Greek ὠσμός, to push) for the phenomenon and observed that the pressure produced appeared to be proportional to the concentration of the solution in contact with the membrane. This was confirmed in 1848 by Karl Vierordt of Karlsruhe. These workers, however, were using animal and parchment membranes permeable to both solvent and solute. Therefore the pressure was only temporary and was related to the relative rates at which the components moved through the membrane.

Semipermeable membranes were developed by Moritz Traube (1826–1894) of Breslau in 1867. With these membranes, which permitted only the transfer of water, it was possible to show that osmotic pressure is permanent and reaches a maximum. A decade later the German botanist Wilhelm Pfeffer (1845–1920) carried on extensive quantitative studies with cane sugar and other solutes, using membranes of hexacyano-copper(II)ferrate(II) formed within the walls of porous ceramic pots. He established the approximately direct proportionality of osmotic pressure and concentration of solute and also observed that osmotic pressure rises with temperature.

The Dutch botanist Hugo de Vries (1848–1935) called these experiments to van't Hoff's attention in 1884 when the latter was developing his theoretical concepts. De Vries not only was familiar with Pfeffer's work but also had been studying the withering of plants. Placed in pure water, they become turgid, but placed in moderately concentrated solutions, they wither. In solutions of intermediate concentration the plant cells remain normal since they have the same osmotic pressure as the external solution. Such isotonic solutions, prepared from various salts, all have the same freezing point. Van't Hoff now recognized an analogy between dilute solutions and gases and applied the second law of thermodynamics to the theoretical understanding of the properties of solutions with respect to concentration. From Pfeffer's data it was obvious that osmotic pressure increases as the volume of water containing one gram of sucrose decreases. Since the product of volume of solvent times osmotic pressure is a constant, the analogy to Boyle's law for gases was obvious. Using Carnot's cycle and Pfeffer's data, van't Hoff also was able to show that osmotic pressure is directly proportional to the absolute temperature, in accordance with the law of Gay-Lussac and Charles. The next step was to extend the relationship to include Avogadro's hypothesis. When this was accomplished, the osmotic pressure of solutions was a new approach to the determination of molecular weights (that is, 22.4 l. of a solution with an osmotic pressure of one atmosphere at 0° C. contains one gram-molecular weight of solute).

Measurement of osmotic pressure did not lend itself readily to molecular weight determinations, but van't Hoff was able to show the relationship between osmotic pressure of a solution and the vapor pressure of the solvent. It had been known since the work of Clemens Heinrich Lambert von Babo in 1847 and Adolf Wüllner in 1856 that the vapor pressure of a solvent is lowered by adding a solute and that the lowering is proportional to the amount of solute added. Even earlier Blagden (1788) had observed the effect of solute in lowering the freezing point[6] and Faraday (1822) had noted the elevative effect of solute on the boiling point.

While such observations existed, they had not been useful in arriving at quantitative relationships because the molecular weight concept was developed slowly and because most work dealing with solute effects

Fig. 15.4. FRANÇOIS MARIE RAOULT.
(Courtesy of the Edgar Fahs Smith Collection.)

was done with salts. In the eighties François Marie Raoult (1830–1901) of Grenoble began publishing his researches on the freezing points of solutions. Since he worked primarily with organic solutes, he was able to show in 1882 that when the freezing point depression for a solution containing one gram of solute in 100 g. of water is multiplied by the molecular weight of the solute, a constant is obtained. He demonstrated how the

---

[6] Knowledge of the effect of solutes on the freezing points of solvents must go far back in history. Fahrenheit made use of it in establishing the low point of his thermometers.

freezing point depression might be used for molecular weight determinations. Two years later he reported that the law failed to hold for salts but applied to the radicals present almost as if the radicals themselves were simply dissolved in water. Raoult also studied the lowering of vapor pressure by solutes in water and in nonaqueous solvents. The determination of molecular weights of dissolved substances by the measurement of freezing-point lowering and boiling-point elevation was pursued in a number of laboratories. Particularly important in this field were James Walker of Edinburgh, Georg Bredig of Heidelberg, Ernst O. Beckmann of Leipzig, and S. Lawrence Bigelow of Michigan; all had studied in Ostwald's laboratory. Beckmann made a major contribution by developing a differential thermometer capable of measuring temperature changes as small as 0.001 centigrade degrees.

Electrolytic Dissociation

Arrhenius studied the anomalous behavior of salts, acids, and bases in connection with electrochemistry. While working for his doctorate at Uppsala under Cleve, he postulated that salt molecules dissociate when they dissolve in water. These submolecular particles he called "ions," using the term Faraday had introduced for the particles discharged at the electrodes during electrolysis.

The idea of dissociation of molecules was not entirely new. Faraday had not suggested it, but others, particularly Williamson (1851) and Clausius (1857), had speculated on such possibilities. At Münster Johann Wilhelm Hittorf (1824–1914) had studied the electrolysis of salts in the 1850's and observed that the current was presumably carried through the solution by ions which moved at different speeds toward the electrodes. Hittorf suggested that the ions could hardly be combined in firm molecules.

Friedrich Kohlrausch (1840–1910) had shown in 1874 that every ion has a characteristic mobility regardless of whatever ion it is associated with in the original salt. The conductivity of a particular salt could now be calculated additively from the mobility of the component ions. Kohlrausch, who was at the time professor of physics at Würzburg, had studied electrolytic processes for some time before introducing the use of alternating current in order to avoid the polarization effects of direct current which had obscured any quantitative relationships in his own work and that of earlier investigators.

Arrhenius not only professed a belief in the existence of ions but also suggested that they are formed when the salt dissolves in water. Thus he disagreed with Faraday, Hittorf, Kohlrausch, and others who believed that ions were produced only when the current began to flow. While pondering the results on the conductivity of extremely dilute solutions, Arrhenius examined the idea of active and inactive molecules developed in preliminary studies by Clausius. He also reviewed Berthelot's thermo-chemical studies and recognized that the strongest acids are also the best conductors.

Arrhenius' thesis, presented in 1883, contained an experimental section dealing with his studies on the conductivity of dilute solutions and a theoretical section developing his theory of ionization. Cleve and the other professors who examined him were not happy with his theoretical speculations and the candidate was passed only with reservations. Since it was obvious that the theory of ionization would receive no support in Sweden, Arrhenius sought it elsewhere by writing to Clausius, Julius

Fig. 15.5. SVANTE AUGUST ARRHENIUS.
(Courtesy of the Edgar Fahs Smith Collection.)

Thomsen, and Ostwald. Only Ostwald was sufficiently impressed by the thesis to make more than a courteous reply. Ostwald visited Arrhenius in Sweden and then invited him to his laboratory in Riga. After working for a time with Ostwald, Arrhenius moved to Kohlrausch's laboratory in Würtzburg where he read van't Hoff's memoir on osmotic pressure in which the work of Raoult was discussed. This led Arrhenius to see that abnormal vapor-pressure lowering by salts could be explained by the ionic theory that was so useful in connection with conductivity. For example, he suggested that the abnormal vapor-pressure lowering observed with sodium chloride is caused by dissociation into two ions, thus creating twice the effect of sucrose which does not split in this manner.

Arrhenius now wrote to van't Hoff, and after a short visit with Boltzmann at Graz worked in van't Hoff's laboratory in Amsterdam. Following further work with Ostwald, who had set up his laboratory in Leipzig, Arrhenius published his theory in the first volume of the *Zeitschrift für physikalische Chemie.*

Arrhenius' theory of ionization was received with mixed emotions. While the charged ions which supposedly formed when a salt, acid, or base dissolved in water were useful in explaining conductivity, osmotic pressure, and vapor pressure phenomena, they also raised difficult questions regarding origin and characteristics and, worst of all, demanded a drastically different mode of thought. The theory was aided immensely by the favorable attitude of Ostwald and van't Hoff, men who were now gaining wide recognition for their activities in the field of physical chemistry. Arrhenius was offered a Giessen professorship in 1891 but chose instead to accept a lectureship in the technical high school in Stockholm where he rose to professor and rector. In 1905 he turned down a professorship at the University of Berlin but was recognized at home by being made director of the newly created Nobel Institute for Physical Chemistry.

COLLOID CHEMISTRY

Colloidal gold, produced by the treatment of gold chloride with essential oils, was known to medical alchemists as "purple of Cassius" and "aurum potabile." Certain other colloidal systems were known in the nineteenth century, but the difference between true solutions and colloidal systems was not recognized until 1844. Francesco Selmi, an Italian toxicologist, investigated sulfur, Prussian blue, and similar colloidal sols, referring to them as "psuedosolutions." He observed no change of temperature when the psuedosolution was produced and reported that the suspended matter was precipitated when salt was added.

A more systematic study was carried on during the next decade by Graham who introduced the term "colloid," meaning gluelike, in 1861. Graham distinguished between "crystalloids," which pass through membranes such as parchment, and "colloids," which are stopped by such barriers. In developing the use of membranes for the separation of crystalloids from colloidal systems, he coined the term "dialysis" to represent such separation. Graham prepared and studied a variety of colloidal systems such as arsenic trisulfide, silicic acid, tungstic acid, and the hydroxides of aluminum, iron(III), and chromium. These fluid systems Graham termed "sols," to distinguish them from semisolid jellies, or "gels." He learned that the addition of even small quantities of neutral salts to the sols causes them to coagulate into flocculent masses. A trace of carbon dioxide causes silicic acid sol to become a translucent gel.

Casual studies on colloids were made in various laboratories following Graham's pioneering efforts, but systematic studies were not pursued until the end of the century. Interest in the subject then grew rapidly.

Although it was known that colloids are coagulated by salts, the effect of salt component valence was not recognized until 1882 when Hans Oscar Schulze of the Freiburg School of Mines reported studies revealing this relationship. The coagulation of colloids by salts was studied systematically around 1900 by W. B. Hardy who recognized the conditions for coagulation, and Herbert Freundlich. The fact that colloidal particles carry electrical charges was clearly established in 1892 when S. E. Lindner and Harold W. Picton showed that colloidal particles move toward one of the electrodes when placed in an electrical field; for example, arsenic sulfide particles migrate toward the anode, ferric hydroxide toward the cathode. They observed a reversal of the charges on the particles when turpentine was used as a suspending medium instead of water. Hardy showed in 1899 that the direction of migration of albumin particles is dependent upon the reaction of the suspending medium. This behavior proved to be characteristic of protein sols in general. The migration of colloids in an electrical field—termed "cataphoresis," or "electrophoresis," —became a valuable technique for the study and purification of colloids.

Davy had known that fine metal particles are torn from metal electrodes when an arc is passed between them. In 1898 Bredig applied this principle to the preparation of metallic sols. A platinum sol was produced by passing an arc between platinum electrodes under water. Iridium, silver, and gold sols were prepared similarly. Svedberg later adapted the technique to the preparation of elemental sols in organic liquids. Using special apparatus and low temperatures, he even prepared sols of alkali metals in such liquids.

The ultramicroscope was developed in 1903 by H. F. W. Siedentopf and Richard Zsigmondy at Jena. With this instrument it became possible to count particles and, with other data, estimate their size. At about this time Peter Petrovich von Weimarn began to suspect that particle size is of considerable significance in connection with the properties of colloidal systems, that such systems are, in fact, intermediate between the molecular and the microscopically visible. It also became apparent that the surface area of such particles is large, thus enhancing adsorptive powers for ions, gases, coloring matters, and noxious odors.

Jean Perrin, Albert Einstein, and others became interested in the Brownian motion of these particles and developed the kinetic theory of molecular bombardment. Perrin even deduced a method for estimating Avogadro's number from such studies.

For some time it was recognized that there are two kinds of sols, those having little affinity for the dispersion medium and those having a great affinity and which frequently lead to gel formation. In 1905 Perrin introduced the terms "hydrophobe" for the first and "hyrophile' 'for the second system.

CONCLUSION

Although chemists had been giving attention to theoretical matters involving chemical substances and chemical reactions from the beginning of the nineteenth century, it was only in the last half of the century that such studies became formalized into physical chemistry. In part, the specialty developed out of physicists' concern with heat phenomena, which led to the first two laws of thermodynamics and the kinetic molecular theory. These physical concepts proved useful in the interpretation of chemical phenomena, particularly in the hands of Horstmann, J. Willard Gibbs, van der Waals, and van't Hoff.

Studies of the dynamics of reactions led to the downfall of earlier affinity concepts and to the recognition of equilibrium principles in chemical systems. Progress in the understanding of dilute solutions followed the insight of van't Hoff in applying gas-law relationships to solutions and Arrhenius' introduction of his theory of ionization. This latter theory also had a significant influence in setting chemists back on a sound path toward the understanding of inorganic compounds, and in developing a fundamental understanding of electrochemistry. The study of colloidal systems also showed steady progress following the distinction drawn by Graham between crystalloids and colloids.

Various physicochemical principles quickly found application in scientific and industrial areas. Success was achieved in the liquefaction of the so-called "permanent gases." The Phase Rule found application in various industrial operations, particularly in interpretation of properties of alloys. The new understanding of equilibrium conditions was to figure significantly in various places; particularly in putting analytical chemistry on a sounder footing and in industrial developments like the Haber process and the separation of complex mixtures like those in the Stassfurt salt deposits.

The acceptance of the concepts of physical chemistry was to encounter resistance in certain quarters, particularly among the organic chemists, but taken as a whole it gave to chemistry a new enthusiasm and a new approach to the solution of troublesome problems.

CHAPTER 16

BIOLOGICAL
CHEMISTRY I.
AGRICULTURAL,
PHYSIOLOGICAL, AND
FOOD STUDIES

Although biological chemistry is in large part a twentieth-century development, the term is used here to include pretwentieth-century work in agricultural and physiological chemistry which provided the foundation for the science. Many of the first biochemical investigations were focused on immediate agricultural or medical problems rather than basic research. Although nineteenth-century investigators were hampered because analytical and organic chemistry had not yet provided essential information, they did contribute much to our understanding of photosynthesis, animal respiration, soil fertility and plant growth, and the composition of biological tissues. And the much wider appreciation of public health problems which resulted directly from this work sparked the movement for sanitation and control of food adulteration.

418

PHOTOSYNTHESIS

Even today the details of the process of photosynthesis are not completely unraveled. The first strides toward understanding were made during the Chemical Revolution, especially by Priestley, Ingenhousz, Senebier, and de Saussure.[1] Jan Ingenhousz (1730–1799), a Dutch physician who practiced in England and Austria, published many experiments dealing with plant physiology; the most important are described in his book *Experiments on Vegetables* (1779). Jean Senebier (1742–1809) was a Swiss minister who pursued botanical studies with such interest that he was able to publish a five-volume *Physiologie végétale* in 1800. Nicolas Theodore de Saussure (1767–1845) was a member of the famous Geneva family of philosopher-naturalists.

The role of air in plant growth appears to have been suggested first by the microscopic examination of plant tissues by Nehemiah Grew (1641–1712) in England and Marcello Malpighi (1628–1694) in Bologna. Both observed the stomata in plant leaves and, since they appeared to be connected with vessels which might serve as air ducts, concluded that these minute pores introduced air and eliminated waste vapors.

In his *Vegetable Staticks* (1727) Stephen Hales further emphasized the importance of air in plant growth. Priestley took up the problem nearly five decades later when he asked why such factors as spoilage by fires and volcanoes and the breathing of animals do not cause air to become permanently vitiated. He observed that plants are able to improve air spoiled by a candle flame or animals. Consistent with his belief in the phlogiston theory, Priestley supposed that animals gave off phlogiston to the air while plants abstracted it. In later experiments he concluded that the improvement of spoiled air confined over water was due to irradiation by sunlight and overlooked the significance of "green matter" (algae) in the water.

Ingenhousz became interested in 1779 and soon established that: (1) plant leaves exposed to light improve the air; (2) Priestley's "green matter" is a plant; (3) air is spoiled by the nongreen parts of plants; and (4) air is also spoiled by green plants in the dark. He explained the phenomena in terms of phlogiston.

Soon thereafter Senebier demonstrated the liberation by plants of "dephlogisticated" air from a solution rich in fixed air and found that the action was restricted to the green portion of plants. De Saussure, an adherent of Lavoisier's ideas, resolved the problem in 1804 when he showed, by growing plants in confined spaces where changes in the air could be studied quantitatively, that in the presence of light carbon dioxide

---

[1] For a good study of the early phases of the photosynthesis problem see Leonard K. Nash, *Plants and the Atmosphere*, Harvard Univ. Press, Cambridge, 1952.

and water are used to form plant tissues and oxygen is simultaneously liberated.

The study of photosynthesis lagged for the greater part of the next century, in part because de Saussure had satisfactorily dealt with the immediate problems, and also because the unanswered questions were so complex that they could not be attacked until considerable progress had been made in the fields of chemistry, physics, and biology.

SOIL CHEMISTRY AND PLANT GROWTH

Since ancient times learned men have been deeply interested in agricultural science. The Romans Pliny and Columella dealt with agricultural matters at considerable length and showed a general realization of the importance of soil fertility. During the Middle Ages the best agricultural methods of antiquity, such as fallowing and crop rotation, were preserved and improved upon. And in the seventeenth century the most eminent natural philosophers began attempting to develop a theoretical agriculture based upon scientific principles. While most of this work was purely speculative, occasionally there were attempts at sound experimentation.

Van Helmont, Bacon, and Boyle accepted water as the primary nutrient of plants, while Mayow and Hales emphasized air. Stahl and his followers stressed phlogiston, while Glauber emphasized the importance of niter. The English physician John Woodward (1665–1728) carried out experiments which persuaded him that earth was the principal nutrient and that water was not as important as others believed. Until Lavoisier's time such alchemical ideas dominated agricultural thought. True, the value of manure, composts, and such minerals as muck, lime, gypsum, and marl was appreciated, but nonempirical speculation reigned.

Serious agricultural experimentation began toward the end of the eighteenth century. On his 300-acre farm near Blois Lavoisier carried out various experiments hoping to improve French agriculture. Even more extensive projects for agricultural experimentation were developed by Albrecht Daniel Thaer (1752–1828) near Celle in Hanover and later at Möglin in Prussia. Thaer's staff chemist, Heinrich Einhof (d. 1808), published extensive analytical data on soils, manures, and fertilizers. Similar studies were made by Sigismund Friedrich Hermbstadt (1760–1833) who sought to correlate the composition of fertilizers with the composition of cereals. In England Davy discussed soil fertility in his lectures on agriculture, which were published in 1813 as *Elements of Agricultural Chemistry*.

Unfortunately the state of analytical chemistry was so rudimentary that much of the work of these chemists had no enduring value. The period between 1820 and 1850 became one of clarification and consolidation. These three decades saw the development of the improved analytical procedures needed for sound research at the same time that the growth of journals made possible the rapid dissemination and criticism of results. However there still remained a great deal of vitalistic and alchemical thinking to be cleared away.

One of the crucial unanswered questions was: Which elements do plants require for nutrition? Ingenhousz had suggested that the only way to determine nutritive essentials was to observe the effect of the absence of an element; the absence of essential elements, he argued, would cause a plant to die. Early attempts at growing plants on synthetic soils—for example, flowers of sulfur (J. C. Schrader, 1800), litharge, extracted peat, sublimed sulfur, granulated lead, or quartz pebbles (H. Braconnot, 1806)—suggested that some transmuting force enables plants to form their own minerals. This hypothesis was not borne out by de Saussure's observations, but the vitalistic concept was not finally abandoned until 1840 when A. F. Wiegmann, a professor at Brunswick, and L. Polstorff, an apothecary, convincingly demonstrated that plants will not grow on a bed of purified quartz, but that similar plants grow normally on the same purified quartz to which minerals and humus has been added. In an even more conclusive experiment the investigator used a platinium crucible containing platinum wire as a supporting "soil." Distilled water was supplied, and the seed was germinated under a bell jar containing a purified atmosphere of nitrogen, oxygen, and carbon dioxide. The seeds sprouted, but the plants turned yellow and died within several weeks. An ash analysis showed that the total ash of the germinated seedlings was no greater than the ash content of the seeds.

Important studies on the mineral needs of plants were made in the 1820's by Carl S. Sprengel (1785–1859), a student of Thaer who taught agricultural chemistry at Göttingen and Brunswick before establishing an agricultural academy and factory for agricultural implements at Regenwald. A careful student of de Saussure's work, Sprengel broke away from Thaer's concepts of transmutation and denied the capacity of plants to form lime from humus. In fact he maintained that humus was valuable only to the extent that its acidity might make other soil constituents more soluble. Its real value, he suggested, lay in its content of potash, lime, and similar minerals. Sprengel analyzed humus by combustion over copper oxide, finding a carbon content of 58 per cent. This study led to the procedure of analyzing soil for humus by first determining the percentage of carbon and multiplying by the factor, 1.724.

Sprengel regarded fifteen elements as essential to plant nutrition— carbon, oxygen, hydrogen, nitrogen, sulfur, phosphorus, chlorine, potassium, sodium, calcium, magnesium, aluminum, silicon, iron, and manganese—and suggested that others might be needed in small amounts. By analyzing good and poor soils he was able to obtain a picture of optimal soil composition. He was the first to realize the importance of the minimum element; that is, plant growth cannot be more luxuriant than permitted by the element present in the soil in least adequate amounts. At the same time he taught that there might be maximal concentrations, beyond which plant growth might be harmed. He disagreed with investigators who believed that one element could serve as a substitute for another and established that lead, arsenic, and selenium are toxic to plants. His teachings emphasized a proper balance of minerals, air, water, and sunshine for optimum plant growth.

Although much of his work reflects a clear insight into soil fertility problems, some of Sprengel's concepts were erroneous and received just criticism. While minimizing the importance of humus, he continued to believe that minerals arising from organic sources were better fertilizers than minerals applied in inorganic form. He emphasized, as did the Swiss botanist A. P. de Candolle (1778–1841), the harm done to vegetation by toxic root excretions of certain plants. Sprengel also stressed the value of burnt clay for its supposed role in the production of ammonia.

One of Sprengel's more noteworthy contemporaries was Jean Baptiste Boussingault (1802–1887). After graduating from the mining school of Saint Étienne in Paris, he spent a decade in South America as a mining engineer for an English firm. On returning to France he obtained through marriage an interest in a large Alsatian estate at Bechelbronn, where he immediately inaugurated extensive experiments in agricultural chemistry which were reported from 1836 forward. In 1839 he was appointed professor of agriculture and analytical chemistry at the Conservatoire des Arts et Métiers in Paris. Here he would teach for a term, then spend the next on his estate.

Boussingault's agricultural experiments were far-ranging. Large-scale field experiments were coupled with laboratory analyses. He studied changes in seeds during germination, the effect of crop rotation, the use of fertilizer, the care of barnyard manure, the assimilation of nitrogen by plants, the nutritive value of forage crops, the feeding of farm animals, and the influence of rations on the yield and composition of milk. In addition to many journal articles he published several important books on agriculture.

Boussingault attacked the problem of the source of nitrogen in plant growth. It had been known from Roman times that plowing-under legumes enriched the soil.[2] In 1838 Boussingault showed that the amount of nitrogen in red clover and peas grown in a soil with no added fertilizer increases during the growth cycle but that the amount in wheat and oats grown under the same conditions does not increase. He could not explain how the legumes acquired their nitrogen. The answer awaited information to be provided by bacteriology, which in 1838 was not yet a science.

[2] Pliny, *Natural History*, **18**, 30.

Fig. 16.1. JEAN BAPTISTE JOSEPH DIEU-
DONNÉ BOUSSINGAULT.
(Courtesy of the Edgar Fahs Smith Collection.)

Boussingault studied the use of minerals by plants by determining the ash composition of plants, as well as by measuring yields under the influence of various fertilizer combinations. He correlated the nutritive value of crops with the nitrogen content, as determined by the Dumas method. Humus was valuable, Boussingault believed, because of its nitrogen content. On the general question of the importance of humus, he took an intermediate position.

Gerardus Johannes Mulder (1802–1880) also figured prominently in the application of science. He obtained a medical degree at Utrecht and practiced for a time before becoming professor of chemistry at Rotterdam and later at Utrecht. He published widely on analytical chemistry and contributed two extensive works on soils and physiological chemistry which attracted considerable attention in translation despite their many mistakes and erroneous speculations.

Mulder coined the word *protein* (Gr., first substance) for those nitrogenous constituents universally found in biological materials. In a day when the "state of the art" made it difficult to determine reliable chemical formulations, he proposed that proteins were basically $C_{48}H_{31}N_{15}O_{12}$ (old equivalents) and that differences between them resulted from multiplication of the primary units in conjunction with sulfur and phosphorus. Thus casein was formulated as $10(C_{48}H_{31}N_{15}O_{12}) + S$, and blood albumin as $10(C_{48}H_{31}N_{15}O_{12}) + S_2P$.

Mulder worked extensively on plant composition, using microscopic and chemical analysis. He distinguished between cellulose proper ànd woody matter (lignin, hemicelluloses), which he considered the source of ulmin, ulmic acid, geic acid, and the other substances it was thought constituted humus.[3] He believed that the humus acids combined with atmospheric ammonia to form salts for synthesis of plant proteins.

Large-scale agricultural experimentation was undertaken in England by John Bennet Lawes (1814–1900) on his estate Rothamsted. His pot experiments with fertilizers begun in 1837 were expanded to field experiments a few years later. In 1843, recognizing the importance of chemical analysis, Lawes hired Joseph Henry Gilbert (1817–1901) who studied under Liebig. The two undertook an experimental collaboration which continued to the end of their lives.

The general plan followed at the Rothamsted Experimental Farm was to grow a crop for many years on soil plots which were untreated, fertilized with barnyard manure, or fertilized with chemical fertilizers. The hundredth consecutive crop of wheat from one of the fields was harvested in 1943. Lawes and Gilbert also studied the effects of fallowing and the influence of crop rotation.

Lawes began experimenting with bone meal very early and found that it is most useful as a fertilizer if first treated with sulfuric acid; otherwise the phosphorus is not of much use to plants. In 1842 he obtained a patent on the acid treatment of phosphates. "Superphosphate of lime" began to be manufactured in a factory at Deptford in 1843, and "J. B. Lawes Patent Manure" became a highly profitable item.

The first English studies on soil bacteriology were conducted at the Rothamsted station. Following the work of Schloesing and Muntz, who in 1877 showed that bacteria turn ammonium compounds into nitrates, Robert Warington at Rothamsted established that soil nitrification is a bacteriological process in which ammonia is first converted to nitrites and these in turn are changed to nitrates. Because of the elementary state of bacteriological techniques he was unable to isolate the responsible organisms. This was accomplished in 1890 by Sergi N. Winogradsky (1856–1953), a Russian-born, chemically trained bacteriologist working at Zurich.

Agricultural chemistry received its greatest stimulus from the activities of Justus Liebig. During the 1830's his many activities in organic and

[3] G. J. Mulder, *J. prakt. Chem.*, **16**, 129, 138 (1839). These terms were in common use among advocates of the humus concept although usage was vague. In 1797 Vauquelin studied a gumlike exudate of the elm tree for which Thomas Thomson in 1813 proposed the name *ulmin* (from the Latin *ulmus* for elm). A supposed similarity between ulmin and the dark-colored alkaline extract from soils led to use of the name for this material as well. Since this material formed compounds with alkalies, the term *Ulmic acid* came into use. Sprengel in 1826 used the term *humic acid* for the same or similar material. Berzelius introduced the terms *gein* (from the Greek word for earth) and *geic* acid in 1832. Meanwhile Braconnot decomposed sugar and starch with hydrochloric and sulfuric acids to obtain a similar material; his work lead to the idea that humus arises from the decomposition of carbohydrates.

analytical chemistry gave him great prestige and caused various groups to seek his advice on practical matters. Needless to say, he was always eager to give it.

In 1840 he appeared at the Glasgow meeting of the British Association for a series of lectures on agricultural chemistry. The lectures provided the basis for his book *Organic Chemistry in its Application to Agriculture and Physiology* which was published in German and in Playfair's English translation. The work was so popular that by 1848 seventeen editions had appeared—four German, four English, two American, two French and one each in Denmark, Holland, Italy, Poland, and Russia. The ninth German edition was published in 1876, three years after Liebig's death. His *Animal Chemistry, or Organic Chemistry in its Applications to Physiology and Pathology* (1842) enjoyed a similar popularity.

The first edition of the *Organic Chemistry* was very largely a critical review of the work done in agricultural chemistry up to 1840. Liebig himself lacked practical farm experience, and his own researches were insufficient to give him more than a piecemeal understanding of the subject. Nevertheless he charged forth with his customary self-assurance. He attacked the botanists and physiologists because they knew no chemistry and asserted that there had been no application of chemistry to agricultural problems since the publication of Davy's book. Then he proceeded to explore the subject, drawing heavily on the work of approximately one hundred authors, a great many of them contemporaries.

The first part of the book dealt with the nutrition of plants, the second with fermentation, putrefaction, and decay. He considered as basic to the growth of crops: (1) carbon and nitrogen: (2) water and its elements; and (3) the soil as a source of inorganic elements. Humus he discounted as unimportant, except as a source of carbon dioxide which he believed absorbable by plant roots. Nitrogen was assumed to come from the atmosphere in the form of ammonia. Contrary to the views of others, Liebig considered soil nitrogen of no importance.

This latter point caused considerable controversy during the next two decades. Liebig overestimated the quantity of ammonia in the atmosphere and underestimated the amount of nitrogen in manures. Boussingault and Gilbert and Lawes established beyond doubt that soil nitrogen has great significance.

Liebig stressed the role of inorganic elements in plant nutrition, emphasizing the need for phosphates, potash, soda, lime, and magnesia. His erroneous suggestion that there was an equivalency among alkalies was attacked by Sprengel who held that these elements had a specific role other than the neutralization of acids. It had been known since the time of Glauber that saltpeter has a highly stimulating effect on plant growth, a point confirmed in pot experiments by Francis Home (1757). Home had also made tests with potassium sulfate, which he called one of the greatest promoters of vegetation. Ordinary salt he reported "an enemy of vegetation."

Some of Liebig's unsound views on plant nutrition resulted from his accepting unreliable ash analyses. Assisted by Will and Fresenius, he developed reliable procedures which were used for very extended mineral analyses in the Giessen Laboratory during the following years. As better data became available, he corrected some of his statements in later editions of his book.

Such revisions were by no means universally made, and certain errors persisted to the end of his life. Liebig was tremendously self-assured, a trait frequently reflected in his students. He readily criticized the ideas and work of others and in turn doggedly held to his position when he found himself under fire. His deserved status as a leading chemist, coupled with his readiness to speak authoritatively on a wide range of subjects, gained for him a devoted following, particularly in England and America. The importance of inorganic nutrition of plants and the Law of the Minimum were considered pronouncements from the highest level. Sprengel's prior work on both of these matters was overlooked and almost completely lost.

Potassium, Liebig believed, should be fused with silicates and applied as a ground glass in order to prevent leaching. This actually placed the potassium into a form that is not readily available. The fallacy of Liebig's reasoning was demonstrated in 1850 by J. T. Way in his studies of the absorptive capacity of soils for salts ordinarily very soluble in water. Liebig also advocated de Candolle's theory of toxic soil excretions.

Despite the shortcomings of his work, Liebig must be counted a major contributor. By emphasizing the importance of inorganic elements, particularly phosphorus and potassium, he stimulated the growth of a commercial fertilizer industry. By criticizing his contemporaries, he forced them to clarify their thinking and improve their experiments. His enthusiasm created an interest in soil chemistry and plant nutrition which resulted in widespread research. And the growth of agricultural experiment stations in Germany and, following the passage of the Morrill Act in 1862, in America resulted in large part from the influence of Liebig and his students.

PHYSIOLOGY AND MEDICINE

Although medicine in the nineteenth century began to consciously adopt the findings and methods of the basic sciences, the fusion of physiology and chemistry was only slowly achieved. Before midcentury the two disciplines hardly spoke the same language, and another quarter century passed before physiological chemistry gained respectability. In part this was because chemists had to reach a position where they knew what they were talking about; and in part because of the traditional reluctance of medical men to realize that another discipline has anything worth exploring from the medical point of view. Nevertheless the growth of physiology through the investigations of Johannes Müller, Du Bois-Reymond, Helmholtz, Magendie, Ludwig, Bernard, and Virchow was bound to create areas of contact with the work of Liebig, Mulder, Hünefeld, Marchand, Maly, Hoppe-Seyler, and Kühne.

From the time of Hippocrates there had been a belief in a single *aliment*, or nutritive component, in foods. Galen wrote about assimilation as a process in which the aliment was separated from the food mixture and used for body needs. This viewpoint was popular into the nineteenth century; it figured in Beaumont's work on digestion (1833).

Galen's theory was attacked in 1834 by Prout in his *Treatise on Chemistry, Meteorology and the Function of Digestion*. Prout suggested three principal constituents of food: *saccharina, oleosa, and albuminosa*, all present in milk which he considered the prototype of the perfect food. The two forms of *albuminosa*, a water-soluble *gelatin* and an insoluble *albumen*, contain nitrogen; Prout obtained them by boiling flesh. Albumen is present in blood and other body fluids, in muscle fibers, in milk, and in plants as gluten. Gelatin is derived from all animal tissues, but skin is particularly rich in it. It is not present in skin as gelatin, but is brought out as such by boiling. Treated with acid, it changes into a form looked upon as a sugar (glycocoll, or glycine), the change being considered analogous to the change of starch into glucose.

In his *Animal Chemistry* Liebig attempted to deal with subjects of physiological importance. In addition to fermentation and putrefaction he treated the nature and role of food materials at considerable lengths. He argued that food components belong to two classes, plastic and respiratory. By *plastic* he meant the nitrogenous substances Mulder had called *protein*. Liebig differed violently with Mulder on details regarding proteins, but he too used elaborate chemical formulas to represent them. As he came to realize the complexity of these compounds, he did not include most of the formulas in later editions of his works. Liebig considered proteins the muscle materials in food which during exercise are converted to urea, uric acid, and similar degradation products for excretion in the urine. His respiratory category included fats, sugars, starch, and alcohol—products which served as a source of heat during oxidation. He was interested in animal heat and attempted to calculate the heat produced by various foods. His view that animals are able to synthesize fat from nonfatty foods was opposed by Boussingault and Dumas and led to a long controversy resolved only when it was shown that pigs and geese fed on vegetable foods essentially free from fat still laid down fat deposits and gained weight.

Digestion

Work in various quarters contributed to the first significant progress in unraveling the digestive process. Many chemists had been interested in the fate of foodstuffs in the body, but the first real opportunity for study came after a shooting fray at a trading post at Fort Mackinac on the United States–Canadian frontier in 1822. Alexis St. Martin, an Indian, came out of the incident with his stomach muscles shot away and his stomach open. The army surgeon at the post, William Beaumont (1785–1853), did his best to patch up the wound and make the patient comfortable; death appeared inevitable. Surprisingly the patient survived, but with a gastric fistula open to the outside world.

Beaumont took advantage of the situation to carry on research on gastric digestion. He tied pieces of meat and other foods to strings, inserted them into St. Martin's stomach, and withdrew them after various intervals to observe the progress of digestive changes. The experiments were continued over a period of years, at intervals when the rebellious subject could be kept tractable. Beaumont studied not only normal gastric digestion but also the influence of alcohol and of the subject's psychological state at mealtime. Samples of gastric juice were collected and sent to Berzelius for examination. The results of Beaumont's experiments were published in 1833 as *Experiments and Observations on the Gastric Juice and the Physiology of Digestion.*

The most important European work on digestion was done by Claude Bernard (1813–1878), the French physiologist at the Collège de France. Bernard studied under François Magendie (1783–1855) who differentiated between nitrogenous and nonnitrogenous foods, recognized nitrogenous foods in plants, and concluded that since animals cannot survive on diets lacking nitrogenous components they are unable to synthesize such substances from other materials. While still a student of this pioneer in experimental physiology, Bernard prepared a thesis dealing with the role of gastric juice in digestion. In later studies he observed the effect of saliva and the influence of pancreatic juice in the digestion of fats. He showed that glucose is at all times present in the blood and it is stored in the liver as glycogen and to a lesser extent as muscle glycogen. He found that a puncture of the medulla oblongata brings about an increase in blood sugar and a consequent elimination of large amounts of blood sugar in the urine. His procedure for creating artificial diabetes made possible the experimental study of the disease.

Bernard's influence on physiology was profound, not only in the realm of digestion but in other areas as well. His book *Introduction à la médecine expérimentale* (1865) has remained a classic on the experimental approach, not only for physiologists but for scientists in general.

Pharmacology

Bernard and his teacher Magendie laid the foundations of pharmacology. Magendie probably more than anyone was responsible for developing an experimental approach to physiology. Most of his predecessors had been contented observers. He however experimented on dogs, cats, and other animals, frequently facing the opposition of antivivisectionists. Magendie, a positivist, helped free biology from the grip of vitalistic doctrines. Toward the end of his career he studied strychnine, morphine, bromides, and iodides and was influential in introducing these substances into medical practice.

Bernard studied the effects of curare, the poison used by South American tribesmen on arrow heads and blowgun darts. His predecessors had taught that toxic substances have a generalized effect on the whole body. Bernard found that curare has a paralytic effect on the motor nerves and demonstrated that poisons generally have a local action on some part of the body. His argument that drugs behave in a similar fashion introduced a new viewpoint to the use of medicines.

RESPIRATION AND ENERGY BALANCES

Interest in respiratory problems was voiced in ancient times and intermittently thereafter. But the understanding we have achieved and the work which continues began with Lavoisier whose collaboration with Laplace has already been mentioned. Lavoisier and Laplace developed an ice calorimeter which measured the heat liberated during combustion and animal respiration. A combustible substance or a small animal was placed in the inner chamber which was surrounded by two concentric outer chambers containing ice. The outermost ice chamber insulated the inner ice chamber from the environment. The amount of heat liberated by the combustible or by the animal was calculated from the amount of water collected from the inner ice chamber during the experiment. Their calorimeter was a refinement of one devised by Black in 1761 when he was arriving at the concept of latent heat of fusion. Black's calorimeter consisted simply of a hollowed block of ice. After drying a substance and noting its temperature, he would place it in the block. The amount of ice melted was measured by soaking up the water in a previously weighed sponge and recording the increase in weight.

In experiments conducted between 1779 and 1784, Lavoisier and Laplace measured the quantities of heat evolved during various chemical changes. They placed a guinea pig in the calorimeter for ten hours; the heat evolved by the animal melted thirteen ounces of ice. Then they determined

Fig. 16.2. LAVOISIER-LAPLACE ICE CALORIMETER.
(From *Œuvres de Lavoisier*, J. B. Dumas, Ed. [1862], vol. 1, Plate 6.)

the quantity of heat produced during the combustion of a known weight of charcoal and obtained the amount of carbon dioxide liberated on burning a weighed quantity of charcoal. From the charcoal experiments they learned how much heat is produced during the formation of a definite quantity of carbon dioxide. Next the amount of carbon dioxide given off by the guinea pig during ten hours was measured; it corresponded to the melting of 10.5 ounces of ice, somewhat less than the amount of heat produced when the animal itself was in the calorimeter. Taking into account probable errors, the experimenters considered the agreement satisfactory and concluded that respiration is a form of combustion.

The very slow combustion was assumed to take place in the lungs where carbonaceous matter of the blood was converted to carbon dioxide, with caloric liberated from the oxygen in the process. The caloric, they believed was absorbed by the moisture of the blood and used to maintain body temperature.

After Cavendish demonstrated the formation of water during the burning of hydrogen, Lavoisier realized that he had failed to take into account the presence of hydrogen in foods and its absorption of oxygen during respiration. He showed that air exhaled by the lungs contains less oxygen than atmospheric air but is enriched in carbon dioxide and moisture.

In later respiration experiments Lavoisier was assisted by Armand Séguin (1765–1835) who frequently served as a human guinea pig. Madame Lavoisier made two sketches of such activities in the Arsenal Laboratory. The experiments were tedious and uncertain, but a good deal of progress was made. However the onset of the Revolution and Lavoisier's involvement with it prevented his concluding the work satisfactorily. Preliminary reports, but no comprehensive summary, were published during his lifetime. Some years later Séguin was able to bring much of the work into print.

Contemporaneous with Lavoisier's earliest work was that of Adair Crawford (1748–1794) in Britain. Crawford devised a water calorimeter in which he measured the heat produced by burning wax or charcoal and by a live guinea pig. He concluded that in the cases of the wax and the animal, heat production is associated with the conversion of pure air into fixed air and water and that charcoal gives only fixed air.

With the rise of the energy concept in the 1840's there came a renewed interest in heat relationships in the living animal. Most of the men involved in the recognition of the Law of Conservation of Energy were interested in biology. While the development of thermodynamics was quickly to become the activity of physicists, many biologists were deeply interested in energy production and utilization.

In 1849 Regnault and J. Reiset devised a closed-circuit apparatus for measuring an animal's oxygen consumption and carbon dioxide elimination. They originated the concept of the *respiratory quotient* (volume of CO_2 exhaled/volume of O_2 used) and measured the oxygen consumption of animals of various sizes, learning that small animals use more oxygen per unit of body weight than large animals; for example, sparrows use more per gram than chickens. They correctly related this observation to the greater surface area per unit weight in small animals which permits greater loss of heat through radiation.

In 1852 Friedrich Bidder and Carl Schmidt of Dorpat showed that the heat loss in fasting animals is the same for animals with the same body volume, surface, and temperature. Bidder and Schmidt also observed the *specific dynamic action of proteins*. A cat's oxygen consumption, they noted, increases strikingly after a large meal of meat. While increased oxygen consumption follows eating any food, it is a marked increase only with proteins, which, they concluded, stimulates metabolic processes.

Extensive research on energy metabolism were carried on at the University of Munich by Karl von Voit (1831–1908) and Max von Pettenkofer (1818–1901) beginning in 1866. Using a Voit-designed respiration calorimeter, they measured oxygen consumption, carbon dioxide and water elimination, and heat production under a variety of circumstances. They followed Liebig's suggestion (1842) that nitrogen elimination in the urine can be used as a measure of metabolized protein. Studies were made on fasting dogs and men and on the same subjects eating various amounts and kinds of foods. They found the respiratory quotient for carbohydrates to be 1, fats to be 0.7, and proteins to be 0.8. In a fasting man they found the R.Q. to be 0.69; hence they concluded that his energy needs were being met by combustion of fat. This conclusion had to be modified to include some burning of protein. Fat was burned during work.

Liebig had supposed that protein metabolism was related to muscular work, and that fats and carbohydrates were consumed by oxygen. Voit showed that muscular work does not increase protein metabolism and that oxygen supply and metabolism are not proportional. He established that, rather than metabolism being caused by oxygen, the rate of metabolism determines the oxygen needs.

Voit and Pettenkofer carried on a very active research program. Many of their students went on to other institutions and initiated similar programs. Particularly active were Rubner at Marburg and Berlin and Atwater in the United States. Max Rubner (1854–1932) measured the heat of combustion of various food materials with the bomb calorimeter and compared these results with the heat liberated by the same food substance when fed to an animal in an animal calorimeter. He found that starches and fats give the same value in each case, but proteins burned in the bomb calorimeter released more heat. This had been expected since proteins are not burned completely in the body, and there is still energy left in the urea formed. Rubner showed that a correction for urea is inadequate since other nitrogenous waste products such as uric acid and creatinine are present and have a different $N:C$ ratio. He also examined the extent to which nitrogen is excreted in the feces.

Rubner's Isodynamic Law formulates the replacement of one food by another on the basis of energy value. Thus 100 grams of fat is equivalent to 232 grams of starch, 234 grams of cane sugar, or 243 grams of dried meat. His standard values for energy content of food components have been in constant use since their promulgation:

1 g. protein	4.1 calories
1 g. carbohydrate	4.1. calories
1 g. fat	9.3 calories

Wilbur Olin Atwater (1844–1907) took a doctorate at Yale under Samuel W. Johnson, a former Liebig student, before he studied in Leipzig and Berlin. When he returned to the United States, he accepted a professorship of chemistry at Wesleyan University in Middletown, Connecticut. Soon thereafter the State of Connecticut legislature established an experiment station at Wesleyan with Atwater as director. His first major project, one which held his attention the remainder of his life, was a study of the composition of American food materials. Up to this time the only comprehensive source of such information was König's *Chemie der Menschlichen*

Nahrungs- und Genussmittel, which dealt only with European foods. Assisted by his own staff and those of other stations, Atwater published extensive data dealing with the moisture, fat, protein, carbohydrate, mineral content and the calorific value of American foods.[4]

After protracted negotiations he received government funds to build a large respiration calorimeter for studying human metabolism. The calorimeter was developed in collaboration with the physics professor Edward Bennett Rosa (1861–1921). It was so efficient that when a weighed amount of alcohol was burned in it, 99.8 per cent of the carbon dioxide was recovered along with 99.9 per cent of the heat. Starting in 1897, human experiments were carried out in collaboration with Francis G. Benedict (1870–1957), also a chemistry professor at Wesleyan. Atwater and Benedict were able to confirm earlier European work and extend respiration studies into new fields. Benedict later became director of the nutrition laboratory established in Boston by the Carnegie Institution. Here he studied metabolism of healthy human beings and of hospital patients. Other American students of metabolism were Graham Lusk (1866–1932) and Eugene F. DuBois (1882–1959) of the Cornell Medical School and the Russell Sage Institute of Pathology and Henry P. Armsby (1853–1921) of Pennsylvania State College. Armsby, who operated a calorimeter large enough to hold an adult steer, contributed much to our knowledge of the energy requirements of livestock and to a more scientific approach to feeding.

FERMENTATION AND PUTREFACTION

Until 1840 fermentation was believed a chemical process. Although yeast was known to be associated with the process, it was regarded as an inert, nonliving substance.

In his *Animal Chemistry* Liebig dealt with the process at great length, formulating a theory strikingly similar to that proposed by Stahl a century earlier. Liebig suggested that a decomposing substance caused a body in contact with it to enter a decomposing state. Sugar molecules, he maintained, were held together by a *vis inertiae*. When they came into contact with a decomposing substance such as yeast, the *vis inertiae* was destroyed by the vibratory motion of the decomposing yeast particles. And yeast, he argued, decomposes relatively readily because of its nitrogen content. Liebig had long been interested in fulminates and he emphasized that nitrogen compounds are particularly prone to undergo decomposition.

Liebig supposed putrefaction to be a process similar to fermentation

[4] W. O. Atwater and C. D. Woods, *The Chemical Composition of American Food Materials,* U.S. Dept. Agr., Off. Exp. Sta., Bull. 28 (1896). An enlarged edition was published under the authorship of Atwater and A. P. Bryant in 1906.

but one which led to a greater variety of by-products. Decay he considered a slow process of the same sort.

A new view of fermentation grew out of discoveries which led to the development of bacteriology. The initial clues were provided by Cagniard de Latour, Schwann, and Kützing. In 1836 Charles Cagniard de Latour, French physicist, observed beer yeast to be composed of microscopic globules probably belonging to the vegetable kingdom. He found similar organisms in wines and showed that drying and freezing did not harm the vitality of the organism. The next year he proposed that the globules were responsible for the formation of carbon dioxide and alcohol.

Shortly after this Theodor Schwann sought to demonstrate that fermentation and putrefraction were caused, not by the oxygen in air, but by a substance in air which is destroyed by heat. The oxygen theory of fermentation was an outgrowth of the success of the French chef Nicolas Appert (*ca.* 1750–1841), in preserving food through a combination of heating and hermetic sealing. Gay-Lussac suggested that the preservative effect was due to lack of contact with oxygen. It was this hypothesis that Schwann set out to disprove. He was able to show that boiled organic substances readily decompose when exposed to common air, but not when the air brought into contact with them is first passed through a red-hot pipe. Schwann recognized yeast as a plant which grows by budding and that after it buds gas and alcohol develop in grape juice.

Friedrich Kützing independently reported the living nature of the yeast plant and associated it with fermentation. He also observed the organism in "mother of vinegar" that causes the conversion of alcohol to acetic acid. Liebig later elucidated the path of acetic fermentation by isolating acetaldehyde as an intermediate step in the change.

Leading chemists such as Berzelius, Wöhler, and Liebig regarded the biological theory of fermentation as unsound, particularly when other investigators failed to verify Schwann's results. In his *Jahres-Bericht* for 1839 Berzelius contemptuously dismissed Schwann's work, arguing that yeast was no more a plant than were various noncrystalline precipitates such as alumina. Kützing's work was similarly treated. Berzelius considered fermentation to be due to a "catalytic force" and looked upon yeast as such a nonliving catalyst. Liebig and Wöhler prepared an anonymous skit which appeared in the *Annalen* under the title "The riddle of vinous fermentation solved." Using a supermicroscope made by Pistorius, Liebig and Wöhler's "researcher" observed that the yeast globules swelled when placed in a sugar solution and gave birth to tiny animals which reproduced with amazing rapidity. Their form differed from that of previously described microscopic organisms, the closest resemblance being to a Beinsdorff still. The animals had a suctorial snout with which they devoured sugar, a stream of alcohol issued from the anus at the same time that carbon dioxide bubbled forth from enormous genital organs.[5]

[5] Anon. (Liebig and Wohler), *Ann.*, **29**, 100 (1839).

Liebig published his serious views on fermentation in Poggendorff's *Annalen der Physik und Chemie* that same year. His argument was essentially the same as that to appear in his *Animal Chemistry* three years later and to which he clung doggedly the rest of his life. The yeast globules or other animalcules seen in fermenting liquids were the result of fermentation, he insisted, not the cause.

Mitscherlich found that yeast globules are large enough to be retained by fine filter paper. In a subsequent experiment he showed that fermentation takes place only in sugar solutions in direct contact with yeast. Despite this he believed that fermentation was brought about by contact with yeast rather than through any metabolic activity of the yeast.

Bloudeau at Rodez studied nonalcoholic fermentations such as lactic, acetic, butyric, urea, and fatty and concluded that they were all caused by vegetative growths.

Louis Pasteur's interest in fermentation which dated from 1854 was an outgrowth of his work on the optically active amyl alcohol (1-methylbutanol-1) associated with the fusel oil produced during fermentation. He studied lactic acid fermentation and discovered the responsible organism. During the next two decades he identified organisms responsible for various fermentations. His studies of butyric fermentation (1861), in which he observed that the organisms function without oxygen, led to his developing the concept of aerobic and anaerobic organisms. Pasteur's conception of fermentation as life without oxygen brought him into another conflict with Liebig who in 1869 still championed his vibratory theory of fermentation, which he had modified in only minor details since its publication in 1839.

Some researchers began asking whether fermentation might be caused by a soluble ferment such as the diastase of malt or the pepsin of gastric juice. Moritz Traube presented such a concept as early as 1858. Numerous attempts including Pasteur's to demonstrate such a ferment failed. Pasteur's idea that fermentation was caused by cell metabolism and was only possible in the presence of living cells was not successfully challenged until 1897 when Eduard Buchner (1860–1917), working at Munich, prepared a yeast juice devoid of living cells. This enzyme, which catalyzed fermentation of sugars without any living cells being present, was named "zymase."

PUBLIC HEALTH

Chemistry came to play an important role in public health problems in many ways. As the science developed so grew the potential to help man by using this knowledge to prevent and control disease, or to harm him by using it, for example, to prepare fraudulent or health-impairing foods and drugs.

The public health movement of the nineteenth century was stimulated not by science but rather by the evils associated with rapid industrialization. Large numbers of factory workers crowded into the slums of rapidly growing cities which had no provisions for sanitary control. Ventilation was bad, sewage and rubbish disposal was primitive or nonexistent, water supplies were polluted, and foods were adulterated in the majority of industrial centers. As a result tuberculosis was endemic, and diseases like cholera, typhoid, typhus, yellow fever, and smallpox spread with violence once they had been introduced into a community. No one knew what caused these epidemics; various hypothesis attributed them to filth, noxious airs, or contagion. It was known that disease was less common in rural communities where fresh air, cleanly conditions, pure water, and pure food were readily available.

Sanitary reform movements originating between 1820 and 1860 were largely associated with the liberal movements of the time. Some of the disciples of Jeremy Bentham, whose doctrine of "the greatest good for the greatest number" gave rise to Utilitarianism, were foremost among the agitators in Britain and America. In Germany Rudolf Virchow concluded that a typhus epidemic in the mill district of Silesia in 1848 had been caused by lack of sanitation. Virchow's report and subsequent activities stressed the importance of healthful living conditions in the prevention of disease. His position, however, actually handicapped his career in the reactionary climate which flourished in Germany after the 1848 revolution. Only after Bismarck's rise to power did the German government give proper attention to public health problems. Sanitary control and medical insurance programs were inaugurated in order to stave off the rise of the socialist movement in Germany.

Max Pettenkofer of Munich was a leader in German hygienic activities. He decided to be an apothecary after he had failed to succeed as an actor. Studying in Liebig's laboratory he discovered the base creatine in meat juice. Liebig undertook researches on meat which led to the development of the dried boullion marketed commercially as Liebig's Extract of Beef. Liebig extolled this product as a health food which supposedly contained the nitrogenous components necessary for muscular repair.

Pettenkofer eventually became professor of pathological chemistry at the University of Munich. Munich then was a typically unsanitary European city. In 1854 it suffered its third cholera epidemic in a generation; epidemics of typhoid and typhus were commonplace. Dr. John Snow in London had recently completed what was to prove a classic investigation in which he related an epidemic to pollution at the Broad Street pump. Pettenkofer failed to find a similar relationship in Munich but developed a ground water theory of cholera and typhus which, while incorrect, led to improvement of the drainage and water supply systems. The incidence of epidemic disease dropped to a very low level.

Pettenkofer's career in public health spanned many years. He was appointed to the first university chair of hygiene in 1866, and Munich established a magnificent Institute of Hygiene for him in 1872. He was widely sought as a speaker and consultant. In 1883 he was one of the founders of the *Archiv für Hygiene*. Despite his insight and accomplishments he refused to accept the germ theory of disease. When Koch discovered the cholera bacillus in 1883, Pettenkofer showed his contempt for the organism by drinking a virulent culture—without harm to himself.

In France progress in public health was achieved more rapidly than in other parts of Europe. The work of René Louis Villermé (1783–1863) and his associates set a pattern followed elsewhere. In Britain the movement had difficulty, in part because of the dominance of the economic philosophy of the Manchester School. In the United States a Sanitary Commission had been set up during the Civil War to deal with military problems, but reform lagged during Reconstruction days, despite abundant evidence of the evils of rapid industrialization in Northern cities.

Water Purification

Sand filtration, a method of water purification introduced in London in 1829 by the Chelsea Water Company, proved valuable and was widely adopted. Results frequently were disappointing because poor civil engineering often resulted in recontamination.

As populations grew cities had to turn to surrounding uplands and moorlands for their water supply. The water obtained, which often was peaty, highly colored, and soft, required flocculation with alum and then rapid filtration through sand. Soft water, which attacked lead pipes and caused lead poisoning, was deacidified with lime. Chlorination was not introduced until 1905 when London was once again faced with a typhoid epidemic.

The problem of sewage disposal was especially serious. The use of the nearest river as a cesspool has always seemed an obvious solution. Surface waters, which have an enormous capacity to purify themselves, could not cope with the increasing pollution with domestic and industrial wastes. The situation in the Thames was so bad in 1855 that Faraday was stirred to write to *The Times* in protest.[6]

Pollution of water with sewage was demonstrated to figure prominently in bacterial epidemics. "Finding" disease germs in water was, however, a difficult job. Chemical analysis proved useful in obtaining presumptive evidence for sewage contamination. The Nessler colorimetric method of determining the ammonia content of water was widely adopted as a means of spotting possible sewage contamination.

---

[6] Michael Faraday letter to *The London Times*, published July 21, 1855. Also see Ernst Cohen in Diergart, *Beitrage aus der Geschichte der Chemie, dem gedächtnis von Georg W. A. Kahlbaum*, Deuticke, Leipzig, 1909.

Late in the nineteenth century a whole body of water- and sewage-analysis procedures were developed—methods for measuring traces of nitrites and nitrates, oxygen consumption, and bacterial count as well as methods for measuring hardness, metals, and other substances of domestic or industrial interest. Official and semiofficial bodies like the American Public Health Association began to publish standard analytical procedures that would stand up in the courts.

Milk similarly came under scrutiny after it had been demonstrated that it might be a carrier of disease. By the end of the century pasteurization was being advocated as a public health measure to insure the safety of city milk supplies. The advocates of pasteurization met the opposition that usually faces proposals of the sort—from dealers who wish to avoid another operation and from consumers who suspect ulterior motives on the part of anyone who proposes change.

Food and Drug Adulteration

During the nineteenth century the problem of pure foods proved increasingly complex. Food adulteration, of course, had always been a problem wherever foods were sold or traded. The Romans had to contend with adulteration of bread, wine, vinegar, and herbs. Medieval records indicate that the problem was an important one, and local laws sometimes dealt severely with transgressors. Prior to the Industrial Revolution producers and consumers were apt to be neighbors if not friends, and this closeness no doubt served to discourage the temptation to adulterate.

Intimate contact between producer and consumer was one of the "victims" first of the rapid growth of industrial centers and later of scientific and technological contributions to food processing. The canning process was introduced early in the nineteenth century. By the end of the century meat, fish, fruits, and vegetables canned in one place were on grocers shelves throughout the world. Sugar refining, already big business at the beginning of the century, prospered with the introduction of improved methods of clarification, crystallization, and evaporation. The discovery that acids will hydrolyze starches to sugars led to the industrial production of commercial glucose. With the introduction of the roller process for flour milling there came a widespread demand for white rather than dark flour. Condensed milk could be used under conditions that prevented the maintenance of supplies of fresh milk. The development of the refrigerator car led to the centralization of meat packing. Oleomargarine was developed in 1869 by Hippolyte Mège-Mouries who sought the prize offered by Napoleon III for a commercial process for a fat as appetizing, nutritious, and stable as butter. Mège-Mouries had observed that starving cattle continued to produce milk, apparently from their own body fat. Out of beef stearin he pressed an "oleo oil" which was made to resemble butter by chilling, working, and coloring. Manufacture of the

product spread to many parts of the world. Vegetable fats frequently
were found in the product sold as genuine butter. This was but one of
many associated with the newly developed industries.

In 1802 the Parisian municipal government created the *Conseil de
Salubrité* (Health Council), an agency which regulated the food trader,
was responsible for controlling food adulteration, epidemic animal diseases,
and unhealthy trades, and governed prisons and public charities. Most
council members came from the medical, veterinary, and chemical professions. Many of the provinces established similar *conseils*. Unfortunately
the *conseils* were not highly effective. Insuring the purity of drugs proved
particularly difficult. A. P. Favre's book dealing with the subject was
published in 1812, and several others appeared before Antoine Bussy
and A. F. Boutron-Charland's definitive treatise in 1829. Several books
on food adulteration and food analysis were published in the forties.
The two-volume *Dictionnaire des altérations et falsifications des substances
alimentaires, médicamenteuses et commericales, avec l'indication des moyens
pour les reconnaître* (1850) of A. Chevallier was a landmark in the field.
Chevallier complained that his many petitions to the government had
been ignored. A general food and drug control law was not passed until
1884.

German experts also began to publish articles and books, but German
programs for food and drug control varied widely from kingdom to kingdom. Only after unification was concerted action taken. An Imperial
Health Bureau was organized in 1876 following outbreaks of trichinosis
from infected pork. The general food law passed three years later was
amended as the need arose. A meat inspection act, prompted by low-
quality imports from the United States, was passed in 1900.

Little was done in England. The nation's political leaders, most of whom
subscribed to laissez faire economic policies, believed government should
interfere with or regulate business as little as possible. They might recognize
that a menace to public health existed, but the notion that the government
should remedy the situation was unacceptable.

Early in the 1800's Frederick Accum (1769–1838), a German chemist,
arrived in London and set up a shop dealing in scientific apparatus.
In 1820 Accum published *A Treatise on the Adulteration of Foods, and
Culinary Poisons*. The title page bore the picture of a cooking vessel with
a Death's Head superimposed. Two serpents were twined around the vessel
which bore the inscription "There is Death in the Pot, 2 Kings C.IV.V. 40."
The book immediately became a best seller. It and the author were soon
nicknamed "Death in the Pot." The book dealt with the adulteration of
foods, drugs, and household needs and gave directions for detecting various
frauds. Accum cited merchants who had been convicted of fraud—including 28 druggists and grocers who had supplied illegal ingredients to
brewers and eighteen publicans who had adulterated beer. The use of

Fig. 16.3. FREDRICK ACCUM.
(Courtesy of H. A. Schuette.)

Fig. 16.4. HARVEY W. WILEY
(Courtesy of the Library of
Congress.)

copper sulfate for coloring pickles and sulfuric acid for fortifying vinegar, cream thickened with starches, spices which contained floor sweepings, and candy colored with vermillion (HgS), which in turn was frequently adulterated with red lead, were some of the evils Accum exposed. The book was widely read but the author was soon forced to leave London following a charge of removing leaves from books in the Royal Institution library. His remaining years were spent teaching in a technical school in Berlin.

For thirty years after Accum the only vigorous demand in England for reform was a highly sensational tract published anonymously in 1830. In 1850 Alphonse Normandy published his well-received *Handbook of Commercial Analysis* which recommended microscopic examination of foods and drugs. Soon thereafter Arthur Hill Hassall reported that with a microscope coffee can be distinguished from chicory, roasted grain, sweepings, and other common adulterants. He was commissioned by Thomas Wakley, publisher of the independent medical journal *The Lancet*, to head an Analytical Sanitary Commission set up to examine other foods. Reports were published regularly. Of 36 samples of coffee, only three were pure; alum was present in all 49 samples of bread. Of 28 samples of Cayenne pepper, only four were pure; thirteen contained red lead and one contained vermillion. Almost all candy samples proved colored with lead chromate, red lead, vermillion, copper arsenate, white lead, or gamboge.

In 1855 Parliament appointed a Select Committee to hear testimony on food adulteration and its effect on health. An Adulteration Act of very limited effectiveness was passed five years later. Finally in 1875 a comprehensive Sale of Foods and Drugs Act was passed. This, with subsequent amendments, proved an effective law. The Society of Public Analysts, an organization of chemists active in food and drug control, was established in 1874. Their journal *The Analyst* played an important role in the development of reliable analytical methods.

In the United States, where the need for legislation was no less urgent, many factors were responsible for delay. The South was in a depressed state during Reconstruction and was in no position to lead a reform movement. And since the South's was still primarily an agrarian economy the sale of fraudulent food was not as commonplace as in the North—although the sale of fraudulent drugs was a problem. Rapid industrialization in the North was characterized by questionable financial manipulations and arrogant business attitudes. Congress and most state governments were controlled by financial interests which openly fought reforms.

This was the state of affairs when Harvey W. Wiley became director of the Bureau of Chemistry in the United States Department of Agriculture in 1880. Trained as a physician and chemist, Wiley was familiar with the opportunities for falsification by food and drug processors. Early in his career he initiated a program of systematic food analysis and in 1884 helped found the Association of Official Agricultural Chemists which still plays an important role in the establishment of reliable methods for the analysis of drugs, foods, feeds, soils, fertilizers, and other agricultural materials.

At the turn of the century Wiley became an outspoken crusader for federal legislation to protect the consumer. In 1902 he began a series of experiments on young men connected with his department. They were fed different foods in order to determine the effect on health of such preservatives as borates, benzoates, salicylates, sulfites, formaldehyde, and coloring materials like copper sulfate. The studies were soon widely known as the "Poison squad experiments"; they helped win popular support for Wiley's points of view, but Congress remained disinterested. Some periodicals were at this time exposing fraudulent practices in the patent medicine industry, and in 1906 the meat industry was jolted by the publication of Upton Sinclair's novel *The Jungle*. While fictional, Sinclair's sensational exposure of packing house practices were sufficiently truthful and convincing to arouse the public. Congress was in effect forced to pass a meat inspection act and a pure food and drug act.

Food and drug control affected the training of scientists. Government departments responsible for law enforcement needed competent analysts, and industry began employing scientists to achieve the kind of processing control necessary to produce an acceptable product. In addition researchers became increasingly interested in probing the composition of foods and the processes of spoilage.

CONCLUSION

Nineteenth-century biochemistry developed very largely out of practical problems connected with agriculture and medicine, except for photosynthesis which had its inception in the work of the pneumatic chemists at the end of the previous century. Early progress was slow because of lack of knowledge of biological constituents and inadequacy of analytical methods.

The application of chemical knowledge to agricultural problems was closely associated with the rise of experimental stations. The early work dealt primarily with problems of plant and animal growth, stressing composition of fertilizers and feeds. This led naturally into studies on composition of biological materials and there gradually evolved a recognition of fats, carbohydrates, and proteins as principal constituents of biological materials. Some attention was given to mineral constituents as recognized by ash determinations but, except for gross constituents like phosphorus and potassium, the role of these elements in biology was not clearly understood.

Although there was progress in the study of digestion and fermentation, the complex nature of cellular metabolism prevented any major understanding of physiology. The role of enzymes in fermentation was only demonstrated in the last decade of the century. Perhaps the most significant progress was made in the study of animal respiration. The deep interest in thermochemical matters combined with the development of large and sensitive calorimeters made this a fruitful area of study.

Chemists figured prominently in the public health movement which developed during the century. The practical need for methods for the detection of pollution of water and adulteration of foods and drugs stimulated the development of organic analysis which in turn led to a new level of understanding of the constituents of biological substances.

CHAPTER 17

CHEMICAL INDUSTRY I. THE NINETEENTH CENTURY

THE RISE OF THE CHEMICAL INDUSTRY

The story of nineteenth-century industrial chemistry is in very large part the story of the Leblanc process, its rise and fall. Nearly all major developments and many minor ones were related to developments in the alkali trade. Although the Leblanc process was introduced in 1791 it did not become important until the nineteenth century was well under way. By the end of the century its survival was threatened by the Solvay process. Today the Leblanc process is only of academic interest.

In 1800 industrial chemical arts differed little from those of 1700 except in the manufacture of sulfuric acid, alkalies, and bleaching chemicals. Even though there was a considerable demand for these items efforts to supply them were disorganized. To a large extent even these chemical products were, as all others, processed in family-type establishments, frequently in association with apothecary shops. The apothecary trade was more compatible with chemical operations than any other. Certain families such as the Roses, the Gmelins, and the Trommsdorffs had carried on such businesses for several generations. Demand was often small, but there was required a broad degree of skill to supply that demand.

In industries like glassmaking, ceramics, metallurgy, sugar-refining, and soap-boiling, operations were sometimes on a sizable scale, but were highly empirical. A few chemists, particularly Black, Cullen, Lavoisier,

Berthollet, and Chaptal, had interested themselves in industrial operations, but for the most part, knowhow was passed on from father to son and master to apprentice without real thought of scientific implications.

Source materials were largely of local origin. Alkalies were chiefly potashes leached from wood ashes and lime burned from limestone. Salt was produced by evaporating seawater or mining rock salt deposits. Saltpeter for gunpowder was prepared from nitrates leached from manure piles or scraped from the wall of stables and burial crypts. Sulfur came from the volcanic deposits in Sicily, charcoal from the forests of Europe.

Naval stores, so necessary in the age of sail, came chiefly from the coniferous forests of Scandinavia and North America. Tanning barks and galls were gathered wherever suitable trees and plants were growing. Dyes likewise came from natural sources. Alum, used as a mordant, was prepared from suitable minerals. White lead for paints was made by exposing lead strips to the vapors of vinegar and fermenting manure or spent tanbark.

By the end of the nineteenth century the picture was a changed one. While some of the simpler operations mentioned above might still be carried out on a family scale, there was a highly developed chemical industry which had the proportions of big business. Acids and alkalies were being produced on an unanticipated scale. The fertilizer industry was making inroads despite the traditional conservatism of farmers. Synthetic dyes and drugs had eliminated certain natural ones from the market. New explosives were about to supplant black gunpowder.[1]

SULFURIC ACID

Sulfuric acid production was a well-established European industry at the beginning of the nineteenth century, and consumer demand was met without particular difficulty. John Roebuck's substitution of lead chambers for glass jars in 1746 stimulated production of the acid. Once restricted largely to the medical and specialty fields because of high price, sulfuric acid was used extensively in textile-bleaching and metal-cleaning when it could be produced less expensively. To satisfy increased demand, larger lead chambers were used, and modifications of these ultimately led to continuous operation.

With the large demand for sulfuric acid resulting from the successful development of the Leblanc soda process, the industry expanded rapidly and the shortcomings of the chamber process became more glaring. Particularly critical was the expense of the nitrate needed to supply

---

[1] For further information about the development of the chemical industry see A. J. Ihde, *Cahiers hist. mondiale* (*J. World Hist.*), **4**, 957 (1958).

the oxides of nitrogen used in the chambers. And the spent gases discharged into the atmosphere created a nuisance. Gay-Lussac suggested absorbing the spent oxides in concentrated acid trickling over coke in a tower. Although proposed in 1827 the scheme was not tried until 1842 and did not come into general use until the Glover tower was introduced in 1859.

John Glover (1817–1902), an English acid producer, showed that the oxides absorbed in a Gay-Lussac tower might be re-used in the chambers if the nitrified acid were passed through a second tower where dilution with water and exposure to sulfur dioxide fumes might take place. Glover's innovation brought about a high percentage of recovery of nitrogen oxides when used in conjunction with the Gay-Lussac tower. At the same time it produced the kind of concentrated acid required by the Gay-Lussac tower. The two towers complemented one another, making it possible to reduce significantly the amount of nitrate used and eliminate pollution of the atmosphere. The innovations were rapidly adopted in European manufacture.

The very small demand for "fuming" sulfuric acid, or oleum, prior to the development of the synthetic dye industry was supplied from 1755 until the 1890's by a single firm, that of Starck located in Nordhausen in Saxony. The acid was prepared by heating green vitriol ($FeSO_4 \cdot 7H_2O$) and dissolving the sulfur trioxide fumes in concentrated sulfuric acid to give a product containing as much as 15 per cent more sulfur trioxide than in pure sulfuric acid.

In 1831 Peregrine Phillips, a Bristol vinegar maker, attempted to prepare sulfuric acid by passing sulfur dioxide and air through red hot platinum tubing. Although the catalyst produced sulfur trioxide, the process was not a commercial success. With the rise of the dye industry there was renewed interest in the catalyzed oxidation of sulfur dioxide.

Rudolph Messel (1847–1920), a German-trained chemist with Squire, Chapman and Co. in England, developed an oleum process in 1875 which utilized the catalytic effect of platinum. He first decomposed sulfuric acid over platinum to obtain pure sulfur dioxide and oxygen which were then passed over platinum to form sulfur trioxide which in turn was dissolved in concentrated acid. Messel had learned in earlier studies that platinum is made ineffective as a catalyst by the impurities present in sulfur dioxide prepared by burning sulfur or roasting pyrites; hence his use of sulfuric acid as his source of sulfur dioxide.

Platinum catalyst poisoning was studied for a quarter-century by Rudolf Knietsch (1854–1906), a chemist with the Badische Anilin und Soda Fabrik. Arsenic was found to be particularly harmful to the catalyst. When sulfur dioxide was carefully purified before being passed over the catalyst, the contact process for producing sulfuric acid became highly successful. The Badische firm pioneered in the production of highly concentrated sulfuric acid in the nineties.

Other Acids

Sulfuric acid was preferred because it is a strong acid, a good oxidizing and powerful dehydrating agent, and relatively inexpensive. The other principal acids, nitric and hydrochloric, were of much less industrial importance. They were produced by the action of sulfuric acid on rock salt and commercial nitrates. Saltpeter provided the nitrate for nitric acid production until 1837 when Chile saltpeter began to be exploited. During the remainder of the century sodium nitrate from the Chilean deposits was a primary source of nitrate, not only for acid production but even for the production of regular saltpeter.

Sodium sulfate, or "salt cake," was a by-product of nitric and hydrochloric acid production. More properly it was the desired product in the first step of the Leblanc process, with hydrogen chloride gas as a by-product. Toward the end of the century, when the Leblanc process was suffering severely from the competition of the Solvay process, saltcake was less in demand and the acids became the principal product.

ALKALIES

Soda Ash

The need for dependable sources of cheap alkali was becoming an acute problem as the eighteenth century drew to a close, in part because traditional sources were being exhausted and in part because the requirements for glass- and soapmaking and other industries were increasing. The forests of Europe and North America supplied potash, but the European forests had been seriously depleted in order to satisfy the demand for charcoal. Kelp was gathered along the Scottish coast for potash, "barilla" was produced from the ashes of saltwort growing along the coast of Spain and soda was imported from the Egyptian natron lakes.

In 1775 France, in a vulnerable position because she depended on overseas supplies, offered through the Académie des Sciences a prize of 2,400 livres for a method for the preparation of alkali from common salt.

Around 1780 several processes, including those proposed by Scheele and by the Abbé Malherbe, were tried commercially with indifferent results. Nicolas Leblanc (1742–1806) modified Malherbe's process, which utilized sodium sulfate prepared by the action of sulfuric acid on sodium chloride, by introducing chalk which served to remove sulfide while also furnishing carbonate. The sodium sulfate was heated with charcoal and chalk to obtain *black ash*, a mixture of calcium sulfide and sodium carbonate. This might be used as such or extracted with water to obtain an impure sodium carbonate (white ash) upon evaporation. If sodium hydroxide were desired, the solution was treated with slaked lime. A plant for producing soda ash was opened at St. Denis, near Paris, in 1791. Because of various conflicting claims the Académie failed to award the prize to Leblanc. As a consequence of the Revolution and technical and financial problems the plant never operated continuously nor successfully, and Leblanc finally committed suicide.

The process was brought into successful operation in 1823 by James Muspratt (1793–1886), an Irish chemical manufacturer who had set up a chamber acid plant in Liverpool in 1822. Josias Gamble (1776–1846) began soda ash production in Glasgow two years later, and soon a number of other producers entered the business. Almost all producers had to contend with irate nearby landowners who were justly irritated by the damage done by sulfurous and hydrochloric acid fumes. In 1836 William Gossage introduced a crude tower for absorbing the fumes with water, but this merely substituted stream pollution for air pollution. Parliament finally passed an Alkali Act in 1863 to bring the nuisance under control.

By the 1860's the bleaching powder industry used sizable quantities of chlorine, and two processes for converting hydrogen chloride into chlorine began to be exploited. Henry Deacon, a chemical manufacturer at Widnes, developed the air oxidation of hydrogen chloride over a cuprous salt catalyst. Walter Weldon developed a scheme for the recovery of manganese dioxide ordinarily lost in the traditional oxidation of hydrogen chloride. These processes led to economies which permitted the Leblanc process to withstand the early competition from the Solvay process.

The reaction of ammonium bicarbonate and salt to give sodium bicarbonate was first studied by Fresnel in 1811. Various investigators sought unsuccessfully to develop the reaction commercially,[2] before the Belgian engineer Ernest Solvay (1838–1922) accepted the challenge. He and his brother Alfred opened a plant in Couillet, Belgium, in 1865. Production was extended to France a few years later, and British rights were obtained by Brunner, Mond and Co. in 1873. By 1890 Solvay process plants were operating in America, Russia, Germany, and several other countries. Competition between Leblanc and Solvay producers was bitter during the last three decades of the century; the latter had the superior process.

Hydroxides

Strong alkalies continued to be produced by the action of calcium hydroxide on potassium or sodium carbonate solutions. Water-slaked lime itself was used where low solubility or presence of calcium was no handicap. For soapmaking and certain other purposes the more soluble

[2] L. F. Haber, *The Chemical Industry during the Nineteenth Century*, Clarendon, Oxford, 1958, pp. 87–88.

Fig. 17.1. Ernest Solvay.
(Courtesy of the Edgar Fahs Smith Collection.)

hydroxides were necessary. Ammonia continued to be produced by the dry distillation of proteinaceous materials, but large quantities were also produced as a by-product of the coal-gas industry.

Toward the end of the century electrolytic processes began to be introduced for the production of sodium hydroxide from salt. Chemische Fabrik Griesheim built a plant in 1890 which utilized a diaphragm cell patented by Matthes and Weber in 1886. The by-product chlorine was at first converted to bleaching powder, but a method was soon devised for liquefying it and shipping it in iron containers. During the next years improved cells were developed in America by Le Sueur and in England by Hargreaves and Bird. A rocking mercury cell was independently invented by Carl Kellner, an Austrian, and Hamilton Y. Castner, an American who developed the process in England. By the end of the century the electrolytic production of alkali and chlorine was firmly established and helped further to wreck the Leblanc industry.

BLEACHING CHEMICALS

During the eighteenth century large areas of land were used as bleach fields near major textile-producing centers such as those of Scotland and Holland. While crude chemical agents such as buttermilk were used to aid the process, sunlight was the main factor in a process which often took months for completion. The lactic acid in buttermilk was helpful in removing mineral impurities which might act as mordants toward dyes. Wood ashes or those of kelp were used to remove greasy substances. When he began manufacturing sulfuric acid, Roebuck introduced the use of sulfuric acid as a substitute for buttermilk in the Scottish bleach fields.

Chlorine solutions were recommended as bleaches by Berthollet after he observed in 1785 that such solutions had a strong decolorizing action on cloth. Some trials were made in France and Scotland, but the unpleasant task of generating chlorine from hydrochloric acid with manganese dioxide kept the bleach from general use. Even the introduction of *eau de Javelle*, made by dissolving chlorine gas in potassium hydroxide solution, did not prove practical because of transportation problems.

Just before the end of the century a Scottish bleacher, Charles Tennant (1768–1838), prepared a bleach liquor by passing chlorine into a sludge of calcium hydroxide. One of Tennant's associates, Charles Macintosh (1766–1843), improved the product to obtain a dry powder which was easily transported and which could be turned into a bleaching agent by merely acidifying with sufuric acid. Manufacture was begun at Glasgow in 1799. The St. Rollox Works founded by Tennant became one of the principal producers of chemicals in Britain during the next century.

FERTILIZERS

The ability of chemical compounds to stimulate plant growth received much attention early in the nineteenth century, particularly in Britain. Books by Kirwan, Dundonald, Erasmus Darwin, and Davy testify to the thought and experimentation in this area. It had long been known that bones promote vegetation, but Dundonald showed that the phosphates in the bones give the beneficial effect. Bone meal however is absorbed very slowly so phosphate fertilizers did not become important until after the 1840's.

The soil fertility researches of Boussingault, Lawes, Liebig, and others created interest in artificial manures and led to the rise of the fertilizer industry. Liebig's enthusiasm and Lawes' practical experimentation stimulated the growth of the superphosphate industry in England. In 1842 Lawes developed the process for making phosphorus more readily available to plants by treating ground bones with sulfuric acid. A factory established at Depford to exploit the process was soon utilizing mineral coprolites and apatites as well as bones for a source of calcium phosphate. The superphosphate industry which grew from this start became the greatest consumer of sulfuric acid. In many cases acid plants were built and operated by fertilizer manufacturers themselves.

While potash had a long history of use as a soil improver, the supply was never adequate for extensive use as a fertilizer. Glauber showed that saltpeter had value in improving plant growth, but the demand for this compound in the production of black gunpowder prevented its general use in agriculture. Dundonald referred to the effectiveness of potassium sulphate, but this too was not generally available. Potassium fertilizers came into practical use only after the development of the Stassfurt salt deposits. These deposits of potassium and magnesium salts were discovered in 1856 while borings for rock salt were being made by the salt works operated by the Prussian government. Adolf Frank, a sugar refinery chemist, developed a commercially practical process for recovering the potassium salts. For sixteen years the deposits were worked by the state. Then private companies were permitted to operate, and separation of sodium sulfate and bromine was begun. By 1900 production of potassium salts, magnesium salts, and bromine was well developed and Germany was supplying the rest of the world with these chemicals.

Nitrogen fertilizers also began to be important around mid-century, even though Liebig believed that plants obtained their nitrogen from the air in the form of ammonia. The importance of the soil ammonia in plant growth was first suggested early in the 1800's. It was established conclusively in 1857 by the American chemist Evan Pugh (1829–1864) on the basis of his researches with Wöhler at Göttingen and Lawes at Rothamsted. Long before this nitrogenous wastes (manures, composts, fish meal) had been prized as fertilizers, but there had been little commercial activity except for the New England fish meal trade. Guano deposits from the offshore islands of Peru and Chile had been used by the Incas from the twelfth century, but regular imports into Britain did not begin until 1845. Starting then there was heavy demand for guano in Britain and the United States. Peru worked the deposits with coolie labor to establish a profitable business. The United States attempted to cut into this monopoly in 1856 by declaring 48 uninhabited and allegedly uncharted islands to be part of the American domain. A skirmish between American sailors and Peruvian soldiers became known as the Guano War.

The guano trade flourished for 35 years, but reckless exploitation of the deposits and slaughter of birds so depleted the sources that the business collapsed. This was true not only of the Peruvian islands but of those off the coast of southwestern Africa and elsewhere. Recently islands where enlightened conservation practices have been adopted have come back into production. The Peruvian islands are now so operated and furnish valuable fertilizer for Peruvian farmers.

The extensive deposits of caliche (Chile saltpeter, $NaNO_3$) began to attract interest in the 1830's, but production was slow in developing. Before 1850 imports into Europe were used principally for the production of nitric acid and potassium nitrate. During the fifties production in

Chile and Peru developed extensively, and use as fertilizer spread. In 1876 the Shank's process—a method invented in 1863 by James Shanks for separating sodium carbonate from black ash—was applied in Chile to the purification of caliche, and exporting of sodium nitrate boomed. The deserts came entirely under Chilean control as the result of the Nitrate War fought by Chile against Peru and Bolivia between 1879 and 1893. British firms dominated production until the beginning of World War I despite German efforts to gain a foothold.

Nitrogen fertilizer, in the form of ammonium sulfate produced as a gasworks by-product, also became important during the nineteenth century.

Other minerals used for centuries as soil improvers were gypsum, limestone, and marl. During the nineteenth century, the value of these minerals began to interest scientists. The enthusiasm of self-trained Edmund Ruffin for the use of marl on the inadequate soils of Virginia and North Carolina was representative of the pioneering occurring in many places.[3] Homer Jay Wheeler's sound, careful studies on soil acidity and its correction, conducted while he was at the Rhode Island Agricultural Experiment Station at the end of the century, represented a new level of the application of science to practical problems.

EXPLOSIVES

Black gunpowder remained a standard military explosive from its introduction in the fourteenth century until the beginning of the twentieth century. For most of this period it was of uncertain quality since the proportions and quality of the sulfur, charcoal, and saltpeter varied greatly. The French product was markedly improved following the saltpeter investigations initiated by Lavoisier after his appointment to the Gunpowder Commission in 1775. By the time that Napoleon rose to power, French powder was the best in the world. While production in France and other countries was a government monopoly, in many countries it remained in the hands of small operators. In the United States, Eleuthère Irénée du Pont de Nemours, who had once been associated with Lavoisier, began production of good quality powder in 1802 on the banks of the Brandywine River in Delaware. The business prospered during the War of 1812, and the du Ponts became a principal producer of American explosives during the following century.

The nitration of cellulose was first carried out by Theophile Jules Pelouze, but the explosive character of the product was first observed by Schönbein who improved the nitration by using concentrated sulfuric and nitric acids. Partially nitrated cotton gave pyroxylin, or nitrocellulose,

[3] A. J. Ihde, *J. Chem. Educ.*, **29**, 407 (1952).

which is inflammable but not explosive. Since it is readily soluble in ether, it came into medical use as collodion and served as a base in early plastics. Highly nitrated cotton (guncotton) explodes sharply with the formation of a large volume of gases when detonated with mercury fulminate, a compound first prepared by E. C. Howard in 1799. Schönbein initiated commercial production of guncotton in England but abandoned his efforts after several explosions resulted in loss of lives.

Commerical production was begun again after 1886 when Paul Vielle of the École Polytechnique prepared a gelatinized, hornlike explosive from the mixture of pyroxylin and guncotton in alcohol and ether. Its smooth-burning properties and lack of smoke made it a suitable propellant, and it rather quickly replaced black gunpowder following the Spanish-American War.

Ascanio Sobrero (1812–1888), an Italian chemist, first prepared glyceryl nitrate, or nitroglycerin, in 1846 by the nitration of glycerol. Commerical manufacture was begun in Sweden by Alfred Nobel in 1862, but sensitivity to shock prevented safe use of the oil as an explosive. In 1866 Nobel learned that it could be more safely handled if it were absorbed on an absorbent solid like kieselguhr, a silicaceous earth. It began to be marketed in stick form as *dynamite*. Since a detonator was now needed to initiate the explosion, it could be transported and handled with comparative safety. Blasting gelatin, a mixture of pyroxylin and nitroglycerin, was introduced by Nobel in 1875. Variations of blasting gelatin contained added nitrates of potassium, sodium, and ammonium and other ingredients such as wood powder. Dynamite and blasting gelatin were very profitable during the era of industrial expansion preceding World War I, and the Nobel enterprises flourished. At the time of his death Nobel left his estate as an endowment for annual awards in chemistry, physics, medicine, literature, and peace. (See Appendix IV.)

Toward the end of the century new explosives in the form of nitrated aromatic compounds began to appear in military circles. Picric acid, or *melinite*, was introduced by the French military in 1885; *lyddite* was the British equivalent. The Germans developed trinitrotoluene (TNT) for the same purpose, although this was not used until World War I. The explosives were liquefied for filling shells which burst with great violence when detonated.

GASEOUS FUELS AND ILLUMINANTS

Except for sporadic trials the illumination of buildings with the gas driven out of coal received little attention before 1800, but it developed quickly during the nineteenth century. The chemical nature of coal gas was revealed by the studies of Frederic Accum and his partner Alexander

Garden, but the practical problems of making the gas acceptable to the public were resolved principally by engineers.

The luminosity of the gas is due to certain trace components in what is largely a mixture of carbon monoxide, methane, and hydrogen. These trace components burn incompletely in what would otherwise be a non-luminous flame. Davy suggested in 1816 that luminosity was due to white-hot solids, in this case unburned carbon, in the flame. Berzelius and other chemists observed that thoria, lime, and certain other oxides incandesce when heated in hot flames. Except for the lime light developed by Thomas Drummond during his work on the Irish survey and a similar lamp perfected by Goldsworthy Gurney, there was no serious adaptation of the incandescent principle before the 1880's. When a gas was of such a composition that it did not give a luminous flame, appropriate compounds were added.

In the eighties the Austrian Auer von Welsbach developed a gas mantle made up of threads of ramie impregnated with the nitrates of thorium, cerium, and traces of other metals. When ignited in a gas flame the organic matter burned away and the nitrates were converted to oxides which incandesced at the temperature of the flame. Gas was now kept free from illuminants and efforts were made to have the temperature of combustion as high as possible.

Other fuel gases also came into use during the century. During the early days of the gas industry various organic materials were destructively distilled in order to obtain a fuel gas. In London, about 1820, a commercial gas was produced from fish oils and sold in cylinders. It was in this gas that Faraday discovered butylene.

Water gas (blau gas) was introduced after 1875 as the result of the work of Thaddeus S. Lowe, an American inventor and aeronaut. It had long been known that an inflammable gas forms when steam is blown onto hot coke. Lowe devised a process for alternately exposing hot coke to steam and air. Tessie du Motay, a French immigrant in America, was developing a similar process at the same time. Water gas, consisting primarily of hydrogen and carbon monoxide, came into use as a fuel gas, either directly or diluted with coal gas.

Producer gas, prepared by passing limited amounts of air through hot coke or coal, was developed as the result of Ludwig Mond's interest in obtaining a better source of ammonia for the Solvay operations of Brunner, Mond and Co. If steam as well as air is passed through the coal fire, some of the nitrogen in the coal is converted to ammonia. This is removed when the gas is passed through sulfuric acid. The stripped gas is low in fuel value but, because of the ease with which it is produced and its cheapness, it came into widespread use as an industrial fuel; it is especially suitable in such furnaces as the Siemens regenerative furnace used for producing open-hearth steel, glass, and phosphorus.

Blast furnace gas began to be used to heat incoming air for such furnaces after Bunsen and Playfair showed in 1845 that while it is too dirty for many purposes it has considerable fuel value.

COAL TAR CHEMICALS AND THE SYNTHETIC DYE INDUSTRY

Coal tar, which is formed as a by-product during the production of coke and coal gas, became a serious industrial nuisance by the mid-1800's.

Fig. 17.2. WILLIAM HENRY PERKIN.
(Courtesy of the Edgar Fahs Smith Collection.)

Demand for it as a wood preservative, road binder, and source of solvent naphtha consumed only a small fraction of the amount produced. Chemists paid little attention to tar before 1840. Garden and Kidd isolated naphthalene in 1819, and Runge isolated aniline, quinoline, pyrrole, and phenol in 1834. Anthracene was separated by Dumas and Laurent, and benzene was isolated in 1842 by John Leigh and again three years later by Hofmann.

Hofmann's careful study of coal-tar chemicals markedly advanced knowledge of their properties beyond the point reached by Runge and Laurent in their earlier work. Charles Mansfield, Hofmann's student at the Royal College, improved the process of fractional distillation to the point where good quality benzene and toluene could be produced; unfortunately he died from burns suffered when vapors around a large still caught fire.

Before Mansfield's work benzene and toluene had been considered only as possible industrial solvents. This attitude suddenly changed after 1856 when William Henry Perkin (1838–1907), a student at the Royal College, sought to synthesize quinine. Perkin reasoned that quinine $(C_{20}H_{24}N_2O_2)$ should result if he started with toluidine (C_7H_9N) and allyl chloride (C_3H_5Cl), let them react to form an intermediate which might dimerize on subsequent oxidation $(2C_{10}H_{13}N + 3O \rightarrow C_{20}H_{24}N_2O_2 + H_2O)$. Upon attempting the synthesis he obtained for his efforts a dirty brown mess. Using a simpler base, aniline, he was rewarded with another dark, sticky mess. While cleaning the reaction vessel with alcohol, Perkin observed a purple solution which yielded dark crystals. The substance was found to hold promise as a dye.

Against Hofmann's advice Perkin quit school and embarked on dye manufacture with his father and brother. Since there was no tradition of fine-chemicals manufacture, it was necessary to develop equipment, processes, and know-how. It was also necessary to overcome the dyers' aversion to any dye from a nonnatural source. The British public did not accept aniline purple at first, but after the color became popular in Paris as *mauve*, or *mauveine*, it became very fashionable in England.[4]

Impressed by the success of mauve, textile industrialists demanded other synthetic dyes. Aniline, toluidine, quinoline, and related compounds were used as starting materials for the synthesis of such new dyes as Fuchsine, Bismark Brown, Imperial Blue, and Quinoline Blue. Dye plants were opened in various places such as Ludwigshafen, Lyons, and Manchester. The creation of new compounds with dye properties stimulated a great deal of highly empirical investigation. Hofmann was one of the few chemists who insisted that it was more important to understand the chemistry of these compounds than blindly to make new ones, yet his Hofmann violets represented a somewhat empirical synthesis whose chemistry was

---

[4] Mauve is often regarded as the first synthetic dye. Actually picric acid was prepared in 1771 by Peter Woulfe, who utilized the action of nitric acid on natural indigo. Picric acid has some use as a dye although it may be argued whether it was synthetic. A method for preparing it from phenol was introduced in 1855. Aurin, or rosolic acid, was the first dye prepared from coal tar. It was discovered in 1834 by Runge, who observed that it produced red colors and lakes with the usual mordants It was not developed commercially.

not entirely understood for a decade. The structure of mauve was established in 1888 by Otto Fischer, after the dye was no longer commercially important.

The first of the triphenylmethane dyes was introduced commercially in 1859 by Renard Frères under the name *fuchsine* and by manufacturers outside France as *magenta*. The chemistry of these dyes was clarified very slowly despite Hofmann's interest in them. He showed in 1862 that *fuchsine* is the salt of an organic base which he called *rosaniline* and learned that the color is intensified by reaction with alkyl iodides, producing compounds which became known as *Hofmann violets*. The chemical structure was finally clarified in 1878 when Emil Fischer and his cousin Otto showed fuchsine to be derived from triphenylmethane, a compound discovered by Kekulé several years earlier. Once the structure was known, new dyes of this class were introduced rapidly. Their brilliance made them popular despite their lack of fastness.

Diazonium salts were prepared in 1856 by Peter Griess (1829–1888), two years before he went to London to become Hofmann's assistant. Subsequent studies showed that, while the salts themselves had no value as dyes, they coupled readily with aromatic amines and hydroxy compounds to form useful dyes. Aniline yellow was produced in 1863, and various other dyes of the class were introduced even before the nature of the azo group was clarified by Otto Witt in 1876. After 1880 the azo class became the most widely produced group of synthetic dyes.[5] Ingrain colors were introduced by Arthur George Green in the 1880's. In this process the cloth is first dyed with a compound containing an amino group which is diazotized on the cloth and coupled with β-naphthol, resorcinol, or a similar compound to intensify the color.[6]

Alizarin, known since 1826 to be the active component of the natural dyestuff madder, came under chemical attack in the sixties in Baeyer's laboratory. Carl Graebe showed it to be derived from anthracene, and he and Carl Liebermann oxidized anthracene to anthraquinone which was brominated and then fused with potassium hydroxide to obtain alizarin. When the bromination step was replaced by treatment with fuming sulfuric acid, the synthesis became commercially acceptable. Manufacture of synthetic alizarin was begun in 1869 by the Badische firm in Germany and by Perkin in England. Within a decade the British textile industries use of alizarin had increased more than fifty-fold, but the farmers of, Provençe, who were dependent on the sale of madder for their livelihood, were ruined.

The period between 1870 and 1900 saw very marked changes in the

[5] Haber, *op. cit.*, p. 83.

[6] R. E. Rose, *J. Chem. Educ.*, **3**, 973 (1926). This paper develops the growth of the dyestuffs industry, including portraits of eleven of the leading figures, structural formulas of the compounds being discussed, and cloth swatches dyed with fifty of the dyes of historical or contemporary interest.

Fig. 17.3. CARL GRAEBE (seated), A. W. HOFMANN, AND
CARL THEODOR LIEBERMANN.
(Courtesy of the Edgar Fahs Smith Collection.)

synthetic dye industry. The early successes in dye synthesis were achieved largely on the basis of a strictly empirical approach. Likely compounds were treated with available reagents to find out what might result. The experimenters generally were well-grounded in chemistry, but they were nevertheless groping since knowledge of structural chemistry and the course of organic reactions was still in a primitive state. Frequently a successful synthesis proved to be a failure when coal-tar chemicals of greater purity came on the market.

The synthesis of alizarin marked a turning point in dye chemistry. The work of Baeyer and his associates revealed that an approach based on a sound knowledge of organic reactions made it possible to duplicate even an important natural molecule. This approach, combined with the application of the new structural ideas, made rapid progress possible during the next two decades.

Witt called attention to the relation of color and constitution, pointing out in 1876 that color was associated with certain characteristic groups which he termed *chromophores*. However the presence of a chromophoric group was not enough to insure a colored compound. Certain associated groups, called *auxochromes*, were found valuable in bringing out color or in varying the shade. Witt also stressed the importance of acidic and basic groups, both as auxochromes and as agents for influencing solubility and attachment to cloth. Absorption spectra were carefully studied in connection with the problem. It was not realized for some time that ultraviolet and infrared absorptions as well as visible absorptions are important in interpreting color and structure. Great progress was made by the dye industry through such application of scientific principles to the synthesis and study of organic compounds.

The nature of indigo commanded considerable attention. Baeyer showed that it was related to isatin, but abandoned his studies out of deference to Kekulé whom he believed interested in the problem. It also has been suggested that Baeyer dropped the problem to protest the pressures exerted by dye manufacturers who wanted a synthetic method for producing the dye. At any rate Baeyer returned to indigo toward the end of the decade after Kekulé had done nothing. A successful synthesis was carried out in 1880, starting with *ortho*-nitrocinnamic acid. The rights were assigned to the Badische and Höchst firms, but the method was without commercial value. Indigo was synthesized from *o*-nitrotoluene by Baeyer in 1882, but this synthesis was also impractical for industry. The final proof of structure was accomplished by Baeyer in 1883.

From this point forward indigo research was financed by commercial dye manufacturers like Badische, Höchst, Kalle and Company, and the Société des Usines du Rhône.[7] Badische had spent the equivalent of one million pounds sterling by the time commercial production was begun in 1897. The synthesis was developed from a reaction carried out by Karl Heumann early in the nineties. His synthesis was based upon the use of anthranilic acid which could be prepared from naphthalene. The first step proved a stumbling block since the traditional oxidizing agent for the conversion of naphthalene to phthalic acid was chromic acid which would make the cost prohibitive. Concentrated sulfuric acid was tried by Sapper, one of the Badische chemists, but proved to be very slow. However the accidental breaking of a thermometer resulted in the introduction of mercury which proved an excellent catalyst for the oxidation. The large quantities of sulfur dioxide formed during the reaction were converted back to sulfuric acid by the newly-developed contact process. By 1900 synthetic indigo production had reached a volume equivalent to the farming of 250,000 acres. The effect on indigo farming in India and elsewhere was devastating.

[7] H. Brunck, *The History of the Development of the Manufacture of Indigo*, Kuttroff, Pickhardt & Co., New York, 1900.

Rise of the German Chemical Industry

By 1875 the synthetic dye industry was becoming concentrated in Germany. During the sixties small German firms were struggling for a part of the dyestuffs market while English firms prospered. The English manufacturers, however, blinded by their early successes with an empirical approach, failed to perceive the necessity for basic research in coal-tar chemistry. Hofmann was lured back to Germany to direct laboratories built for him first in Bonn and almost immediately thereafter in Berlin. Nearly all the chemists connected with English dye manufacture had been trained by Hofmann. With the exception of Perkin, Nicholson, and Medlock most of them were Germans; they began to drift back to Germany to take important positions with German producers who needed the know-how they had acquired in England. Caro, Martius, and Witt were the most important repatriates, but there were others. By 1880 the drift of the industry to Germany was virtually complete.

In England Nicholson had sold his business in 1868. Perkin retired in 1874, a wealthy man determined to return to research in basic organic chemistry. Greville Williams sold his property in 1877. Only Ivan Levinstein, who had begun the manufacture of synthetic dyes in Manchester in 1864, persisted in the face of the growing German competition. British capitalists had little real interest in the chemical industry, and the leading figures in England at the end of the century were of foreign origin—Messel (German), Castner (American), Brunner (Swiss), and Mond (German).

In Germany, on the other hand, there was real enthusiasm for the chemical industry. German politicians and businessmen realized that England had gained a dominant position in textiles and steel during the Industrial Revolution. Germany, an ambitious nation under Bismarck, hoped to become preeminent in industries closely allied with science. While its chemical firms arose from small beginnings, the economic climate was such that they could become veritable giants by the end of the century. Adequate financing, vigorous approach to sales, farsighted management, a favorable patent policy, and mutually agreeable competitive arrangements enabled German industry to flourish. Scientific education was given sympathetic support by state and industry. Well-trained chemists were available to carry on the research necessary to master the indigo problem and to expand greatly the knowledge of useful compounds. By 1897 there were 4,000 chemists active outside the academic field in Germany.[8]

The famous Badische firm was founded in 1861 as the Chemische Fabrik Dyckerhoff, Clemm & Co. of Mannheim. When it moved to Ludwigshafen in 1865 it was incorporated as the Badische Anilin und Soda-Fabrik. After Heinrich Caro became chief chemist in 1868, the firm naturally pioneered in the synthesis of alizarin. During the following years

[8] Haber, *op. cit.*, p. 188.

the firm, by internal expansion as well as by the acquisition of other firms, achieved such integration of its operations that it was distilling tar and producing intermediates as well as producing finished dyes. Its laboratory staff concentrated so diligently on the indigo problem during the eighties and nineties that diversification of dye manufacture was neglected and competitors gained a lead in the azo dye field. However René Bohn's discovery of the indanthrene type of vat dyes in 1901 opened up a new field. Badische's growth and integration parallelled very much the rise of Heinrich von Brunck (1847–1911) in the firm. Brunck, who had studied with Kekulé at Ghent, joined Badische in 1869 and rose through the production ranks to become technical manager in 1883. Later, as chairman of the board he had a great deal to do with encouraging the broad integration of the company, the development of the process for synthetic indigo, and in the twentieth century the development of the Haber process.

In the Frankfurt area four firms rose to prominence during this same period. Leopold Casella & Co., which had existed since 1812 as a dealer in vegetable dyestuffs, entered the aniline color field in 1867, specializing at first in triphenylmethane dyes, but later developing direct cotton and sulfur dyes. Otto N. Witt was chief chemist.

At Offerbach the firm of K. G. R. Oehler, long a dealer in dyewoods, bought a tar distillery in 1856 and began the production of picric acid, mauve, and fuchsine, the first synthetic dyes manufactured in Germany. This firm later specialized in intermediates. It was purchased by Griesheim Elektron in 1905. Kalle and Co. was founded in 1863 by Dr. Paul Wilhelm Kalle for the production of specialty dyes.

The chemist Dr. Eugen Lucius joined with two merchants in founding Meister, Lucius & Co. in 1862. It was renamed Meister, Lucius, and Brüning four years later, but became generally known as Farbwerke Höchst, or simply Höchst. after the city where it operated its plant. The firm varied its production to become a highly integrated one.

On the lower Rhine the firm of Bayer, founded in 1861 by Friedrich Bayer (1825–1880) and J. Westkott for the production of fuchsine, became important. It carried on no research and manufactured only a limited number of standard dyes. In 1880 it was reorganized as the Farbenfabrik vormals Friedrich Bayer & Co., and three years later Carl Duisberg (1861–1935) joined the firm. Duisberg had studied at Göttingen and Jena and was sent by the firm to Strassburg for postgraduate study under Fittig before taking over his duties in the azo department.[9] He

---

[9] The Bayer company hired three young chemists at this time and, in order to keep them from being diverted to production problems, sent each to a university where a good laboratory and expert guidance were available for postdoctoral experience. Duisberg worked with Fittig at Strassburg; Oscar Hinsberg worked at Freiburg; Wilhelm Herzberg at the Technische Hochscule in Munich. See J. J. Beer, *Isis*, **49**, 128 (1958).

Fig. 17.4. CARL DUISBERG.
(Courtesy of the Edgar Fahs Smith Collection.)

became chief of the research division in 1891 and rose to a management position in the firm before the century ended. Duisberg was a dominant figure in the growth of the Bayer company, taking the lead in the introduction of new dyes and pharmaceuticals. His planning of the new Bayer plant at Leverkusen resulted in an installation that served as a model far into the twentieth century.

The Aktiengesellschaft für Anilinfabrikation (AGFA) was organized in 1867 by Paul Mendelssohn-Bartholdy, son of the composer, and Carl Alexander Martius. The latter had worked with Hofmann and Griess in London and then had returned to Germany with Hofmann.

All of these firms rose to positions of great power in the chemical industry. By encouraging vigorous research in their own laboratories and utilizing the services of academicians as consultants, they were able to develop their industry on truly scientific foundations. At the same time leaders in the industry did not hesitate to enter into mutually advantageous economic arrangements which enabled them effectively to stifle competition in other parts of the world. Only a few Swiss specialty firms like Gesellschaft für chemische Industrie Basel (CIBA) and Geigy and the English firm of Levinstein were able to survive the vigorous and sometimes cutthroat practices of the German industry.

PHARMACEUTICALS

It was natural that chemists working with coal-tar compounds should seek uses for derivatives unsuitable for dyes. The drug field was an obvious one for exploitation, particularly in view of the recent development of the germ theory of disease based on the work of Pasteur and Koch. Lister had successfully used phenol as a spray in operating rooms to reduce infections, the plague of surgery. The toxicity of phenol toward bacteria naturally prompted the search for germicides that are not as violently toxic when taken internally.

Salicylic acid was accepted by the medical profession in large part because of the enthusiasm of Kolbe who developed a synthesis in 1853. It was found effective as a preservative for meat, milk, beer, and other foods, and tests for medical uses were promising. Kolbe's former student, Friedrich von Heyden, undertook its commercial manufacture. However, researchers finally determined that while salicylic acid was effective in reducing fever, it failed completely in preventing or curing illness. Kolbe's theory that the acid slowly broke down in the blood stream to phenol was without foundation. Its preservative action in foods is due to the acid itself; in the body it is effective only in reducing fever and relieving pain. Traditional remedies for this purpose, such as willow bark and oil of wintergreen, are natural sources of salicylates. Salicylic acid proved useful in the treatment of rheumatism and for reducing fever and pain, but its irritant action in the digestive system limited its application.

Quinoline was obtained by Gerhardt in 1842 when he treated quinine with alkali. After Skraup synthesized quinoline from aniline and glycerol in 1880, chemists sought to synthesize quinine or modify quinoline to obtain a quinine substitute. Otto Fischer prepared many compounds; some proved effective in reducing fever but ineffective against malaria and too toxic for general use. Several other derivatives of aromatic compounds were introduced as antipyretics by 1895, but none of them survived the competition of aspirin, introduced by the Bayer company in 1899. This was developed as a result of Felix Hoffmann's research for a variant of salicylic acid without its undesirable side reactions. The acetyl ester, originally prepared by Gerhardt in 1853, was only hydrolyzed in the alkaline juices of the intestine and hence did not cause the stomach upsets so common with salicylic acid. The word "aspirin" was derived from *spiric acid*, an alternative name for salicylic acid, which is found in plants of the *Spirea* genus; the a- stands for acetyl. It was marketed as the sodium salt. Vigorously promoted among members of the medical profession, aspirin quickly attained widespread use and became a very profitable item for the Bayer company. When other companies began manufacturing aspirin upon expiration of the patent, Bayer sought to maintain its monopoly by arguing the word "Aspirin" to be a trademark but was unsuccessful in upholding the claim in the courts.

Research also began on local anesthetics after 1884 when Carl Koller, working in Sigmund Freud's clinic, used cocaine in eye surgery. Cocaine was isolated from coca leaves by Albert Niemann, one of Wöhler's students, in 1860. The leaves, and later cocaine itself, were used to deaden sensitivity to pain for some time before the dangers of addiction were fully appreciated. Even before the structure of cocaine was established by Richard Willstätter in 1898, there were efforts to create a non-habit-forming substitute. Of several promising aminobenzoic acid esters, *novocaine*, synthesized by Alfred Einhorn, proved most statisfactory.

Studies on sleep-inducing drugs moved from bromides, to the dangerous chloral, to organic sulfur compounds such as *sulfonal*, and to the barbiturates, which were first studied by Emil Fischer when he condensed urea with various substituted malonic acid derivatives. The first such hypnotic, *veronal*, was introduced commercially in 1903.

OTHER ORGANIC FINE CHEMICALS

Although dyes, and to a certain extent pharmaceuticals, accounted for the major production of the organic chemicals industry, a variety of lesser products were profitably produced and marketed. Among these were food adjuncts such as colors, preservatives, flavors, and sweeteners and perfumery and photographic chemicals. Many of these chemicals were so closely related to the aromatic compounds common around the dye factory that their production was a natural consequence of the business.

Saccharin, the leading synthetic sweetener, was discovered at Johns Hopkins by Ira Remsen and Constantin Fahlberg. The latter, without consulting Remsen, obtained a patent and granted manufacturing rights to *CIBA* in Switzerland. Coumarin, first prepared by the elder Perkin, was used as a vanilla substitute for many decades before its danger to health was recognized. It was also used in perfumery, being the first natural perfume to be synthesized from coal tar. Early investigations had revealed the resemblance of various aliphatic esters to the odors and flavors of fruit. With the growth of the chemical industry various esters of this sort were marketed as synthetic fruit flavors. Some were truly present in the natural fruit juices; others merely had a superficial resemblance to the natural.

A natural outgrowth of the intense interest in terpenes during this era was the partial duplication of natural perfumes. Ferdinand Tiemann succeeded in isolating irone from the rhizome of orris (Florentine iris) in 1893. This apparently corresponded with the principal odor in violet essence, one of the most expensive and most prized perfumes. Tiemann was unable to establish the structure but attempted a synthesis from citral, the major component of oil of lemon grass, obtaining a mixture of α- and β-ionone. These compounds have a strong violet odor, but do not correspond to irone. The ionones came into use in the preparation of perfumes approximating that from the violet plant.

Terpene studies uncovered other compounds which were either trace components of the petals of flowers or resembled them in odor. A synthetic lilac essence was prepared from turpentine by G. Bouchardat and R. Voiry. Francis D. Dodge, an American student of Victor Meyer, isolated citronellol in 1889 and showed it and geraniol to be the major constituents of rose oil. He synthesized citronellol from inexpensive oil of citronella. Returning to New York in 1891, Dodge became a member of the family firm of Dodge and Olcott, essential oil importers founded in 1789. Young Dodge expanded the operations of the firm to produce concentrated natural and converted terpenes for the perfumery and flavor trade. Similar efforts elsewhere to isolate and convert terpenes led to the partial substitution of synthetic chemicals for the very expensive natural oils extracted from flowers.

Photographic Chemicals

During the early years after William Henry Fox Talbot introduced his type of photography, developing was done with gallic acid, which later was supplanted by pyrogallol as the result of Regnault and Liebig's independent researches. During the eighties German research led to the introduction of improved phenol and aromatic amine developers. Improved sensitivity to yellows and oranges resulted from the work of H. W. Vogel of Berlin, who showed in 1873 that the presence of certain dyes in the silver halide emulsion brought about a more faithful reproduction of the object. Dyes fully sensitive to reds were not introduced until 1906. The German industry held a virtual monopoly until World War I as a supplier of photographic chemicals, optical glass, and high quality cameras.

GLASS

Glassmaking remained a highly empirical chemical art until the twentieth century, except for the development of optical glass and chemical glassware during the 1800's. During the early part of the nineteenth century chemical glassware was chiefly of local origin; it usually was ordinary glass fabricated into shapes desired by the customer. A company started by Gay-Lussac and the glassblower Collardeau was an important supplier of specialized glassware such as burettes, vapor density apparatus, alcoholometers, and similar laboratory devices. This was chemical glassware only in the sense of being designed for chemical studies; the glass was merely of standard composition and was subject to attack by various chemicals.

As Germany became preeminent in training and research in chemistry, its leadership extended to the production of chemical glassware also. Liebig, Bunsen, Mohr, Fresenius, Hempel, Lunge, and Geissler were all active in the design of new apparatus. Chemists soon recognized that glassware from the Thuringian region was more resistant than any other to chemical attack. They did not know at the time that this was due to alumina present as an impurity in Thuringian glass sand.

In 1860 Stas realized that the accuracy of his atomic weight work was handicapped by contamination from glassware. He developed a high-silica, high-potash variant of soda-lime glass which proved of superior quality.

A widely used chemical glassware was also produced by the works founded in 1837 by Francis Kavalier in Bohemia. His sons carried on the business, which was a leading supplier until the development of the works at Jena. Kavalier hard glass had an excellent reputation for combustion tubing.

Jena glassware was developed as a result of F. O. Schott's experiments sponsored by the Prussian state. The Jena Glass Works, founded by Schott and Genossen in 1884, quickly acquired a reputation for the quality of its glassware, which was highly resistant to chemicals and thermal shock.

METALLURGY

At the time of Lavoisier metallurgy depended on processes that had been described by Agricola two centuries earlier and probably differed little from those used in antiquity. Depletion of forests by the insatiable demand for charcoal in mining centers had led Abraham Darby to introduce coke in the smelting of iron in 1735. Coal and coke had been tried earlier, but after Darby's experiments the use of coke began to spread. As a result the iron industry moved from wooded sections of Britain to the coalfields of Scotland, the Midlands, and South Wales. The English iron industry expanded rapidly toward the end of the eighteenth century, being stimulated by, and in turn stimulating, the Industrial Revolution. Steam engines, pumping machinery, transportation equipment, and mill machinery all made heavy demands on the industry. This demand was accelerated during the whole of the next century with the rise of railways, metal ships, bridges, as well as minor articles. The industry could no longer satisfy demand with centuries-old operations. Practical inventors and the scientists began to change the art of metallurgy.

In 1786 Monge, Vandermonde, and Berthollet established that the difference between iron and steel is due to carbon. The blast furnace came into use during the eighteenth century for the production of pig iron and was improved in 1828 when James Neilson introduced the hot blast, thereby reducing coke consumption. The puddling furnace for the production of wrought iron was developed by Henry Cort in 1784. Cort utilized a reverberatory furnace in which molten iron could be stirred while carbon and other impurities were being burned out. The high purity iron which resulted could be used directly for tough, malleable products, or it could be converted into steel by incorporating charcoal in the puddling furnace.

Steelmaking was drastically modified shortly after 1850 with the development of the Bessemer and the open-hearth processes. The process of Henry Bessemer, which was being independently developed in America by William Kelly, burned out the impurities in molten pig iron through the use of an air blast. The invention drastically cut the time and labor required for steelmaking, lowered costs, and made steel abundant for the first time.

The open-hearth process was developed by the Siemens brothers in England and the Martin brothers in France. William and Frederick Siemens were brothers of the founder of the electrical firm of Siemens and Halske in Germany, but they found patent conditions in England more encouraging to engineers and inventors. In 1856 Frederick developed a regenerative furnace in which hot, spent gases were used to heat networks of brick which in turn heated incoming air and fuel gas, thus conserving fuel and making higher temperatures possible. The furnace was adopted immediately by the glass industry and soon thereafter by the steel industry. In 1861 William Siemens developed a gas producer which provided a cheap gaseous fuel for the regenerative furnace.

In France Pierre and Émile Martin made steel at Sireuil in 1864, melting pig iron and adding wrought iron until the carbon content of the mixture was reduced to the desired level, primarily by dilution. Meanwhile the Siemens brothers were experimenting in a steel plant at Birmingham, utilizing iron ore as an agent for the oxidation of impurities in molten pig iron. By 1868 the essential features of the open-hearth process were worked out: regenerative heating, gaseous fuel, siliceous hearth lining, and additives like iron ore, scrap iron, and limestone for removal of impurities. The long melting and purifying period made careful control possible and resulted in the mass production of high quality steel from materials the Bessemer converter could not handle. Such steel, when made into boiler plate, permitted the use of high-pressure steam with corresponding increases in the efficiency of engines.

Since phosphorous makes steel brittle, the phosphorus-rich iron ores in parts of England, France, Luxemburg, Belgium, and Germany remained useless until the invention of a process for phosphorus removal by Sidney Gilchrist Thomas and his cousin Percy Carlyle Gilchrist. These workers substituted a magnesia lining for the silica firebrick in the Bessemer converter, thus creating basic conditions favoring the removal of phosphorus oxide from the pig iron. The principle of a basic lining was extended to the open-hearth process by J. H. Darby in 1884 at Brymbo, Wales. Gradually the basic open-hearth process superseded the basic

Bessemer process because of the better control which could be maintained over the quality of the finished steel.

Nonferrous Metals

Gold and silver were extracted by the amalgamation process until the cyanide process was introduced in 1887 by John S. MacArthur and the Forrest brothers in England. The use of cyanide made it economically possible to extract the precious metals from ores containing mere traces and from tailings. The process, which quickly came into use in major gold-producing areas, created a great demand for sodium cyanide. Synthetic methods for preparing cyanides were developed during the nineties.

Methods for smelting copper did not change significantly until the end of the century. Then the demand for very pure copper by the electrical industry led to an enlarged scale of operations using water-cooled blast furnaces and Bessemer converters. Electrolytic refining of the copper matte from the converters was necessary to obtain the high state of purity necessary in conductors.

Lead smelting followed tradition until 1829 when Pattinson's process for removing lead from silver was introduced. The molten lead was cooled slowly, pure crystals of lead forming on the surface. These were skimmed off as they appeared. When the silver concentration of the residual molten lead became high enough, it was separated by the ancient method of cupellation. In 1852 Parkes introduced a more efficient method utilizing molten zinc as a partition solvent for silver. Molten lead and zinc are immiscible, and silver is most soluble in zinc. The zinc can then be distilled to recover the silver.

New Metals

Aluminum was isolated in a highly impure form by Hans Christian Oersted of Denmark in 1825, and a better preparation was made two years later by Friedrich Wöhler. Both separated it from its chloride through the action of potassium. The metal attracted considerable attention because it was light and mineralogically abundant, but the use of expensive potassium metal for its preparation caused it to remain a laboratory curiosity. In 1854 Henri Sainte-Claire Deville devised a process for lowering the production cost of sodium and found that this metal could be substituted for potassium in the preparation of aluminum. He devised a process for producing aluminum metal from bauxite ore found near Baux in southern France. This process greatly lowered the price, but not to a point permitting widespread use.

In 1886 two young men independently developed an electrolytic method for producing the metal. Paul L. V. Héroult in France and Charles Martin Hall in the United States obtained patents in their respective countries for the electrolysis of alumina (aluminum oxide) dissolved in fused cryolite

Fig. 17.5.   CHARLES MARTIN HALL.
(Courtesy of the Aluminum Company of
America.)

(sodium aluminum fluoride). Production was begun in America by the
Pittsburgh Reduction Co., which became the nucleus of the Aluminum
Company of America. In France Hérault's patents were exploited vigor-
ously and the price of aluminum dropped significantly.

Nickel, used in electroplating, was not produced in quantity until
the Mond process was introduced in 1890. Ludwig Mond (1839–1900) was
educated in Germany and held several positions in continental chemical
firms before settling in England where he and John Brunner introduced
the Solvay process. Mond devised a number of unique chemical processes,
including one for manufacturing producer gas. He observed about 1890
that nickel valves are corroded by hot gases containing monoxide. Further
studies with Carl Langer led to the isolation of a colorless liquid with
the formula $Ni(CO)_4$. A volatile iron carbonyl, $Fe(CO)_5$, was discovered
soon there-after by Berthelot and by Mond and Quincke, and various
other metal carbonyls were reported subsequently.[10]

Nickel carbonyl was readily volatilized. Mond utilized this property
for the commercial separation of nickel from cobalt and other metallic

[10] L. Mond, C. Langer, and F. Quincke, *J. Chem. Soc..* **57**, 749 (1890); Mond, *J. Soc.
Chem. Ind.*, **14**, 945 (1895); M. Berthelot, *Compt. rend.*, **112**, 1343 (1891); Mond and
Quincke, *Ber.*, **24**, 2248 (1891), *J. Chem. Soc.*, **59**, 604 (1891) *Chem. News*, **63**, 301
(1891).

contaminants. The Mond Nickel Co. was formed to exploit the process; a plant was built at Sudbury, Canada, and brought into successful operation.

ELECTROCHEMISTRY

Cells

Early electrochemical cells were primarily research tools and as such were expensive to operate and unstable with respect to voltage. The instability was overcome in a cell introduced in 1836 by John Frederic Daniell that utilized copper and zinc electrodes submerged in copper and zinc sulfate solutions respectively. The solutions were separated by a porous diaphragm, or, in the gravity cell modification, the zinc sulfate solution was layered over the more dense copper sulfate solution. This cell was widely used in telegraphy, electroplating, and research. Research-type cells were also developed by William R. Grove in England, and by Bunsen.

In 1867 Georges Leclanché introduced a cell having a zinc electrode in ammonium chloride solution and a carbon electrode immersed in manganese dioxide, the latter serving as a depolarizer. The cell proved useful when demands for a convenient cell were rising. It was later replaced by the "dry cell" introduced by Gassner in 1888. Moist ammonium chloride was substituted for the solution in the Leclanche cell, and the zinc electrode also served as a container.

Efforts to produce a rechargeable cell were being made simultaneously. Gaston Planté used lead sheets in a sulfuric acid solution in 1859, the charge being developed in the cell by a tedious "forming" process. The difficulty of this operation together with the lack of a practical generator caused the cell to be a commercial failure. A more practical accummulator, or storage cell, was invented in 1881 by Camille A. Faure who used lead oxides on lead plates but still had difficulty with the charging operation. Volckmann, Swan, Correns, and others introduced improvements, but the cell did not become truly practical until the new century began.

Electrolytic Industries

Except for small-scale electroplating operations dating from 1840 in the silver and gold industries, commercial electrochemistry remained underdeveloped until the last two decades of the century. With the advent of practical generators after 1870, electricity became more widely available, and chemical operations involving heavy use of electric current became practical in Scandinavia and other places where hydroelectric power was developed. Production of chlorine and sodium hydroxide developed rapidly in the vicinity of Niagara Falls where salt and power were plentiful. Germany also made rapid advances in the field since the Stassfurt deposits provided a suitable source of salts. Nickel plate became a very popular protective coating at this time.

Electric Furnace Products

The electric furnace, invented by Henri Moissan in 1892, who used it for the study of metal carbides, metal nitrides, the reduction of refractory metal oxides, and, with Samuel Smiles, silicon hydrides. In places where electricity was cheap the furnace was soon used for the production of calcium carbide, artificial graphite, and silicon carbide.

In 1862 Wöhler had prepared calcium carbide by heating carbon with an alloy of zinc and calcium. He found that it gave acetylene with water, but the carbide had no industrial importance until Moissan showed that the electric furnace made its production from lime and carbon practical. This process was independently discovered by Thomas L. Willson, a Canadian. The acetylene produced from calcium carbide and water was promoted as an illuminant, but because of its sooty flame it failed to make inroads into the conventional market. However it was readily prepared for immediate burning in canisters containing carbide and water, and so was suitable for portable lamps such as those used by miners and on early bicycles and automobiles. Acetylene proved most useful when burned with oxygen in welder's torches; still later the gas was employed in a large variety of organic syntheses.

In 1898 Fritz Rothe of Germany learned that when nitrogen is passed over hot calcium carbide, calcium cyanamide is formed. This compound has direct value as a fertilizer and is useful in the production of ammonia and of cyanides.

Silicon carbide was discovered by the American Edward G. Acheson in 1891. Its abrasive properties were quickly recognized and it began to be marketed under the trade name "Carborundum." Acheson also developed the process for preparing artificial graphite in the electric furnace. Electrodes made from this material played an important role in the further development of electrochemical processes.

CONCLUSION

The rise of chemical industry in the nineteenth century was a prelude to the influence which basic science would soon reflect on economic, political, and social affairs. In 1800 chemistry was hardly evident in the operations of the family-type businesses which dealt in soap, drugs, pigments, and natural dyestuffs. To be sure, such chemists as Berthollet, Chaptal, and Black were making pronouncements about the potential contributions science might make to industry and agriculture but there were few signs that such contributions would be of more than trivial importance. By 1900 such firms as Badische Anilin and Soda Fabrik and Brunner, Mond and Co. were industrial giants with sales even in distant parts of the world. And synthetic alizarin and indigo had wrecked the agricultural communities dependent on their production.

The development of the chemical industry, particularly the fine chemicals segment, revealed that basic chemistry had now reached the point where it could contribute in a major way to the development of new products and processes.

IV

THE CENTURY OF THE ELECTRON*

$\mathbf{D}$ISCOVERY OF THE ELECTRON JUST before the twentieth century began was ultimately to have a profound influence, not only on chemistry but on all of science. The nineteenth century may be characterized as the century of the atom, the century in which the atomic particle made its serious entry into science. Much of the chemical activity of the century was expended toward understanding the concept and making it useful in chemistry. The twentieth may be looked upon as the century of the electron, a subatomic particle which related the atom to spectroscopy and led to a new understanding of chemical combination.

Proof of the existence of a particle distinctly less massive than the hydrogen atom shocked the scientific community even though prior discoveries in chemistry and physics had been preparing the ground. The electron proved to be a part of every kind of atom and the principal theoretical problems of chemistry and physics proved to be associated with the role of electrons in the structure of atoms, in chemical reactions,

---

* In a book of this type it is obviously impossible to deal with twentieth-century developments in the detail given previous centuries. The enormous literature associated with even a narrow field such as sex hormones, organometallic compounds, or infrared spectroscopy makes it utterly impossible to trace the detailed history or give credit to the work of individual persons in a consistent manner. Hence it will be necessary in these chapters to generalize to a high degree, even though generalizations are always faulty oversimplifications. An effort will be made to point out major developments and trends, even though this is dangerous when the perspective of time is unavailable.

and in electrical and optical phenomena. The understanding of the role of electrons in electrical phenomena led to the development of a myriad of electronic instruments which, in turn, found application in chemical research and industrial control.

As a consequence of the build-up of scientific knowledge the twentieth century has witnessed a scientific "explosion." Factual and theoretical knowledge have borne fruit too rapidly for man to deal adequately with the consequences. This explosion draws its power not only from chemistry, but from mathematics, physics, geology, astronomy, and biology as well. Industry, agriculture, and medicine have been revolutionized; the nature and consequences of warfare have been transformed; indeed the whole pattern of living in the Western World has changed, and the East is being affected profoundly.

The scientific "explosion" has also changed the organization of science. Although research activity is greater than ever in the universities, finance and direction come from government and industry to a large degree. At the beginning of the century research was largely an avocation pursued by professors between classes; after World War I it became an integral part of university life, sometimes to the point where classes became a necessary evil between experiments. Financed at first principally by universities, research gradually attracted the greatest fraction of its support from foundations, governmental agencies (especially agricultural, medical, and military), and industrial firms. Industrial research, almost non-existent at the beginning of the century, grew to enormous proportions as the century progressed.

The century's two world wars and major economic depression stimulated science and undoubtedly had a marked effect on the nature of the programs undertaken. War channeled science into areas of actual or potential military application. When poison gas was used for the first time as a major weapon in World War I chemists were called upon to develop protective devices and new poisons. Science truly was "unleashed" during World War II, leaving mankind with the power to destroy civilization in an irresponsible moment.

CHAPTER 18

RADIOCHEMISTRY I.
RADIOACTIVITY
AND ATOMIC
STRUCTURE

SHORTCOMINGS OF THE DALTONIAN ATOM

While the small indivisible particle was distinctly useful to the organic chemist, many aspects of the Daltonian atom remained unsatisfactory and in fact caused some physical scientists to question the value of the whole concept. The physicists were unconvinced that the atom or the molecule really had a place in science. Their skepticism grew out of their attempts to build a theoretical system consistent with dominant physical principles. Unlike the chemists they were more concerned with space than with discrete matter and hence sought to treat atoms, when they recognized them at all, as heterogeneities in a continuum. The philosophy of continuity had always had a strong appeal for physicists. Continuity was consistent with mathematics, and a continuum such as an ether seemed essential to the wave theory of light.

As early as 1849 the Scottish physicist William John Macquorn Rankine proposed a vortex theory to derive equations of elasticity and thermo-dynamics. His vortex molecules were elastic atmospheres revolving about a central nucleus. Rankine found the concept useful for mathematically dealing with certain physical phenomena, though he considered its objective reality incapable of proof.

In 1868 Helmholtz developed a theory of vortex motion in a perfect fluid (one lacking in viscosity) showing that once formed in such a fluid vortices should permanently undergo vortex motion and retain their identity. In 1867 William Thomson (Kelvin) suggested a vortex atom in a perfect fluid, presumably the ether. With this atomic theory he was able to preserve the continuum and at the same time provide for discreteness. The properties of the vortices could be derived mathematically, whereas the chemists' atoms were assigned properties quite arbitrarily. The vortex atoms were perfectly elastic, and Thomson felt that a rigorous kinetic theory could be derived from such atoms, that the thermal expansion coefficient could be derived from swelling of the vortex with increasing kinetic energy, and that spectral lines could be reconciled with vortex vibrations. Peter Guthrie Tait recognized an analogy with the behavior of smoke rings.

Chemists never became greatly excited about vortex atoms, although Roscoe suggested in 1884 that the hydrogen atom might correspond to a single vortex and that the molecule might consist of two enchained vortices. Most other chemists preferred to conceive of a less nebulous atom though such atoms strained the credulity of the physicists.

Physicists themselves were only lukewarm about the vortex atoms, but for a different reason. They were generally not convinced that dealing with physical phenomena required a discontinuous philosophy. Even when they started with a continuous ether, they found it soon acquired particulate qualities. Maxwell considered the vortex atom a possible representation of reality, but pointed out in 1875 that atoms need not be assumed when mathematicians used variables to represent positions for molecules. Willard Gibbs developed his thermodynamics and statistical mechanics without concern for the ultimate structure of matter.

Ernst Mach (1838–1916), the Viennese physicist and philosopher who suggested theoretical models, attacked atomic theories (and also ethers). He noted that models were not employed in thermodynamics and concluded that it was the role of the scientist to observe facts and phenomena. These, he agreed, might be generalized through laws and rules, but he argued that such laws and rules served merely as memory devices and had no physical significance. Theorization, he believed, could serve no useful purpose, and might prove harmful by being misleading. Mach's negative attitude toward the atomic theory was opposed by Boltzmann who in 1899 pointed out that a theory which was consistent with so many physical, chemical, and crystallographic facts should be further developed rather than combated. Nevertheless Mach was supported by influential followers, notably Ostwald who deliberately avoided particle concepts in his widely used *Grundlinien der anorganische Chemie* (1900).[1] However most chemists found the atomic concept useful despite the obvious shortcomings of a small indivisible particle.

Spectral Problems

Chemists and physicists long were baffled by spectral lines which suggested some sort of harmonic oscillation in atoms. The source of this oscillation remained a mystery for many years. Immediately following the introduction of the spectroscope in 1859 spectral studies produced a mass of information about the lines of various elements. Theoretically minded physicists were quickly overwhelmed by the data on wavelengths, and no real progress was made before 1885.

In that year Johann Jacob Balmer (1825–1898), a Swiss schoolmaster, published a paper revealing an empirical relationship between the four prominent lines of the hydrogen spectrum. Utilizing the wavelengths published earlier by the Swedish physicist and astronomer Anders Jonas Ångström, he showed that these agreed with wavelengths calculated by the formula

$$\lambda = 3645.6 \times \frac{n^2}{n^2 - 2^2},$$

where n is a whole number, 3 for the red line, 4 for the green, 5 for the blue, and 6 for the violet. Without having seen such a line, he suggested that there should be one at the edge of the visible violet. A colleague informed him that such a line, and several more, had been detected in the ultraviolet. These corresponded in wavelength almost exactly with those calculated from Balmer's formula. These lines crowded together as the formula demanded to suggest a series convergence limit at 3645.6 Å if n were taken as infinity. In his paper Balmer suggested that there might be other series of lines in the hydrogen spectrum corresponding to wavelengths calculated by replacing the 2^2 in his formula by 1^2, 3^2, 4^2, etc.

The Balmer formula was revised by Johannes Robert Rydberg[2] (1854–1919) of the University of Lund in 1890. Rydberg introduced the concept of wave numbers $(1/\lambda)$ and expressed the wave numbers according to the formula

$$\text{Wave no.} = \frac{10^8}{\lambda} = n = n_0 - \frac{N}{(m + \mu)^2},$$

where n represents the number of wavelengths per centimeter, m is any positive integer, n_0 and μ are constants peculiar to the series, and N is

[1] Wm. Ostwald, *Grundlinien der anorganische Chemie*, Engelmann, Leipzig, 1900, p. 92.

[2] J. R. Rydberg, *Phil. Mag.*, [5] **29**, 331 (1890).

a constant common to all series and all elements. The equation came to be commonly expressed in the form

$$\frac{1}{\lambda} = R\left(\frac{1}{2^2} - \frac{1}{n^2}\right),$$

where R is a constant known as the Rydberg constant, and n has the values 3, 4, 5, etc. When wavelengths were expressed in centimeters, the Rydberg constant had the value 109,720 cm^{-1} (recent value = 109,677.58 cm^{-1}). Rydberg also adopted Liveing and Dewar's series classification as:

(1) *principal*, including lines of great intensity,
(2) *diffuse*, including lines of intermediate intensity,
(3) *sharp*, including the weak lines.

In 1908 a fourth group of lines was discovered by Arno Bergmann of Jena who showed that these lines have properties much different from those of the three earlier groups (*fine*).

If the Rydberg formula is generalized, following Balmer's speculation, it may be written

$$\frac{1}{\lambda} = R\left(\frac{1}{n_f^2} - \frac{1}{n_i^2}\right),$$

where n_f represents an integer which remains fixed for a given series, while n_i represents integers having the values $n_f + 1$, $n_f + 2$, . . . for the successive lines of the same series. Until 1908 there was no evidence that such series, other than the Balmer series (where $n_f = 2$), actually existed, but in that year the German physicist Friedrich Paschen (1865–1947) found two lines in the infared whose wave numbers were correctly given by the Rydberg formula if $n_f = 3$ and $n_i = 4$ and 5.

Other series were encountered as improvements in spectroscopy made it possible to explore the infrared and ultraviolet. Between 1906 and 1914 Theodore Lyman (1874–1954) at Harvard discovered in the ultraviolet the hydrogen lines corresponding to $n_f = 1$. In the infrared Frederick Summer Brackett discovered the $n_f = 4$ series in 1922, and August H. Pfund discovered the $n_f = 5$ series two years later; both studies were made at Johns Hopkins.

GAS DISCHARGE TUBES AND THEIR CONSEQUENCES

Still another troublesome problem for atomic theory arose out of the electrical behavior of matter. We have already noted the growth of the theory of electrolytic dissociation. While this theory adequately explained electrolysis and vapor pressure phenomena of certain solutions, it raised the new question of the origin of the charged ion. Problems raised by behavior of gases in electrical discharge tubes were even more troublesome and took longer to resolve.

Early experiments indicated that electricity was conducted through solids without apparent change in the conductor; that through liquids conduction caused decomposition; and that through gases electricity was conducted only at high voltage in the form of violent sparks. In 1821 Davy learned that air becomes a better conductor when pressure is reduced. The numerous experiments on electrical discharge through gases carried out between 1821 and 1860 by Faraday, Snow Harris, von Marum, Geissler, and others produced little more than a mass of confusing information. There was much interest in the colored glow produced in the tubes, but little attention was paid to the fluorescence of the glass in the tubes.

In the fifties the development of improved tubes by Heinrich Geissler (1814–1879) made new advances possible. Geissler, an expert mechanic and glassblower who founded a shop in Bonn for making scientific instruments, introduced improved methods for producing low pressures, particularly with his mercury vapor pump. His gas discharge tubes were so much better than those most physicists could make that Geissler tubes were eagerly sought. With these tubes, whose gaseous pressure was frequently three or four millimeters and sometimes lower, new studies could be undertaken.

Julius Plücker (1801–1868), a professor of physics and mathematics at Bonn who carried out many experiments with Geissler tubes, observed that the fluorescent spots associated with the operation of the tubes could be moved by a magnet. This suggested that the beam causing the fluorescence was electrically charged.

Hittorf, a former student of Plücker, not only confirmed the influence of the magnet, but also discovered in 1869 that the radiations in the tube cast shadows when a solid body is placed between a pointed cathode and the glass wall of the tube. This observation was confirmed in 1876 by Eugen Goldstein (1850–1930), the German physicist who named the radiation which originates at the cathode and resembles light *Kathodenstrahlen*.

William Crookes (1832–1919) reported the results of extensive research in 1879. He not only verified the results of earlier workers, but also extended the knowledge of cathode rays into new areas with cleverly designed and highly evacuated tubes. With a tube containing a metal Maltese cross he verified the shadow-producing effect of the rays. The anode was placed at the side, showing that the position of the anode has no bearing on the production of rays from the cathode. He also discovered fluorescence fatigue, noting that glass exposed to the rays loses its glow with continued operation of the tube. He found that the rays cause minerals such as zinc sulfide to fluoresce and that they too show fatigue.

Using a tube containing a paddlewheel on a glass track, Crookes demonstrated that the cathode rays would roll the paddlewheel toward the anode. From this he deduced that the rays were really particles with a definite mass, rather than being light waves as most German physicists supposed. (Two decades later J. J. Thomson established that the masses are too small to move paddle wheels. Thomson attributed the effect to the heating produced by the greater recoil of colliding molecules striking the heated side of the vane. This explanation has since been confirmed.)

In another refinement of a Goldstein experiment Crookes showed that a narrow beam of cathode rays is deflected by a magnet in the direction that indicates a negative charge. His conclusion that the cathode rays were negatively charged atoms streaming from the cathode did not meet with the approval of German physicists who argued that a beam of negative particles must be accompanied by an oppositely directed beam of positive particles.

Partial evidence for this was obtained in 1886 by Goldstein, using a tube with a large-diameter cathode pierced by several holes. Rays were visible beyond the holes in the cathode. Named *Kanalstrahlen*, they differed from cathode rays in being self-luminous. Cathode rays have no self-luminosity, they only become apparent when they strike an object which fluoresces (e.g., air or glass). The luminosity of the canal rays is characteristic of the gas in the tube, being yellowish in air, rose-colored in hydrogen, yellowish-rose in oxygen, and greenish-gray in carbon dioxide. Cathode rays behaved the same regardless of the gas. Goldstein was unable to observe any deviation in a magnetic field, so the canal rays created a new problem without really settling anything. They attracted much less attention during the next decade than the cathode rays.

In 1892 Heinrich Rudolph Hertz (1857–1894), a professor at Bonn, learned that cathode rays easily penetrate gold or aluminum foil placed in their path in the tube. Later his pupil, Philipp Lenard (1862–1947), placed an aluminium-foil window in the tube and observed that the rays not only penetrated the window, but also caused the air to fluoresce up to two centimeters beyond the window. The discovery of "Lenard rays" reinforced the viewpoint of German physicists who held that the cathode rays were waves moving in ether rather than being charged atoms, since atoms should have been stopped by the metal foil. The particle position was defended by the French physicist Jean Baptiste Perrin (1870–1942); in 1895 Perrin detected a negative charge in an aluminum-collecting vessel placed in the path of cathode rays.

Joseph John Thomson (1856–1940), long associated with the Cavendish Laboratory of Cambridge University, now appeared on the scene in support of the particle hypothesis. He repeated Perrin's experiment, using a collector placed beyond a perforated anode and not in line with the cathode and anode. The cathode rays penetrated the anode and produced a fluorescent spot on the opposite glass wall. This spot could be moved about by a magnet. When the spot was moved over the opening of the metal collector-box, an attached electroscope immediately indicated a strong negative charge. When moved beyond, the charge on the box immediately fell off. The experiment showed that fluorescence and charge cannot be separated.

Fig. 18.1. J. J. THOMSON.
(Courtesy of the Edgar Fahs Smith Collection.)

Thomson next set out to demolish the argument that the cathode rays are not deflected by an electrostatic field. Since cathode rays cause ionization of the gas in the tube, this ionized gas, he reasoned, would be attracted to the plates and would neutralize the electrostatic charge. He also hypothesized that in a highly evacuated tube the production of gaseous ions would be reduced to a point where the electrostatic charge would be retained and deflection would be observed. Experiment showed Thomson's reasoning to be correct. Cathode rays could be deflected by an electrostatic field into a parabolic path as theory demanded.

The evidence in favor of a "corpuscle," as Thomson called it, now appeared to be convincing, but information regarding the magnitude of the charge and of the mass was still needed. There was no apparent method for determining either of these values, but Thomson devised a method for determining the ratio of charge (e) to mass (m). The results were variable, the best values for e/m lying between 10 and 31 million electromagnetic units per gram (present value, 1.7×10^7 emu/g or 5.1×10^{17} electrostatic units/g). These e/m values were startling since they implied a specific charge on the "corpuscles" at least a thousand times as great as that on the hydrogen ion (2.87×10^{14} esu/g). Thomson suspected the charge to be equal but opposite to that on hydrogen ions and that the mass of the "corpuscle" must be a thousandth that of the hydrogen ion. These results, published in 1897, were received with deep skepticism. Thomson's mass value ultimately proved to be too high.

The name *electron* came into use despite Thomson's objections. The term had been used earlier by the Irish physicist George Johnstone Stoney to designate charge units in atoms without reference to mass.

At the same time Thomson was conducting his studies, negative charges were obtained in unrelated experiments. Hertz learned that a negatively charged metal such as zinc gives up negative charges when exposed to ultraviolet light. It was also found that negative charges are emitted by incandescent filaments of metal or carbon. Thomson examined these charges and was able to demonstrate that they consist of particles with e/m values similar to the particles in cathode rays. He also showed that the e/m values of electrons are independent of the kind of materials present in filaments, cathodes, glass, or residual gas. The electrons were evidently universal constitutents of matter which when removed from an atom left it as a positively charged ion. This knowledge led Thomson to an explanation of the ionization of gases.

It was now essential to determine the mass, or charge, of the electron in order to learn if the assumed tiny mass was correct. The solution to this problem was not readily obvious, but through the application of seemingly unrelated principles it was ultimately possible to arrive at reliable values. The solution came through the production and study of fogs.

John Aitken of Edinburgh discovered in 1880 that a fog is produced when air saturated with water vapor is cooled by a slight expansion of volume. Thomson's deduction that a particle of atomic size should serve as a nucleus for fog formation if it is charged was confirmed in 1897 at the Cavendish Laboratory by Charles Thomson Rees Wilson (1869–1959) who formed fogs in dust-free air ionized by x radiation. These findings led to the development of the Wilson cloud chamber which proved such an effective tool in studies of radiation after 1912.

Thomson set out to determine the charge on the electron. Using a cloud chamber so adjusted that fog droplets formed on the electrons and the formula developed in 1849 by George Stokes for the rate of fall of spherical bodies through a fluid (e.g., raindrops through air), he arrived at a value equal to 3×10^{10} esu. The value was refined by H. A. Wilson in 1903 who measured the voltage necessary to prevent droplets from settling between two charged plates.

The best work on determination of the charge on the electron was done by Robert Andrews Millikan (1868–1953) at the University of Chicago between 1908 and 1917. Millikan used oil droplets for his charge carriers, since oil is sufficiently nonvolatile that the mass of the droplet does not change during the experiment. His apparatus was so designed that the movement of an oil droplet between two plates could be studied with a microscope set at right angles to the source of illumination. The rate of fall of an oil droplet between uncharged plates was first studied in order to calculate the mass from the Stokes' formula. Then a charge was placed on the horizontal plates, positive on the upper and negative on the lower. An oil droplet carrying one or more ions would now be caused to rise and the rate of ascent measured. Millikan observed that the rate of ascent frequently changed, but always by finite amounts, all of which were multiples of a minimum value. This behavior was attributable to loss or gain of ions, and the measurements showed the charge on the electron to be a finite unit of electricity. Millikan also examined the Stokes' formula critically and made certain improvements. His final value for e was 4.77×10^{10} esu. This value was recalculated to 4.8020×10^{10} esu in 1941 by the Australians Hopper and Laby on the basis of an improved value for the viscosity of air in the Stokes' formula.

Modern determinations of e/m show the ratio to be close to 1.7×10^7 esu/gm. This places the mass of the electron at 1/1837 of that of the hydrogen atom. This of course is a rest mass. Einstein showed in 1905, in his special theory of relativity, that the ratio e/m decreases with velocity because of a gain in mass. The velocities of electrons attained by Thomson were not sufficiently high to affect his results significantly, but the high velocities attainable in modern work, approaching within a few per cent of the speed of light, cause the ratio of e/m to drop below the rest ratio.

Discovery of X Rays

While work on the electron progressed, x rays, another outgrowth of the operation of discharge tubes, also were being studied. In 1895 Wilhelm Konrad Röntgen (1845–1923), a professor of physics at Würtzburg, was studying emission of ultraviolet light by discharge tubes, using crystals of barium platinocyanide, $BaPt(CN)_4$, which fluoresces under ultraviolet light, as a detector. One day he observed that the crystals fluoresced at some distance from the tube, even when the tube was wrapped in black paper. The radiation, which caused fluorescence on a screen several feet away, was found to originate from the glass at the end of the discharge tube where the cathode ray beam struck.

Röntgen studied the passage of the rays, which he designated *x rays*, through various substances. In general he found that the radiation is absorbed by dense and thick materials. Metals are fairly opaque to the radiation, especially those of high density like lead. Bones were found to be highly absorbent, whereas the soft tissues of the body transmit the radiation readily. It was also learned that the rays cause exposure when they strike a photographic plate and that unlike ordinary light they are not stopped by the black wrappers. The medical profession began using x rays a few months after their discovery. Unfortunately the harmful effect on living tissues was not recognized until many operators of x ray tubes were severely overexposed and several died from effects of exposure.

For more than a decade physicists debated vigorously whether x rays were waves or particles. Since they could not be deflected by a magnet they could not be charged particles like electrons. On the other hand they could not be reflected or refracted like ordinary light, nor could they be diffracted by even the most closely ruled diffraction gratings. In 1897 George Stokes theorized that if a charged particle were suddenly stopped, an electrical and magnetic disturbance should spread out from the point with the speed of light. Part of the energy of the charged particle would be dissipated in this fashion; the rest would become heat.

Since the penetrating power of x rays increases with the voltage used in their production, Stokes suggested that the radiations are light rays of very short wavelength. Because sufficiently fine diffraction gratings

Fig. 18.2. WM. LAWRENCE BRAGG AND WM. HENRY
BRAGG.
(Courtesy of the Edgar Fahs Smith Collection.)

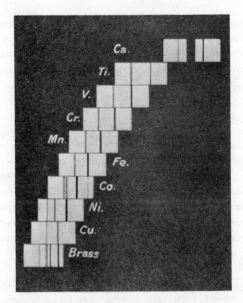

Fig. 18.3. X-RAY SPECTRA OBTAINED BY
MOSELEY FOR SEVERAL METALS.
(From *Philosophical Magazine* [1913], with per-
mission of Edward Arnold, Publishers.)

were not available, the wavelength of x rays could not be measured.
In 1912 Max von Laue (1879–1950) of Zurich reasoned that the distance
between atom layers in crystals might be of the proper order of magnitude
for use as a diffraction grating. W. Friedrich and P. Knipping proved
Laue's deduction correct. By passing an x ray beam through a zinc sulfide
crystal they obtained a symmetrical pattern of spots on a photographic
plate.

The interpretation of the spots obtained from x ray diffraction by
crystals was a difficult job. William Henry Bragg (1862–1942), professor
of chemistry at Leeds and later at the University of London and (from
1923) at the Royal Institution, and his son William Lawrence Bragg
(1890–1971) resolved the problem by treating the crystal as a three-
dimensional diffraction grating. These studies made it possible to establish
the character of x rays as a form of light with wavelengths ranging from a
fraction of an Ångström unit to several hundred Ångströms.

Of particular significance to chemistry was the study made in 1913 by
Henry Gwyn-Jeffreys Moseley (1887–1915). Moseley used various metals
as the target in x ray tubes and measured the wavelengths of the two
principal radiations produced.[3] The wavelengths revealed a systematic

[3] H. G. J. Moseley, *Phil. Mag.*, [6] **26**, 1024 (1913).

decrease as the target metal was changed from one element to the next in the periodic table. Figure 18.3 shows the results of Moseley's findings for the elements between calcium and zinc in the periodic table. By plotting square root of frequency against order of the elements in the periodic table (atomic number), straight lines were obtained. These results led to the conclusion that the order of the elements is not mere happenstance, but is based on some fundamental principle of atomic make-up. The research was also important because it showed that despite the atomic weight inversion cobalt is properly placed ahead of nickel, and that potassium immediately precedes calcium, with a gap between chlorine and potassium as the proper place for argon. Subsequent studies verified the blank spaces in the periodic table for elements between molybdenum and ruthenium and tungsten and osmium. Moseley was also able to show a missing element between neodymium and samarium in the rare earth series. A year later he was killed during the British storming of Gallipoli.

Positive Rays

Goldstein's canal rays were examined in 1898 by the German physicist Wilhelm Wien (1864–1928). By using a very powerful magnetic field, he was able to deflect the canal rays and determine from the direction of deflection that they carry a positive charge. At low pressures it was also possible to deflect them with an electrostatic field.

Wien next sought to determine the ratio of charge to mass; his techniques were similar to Thomson's. He used a tube containing an iron cathode in the form of a cylindrical plug 3 cm long and containing an axial hole 2 mm in diameter. The canal rays formed in the tube streamed toward the cathode, some of the rays passing through the hole to form a beam on the opposite side. This beam could then be passed through a magnetic or electrostatic field to impinge on the flattened end of the tube which served as a fluorescent screen. In order to compensate for velocity differences of particles having the same mass and charge, Wien placed the electrostatic and magnetic fields parallel to each other. Thomson had used them at right angles so they would counteract each other. Wien's arrangement caused all particles with the same e/m ratio to fall somewhere on the same parabolic curve, regardless of velocity. Particles with a different e/m fell on a different parabolic curve.

Wien's results, published in 1902, showed that the e/m values vary with the nature of the residual gas in the tube. His results were highly inaccurate, but despite this it was evident that the value obtained when hydrogen was present in the tube compared favorably with the value of e/m calculated for the hydrogen ion from electrolytic evidence. With oxygen gas the value was much less. The evidence clearly indicated that canal rays were made up of gaseous ions.

RADIOACTIVITY

Learning of Röntgen's discovery of x rays, it occurred to Antoine-Henri Becquerel (1852–1908) that, since the rays emanated from the fluorescent spot on the tube, there might be a relationship between x radiation and fluorescence. He immediately tested the hypothesis, using various fluorescent and phosphorescent materials available in his laboratory at the Sorbonne. Crystals of material were placed on well-wrapped photographic plates and exposed to the sun for several hours. Several substances gave negative results, but a salt of uranium gave off penetrating radiation which fogged the wrapped plate. Then his work was interrupted by several cloudy days during which his prepared plates and minerals lay in a dark drawer. These plates were developed without the accompanying minerals having been exposed to sunlight and therefore never having fluoresced in the presence of the plates. To his amazement halos produced by the salts appeared on the plates. Further experiments revealed that the radiation had no connection with fluorescence, but was a characteristic of uranium itself. Becquerel found that the radiations were like x rays in their penetrating ability, their effect upon a photographic plate, and their ability to ionize the air and render it conducting. The intensity of the radiation was in direct relationship to the quantity of uranium present in the sample.

Not long thereafter the subject began to be investigated by Marie Sklodowska Curie (1867–1934) as research toward her doctorate. Marie Sklodowska had come to Paris from Poland several years before to take a degree in mathematics and physics as a student of Gabriel Lippmann. During her period of study she met and married Pierre Curie (1859–1906), a member of the physics department who with his brother Jacques had discovered piezo-electricity.

Marie Curie set out to learn the extent of the new phenomenon, testing salts and minerals of a great many elements with her husband's piezo-electric quartz electrometer. Her desire to examine all known elements was aided by loans from Demarçay and others. Thorium compounds also gave off the ionizing radiation, a fact independently discovered in Germany by Gerhardt Carl Schmidt at Erlangen. No other elements showed the property, which appeared to be an atomic phenomenon. The nature of the chemical combination of uranium and thorium had no influence; neither did heat or other physical effects. Testing the mineral pitchblende, which is composed of about 80 per cent U_3O_8, she observed that its ionizing power is several times that of the same weight of pure uranium.

Madame Curie attributed the unexplained intensity to a new element which she and her husband undertook to isolate. Pitchblende ore was dissolved with acid and put through a separation process resembling the scheme used by freshmen students in qualitative analysis. The direction taken by the radioactive material was followed at every step with the electrometer. After months of tedious work with primitive facilities, they obtained a high radioactivity in the bismuth concentrate. They concluded that this contained a new element; Madame Curie named it *polonium*, in honor of her homeland. The announcement was made in July 1898.

Fig. 18.4.   MARIE SKLODOWSKA CURIE.
(Courtesy of the Richard Fischer
Collection.)

There was still intense radioactivity in another fraction, that of the alkaline earth precipitates. Hence it was necessary to fractionate the barium residues in order to track down still another element. When they fractionated barium chloride they found that the greatest activity accompanied the most insoluble portion. In a sample having 60 times the specific activity of uranium the spectroscopist Demarçay detected a faint new line among the barium lines. Further fractionation to a specific activity 900 times that of uranium caused an intensification of the new line. On the basis of this evidence the Curies announced their second new element in December 1898. The white salt, which was still grossly contaminated with barium chloride, glowed in the dark and hence was named *radium*.

The Curies now set out to purify the new element. Through the help of the Vienna Academy of Sciences they were able to obtain, as a gift of the Austrian Government, a ton of pitchblende residues from which the uranium had been extracted for use in glassmaking. The ore was mined

Fig. 18.5. PIERRE CURIE.
(Courtesy of the Edgar Fahs Smith
Collection.)

near the Bohemian city of Joachimstahl, which was at that time under Austrian control. By July 1902 several tons of residues had been dissolved and fractionated to obtain 0.1 gram of radium chloride sufficiently pure to show no spectroscopic evidence of barium. The atomic weight was determined as 225, on the assumption that radium, like barium, is divalent. The Curies and Becquerel were voted the Nobel Prize in Physics for 1903. Elemental radium was not isolated until 1910, when Marie Curie and André Debierne (1874–1949) succeeded in electrolyzing fused radium chloride. (Her husband had been killed in a traffic accident in 1906.) She was awarded the Nobel Prize for Chemistry in 1911; she remained the only person to be awarded two Nobel prizes until 1963 when Linus Pauling received his second.

At first Becquerel believed the radiations from uranium to be x rays. As more intense sources of radiation became available, however, it was recognized that the radiation was not homogeneous. In 1899 Becquerel and, in Germany, F. Giesel learned that part of the radiation is deflected by a magnet in the same direction as cathode rays. Velocity determinations gave a speed approximately half that of light, and e/m determinations suggested the particles to be electrons. Early in the same year Rutherford observed that uranium gave off a highly penetrating radiation which he named *beta* and an easily absorbed radiation which he named *alpha*. The α-radiations were effectively stopped by thin metal foils (e.g., 0.003 cm of Al), whereas the β-radiations were roughly 100 times more penetrating. The Curies also studied magnetic deflections and showed, in 1900, that a negative charge is associated with the β-radiations. On the basis of available information Becquerel concluded that the β-rays were similar to cathode rays (electrons).

At first the α-rays, which Becquerel and others were unable to deflect in a magnetic field, were considered to be easily absorbed x rays, perhaps produced when the particles were stopped. Rutherford, however, was struck by the fact that helium was always found associated with uranium and thorium minerals. He was able to deflect α-rays in 1903 by using powerful magnetic and electric fields. From the direction of the deflection it was apparent that the rays were positively charged. The e/m was determined to be about 2×10^{14} esu/g, comparing in order of magnitude to the atomic particles in positive ray beams rather than to electron beams. More careful e/m measurements made in 1906 gave a value of 1.5×10^{14} esu/g, approximately half the e/m value for hydrogen ions.

Such a ratio corresponds to that of a molecule of hydrogen with one electron missing or an atom of helium with two electrons missing. Rutherford leaned toward the latter hypothesis since W. Ramsay and F. Soddy had shown (1903) helium gas to be steadily liberated from radium chloride. Helium was also found in association with other salts of radioactive elements.

Rutherford and Hans Geiger (1882–1945) studied the total charge on the particle in 1908, obtaining a value of about 9.6×10^{10} esu, equivalent to twice the charge on the electron. E. Regener obtained a similar value in Germany. The association of α-particles with helium was clearly demonstrated in 1908 by Rutherford and Thomas D. Royds (1884–1955). Utilizing the recently discovered gas radon, which had been shown to be an α-emitter, they placed the gas in a glass bulb with walls so thin that α-particles passed through, but sufficiently strong to resist atmospheric pressure. This bulb was placed inside a tube which had one end drawn

Fig. 18.6. ERNEST RUTHERFORD.
(Courtesy of the Edgar Fahs Smith Collection.)

into a capillary with wires sealed into the ends of the capillary. The outer
tube was evacuated. At intervals any gas in the outer tube was forced
by mercury into the capillary, whereupon a discharge was passed between
the wires and the spectrum was examined. After four days the yellow and
green lines of helium became visible, and in six days the whole spectrum
of helium was clearly evident. A control experiment using helium gas
inside the thin-walled bulb showed no evidence of a helium spectrum in
the capillary after many days. Clearly the source of helium in the first
instance was the α-particles from radon.

A third type of radiation, unresponsive to a magnetic field but with
marked penetrating power, was recognized and studied by P. Villard in
France. This radiation came to be termed *gamma*.[4] Rutherford and
Charles G. Barkla (1877–1944) believed these rays to be similar to x rays.
Others, notably F. Paschen and W. H. Bragg, looked upon them as high-
speed particles. Rutherford and A. N. da C. Andrade established the
radiation hypothesis in 1914 when they succeeded in diffracting γ rays

[4] S. Glasstone, *Sourcebook on Atomic Energy*, Van Nostrand, Princeton, 2nd ed.,
1958, p. 56n, questions the usual attribution of the naming of γ-rays to Villard.
Glasstone believes that Rutherford, or possibly Becquerel, was responsible for the
name. At any rate the term came into general use in 1903.

with a crystal. Wavelength determinations showed correspondence with very short x rays.

Additional Radioactive Elements

As early as 1899 evidence began piling up for still other radioactive elements. In that year André Debierne, working in the Curies' laboratory, obtained an active substance in an iron precipitate which he separated from pitchblende. This substance was named *actinium*. When Crookes purified some uranyl nitrate, he found that the uranium was left in an apparently inactive state, but an impurity, named *uranium X*, was strongly radioactive. Crookes consequently suggested that the apparent radioactivity of uranium might be due to the contaminant. Becquerel also observed that uranium lost its activity when barium sulfate was precipitated from a solution containing uranium salts. However he set aside his materials for later examination. This revealed that the barium sulfate preparation soon lost its activity, whereas the uranium salt recovered its full activity within 18 months.

Rutherford and Soddy reported similar results with thorium in 1902. They were able to separate an active material, named *thorium X*, while leaving the thorium apparently inactive. Upon standing, the thorium X lost most if its activity while thorium became as active as it had been initially. Rutherford and Soddy plotted decay and recovery curves which suggested the establishment of an equilibrium state composed of slowly decaying thorium mixed with a rapidly decaying product.

A radioactive gas began to be observed in laboratories working with radioactive substances. The Curies reported in 1899 that materials placed near radium preparations acquired an "induced" radioactivity. Robert B. Owens (1870–1940), working with Rutherford at McGill, observed that air currents affect the activity of thorium. Rutherford showed in 1900 that thorium compounds continuously emit a radioactive gas which he named *emanation* (or *thorium emanation, thoron*). This gas disappears rapidly but leaves a radioactive residue on objects with which it comes in contact. At the same time Friedrich Ernst Dorn (1848–1916) at Halle isolated a gaseous emanation from radium. This inert gas was named *niton*, later *radon*. Debierne discovered a similar emanation in actinium preparations in 1903. Soddy and Rutherford showed that the emanation of thorium could be liquefied. Ramsay and Robert Whytlaw Gray made density determinations in 1910, showing it to be the heaviest known gas, with an atomic weight of 222.

The study of radioactivity attracted interest in many quarters but was pursued with particular activity in Paris, in parts of Germany, and in Rutherford's laboratories, first at McGill University in Montreal (1898–1907) and then at Manchester (1907–1919) and Cambridge. A New Zealander, Ernest Rutherford (1871–1937) studied at the Cavendish

Laboratory before he began teaching at McGill. Many of his associates, particularly Soddy, Geiger, Hahn, Fajans, Bohr, and von Hevesy, made notable contributions to the field. In Germany the Berlin laboratory of Otto Hahn (1879–1968) was particularly active. Lise Meitner (1878–1968) became closely associated with Hahn's researches until forced to flee Germany with the rise of Naziism. B. B. Boltwood (1870–1927) who worked at Yale also made notable contributions, particularly in discovering *ionium* and showing it to be the immediate precursor of radium. Ionium was independently discovered by Hahn and by Willy Marckwald in Germany.

By 1912 approximately thirty radioactive elements had been reported. The detailed sequence is confusing and is hardly worth reciting here. (See Appendix II.) Suffice it to say that it soon became evident, as Rutherford and Soddy pointed out in 1902, that radioactive elements are composed of unstable atoms which spontaneously decay with α, β, or γ emission into new elements. Radium was found to be part of the uranium decay series, being immediately preceded by ionium and followed by radon. Radon in turn decays into several successive elements, named *radium A* to *radium G*, before a stable state is reached. Somewhat similar decay patterns were observed with actinium salts and thorium salts (see Appendix III). The concept of decomposing atoms conflicted in every detail with the old belief in the atom as an indivisible particle, but there appeared to be no other interpretation.

Decay Rates

At the same time it was being observed that a particular element's rate of decay decreases according to a particular formula. For example, it was learned that the rate of activity of a specific amount of radon gas is high when first removed from radium but rapidly decreases. The rate of decay was found to follow an exponential curve according to the equation:

$$I_t = I_0 e^{-\lambda t},$$

where I_t is the activity at the end of a time interval t, I_0 the initial activity, e the base of natural logarithms, and λ a radiactive constant which represents the fraction of the substance undergoing decay in a unit of time. Thus the decay of a particular substance can be expressed in terms of the half life, the time during which half of a given mass will undergo radioactive change. Half lives of radioactive elements vary from fractions of a second to billions of years.

Recognition of Isotopes

Early in the study of radioactive substances it became evident that the element radium D closely resembles lead. When the two elements are mixed they can not be separated. J. Elster and H. Geitel showed that lead deposits, particularly those associated with uranium minerals, are often radioactive. Lead ores from old rocks uncontaminated with uranium are not radioactive.

Several investigators reported different atomic weights for lead from different sources. Fajans sent Max E. Lembert, his assistant, to Harvard to make determinations with T. W. Richards. In 1914 they reported that the atomic weight of lead from radioactive ores varied from the standard value of 207.2. Atomic weights as low as 206.4 were found for lead from pitchblende and as high as 208.4 for lead isolated from thorite. Similar results were reported independently from the laboratories of Hönigschmidt and Maurice Curie (nephew of Pierre).

Other similarities between particular radioactive elements were becoming evident. The three known emanations all had the properties of a radioactive rare gas. Boltwood discovered ionium which has properties similar to those of thorium. The two could not be separated if mixed, and both appeared to have the same spectral lines.

Scientists began to doubt that the atoms of a particular element were all alike, as Dalton had supposed. As early as 1907 H. N. McCoy and W. H. Ross at Chicago were groping toward an isotope concept, but it was Soddy who clearly enunciated the idea in 1913 when he, Alexander S. Russell, and Fajans independently developed the Radioactive Displacement Law.

In connection with his clarification of the nature of mesothorium in 1911, Soddy had pointed out that when an α-particle is lost the resulting element corresponds to the one two places left in the periodic table. About the same time Russell found that following β-decay the product corresponds to the next element in the periodic table. Soddy, Russell and Fajans generalized these specific findings to hold for all radioactive changes. Application of the rule meant that atoms of the same element might differ in weight—in fact, Soddy predicted that lead from uranium ores would have atomic weight 206 and that from thorium ores 208 before Richards and Hönigschmidt reported their results. Soddy applied the term *isotope* (Greek, meaning "same place") to radioactive elements that appear to have the same place in the periodic table but come from different sources. He used the term only in connection with the elements encountered in radioactive decay and did not associate it with elements in general.

Mass Spectra and Stable Isotopes

Early in the century J. J. Thomson set out to examine seemingly incongruous results obtained by Wien in his e/m determinations of positive rays. Wien had obtained, not the sharp parabolas called for by atomic theory, but feathery parabolas which suggested that either the constancy of atomic weight was invalid or the magnitude of the electrical charge was variable. Thomson concluded that Wien's experiments were faulty, due to collisions between positive rays and residual gas molecules in the sorting tube. He designed a positive-ray tube operating at very low pressures and obtained the sharp parabolic lines that theory demanded. With hydrogen there are two parabolas, at positions which give e/m values corresponding with H^+ and either H^{++} or H_2^+. Thomson rejected the idea of hydrogen losing two electrons in the tube and considered the parabola to represent H_2^+. With oxygen three lines correspond with O^+, O^{++}, and O_2^+. Here, incidentally, was evidence from a new direction for diatomic gas molecules.

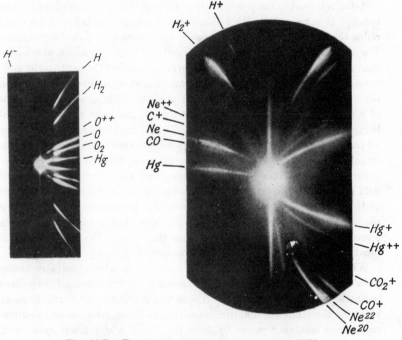

Fig. 18.7. PARABOLIC MASS SPECTRA OF VARIOUS ELEMENTS.
(From F. Aston, *Mass Spectra and Isotopes* [1942], with permission of Edward Arnold, Publishers.)

Thomson also observed that with some gases he obtained very faint inverted parabolic lines suggesting the presence of negatively ionized gases. These negative ions were observed for hydrogen, oxygen, and chlorine, but never for neon, helium, mercury, or nitrogen.

With neon in his tube Thomson obtained a parabola corresponding to atomic weight 20, but another one appeared corresponding to atomic weight 22. The atomic weight of neon according to Ramsay's work was 20.2. The second line was difficult to explain. It was suggested that it might be due to an unknown hydride of neon, NeH_2, or to doubly charged carbon dioxide. Thomson believed that an element closely related to neon might be responsible for a relationship similar to that of iron, cobalt, and nickel. His assistant, Francis William Aston (1877–1945), sought to separate these elements. Since neon and its supposed companion were chemically inert, he resorted to differential diffusion through fine orifices and to differential adsorption of the gases on charcoal at liquid air temperatures. He had not succeeded by the time World War I forced him to put this work aside.

Aston returned to the problem after the war. He was aware of the isotope concept in connection with radioactive elements and believed he could refine the Wien-Thomson technique for separating ions according to mass. Wien and Thomson had utilized electrical and magnetic forces so oriented that one force threw the ion upward while the other threw it to the side, causing the ions with the same e/m to fall somewhere on the same parabola, the exact position depending on their velocity. This meant that all ions of a given e/m value were spread out over a parabolic curve instead of being concentrated in one spot. Since it was impractical to obtain ions all having the same velocity, Aston utilized another technique for concentrating them. By passing the positive-ray beam first through an electrostatic field oriented to turn them down and then through a magnetic field oriented to turn them up, ions with the same e/m could be focused at the same spot regardless of their velocity. By placing a photographic plate in line with the focal points of the ions, Aston was able to obtain a record of the particles in the gas according to their masses. The instrument became known as a *mass spectrograph*.

When neon gas was placed in the instrument, which had a resolving power ten times greater than Thomson's, Aston obtained two lines indicating masses of 20 and 22. There was no line at 20.2, the Ramsay value for the atomic weight of neon, although the instrument had a sufficiently good resolving power to show the line if a significant number of particles of such a mass were present. Neon gas was shown to be a mixture of two isotopes, 90 per cent mass 20 and 10 per cent mass 22. Aston thus extended the isotope concept beyond the radioactive area to include stable elements of low atomic weights.

Aston studied other elemental substances, obtaining evidence for three isotopes of sulfur (32, 33, 34), two of chlorine (35, 37), three of silicon (28, 29, 30). By placing solid salts in the path of the electron beam, he was able to obtain the lines of the component elements, e.g., sodium chloride gave sodium as well as chlorine lines. In this manner the masses of nongaseous elements were reexamined. Compared to oxygen with a value of 16, the masses uniformly appeared to be whole numbers, indicating that Prout's hypothesis might be a useful concept after all. Aston proposed the *whole number rule*.

Improved instruments for measuring atomic masses were built in the laboratories of Aston, Arthur J. Dempster at Chicago, Kenneth T. Bainbridge at Bartol Research Foundation and Harvard, and A. O. Nier at Minnesota. The instrument Aston developed in 1927 could measure mass with an accuracy of one part in 10,000. Optical spectra of incandescent gases and vapors also proved useful in determining isotopic masses. In 1929 William F. Giauque and H. L. Johnson at the University of California determined from optical spectra that oxygen consists of three isotopes of masses 16, 17, and 18. Chemists continued to base their atomic weight tables on natural oxygen as 16, but physicists introduced a table where $^{16}O = 16$.

Reexamination of hydrogen suggested to several investigators that there might be a heavy isotope. In December 1931 Harold Urey and George M. Murphy at Columbia and F. G. Brickwedde of the United States Bureau of Standards reported experimental evidence of such an isotope: from a 1 ml. residue remaining after evaporation of 4 liters of liquid hydrogen by Brickwedde, Urey and Murphy observed in the optical spectrum a line calculated to be due to heavy hydrogen. The next year Hugh Taylor, Henry Eyring, and A. A. Frost at Princeton prepared a 99 per cent concentrate of heavy water, using an electrolytic technique which produced 82 ml. of heavy water from 2800 liters of ordinary water. This hydrogen isotope was named *deuterium* by Urey and his colleagues.

Departure from Whole-Number Masses

Mass spectrographic studies of the isotopes of the elements, particularly Aston's with his 1927 instrument, soon revealed that the masses relative to $^{16}O = 16$ are close to, but not exactly, whole numbers. In general the exact masses of the elements through fluorine run a trifle above the nearest whole number (known as the *mass number*) and those from neon through osmium have exact masses a trifle below the mass number. From iridium through the rest of the periodic table the exact masses steadily rise above the mass number.

The departure of exact masses from whole numbers apparently undermined Prout's hypothesis, which otherwise seemed so nicely in accord with the isotope concept. Of course chemists had long known that the best atomic weights of hydrogen and oxygen do not bear a whole number relationship (1.008:16 or 1:15.873). Aston's mass spectrograph of 1919 did not have sufficiently good resolution to determine whether the mass ratio was 16/1 or 16/1.008. It also was believed possible that hydrogen might contain a heavy isotope in the proportion of one in every 130 atoms, but the mass spectrograph could not resolve this question because of interference from hydrogen molecules. Aston solved the problem in 1920 by

Fig. 18.8. ASTON'S FIRST MASS
SPECTROGRAPH.

(From F. Aston, *Mass Spectra and
Isotopes* [1942], with permission of
Edward Arnold, Publishers.)

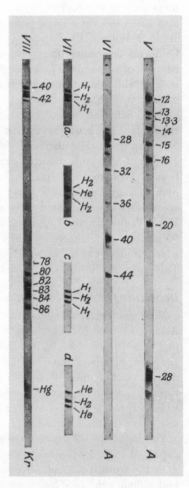

Fig. 18.9. TYPICAL MASS SPECTRA
OBTAINED WITH THE FIRST MASS
SPECTROGRAPH.

(From F. Aston, *Mass Spectra and
Isotopes* [1942], with permission of Edward
Arnold, Publishers.)

bracketing the line for the hydrogen molecule (H_2^+) between two helium
lines formed at variable intensities of the electric and magnetic fields.
Since the line was off center between the two helium lines, he was able
to calculate its exact mass as corresponding to 1.008.[5] (See Fig. 18.9,

[5] F. W. Aston, *Isotopes*, Arnold, London, 2nd ed., 1924, p. 71 and plate III (item
vii).

spectrum VIId.) Actually this finding, rather than eliminating Prout's hypothesis, reinforced it at a new level of sophistication since it is consistent with Einstein's theory of relativity which suggests a small mass decrease in the building up of elements from hydrogen.[6]

In 1905 Albert Einstein (1879–1955) indicated the equivalence of mass and energy in his special theory of relativity. According to the formula $E = mc^2$ (where c represents the speed of light) the destruction of one gram of matter will result in the production of 9×10^{20} ergs of energy. Simple combination of hydrogen atoms, as suggested by Prout's original hypothesis, would merely multiply without providing for binding them. If the combination were accompanied by a release of energy, as in an exothermic chemical reaction, then a stable atom might be expected. Einstein's theory suggests that such energy release may take place with the expenditure of mass. Since the mass ratio of oxygen to hydrogen is 16/1.008, not 16.128/1.008 (16/1), Prout's hypothesis is consonant with Einstein's special theory of relativity.

In his study of exact masses of isotopes Aston introduced the term *packing fraction* to designate the difference between the exact mass and the closest whole number (mass number), using $^{16}O = 16$ as the reference isotope.

$$\text{Packing fraction} = \frac{\text{Exact mass} - \text{Mass no.}}{\text{Mass no.}} \times 10,000$$

He plotted packing fractions against mass numbers to obtain a curve in which the elements preceding oxygen—16 have positive values dropping off sharply following hydrogen. The succeeding elements show negative values; after a plateau in the vicinity of iron the curve gradually rises, crossing the zero value in the neighborhood of mercury. Helium, carbon, and oxygen, which Rutherford observed to be very stable elements, lie below the main curve. A correlation between packing fraction and stability is apparent, those elements lowest on the curve being of greater stability than those at the extreme ends. Rutherford pointed out that the gradual slope of the curve in the region of the high atomic weights is consistent with the fact that α-particles are eliminated rather than protons.

THE STRUCTURAL ATOM

By the turn of the century the Daltonian atom was doomed by experimental evidence incongruous with an indivisible particle. Of course there had often been a tendency to surround the atom with a capsule—*caloric*

---

[6] In the reasoning taking place at this time, scientists thought not in terms of the original Prout Hypothesis but rather in terms of combination of protons, neutrons, and electrons. However, the following paragraph is treated in terms of the simple combination of hydrogen atoms to form heavier atoms.

a century earlier, an *electrical fog* in the late 1800's. Physicists were beginning to ask whether the atom might have negative particles which might be lost under certain circumstances, leaving a positive ion. Of course they were uncertain about the nature of the relationship between the negative particle and the positive ion.

Those groping in this direction were provided some clues by the discovery of the electron. The production of electrons in different ways from different materials indicated the significance of the electron as an apparently universal constituent of atoms. The probability of an electrical structure of matter was recognized by 1895 by H. A. Lorentz of Leyden and Joseph Larmour of Cambridge.

In 1903 Lenard undertook one of the first attempts to create a model of a structural atom. He had brought cathode rays through an aluminum window in 1894, thus obtaining evidence for the emptiness of atoms. Hence when he sought to devise a structure, he postulated a shell largely empty except for the presence of *dynamids* at the center. A single positive charge connected with a single negative charge made up a dynamid. The dynamids, their number proportionate to the atomic weight of the element, lay as a cluster in the center of the atom where they formed an impenetrable barrier. Since Lenard's atom was primarily an empty shell, electrons might easily pass through with little chance for obstruction by the tiny dynamid cluster.

The Japanese physicist Hantaro Nagaoka postulated a saturnian atom. His atom was a positive sphere with a halo of mutually repellent electrons; it resembled the planet Saturn. This model more readily explained the loss of electrons with the formation of positive ions than did Lenard's. It also sought to account for spectral lines by suggesting that the electrons might vibrate or quiver.

A third model, sometimes referred to as the "plum pudding," or "raisin muffin," atom, was based upon a suggestion of Kelvin elaborated by J. J. Thomson in 1904. The Thomson model, which consisted of a uniformly positive sphere of matter in which electrons were imbedded, did not account for the ease with which an atom is penetrated or the ease with which gaseous atoms are ionized. Thomson suggested that when the electrons were moving too fast within the positive matrix, radioactive particles were ejected and also made a vague attempt to account for radiation and chemical combination. Beyond noting that it must be large, he did not concern himself with the number of electrons that might be present in an atom.

In 1906 Rutherford suggested that there were 1000 electrons in the hydrogen atom; his notion derived from the then current supposition that the mass of the electron was 1/1000 of that of the hydrogen atom. We cannot be sure, but Rutherford may have considered half these electrons to be positively charged. At the same time Oliver Lodge was looking upon the atom as a complex of interrevolving positive and negative electrons.

Scientists needed a positive particle, but no particle of less than atomic mass had been encountered experimentally. By 1906 Thomson, who had begun clearing out the particles of the atom, believed the hydrogen atom contained one electron which revolved in some sort of nebulous positive jelly. He also held, on the the basis of experiments on the dispersion of light and scattering of x rays, that the number of corpuscles was not appreciably different from the atomic weight.

Ramsay in 1908 sought to treat the electron as an "element" in connection with the ionization phenomenon. He supposed that in the ionization of hydrogen chloride the hydrogen atom acquired its positive charge by losing an electron to the chlorine atom, which thus acquired the charge of the chloride ion. According to Ramsay the chloride ion in the neutral HCl molecule forced the extra electron back into the hydrogen ion. The association of the electron with valence phenomena was treated in the concepts of Lodge, Thomson, and Ramsay only at a vague, unsophisticated level as rods and slots or as lines of force.

An electron gas theory of electricity was also considered. E. E. Fournier d'Albe (1868–1933) sought to treat electrons as molecules making up the mutually repellent particles of an "electron gas" and suggested that electrons roaming among the bound electrons in metals were responsible for conductivity. D'Albe held that there were no positive electrons and thus created a theoretical structure somewhat related to Franklin's one-fluid theory of electricity.

In 1911 the problem of atomic structure was taken out of the realm of broad speculation by Rutherford on the basis of his experiments dealing with the scattering of α-particles in gases and by thin metal foils. With the refinement of C. T. R. Wilson's procedure for vapor condensation on ion tracks, the tracks formed by α-particles moving through air could be photographed. It was observed that in most cases the path is a straight line, but that occasionally the path terminates with a sharp turn suggesting an encounter. Rutherford also observed in 1906 that whenever a metal foil was placed in the path of α-particles the particles penetrated the foil but were spread out into a diffuse spot on a photographic plate behind the foil.

In 1908 Rutherford and Hans Geiger observed a sharp deflection of a small number of α-particles when a beam was passed through a thin metal foil. During the next year the phenomenon was studied carefully in the Manchester laboratory by Geiger and Ernest Marsden, who found that while the greatest number of particles could be detected within a few degrees of the spot behind the foil the beam would have struck had there been no foil in its path, a significant number of particles were sharply deflected in passing through the foil. The scattering was sufficiently great to suggest that the particles had suffered deflections by a powerful force. A few of the particles were deflected more than 90 degrees, into the direction from which they had come. Many years later Rutherford described the unexpected results as follows: "It was about as credible as if you had fired a 15-inch shell at a piece of tissue paper and it came back and hit you."[7] Rutherford's hypothesis that the α-particles were deflected by a close approach to a center of highly positive electricity was consistent with Geiger and Marsden's scattering distribution studies in which calculations were based on the effects of coulombic repulsion between α-particles and positive centers in the metallic atoms.

In 1911 Rutherford postulated an atom which consisted of a central region (called the *nucleus* in 1912) containing nearly all of the mass of the atom along with the positive charge. One or more electrons, dependening upon the nuclear charge, revolved around the nucleus at a comparatively great distance. The atom was thus visualized as mostly empty space, a view consistent with the penetrability of atoms by α-particles. And the scattering was consistent with the presence of a highly positive nucleus occupying only a tiny fraction of the total "volume" of the atom.

The magnitude of the charge on the nucleus puzzled scientists for a time. In 1913 Geiger and Marsden, arguing from their scattering studies, suggested it to be one-half the atomic weight. On the basis of x ray scattering studies made between 1903 and 1911, Charles Glover Barkla had arrived at this same conclusion for the number of electrons in various atoms.

Moseley's x rays studies in 1913 contributed to the problem's solution. His bombardment of many of the elements between aluminum and gold in the periodic table led to the establishment of the proper atomic numbers. Rutherford recognized that Moseley's work correctly identified the number of positive charges of the nucleus and with it the number of planetary electrons.

The problem of the difference between the atomic weight and the atomic number remained. In 1913 the Dutch physicist A. van den Brock suggested that there might be electrons in the nucleus. By 1914 Rutherford had definitely accepted electrons in the nucleus. The nucleus thus was believed to have positive charges equal in number to the atomic weight, with enough compensating electrons to bring the net nuclear charge down to the atomic number. The nucleus of the hydrogen atom was thus apparently the positive particle, or positive electron, theory seemed to require, although its actual discovery and naming had to wait the termination of World War I.

The Bohr Atom

The Rutherford atom still had a grave shortcoming. According to classical theory, which suggested that electromagnetic waves would be

---

[7] E. Rutherford in his 1936 lectures on the *Background to Modern Science*. Quoted from Glasstone, *op. cit.*, p. 93.

radiated by an oscillating charge, the Rutherford atom should radiate energy continually since the electron travelling around its orbit changes direction continually and is therefore accelerating. If the electron loses energy on every revolution, it should quickly spiral into the nucleus. This of course does not happen, so either the model or classical theory was in error.

The problem was attacked by Niels Bohr (1885–1962) of Copenhagen who had recently completed a period of study with Rutherford. Bohr decided to treat the structural atom from the standpoint of the quantum theory, and in doing so to assume that energy emission fails to take place as long as the electron travels about a proper orbit, but does take place when it leaves the orbit to drop to another.

Fig. 18.10. Niels Bohr.
(Courtesy of the Edgar Fahs Smith Collection.)

Quantum theory, a somewhat radical innovation in physics, had been introduced in 1900 by Max Planck (1858–1947). Planck had been searching for a sound mathematical treatment of black body radiation when he found it necessary to assume a discontinuous emission of energy. That is, if the energy is emitted in discrete units, or *quanta*, a logical mathematical treatment is possible. The energy of a quantum of light was considered equal to hv; h is Planck's constant and v frequency of the light. The concept of discrete energy states and quanta of energy represented such a marked departure from classical ideas in physics that the theory met much opposition. The theory, however, satisfactorily explained the experimental observations connected with thermal radiation and was useful in explaining other phenomena such as the photoelectric effect.

Bohr proceeded to design a theoretical model of the hydrogen atom, assuming like Rutherford a central nucleus of single positive charge and a single electron revolving about the nucleus. By dealing with the energy of a harmonic oscillator, Bohr was able to formulate the kinetic energy of the electron in such a manner that the energy changed in discrete amounts rather than continuously. He derived the formula:

$$\nu = \frac{2\pi^2 m e^4}{h^3} \left(\frac{1}{n_1{}^2} - \frac{1}{n_2{}^2} \right).$$

This is easily recognizable as Rydberg's formula for the Balmer lines of hydrogen if n_1 is replaced by 2 and if the ratio $2\pi^2 m e^4/h^3$ is identified with the Rydberg constant. When Bohr substituted the appropriate values for mass, charge, and Planck's constant, the result was surprisingly close to the value of the constant arrived at empirically by Rydberg. The whole set of Balmer lines could be correlated with the formula when the n_2 term was given values of 3, 4, 5, 6, etc. The recently discovered Lyman series in the ultraviolet coincided with $n_1 = 1$ and $n_2 = 2$, 3, 4, etc., while the infrared Paschen series took $n_1 = 3$ and $n_2 = 4$, 5, 6, etc.

Bohr next developed a theoretical model of the hydrogen atom, supposing the electron capable of revolving about the nucleus in certain permitted orbits, or energy levels, where no emission of energy takes place. In moving from lower to higher energy levels, and vice versa, energy is absorbed or emitted in finite amounts, corresponding to spectral lines. Bohr also identified certain hitherto puzzling stellar spectrum lines with those calculated for singly ionized helium. This introduced the atomic number Z into his equation, it had been overlooked in the case of hydrogen where it is 1. Bohr now had an equation with which wavelengths of the spectral lines of all one-electron atoms (H, He^+, Li^{++}, Be^{+++}, etc.) could be calculated. Later such ions for all elements up to oxygen were prepared in violent electric discharges and their spectral lines were found in agreement with Bohr's formula.

Extension of the Bohr Theory

As a first attempt the Bohr model naturally was found to have flaws, especially when it was extended to multielectron systems. Hence the development of the concept of the atom was turbulent. New minds began to attack the problem and fresh views were introduced. World War I delayed progress, but once the conflagration ended, progress was rapid. Even while the war was in its early stages, new suggestions were made.

In 1915 the German physicist Arnold Sommerfeld (1868–1951), introduced the use of elliptical orbits. These had not been excluded by Bohr, but he had restricted himself to circles. William Wilson (1875–1965) was independently active in the mathematical development of elliptical orbits. Into the twenties these men and Bohr continued to make the major contributions to the further refinement of atomic theory.

The Bohr theory also had to be applied to the structure of multielectron atoms. This introduced many complications since the mathematics of one-electron atoms was by no means simple. Moseley's discovery of atomic numbers was opportune since the high frequencies of the x ray spectra of the elements were of particular significance in understanding the displacement of inner electrons.

The elucidation of the structures of atoms like lithium and sodium progressed most rapidly. To a degree these atoms resembled one-electron systems. It was reasoned that the third electron in lithium and the eleventh electron in sodium traveled in a highly elliptical orbit, mostly outside the region occupied by the inner electrons. These inner electrons might even be treated as shielding electrons which cut down the magnitude of the positive nuclear attraction for the outermost electron. Bohr found it possible to treat the sodium and the hydrogen atom in a similar manner if the nucleus and the ten inner electrons of the sodium atom were regarded as a somewhat enlarged nucleus with a charge of $+1$.

Bohr next devised an arrangement of the elements in periods of 2, 8, 8, 18, 18, and 32. His periodic table arranged the elements in this fashion: Elements 21 and 28 were boxed to set them off from the rest of the fourth period. The corresponding elements in the fifth period were similarly treated, while the sixth and (incomplete) seventh periods were placed in a long box which included an inner box containing the rare earth elements.

Bohr recognized that the electrons in the second shell are not all similar in energy level, so a subshell was created. The third and fourth shells were subdivided into three and four subshells, respectively. Bohr was not sure of the distribution of electrons in the subshells, and until 1920 he divided them evenly among the available subshells. The second shell of neon consisted of two subshells containing four electrons each; the third krypton shell had three subshells with six electrons each; the fourth niton (radon) shell had four subshells with eight electrons each. This arrangement proved inconsistent with chemical and spectral properties, and later workers found it necessary to redistribute electrons into subshells containing two, six, ten, and fourteen electrons.

Bohr's thinking was influenced primarily by spectral behavior of the atoms, as was that of Sommerfeld and William Wilson. Charles R. Bury, at about the same time as Bohr, was arriving at the same conception of multielectron atoms on the basis of chemical considerations. Since 1916 chemists had been considering a structural atom which might be useful in explaining chemical properties. Walther Kossel (1888–1956) in Berlin and G. N. Lewis and Irving Langmuir in America devised a static atom model which was somewhat successful in explaining ionic and covalent combination. These models will be considered in more detail in Chapter 20.

Despite the fruitfulness of Bohr's theory of atomic structure, many details of chemistry and physics remained outside its scope or were inconsistent with it. During the twenties there arose a group of theoretical physicists who introduced new schemes of dealing with the atom mathematically. The revolution leading to the new quantum mechanics was initiated in large part by Louis Victor de Broglie (b. 1892) in France and Werner Heisenberg (1901–1976) in Germany.

De Broglie treated electrons, not as particles following orbits in Newtonian fashion, but as entities whose motion was governed by the pattern of their standing waves. All matter was thus a wave. This concept has no implications as far as the motion of gross objects is concerned but is of real significance with regard to objects of the magnitude of the electron. The *wave mechanics* suggested by de Broglie was developed largely through the systematic work of the German physicist Erwin Schrödinger (1887–1961) in 1926.

Heisenberg's *matrix mechanics* was developed in 1925. The German argued that it was necessary to deal only in observables. The atom cannot be observed when it is in a steady state. When an electron jump takes place it is the frequency of the radiation that is observed. Hence the atom must be described in terms of frequencies and energies. These frequencies of spectrum lines, or energies of state, were listed in tabular form (matrices). Heisenberg developed the algebra of matrices necessary to describe the properties of atoms. This algebra was extended by Max Born (1882–1970), another German.

At first de Broglie and Schrödinger's wave mechanics and Heisenberg's matrix mechanics bewildered some scientists; they appeared to attain the same results but utilized drastically different mathematical approaches. This dilemma was resolved when Schrödinger revealed an inner relationship between them.

Heisenberg's work led to the Uncertainty Principle which postulated the impossibility of precisely determining both the position and the velocity of the electron. If position is accurately determined, the velocity becomes uncertain, and vice versa. This is not caused by inadequacies in measuring devices; rather it is a fundamental restriction of nature. The error in measuring position multiplied by the error in measuring momentum is of the order of magnitude of Planck's constant.

The concepts of the new quantum theory were developed in a coldly logical, elegant form by the English physicist Paul A. M. Dirac (b. 1902) in 1927, Dirac integrated both the matrix mechanics and the wave mechanics into a new mathematical system. Important contributions to the problem were also made by the French physicist Pascual Jordan.

The over-all effect of the work of these men was to make the concept of the structural atom subtle by substituting for Bohr's readily visualizable orbital model a nonvisualizable mathematical model. This mathematical model, consisting of wave functions, satisfied the theoretical physicist but was too abstract to be useful in the day-to-day work of most chemists. "Pictures" of electrons in orbits continued to be used by chemists for many years, even though many physicists looked askance at such models.

Actually, the theoretical "model" used in science is usually determined by the practical problem immediately at hand. Many times a rigorous mathematical treatment is too complex for use in handling a particular problem. A simple model will frequently serve the scientist satisfactorily if his interest is in the use of theory as a tool rather than in developing theory for its own sake. The model that gets the job done fastest and with reasonable satisfaction is often utilized, even when it is open to criticism with respect to mathematical rigor.

CONCLUSION

As the nineteenth century was drawing to a close the Daltonian atom became less and less adequate to explain the facts of chemistry and physics. Solution of the problem was to come, not directly from the field of chemistry but from parallel studies in physics—interpretation of spectral lines, the behavior of gases in discharge tubes, and radioactivity. Early in the twentieth century it was becoming evident that atoms could no longer be looked upon as permanently indivisible particles but as particles in which electrons were an important component.

Studies in mass spectrography converged with studies of radioactive decay products to reveal the existence of isotopes, thus demolishing the Daltonian assumption that all atoms of a particular element are alike.

The α-scattering experiments carried out in Rutherford's laboratory provided the clue for the nuclear atom, but it was necessary for Bohr to depart from classical physical concepts and introduce quantum interpretation to arrive at a concept for the hydrogen atom which was scientifically satisfying. In doing so he was able to clarify the physical significance of Rydberg's constant, introduced on empirical grounds two decades earlier. The Bohr concept proved inadequate to deal with more complicated atoms and the 1920's saw a high degree of mathematical refinement of atomic theory in the work of de Broglie, Schrödinger, Heisenberg, and Dirac.

CHAPTER 19

RADIOCHEMISTRY II.
THE NUCLEAR AGE

TRANSMUTATION

When Rutherford and others realized that atoms spontaneously undergo decay into atoms of different elements, the concept of the element as a permanent substance, the basis of chemical progress since Lavoisier, was badly shaken. The change of one element into another was the goal alchemists had vainly sought for a millenium and a half. It was now natural for the question to arise: "Can such a transmutation be directed by man?"

In 1905 W. H. Bragg and R. Kleeman observed that α-particles emitted by a particular substance have a characteristic range in air, the ionization produced by them ceasing abruptly after the particles travel a certain distance. And in 1914 E. Marsden working in Rutherford's laboratory learned that the α-radiations from radium C produce a few long-range particles when passed through hydrogen gas. These particles have a longer range than α-particles and were presumed to be the nuclei of hydrogen atoms.

Rutherford planned to examine this hypothesis experimentally, but World War I began and he spent the war years doing research on submarine detection. As soon as the war ended he turned again to his work on radioactivity and in 1919 reported the results of experiments in which he studied the effect of α-particles on different gases. His source of α-particles was contained in a cylinder with a thin metal foil at one end capable of stopping the α-particles. The scintillations of a zinc sulfide screen just beyond the metal foil indicated that particles were penetrating the foil. When hydrogen gas was present in the cylinder, many scintillations were observed, indicating that the α-particles were producing high-speed hydrogen nuclei by collision with the gas. When oxygen or carbon dioxide was in the cylinder, hardly any particles registered on the screen.

When air was used, however, the number of scintillations increased. Neither oxygen, carbon dioxide, not water vapor of the air was responsible for this effect; it was found to be associated with the nitrogen. Rutherford reported, "It is difficult to avoid the conclusion that these long-range atoms arising from the collision of alpha particles with nitrogen are not nitrogen atoms but probably atoms of hydrogen. . . . If this be the case, we must conclude that the nitrogen atom is disintegrated under the intense forces developed in a close collision with a swift alpha particle."[1]

During the years that followed, Rutherford and his collaborators assiduously studied the effect of α-particles on various elements. With James Chadwick (1891–1974) he observed the deflection of the long-range particles in a magnetic field and showed that they are positively charged hydrogen nuclei, or protons. Under α-bombardment protons were obtained from all the atoms from boron to potassium except carbon, oxygen, and possibly beryllium. The energy of the protons was sometimes greater than that of the impinging particle, showing that energy is sometimes acquired by nuclear rearrangement. The source of the positive charge of the atomic nucleus was clearly identified with the proton, or hydrogen nucleus.

It was not known until 1925 whether the proton was dislodged from the nucleus by a severe collision with an α-particle or whether the α-particle actually combined with the nucleus, which then rearranged with the ejection of a proton. If the nature of the resulting nucleus were known, this question could have been answered, but there was no way of identifying this nucleus. The problem was resolved by P. M. S. Blackett (1897–1974) in England and William D. Harkins at Chicago as the result of their examination of thousands of cloud chamber photographs. Cloud trails found in a few cases revealed that the α-particle trail ended abruptly and a forked trail began, one part of which was short and heavy, the other long and thin. The long track was that of a proton, the short one that of a recoil nucleus of considerable mass. This clearly showed that the α-particle was captured, since otherwise there should have been a third fork corresponding to a deflected α-particle. The reaction was,

$$^{4}_{2}\text{He} + ^{14}_{7}\text{N} \rightarrow ^{17}_{8}\text{O} + ^{1}_{1}\text{H}.$$

Blackett took over 20,000 photographs of more than 400,000 α-particle trails in nitrogen. He found eight forked trails.[2]

[1] E. Rutherford, *Phil. Mag.* [6] **37**, 586 (1919).
[2] P. M. S. Blackett, *Proc. Royal soc.*, **107**, 349 (1925).

By 1926 many elements had been bombarded with α-particles, and protons had been obtained from most of those of low atomic number. The heavier elements merely scattered the α-particles, however. The repulsion of the positive nuclei for the positive α-particle presented too great a barrier for penetration. From wave mechanics George Gamow calculated that the prospects for successful penetration of the potential barrier were better for protons than for α-particles of the same energy. Researchers then undertook to produce protons of such energies, which were not yet available.

The first atomic transmutation utilizing protons was achieved in 1932 by J. D. Cockroft and E. T. S. Walton, who were associated with Rutherford's laboratory at Cambridge. They bombarded lithium oxide with protons prepared from hydrogen in a discharge tube and accelerated by a voltage multiplier. Analysis of the evidence from cloud chamber photographs revealed the production of α-particles emerging in opposite directions from the lithium target. Further work in Rutherford's laboratory by M. L. E. Oliphant, E. S. Shire, and B. M. Crowther, who used lithium isotopes separated in a mass spectrograph, revealed that it was the ^{7}Li isotope which was being transmuted into helium.

$$_1^1H + {}_3^7Li \rightarrow {}_2^4He + {}_2^4He.$$

The voltage multiplier employed by Cockroft and Walton was an adaptation of a device used earlier for acclerating electrons. Consisting of condensers, vacuum tube valves, and a transformer, this system has the effect of multiplying a high potential to a much higher one [up to 0.8 million electron volts in Cockroft and Walton's multiplier, up to 3 Mev in later models]. Another form of accelerator was the electrostatic generator designed in 1931 by Robert Jemison Van de Graaff (1901–1967) at the Massachusetts Institute of Technology. Based on the principle of the classical static machine, it consisted of a large hollow sphere given a large positive charge by transferring electrons out of the sphere by means of a moving belt. A potential of 1.5 million volts was attainable with the original generator.

Linear accelerators were being developed at the same time, based on principles introduced by R. Wideröe of Germany in 1929 and extended by D. H. Sloan and E. O. Lawrence in the United States. A series of progressively longer, alternately charged cylinders was used. As a positive ion leaves a negative cylinder, the charges are alternated and the particle is accelerated in the gaps between cylinders. The linear accelerator, originally used to accelerate heavy ions, ultimately was refined so it could be employed to accelerate protons.

The principle of successive accelerations of positively charged particles was significantly advanced by the cyclotron developed by Ernest Orlando

Fig. 19.1. ERNEST O. LAWRENCE (right) AND M. STANLEY LIVINGSTONE
WITH THE 27-INCH CYCLOTRON IN 1934. The magnets weighed 74 tons. Energy:
5 Mev. deuterons.
(Courtesy of the Lawrence Radiation Laboratory, University of California,
Berkeley.)

Lawrence (1901–1958) and M. S. Livingston at the University of California
in 1931. In this accelerator the electrodes are designed as two flat, semi-
circular boxes known as *dees*, or "D's" because of their shape. These
electrodes, with their open diametric edges facing each other, lie horizontally within a chamber containing hydrogen or helium at low pressure.
The poles of an electromagnet above and below the chamber cause
charged ions, formed by electron bombardment of the gas at the center
of the chamber, to move in a circular path. The ions are accelerated in
the gap between the electrodes by rapid alternation of the charges on the
electrodes. As a result of successive acclerations the protons move in

an ever-enlarging spiral until they reach the periphery of the electrodes where the ion beam is deflected to bombard an appropriate target. Lawrence's original cyclotron produced protons with energies of 80,000 electron volts. A 1949 modification produced 350 Mev protons.

The development of high-energy proton accelerators initiated a decade of research in which high-energy particles were hurled at all sorts of elements in the hope of achieving transmutations. Many successful

Fig. 19.2.   First two chambers in which the cyclotron principle was tested in 1930, using a 4-inch magnet.
(Courtesy of the Lawrence Radiation Laboratory, University of California, Berkeley.)

transmutations were achieved, isotopes of many elements became available for study, and much was learned about the character of the nucleus and of nuclear changes. Before 1932 about eleven transmutations had been achieved, all by the use of α-particles from natural sources and all involving the ejection of a proton from the bombarded nucleus (an α, p reaction). Since 1932 there have been dozens of proton-induced transmutations.

Following Urey's discovery of deuterium in 1932, the heavy hydrogen nucleus, or deuteron, was quickly put to use as a projectile. Since the deutron has double the mass of the proton but the same charge it proved to be superior projectile for many purposes. With the discovery of the neutron an even more useful projectile became available.

An uncharged particle with the mass of the proton was postulated as early as 1920 by William Harkins in America, Orme Masson in Australia, and Rutherford in England, whose staff tried but failed to isolate such a particle. The actual production of neutrons, when it did occur, went unrecognized for two years.

In 1930 the German scientists W. Bothe and H. Becker observed a very penetrating radiation associated with the α-particle bombardment of beryllium, boron, and lithium. They believed it to be very high-energy gamma radiation. Repeating these experiments two years later, Irène Curie and her husband Frédéric Joliot, observed that high-energy protons were ejected when paraffin or some other hydrogen-rich material was placed in the path of this radiation. The Joliots intepreted this as a new mode of interaction of radiation with matter, by means of which high-energy gamma radiation was able to transfer great momentum to light atoms. Their hypothesis, however, was not consistent with the classical laws of physics. H. C. Webster reported similar experimental observations.

James Chadwick examined these experiments and concluded that the results could be explained if the radiation were considered to be made up of chargeless particles with the mass of a proton. Chadwick's explanation was consistent with the great penetrating power of the particle, which, since it carries no charge, leaves no ionization in its path and thus leaves no trail in a cloud chamber.

The neutron proved useful because it could now be considered a part of the nucleus. We remember that scientists had been postulating electrons in the nucleus in order to account for the difference between atomic weight and atomic number. The neutron appeared to be a general component of nuclei, for it was also produced by α-particle bombardment of nitrogen, fluorine, sodium, magnesium, and aluminum. The difference between isotopes was now merely a matter of the number of neutrons.

Because the neutron has no charge, it is not repelled by the nucleus, and it was quickly utilized as a projectile for atomic transmutations. Neutrons are produced by an (α, n) reaction with one of the light elements, and once formed, they cannot be accelerated. Nevertheless they are a useful projectile. A common procedure for obtaining neutrons is to mix beryllium with an α-source such as radium, polonium, or radon (and later, plutonium). With the cyclotron they may also be produced by bombarding beryllium or lithium with deuterons (d, n).

ARTIFICIAL RADIOACTIVITY

Investigating the effects of α-particles on such light elements as boron, magnesium, and aluminum, the Joliots found that the emitted particles include neutrons and positrons as well as protons. (The positron, the positive electron, had been discovered in 1933 by Carl Anderson of the California Institute of Technology in connection with cosmic ray studies.) The presence of neutrons and positrons was not considered unique at first, since the two particles taken together might be considered the equivalent of the proton. Early in 1934 the Joliots observed, however, the emission of protons and neutrons from aluminum ceased when bombardment with α-particles stopped, but that the positron emission continued. Evidently the bombarded material was radioactive. This behavior was also observed after α-bombardment of boron and magnesium.

The Joliots concluded that the transmutation was of the (α, n) type and that the isotope formed thereby was not a natural one and therefore might be expected to undergo radioactive decay. They hypothesized that bombardment of aluminum formed phosphorous of mass 30 thus:

$$\,^4_2\text{He} + \,^{27}_{13}\text{Al} \rightarrow \,^{30}_{15}\text{P} + \,^1_0\text{n},$$

and that the phosphorus isotope then decayed into silicon by loss of a positron:

$$\,^{30}_{15}\text{P} \rightarrow \,^{30}_{14}\text{Si} + \,^{0}_{+1}\text{e}.$$

This was confirmed by bombarding aluminum foil with α-particles and then dissolving the foil in hydrochloric acid. The hydrogen gas evolved was radioactive, presumably due to the presence of phosphine $^{30}\text{PH}_3$. Further experimentation established that aluminum had been transmuted into a radioactive isotope of phosphorus. Similarly, α-particles converted boron into radioactive nitrogen (^{13}N) and magnesium into radioactive silicon (^{27}Si). Each of these elements decayed according to a characteristic half life pattern, and they were quite clearly examples of artificial radioactivity.

Bothe suggested using mass number as a superscript to designate the isotope and symbol of the projectile and emitted particle in parentheses to designate the type of nuclear transformation involved. Thus ^{27}Al (α, n) ^{30}P describes the transmutation which led to the discovery of artificial radioactivity.

Great interest developed at once in artificial radioactivity, both for its own sake and because of the many uses that could be made of radioisotopes. The latter aspect will be considered in subsequent chapters. Immediate attention had to be given to radioisotope production and identification and to the determination of half lives and decay products. In addition there were a great many theoretical problems to be attacked, such as those concerned with likelihood of projectile capture, range of n/p ratios which are stable, and decay characteristics.

Fortunately artificial radioactivity was discovered just when effective accelerators and radiation detectors had been developed. Ionization chambers, proportional counters, and Geiger-Müller counters were all in a fairly acceptable state of development by 1935 and were continually improved as the years moved on.

The Geiger-Müller counter proved particularly popular. The principle of this counter was used by Rutherford and Hans Geiger in 1908 while they were counting α-particles in order to determine their charge. Geiger improved the tube in 1913, but it was not until 1928 that he and W. Müller, in Germany, developed the highly sensitive tube which became so universally useful. Further refinement of counting tubes has made them highly sensitive and also highly specialized. Improvements in amplifiers and recorders also have contributed significantly to the development of counters.

In 1903 Crookes and J. Elster and H. Geitel observed that the fluorescence caused by radiation on a zinc sulfide screen is made up of individual flashes (caused by α-particles). This knowledge provided the basis for the development in recent years of the scintillation counter. The cloud chamber still is useful for certain purposes as well as the photographic plate.

By 1930 around 200 stable isotopes and about 40 radioactive isotopes were known. Now more than 900 radioactive isotopes,[3] including representatives of every element in the periodic table, are known. Many of these were prepared in the thirties by the bombardment of elements with protons, deuterons, α-particles and neutrons. Still others were found in the forties among the fragments produced in nuclear fission. It is not always easy to identify isotopes produced in transmutation experiments because the quantity produced is too small for ordinary chemical analysis and even for identification with the mass spectrometer. The unknown isotope is usually identified by association with a suitable carrier, that is, a compound which contains the same element or a closely related element in a stable form. Upon precipitation of the carrier, the radioisotope, if it is the same or a closely related element, will be coprecipitated. If the carrier is composed of an element markedly different from the radioisotope, the radioactivity will not be present in the precipitate and a different carrier must be tried. Once the radioisotope is identified with a particular element it is still necessary to determine its mass number. This often has to be done by cross-bombardment, a scheme whereby the isotope is produced from more than one element, thus making it possible to eliminate various isotopic masses to arrive at the correct one.[4]

[3] According to a count made from the General Electric *Chart of the Nuclides* (Knolls Atomic Power Laboratory, 5th ed., April 1956), there were then known 269 stable isotopes of natural occurrence, 55 radioactive isotopes of natural occurrence (some of them existing only momentarily as part of a decay series), and 872 radioactive isotopes of artificial origin. Of the total, 156 isotopes (natural and artificial) have isomeric transition states.

[4] These processes are described, with examples, in S. Glasstone, *Sourcebook on Atomic Energy*, Van Nostrand, Princeton, 2nd ed., 1958, pp. 294 ff.

NUCLEAR FISSION

Almost immediately after its discovery scientists were investigating the neutron's potentialities as an atomic projectile, especially because it is chargeless and hence the coulombic repulsion between nuclei and neutron projectiles is inconsequential. Among the first to undertake neutron bombardment of elements was Enrico Fermi (1901–1954), physics professor at the University of Rome. His neutron source was a radon-beryllium mixture; the radon was furnished by Guilo Cesare Trabacchi, director of the physics laboratory of the bureau of public health, which

Fig. 19.3. Enrico Fermi.
(Courtesy of the University of Chicago.)

was housed in the same building. The bureau owned a gram of radium from which the radon was extracted. There were no detectable results from bombardment of the first elements in the periodic table, but fluorine formed a radioactive substance. Fermi and his associates—Franco Rasetti, Emilio Segrè, Edoardo Amaldi, and the chemist Oscar D'Agostino —were repeatedly successful in inducing artificial radioactivity in elements beyond fluorine. They examined every element they were able to procure and observed the formation of isotopes, frequently of the element of next higher atomic number.

Neutron bombardment of uranium, the last element in the periodic table, resulted in activity which suggested the production of more than one element. Beta-decay of isotopes having half lives of ten seconds, forty seconds, and thirteen minutes was detected. Since uranium-239 was a possible isotope if an (n, γ) reaction took place, the theory of β-decay suggested to Fermi and his colleagues that uranium-239 should change to element 93, which in turn should be transmutated to element 94 by beta emission and that elements of higher atomic number might be formed. Believing that elements 93 and 94 might be congeners of the metals in periodic table groups seven and eight, they used manganese, rhenium, and platinum salts as carriers and learned that a manganese precipitate contained some of the thirteen-minute activity and that a rhenium precipitate carried more than 50 per cent of this activity, suggesting the formation of an eka-rhenium (element 93). A report of the probable preparation of transuranium elements was prematurely announced in a speech by Senator Orso Mari Corbino, Fermi's chief in the physics department. To Fermi's consternation[5] the Fascist press gave the "contribution of Italian scientists" great publicity which soon attracted worldwide attention among scientists. During the next three years Fermi and his colleagues sought to identify their active products more definitely. Carrier experiments appeared to rule out lead, bismuth, and elements 86 through 91 and suggested the probable correctness of the transuranium hypothesis. Other investigators, however, were unable to confirm the presence of such elements, and the whole matter became highly confused.

The German Ida Noddack[6] (b. 1896) criticized the Fermi group for not considering other elements, since she found that a manganese carrier precipitate showed evidence of numerous heavy elements, perhaps brought down by occlusion. Her suggestion that the uranium atom may have been split during neutron bombardment was not well received among workers in the field, and her experimental results were explained in other ways.

Experimenting at the Kaiser Wilhelm Institute at Berlin, Otto Hahn (1879–1968) and Lise Meitner (1878–1968) also reported results at variance with Fermi's. In their early work Hahn and Meitner also associated the thirteen-minute activity with protactinium. In their later work they eliminated this element as a possibility and excluded all elements from mercury to uranium (except 85) as possible isotopes. A platinum carrier separated the thirteen-minute activity and also a ninety-minute activity. The latter was subsequently found to be heterogeneous. Hahn and Meitner suggested that the isotopes probably represented platinum-type metals beyond element 93.

[5] See the remarks of Mrs. Laura Fermi in the biography of her husband, *Atoms in the Family*, University of Chicago Press, Chicago, 1954, pp. 90–93.

[6] Ida Noddack, *Angew. Chem.*, **47**, 653 (1934).

In the meantime Fermi's group bombarded uranium with slow neutrons and obtained activities they considered members of a β-decay chain of transuranium elements. Hahn's group obtained variant results. They detected a four-minute thorium activity and postulated an (n, α) reaction (to give ^{235}Th from ^{238}U) followed by three successive β-decays to give $^{235}_{91}$Pa, $^{235}_{92}$U, and $^{235}_{93}$Eka-Rh. However G. von Droste failed to detect any α-particles during neutron bombardment of uranium, so this hypothesis had to be abandoned.

Irène Curie-Joliot and Paul Savitch observed a 3.5-hour activity in uranium subjected to long bombardment with either fast or slow neutrons and ultimately showed that the activity was precipitated with a lanthanum carrier. Since they considered lanthanum an improbable radioisotope, they proposed that the isotope was actinium. Lawrence Quill, at Ohio State University, suggested that the reaction might involve capture of an electron from the K shell as well as emission of an α-particle.

$$^{1}_{0}n + {}_{92}U + K \text{ capture} \rightarrow {}_{89}Ac + {}^{4}_{2}He$$

Curie-Joliot and Savitch then carried out a careful fractionation in order to verify the actinium hypothesis. The activity was precipitated with lanthanum and actinium carriers. Separation of the carriers showed that the activity stayed with the lanthanum rather than the actinium. Further fractionation of the lanthanium produced an inactive and an active fraction, which suggested that the activity was caused by a new transuranium element. It was later learned that the lanthanum salt used by Curie-Joliot and Savitch was contaminated with another rare earth. The inactive fraction, which they believed lanthanum, was really this impurity, and the active fraction was lanthanum.

Meanwhile Hahn, Meitner, and Fritz Strassmann (b. 1902) irradiated 20 g. of uranium with neutrons for eighty days. They detected activities associated with barium, lanthanum, and zirconium carriers. The barium activities they considered radium isotopes, and to explain their origin, they postulated an (n, 2α) mechanism leading to a β-decay chain giving actinium and thorium isotopes.

$$^{1}_{0}n + {}^{238}_{92}U \rightarrow {}^{231}_{88}Ra + 2{}^{4}_{2}He$$

$$^{231}_{88}Ra \xrightarrow[25 \text{ m.}]{-\beta} {}^{231}_{89}Ac \xrightarrow[40 \text{ m.}]{-\beta} {}^{231}_{90}Th$$

The German investigators also noted activities ranging from 110 minutes to several days in their barium carrier. These were believed isomeric forms of radium-231 which also decayed into actinium and thorium.

Between 1934 and 1939 researchers reported similar results for neutron bombardment of thorium. As with the uranium studies, the thorium investigators were blinded by the notion that neutron bombardment of an element should result in no more than a minor change in atomic number. Physicists maintained that particles ejected during a transmutation were unlikely to be other than protons, deuterons, α-particles, or neutrons. Hence everyone sought to identify radioisotopes with atomic numbers in the vicinity of 92. Since the chemistry of transuranium elements was completely open to speculation, and since no such element was available as a carrier, it was easy to be misled into thinking that activities carried by manganese and platinum salts were transuranium elements, while activities carried by barium, lanthanum, and zirconium salts were elements near uranium, but with lower atomic numbers. In 1938 the idea became current that bombardment of uranium with neutrons resulted in two types of product: (1) neutron capture followed by successive β-decays to give isotopes of elements 92 to 96; and (2) (n, 2α) reactions resulting in the formation of isomeric radium isotopes which then underwent successive β-decays.[7]

Hahn and his associates were the first to recognize that everyone had been misinterpreting these studies and overlooking the fission of the uranium atom. Since the beginning of the century Hahn had been in the forefront of radiochemical research. In 1938, following the Austrian Anschluss, one of Hahn's long-time colleagues, the physicist Lise Meitner, fled Germany. Hahn and Strassmann continued their researches and near the end of the year recognized that they were dealing with barium, lanthanum, and cerium (elements 56, 57, and 58). What they had believed to be radium could not be separated from the barium carrier and in fact could be separated from a radium carrier. A paper published early in January 1939[8] cautiously reported the unexpected presence of barium, lanthanum, and cerium among the radioisotopes formed during the bombardment of uranium with neutrons. The authors, chemists, realized that physicists would not believe the results and therefore refrained from positive claims that the uranium nucleus had been split. However the presence of isotopes with atomic numbers in the fifties spoke for itself.

Since barium, with a mass of about 140, accounted for a little more than half the total mass of the uranium atom, Hahn and Strassmann reasoned that another fragment with a mass around 100 should be present among the fission fragments. They started to search for isotopes having masses between 80 and 100, realizing that several neutrons might be liberated during the fission process. Strontium was the first isotope in this range to

[7] An excellent review of the numerous papers published before the discovery of nuclear fission was prepared by Lawrence Quill, *Chem. Rev.*, **23**, 87 (1938). Four years later the subject was brought up to date by Robert A. Staniforth, *Nuclear Fission and the Transuranium Elements*, M. S. thesis, Ohio State University, Columbus, 1942. This thesis, unfortunately, was never published, although duplicated manuscript copies were made available to certain persons connected with the Manhattan Project. I am obligated to Edwin M. Larsen for calling my attention to this thesis.

[8] O. Hahn and F. Strassmann, *Naturw.*, **27**, 11 (1939).

Fig. 19.4. OTTO HAHN.
(Courtesy of J. W. Williams.)

be identified. Subsequent studies revealed the presence of krypton and rubidium.

Meitner had been informed immediately by letter of the presence of barium in the residues produced during the neutron bombardment of uranium. She discussed the findings with her nephew Otto R. Frisch (b. 1904) during the Christmas holidays in Sweden. Frisch, also a prominent radiophysicist and refugee from Germany, was associated with Bohr's laboratory in Copenhagen. They accepted the splitting of the uranium atom reluctantly, as the only possible explanation for Hahn and Strassmann's findings.

Because of the difference in packing fraction between uranium and the middle elements, Meitner and Frisch suspected that the large energy production associated with fission causes strong ionization by the fragments. This was confirmed experimentally by Frisch on January 15 when he bombarded uranium with neutrons from a radium-beryllium source and observed intense electrical pulses on an oscilloscope screen. Frisch's was essentially the same experiment carried out by Fermi in 1934, but Fermi had used a thin aluminum foil to shield his instrument from short-range α-particles from uranium while permitting long-range radiations from transuranium elements to be recorded. The foil was sufficiently thick to prevent the recording of the great energies liberated during fission.

On January 16, 1939, Meitner and Frisch posted a letter to *Nature* in which they suggested that on the basis of chemical evidence the uranium atom must undergo fission under the impact of neutrons. They hypothesized further that because of a high ratio of neutrons to protons the fission fragments are radioactive and undergo a series of decays before a stable isotope is formed.

Niels Bohr brought news of the fission of uranium to the United States in January 1939. He passed the information on to his former student John A. Wheeler, now at Princeton, and from there word reached Fermi, a refugee from Fascism at Columbia University. On January 26 Bohr and Fermi made a verbal report at a conference on theoretical physics in Washington, D.C. Several physicists left the session before it closed and, working in laboratories at the Carnegie Institution of Washington and at Johns Hopkins, that day confirmed the fission of uranium. No one in Washington knew that Frisch and, the night before, John R. Dunning, Francis G. Slack, and Eugene T. Booth at Columbia had observed the high energies. The extreme recoil of atoms formed during neutron bombardment of uranium was reported from the laboratories of F. Joliot in France, Meitner and Frisch in Denmark, and E. M. McMillan in California. McMillan reported fission fragments with ranges up to 2.2 cm. in air.

Several elements of medium atomic number were identified as fission products in the laboratories of I. Curie-Joliot in Paris, N. Feather in England, P. Abelson at California, Hahn in Berlin, and F. A. Heyn, A. H. W. Aten, and C. J. Bakker in Holland. These included bromine, krypton, rubidium, strontium, molybdenum, antimony, tellurium, iodine, xenon, cesium, and barium. The energy associated with the fission of uranium was shown by Meitner and Frisch and by others to be of the order of 200 Mev, about ten times the most energetic transmutation previously studied—Li (d, α) He = 22.2 Mev.

The liberation of neutrons during fission was demonstrated by H. von Halban, F. Joliot, and L. Kowarski in France and by two American groups (Herbert L. Anderson, Fermi, and H. B. Hanstein and Leo Szilard and Walter H. Zinn). This possibility had been suggested by Fermi in his conversation with Bohr at the Washington conference; Fermi pointed out that the total number of neutrons present in uranium was probably too large for them all to remain, even momentarily, in the fission fragments. Frantic investigation of the fission process was undertaken during the year. When L. A. Turner of Princeton wrote his review article on the subject for the *Reviews of Modern Physics*,[9] nearly one hundred papers were covered.

In accordance with Einstein's equation for the equivalence of mass and energy ($E = mc^2$), the possibilities for large energy release accompanying fission were readily apparent. It was reasoned that because the binding

[9] L. A. Turner, *Rev. Mod. Phys.*, **12**, 1 (1940).

energy of the nuclei formed during fission is high compared with that of the uranium nucleus, there should be significant conversion of mass into energy during fission.

NEPTUNIUM AND PLUTONIUM

After Fermi's claim for the discovery of transuranium elements was discredited researchers became cautious when reporting on work in this area. During their investigation of the isotopes associated with neutron bombardment of uranium, Hahn, Meitner, and Strassmann observed one with a 23-minute half life which they believed might be ^{239}U. In 1940 McMillan found that when thin uranium foil is bombarded with neutrons, the fission fragments are ejected by their recoil energies while the 23-minute activity remains in the foil. Associated with it is a 2.3-day activity. The 23-minute activity turned out to be ^{239}U. The longer activity was suspected to be an isotope of element 93.

McMillan and Abelson now proceeded to identify the 2.3-day isotope. Segrè had already shown that it does not resemble rhenium, its expected homologue, but is more like the rare earths in its properties. McMillan and Abelson showed that it is carried by neither uranium nor the rare earths, but has a character uniquely its own, with oxidation states of IV and VI. The name *neptunium* was proposed since it is the element beyond uranium, just as Neptune is the planet beyond Uranus. The existence of the element was confirmed in Germany by K. Starke and also by Strassmann and Hahn.

Shortly thereafter McMillan left Berkeley to undertake war research; the study of neptunium was continued by Glenn T. Seaborg. Seaborg and his associates pursued the subject vigorously; but their papers remained unpublished until 1946. By mid-1944, 45 micrograms of neptunium had been obtained from cyclotron bombardment of uranium and from nuclear pile operation; hence chemical studies were possible. Actually the neptunium-237 isotope was studied, since it has a long half life of 2.2×10^6 years. An (n, 2n) reaction of ^{238}U takes place to some extent in the nuclear pile, resulting in the formation of ^{237}U which decays by β-emission with a half life of 6.75 days to form ^{237}Np.

With the discovery of neptunium-239 McMillan realized that β-decay of this element should lead to the formation of element 94. However the scale of the original experiments prevented its detection, and Seaborg's group had to devise another approach. They bombarded uranium with deuterons to obtain:

$$^{238}_{92}U + ^{2}_{1}H \rightarrow ^{238}_{93}Np + 2n,$$

$$^{238}_{93}Np \xrightarrow[2.1\ d]{-\beta} \ ^{238}_{94}Pu \quad \text{(α-emitter, 90 year half life).}$$

Tracer studies were feasible with this isotope, and chemical studies were pursued by Seaborg, Wahl, and J. C. Kennedy.

Knowledge of the chemistry of neptunium made it possible to detect and study element 94. Large amounts of neptunium-239 were prepared by neutron bombardment of uranium-238, the uranium-239 formed undergoing β-decay with a 23-minute half life. The neptunium 239 in turn undergoes β-decay (2.3-day half life) to form element 94, an α-emitter with a half life of 24,300 years. The name *plutonium*, after the planet Pluto, seemed an obvious choice. Kennedy, Seaborg, Segrè, and Wahl showed it to be fissionable with slow neutrons.

EXPLOITATION OF NUCLEAR ENERGY

Once it was known that neutrons are emitted during fission, the possibilities of a chain reaction were grasped—as were the military possibilities of nuclear energy. The foreign-born scientists who now worked in the United States were particularly concerned and almost immediately sought to restrict publication in the field and to interest the government in the problem. Albert Einstein was persuaded by Leo Szilard and Eugene Wigner to sign a letter, dated August 2, 1939, to President Roosevelt, calling attention to the nature of the problem and advising further investigation. The United States, together with Britain and Canada, ultimately became deeply involved in research and development which resulted in: (1) the successful demonstration by Fermi and his associates at the University of Chicago that a chain reaction of fissioning uranium can be sustained in a nuclear reactor (pile); (2) the successful separation of uranium-235 from the natural isotopes on a commercial scale by use of gaseous diffusion of uranium hexafluoride; (3) the successful production on a commercial scale of plutonium-239; (4) the design and production of fission bombs, which were used on Japanese cities by August 1945.[10]

The development of nuclear energy for military use was carried out under extreme secrecy, despite the ultimate involvement of 125,000 scientists, administrators, and workmen. It was feared that the Germans would develop and exploit nuclear energy before the Manhattan project

---

[10] The details of the wartime effort cannot be recounted here, but see Henry D. Smyth, *Atomic Energy for Military Purposes*, Princeton University Press, Princeton, 1945; Arthur Holly Compton, *Atomic Quest, A Personal Narrative*, Oxford University Press, New York, 1956; Ralph E. Lapp, *Atoms and People*, Harper, New York, 1956; Laura Fermi, *Atoms in the Family*, University of Chicago Press, Chicago, 1954; Wm. L. Laurence, *Men and Atoms*, Simon & Schuster, New York, 1959; Michael Amrine, *The Great Decision, The Secret History of the Atomic Bomb*, Putnam's, New York, 1959; Leonard Bertin, *Atomic Harvest: A British View of Atomic Energy*, Freeman, San Francisco, 1957; John M. Fowler, ed., *Fallout: A Study of Superbombs, Strontium 90, and Survival*, Basic Books, New York, 1960; and Erwin N. Hiebert, *The Impact of Atomic Energy*, Faith and Life Press, Newton, Kansas, 1961.

Fig. 19.5. NUCLEAR BOMB OF THE TYPE EXPLODED OVER HIROSHIMA. The bomb is 28 inches in diameter and 120 inches long, weights 4½ tons and has explosive power equivalent to about 20,000 tons of high explosive.
(Courtesy of the Los Alamos Scientific Laboratory.)

had succeeded. As the war came to an end in Europe, it was learned that the Germans, directed by Werner Heisenberg, had made little progress.[11] Ironically, such security leaks as occurred, leaked not to the German enemy but to the Soviet ally.

Since the war the moral and security aspects of nuclear weapons have been discussed extensively, at times hysterically. Russia was able to explode a nuclear bomb in August 1949. The hazard from fallout of radioactive materials became more and more a matter for concern, particularly in view of the possible genetic consequences. Nevertheless research and testing continued.

A somewhat subsidiary consequence of the developing knowledge of nuclear energy taking place in the 1930's pointed to a solution of the long mystery of the source of energy output by the sun and stars. Early hypotheses could not account for the age of the solar system, but with increased understanding of the enormous energies involved in nuclear processes the source of solar energy became evident. In 1939 Hans Bethe of Cornell proposed a mechanism involving a carbon cycle in which four protons were fused into a helium atom with a mass-to-energy conversion yielding 27 Mev per helium nucleus formed. A similar mechanism was advanced by C. F. von Weizsäcker in Germany. Later studies revealed that a proton chain might be even more important in the sun and cooler stars.

[11] Samuel Goudsmit, *Alsos*, H. Schuman, New York, 1947, pp. 128 ff.

Since temperatures in excess of a million degrees Kelvin are reached in nuclear explosions, scientists recognized that such temperatures might be utilized to trigger man-made hydrogen-to-helium conversions on earth. The United States embarked on a development project in 1949, utilizing a principle introduced by Edward Teller. Through concentration of deuterium and the production of tritium, and using deuterium and tritium in conjunction with lithium isotopes, a successful bomb test was carried out in November 1952. A Soviet bomb was tested three years later.

Peaceful uses of nuclear energy have not been ignored. Heat from a reactor was used to heat buildings in 1951 at the Harwell Atomic Energy Establishment in England. The British brought a nuclear power plant into commercial operation at Calder Hall in 1957.[12] A submarine powered by a nuclear reactor was launched by the United States in 1954. The development and utilization of nuclear power reactors has proceeded rapidly since 1954 in a number of countries. Scientists also are concerned with the practical utilization of fusion energy. Here the containment of enormous temperatures within magnetic fields to avoid vaporization of apparatus is a major problem.

PROBLEMS OF NUCLEAR STRUCTURE

Naturally the theoretical exploration of the atom is of no less importance than the development of practical applications of nuclear energy. Nuclear reactors serve as excellent neutron sources for the preparation of isotopes which can be studied themselves or used as tools for a great variety of chemical, medical, agricultural, and engineering research projects. Some of these uses will be discussed in subsequent chapters.

Of greatest theoretical significance, however, are the results attained with the powerful accelerators now available. Unlike earlier models these instruments, which are of huge dimensions, are designed to adjust for the mass increase of high velocity particles. The new instruments can hurl ions with energies above the billion electron volt (Bev) range. They can accelerate not only protons and deuterons but also ions as large as carbon, nitrogen, oxygen, and fluorine.

The twentieth century may be considered the century of the subatomic particle; except for the electron all such particles were first encountered after 1900, with the majority of the discoveries coming after 1930. In some cases the particles were postulated prior to their actual experimental observation.

Early investigators of radioactivity encountered problems with the

---

[12] According to P. Semenovsky in *Conquering the Atom* (Moscow, 1956, pp. 47–59), a Russian power plant of 5000 kw capacity began operating in June 1954.

spontaneous discharge of electroscopes, first reported in 1900 by C. T. R. Wilson in England and by J. Elster and H. Geitel in Germany. Until 1909 this discharge was attributed to terrestrial radiation, but within a few years it was found the rate of discharge of well-insulated electroscopes increases with altitude. After World War I Robert Millikan began systematic altitude studies with instruments in free balloons and confirmed that rate of discharge increases as altitude increases. In 1925 Millikan termed the responsible radiation *cosmic rays*. The intense energy of these rays was shown by their effect on electroscopes at the bottom of deep lakes and deep mines. Millikan considered the rays to be a very high-frequency electromagnetic radiation, but Arthur H. Compton showed that their distribution varied with the earth's magnetic field, and that hence they must be high-energy charged particles.

The nature of cosmic rays has been a controversial problem because it has been difficult to separate cause from effect. There is now some agreement that they are high-energy protons moving into the earth's atmosphere from outer space. However they are studied mainly from the effects they produce when they collide with components of the atmosphere. These secondary particles have been largely identified in the past twenty years as high-energy photons, protons, heavier nuclei, neutrons, electrons, positrons, and particularly mesons.

In 1935 Hideki Yukawa postulated the meson as a subatomic particle in connection with work on the theory of intranuclear forces. He estimated it to have a mass about 140 times that of the electron and a mean life of about 10^{-6} second. Such particles were discovered in cosmic radiation two years later by Carl D. Anderson and Seth H. Neddermeyer of the California Institute of Technology and by J. C. Street and E. C. Stevenson of Harvard. Confirmation quickly came from various sources for positive and negative short-lived particles having masses about 200 times that of the electron. The name *mesotron* (Gr., *meso*, intermediate) was proposed by Anderson and Neddermeyer, but the shortened form *meson*, proposed by the Indian physicist H. J. Bhabha, was adopted by the International Union of Physics in 1947.

In 1947 C. F. Powell and G. P. S. Occialini and coworkers at Bristol established the existence of mesons of different masses. The heavier meson, termed π (for primary), was found to have a mass of about 273. It may be either positive or negative and decays within about 2×10^{-8} seconds into lighter (207) μ mesons. Since the μ mesons have the same charge as the parent π mesons, the emission of a neutral particle of mass 66 the *neutretto*, has been postulated. Negative π mesons are commonly captured by atomic nuclei before decay can take place. When the negative μ meson is captured, a multiple track or star appears on a photographic plate, indicating that capture probably is accompanied by explosion of the nucleus.

Mu-mesons decay with a mean life of 2.15×10^{-6} seconds. The positive μ mesons decay with the emission of a positron and some sort of neutral particle or photon. Negative μ mesons are usually captured by atomic nuclei, but the nature of the capture is uncertain. Such mesons captured by heavy nuclei are termed rho (ρ) mesons.

Claims for still other charged and uncharged particles have been advanced. Neutral π-mesons are taken seriously; it is believed that they decay into γ-ray photons leading to the γ-ray showers observed on cosmic ray photographs. There is also evidence for neutral and charged particles with a mass about 1000 times that of the electron. These, named tau (τ) mesons, were first encountered by Powell's groups on a plate exposed under a twelve-inch block of lead atop the Jungfrau.

Mesons have been produced during bombardment experiments with high-energy accelerators. Eugene Gardner at Berkeley, collaborating with the Brazilian physicist C. M. G. Lattes, bombarded various targets with 380 Mev alpha-particles and obtained photographic tracks of π mesons. By moving the photographic plates with reference to the magnets it was also possible to detect π^+, μ^+, and μ^- mesons.

The picture rapidly became more chaotic as larger accelerators with higher energy potentials were developed. More and more "particles" were encountered from these high-energy accelerator studies and more refined cosmic ray studies. Some of the discoveries, like that of the *neutrino*, verified earlier hypotheses. Others just compounded confusion.

The neutrino theory, proposed in 1934 by Pauli and elaborated by Fermi, sought to explain a hitherto unaccountable energy loss during β-decay. The neutrino was postulated as a neutral particle with essentially no mass but with angular momentum, or spin. Because of its neutrality it would pass through matter for great distances without leaving tangible evidence. Physicists despaired of detecting the particle, although it was necessary in their equations in order to conserve energy and spin. When large reactors became available researchers realized that large neutrino production should accompany the β-decay of fission products and that the neutrino might be detected by reversing β-decay with the aid of a huge scintillation counter rich in a proton-containing liquid. A tentative identification was made in 1953 by the Los Alamos physicists F. Reines and C. L. Cowan who were using one of the Hanford reactors. Three years later Reines and Cowan and their colleagues at the Savannah River plant of the Atomic Energy Commission[13] confirmed the identification.

In a sense the positron is a form of antimatter. The counterpart of the electron, it served no necessary role in the stable atom. Its property of

[13] C. L. Cowan, Jr., F. Reines, F. B. Harrison, H. W. Kruse, A. D. McGuire, *Science*, **124**, 103 (1956). A tentative identification was reported by Reines and Cowan, *Phys. Rev.*, **90**, 492 (1953), **92**, 830 (1953). Also see P. Morrison, *Sci. Amer.*, **194**, no. 1, 58 (1956).

annihilation of itself and an electron with the emission of gamma radiation showed that it lacks a stable role in earthly matter. However its discovery lent weight to Dirac's theoretical calculations regarding the properties of the electron. Dirac's equation, based upon general principles of relativity and quantum mechanics, was in agreement with the electron's behavior in the hydrogen atom. He learned, however, that the equation demanded the existence of positive and negative electrons. Anderson's discovery of the positron was a triumph for Dirac's equation.

Slight modification of the equation showed that it might be applied to the proton as well as to the electron. In this case too it predicted an antiparticle—a negative proton. It was not possible to search for this particle until the California bevatron was in operation. This instrument, which cost \$9,550,000 provided by the AEC, can accelerate protons to energies of 6.2 billion electron volts (Bev). A team made up of Owen Chamberlain, Segrè, Clyde Wiegand, and Thomas Ypsilantis, hurled 6.2 Bev protons at a copper target where energy was converted into a proton-antiproton pair during collision with neutrons.

$$_1^1H + {_0^1}n + \text{energy} \rightarrow {_1^1}H + {_0^1}n + {_1^1}H + {_{-1}^1}H$$

The antiprotons were identified by measuring the velocity and momentum of the particles coming from the target. Another team, led by E. Amaldi of the University of Rome, located evidence of antiprotons on photographic plates sent from California. The results were announced late in 1955. The antiproton, a short-lived particle, is annihilated in a shower of mesons and other radiation upon collision with a proton or neutron. It has been postulated that there should also be an antineutron, but discovery of this would pose some new and difficult problems.

More than thirty subatomic particles have been discovered or postulated. The chaos they have created for nuclear scientists is reminiscent of the chaos that prevailed a century ago with respect to the formulation of chemical compounds. Only after Kekulé, Cannizzaro, and Lothar Meyer began their work did order begin to appear. Today nuclear studies are similarly confused, with newly discovered particles creating continually greater problems. The main saving grace appears to lie in the possibility that most of these particles are not regular components of the nucleus, but are formed fleetingly when the nucleus is subjected to great energy changes.

Nuclear theorists, therefore, have been able to proceed on the assumption that the nucleus is made up of neutrons and protons. Evidence from a number of sources—such as isotope stability, electron scattering, and magnetic data—give clues that the nucleus is not a homogeneous body but has a distinct structure which is subject to quantum treatment. Studies of the nucleus reveal that there is a variation of the density outward from the center. Evidence for shell structure is strong, although the distribution of nuclear particles (neutrons and protons, collectively called *nucleons*), is different from that in electron shells.

Examination of a table of abundance of the elements in the universe shows great differences, but these differences are readily associated with atomic numbers and masses. The elements of even atomic number are, except for hydrogen, more abundant than those of odd atomic number, e.g., $_2$He, $_6$C, $_8$O, $_{10}$Ne, $_{12}$Mg, $_{16}$S, $_{20}$Ca. Six isotopes—$^{16}_8$O, $^{24}_{12}$Mg, $^{28}_{14}$Si, $^{40}_{20}$Ca, $^{48}_{22}$Ti, and $^{56}_{26}$Fe—make up about 80 per cent of the earth's crust.

Examination of a table of stable isotopes reveals the following:

Even at. no., even no. of neutrons	162
Even at. no., odd no. of neutrons	56
Odd at. no., even no. of neutrons	52
Odd at. no., odd no. of neutrons	4 ($^2_1$H, $^6_3$Li, $^{10}_5$B, $^{14}_7$N).

When the atomic number is odd, there are never more than two stable isotopes, and these are always two mass numbers apart, e.g., $^{35}_{17}$Cl and $^{37}_{17}$Cl.

Whereas the numbers 2, 8, 18, 32 have particular significance with regard to the distribution of electrons, the nuclear magic numbers are 20, 50, 82, and 126. Elements with these numbers of total protons or neutrons have peculiar stability and are inordinately abundant as compared with neighboring elements.

Thus the elements where the atomic weight equals 4, 16, 20, 40, etc., are especially stable (^{4}He, ^{16}O, ^{20}Ne, ^{40}Ca). Fifty-neutron elements like $^{90}_{40}$Zr and $^{88}_{38}$Sr are particularly stable. Tin, with atomic number 50, occurs to the extent of 40 ppm. in the earth's crust and has 10 stable isotopes. Lead-208 has two magic numbers with 82 protons and 126 neutrons.

Nucleons are treated as having angular momentum. Therefore the pairing of spins will create greater stability. The shells and subshells are treated in quantum fashion, with the number of nucleons per subshell corresponding to 2. A pair of protons or of neutrons may be considered to make up a subshell having great stability.

CONCLUSION

The discovery of radioactivity had many physical and chemical consequences, most of which came into fruition after World War I. When Rutherford bombarded nitrogen with α-particles in 1919 he not only achieved a successful transmutation but also discovered the proton. Subsequent refinement of bombardment methods led to the discovery of the neutron and of artificial radioactivity. The neutron and proton made possible a more meaningful interpretation of the nucleus. The neutron also provided a projectile for studying transmutation of the heavy elements, leading to the release of nuclear energy and the production of transuranium elements. Artificial radioactivity made available a variety of isotopes suitable for inquiry into a multitude of scientific problems.

As high-energy accelerators became available a host of new subnuclear particles were discovered, leading to complication of nuclear theory.

CHAPTER 20

PHYSICAL CHEMISTRY II.

MATURITY

At the beginning of the twentieth century physical chemists were very much concerned with the properties of dilute solutions, molecular weights and the relation of physical properties to chemical constitution. In fact the obsession for precise measurement was such that unsympathetic workers in other fields jibed that physical chemistry was nothing more than the very exact measurement of the properties of very slightly impure substances. Organic chemists were fond of pointing out that physical chemists proved mathematically what empirical organic chemists had known to be true for a quarter century.

Physical chemists were admittedly feeling their way and during the formative period they needed highly accurate data to determine whether their hypotheses were valid. Hence tedious measurements were essential if the field were to develop into a sound and useful part of chemical science. Unfortunately they sometimes extended their claims beyond the measure of their data, thus detracting from the reputation of the discipline.

By 1925 the discipline had developed enough to be useful in many areas of industrial and pure chemistry. Industrial planning particularly was helped by applications of the Phase Rule, equilibrium constants, reaction rates, and catalysts. And knowledge of thermodynamics was especially important in helping the fast-growing petroleum industry solve refining problems. By midcentury, physical chemistry was beginning to supplant the always popular organic field as the most attractive area for graduate study.

While physical chemistry has contributed much to other fields of science and technology, it also has borrowed heavily, particularly from physics.

The techniques of radiochemistry, mass spectrography, x ray crystallography, and electron diffraction, for instance, have proven invaluable to chemists. The development and refinement of the electron tube has contributed instruments which play a major role in all types of physico-chemical research.

CHEMICAL THERMODYNAMICS

Thermodynamics at first seemed almost more important to steam engine operators than to chemists. Actually knowledge of thermodynamics was useful, particularly to researchers concerned with gaseous equilibria and equilibria in solutions. Van't Hoff's work and the equation developed by Willard Gibbs and Helmholtz effectively demonstrated the relation between heat of reaction, temperature, and equilibrium constant which also was expounded by Fritz Haber in his *Thermodynamics of Technical Gas Reactions* (1908).

Although van't Hoff's generalization of the Clausius-Clapeyron equation permits the calculation of the effect of temperature change on equilibrium from thermal data, it fails to give the value of the equilibrium constant unless this is already known experimentally for one set of conditions. The integration of equation

$$\frac{d \ln K}{dT} = \frac{Q}{RT^2}$$

gives

$$\ln K = -\frac{Q}{RT^2} + c,$$

where the integration constant c is of unknown value. This obstacle was finally overcome in 1906 when Walther Nernst (1864–1941), developing an idea suggested by T. W. Richards, arrived at the third law of thermodynamics. This concept, which holds that the entropy of a pure crystalline material at absolute zero is zero, has yet to be proved conclusively, but it has been checked in practice and has been found capable of widespread application. With the Nernst heat theorem it is possible to calculate the chemical equilibrium for a given system from a few physical constants.

Physical chemists have applied the third law to a wide variety of chemical problems. Nernst, Abegg, Haber, Sackur, and G. N. Lewis were particularly active in this work, which, as any application of thermodynamics, depends on highly accurate thermochemical measurements. Since the early 1900's a large number of chemists have devoted themselves to this problem. Frequently the apparent failure of the thermodynamic approach has been shown to result, not from the failure of thermodynamics, but from the use of inaccurate data.

Fig. 20.1. HERMANN WALTHER NERNST.
(Courtesy of the J. H. Walton Collection.)

Nernst demonstrated the value of the third law by calculating the temperature at which transition would take place from the specific heats of rhombic and monoclinic sulfur and from the heat of transformation at constant volume. J. T. Barker, working in Nernst's laboratory, used the heat theorem to calculate the temperature of fusion of single substances, and Julius Thomsen used it in his investigation of salt hydration. The electromotive force of voltaic cells was studied in several laboratories, as was the application of the law to various homogeneous and heterogeneous equilibria.

Thermodynamic treatment has been exceedingly fruitful in the study of industrial systems, especially when gaseous reactions are involved. Haber used it very effectively in developing his process for fixing nitrogen by combining hydrogen and nitrogen. Also experiments confirmed the hypothesis based on calculations that significant quantities of nitric oxide are formed from the elements only at high temperatures.

From Thomsen's experimental determination of the heat of formation of hydrocarbons from the elements, it was calculated that if hydrogen is passed over carbon at 500° C. and atmospheric pressure, the product gas at equilibrium should be 70 per cent methane. Experimental results were in substantial agreement, showing that the prospect of obtaining gasoline-size molecules was poor. The higher hydrocarbons were found to be thermodynamically unstable at ordinary temperatures, tending to change into carbon and methane. However the velocity of this reaction at ordinary temperatures is so low that an observable change does not take place. If such molecules are heated to a high temperature, the "cracking" of the unstable molecules is accelerated, and coke and smaller hydrocarbon molecules are formed. If the heating is done at atmospheric pressure, the large hydrocarbons decompose almost entirely to coke and methane. High pressures, however, tend to prevent the formation of coke and methane and favor the cracking into molecules of higher molecular weight, such as those present in gasoline.

These experimental observations confirmed the Nernst equation which has proved invaluable in suggesting favorable conditions for carrying out desired conversions of hydrocarbons. As a result, an industry that initially separated petroleum fractions by simple fractional distillation has reached a point where large-scale tailoring of hydrocarbon molecules is commonplace.

Interest in thermodynamics and its application was greatly stimulated by G. N. Lewis and Merle Randall's *Thermodynamics and the Free Energy of Chemical Substances* (1924). At the University of California Lewis built up a strong department of chemistry which trained a large number of physical chemists. Arthur Amos Noyes (1866–1936) directed work in physical chemistry at M.I.T. until 1920 when he moved to the California Institute of Technology. Edward W. Washburn (1881–1937), an M.I.T. graduate who proposed the concept of thermodynamic environment, taught at the University of Illinois until his appointment as director of the United States Bureau of Standards in 1922. As editor-in-chief of the *International Critical Tables* he was instrumental in making readily available in French, German, and English the mass of physical data essential to those working in physics and chemistry.

Considerable research on the energy relations in chemical reactions has been stimulated by theoretical and practical needs. Richards and his students contributed much to our knowledge of the temperature coefficients of galvanic cells. The work of Lewis, George Ernest Gibson (1884–1959), and Wendell M. Latimer (1893–1955) was particularly important in extending the use of the third law. The California group devoted much of their time to the energy changes of substances at temperatures approaching absolute zero. Their work has been extended in subsequent years by William F. Giauque (b. 1895).

KINETICS

Interest in chemical kinetics, considerable during the 1890's, steadily declined from 1900 until shortly after World War I, perhaps because the field lacked the stimulus of a theoretical development that suggested experiments and sparked debate. In 1918 the radiation hypothesis provided such a stimulus. According to this hypothesis, decomposing molecules in monomolecular reactions receive their activation energies from radiation from the walls of the container, rather than from intermolecular collisions. It grew out of the work of Jean Perrin in France, M. Trautz in Germany, and William C. McC. Lewis at Liverpool. Perrin, who was groping toward the hypothesis as early as 1913, was impressed by the observation that frequency of collision in a monmolecular reaction does not determine reaction rates. Lewis and Trautz were active in developing the concept after 1918.

At first the hypothesis was received enthusiastically and stimulated considerable experimentation—which soon revealed flaws in the hypothesis. F. Daniels at Wisconsin and Hugh Taylor at Princeton reported inactivity of infrared radiation in the decomposition of nitrogen pentoxide. Experiments on the decomposition of carbon dioxide and hydrogen chloride and alcohol vapor oxidation pointed to similar conclusions, and the radiation hypothesis was rejected.

F. A. Lindemann urged that the collision hypothesis be reconsidered, noting that since a time interval occurred between collision and dissociation, the activation due to collision might be overlooked. Beginning in the midtwenties, Worth H. Rodebush, Cyril N. Hinshelwood, Eric K. Rideal, O. K. Rice, Louis S. Kassel, and others investigated the collision hypothesis exhaustively. It soon was established that before a monomolecular reaction can occur the activation energy must be concentrated in a particular bond.

The chain reaction mechanism, suggested by Bodenstein in 1913, continued to attract attention, especially because of the work of S. C. Lind, Michael Polanyi, and N. N. Semenov. Applying this concept, especially to many reactions in organic chemistry, required the postulation of the momentary existence of free radicals such as CH_3 and C_6H_5.

Although this approach was out of accord with traditional valence concepts, there was experimental evidence for such free radicals. Triphenyl methyl radicals had been reported by Moses Gomberg[1] of Michigan at the beginning of the century. Other large organic radicals of this sort had been observed, but the idea of such radicals as common participants in chemical reactions was revolutionary when Hugh Taylor made the suggestion in 1925, in connection with his studies of the hydrogenation of ethylene under the influence of the light of a mercury vapor lamp.[2] Paneth in Germany and later Francis Owen Rice at Johns Hopkins showed by the removal of metallic mirrors that certain organic compounds are apparently broken up in hot tubes to form free radicals. Their conclusion was based upon the removal of metallic mirrors from the inside of the tube in the region where free radicals might form, the free radical apparently reacting

[1] M. Gomberg, *J. Am. Chem. Soc.*, **22**, 757 (1900).
[2] H. S. Taylor, *Trans. Faraday Soc.*, **21**, 560 (1925).

with the metal of the mirror and carrying it onward. Other chemists reported results that could most easily be explained by assuming the formation of free radicals. Of course spectroscopists had long been interpreting certain spectral bands as due to free radicals such as CN, TiO, OH, SiO, C_2, and CH_2. Calculations from quantum mechanics indicated that such radicals should be stable, although low activation energies would make them highly reactive. In 1934 Rice and Herzfeld used the free radical hypothesis to explain the decomposition of organic vapors and the hypothesis has had considerable favor since that time. Morris Kharasch at Chicago became a leader in the application of free radical mechanisms to organic reactions.

A significant body of experimental and theoretical work has been accumulated, although it can hardly be said that all major kinetic problems are close to being solved. Even reactions that at first appear simple generally turn out on investigation to be highly complex. In addition to those mentioned before, other principal and still active contributors to our understanding of kinetics include F. Daniels, L. Farkas, H. Eyring, G. Kistiakowsky, E. A. Moelwyn-Hughes, J. Hirschfelder, G. Scatchard, and W. A. Noyes, Jr. Harvard's Kistiakowsky has become a leader in the study of the kinetics of explosives and propellants, a field which acquired great prominence during and after World War II, and Hirschfelder has made important contributions to the theory of flames.

CHEMICAL BONDING

In the previous chapter the atom was treated principally from the standpoint of physical phenomena. If an atomic concept is to have universal acceptance, it must be capable of explaining chemical phenomena as well. Even before Rutherford proposed a nuclear atom, the chemist Ramsay was thinking of chemical combination in terms of electrons. There was no further clarification of the matter until after 1913, the year Bohr announced his atom.

Within a few years,[3] G. N. Lewis,[4] Walther Kossel,[5] and Irving Langmuir[6] arrived at a rather similar scheme for dealing with chemical bonding. Kossel (1888–1956) was professor of physics at the technical high

[3] G. N. Lewis seems to have speculated about electrons and their significance in atomic structure as early as 1902 according to a memorandum made at that time. See Fig. 20.2, taken with the permission of the Reinhold Publishing Company from G. N. Lewis, *Valence and the Structure of Atoms and Molecules*, Chemical Catalog Co., New York, 1923.

[4] G. N. Lewis, *J. Am. Chem. Soc.*, **38**, 762 (1916); also see his book on the subject.

[5] W. Kossel, *Ann. Physik*, **49**, 229 (1916).

[6] I. Langmuir, *J. Am. Chem. Soc.*, **41**, 868 (1919); **42**, 274 (1920).

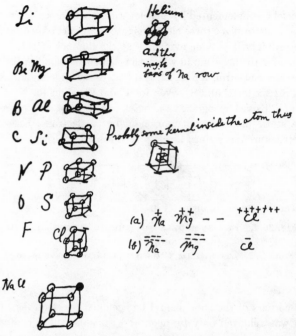

Helium

And this may be base of Na row

Probably some kernel inside the atom thus

(a) $\overset{+}{Na}$ $\overset{++}{Mg}$ -- $\overset{+++++++}{Cl}$

(b) $\overset{=--}{Na}$ $\overset{=--}{Mg}$ $\overline{Cl}$

Fig. 20.2. G. N. LEWIS MEMORANDUM OF 1902 SHOWING SPECULATION
REGARDING THE ROLE OF ELECTRONS IN ATOMIC STRUCTURE.
(From G. N. Lewis, *Valence and the Structure of Atoms and Molecules* [1923], with
the permission of Reinhold Press.)

school in Danzig at the time. While Lewis and Kossel's studies were
reported before Langmuir's, the consequences of the concept were refined
and extended by the latter. Langmuir was with the General Electric
Company.

The theory of chemical bonding could be developed only after the
number of electrons outside the nucleus was known. Bohr's work and that

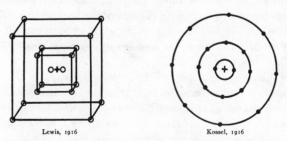

Lewis, 1916   Kossel, 1916

Fig. 20.3. TWO 1916 MODELS OF THE ARGON ATOM.
(From G. N Lewis, *Valence and the Structure of Atoms and Molecules* [1923], with
permission of the Reinhold Press.)

of Sommerfeld, supplemented by Moseley's atomic number determination, made it possible to place these electrons into some sort of orbital arrangement. Accordingly it became evident that the number of outer electrons corresponded with the group in which the element is found in the periodic table: one outer electron for H, Li, Na, and K; two for Mg and Ca; four for C and Si; six for O and S; seven for F and Cl; eight for Ne and Ar.

Lewis and Kossel recognized an octet relationship and proceeded to develop a static model based upon eight outer electrons. Kossel placed these electrons around a ring; Lewis placed them at the corners of a cube.

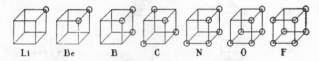

Li   Be   B   C   N   O   F

Fig. 20.4. LEWIS MODELS OF THE ELEMENTS OF THE SECOND PERIOD.
(From G. N. Lewis, *J. Am. Chem. Soc.* [1916], with permission of the editor.)

While all corners of the cube should be occupied by electrons, this ideal condition was achieved only by neon and the subsequent rare gases. In the halogen atoms one corner of the cube was unoccupied; in oxygen and sulfur two corners were empty; and so on to the left across the periodic table to the first group where lithium, sodium, and potassium had but a single electron in the outer cube. Chemical combination was a consequence of the unoccupied corners of the outer cubes, since, by transferring or sharing electrons, it was possible to fill up outer cubes.

This mechanism not only explained the sort of combination taking place between metals and nonmetals but also indicated that ions are formed at the moment of compound-formation. The source of ions in ionic combination thus became evident. Also, the Braggs' work was revealing that there is no molecular structure in salts, merely a lattice of ions surrounded by oppositely charged ions. The combination of divalent and trivalent atoms with their opposite numbers was, of course, obvious, as was the possibility for combination of atoms not having the same number of outer electrons and empty corners, i.e., $MgCl_2$, K_2O, Al_2O_3.

The Lewis-Kossel-Langmuir concept also could explain the combination of nonmetals with nonmetals and at the same time deal with hydrogen, an element which created an anomaly if treated as a metal. Since a hydrogen atom consisted merely of an electron surrounding a nucleus, transfer of the electron to a nonmetal would not expose a stable cube, but would expose the nucleus itself. Therefore hydrogen had to be treated as an atom with a satisfactory cube when two electrons were present, not as an electron-losing atom. This could be done by having the hydrogen cube share an edge with an electron-accepting cube.

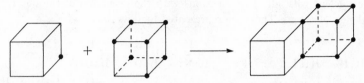

By sharing a pair of electrons, the hydrogen atom controlled as many electrons as the rare gas helium, while the chlorine atom controlled as many as the rare gas argon. Both atoms achieved a stable rare gas outer cube, but in doing so became bonded together by a shared electron pair. In the same fashion, two hydrogen atoms could share an edge to become a stable H_2 molecule and two fluorine or two chorine atoms could share a pair of electrons along an edge of their respective cubes to form molecules of F_2 and Cl_2.

The old question raised by Berzelius against the Avogadro hypothesis was being answered. It was suggested that divalent oxygen atoms might similarly share two pairs of electrons by having a face in common.

The molecular structure of all manner of compounds, such as HCl, H_2O, NH_3, CH_4, CCl_4, and CO_2, could be explained by means of shared edges and shared faces, very much in conformity with valence principles.

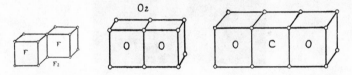

Fig. 20.5. LANGMUIR MODELS OF SEVERAL MOLECULES.
(From I. Langmuir, *J. Am. Chem. Soc.* [1919], with permission of the editor.)

This was, however, a static concept, and limitations were immediately apparent. The N_2 molecule, for example, could not be incorporated into the system. Besides, the physicist could not accept an atom with stationary electrons. Though obviously doomed to a short life, the model called attention to two useful working rules: the importance of the octet of outer electrons as a stable structure and the value of the electron pair as a basis for bonding between atoms.

The formation of salts by electron transfer obviously required little attention, even though the static model has been discarded. This type of combination came to be termed *electrovalent,* or *ionic,* in distinction from *electron pair,* or *covalent,* bonding found in combinations of nonmetallic atoms. The term *covalent* was introduced by Langmuir.

Lewis[7] developed the symbolism for electronic bonding in compounds, letting the ordinary symbol stand for the kernel of the atom, that is, the nucleus with all of its electrons with the exception of the outer, or valence, electrons. These are represented by dots thus:

$$
\begin{array}{ccccccc}
 & & & \text{H} & \text{H} & \text{H} & :\text{N}:::\text{N}: \\
\text{H}:\text{H} & \text{H}:\ddot{\text{C}}\text{l}: & :\ddot{\ddot{\text{F}}}:\ddot{\ddot{\text{F}}}: & \ddot{\text{H}}:\ddot{\text{N}}: & \text{H}:\ddot{\text{O}}: & \text{H}:\ddot{\text{C}}:\text{H} & \\
 & & & \text{H} & & \text{H} & \text{H}:\text{C}:::\text{C}:\text{H} \\
\end{array}
$$

This treatment made the electron pair bond equivalent to the usual dash line in structural formulas.

Fig. 20.6. NEVIL VINCENT SIDGWICK.
(Courtesy of *J. Chemical Education.*)

Although it worked very well for a large number of compounds, electronic formulation of this sort still had inadequacies. Extremely important contributions to the solution of these problems were made in 1923 by Nevil Vincent Sidgwick[8] (1873–1952) at Oxford who showed that in some compounds the electrons of a shared pair are contributed by the same atom, a situation he termed *coordinate covalency*. In some such molecules a polarity is created, and the donor atom becomes slightly positive and the acceptor atom becomes slightly negative. Sidgwick represented such a linkage by a directed bond (→), following earlier usage suggested by

[7] G. N. Lewis, *Valence and the Structure of Atoms and Molecules*, Chemical Catalog Co., New York, 1923.

[8] N. V. Sidgwick, *The Electronic Theory of Valency*, Clarendon Press, Oxford, 1927; *Ann. Repts. Chem. Soc.*, **30**, 112 (1933); W. A. Noyes, *Chem. Revs.*, **17**, 1 (1935).

Lowry. This sort of electron-sharing was necessary to explain the combinations present in many molecules, particularly those around a central atom as found in the perchlorate ion and the sulfate ion.

$$\left[\begin{array}{c} :\!\overset{\cdot\cdot}{O}\!: \\ :\!\overset{\cdot\cdot}{O}\!\overset{*\,*}{:}\!Cl\overset{*}{:}\overset{\cdot\cdot}{O}\!_\circ \\ :\!\overset{\cdot\cdot}{O}\!: \end{array} \right]^{-} \quad \left[\begin{array}{c} :\!\overset{\cdot\cdot}{O}\!: \\ {}^\circ\!\overset{\cdot\cdot}{O}\!\overset{*\,*}{:}\!S\overset{*}{:}\overset{\cdot\cdot}{O}\!_\circ \\ :\!\overset{\cdot\cdot}{O}\!: \end{array} \right]^{=}$$

In each case the * electrons originate from the Cl or S, the · electrons from the O, the ° electrons from the element forming the accompanying positive ion.

The concept of coordinate covalency was particularly useful in dealing with complex Werner-type compounds. When Werner proposed his ideas on coordination complexes in the nineties, he could not explain the nature of the bonding which took place, since it was not within the scope of ordinary valence theory. The concept of electron pair bonds permitted a more convincing explanation of such compounds.

In 1922 Maurice L. Huggins pointed out that nitrogen in ammonia has one unshared pair of electrons. Sidgwick suggested that in the formation of the hexammine cobalt ion, each of six molecules of ammonia contributes its unbonded pair of electrons to the cobalt ion which thus acquires an outer shell of twelve electrons. Even though the electrons are shared with six nitrogen atoms, the cobalt ion is in a greater state of stability than as the bare ion. Sidgwick was also successful in extending the concept of the coordinated electron pair bond to many other types of inorganic and organic compounds, for example, the chelation complexes.

Quantum Mechanics

The quantum mechanics developed in the middle twenties was quickly applied to the problems of chemical bonding and structure. Greater understanding of electron distribution in atoms resulted from spectral studies on the one hand and chemical correlation on the other.

We have observed that J. J. Thomson proposed an atom in which electrons revolved inside a positive shell. In 1904 he further postulated that when a certain number of electrons occupied a certain orbit, mutual repulsion would prevent more electrons from coming into the same orbit. His postulate was based in part on the behavior of floating magnets. Thomson stuck magnetized needles through corks which were placed in a dish of water with the north poles up. A strong magnet with its south pole up was placed in the center of the dish. Only a limited number of magnets would revolve about the central magnet in an innermost circle. An additional magnet placed in this circle would immediately be ejected, or one of the other magnets in the circle would leave. If a number of additional magnets were placed in the dish, they formed an outer circle concentric with the first. In the same year Abegg remarked on the stability of an eight-electron group.

Very soon after embarking on his task of elucidating the structural nature of the atom, Bohr began arranging the electrons in levels of two, then eight, and then eighteen—in order to be consistent with energies revealed by spectral lines. He quickly recognized the need for subdividing the electrons in the second, third, and fourth levels. Guided by Rydberg's 1914 postulate, but not blinded by it, he moved toward periodic groups containing 2, 8, 8, 18, 18, and 32 elements, with electron levels containing 2, 8, up to 18, and up to 32 electrons in the K, L, M, and N shells. In the case of boron, atomic number five, he treated the fifth electron, that is, the third in the L shell, differently than the first two in that shell, reasoning by analogy with aluminum which gives spectral lines indicating that not all of its M electrons behave in the same manner.

Bohr at first devised a subshell arrangement in which he divided the total number of electrons in the main shell by the shell number to arrange the electrons (2), (4, 4), (6, 6, 6), and (8, 8, 8, 8). He quickly recognized that this arrangement was unsound, and next arranged the subshells on the basis: (2), (2, 6), (2, 6, 10), and (2, 6, 10, 14).

Early analysis of spectral lines revealed characteristics ultimately designated *strong, principal, diffuse,* and *fundamental.* The initial letters of these terms (s, p, d, f) were incoporated into quantum symbolism to represent the various subshells occupied by electrons in the atom. Spectroscopists had also observed that high dispersion resolves the fine structure found in many lines into doublets, triplets, and multiple lines. Some lines are split by a magnetic field, the Zeeman effect, while others are split by an electrostatic field, the Stark effect. These lines were used as clues to the energy state of an electron in the atom. Each electron thus came to be described according to four quantum numbers: principal (n), orbital (l), magnetic (m_l), and spin (m_s).

Wolfgang Pauli (1900–1958) introduced the Exclusion principle in 1925, pointing out that no two electrons in a given atom can have all four quantum numbers the same. Thus there might be two, but no more, s electrons for each principal quantum number. Six different p electrons would be possible, and there might be ten of the d and fourteen of the f variety. Consequently quantum consideration became consistent not only with spectral behavior but also with the arrangement of elements into periods.

That two electrons can have their first three quantum numbers the same but differ in their spin quantum numbers, suggested that electron pairs represent a stable state within the atom and clarified the significance of the electron pair bond in covalent compounds. It was also consistent with the magnetic character of chemical substances and with color. As G. N. Lewis pointed out in 1923, paramagnetism and color can usually be associated with unpaired electrons. The theory of the electron pair bond was developed quantum mechanically after 1927; W. Heitler and Fritz

London made especially important contributions. This was a natural outgrowth of G. Uhlenbeck and Samuel Goudsmit's work on the electron's spin properties in the immediately preceding years. Others who played a role in the quantum mechanical development of valence theory were Max Born, Herbert Weyl, Linus Pauling, and John C. Slater. The concept of the orbital was introduced to refer to the wave function associated with the orbital motion of an electron.[9] The electrons with spins opposed occupy an orbital and, if the electrons are contributed by two different atoms, serve to bond the atoms together.

The resonance concept grew out of Heisenberg's 1926 study of the quantum states of the helium atom. It was soon extended to deal with structural problems in compounds such as nitrous oxide where no written structure was entirely consistent with observed properties. Employing quantum mechanical treatment, investigators were able to treat the structure as a nonvisualizable stable state (resonance hybrid) lying somewhere between two or more visualizable structures. Resonance showed a certain kinship with organic theories of an earlier day, for example, Thiele's theory of partial valences and the ideas on mesomerism being developed by British chemists in the twenties. Leaders in the development of resonance theory were Slater and Pauling in America and Hückel in Holland.

The concept of partial ionic character was developed to explain the structure of compounds which have properties intermediate between purely ionic and purely covalent compounds. Many instances were evident where salts such as metal halides could not be considered 100 per cent ionic. On the other hand, compounds such as HF and HCl had to be assigned some ionic character since the electrons constituting the bond spent a greater fraction of their time in the vicinity of the halogen atom than in that of the hydrogen atom. Pauling contributed a great deal to this field in the early thirties, introducing the useful concept of electronegativity to represent the ability of an atom in a molecule to attract electrons.

The hydrogen bond was another development of this period, although T. S. Moore and T. F. Winmill had speculated about such a linkage in 1912 when they sought to explain the weakly basic character of trimethyl ammonium hydroxide. P. Pfeiffer used the hydrogen bond in organic chemistry a year later, but its real value was not recognized until 1920 when Latimer and Rodebush employed it to explain association in such liquids as water, hydrofluoric acid, ammonium hydroxide, and acetic acid. Sidgwick's spectroscopic studies and crystal analysis revealed many other instances where the bond appeared to be important.

At first it was supposed that the hydrogen atom accepted a second pair of electrons, but Pauling showed by quantum mechanics that this hypothesis was untenable, since only the $1s$ orbital had any importance in

[9] L. Pauling, *The Nature of the Chemical Bond*, Cornell Univ. Press, Ithaca, 1939, p. 24n.

Fig. 20.7. Linus Pauling.
(Courtesy of the California Institute
of Technology.)

hydrogen atoms. The hydrogen bond, researchers learned, is present only where the hydrogen atom is held between ionic centers. Electronegative atoms such as fluorine, oxygen, nitrogen, and chlorine are almost the only ones where these bonds are found, with the tendency greatest in fluorine and least in chlorine. Any factors which increase the electronegativity of an atom increase the tendency for hydrogen bonding, i.e., phenols are more prone to bond than aliphatic alcohols. Through the hydrogen bond it became possible to explain the anomalous physical properties of water, hydrogen fluoride, and ammonia.

Another problem was posed by the metals, which are characterized by their malleability, ductility, and high conductivity. H. A. Lorentz advanced a theory of the metallic state in 1916 which treated the metal atoms as spheres closely packed with free electrons in the interstices between spheres. Pauli developed this further in 1927, setting up energy levels for the free electrons. Subsequently the theory of metal structure was developed in accordance with quantum mechanics by Sommerfeld, Bernal, and others. The model ultimately developed consisted of the kernels of the metallic atoms in a closely packed arrangement, with the valence electrons forming an atmosphere of electrons bonding the atoms together in a resonating system.

Intermetallic compounds naturally were considered in this connection, with important contributions being made by G. Tammann, W. L. Bragg, H. A. Bethe, H. Jones, R. Peierls, and W. Hume-Rothney who pointed out as early as 1926 that certain intermetallic compounds with similar structures, but unrelated stochiometric composition, have the same ratio of valence electrons to total atoms. For example, in such compounds as the following the ratio is 3/2: CuZn $(1 + 2)/2$; Cu$_3$Al $(3 + 3)/4$; Cu$_5$Zn $(5 + 4)/6$; AgZn $(1 + 2)/2$; AgMg $(1 + 2)/2$; Ag$_3$Al $(3 + 3)/4$. In another series the ratio is 21/13: Cu$_5$Zn$_8$ $(5 + 16)/13$; Cu$_9$Al$_4$ $(9 + 12)/13$.

Interesting studies were made of metals combined with boron, carbon, and nitrogen. Carbon, of course, had been a standard component of steel for centuries and had been thoroughly investigated from the standpoint of the Phase Rule in Roozeboom's laboratory late in the nineteenth century. Now, x ray crystallography and quantum mechanical studies were invaluable in the elucidation of the nature of steel and other alloys, especially in the hands of G. Hagg and K. Friedenhagen.

SOLUTION THEORY

Despite the enthusiasm with which Ostwald, van't Hoff, and their followers accepted Arrhenius' theory of electrolytic dissociation, some chemists were only lukewarm and a few openly hostile. Critics asked how chlorine could be present in a colorless solution without oxidizing power, or how sodium could exist in water without liberating hydrogen. They of course missed the distinction between ions and atoms. The source of the charge, however, was a more difficult question to deal with and was only answered satisfactorily when the nature of ionic compounds began to be revealed.

Particularly hostile to the theory were Mendeleev, Henry Armstrong (1848–1937) at South Kensington, and Louis Kahlenberg (1870–1941) at Wisconsin. Kahlenberg had taken his doctorate under Ostwald and returned to America enthusiastic for the theory. After setting up his own research program, he soon encountered troublesome questions, especially in connection with studies of nonaqueous solutions. Arrhenius had developed his theory without considering solvents other than water. Kahlenberg showed in 1903 that the dissociating power of the solvent is related to the dielectric constant. During these same years Armstrong caustically attacked the theory, insisting that the solvent must play a role in the properties of the solution.

The criticisms of Arrhenius' theory were answered vigorously by the rising group of young physical chemists, most of them from the Ostwald school, in debates which often resembled a barroom brawl, but which served a useful purpose. The critics raised points on which the original theory was vulnerable. Renewed research was stimulated and ultimately a much improved understanding of solutions was achieved.

Almost from the start it was necessary to concede that Arrhenius's theory was only applicable to very dilute solutions. Also the source of the ionic charge remained a mystery. A more comprehensive theory of solutions was developed in 1923 by Peter Debye (1884–1966) and Erich Hückel[10] (1896–1981) at Zurich. According to Arrhenius' concept, an electrolyte was molecular until dissolved, at which time it dissociated to a lesser or greater extent, depending upon the nature of the solute. The extent of dissociation might be determined by observing the conductivity. Even strong electrolytes were found to be only about 90 per cent dissociated, except at very high dilution where 100 per cent dissociation was approached.

In the early twenties, when the nature of ionic combination began to be understood, it became apparent that salts and metallic hydroxides

Fig. 20.8. PETER DEBYE.
(Courtesy of Peter Debye.)

are ionized in the solid state and therefore should be 100 per cent ionized in solution. It was demonstrated in 1920 by C. Tubandt and S. Eggert that fused salts are excellent conductors—since melting permits mobility of the ions already present in the solid. Debye and Hückel attributed the failure to obtain 100 per cent ionization values by conductivity or freezing point measurements to decreased ion mobilities. They postulated that an ion will attract unto itself an atmosphere of oppositely charged ions. Hence in an electrical field the central ion will be attracted to the oppositely charged electrode, but its progress will be retarded by the "atmosphere" associated with it and pulled in the opposite direction. The total reduction

[10] P. Debye and E. Hückel, *Physikal. Z.*, **24**, 305 (1923); **25**, 145 (1924); *Trans. Faraday Soc.*, **23**, 334 (1927).

of mobility was shown to be proportional to the square root of the concentration. The Norwegian physicist Lars Onsager improved the equation in 1926 by taking the Brownian motion of the ion into account.

These innovations resulted in a reasonably satisfactory theoretical treatment for dilute solutions. The problem of concentrated solutions is still unresolved since solvation of the ions defies the estimation of concentration and introduces various chemical and physical complications. In addition, reaction between ions to form complex ions very often results in the presence of several species.

The Debye-Hückel-Onsager treatment lends itself to nonaqueous solutions, but here again all of the problems involved in aqueous concentrated solutions are encountered.

Acid-Base Theory

Arrhenius's theory assumed the presence of hydrogen ions in acid solutions, the strength of the acid being proportional to the degree of dissociation into ions and therefore to the concentration of hydrogen ions. Basic solutions contained hydroxyl ions in proportion to the degree of basicity. Neutralization was simply combination of hydrogen and hydroxyl ions to form water. The soundness of this concept was illustrated by the uniformity of the heats of neutralization whenever equivalent solutions of any strong acid and any strong base neutralize one another.

Because of the very slight conductivity of very pure water, it was necessary to suppose that water itself is slightly dissociated into hydrogen and hydroxyl ions. After the ion product for water was shown to be 1×10^{-14}, the concept of hydrogen ion concentration became a useful way to express the effective acidity of an acid or base solution. In 1909 S. P. L. Sørensen (1868–1939), Kjeldahl's successor at the Carlsberg Laboratory, introduced the convenient pH concept to express hydrogen ion concentration. The use of the negative logarithm was later extended to the expression of equilibrium constants of acids and bases in the form of pK_a and pK_b values.

Brønsted Theory. Arrhenius' theory of acids and bases, based upon hydrogen and hydroxyl ions, explained acid and base phenomena in aqueous systems, although it failed to account for dissociation of acids, compounds belonging to the covalent class. Many researches on nonaqueous solutions indicated instances of behavior similar to neutralization even though hydrogen ions were not present. Edward C. Franklin (1862–1937), reported that in liquid ammonia the reaction $NH_4Cl + NaNH_2 \rightarrow 2NH_3 + NaCl$ showed a typical neutralization curve during electrometric titration. In glacial acetic acid a similar phenomenon was observed: $HCl + NaC_2H_3O_2 \rightarrow HC_2H_3O_2 + NaCl$.

A completely enlarged viewpoint of acids and bases was developed from the proposals made in 1923 by Thomas Martin Lowry (1874–1936) in

England and independently by J. N. Brønsted (1879–1947) and Niels Bjerrum (1879–1958) in Denmark. These investigators proposed a definition based upon the role of the hydrogen ion in acid-base systems. A base was defined as a hydrogen ion acceptor, and acid as a substance capable of losing hydrogen ions. Thus Brønsted looked upon salts as ionic systems defined by the theories of Debye, Onsager, and others, whereas acids and bases were characterized by their capacity to participate in the transfer of hydrogen ions:

$$\text{acid} \rightleftarrows \text{base} + \text{hydrogen ion}$$

However, this reaction could go to the right only if a hydrogen ion acceptor were present; hence

$$\text{acid} + \text{base} \rightleftarrows \text{new acid} + \text{new base.}$$

The reaction might be expected to go in the direction favoring the formation of the weakest acid and base. According to this concept, a neutralization reaction formed a new acid and base, not a solvent, the solvent itself figuring as an acid or base. It was further realized that the hydrogen ion was unlikely to be a naked proton but might be solvated. Thus the dissociation of covalent molecules might be explained, for example, $HCl + H_2O \rightleftarrows H_3O^+ + Cl^-$.

The Brønsted theory also provided a logical explanation for buffer action and common-ion effect and revealed that a particular ion might act as an acid toward one reagent and as a base toward another. It thus broadened the scope of acid-base systems but was dependent on the presence of a hydrogen-containing molecule or ion to serve as a proton donor.

However there were nonprotonated systems where investigators could also see evidence of acid-base behavior. Germann suggested in 1925 that liquid phosgene might be considered to dissociate into positive and negative ions, that is, CO^{++} or $COCl^+$ and Cl^-. Paul Walden (1925) and Gerhart Jander (1936) treated liquid sulfur dioxide as a sytem which dissociated into SO^{++} and $SO_3^=$.

G. N. Lewis suggested a scheme for handling acids and bases within a broader system in 1923 and developed it further in 1938. Sidgwick also contributed to the elaboration of the concept. Lewis supposed an acid to be any molecule or ion with an incomplete electron grouping around one of its atoms. This atom was able to accept an electron pair from another atom, the donor atom being present in an ion or molecule termed a *Lewis base*. Representative acids in this sytem were: O, HCl, SO_3, BCl_3, H^+, Ag^+, $SnCl_4$, NH_4^+, Fe^{+++}, and Cu^{++}; representative bases were: tertiary amines, CN^-, OH^-, triphenyl methyl, ether, and $O^=$. Sidgwick called these *acceptor* and *donor* units, but Lewis used more conventional names. The reaction between Lewis acids and bases was an addition reaction, although this might be followed by rearrangement to a more stable product. The concept of acids as electron pair acceptors made it possible to include such nonhydrogen containing molecules as SO_3 and SO_2, constituting in a sense a return to the earlier day when the nonmetal oxide was considered an acid. Lewis argued: "To restrict the group of acids to those substances which contain hydrogen interferes as seriously with the systematic understanding of chemistry as would the restriction of the term oxidizing agent to those substances containing oxygen."[11]

Fig. 20.9. GILBERT NEWTON LEWIS.
(Courtesy of Farrington Daniels.)

Still more all-inclusive systems were proposed, like that of M. Usanovich[12] of the Central Asiatic University at Tashkent, which would essentially bring oxidation-reductions within the system. The acid-base concepts have had a varied reception and illustrate again that the value of a theory lies in what one wishes it to do for him. The hydrogen ion concept of an acid is entirely adequate for aqueous systems, but the proton donor concept is necessary if liquid ammonia and glacial acetic acid systems are to be considered. The electron pair accepter concept of Lewis and Sidgwick represents a still broader point of view which has been useful in explaining certain reactions for which the other theories are inadequate.

[11] G. N. Lewis, *J. Franklin Inst.*, **226**, 297 (1938).
[12] M. Usanovich, *J. Gen. Chem.* (U.S.S.R.), **9**, 182 (1939). See N. F. Hall, *J. Chem. Educ.*, **17**, 127 (1940).

CRYSTALLOGRAPHY (STRUCTURAL ANALYSIS)

After it was learned that crystals can be used as a three-dimensional diffraction grating for the measurement of x ray wavelength, it became obvious that x rays might be used as a tool for the elucidation of crystal structure. Laue developed the mathematical interpretation necessary to reconstruct, from the diffraction spots on a photographic plate, the arrangement of atoms within the crystal which produced the pattern of spots.

In 1912 W. H. and W. L. Bragg simplified the analysis somewhat by using x rays of definite wavelength. The x rays were reflected from layers of atoms lying in planes parallel to the crystal face and were picked up by an ionization chamber. By obtaining measurements of the angles of reflection from the various systems of planes it was possible to calculate the distance between layers of atoms and the position of atoms within the crystal. Fourier analysis was introduced into crystal analysis by the Bragg's in 1915.

A third procedure was introduced in 1916 and 1917 by Debye and Paul Scherrer at Göttingen and Albert W. Hull of the General Electric Co. in the United States. The Laue and Bragg methods required well-formed crystals of fair size, which gravely limited these procedures since suitable crystals were not obtainable for many substances. The new method of Debye, Scherrer, and Hull used powder. The reflected x rays are photographed on film mounted in a cylindrical camera, where lines correspond to particular atomic planes. Although the powder contains tiny crystals oriented in many directions, some are located so as to reflect from one plane and some from another in order to give an interpretable pattern of concentric cones which show on the film strip as lines.

Application of these techniques led to new levels of understanding of the structure of substances in the solid state. Sodium chloride crystals were shown to have a simple cubic arrangement with chloride ions alternating with sodium ions. Beneath the surface each sodium ion is surrounded by six chloride ions and each chloride ion by six sodium ions. The concept of molecules had to be abandoned in crystals of this sort. The distance between atomic centers was determined to be 2.817 Ångström units (Å).

Examination of the diamond showed that each carbon atom appears to be at the center of a tetrahedron surrounded by four equally spaced carbon atoms, all carbon atoms being 1.54 Å apart. In graphite, however, carbon atoms are joined together in a continuous plane of hexagons, each carbon being combined with three others at a distance of 1.42 Å. Planes above and below are at a distance of 3.40 Å. The hardness of diamond can be correlated with the uniformly close bonding of the carbon atoms;

the softness of graphite can be correlated with the large distance and the lack of strong bonding between planes.

There was a natural interest in the structure of benzene, but since it is a liquid at convenient working temperatures the earliest studies were made on related solids such as diphenyl and hexamethyl benzene. Kekulé's six-membered planar ring was confirmed with carbon atoms spaced 1.42 Å apart as in graphite. J. M. Robertson at Glasgow investigated napthalene and anthracene, showing that the conventional structures with two and three fused rings are apparently sound. Studies of solid hydrocarbons and of fatty acids revealed a zig-zag orientation of carbon atoms in the long chains.

The Hull-Debye-Scherrer method proved applicable to the study of alloys since these consist of tiny metallic crystals in random arrangement. X ray crystallography is also valuable in unraveling the structure of the silicates. High-speed calculators have enabled researchers to increase our understanding of crystalline organic and inorganic materials. Since electrons are responsible for the scattering of x rays, crystallographers can mathematically calculate and plot the electron densities within the unit cell and obtain, as it were, a "picture" of complex molecules. Molecular dimensions, bond lengths, and bond angles all are susceptible to estimation through crystallographic analysis.

An example of the power which x ray crystallography has attained is evident in the work for which Max F. Perutz and John C. Kendrew received the Nobel Prize for Chemistry in 1962. Perutz began working at Cambridge on the structure of horse hemoglobin in the late 1940's. Soon thereafter Kendrew joined the laboratory and began work on sperm whale myoglobin, the pigmented protein of muscle. Hemoglobin has a molecular weight of about 65,000 and each molecule has about 6,000 atoms capable of scattering x rays (hydrogen atoms are ineffective); myoglobin is much simpler but still has a molecular weight of about 17,000. Perutz found that improved results could be obtained by introducing heavy atoms like iron or mercury into the protein crystals. In 1958. Kendrew was able to announce an image of the myoglobin molecule, a wormlike structure suggestive of the alpha helix arrangement for proteins proposed by L. Pauling, Robert B. Corey, and H. R. Branson of the California Institute of Technology in 1951. In 1960 Perutz completed a similar study on hemoglobin, in which it became evident that the molecule is composed of four subunits similar in structure to myoglobin.

The 1962 Nobel Prize for Medicine and Physiology which was awarded to Francis H. C. Crick, James D. Watson, and Maurice H. F. Wilkins also had crystallographic connotations. In 1953 Watson and Crick suggested that deoxyribonucleic acid (DNA), important in the storage and transfer of genetic information, has the form of a double helix. In part, their deduction was based upon analysis of x ray diffraction photographs made in various laboratories. Even better photographs of DNA fibers, obtained by Rosalind Franklin and by Wilkins, substantiated the double helix structure. This released a flood of research into the chemical nature of DNA and the functionally related types of ribonucleic acid (RNA).

The first nucleic acid was discovered in 1869 by Friedrich Miescher, a student of Hoppe-Seyler, who isolated his material from pus-forming bacteria. Later on, yeast and thymus glands became the favorite source of the acids. Albrecht Kossel in Germany and P. A. Levene in the United States established the presence in nucleic acids of purines (adenine and guanine), pyrimidines (cytosine, thymine, and uracil) in a repetitive unit containing phosphoric acid and pentose sugars (ribose and deoxyribose). There was much confusion regarding the nature and function of RNA and DNA. In 1944 O. T. Avery, Colin M. MacLeod, and Maclyn McCarty of Rockefeller Institute produced evidence that the transformation of pneumococcal types is due not to transmission of genetic information by proteins but to transmission by a specific DNA. Soon thereafter Erwin Chargaff and associates at Columbia published analytical data on on the proportions of various bases in DNA from various animals and microorganisms. These data showed a striking equality between the numbers of thymine and adenine groups and between the cytosine and guanine groups. Then as the chemistry of the Watson-Crick double helix developed, the studies of Wilkins confirmed the system of bases hydrogen bonded to one another to provide an attachment between helices. Since, adenine bonds to thymine and cytosine to guanine, the unraveling of the helices provides two strands, each capable of serving as a template directing the synthesis of its complement.

We can even study seemingly noncrystalline substances with x rays, for example, glass, rubber, and fibers such as wool, silk, cellulose and its derivatives, and synthetics. X rays have also been used to detect structure in liquids. Debye and Scherrer showed in 1916 that liquids such as water, benzene, hexane, methanol, and ethanol produce a set of diffuse rings on a photographic plate when exposed to a beam of monochromatic x rays.

Electron Diffraction

Louis de Broglie predicted in 1923 that moving electrons would show wave properties that lead to interference, diffraction, and other properties of light waves. The prediction was experimentally verified in 1927 by Clinton Joseph Davisson and Lester Halbert Germer at the Bell Telephone Laboratories, and independently by J. J. Thomson's son George Paget Thomson at Aberdeen.

A few years later the German physicist R. Wierl showed that the diffration of electrons provides a tool for the study of gases and vapors. Electron beams at low voltages are less penetrating than x rays and thus are useful for studying the structure of gases, adsorbed gases, and surface layers.

Bond distances and bond angles can be calculated from the intensities and distances shown on photographs of diffracted electrons.

SPECTROSCOPY

With the rise of structural theory spectroscopy could be developed from a highly empirical technique to one with a strong theoretical basis. Through quantum theory certain energy absorptions or emissions could be associated with certain electron or molecular displacements. At the same time progress in instrumentation made it possible to extend the range of the useful spectrum well out into the long wavelengths.

At the beginning of the century visible emission and absorption spectra were widely used in connection with identification problems. There was even some study in the ultraviolet, but this region was not extended and made especially useful until instrumentation was improved. The greatest progress has been made in the regions beyond the visible red where only the most elementary explorations had been made by 1920.

When infrared light passes through a compound, energy is absorbed at wavelengths in the shorter infrared because certain wavelengths have energies corresponding to certain atomic vibrations and rotations. Wavelengths in the longer infrared are associated with certain molecular rotations. Beginning with A. Kratzler's work in 1920 investigators have had considerable success in combining theory and practice. In many cases it is possible to identify functional groups, to distinguish normal organic compounds from iso-compounds, *cis* from *trans* compounds, and polymers from monomers, and to calculate bond angles.

The technique of microwave spectroscopy, an extension of spectroscopy into the very far infrared and short Hertzian range of wavelengths, became available with the development of generators of microwave frequencies such as the Klystron tube used in connection with radar during World War II. With these generators compounds can be exposed to radiation of 0.1 to 30 cm. wavelength, energies which correspond to those involved in the rotation of molecules. Absorption is dependent upon the moment of inertia of the absorbing compound and can be interpreted in terms of masses of atoms, interatomic distances, and angles of symmetry in the molecules. E. Bright Wilson of Harvard has been particularly active in the field; the bulk of his work involves fairly simple gases.

Still another fruitful approach to constitutional problems has been the use of Raman spectroscopy. Adolf Smekal at Halle predicted in 1923 that a transparent substance would scatter monochromatic light and that in the scattered light there would be some wavelengths differing from that of the incident beam. The prediction was verified by Chandrasekhara Venkata Raman (1888–1970) and K. S. Krishman of the University of

Calcutta in 1928. The wavelength and intensity of the scattered light is a source of information regarding the molecular structure of the compound causing the scattering.

PHOTOCHEMISTRY

The influence of light on chemical combination was noted early in the nineteenth century by Dalton and by Cruikshank. In 1841 John W. Draper (1811–1882) of New York indicated that only those rays of light actually absorbed are effective in producing a chemical change, a generalization noted in 1818 by Grotthuss. Bunsen and his student Henry Roscoe undertook a careful investigation of photochemical reactions in 1854, exposing hydrogen and chlorine to light of measured intensity for various periods of time. They found that the amount of hydrogen chloride formed was proportional to the intensity of light and the time of exposure, or more exactly, the amount of chemical change produced was proportional to the quantity of light absorbed. Bunsen also noted that a period of preliminary exposure was necessary before the reaction reached a constant velocity. He called this the *period of photochemical induction*.

Subsequent nineteenth-century studies confused rather than clarified the nature of photochemical reactions. Early twentieth-century research showed that the induction period is attributable to the presence of impurities such as ammonia, rather than being a characteristic of photochemical reactions. This confirmed again the opinion expressed by van't Hoff in his *Études* that induction or acceleration of reactions is caused by interfering reactions and is not a characteristic of the main reaction.

In 1912 Einstein, at Prague, entered the field with his law of photochemical equivalence based upon Planck's quantum theory. Einstein suggested that a molecule absorbs one quantum of energy when it undergoes photochemical change. The total energy absorbed by one mole would be expressed by:

$$U = N \cdot h\nu,$$

where U represents the energy absorbed by one mole, N is the Avogadro number, h is the Planck constant (6.62×10^{-27} erg-seconds), and ν is the frequency of the absorbed light. Since frequency varies inversely with wavelength, it is obvious that the energy absorption per quantum is greatest with short wavelength light. Thus ultraviolet is often more effective than visible light since it supplies a greater quantity of energy. Experience indicated that while the general law was sound it was frequently complicated by secondary reactions.

Subsequent studies on photochemistry revealed that such reactions are seldom simple. Much work has been carried out on the photosynthesis reaction, the conversion of oxygen to ozone, and the conversion of anthracene to dianthracene, in which there is an increase of free energy in the products. In certain other reactions, in particular, the formation of hydrogen chloride from the elements, a small quantity of light produces a very pronounced effect. In 1913 Max Bodenstein observed that a single photon might initiate the reaction of millions of molecules. The mechanism he postulated to explain this was expanded in 1917 by Nernst. It was assumed that the photon caused dissociation of a chlorine molecule into atoms which in turn reacted with hydrogen molecules to form HCl and H, the latter reacting with chlorine molecules, and so on. In this fashion, the reaction might go on at great length, being stopped only by the reaction of free atoms to form Cl_2 or H_2. While late research showed that the reaction is actually more involved, the validity of the chain mechanism was not undermined. The theory of the chain reaction has been extended to other reactions, not all of them photochemical. It has had some success in the theory of negative catalysis where the catalyst may play a role in the breaking of the chain.

In 1873 Hermann Wilhelm Vogel (1834–1898) of Berlin observed that light may be absorbed by a secondary substance which acts as a sensitizer. This finding provided the basis for the improvement of photographic film. Silver halides are sensitive to ultraviolet and the shorter wavelengths of the visible spectrum. Panchromatic film was produced by the addition of dyes sensitive to the orange and red portions of the spectrum. These dyes absorb low energy photons, passing the accumulated energy on to sensitize silver halide, thus giving the more satisfactory results of panchromatic film.

COLLOID CHEMISTRY AND HIGH POLYMERS

The science of colloidal systems has become a very attractive field for investigation. At the turn of the century the study of lyophobic systems, such as the metal sols, sulfides, and hydrated oxides was emphasized. Interest shifted rather quickly to the study of large molecules, in part because of the biochemists' concern with proteins and carbohydrates, and before 1950 researchers were also investigating synthetic molecules of colloidal dimensions.

The molecular weights of proteins and related natural colloids are so high that they can not be determined by methods available in the early 1900's. Hence colloid chemists sought new theoretical approaches to the problem. Einstein, for example, showed that the scattering of light by solutions was related to the change of osmotic pressure and refractive index produced by the solute. Investigators derived a formula with a form similar to that of Beer's law, whereby turbidity could be calculated from the light scattered. Turbidity, which is related to molecular weight, provided a measure of the latter value.

Chemists also undertook to measure molecular weights through osmotic pressure, viscosity, and sedimentation velocities. Developments in the latter field were stimulated by The Svedberg's (1884–1971) invention of the ultracentrifuge. With refined versions of this instrument a colloidal suspension can be whirled in an oil-turbine driven rotor at speeds above 50,000 r.p.m. to achieve sedimenting forces as much as 240,000 times that of the earth's gravity. Under such forces even relatively small molecules like sucrose are caused to settle. The settling of colloids in the ultracentrifuge can be studied by means of an optical system which measures the change in refractive index as the boundary between the colloidal solution and pure solvent moves downward. With colloidal mixtures such as those present in blood plasma the different sedimentation rates of several components can be observed. Approximations of molecular weights can be made using equations dealing with fall of particles through fluids of known viscosity.

This area of research was pursued at Uppsala by Svedberg and his collaborators and at Wisconsin under the direction of John W. Williams. Svedberg began working on the instrument in the early twenties while he spent a term at Wisconsin. His ultracentrifuge has been replaced by less costly, air-driven ultracentrifuges developed by Jesse Beams at the University of Virginia and electrically driven ultracentrifuges developed at the Rockefeller Foundation by Edward G. Pickels.

The use of electrophoresis in the study of colloids was advanced notably in 1937 by the development of apparatus in which the density changes due to internal heating by the current are largely nullified at the temperature of maximum density. This technique was developed by Arne Tiselius (1902–1971) in Svedberg's laboratory.

Researchers also have been interested in surface phenomena. Work on adsorption phenomena was considerably advanced by the adsorption isotherm which relates the adsorption to the adsorbent, adsorbate, and the nature of the solution. Herbert F. Freundlich (1880–1941) and Irving Langmuir (1881–1957) were among the foremost investigators in this area.

In 1899 Rayleigh suggested that the thin films formed by oily substances are a single molecule in thickness if there is sufficient area for them to spread out freely. William D. Harkins of Chicago called attention to the fact that substances which spread out in this way contain a polar group such as OH or COOH which tends to be water soluble. In the same year, 1917, Langmuir at General Electric and N. K. Adam in England studied such films and used the measurements to calculate film thickness and length of molecules. They proceeded on the supposition, advanced by Langmuir and independently by William Hardy of Cambridge, that

Fig. 20.10. THE SVEDBERG.
(Courtesy of J. W. Williams.)

molecules such as oleic acid will align themselves with the long chains parallel and the polar ends attracted to the water surface. Studies of monomolecular films by Eric K. Rideal and his group at Cambridge were important in elucidating the nature of cell walls, the behavior of toxins, and the nature of staining reactions.

With the development of the electron microscope by Ernst Ruska (b. 1906) and Max Knoll (b. 1897) in 1931 minute bodies, of the size range of large protein molecules, could be studied more effectively. The electron microscope followed very rapidly the discovery of the wave behavior of the electron. By accelerating electrons at a voltage of 50,000 researchers can obtain wavelengths of 0.05 Å capable of resolving objects having diameters of 20 Å. The electrons are focused by electrical and magnetic fields which fulfill the role of lenses in the optical microscope. In this manner viruses, fibers, colloidal particles, bacteria, and surface details of crystals can be examined.

CONCLUSION

Physical chemists have been strikingly successful in advancing their science since 1900 both in theoretical and practical directions. Thermodynamics held a particularly important position, especially during the earlier decades of the century. Quantum mechanics took on great significance in connection with problems of atomic structure and chemical bonding.

Instrumental techniques opened up new ways of gaining information about molecules. X ray diffraction proved to be a tool of remarkable power in the study of crystal structure, particularly after the advent of high-speed computers. Spectroscopy also took on new importance after developments in instrumentation led to routine use of infrared, microwave, and Raman techniques.

Solution theory underwent a series of extensive changes after the shortcomings of Arrhenius' theory became evident. However, even by 1950 there were no adequate theoretical structures for dealing with the properties of concentrated solutions. Kinetics continues to be another area of study without an adequate theoretical structure.

Important progress is being made in the chemistry of huge molecules. Staudinger's concept of macromolecules is important in providing a useful theoretical approach to the problems of colloid chemistry, and instrumental developments make it possible to make meaningful measurements. The success of this branch of chemistry is reflected in the growth of the high polymers industry from a highly empirical state to one with a sound scientific foundation.

CHAPTER 21

ANALYTICAL
CHEMISTRY II.
EXPANSION

During the first three decades of the century analytical chemistry was a rather unspectacular handmaiden of the other fields of chemistry. For the most part analysts used the methods of gravimetric and volumetric analysis developed during the 1800's. True, these had been refined in the light of theoretical and experimental knowledge, but they were still tedious and painstaking enough to impede progress in chemical investigation. Some researchers employed micro methods, particularly in organic analysis, and unusual organic reagents occasionally were used to detect and estimate trace amounts of particular ions.

Since 1930 the whole complexion of analytical chemistry has changed, although the full impact of that change has not yet been felt. Elementary courses in analytical chemistry differ little from those of thirty years ago, except perhaps for greater emphasis on theoretical factors involved in experimental operations. Many routine analyses are still carried out by traditional methods, but often with speedier balances, improved drying ovens, and other aids.

Analytical chemistry depends more and more on nonstoichiometric methods based on instrument-measured properties, for example, color, spectrum, refractive index, mass distribution, radioactivity, and the like. Instruments also are being utilized in large industrial operations to achieve automatically controlled systems. In the petroleum industry this application of instrumentation is ideally suited to the continuous analysis of gaseous mixtures. Whenever the composition falls outside certain limits, automatic adjustment of temperature, pressure, flow rates, or composition of feed stock is instituted. This degree of automation requires careful planning, heavy capital investments, and intelligent supervision, but once set up it is capable of amazing results.

This trend in analytical chemistry derives primarily from developments in physics, mathematics, and electronic technology. The principal tools have been improved amplifying devices developed by the radio industry and the high-speed computers created to fill the needs of mathematicians, the military, and industry. Research in inorganic and organic chemistry also have benefited from these developments. Very frequently, for instance, a reaction mixture can be examined without further purification to learn if a certain compound or group is present.

DEVELOPMENTS IN TRADITIONAL QUANTITATIVE ANALYSIS

Perhaps the most important aspect of quantitative analysis in the early 1900's was the refinement of gravimetric and volumetric techniques which resulted from the application of theoretical concepts to analytical problems. Of course this work had commenced in the 1800's, but without the consciousness of physical chemistry essential to the advances made after the turn of the century.

The new knowledge of equilibrium principles—the common ion effect, solubility product, pH, buffer action, and complex ion formation—helped the analyst better understand his methods. For example, the precipitation of hydroxides and carbonates was carefully studied by W. F. Hillebrand and Gustav Lundell who established the importance of careful control of hydrogen ion concentration.

Refinement of theory resulted in more effective use of indicators. In 1881 Lunge introduced methyl orange for the titration of alkaline carbonates purely on empirical grounds. This empiricism was still present in P. T. Thomson's study of commonly used indicators three years later. Nevertheless Thomson was able to provide a set of practical criteria for indicator use. Ostwald's indicator theory introduced in 1891, though unsatisfactory, paved the way for A. Hantzsch's concept of indicators as pseudoacids and -bases. E. Salm, J. Thiel, and other investigators examined the relation of hydrogen ion concentration to color change. Salm studied 28 indicators in a variety of solutions. His work laid the foundation for Sørensen's introduction of the pH concept in 1909. Henry T. Tizard (1885–1959) studied the sensitivity of indicators in 1911. Three years later Niels Bjerrum brought out his monograph on indicator theory,[1] which provided a very fine treatment of the hydrolysis of salts and also emphasized the importance of titrating to a specific pH, a goal ultimately achieved by using an indicator with its change point at the proper pH or by employing potentiometric titration. Bacteriologists found pH indicators of great significance in the preparation of culture media. A very careful study of dyes suitable for this purpose was carried out around 1915 by Wm. Mansfield Clark and Herbert A. Lubs, chemists with the United States Department of Agriculture.

The sulfonphthalein class of indicators was introduced largely through F. S. Acree's work in 1916 and after. Thymol blue proved of particular interest because of its two color changes, later explained as resulting from its ability to act as a dibasic acid.

Investigators also were concerned with the significance of the ionization constant of acids and bases. Tizard and Boeree in 1921 showed how the titration of dibasic acids must be handled.

The titration of amino acids was sufficiently important to command attention. In 1907 Sørensen showed that this might be satisfactorily done if the basic amino group were first blocked by formaldehyde. An alternative procedure was suggested early in the twenties by F. W. Foreman and independently by R. Willstätter and E. Waldschmidt-Leitz after they observed the effect of ethanol in lowering the strength of the amino group disproportionately to that of the carboxyl group.

A key advance in precipitation titrimetry was the introduction of adsorption indicators by Kasimir Fajans in 1923. Fajans found that fluorescein and its derivatives are able to indicate clearly the end point of titrations of halide samples with silver ion solutions. Later tartrazine and phenosafranine were shown to be effective for titration in acid solutions. It also was demonstrated that if two indicators are used, iodide and chloride are titratable in the same solution.

New reagents were also introduced for oxidation-reduction titrimetry; the ceric sulfate methods developed by N. Howell Furman of Princeton and Hobart H. Willard of Michigan are particularly important. Research on oxidation potentials also led to new indicators. Knop, for instance, suggested diphenylamine as an internal indicator in the dichromate method for iron, but in 1931 I. M. Kolthoff and L. A. Sarver showed that diphenylamine sulfonic acid is preferable.

Many of the research projects focused on the less familiar elements contributed to the development of specialized procedures for these elements. Certain dibasic organic acids, such as tartaric and oxalic, were useful in some of these studies because of their ability to form complexes with various ions. In the twenties Schoeller and his associates found tannin

---

[1] N. Bjerrum, *Die Theorie der alkalimetrischen und azidimetrischen Titrierungen* Enke, Stuttgart, 1914.

Fig. 21.1. KASIMIR FAJANS.
(Courtesy of K. Fajans.)

to be a useful agent for precipitating tantalum and niobium from such complexes. L. Tschugaeff (1873–1922) observed the reaction of dimethyl glyoxime with ammoniacal solutions of nickel salts in 1905. This reagent was applied in qualitative analysis by H. Kraut, and a gravimetric procedure was published by O. Brunck in 1907.

Balances

The analytical balance remained a standard instrument for many years. Time-saving changes—for example, damping devices to bring the beam to rest quickly, the gold-plated chain to eliminate the use of the rider and small weights, and a keyboard-operated lever system for adding and removing weights—were long rejected by chemists fearful that such innovations would impair the accuracy of the balance. Demands of time and convenience slowly overcame this resistance. By 1950 apparatus makers developed electrically operated, direct-reading balances. Even the weight pan is eliminated in the more advanced models.

Water Determination

An interesting development in 1935 was the introduction of a new approach to the determination of water by Karl Fischer (1899–1958) of the Edeleanu Gesellschaft. His procedure was based upon a reaction observed by Bunsen in 1853, in which iodine reacts with water in the presence of sulfur dioxide to form hydrogen iodide and sulfuric acid. The reaction was soon utilized by Ferdinand Reich for the determination of sulfur dioxide in kiln gases, and in later years it was adapted to the determination of sulfur dioxide in acid chamber gases. Fischer adapted the principle to the determination of water in sulfur dioxide but only after an extensive investigation into the nature of the reaction. He found that the erratic results obtained in quantitative use of the reaction for the determination of water were caused in part by an increased concentration of acidic substances which might actually cause a reversal of the reaction. Fischer used pyridine to suppress the unfavorable effect of the acids, and a practical titrimetric method for the determination of water was evolved.

The method proved applicable not only to samples containing free moisture but also to the determination of a variety of substances which form water as a product in certain reactions as well as of substances which react with water. The post-1935 decade produced a steady stream of papers dealing with the use of the Karl Fischer reagent for the determination of alcoholic hydroxyl, phenols, carboxylic acids, acid anhydrides, carbonyl compounds, amines, nitrites, peroxides, and other compounds.[2]

INSTRUMENTATION

Today great emphasis on instrumental approaches to analytical work is the consequence of decades of developing ever more refined and useful tools. We considered the introduction of instrumental methods in Chapter 11. The refractometer and the polarimeter were exceedingly helpful tools in a few fields of analysis, and by the end of the nineteenth century electroplating methods for certain metals were proving useful. Emission spectroscopy was an imporant tool in qualitative analysis, and absorption spectroscopy was somewhat useful in quantitative and qualitative analysis. The microscope also was employed, particularly in food and drug analysis.

Émile Chamot (1869–1950) and his student Clyde W. Mason (b. 1898) contributed significantly to the development of methods for using the microscope to study chemical problems. Before Chamot began his lengthy researches at Cornell, the microscope had rarely been used to identify pure compounds. He adapted the instrument to the identification of many crystalline substances and developed procedures for carrying out precipitation reactions on the stage of the microscope. For a time Chamot's work was influential, but because its successful application required careful training and experience it lost ground to simpler methods.

Electrochemical methods attracted considerable attention in the early 1900's. But except for a few electroplating procedures, the approach was not found particularly fruitful, and by 1925 electrochemical apparatus was generally relegated to the attic or the museum. Potentiometric titrations required apparatus that was frequently temperamental and, in most titrations, offered no particular advantage over indicators. Conductimetric titrations also required special apparatus, and in most cases the changes in conductivity had to be plotted in order to locate the place where the solution showed a discontinuity in conductivity. However, after electrodes and current-measuring devices were improved and when methods suitable for special types of samples were required, these procedures began to find new favor in certain quarters.

The success of potentiometric titrations was dependent upon a convenient system for measuring potential, which related, of course, to developments in the measurement of pH. Despite its large range and its freedom from salt errors, the hydrogen electrode, which was established as a reference electrode, was not a convenient electrode for everyday analytical use. The quinhydrone electrode was of limited range and, like the hydrogen electrode, could not be used in the presence of oxidizing and reducing agents. The antimony electrode was sufficiently rugged for industrial purposes and had certain other advantages, but it could not be used in the presence of oxidizing agents and was readily poisoned by certain ions. The glass electrode, which had been suggested in 1909 by Haber and Klemensiewicz, was finally developed into a practical electrode in the thirties.

With the glass electrode, portable pH meters convenient for both laboratory and factory could be produced. It also revived interest in potentiometric titration. The technique lends itself very well to the determination of acidity or alkalinity in turbid or colored solutions, for example acid value of fatty oils; acidity of tannin-conditioned boiler feed waters. It can be used for the titration of a mixture of two acids, provided their dissociation constants differ by at least 10^4 when concentrations are close to equal. Potentiometric titrations are also possible for oxidation-reduction reactions and for precipitations in certain cases, for example, the determination of vitamin C in colored juices or the determination of chloride by titration with silver ion, using a silver wire as the indicator electrode.

The Polarograph

Polarography was an outgrowth of the work of Jaroslav Heyrovsky (1890–1967) in Prague. Heyrovsky subjected a dropping mercury cathode and an anode consisting of a pool of mercury to a continuously increasing negative potential. He observed that the current increased in steps, not steadily, if the solution in contact with the electrodes contained reducible ions or groups.

Although Heyrovsky enunciated the principle of the polarograph in 1925, many years passed before it began to take on practical significance. Books on the subject by Hohn, Heyrovsky, and Kolthoff and Lingane were influential in attracting interest. Now, commercially available

Fig. 21.2.   JAROSLAV HEYROVSKY AND CHANDRESEKHARA
VENKATA RAMAN.
(Courtesy of Philip J. Elving and *J. Chemical Education.*)

instruments permit the determination of substances at low concentrations
in mixtures, with a minimum of preliminary separations. The polarograph
is useful for the determination of cations, certain anions such as chromate,
nitrate, and bromate, and organic compounds with electroreducible
groups. The instrument is also useful in physical chemical studies on
the nature of solutes and on the mechanism of reactions in solution. The
polarograph is rarely applicable to casual analyses, but it can be very
useful in particular problems after the necessary research has been done.

Colorimetry

Colorimetric analytical methods have had a long history. The deep
blue color of the tetrammine copper ion was used for the estimation of
copper around 1830. Nessler's method for ammonia dates from 1852,
and thiocyanate was used for the analysis of iron about the same time.
The yellow color produced by the reaction of titanium salts with hydrogen
peroxide was reported by Schönn in 1869 and developed into a colorimetric method for titanium by Weller in 1882. Vanadium undergoes a
similar reaction with peroxide, forming an orange complex. In 1912
Mellor developed a method for the colorimetric determination of both
elements, using a reaction discovered by Fenton in 1908 (orange-yellow
color of dihydroxymaleic acid with titanium, none with vanadium) on
one portion, peroxide on another.

Color-absorption techniques provide a good example of a nonstoic-hiometric approach. The light absorption of a colored compound varies with the wavelength of the radiation used. Hence early colorimetry was highly dependent on empirical comparisons of the unknown with standard solutions having approximately the same concentration, as, for instance, in Nessler's method for ammonia. Colorimeters, such as the Duboscq, for matching the color of the unknown with the color of a standard by varying the depth of the transmitting solutions and using Beer's law were not satisfactory for all colored compounds and were highly empirical at best.

In 1729 P. Bouguer observed that the fraction of incident light absorbed by a medium is proportional to the thickness of the medium. This was later rediscovered by Johan Heinrich Lambert (1728–1777) whose treatment of the absorption of monochromatic light led to the relation

$$\frac{-dI}{I} = a \cdot dx,$$

where I is the intensity of light passing through a medium of thickness x and a is an absorption coefficient. Integration, with the boundary condition $I = I_0$ at $x = 0$, gives

$$I = I_0 e^{-ax}.$$

August Beer (1825–1863) showed in 1852 that for many solutions the absorption coefficient a is proportional to the concentration of the solute c. Although Beer did not himself formulate the exponential absorption law in which concentration and thickness appear as symmetrical variables, the relationship

$$I = I_0 e^{-acx}$$

is referred to as Beer's law. The term appears first to have been used in 1889.[3]

Until 1940 colorimetry was mostly visual and frequently at a high level of empiricism—depending as it did on Nessler tubes, the Duboscq colorimeter, and Lovibond tintometer. The tintometer utilized super-imposable colored glass disks as the basis for color comparisons. Some determinations were even carried out by comparing color with colored paper or colored glasses. The nephelometer, used for measuring light-scattering by slightly turbid solutions, was developed by T. W. Richards in connection with the determination of silver halides.

Spectrophotometers came into widespread use beginning around 1940. Several commercial instruments of high quality and simplicity caused

---

[3] H. G. Pfeiffer and H. A. Liebhafsky, *J. Chem. Educ.*, **28**, 123 (1951). These authors found B. Walter, *Ann. Physik*, **36**, 502 (1889) apparently to have been the first to have used the term "Beer's Law."

colorimetry to acquire a new popularity. The best-known instruments, the Zeiss-Pulfrich, Hilger, Spekker, Beckman, and Coleman employ filters, prisms, or gratings to provide illumination within a narrow range of wavelengths. Absorption is generally measured by photocells.

Representative of colorimetric reagents is diphenyl-thiocarbazone, commonly known as dithizone, which was discovered by Emil Fischer in 1882. He observed the marked tendency of dithizone to form colored compounds with metallic ions but did not pursue the work. In 1926 Hellmuth Fischer studied the compound and reported its analytical possibilities, which were exploited to the fullest during the thirties. The colored chelates formed by this reagent with a large number of cations are extremely soluble in organic solvents such as chloroform. Hence the complex can be extracted from large quantities of water into small quantities of solvent, making the method very sensitive to trace amounts.

With instruments, colorimetry can be applied in the ultraviolet to wavelengths as short as 2000 Å. Further extension into the ultraviolet is not possible because of the absorption of containers and prisms and of the air itself. Recording, at first primarily photographic, was markedly improved with the development of practical photocells. Ultraviolet spectro-photometry is particularly valuable in the determination of aromatic compounds, for example, phenols, anthracene, styrene.

Ultraviolet absorption also figures in studies of the structure of organic compounds. Ultraviolet absorption is associated with loosely bound electrons, such as those found in double bonds. The unsaturated bonds in ethylene, acetylene, carbonyl groups, and cyanides absorb below 2000 Å and hence are outside the range of the ultraviolet spectrophotometer. Substitution around the unsaturated bond tends to move the absorption into the longer wavelength region but hardly far enough for practical purposes. Azo, nitro, nitrite, nitrate, and nitroso groups absorb between 2500 and 3000 Å. Conjugation of unsaturated linkages tends to intensify the absorption and shift it into longer wavelengths. Aromatic rings have a characteristic absorption which can be of diagnostic value.

Infrared. Perhaps the greatest advance in the use of absorption of radiant energy as an analytical tool has been in the field of infrared spectroscopy. Instruments utilizing the portion of the spectrum from 8000 Å to several tenths of a millimeter were available before 1920, but the real progress in infrared work occurred after 1940. This region of the spectrum contains frequencies such as those involved in vibrations of the molecular structures. Such factors as the masses of atoms, strength of bonds, and configuration of molecules are important in connection with the energy absorbed. Thus certain bands are readily identified with certain groups such as OH, NH, $C=C$, and $C=O$.

The rise of infrared spectroscopy has depended on progress in the development of thermopiles and bolometers, amplifiers, and recorders

For many years the optical parts of such instruments were far more satisfactory than the detection and recording mechanisms.

Infrared spectroscopy, while primarily a qualitative tool, can be used for quantitative analysis if numerous routine analyses—for example, those frequently necessary in industrial practice—justify the research required to perfect the procedure for such work. Quantitative I-R spectroscopy has been used in the analysis of nitroparaffin mixtures, cresol mixtures, and benzene hexachloride isomers. In the latter case the *gamma* isomer is active as an insecticide. I-R spectroscopy has been a useful tool in the determination of γ-benzene hexachloride when contaminated with its closely related isomers. In qualitative analysis I-R spectroscopy is exceedingly valuable since a wealth of information is provided by the position and intensity of absorption. A great deal of work has been done in mapping the nature of absorption by various bonds and groups so that this data can be quickly applied to the determination of structure in new compounds.[4] Industrially, I-R is useful in following the progress of polymerization since the absorption bands of monomer and polymer differ from one another.

Mass Spectrometer

The mass spectrometer was used for two decades primarily as a research instrument before it began to be considered seriously as an analytical tool. At first the mass spectrometer was a highly tempermental instrument which gave reliable results only for its builder. Early investigators found themselves so occupied with the measurement of exact atomic masses and the distribution of isotopes that they were not actively seeking new uses for the instrument.

With the onset of isotopic tracer studies, the utility of the mass spectrometer for analysis rapidly became evident. Biochemical studies on nitrogen metabolism in plants required the use of ^{15}N as a tracer. Since this is a stable isotope, and since it is not accurately determined by density measurements, the mass spectrometer was the obvious analytical instrument. It also served a useful purpose in tracer studies using stable ^{13}C and in other work based upon identification of stable isotopes. Commercial instruments of standard types have been available for about fifteen years.

During the forties the petroleum industry began to employ the mass spectrometer in connection with the analysis of hydrocarbon mixtures. Although the quantitative interpretation of such mass spectrograms poses a formidable calculation problem, this instrument proves significantly successful in industry when high-speed calculators are available.

---

[4] The I-R absorption of characteristic organic groups can be seen in summary form in C. R. N. Strouts, J. H. Gilfillan and H. N. Wilson, *Analytical Chemistry*, Clarendon, Oxford, 1955, vol. 2, pp. 826–827.

Emission Spectroscopy

Emission spectroscopy has developed more rapidly than any other spectroscopic fields. Indeed as early as 1920 it had reached a high level of usefulness. Practical procedures have been developed for the identification of the metals and metalloids, which means that only the elemental gases, sulfur, and the halogens are outside the scope of ordinary operations. The technique is sensitive to a few parts per million and thus is very useful for the identification of trace contaminants.

Quantitative analysis became feasible with the standarization of photographic plates, instruments, and procedures. The quantity of each element present is determined by comparing the density of certain emission lines of the different elements present. Such comparisons were made visually at first, but with microphotometers the density of silver deposits on the photographic plate can be measured accurately. Quantitative and qualitative emission spectroscopy are best suited to large-scale, routine operations. Small-scale and nonroutine spectra present such difficult interpretation problems that costs are prohibitive. It is better and cheaper to use some other analytical techniques than to pay the interpreters.

Large industrial spectroscopic laboratories are turning more and more to completely automatic interpretation and recording of results. In spectrographs of high resolution, photocells are placed at the positions where the important spectral lines of the expected elements will fall. By appropriate amplification, calculation, and recording devices, a printed record of the results is available within seconds after the emission is passed into the instrument. This degree of automation naturally involves a large capital investment and well-qualified maintenance and is restricted in the number and kind of elements it will analyze without extensive alteration. However in industrial control work, where many batches of similar alloys are produced every day, the saving in time and labor justifies the large investment.

Arc spectra, made up chiefly of lines from neutral atoms, and spark spectra, made up chiefly of lines of ions, have long been satisfactory spectra for analytical work. Flame spectra, although comprised of a smaller number of lines of low energy transitions, are completely adequate for many analyses, particularly of alkali and alkaline earth metals. But since their lines are weak and transient flame spectra are not as generally suitable for quantitative work.

Two procedures developed since 1929 extended the usefulness of flame spectra. Ramage introduced a flame method for the analysis of soil extracts and plant ashes which utilizes an oxy-coal gas flame in a silica burner, so designed that a filter-paper coil impregnated with the sample can be fed into the flames. About the same time Lundegardh developed an apparatus where a fine spray of solution is introduced into an acetylene-compressed air flame. Lundegardh claimed success in the quantitative determination of 32 metals. Neither of these methods is sensitive for metals such as lead, zinc, and mercury. In recent years flame photometers, instruments based on the Lundegardh principle, have become commercially available. They are especially useful for the routine determination of sodium and potassium.

Nuclear Magnetic Resonance

The technique of studying the magnetic moment which results from the intrinsic angular momentum, or spin, possessed by many atomic nuclei has been developed since 1940. This work is dependent on the fundamental studies of nuclear theory contributed by physicists. The first rough measurements of the magnetic moment of nuclear particles were made by the Germans Otto Stern and I. Estermann in 1933, and significant progress in the field was achieved in the laboratory of Isidor Issac Rabi (b. 1898) at Columbia. These studies have been valuable in the development of nuclear theory.

When a beam of atoms exposed to a strong magnetic field has a weak, oscillating field of known frequency superimposed upon it, energy of certain frequencies is taken up by the nuclei as they undergo transition to higher magnetic sublevels. By measuring the beam strengths of a magnetic field of steadily varying frequency, the frequency at which nuclei absorb can be measured. This technique, which was first used for gaseous substances, was extended to liquids and solids through the work of Felix Bloch (b. 1905) at Stanford and Edward M. Purcell (b. 1912) at Harvard. Block's group first measured the resonance induction for protons in water, Purcell's the resonance absorption for protons in solid paraffin. Since the publication of these studies in 1946 the field has developed rapidly. While the technique serves physicists investigating nuclear properties, it is valuable as well to chemists concerned, for example with identification and analysis, following the course of chemical reactions, or studying complexes, hindered rotation, or defects in solids. In 1949 W. D. Knight showed that the resonance frequency for particular nuclei in a specific field is sometimes dependent on the chemical form in which the atom is present. For example, three independent peaks are seen for the protons in ethanol, corresponding to binding in CH_3, CH_2, and OH. This so-called chemical shift is correlated with the shielding effect shown by the valence electrons on the magnetic field in its action upon the nucleus.

CHROMATOGRAPHY

Analytical chemists usually classify chromatography as an instrumental method, but it will be given separate treatment here since it differs quite distinctly from the methods described above in objective and principle. Utilized for the most part in the separation of complex mixtures, it has only limited applicability to quantitative measurements. The individual components of a mixture separated by chromatography are ordinarily identified and measured by traditional procedures.

Chemists have long known that many substances are adsorbed from gases and solutions onto the surface of solid materials. Scheele described the adsorption of gases on charcoal in 1773, and before 1800 the principle was applied to industry—in the use of charcoal for the clarification of sugar solutions. Organic chemists have used boneblack to decolorize their preparations for at least 150 years. And in the dye industry workers frequently tested the quality of dye baths by placing a drop of solution on paper or cloth and noting the pattern of concentric rings formed by the dye components as the solvent moved outwards due to capillary action. The German dye chemist F. F. Runge recognized the general applicability of adsorption as a basis for separating, not only dyes, but colored substances in general. He showed that inorganic cations can be separated as a result of their different rates of migration through paper or other porous materials. His method and results are described in two books published in the 1850's.[5] The later work was illustrated by actual paper chromatograms pasted in by hand.

Runge's work influenced further investigation of capillary separation undertaken by Schönbein and his pupil F. Goppelsröder around 1861. Goppelsröder wrote several books on capillarity and made extensive investigations of the rate at which dissolved materials advance on porous substances. He tested adsorptive powders but found paper and textiles more satisfactory and experimented with columns of Kieselguhr and other fine powders and observed some separation, as did Matteucci in Italy. About this same time the English soil chemists H. S. Thomson and J. T. Way, observed that when solutions of minerals are poured through columns of earth, the various minerals become concentrated at different levels. In 1886 C. Engler and M. Boehm noted that when hydrocarbons are passed through charcoal columns the unsaturated compounds are retained while the saturated hydrocarbons remain in the fluid leaving the column. This observation was applied in Germany in the production of petroleum jelly. Although several investigators reported similar observations on dyes, minerals, and other materials, interest in the technique was not widespread or infectious during the nineteenth century.

In 1897 David T. Day, a petroleum chemist with the United States Geological Survey, hypothesized that petroleum from different parts of the world differed in composition because of the influence of rocks in

[5] F. F. Runge, *Farbenchemie*, 1850; *Der Bildungstrieb der Stoffe*, 1855; See T. I. Williams, *The Elements of Chromatography*, Philosopical Library, New York, 1953, p. 2 for an illustration of these chromatograms.

adsorbing out various constituents as the petroleum seeped through. Day passed crude petroleum through columns of fuller's earth and other adsorbents and observed that the composition changed while passing through the columns and that different components of the petroleum were to be found at different levels in the column. He reported his results in Paris at the First International Petroleum Congress held in 1900. Day realized the importance of column adsorption, not only in petroleum research, but in chemical analysis in general. The method was soon put to work by petroleum chemists in the United States, Russia, and Germany. Engler and E. Albrecht made extensive studies, utilizing a column with ports at various levels from which samples might be withdrawn and studied to observe changes in boiling point, density, and viscosity. In Russia S. K. Kvitka was awarded a certificate in 1900 by the Baku Technical Committee for a fractionation procedure involving adsorption.

Michael Tswett (1872–1919) was responsible for developing chromatography as a method for the separation of plant pigments. Tswett was born in Italy, the son of an Italian mother and a Russian father. He studied in Switzerland, and his research and teaching activities were carried out in Warsaw and Russian educational centers. As a botanist he was interested in plant pigments, but it was his training in physical chemistry which led him to utilize adsorption as a separation technique. He was struck by the fact that, although plant pigments are readily soluble in petroleum ether, the solvent is very unsuitable for extracting the pigments from plant tissue. In a little-known paper published in 1903[6] he reasoned that the pigments are strongly adsorbed to the plant tissues and hence are not easily extracted by petroleum ether. The extraction, he believed, could be made by alcohol or even petroleum ether with alcohol added.

Tswett supported his adsorption hypothesis by showing that extracted pigments are strongly adsorbed from petroleum ether by filter paper. Petroleum ether fails to extract a significant amount of the pigment from the filter paper until a small quantity of alcohol is added. Thus the paper holds the pigment in the same manner as plant tissue. Tswett then tested the ability of more than 100 chemical substances to adsorb chlorophyll extracts. He also studied the adsorption behavior of certain natural and synthetic dyes and of lecithin. Tswett's investigations appear to have been influenced by Goppelsröder but not by any of his contemporaries' work.

Between 1906 and 1914 Tswett carried on extensive research on chlorophyll and the associated xanthophylls and carotenes. In his well-known paper of 1906 he separated the pigments in a petroleum ether extract by

[6] M. Tswett, *Proc. Warsaw Soc. Nat. Sci.*, *Biol. Sect.*, **14**, minute no. 6 (1903). German and English translation were recently published by G. Hesse and Herbert Weil, *Michael Tswett's First Paper on Chromatography*, Woelm, Eschwege, 1954.

passing the solution through a column of calcium carbonate. The two chlorophylls, the carotene and the xanthophyll, appeared as colored bands which could be separated and eluted with alcohol. Tswett was highly enthusiastic about column chromatography but was able to evoke little interest. Dhéré and Vegizzi used it in their study of chlorophyll in 1916, and L. S. Palmer at Minnesota made some use of the technique around 1920 in his extensive study of the carotenoids.

Fig. 21.3. Mikhail Semenovich Tswett.
(Courtesy of *J. Chemical Education.*)

In 1931 adsorption chromatography was recognized for its real worth as a consequence of the work of Richard Kuhn and his associates at the Kaiser Wilhelm Institute in Berlin. Their use of the tool in their studies on carotenes and xanthophylls brought it to popular attention in a period when interest in vitamins and other biologically active substances was high. They showed activated alumina to be a widely useful adsorbent. Karrer, Windaus, Ruzicka, Winterstein, Heilbron, Zechmeister, and Willstätter are a few of the organic and biochemists who were prominent in developing chromatography as a useful separatory tool during the thirties. Its effectiveness in dealing with colored compounds was quickly evident, and it was applied also to the resolution of mixtures, concentration of trace substances, and checking of homogeneity. In time, various techniques were developed for dealing with colorless substances.

Frontal Analysis and Displacement Development

Frontal analysis and displacement developments are the chromato-graphic techniques introduced by Arne Tiselius and his associates, particularly Stig Claesson, in the early forties. In frontal analysis the refractive index of the liquid leaving the column is continuously measured. The curve of refractive index versus volume then gives a record of the movement of components through the column. At first the refractive index will be that of the solvent since there is sufficient solid in the column to adsorb all of the solutes. When the refractive index changes, the most poorly adsorbed solute is being eluted from the adsorbent and is coming through, and so on. The method has been applied not only to the separation but also to the quantitative analysis of such mixtures as fatty acids and their esters and aliphatic alcohols. Tiselius and his group originally applied the method to the separation of sugars and of protein cleavage products on columns of carbon.

Displacement development, while similar in principle to frontal analysis, involves development with a solution containing a solute more avidly adsorbed than the substances already on the column. The bands move down in juxta-position as the column is developed. The emerging solution is analyzed as in frontal analysis. The technique has the advantage that successive zones come through with less contamination with previous and later zones than is the case in frontal analysis.

Ion Exchange

Ion exchange materials, long known in the form of naturally occurring silicates such as the greensands, were used in the earliest types of regenerative water softeners. Since 1930 chemists have had marked success in producing synthetic resins with ion exchange properties. Such resins are often used in ion exchange columns, which are highly effective in bringing about certain separations.

Ion exchange techniques were utilized very effectively during World War II in the separation of fission products from nuclear reactors. Many of these fission products are in the rare earth portion of the periodic table where separations are extremely tedious. The use of cation exchange resins greatly enhanced the separations. The results of this work were published in a series of papers by Edward R. Tompkins, Frank H. Spedding, G. E. Boyd, Lawrence S. Myers, John A. Ayres, and collaborators in 1947.[7] The use of citrate, which complexes with rare earth ions, and carefully controlled pH values enabled them to obtain good separations. Ion exchange methods are employed for large-scale separation of rare earth elements by Spedding's group at the Institute for Nuclear Studies

---

[7] Fourteen papers on the subject will be found in *J. Am. Chem. Soc.*, **69**, 2679–2879 (1947).

at Iowa State College. Commercial production by the Lindsay Chemical Co. near Chicago utilized a similar approach.

Amino acids were separated into basic, neutral, and acidic types by Tiselius, Drake, and Hagdahl on ion exchange resins in 1947. Even earlier, Block successfully made similar separations on Amberlite I. R. 100, a sulfonated phenolformaldehyde resin with cation exchange properties.

Partition Chromatography

Partition chromatography was introduced in 1941 by Archer John Porter Martin (b. 1910) and R. L. Millington Synge[8] (b. 1914). Synge had observed in 1938, while working with acetylated amino acids, that their partition coefficients between chloroform and water differ markedly. Martin had previously developed a countercurrent extractor for vitamin purification. As collaborators their attempts to separate amino acids by partitioning between two liquid solvents moving countercurrently were unsuccessful. They then found that if the water phase is held stationary by adsorbing it on silica gel, while permitting the chloroform phase to flow over it, the partitioning proceeds satisfactorily. And they learned by using a chloroform-insoluble indicator such as methyl orange it is possible to detect the bands of acetyl amino acids on the column.

The technique was soon applied by Isherwood[9] to the separation of fumaric, succinic, oxalic, malic, citric, and tartaric acids, using chloroform-butanol as the mobile phase and dilute sulfuric acid as the static phase. Levi and colleagues used it for the separation of penicillins. Ramsay and Patterson[10] of the United States Food and Drug Administration applied partition chromatography to the separation of isomers of hexachlorocyclohexanes and of C_5—C_{10} fatty acids. They pioneered in the development of nonaqueous systems, using nitromethane on silica and n-hexane in the first case, methanol on silica and trimethylpentane in the second.

In 1943 A. H. Gordon, Martin, and Synge encountered difficulty in separating certain amino acids on silica. They turned to other water-adsorbing substances such as starch and cellulose. Using filter paper in an enclosure saturated with water vapor, they were able to separate various amino acid mixtures and detect the position of the amino acids by spraying with ninhydrin and heating.

Paper-strip chromatography developed rapidly from this point. In 1944 R. Consden with Gordon and Martin developed a two-dimensional technique for bringing about separation of resistant combinations.

[8] A. J. P. Martin and R. L. M. Synge, *Biochem. J.*, **35**, 1358 (1941); **37**, 79, 313 (1943); with A. H. Gordon, **38**, 65 (1944).

[9] F. A. Isherwood, *Biochem. J.*, **40**, 688 (1946).

[10] L. L. Ramsey and W. I. Patterson, *J. Assoc. Off. Agr. Chem.*, **29**, 337 (1948); **31**, 139, 164 (1948).

According to this method, a square of filter paper is spotted with the test substances, for example, a protein hydrolysate, in one corner and developed with a suitable solvent. The various components are thus spread along the edge of the paper which is then dried, turned 90 degrees, and developed with a second solvent. As a result components not separated by the first solvent are moved apart. Consden and his colleagues located the position of over twenty amino acids by development, first with phenol and water, then with collidine and water.

Partridge applied the paper-strip technique to the identification of reducing sugars, bringing out the zones with ammoniacal silver nitrate solution. Goodall and Levi utilized paper chromatography to separate penicillins, locating the position of each active compound by placing the paper on nutrient agar inoculated with spores of B. subtilis. This method was refined into a quantitative test by observing the extent of growth inhibition in the vicinity of the penicillin spot. Paper chromatography has also found a variety of other applications, including the analysis of mixtures of inorganic cations and anions. It is useful primarily in qualitative work.

Gas Chromatography

Gas chromatography, or vapor-phase chromatography, has been very rapidly applied since its introduction by A. T. James and A. J. P. Martin of England in 1952. It is actually a variant of partition chromatography, with one of the phases being gaseous. The chromatographic adsorption of gaseous components had been utilized in 1941 by G. Hesse in Germany, but it was James and Martin who developed vapor-phase chromatography into an effective analytical tool. They used a liquid phase consisting of silicone oil and stearic acid supported on kieselguhr for the separation and analysis of volatile fatty acids. Acetic, propionic, isobutyric, n-butyric, β-methylvaleric, α-methylvaleric, and n-valeric acids separated in that order from nitrogen as the carrier gas. The quantity of each acid brought through by the effluent gas was determined by means of an automatic recording burette.

James and Martin extended the method to the determination of ammonia and the methylamines, the volatile aliphatic amines, and the homologues of pyridine. The technique was quickly adopted by other laboratories. It is particularly valuable in the analysis of hydrocarbon mixtures of high volatility and has become an important analytical tool in the petroleum industry. With the development of high temperature columns the technique can be extended to mixtures of low volatility, such as fatty acid esters.

Broadening of their applicability followed the introduction of new detection devices. J. Janak of Hungary utilized the nitrometer. The thermal-conductivity cell has had the most extensive use. S. Claesson of

Hungary introduced the use of the gas-density balance, which was further developed by James and Martin. A. E. Martin and J. Smart introduced the infrared gas analyzer. Other techniques involve the detection of ionization in a hydrogen flame, surface potential, radioactivity, flow-impedence, and specific heat. Changes in the effluent gas are recorded automatically.

MICROANALYSIS

While analytical methods sensitive to very small amounts had been in use for many years—for example, the Marsh test for arsenic, the Nessler method for ammonia—microanalysis as a standard analytical procedure is a twentieth-century development. It was established very largely as a result of the work of Pregl and Emich, both of the University of Graz, in Austria. To a very large degree the methods of quantitative microanalysis represent a high degree of refinement of traditional macromethods, although in some cases entirely original procedures have been introduced.

Friedrich Emich (1860–1940) was in a sense the founder of quantitative microanalysis during his long career at Graz. His contributions included methods in both the organic and inorganic fields, with greater emphasis on the latter. One of Emich's most renowned pupils, Anton A. Bendetti-Pichler (1894–1964), made many contributions to microanalysis, both while associated with Emich at Graz and after 1929 at New York University and Queens College in New York City.

Fritz Pregl (1869–1930) was trained in medicine at Graz where, because of his interest in physiological chemistry, he was appointed to the physiological institute as an assistant to Rollet. He remained there the rest of his life, except for a year of study with Ostwald and Emil Fischer and from 1910 to 1913 when he was on the faculty at Innsbruck.

In his researches on bile acids Pregl obtained a very small amount of a compound for which he needed an analysis. Traditional methods required several tenths of a gram—and months more of tedious work. Pregl chose instead to refine the method so that a few milligrams would be sufficient. Following the pattern set by Emich, he scaled down apparatus and introduced innovations which made this possible. An expert glass blower and mechanic, he was able to fabricate the necessary apparatus himself.

The Hamburg instrument maker W. H. Kuhlman had produced a microbalance sensitive to 0.01 mg. Emich had already shown that satisfactory analyses could be made with such a balance. Encouraged by Pregl, Kuhlmann refined the balance, increasing its sensitivity tenfold while permitting a maximum load of twenty grams.

Over three years Pregl literally revolutionized organic analysis. Liebig's combustion method for carbon and hydrogen was shrunk down in size.

Fig. 21.4. FRITZ PREGL.
(Courtesy of the Edgar Fahs Smith Collection.)

The *kali apparat* was replaced by a small tube containing soda lime. Water was absorbed in a similar tube containing phosphorus pentoxide. A public demonstration of the method was made in Vienna before the Naturforscherversammelung in 1910. Pregl similarly reduced the Dumas' method for nitrogen, the Zeisel method for methoxyl groups, and the Carius' methods for sulfur and for halogens to micro scale. Apparatus was developed for micro melting-point determinations and for the measurement of freezing-point depressions and boiling-point elevations.

Pregl made it possible to carry out a complete analysis with only a few milligrams of material. Publication of the methods quickly brought them into use, not only for traces of material but also for abundant materials, since the saving in time, space, and reagents was a clear advantage. Pregl's laboratory in Graz was sought out by chemists who wished to gain experience with microanalysis. In 1916 Pregl responded to many demands by publishing a text, setting down in detail the methods he had perfected.

Others have contributed to the improvement of microchemical analysis, notably Joseph B. Niederl (b. 1899) of New York University, who was born in Graz and studied under Pregl. His teaching, textbook, and research have had an important role in the use of micromethods in America.

The determination of active hydrogen is now highly standardized through the development of the Grignard "machine," the procedure being based upon the liberation of methane when methyl magnesium iodide comes in contact with active hydrogen. The micromethod, developed to a high state of reliability by A. Soltys,[11] is based upon the macromethod introduced in 1902 and perfected by T. Zerewitinov during the following decade.

Spot Tests

Spot tests play an important role in qualitative microanalysis, organic reagents serving a wide range of usefulness for the recognition of both organic and inorganic compounds. Although generally ignored in formal instruction, they are nevertheless widely used in educational and industrial laboratories. Such tests have been introduced intermittently over the years; a few, such as the nutgall test for iron, go back to antiquity.

While Emich contributed significantly to the systematization of these spot tests, Fritz Feigl (1891–1971) was primarily responsible for making such tests convenient and reliable. Feigl received his doctorate in analytical chemistry under Wilhelm Schlenk at Vienna. During his doctoral study, Feigl neglected the phosphate problem which Schlenk assigned him and concerned himself with spot tests instead. He had become interested in Goppelsröder's work in this field and ended up with a thesis on the subject. He remained at Vienna until 1938 when the Nazi Anschluss caused him to take a research position in Ghent. Here he solved the problem of making the gas mask protective against arsine. Utilizing one of the reactions described in his book, he showed that arsine might be oxidized to an adsorbable form by using manganese dioxide on silica gel. When Germany invaded Belgium, Feigl escaped to Brazil where he was given a research position in the Ministry of Agriculture.

Biological Methods

The use of living organisms as analytical reagents rose to great prominence between 1900 and 1940 in connection with the nutrition studies which attracted the attention of biochemists during that period. With the recognition of the importance of vitamins and minerals in the diet, animal assays took on major importance. Until 1930 the chemical nature of the vitamins was unknown, and only a few empirical chemical tests, such as the antimony trichloride test for vitamin A and the ferric chloride test for vitamin E, had been developed. The most reliable assays were those using living animals. This was less true in the realm of minerals where chemical analyses were well developed, but even here animal assays were of greater reliability since the total amount of a particular mineral in the food is not always available to the animal.

Although several had carried out dietary experiments with animals by 1912, E. V. McCollum probably more than anyone, directed attention

[11] A. Soltys, *Mikrochemie*, **20**, 107 (1936).

to the value of the white rat as a test animal. The white rat had been used many times in experiments, but it was McCollum who recognized its real virtues. He had been associated with cattle-feeding experiments at Wisconsin where he came to recognize the value of a small animal's limited space requirements, low food consumption, rapid maturation period, short gestation period, short life span, ease of handling, and prolific birth rate. In addition the rat's omnivorous food habits made it ideal for studying the wide variety of foods used by human beings. It was subject to the major deficiency diseases except scurvy, and here the guinea pig proved a satisfactory substitute.

By using diets adequate in all nutritional components except for the vitamin or mineral component under assay, the animal soon shows symptoms of the deficiency disease. The test food is then supplemented to find the level at which a cure can be attained or the test food is added to the deficiency diet at several levels to find the level that is just protective.

Assay methods for drugs are also frequently carried out biologically —for example, in the estimation of penicillin by observing inhibition of growth of bacteria—especially during the early period in the development and introduction of a new drug. In some cases bioassay is superior to any chemical or physical procedure at all developmental stages.

Other microbiological assays were introduced around 1940 as an outgrowth of studies on the nutritional requirements of microorganisms. It was learned that some species of bacteria and molds cannot synthesize certain vitamins or amino acids and that therefore their growth is very sensitive to the level of such compounds in their nutrient medium. This was observed in various laboratories, but particularly in the fermentation laboratory of William H. Peterson (1880–1960) at Wisconsin. In Peterson's laboratory and later at Texas, Esmond E. Snell studied the nutritional requirements of *Lactobacillus casei* and introduced the use of the organism for the assay of riboflavin. The assay material was added at various levels to a nutrient solution containing all nutritional requirements except riboflavin. Since growth and lactic acid production are related to the amount of riboflavin available in the assay material, Snell could assay the riboflavin by measuring the turbidity caused by bacterial growth, or by titrating the lactic acid produced.

The same organism was found by Frank Strong to require pantothenic acid, and it could therefore be used for the microbiological assay of this vitamin. Strong utilized *L. arabinosus* for the bioassay of nicotinic acid. The mold *Phycomyces blakesleeanus* was used for the determination of thiamin, the weight of dried mycelium being proportional to the concentration of thiamin in the assay material.

The red bread mold *Neurospora crassa* proved a veritable treasure for bioassay methods. The natural mold grows well on nutrient medium containing only mineral salts, sugar, and biotin. Thus it can be used for biotin assays. In 1940 the geneticist George Wells Beadle (b. 1903) and the chemist Edward Tatum (1909–1975) at Stanford began irradiating the mold with x rays to produce mutations. They first obtained a mutant which no longer possessed the ability to synthesize pyridoxine (vitamin B_6). Further work of this sort led to a great variety of molds which had lost the capacity to synthesize a particular vitamin or amino acid. Such molds, of course, have obvious uses as assay organisms.

ISOTOPES IN ANALYTICAL CHEMISTRY

As early as 1913 G. Hevesy and F. A. Paneth in Germany used radium D (^{210}Pb) as an analytical aid in determining the solubility of lead salts. The limited number of radioactive elements available at that time strictly limited the further application of the technique. Since many isotopes are now available, the use of isotopes as tracers in analytical problems has become widespread. The modes of application fall into three categories: isotope dilution analysis, activation analysis, and isotope derivative analysis. Both stable and radioactive isotopes are used, the latter being preferred since they can be determined without a mass spectrograph.

Classical analytical methods are traditionally burdened by the need to isolate the sought-for-substance in high purity and high yield before the determinations can be completed by weighing, titration, or measurement of an appropriate physical property. The requirement of both high yield and high purity has been the stumbling block of quantitative analysis throughout its history. It is generally not too difficult to obtain a substance in a high state of purity if yield is of no consequence. Conversely high yields of a badly contaminated substance are usually not hard to obtain. Isotopic tracer techniques make it possible to concentrate on one objective without being too concerned about the other. Advantage is taken of the fact that the tracer isotope behaves chemically like the same element in the sample but is readily detectable because of its radioactivity.

Isotope Dilution Analysis

Isotope dilution analysis is particularly applicable to samples where the concentration of sought-for substance is sufficiently high for chemical measurement but the presence of interfering substances makes isolation in high yield difficult. A measured quantity of tracer isotope, in the form of a suitable compound, is added to the sample which is then processed so as to recover the sought-for-substance in high purity and measurable form. This product is then measured chemically and by counting. The count, compared to the count of the total added tracer, enables the analyst to calculate his chemical yield, after which the measurement of product recovered can be equated back to total quantity in the sample. Accurate analyses are possible even if 90 per cent of the sought-for-substance is lost in processing—truly the answer to the prayers of the sloppy chemist! The technique has been effectively applied in the analysis of organic mixtures where quantitative separation is virtually impossible, for example, vitamins, antibiotics, insecticides, herbicides, steroids.

Activation Analysis

Activation analysis is frequently used in analyses where the element to be determined is present in such low concentration that chemical isolation in uncontaminated measurable form is virtually impossible. The sample is irradiated with thermal neutrons in a reactor, and the active isotope can then be counted. The method is applicable to qualitative as well as quantitative analysis, since by identification of half lives and energies the presence of particular isotopes can be detected. Because the neutron irradiation generally activates more than a single element in the sample, it is usually necessary to isolate the element being analyzed. This can be done by a reverse isotopic dilution, adding a suitable compound containing the desired element in inactive form and processing to recover the element uncontaminated with other activated elements. Recovery need not be quantitative since the difference between amount of element added and amount recovered indicates the amount of activated isotope recovered. Activation analysis has been used for the determination of arsenic in sea water ($2 \ \mu g/l$), semiconductors, and biological materials; the analysis of meteorites for gold, gallium, palladium, and rhenium (0.1–0.01 ppm); the determination of impurities in high purity materials; the determination of hafnium in zirconium, of rare earths in rare earth mixtures (very difficult by ordinary chemical methods) and of trace elements in biological materials (for example, gold in arthritic tissue). About two-thirds of the elements can be determined at a level of one microgram or less. A few elements can be determined at a level below $10^{-4} \ \mu g$.

Isotope Derivative Analysis

Typical organic compounds do not lend themselves to activation analysis since carbon, hydrogen, nitrogen, and oxygen give radioisotopes whose half lives are too short for practical analysis. Molecules containing sulfur, halogen, or phosphorus are changed during activation, so even these elements do not lend themselves to such analysis. In such cases isotope derivative analysis is sometimes applicable. The sought-for-compound is converted to a suitable derivative, using a labeled reagent. After excess reagent is removed, a stable derivative is added as a carrier and the sample is processed to recover the derivative in pure form. The procedure has been used very effectively by Keston, Undenfriend, and coworkers in the analysis of protein hydrolysates. By preparing *pipsyl* derivatives, they have been successful in determining up to twelve amino acids in milligram samples, with some of the acids present at the μg level.

$$^{131}\text{I}\left\langle\bigcirc\right\rangle-\text{SO}_2\,\text{Cl} \;+\; \text{H}\overset{\text{H}}{\underset{}{\text{N}}}-\overset{\text{R}}{\underset{}{\text{CH}}}\cdot\text{COOH} \rightarrow$$

$$^{131}\text{I}\left\langle\bigcirc\right\rangle\text{SO}_2-\overset{\text{H}}{\underset{}{\text{N}}}-\overset{\text{R}}{\underset{}{\text{CH}}}\cdot\text{COOH} + \text{HCl}$$

ISOTOPES AND THE MEASUREMENT OF TIME

The study of mineral age by radioactive means was suggested in 1907 by Boltwood. He believed that the age of radioactive rocks might be

Fig. 21.5. WILLARD F. LIBBY.
(Courtesy of University of California in Los Angeles.)

estimated from the half life of uranium and the quantity of accumulated helium. An improved approach was developed when it was learned that the end product of the decay of uranium-238 is lead-206. The age of the oldest rocks has been found to be in the vicinity of 4.5×10^9 years.

The age of carbonaceous materials which lived within the last 50,000 years became datable with the recognition of the presence of carbon-14 in the atmosphere. It was suggested in 1937 by Aristid V. Grosse that radioactive isotopes might be formed during cosmic ray collisions with atoms. Willard F. Libby (1908–1979) of Chicago showed in 1946 that living matter contains a small and constant quantity of carbon-14, derived as a consequence of collision of cosmic ray neutrons with atmospheric nitrogen to give an (n, p) reaction. With a half life of 5,600 years there is ample opportunity for oxidation to carbon dioxide and incorporation into living organisms by photosynthesis. Libby and his associates showed the level of carbon-14 to be constant in living animals and plants but to fall off in proportion to age after death. The reliability of the technique was established by comparison of radiocarbon dates of archeological objects with dates established by other methods.

CONCLUSION

After three inauspicious decades analytical chemistry made noteworthy strides forward. Speed, sensitivity, and selectivity of analytical methods were all improved through greater application of theoretical principles, use of instruments, and the availability of radioactive isotopes as analytical aids. There was a marked trend toward the general use of micro methods and, in the field of radiochemistry, of ultramicro methods.

Chromatography proved to be a particularly important development. In its early period, the technique was used principally for separations, but in time it proved adaptable to qualitative and even quantitative analyses. The introduction of vapor phase chromatography in the 1950's made possible the rapid analysis of mixtures which had been analyzed with the greatest difficulty before. For example, the volatile compounds responsible for the flavor of fruits and vegetables had been known incompletely and sometimes even erroneously before gas chromatography came into use. Now it became possible to study the balance of the numerous compounds responsible for the flavor of onions and strawberries, for example. Studies of a great many natural substances and industrial mixtures became commonplace with the introduction of various types of chromatography.

CHAPTER 22

INORGANIC
CHEMISTRY II.
DECLINE AND RISE

From 1900 until the onset of World War II inorganic chemistry was stagnant. Compared with organic chemistry it lacked system; with physical chemistry it lacked rigor and logic. Many believed that inorganic chemists concerned themselves solely with the preparation of insignificant compounds, and the whole discipline appeared to be a dull study of unconnected facts. Such compilations as J. W. Mellor's *Comprehensive Treatise on Inorganic and Theoretical Chemistry*, and Gmelin-Kraut's *Handbuch der anorganischen Chemie* had the character of ponderous collections of information. While the periodic table was used as a unifying device, behavior of individual elements was sufficiently irregular to decrease its value as a generalizing tool.

Werner's contributions should have served as a rallying point for inorganic chemists, but they failed to have that influence during his lifetime. On the one hand, his contributions were so comprehensive that many erroneously believed that little remained to be done; on the other, the complex nature of coordination compounds caused others to despair that such compounds could be dealt with in a grand, systematic manner. This attitude was intensified by the problems created in explaining primary and secondary valences before electron concepts were available. The lack of a good theory of chemical bonding was a serious handicap to inorganic chemists. Organic chemists, dealing with a multitude of compounds containing only a limited number of elements, were quite successful in dealing with the structure of their molecules. Inorganic chemists, dealing with compounds containing a wide variety of elements, had very limited success in evolving a useful structural theory.

The introduction of x ray diffraction techniques for the study of crystal structure provided an important tool for inorganic chemists, but the large amount of work involved in mathematical analysis of the spectra limited application mostly to the simpler compounds. Only with the introduction of high-speed computers did it become practical to undertake extensive structural analysis of complex compounds.

The electronic concepts of chemical combination permitted great progress in work on inorganic combinations, particularly when applied in combination with the quantum mechanical principles of Pauling and others. For instance, magnetic susceptibilities had been measured by Faraday in the 1850's, but the theoretical treatment of the phenomenon was unsatisfactory until quantum mechanics was used to relate magnetic susceptibilities with the number of unpaired electrons.

A new interest in inorganic chemistry was apparent in the thirties, but it was the stimulus of wartime problems that really demonstrated the importance of this field of chemistry. The development of nuclear energy raised questions regarding properties of many of the elements. Answers were provided by the intensive studies made in connection with the Manhattan Project.

Since World War II, interest in inorganic chemistry in the universities and in industry has not abated. Before the war most chemists received no more instruction in inorganic chemistry than was included in the introductory college course in general chemistry; now advanced courses in the field are required. These courses tend to approach the subject from the viewpoint of structure (bonding, stereochemistry) and reactions (products, thermodynamics, kinetics).

In the 1930's researchers began to consider compounds representing rare and unknown oxidation states of many of the elements as well as traditional compounds. It was shown that oxidation states of one and zero are fairly common among transition elements. Coordination complexes with water, ammonia, and amine molecules and with chloride, cyanide, and thiocyanate ions as ligands were extensively studied by Werner and his followers. Later workers successfully prepared complexes with such ligands as ethylene, cyclopentadiene, and benzene, as in the metal sandwich compounds. Chelated combinations not only gave an insight into the nature of such natural substances as chlorophyll and hemoglobin, but also introduced a linkage which had practical significance in the softening of water and the removal of undesirable ions from industrial liquids.

The chemistry of various little-known elements began to receive considerable attention after 1900. Much work has been done on silicon, both in the inorganic state and in combination with organic radicals. Stock,

Schlesinger, and others concentrated on the hydrides of silicon and boron; fluorine and its compounds also have received extensive study. The availability of radioactive isotopes provided a real stimulus to inorganic research. Radioisotopes and stable isotopes such as deuterium made it possible to pursue the study of structure, equilibrium processes, and reaction mechanisms. (While a large part of radiochemistry is a part of inorganic, the major part of the subject has been given separate treatment in Chapters 18 and 19.)

NOMENCLATURE

The system of inorganic nomenclature introduced by Lavoisier and his associates and extended by Berzelius served chemistry well for more than one hundred years but began to prove inadequate as the twentieth century advanced. The multitude of coordination compounds, the multiplicity of oxidation states, and the clarification of the composition of many compounds all served to create a need for a more expansible and systematic nomenclature.

In 1902 Bohuslav Brauner proposed a system in which the valence of the element would be indicated by a specific suffix; *a* for 1, *o* for 2, *i* for 3, *e* for 4, *an* for 5, *on* for 6, *in* for 7 and *en* for 8. Thus $KAg(CN)_2$ became potassium dicyanoargentate, $Na_2S_2O_3$ became sodium thiotrioxosulfuronate, and K_2PtCl_6 became potassium hexachloroplatenate. There was little enthusiasm for Brauner's system, probably because it required a break from traditional nomenclature which had not yet proved too inadequate. Seven years later A. Rosenheim and I. Koppel suggested the use of Arabic numerals to indicate the number of atoms in a compound; thus Fe_3O_4 would be called 3-iron 4-oxide. The proposal was considered too cumbersome by Alfred E. Stock and others. After World War I there were various proposals for nomenclature reform; the most useful ideas were based upon Werner's earlier suggestions as subsequently developed by Stock.[1] Different groups sought to deal with the problem, but international action was not taken until 1938 when the International Union of Pure and Applied Chemistry established the Committee on the Reform of Inorganic Nomenclature, under the chairmanship of W. P. Jorissen of Leyden and with H. Bassett of Reading, A. Damians of Paris, F. Fichter of Basle, and H. Rémy of Hamburg as members. After meeting in Berlin and Rome, the committee released recommendations for the naming of inorganic compounds in 1940.[2]

---

[1] A. Stock, *Z. angew. Chem.*, **32**, 1, 373 (1919); **33**, 1, 79 (1920); *Angew. Chem.*, **47**, 568 (1934).

[2] Jorissen, *et al.*, *J. Chem. Soc.*, **1940**, 1404; *Ber.*, **A73**, 53 (1940); *J. Am. Chem. Soc.*, **63**, 889 (1941); J. Scott, *Chem. Revs.*, **32**, 73 (1943).

The rules were based very largely on recommendations presented by Rémy as the outgrowth of a study made by a committee for the Deutschen Chemischen Gesellschaft. This committee included Stock of Karlsruhe, Germany's leading inorganic chemist. Stock had long been interested in nomenclature and many of his suggestions were incorporated into the proposed rules.

Table 22.1. COMPARISON OF TRADITIONAL AND STOCK NOMENCLATURE

Formula	Traditional Name	Stock Name
FeO	Ferrous oxide	Iron(II) oxide
Fe_2O_3	Ferric oxide	Iron(III) oxide
Fe_3O_4	Magnetic oxide, black oxide	Iron(II, III) oxide
N_2O	Nitrous oxide	Nitrogen(I) oxide
NO_2	Nitrogen dioxide	Nitrogen(IV) oxide
N_2O_4	Nitrogen tetroxide	Dimer of nitrogen(IV) oxide
N_2O_5	Nitrogen pentoxide	Nitrogen(V) oxide
$Fe(CN)_6^{-3}$	Ferricyanide ion	Hexacyanoferrate(III) ion
$Fe(CN)_6^{-4}$	Ferrocyanide ion	Hexacyanoferrate(II) ion

There was no formal activity on nomenclature during the war. In 1948 work was resumed. Stock's proposals had been worked with for years and, except for certain cases requiring revision and extension, were found to be sound. W. Conrad Fernelius and his associates[3] submitted extensive proposals in 1948, some of which have come into general use. The Fernelius group was concerned primarily with coordination compounds, as were Ewens and Bassett.[4] Although the problem has received continuous study, many decisions remain to be made. The large amount of research on the chemistry of silicones and boron and fluorine compounds creates problems transcending the inorganic and organic fields.

FILLING THE PERIODIC TABLE

At the beginning of the century the periodic table was generally accepted as a useful tool, but it suffered from obvious vacant spaces as well as un-certainties of a more fundamental sort. The rare earth mess was still unresolved; no one knew how many rare earths there should be. The relation of the transition elements to the main groups was not entirely

---

[3] W. C. Fernelius, *Chem. Engr. News*, **26**, 161 (1948); Fernelius, E. M. Larsen, L. E. Marchi, and C. L. Rollinson, *ibid.*, p. 540; Fernelius, *Advances in Chemistry Series* [8], 9 (1953). Also see J. C. Bailar, Jr., *Coordination Compounds*, Reinhold, New York, 1956, and T. Moeller, *Inorganic Chemistry*, Wiley, New York, 1952.

[4] R. V. G. Ewens and H. Bassett, *Chemistry and Industry*, **1949**, 131.

settled. And the rapid discovery of radioactive elements made the correctness of the last part of the table open to serious doubt.

With Moseley's work on atomic numbers, the problem of the total number of elements and the number of rare earths approached resolution. His research covered a great many of the elements between aluminum (13) and gold (79). In Bohr's periodic table of 1920, uranium was listed as element 92. Positions 43, 61, 72, 75, 85, and 87 remained unfilled. Moseley's study of the x ray spectra of the rare earths revealed a missing element between neodymium (60) and samarium (62) and another following lutetium (71). His work served to renew the search for the missing elements, both by pointing out where they occurred in the periodic table and by providing a new means for identification.

Hafnium

Soon after he learned of Moseley's research, Georges Urbain took some of his rare earth preparations to Oxford for examination. Moseley confirmed the results of Urbain's many years of work in a few days by identifying the lines produced by erbium, thulium, ytterbium, and lutetium. At the same time he showed that the material Urbain believed to be a new element (*celtium*) on the basis of the arc spectrum, showed x ray spectrum lines corresponding only to ytterbium and lutetium.

After World War I ended Urbain resumed the search for element 72 in the rare earth residues he and Moseley had examined earlier. Using an improved method of x ray analysis developed by de Broglie, he believed that he observed two faint lines which nearly coincided with those predicted for element 72. However he was unable to achieve further concentration—because he was looking for the element in rare earth residues, an unlikely place! Meanwhile a more successful search was going on elsewhere.

Arguing from his quantum theory, Bohr supposed that element 72 must be quadrivalent rather than trivalent and therefore would not be found among the rare earths. Hence he refused to accept Urbain's claims for celtium, insisting that the rare earth series should end with element 71 and that number 72 should be related to zirconium. Actually, it had long been suspected that zirconium was not a homogeneous element, and on several occasions claims had been made for the discovery of a new element associated with zirconium. None of these claims had proved reliable. Bohr now advised his Hungarian associate Georg von Hevesy (1885–1966) to examine zirconium ores for the element.

The search proved successful. Early in 1923 Hevesy and Dirk Coster, professor of physics and meteorology at Copenhagen, announced the discovery of the element with the aid of the x ray spectrum, a research technique in which Coster specialized. The name *hafnium* was given to the new element, after Copenhagen.

Fig. 22.1. GEORGES URBAIN.
(From H. M. Smith, *Torchbearers of Chemistry*, with permission of Academic Press, Inc.)

Fig. 22.2. GEORG VON HEVESY.
(Courtesy of *J. Chemical Education.*)

Rhenium and Technetium

Elements 43 and 75 were announced in 1925 by Walter Noddack (1893–1960), Ida Tacke (b. 1896), and Otto Berg, an x ray spectroscopist in the laboratory of the Werner-Siemens company. Noddack and Fraulein Tacke, who later became Frau Noddack, were in the Physico-Chemical Testing Laboratory in Berlin. They began searching for elements 43 and 75 in 1922. First they surveyed the elements, observing that those of odd atomic number occur less commonly. Then they examined platinum ores which commonly contain the elements between atomic numbers 24 to 29, 44 to 47, and 76 to 79 and columbite ores which commonly contain elements 39 to 42 and 72 to 74.

Since elements 43 and 75 occur under manganese in the periodic table, their properties, it was believed, should resemble manganese. Before their discovery they were commonly referred to as eka-manganese and dwimanganese. Noddock and Tacke searched for such elements and finally, after a 100,000-fold concentration of a gadolinite fraction, obtained a material which showed x ray spectral lines corresponding to element 75. They called the new element *rhenium*, naming it for the Rhineland.

Noddack, Tacke, and Berg also advanced claims in 1925 for faint x ray lines corresponding to element 43 in their rhenium concentrate. They proposed the name *masurium*, after the Masurenland in East Prussia.

There was a strong dose of nationalism in the names selected by Noddack; the Rhine and the Masurian Marshes represented barriers to the West and East of Germany never crossed by the Allied armies during World War I.

The claims for element 43 were not substantiated. The Noddack team never succeeded in concentrating the element, and, although the symbol Ma was to be found in periodic tables of the thirties, there was less and less confidence that it really belonged there.

In 1936 Emilio Segrè of Palermo reopened the subject of the forty-third element, this time from a radiochemical standpoint. He and C. Perrier obtained from E. O. Lawrence a sample of molybdenum which had been bombarded with deuterons in the California cyclotron for several months. Segrè and Perrier found that the radioactivity was not associated with niobium, zirconium, or molybdenum, but that it was associated with manganese and rhenium carriers. Separation from the carriers was achieved by volatilization in a stream of hydrogen chloride gas. The amount of element 43 used in their experiments was unweighable, being of the order of 10^{-10} grams. In 1940 Segrè and C. S. Wu[5] encountered element 43 among uranium fission products. After F. A. Paneth suggested in 1947 that the first producer of an artificial element be entitled to the same naming privileges as the discoverer of a natural element, Perrier and Segrè introduced the name *technetium* (*artificial*), since it was the first element, previously unknown on earth, to be prepared artificially.

Elements 85 and 87

The search for elements 85 and 87 was an active one, particularly after Moseley's work. Wide publicity was given Allison's claims in 1931 for the discovery of both elements. Fred Allison (1882–1974), who was a professor of physics at Alabama Polytechnic Institute, developed a magneto-optical method for the analysis of chemical elements. He and one of his students believed that with this device they had obtained evidence of intensities corresponding to the missing elements, which Allison named *virginium* (87) and *alabamine* (85) after the states of his birth and his career. Other investigators were unable to confirm his results, even when using his instrument. Other claims for discovery proved equally impossible to verify, for example, russium, alcalinium, and moldavium for element 87. The conviction gradually grew that these elements would be radioactive and therefore not naturally occurring, unless they were part of some element's decay series.

In 1939 Mlle. Marguerite Perey of the Curie Institute encountered evidence for element 87 among the decay products of actinium-227. The β-decay of this element had long been known. Perey found that, to a slight extent (1%), it also undergoes α-decay, forming the mass-223 isotope of

---

[5] E. Segrè and C. S. Wu, *Phys. Revs.*, **57**, 552 (1940). Also see Segrè, *Nature*, **143**, 460 (1939).

element 87. She first referred to this decay product as actinium K, and was able to follow up with purification and study of the element, which confirmed its expected behavior as an alkali metal. The name *francium* was proposed by Perey in 1946.

The first isotope of element 85 was discovered in 1940 by D. R. Corson, K. R. Mackenzie, and Segrè at California where they used the cyclotron to bombard bismuth with α-particles. They were able to isolate an isotope of 7.5 hour activity which could be separated from all adjacent elements and showed metallic properties. Further investigation was deferred until after the war when they studied the chemistry of the isotope by refined tracer techniques on account of the very short half life. The element fits into the halogen group but shows considerable metallic character; this is consistent with the increase in electropositive character as one moves down group VII in the periodic table. It is less volatile than iodine but shows a strong affinity for silver and, like iodine, tends to concentrate in the thyroid gland of animals. The discoverers proposed the name *astatine* (*astatos*, unstable) since it is the only halogen without a stable isotope.

Promethium

Arguing from the discontinuity of properties between neodynium and samarium, Brauner predicted in 1902 that there should be an element between them. This prediction was validated by Moseley, who showed that there is a discontinuity of x ray lines at position 61 in the rare earth series. After World War I ended, chemists in several laboratories undertook a careful search for the missing element. It was felt that monazite sand, which contains the two adjoining elements, should also yield number 61.

Luigi Rolla and Rita Brunetti of Florence deposited a sealed package with the Reale Accademia dei Lincei in 1924, claiming discovery of the element which they named *florentium*. In 1926 Charles James (1880–1928) and H. C. Fogg at New Hampshire prepared a concentrate which, when examined in the x ray laboratory of J. M. Cork at Michigan, appeared to show evidence of the element. Shortly before this B. Smith Hopkins, J. Allen Harris and L. F. Yntema at Illinois announced discovery of the element in rare earth residues furnished by the Lindsay Light Co. and the Welsbach Mantle Co. Hopkins proposed the name *illinium* for the element. Rolla at this time revealed the content of the sealed package.

Rolla, James, and Hopkins had all done a great deal of good work on the chemistry of the rare earths, and all of them had the cooperation of competent x ray spectroscopists. There was little reason to question their results, but the laboratories of W. Prandtl and of the Noddacks were unable to confirm the findings.

When the Ohio State cyclotron began operating in 1938, L. L. Quill and his associates bombarded neodymium and samarium with various pro-

jectiles. They obtained a number of activities, some of which were believed
due to element 61 for which they proposed the name *cyclonium*. Their results were based on limited chemical separations and not all of their
conclusions were correct, although it is highly probable that their materials
contained isotopes of the element.

A conclusive separation and identification was made later by J. A.
Marinsky and L. E. Glendenin when they were working with Charles D.

Fig. 22.3.   LEFT TO RIGHT, JACK A. MARINSKY, LAWRENCE
E. GLENDENIN, HAROLD G. RICHTER, AND CHARLES D.
CORYELL.
(Courtesy of C. D. Coryell.)

Coryell's group on fission product identification during World War II.
They precipitated cerium from the rare earth fraction as the iodate,
removed yttrium, samarium, and europium by digestion with carbonate
solution, and passed the residue through an ion exchange column containing an Amberlite resin. Praseodymium, neodymium, element 61, and
yttrium were adsorbed at different levels. Elution with ammonium citrate
at proper pH gave fractions whose activity could be studied. Element 61
was associated with a β-emitter of a 3.7-year half life. A similar half life

was associated with the eleven-day β-decay of $^{147}_{60}\text{Nd}$. The identification of $^{147}61$ was verified by the mass spectrograph. Other isotopes have been prepared, one of them with a 25-year half life. The name *prometheum*, proposed by Marinsky and Glendenin, is based upon the myth in which Prometheus, the Titan, stole fire from the gods for the benefit of man. The name was suggested by Grace Mary Coryell and was adopted by the International Union of Chemistry in 1949, with the spelling *promethium*. The International Union thus recognized the claim of Marinsky and Glendenin to priority in the identification of the element. The claims of the twenties can hardly have been valid since none of these investigators was working with radioactive materials and since no stable isotopes have ever been encountered.

Transuranium Elements

The unquestioned recognition of the transuranium elements began in 1940 with the discovery of element 93 by McMillan and Abelson and of element 94 by Seaborg and his associates in 1941. These discoveries have been treated in Chapter 19 and will receive no further discussion here beyond noting that plutonium was shown to be present in natural uranium by Seaborg and his associates[6] in 1942. Since this plutonium has been shown to be ^{239}Pu, it is presumably a product of neutron reaction with ^{238}U in mineral deposits; the half life is too short for it to have been a primordial element. The elements from number 95 on have not been considered up to now and deserve mention.

With the discovery of elements 93 and 94 it was realized that the prospect of synthesizing still other transuranium elements was good. Transmutation methods were well established and techniques for identification of isotopes had reached an advanced stage of effectiveness. That transuranium elements are related to the rare earths was to be expected, since $5f$ electrons should begin to take on importance at this point, but had been generally overlooked for many years. Chemists were inclined to consider thorium, protactinium, and uranium as congeners of hafnium, tantalum, and tungsten, rather than of the rare earths. The discovery of protactinium may have been delayed by the belief that it would closely resemble niobium and tantalum, and the tendency to relate uranium to chromium, molybdenum, and tungsten was responsible in part for errors that handicapped the atomic energy program in its early days. Bohr was one of the few who recognized a probable relationship between the elements beyond radium and a second rare earth series. In mid-1944 Seaborg began working on the principle that new elements in the transuranium series should behave like rare earths. This principle was important because

[6] C. A. Levine and G. T. Seaborg, *J. Am. Chem. Soc.*, **73**, 3278 (1951); Seaborg and M. L. Perlman, *ibid.*, **70**, 1571 (1948); C. S. Garner, N. A. Bonner and Seaborg, *ibid.*, **70**, 3453 (1948).

thorium, protactinium, and uranium show just sufficient analogy to the transition elements to obscure the "rare earth" character. With elements like number 95 and beyond, the rare earth character becomes clean-cut.

In 1944 Seaborg, R. A. James, and A. Ghiorso obtained evidence for element 96 after bombarding plutonium-239 with helium ions in the Berkeley cyclotron. Late in the year neutron bombardment of ^{239}Pu produced evidence for element 95. Work on this element was also carried on at the Metallurgical Laboratory by Seaborg's group which included James,

Fig. 22.4. GLENN T. SEABORG.
(Courtesy of F. Daniels.)

L. O. Morgan, and Ghiorso. B. B. Cunningham prepared a pure compound of element 95 late in 1945, but it was two more years before L. B. Werner and I. Perlman were able to do this with element 96. By analogy with the rare earth *europium*, element 95 was named *americium* for the Americas. And by analogy with *gadolinium*, which was named for Gadolin, the pioneer in rare earth studies, element 96 was named *curium* for the Curies, pioneers in the study of radioactive elements.[7]

If theory is correct, as it ultimately appeared to be, with each increase

[7] K. Street, Jr. and Seaborg, *J. Am. Chem. Soc.*, **72**, 2790 (1950); Street, U.S. Patent 2,711,362. R. A. Glass, *J. Am. Chem. Soc.*, **77**, 807 (1955); G. R. Choppin, B. G. Harvey, and S. G. Thompson, *J. Inorg. Nuclear Chem.*, **2**, 66 (1956).

in atomic number the added electron will go into the $5f$ subshell and therefore will not greatly alter the properties of successive transcurium elements. Since separation of such elements is exceedingly difficult when conventional chemical separations are used, it was fortunate that ion exchange techniques were developed before these studies began.

In 1949 S. G. Thompson, A. Ghiorso, and Seaborg bombarded ^{241}Am with 35 Mev helium ions in the cyclotron to produce the first isotope of element 97 by the (α, 2n) reaction. Since the element is the homologue of *terbium*, named after the town of Ytterby which provided the source of rare earths, it was named *berkelium* after Berkeley, California, where the cyclotron provided the source of new actinide elements.

Element 98 was discovered shortly after berkelium by Thompson, Street, Ghiorso, and Seaborg who bombarded curium-242 with 35 Mev helium ions in the cyclotron to produce an isotope which decayed by orbital electron capture with a half life of 45 minutes. It was found to be the isotope of mass 245 and also to decay to the extent of 34 per cent by α-emission. The name *californium* was chosen in honor of the university and the state. There is no analogy to the name of the rare earth homologue dysprosium (hard to come by).

Research groups at California, the Argonne Laboratory, and the Los Alamos Laboratory identified elements 99 and 100 in heavy-element samples from the "Mike" thermonuclear explosion in November 1952. The results remained classified, however, until 1955 when isotopes were prepared and studied at California and by the Argonne Laboratory group working at Arco, Idaho.[8] Element 99 was named *einsteinium* in honor of the noted physicist and humanitarian, and 100 became *fermium* in honor of the father of the nuclear age.

The large accelerators built in the late forties and early fifties have made it possible to use heavier ions as projectiles. While helium ions continued to be used, ions of beryllium, carbon, nitrogen, oxygen, and even fluorine can be accelerated. Atomic number and mass hence can be jumped by a significant increment. For example $^{250}_{100}$Fm was prepared from $^{238}_{92}$U by the ($^{16}_{8}$O, 4n) reaction, $^{246}_{99}$Es from $^{238}_{92}$U by the ($^{14}_{7}$N, 6n) reaction, and $^{255}_{100}$Fm by the ($^{9}_{4}$Be, α 2n) reaction on $^{252}_{98}$Cf. The high neutron flux of nuclear reactors which produces many isotopes of high atomic number because of multiple neutron capture, also is utilized.

The identification of seventeen atoms of element 101 was announced in

[8] The work on heavy element samples from the 1952 explosion was reported by A. Ghiorso, S. G. Thompson, G. H. Higgins, G. T. Seaborg (California); M. H. Studier, P. R. Fields, S. M. Fried, H. Diamond, J. F. Mech, G. L. Pyle, J. R. Huizenga, A. Hirsch, W. M. Manning (Argonne); C. I. Browne, H. L. Smith, and R. W. Spence (Los Alamos), *Phys. Rev.*, **99**, 1048 (1955). Reports of other work of the California group are in *ibid.*, **93**, 908, 1129 and **94**, 1080 (1954), of the Arco group in *ibid.*, **93**, 1428 (1954). Also see *Chem. Engr. News*, **33**, 1956 (1955).

1955 by Ghiorso, Harvey, Choppin, Thompson, and Seaborg.[9] They bombarded einsteinium-253 with 41 Mev helium ions and detected activity on an ion exchange column at the position which should belong to element-101. The name *mendelevium*, honoring the Russian founder of the periodic system, was introduced for the element, which was positively identified in a series of experiments.

In the spring of 1957 claims were advanced for the production of an isotope of element 102 at the Nobel Institute of Physics in Stockholm. Curium-244 was bombarded with 90 Mev $^{13}C^{+4}$ ions to give what was supposed to be isotope 251 or 252, decaying by α-emission with a half life of about ten minutes. The work was done by an international group made up of P. R. Fields and A. M. Friedman of the Argonne Laboratory, J. Milsted and A. B. Beadle of the Atomic Energy Research Establishment at Harwell, England, and H. Atterling, W. Forsling, L. W. Holm, and B. Åstrom of the Nobel Institute. The name *nobelium* was proposed, in honor of Alfred Nobel and his contributions to the advancement of learning and of peace. Workers at the California Radiation Laboratory failed to confirm the findings. In April 1958 Ghiorso, T. Sikkeland, J. R. Walton, and Seaborg bombarded curium-246 with carbon-12 ions and reported the identification of isotope 254 of element 102. During 1957 and 1958 G. N. Flerov and his colleagues in Russia produced an isotope of element 102 by bombarding plutonium with oxygen ions.

In the spring of 1961 Ghiorso, Sikkeland, Almon E. Larsh, and Robert M. Latimer announced the preparation of element 103 by bombardment of californium with boron nuclei. The name *lawrentium* was proposed in honor of Ernest O. Lawrence.

It is possible that as this is being revised (September 1963) still additional elements are being produced. The work becomes difficult as the atomic number goes up since the instability of the nuclei formed leads to extremely short half lives. The detection of such isotopes poses serious problems since stable and semistable starting materials become less and less available with increasing atomic number. However the more powerful particle accelerators are now capable of hurling ions up to neon in mass, and improvements are still foreseeable. Atomic-number jumps of considerable magnitude may be successful.

The preparation of several more elements will certainly be pursued with enthusiasm. With element 103 the $5f$ subshell will be filled with electrons. This will terminate the actinide, or second rare earth, series. Elements 104, 105, and 106 should represent eka-hafnium, eka-tantalum, and eka-tungsten since the next electrons presumably will enter the $6d$ subshell. Naturally, chemists are eager to show that the supposed correspondence is real—or if it is not real, to find out why.

[9] A. Ghiorso, B. G. G. Harvey, G. R. Choppin and G. T. Seaborg, *Phys. Rev.*, **98**, 1518 (1955).

COORDINATION COMPOUNDS

While Werner's ideas on coordination compounds were the first to fit into the experimental observations, his notions of primary and secondary valences were not well received during his lifetime, to a large extent because there appeared to be no reasonable system that would account for the secondary valences. With the rise of the electronic theory of valence this obstacle was overcome. Following the treatment given the subject by G. N. Lewis, W. Kossel, N. V. Sidgwick, and others, there was rapid acceptance of Werner's theories. And confirmatory evidence was supplied by x ray diffraction data.

In 1920 Kossel postulated an electrostatic model for coordinated ions whereby it became possible to correlate stability with the ratio of charge to size of the central ion. Complexes were supposedly held together by the electrostatic attraction of charged ions on dipolar molecules. Ions with a high charge and small radius would form highly stable complexes. It already had been observed by De that metals with the greatest coordinating ability (Cr, Fe, Co, Ni, Cu, Ru, Rh, Pd, Os, Ir, Pt, Au) have small atomic volumes. These ideas had been vaguely anticipated as early as 1899 by Richard Abegg and Guido Bodlander.

It was readily apparent to students of the subject that Kossel's model represented an oversimplification. Fajans, in particular, developed ideas about ion deformation and interpenetration of electron clouds. His suggestion that high colors of coordination complexes result from the deformation of electron clouds was amplified in 1941 by Kenneth Pitzer and Joel Hildebrand.

Of the various contributions by M. L. Huggins, T. M. Lowry, Main-Smith, D. M. Bose, Butler, and Sidgwick, the concept of the *Effective Atomic Number* (E.A.N.) was the most useful. Sidgwick, who was primarily responsible for the concept, pointed out that the total number of electrons controlled by the central atom of a complex, including those shared in coordination, is frequently equal to the atomic number of the next inert gas in the periodic table. For example, in $PtCl_6^=$ the Pt^{+4} ion contributes 74 electrons; twelve more are introduced by the six coordinated Cl^- ions, making a total of 86—equivalent to the electrons in the radon atom. An E.A.N. of 36 (krypton) is observed in $Fe(CO)_5$, $Fe(CN)_6^{-4}$, $Co(NH_3)_6^{+3}$, and $Ni(CO)_4$; of 54 (xenon) in $RhCl_6^{+3}$, and $Pd(NH_3)_6^{+4}$. The concept was useful in work on certain complexes, particularly the carbonyls and nitrosyls, but did not explain many other stable complexes. For example, $Cr(NH_3)_6^{+3}$ has an E.A.N. of 33, $Fe(CN)_6^{-3}$ has 35, $Co(NO_2)_6^{-4}$ has 37, $PdCl_4^{-2}$ has 52, and $Pt(NH_3)_4^{+2}$ has 84, yet all of these are stable complexes.

With the introduction of quantum mechanics, a new level was reached in the explanation of coordination compounds. It is supposed that in the formation of complex ions, electron pairs from a ligand such as ammonia serve to occupy and fill available orbitals at particular sublevels. For example, according to Pauling[10] the electron distribution in cobalt and its ions may be explained as follows (in all cases the earlier subshells contain $1s^2$, $2s^2$, $2p^6$, $3s^2$, $3p^6$ electrons):

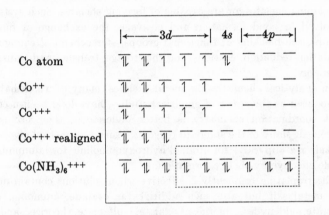

The complex ion can form as the result of the pairing of the four unpaired electrons in $3d$ orbitals, thus emptying two of the $3d$ orbitals. These, plus the $4s$ and the three $4p$ orbitals, are now available for the twelve electrons introduced with the six ammonia molecules. These six orbitals are now spoken of as d^3sp^4 hybrid orbitals.

Coordination concepts have proved valuable in work on a variety of inorganic phenomena such as the stabilization of unusual oxidation states, acid-base theory, the nature of amphoteric hydroxides and hydrated oxides, the structure of the polyacids, analytical implications of metal complexes, and the industrial use of complexing agents.

Chelation

The term *chelate* was introduced in 1920 by Morgan and Drew for coordination compounds whose central metal ion is attached to an organic molecule by two or more bonds. The term is derived from the Greek word *chela* referring to the "claw" of a crab or lobster. The stability of chelates is greater than compounds coordinated by single linkage, for example, $Ni(NH_2CH_2CH_2NH_2)_3$ is more stable than $Ni(NH_2CH_3)_6$. Beta-diketones such as acetylacetone, $CH_3 \cdot CO \cdot CH_2 \cdot CO \cdot CH_3(acac)$, enolize to form very stable six-membered rings with metal ions. Schwarzenbach[11] showed that

[10] L. Pauling, *J. Am. Chem. Soc.*, **53**, 1386 (1931).
[11] G. Schwarzenbach, *Helv. chim. acta*, **35**, 2344 (1952).

Be(acec)$_2$ distills at 270° C. and Al(acac)$_3$ at 314° C., both without decomposition. Such chelate rings have stabilities of the same order as aromatic rings. Three-ring chelate systems which decompose only at red heat also have been prepared.

Such compounds have great significance in biology, chlorophyll and hemin being porphyrin-ring chelates with magnesium and iron, respectively. Many of the enzymes involved in biochemical reactions are metal-containing chelates, or are capable of forming chelates. Such systems are involved in bond formation and cleavage, the exchange of functional groups, the blocking of functional groups, stereochemical configuration, oxidation-reduction reactions, and the storage, transfer, and transmission of energy.

In analytical chemistry the metal chelates, many of which have been introduced as the result of empirical studies, have long been used. Now that coordination chemistry is better understood, their use is being extended. Many of the standard reagents—such as dimethyl glyoxime for nickel, α-α'-dipyridyl for iron, 8-hydroxyquinoline for aluminum—are made effective by chelate formation.

Electroplating frequently is effective when solutions contain not only the metal salt but also such additives as cyanide, ammonia, citrate, tartrate, aldehydes, fluorides, oxalate, sulfamate, licorice, and other puzzling substances. The effectiveness of these additives is interpretable only in the light of coordination chemistry.

Mordants in the form of metal salts have been used in dyeing since antiquity. In 1908 Werner called attention to the role of coordination in their effectiveness. The chemistry of their action has been extensively developed, particularly as the result of the work of Gilbert T. Morgan and his colleagues at Birmingham.

In the field of water softening, chelating agents have assumed an important place during the past thirty years. The term *sequestration* was introduced for tying up magnesium and calcium ions into a soluble complex of such stability that precipitation does not subsequently take place while the water is being used. The polyphosphates, both in the form of chains and rings (metaphosphates) have proven highly effective. The other principal types of sequestering agents are the polyamine acids (represented by triglycine) and ethylenediaminetetraacetic acid (Versene, Sequestrine, Nullapon).

Clathrates

Clathrates, sometimes referred to as *molecular* compounds, are frequently completely organic in character, but will be discussed here since they represent coordination types. The clathrates are compounds in which formal bonding between certain components is missing; one component becomes wrapped or encased in another. Three general types are known.

In the first, a packing phenomenon occurs in which molecules of two compounds make up a crystal lattice in a certain ratio, as is the case with β-naphthylamine and 1,3,5-trinitrobenzene.

In the second type, one of the components forms a channel in which molecules of the second substance become enclosed. Typical examples are the choleic acids and the urea adducts. The choleic acids, discovered by Heinrich Reinboldt in 1927, are formed by the combination of certain bile acids (desoxycholic is the best) with fatty acids, esters, long chain hydrocarbons, or similar compounds. The number of desoxycholic acid molecules coordinated by a fatty acid molecule is approximately the same as the number of groups around a metal ion, for example, fatty acids with four to eight carbon atoms coordinate four molecules of desoxycholic acid. In the urea and thiourea adducts, first prepared in 1949 by W. Schlenk in Germany, a similar channel is formed by the urea, with the fatty acid lying within the channel.

In the third type, one component is constant and forms a cage in which molecules or atoms of proper size may be entrapped. The trapped units,

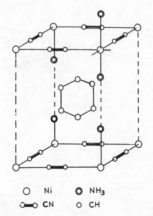

○ Ni   ○ NH₃
○—○ CN   ○ CH

Fig. 22.5. Example of a cage compound containing benzene trapped within a lattice of nickel, ammonia, and cyanide groups.
(From J. C. Bailar, Jr. (Ed.), *Chemistry of Coordination Compounds* [1956], with permission of Reinhold Press.)

which show no chemical relationships, are merely small enough to fit into the interstices of the cage, but large enough so that they can not escape. For example, hydroquinone forms a cage which encloses methanol but not water or ethanol. Molecular compounds of hydroquinone have been prepared with H_2S, SO_2, HCl, HBr, HCN, CO_2, C_2H_2, and even the rare gases argon, krypton, and xenon. Nickel cyanide was found to form a clathrate with ammonia and either benzene or thiophene. Such cage compounds were extensively studied by x ray analysis in the laboratory of H. M. Powell[12] at Oxford.

π-*Complexes*

It has long been known that unsaturated hydrocarbons such as ethylene form complexes with the halides of some of the platinum metals and with copper, silver, and mercury. Even a few cases of complexes of benzene and its homologues—for example, Menschutkin's complexes of benzene with the antimony halides—were known. The nature of the bonding in these complexes remained in dispute until the discovery of dicyclopentadienyl iron [ferrocene, $Fe(C_5H_5)_2$] in the early fifties[13] stimulated widespread interest in these compounds. Investigation of properties and structure, particularly in the laboratories of G. Wilkinson at Harvard and E. O. Fischer at Munich, established that conventional bonding explanations

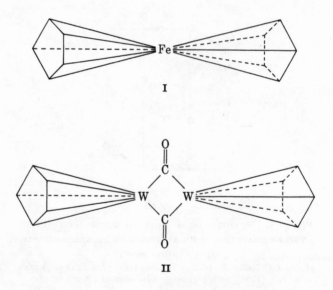

[12] H. M. Powell, *et al.*, *J. Chem. Soc.*, **1947**, 208; **1948**, 61, 571, 817; **1950**, 298, 300, 468; **1952**, 319; *Endeavour*, **9**, 159 (1950).

[13] The discovery is announced independently by T. J. Kealy and P. L. Pauson, *Nature*, **168**, 1039 (1951) and S. A. Miller, J. A. Tebboth, and J. F. Tremaine, *J. Chem. Soc.*, **1952**, 632. On structure see G. Wilkinson, M. Rosenblum, M. C. Whiting, and R. B. Woodward, *J. Am. Chem. Soc.*, **74**, 2125 (1952) and E. O. Fischer, *Z. Naturforsch.*, **96**, 619 (1954).

were untenable. A "sandwich," or "double cone," structure, in which the iron atom lay between the two cyclopentadiene ions (I) was proposed in 1952. Utilizing the concept of π-orbitals being developed by organic chemists in connection with double bonds, Wilkinson and Fischer reasoned that the iron was bonded to the cyclopentadiene molecules by two quintets of π-electrons, one quintet from each hydrocarbon.

Wilkinson's group prepared similar compounds of ruthenium, cobalt, nickel, molybdenum, tungsten, titanium, vanadium, and chromium. Related tungsten and molybdenum complexes with cyclopentadiene and carbon monoxide were interpreted as having a "doubledecker" structure (II). (See p. 602.)

Addition compounds of olefins with metal ions, for example, ethylene with silver perchlorate, are now considered complex ions in which the metal sits on the two π-electron orbitals of the double bond rather than being bonded to a particular carbon atom.

OPTICAL ACTIVITY

As the study of optical activity developed, there appeared to be no reason to suspect that optical activity need be restricted to carbon compounds and asymmetric carbon atoms, provided that a structural asymmetry is present in the molecule. Early efforts of Le Bel and others to prepare optically active compounds with an asymmetric nitrogen atom were unsuccessful. Finally, in 1891, Le Bel exposed a solution of methyl ethyl isopropyl isobutyl ammonium chloride to the action of the mold *Penicillium glaucum*. The specific rotation of the solution ultimately dropped to -7, indicating destruction of a *dextro* isomer. In 1899 William J. Pope and S. J. Peachey prepared racemic phenyl benzyl methyl allyl ammonium salts and succeeded in resolving the isomers. After this, numerous quaternary ammonium salts were resolved. In 1925 a spirane type compound was resolved, showing the tetrahedral nature of the nitrogen atom. Optically active amine oxides were prepared by J. Meisenheimer.

Optically active sulfur compounds were first encountered in 1900 when successful preparations were made by Pope and Peachey and independently by J. Smiles. In these compounds the sulfur carries three constituents; it was recognized later that the unshared electron pair serves the role of a fourth constituent. Active compounds of the sulfur type were also prepared with selenium, tellurium, tin, and germanium. Active phosphorus compounds were prepared and resolved by Meisenheimer and by F. S. Kipping. Other elements found to form active compounds include boron, beryllium, palladium, and platinum.

The preparation of optically active isomers of coordination complexes has been examined in Chapter 14. The studies of Werner, Pfeiffer, Jaeger, and others showed that the central atom has an octahedral coordinating power which, with bivalent coordinating groups, may give rise to mirror image structures. Various compounds of aluminum, arsenic, cobalt, chromium, iron, platinum, rhodium, iridium, and zinc have been resolved.

Early in the study of complex ions it was recognized that certain four-coordinated metals might have a planar structure as have bivalent nickel, platinum, and palladium. Werner proposed a planar configuration for such compounds and predicted *cis-trans* isomerism. This was difficult to establish; finally in 1922 x ray analysis showed the metal atom to be at the center of a plane surrounded by four chloride ions in such compounds as K_2PdCl_4, $(NH_4)_2PdCl_4$, and K_2PtCl_4. A decade later Pauling developed the probable distribution of groups in such planar compounds from wave mechanical considerations. Samuel Sugden (1892–1950) confirmed the prediction of *cis-trans* isomerism in planar molecules when he isolated geometric isomers of nickel bis-benzylmethylglyoxime. The compounds proved diamagnetic, whereas a tetrahedral arrangement around the nickel atom should have led to paramagnetic optical isomers. Corresponding compounds of palladium were prepared by F. P. Dwyer and D. P. Mellor. Evidence for the planar configuration of four-covalent platinum was obtained when William H. Mills and Thomas Quibell prepared *meso*-stilbenediamino isobutylenediamino platinum(II) salts. Since the substance could be resolved into isomers with opposite rotations, the evidence for planar rather than tetrahedral structure was complete.

SILICONES

Early in the century Frederic Stanley Kipping (1863–1949) of University College, Nottingham, showed that Grignard reagents react with silicon chloride to form organochlorosilanes.

$$RMgBr + SiCl_4 \rightarrow RSiCl_3 + MgBrCl$$
$$RMgBr + RSiCl_3 \rightarrow R_2SiCl_2 + MgBrCl$$
$$RMgBr + R_2SiCl_2 \rightarrow R_3SiCl + MgBrCl$$

The resultant compounds hydrolyze readily to form silanols.

$$RSiCl_3 + 3H_2O \rightarrow RSi(OH)_3 + HCl, \text{ etc.}$$

These silanols easily lose water, forming what were at first believed to be the equivalent of ketones. Hence they were named *silicones*. They are actually condensation products of varying complexity. Kipping investigated these compounds for more than three decades.

In 1941 Eugene G. Rochow (b. 1909) and William F. Gilliam of the

General Electric Co. polymerized alkyl chlorosilanes to a rubbery plastic which gave promise of commercial possibilities. Subsequent studies have led to a variety of polymers useful as varnishes, greases, "rubber," and plastics. The great virtue of these polymers resides in their water resistance and ability to maintain their properties over large temperature ranges. The presence of silicon in the molecule results in stability at temperatures where completely organic molecules decompose.

Commercial production in the United States is carried on principally by General Electric and the Dow Corning Corporation, which was formed by Dow Chemical and Corning Glass in 1943 to produce a silicone for sealing ignition systems of high-altitude aircraft.

Rochow, now at Harvard, and others have extended the study of metal-organic polymers to include compounds of germanium and tin.

BORON AND SILICON HYDRIDES

Silane SiH_4 was first prepared in 1858 by Wöhler. He had prepared magnesium silicide Mg_2Si by heating the elements in a crucible. The silicide reacted with hydrochloric acid to give magnesium chloride and a mixture of gaseous silicon hydrides in which silane probably predominated. There was a general interest in the silicon hydrides since the periodic relationship between silicon and carbon suggested the possible existence of silicon compounds analogous to the hydrocarbons. However the silicon hydrides are volatile and spontaneously combustible in air, so their study was sufficiently difficult and unpleasant to discourage careful analysis. While a few chemists prepared silicon compounds carrying organic radicals, there was no real success in studying the hydrides until the twentieth century when Stock developed vacuum techniques for preparing and handling such compounds in closed-system glass apparatus.

Employing his fractionation techniques Stock studied SiH_4, Si_2H_6, Si_3H_8, Si_4H_{10}, and higher hydrides. These compounds are all highly reactive so the parallelism with paraffin hydrocarbons is limited. Efforts to prepare compounds with $>Si=Si<$ and $>Si=C<$ bonds failed.

Stock's techniques also proved useful in studying the hydrides of boron. Boron hydrides were first prepared from magnesium boride and hydrochloric acid in 1879 by F. Jones. The gas evolved in this reaction was shown by Ramsay and Hatfield to be a mixture of hydrides and condensable with liquid air. The simplest hydride proved to be diborane B_2H_6, the monomer BH_3 being unknown. Stock and his students studied the boron hydrides for several decades, contributing much to our knowledge of these unusual compounds.[14]

[14] A. E. Stock, *The Hydrides of Boron and Silicon*, Cornell Univ. Press, Ithaca, 1933.

Fig. 22.6. ALFRED STOCK.
(From A. Stock, *Hydrides of Boron and Silicon*
[1933], with permission of Cornell Univer. Press.)

The structure of diborane provided a puzzle which led to many polemics.[15] There are not enough electrons to form a structure analogous to
ethane C_2H_6. Early proposals suggested an ionic distribution of electrons,
but Stock's work showed this to be untenable. Other authors suggested
participation of the K electrons of boron in the bonding, but this was a
phenomenon not characteristic of other compounds. Other unconventional
bonds such as the one-electron bond (Sidgwick, Pauling) and no-electron
bond (Lewis) were proposed. In 1921 W. Dilthey proposed a vague form
of bonding through two of the hydrogen atoms without paying attention
to disposition of electrons. Still another theory suggested a double bond
between the boron atoms.

All of the proposals had shortcomings of one sort or another. Experimental studies revealed that two of the hydrogen atoms must be bonded
in a manner somewhat different from the others. As a consequence of this,
Stock's former student E. Wiberg suggested that the boron hydrides
might be unsaturated polybasic acids. H. C. Longuet-Higgins suggested a
resonating structure involving hydrogen bridges. This reasoning was
extended to beryllium borohydride $Be(BH_4)_2$, beryllium hydride $(BeH_2)_x$,
and aluminum hydride $(AlH_3)_x$.

[15] Stock, *ibid.*, p. 154 ff has a good discussion of the ideas presented up to 1931.

$$
\begin{array}{ccc}
H & H & H \\
\diagdown & \diagup \diagdown & \\
B & & B \\
\diagup & \diagdown & \diagdown \\
H & H & H
\end{array}
\qquad
\left[\begin{array}{cc}
H & H \\
\overset{..}{B}{}^- &:: \overset{..}{B}{}^- \\
\overset{..}{} & \overset{..}{} \\
H & H
\end{array}\right] 2H^+
$$

Dilthey (1921)   Wiberg (1930)

$$
\begin{array}{cccccc}
H & H & H & H & H & H \\
\diagdown & \diagup \diagdown & \diagup & \rightleftharpoons & \diagdown \diagup & \diagdown \\
B & & B & & B & B \\
\diagup & \diagdown & \diagup \diagdown & & \diagup \diagdown & \diagdown \\
H & H & H & H & H & H
\end{array}
$$

Longuet-Higgins (1943)

In 1945 Kenneth S. Pitzer introduced the concept of the "protonated double bond" where a normal covalent double bond has two protons embedded in the electron cloud while the boron atoms each carry a negative charge.

$$
\begin{array}{cc}
H & H \\
\diagdown & \diagup \\
\bar{B} \underset{H^+}{\overset{H^+}{=\!=}} \bar{B} \\
\diagup & \diagdown \\
H & H
\end{array}
$$

Pitzer (1945)

In his theoretical treatment of electron deficient bonds Robert E. Rundle[16] suggested a boron-boron bond in which two hydrogen atoms each form a half-bond to each boron atom. The concept resembles Pitzer's without assuming the existence of protons within the bond and that of Longuet-Higgins without assuming resonance. The concept has since been extended to other boron compounds, to $Al_2(CH_3)_6$, and to certain borohydrides.

The bonding problems raised by the boron hydrides led Hermann I. Schlesinger (1882–1960) of Chicago to a study of these compounds in the thirties. His group developed improved synthetic methods for boron hydrides and their derivatives. When it became known that trimethyl aluminum exists as a dimer, an attempt was made to prepare a compound containing H_3B and $Al(CH_3)_3$. The projected compound did not materialize, but aluminum borohydride $Al(BH_4)_3$ was obtained instead. Beryllium and lithium borohydrides were soon prepared, the latter being a stable, water-soluble, saltlike compound. The aluminum and beryllium compounds are the most volatile of the compounds of these elements.

Herbert C. Brown (b. 1912) at Chicago and Purdue was particularly active in studying the properties and application of the borohydrides and aluminum hydrides. Lithium aluminum hydride proved a valuable reagent

[16] R. E. Rundle, *J. Am. Chem. Soc.*, **69**, 1327, 1719 (1947); *J. Chem. Phys.*, **17**, 671 (1949).

for carrying out reductions of organic compounds which are ordinarily troublesome, for example, COOH to —CH$_2$OH, —CH$_2$Cl to —CH$_3$, —CN to —CH$_2$NH$_2$. Sodium borohydride is also useful, although the smaller size of the boron ion causes the hydrogen to be held more tenaciously.

In 1941, when there was an interest in volatile uranium compounds, Schlesinger prepared uranium(IV) borohydride, which proved the most volatile compound of quadrivalent uranium. However it is highly flammable and was not used in the gaseous diffusion process for separating uranium isotopes when uranium(VI) fluoride proved a satisfactory gaseous uranium compound.

Since lithium and sodium borohydrides liberate hydrogen from water, they were used during the war as solid sources of hydrogen; industrial production was initiated very soon after their discovery. The compounds also proved excellent reducing agents, particularly for unsaturated aldehydes and ketones since they do not hydrogenate the double bond. The related lithium aluminum hydride LiAlH$_4$ and sodium aluminum hydride NaAlH$_4$ were soon prepared and took their place as reducing agents, not only in the laboratory but in industrial operations as well.

Within the past decade the boron hydrides have assumed great significance. Their high energy yields on combustion give them a role in rocket fuels.

Boron-Nitrogen Compounds. When boron and nitrogen atoms are bonded together, the same number of electrons are present as when two carbon atoms are bonded. A number of compounds which represent boron-nitrogen analogs of carbon compounds have been prepared. Boron nitride, (BN)$_x$ has a structure resembling that of graphite and is inert to most reagents. Borazole B$_3$N$_3$H$_6$ prepared by heating a primary amine with diborane, has physical properties resembling those of benzene, although its chemical properties are modified by the polarities of B—N bonds.

Borazole   Benzene

FLUORINE CHEMISTRY

Following the preparation of elemental fluorine, there was only limited investigation of its compounds because of its high degree of reactivity and toxicity. F. Swarts in Belgium, O. Ruff in Germany, and A. L. Henne in the United States were the only workers to carry on extensive research on the element during the first quarter of the century. After 1925 fluorine compounds took on new importance and fluorine research was carried on vigorously, especially by E. T. McBee at Purdue, J. H. Simons at Penn State and later at Florida State, and in certain industrial laboratories.

The first of the industrially important organic fluorides were the Freons, developed by Thomas Midgley and Henne as nontoxic gases for mechanical refrigerators. In 1907 Swarts prepared the first such compound —CCl_2F_2 (Freon 12)—by treating carbon tetrachloride with antimony

Fig. 22.7. FREDERIC SWARTS.
(Courtesy of *J. Chemical Education.*)

trifluoride. Midgley and Henne prepared a whole group of fluorinated methanes and ethanes using the Swarts reaction.

Fluorine was in heavy demand during World War II for the production of uranium hexafluoride in the gaseous diffusion process. It was produced electrolytically, as in Moissan's method, but using steel cells with carbon electrodes. An electrolyte of potassium fluoride and anhydrous hydrogen fluoride was used in all procedures, although the ratio varied; a German process, for example, employed equimolecular proportions at a temperature near 250° C. The gas is compressed and transported in steel cylinders; the metal is not destroyed as long as water is absent. The demand for fluorine in the United States reached a level by 1942 where industrial production could be scaled up. For some purposes fluorine is converted to chlorine trifluoride which is liquid at slight pressure and is nearly as effective a fluorinating agent as fluorine itself.

Fluorocarbons, organic compounds which have their hydrogen replaced by fluorine, have been studied intensively since the 1930's; some are produced industrially for a variety of purposes. When all of the hydrogen is replaced by fluorine the resulting compound is highly inert and even resists oxidation. Such products are excellent electrical insulators and are valuable as lubricants. Teflon, a plastic produced from tetrafluorethylene, can be used under conditions of heat and chemical attack where nearly all other substances break down. Sulfur hexafluoride, a completely fluorinated sulfur compound, shows surprising chemical inertness and is used in gaseous form as an electrical insulator.

Besides the Swarts' reaction, fluocarbons are prepared by the catalytic fluorination with cobalt(III) fluoride, direct addition of hydrogen fluoride to olefins, and electrofluorination The latter process was developed in Simons' laboratory at Penn State. It involves dissolving the organic compound in liquid hydrogen fluoride and exposing the solution to a low voltage.

RARE GAS COMPOUNDS

At the time of their discovery in the 1890's the rare gases appeared to be unreactive toward other elements and compounds. With the development of knowledge of electronic structure of the elements, it became evident that each rare gas beyond helium possessed an octet of electrons in its outermost shell. Chemists generally assumed that these elements are therefore unreactive, a conclusion consistent with experimental observations.

In 1933 Pauling suggested the possibility of a reaction between xenon or krypton and fluorine and predicted xenon and krypton hexafluorides. Don M. Yost and A. L. Kaye sought unsuccessfully to obtain a reaction between krypton and fluorine and xenon and fluorine in an electrical discharge. There was little further work until 1962 when Neil Bartlett of the University of British Columbia prepared xenon hexafluoroplatinate $XePtF_6$. Soon thereafter H. H. Claassen, H. Selig, and John G. Malm of the Argonne Laboratory prepared xenon tetrafluoride by heating xenon and fluorine to 400° for one hour. Their success in preparing a stable crystalline compound led to vigorous experimentation in other laboratories. Within the next year stable compounds of fluorine and oxygen with xenon, krypton, and radon were reported. The discoveries stimulated a great deal of research into the crystal structure of these compounds, leading to the belief that the bonding problems presented do not demand the introduction of new concepts in quantum chemistry.

GEOCHEMISTRY

Although the term was used by Schönbein as early as 1838, geochemistry is primarily a twentieth-century development. Earlier work was largely restricted to the analysis of minerals, rocks, waters, and gases. By 1850 a large number of mineral analyses had been accumulated, especially as the result of Berzelius' work. Carl Gustav Bischof of Bonn attempted to organize and interpret this data in his *Lehrbuch der physikalischen und chemischen Geologie* (1879–1893). The bulk of the data reported in these works was of European origin.

With the formation of the United States Geological Survey in 1884, chief chemist Frank Wigglesworth Clarke initiated an extensive chemical study of the earth materials of the North American continent. Besides collecting factual information, Clarke was deeply interested in the significance of the data. In 1908 he published his *Data of Geochemistry*, a Geological Survey bulletin which sought to correlate and interpret the mass of information which had been accumulated. This work was repeatedly revised, the last edition appearing in 1924.

In this publication Clarke attempted to work out estimates of the abundance of elements in the earth's crust. The first such effort had been made by Döbereiner. Clarke's estimates are still basic in current tables of abundance. His last compilation was made with the aid of Henry S. Washington. Confirmation of the Clarke-Washington conclusions was provided by Goldschmidt, who argued that the glacial drift present over much of southern Norway is representative of the average composition of the earth's crystalline rocks.

An important school of geochemistry was developing in Norway in the first decade of the twentieth century under the leadership of J. H. L. Vogt and W. C. Brøgger. Its most important investigator was V. M. Goldschmidt (1888–1947) who graduated from Oslo in 1911. In his doctoral thesis he applied the phase rule to the changes induced during metamorphism in shales, marls, and limestones, showing that these changes might be treated in terms of chemical equilibria. These studies on metamorphism were continued, stimulating phase rule interpretations in other Scandinavian laboratories. Goldschmidt's laboratory became the principal center for geochemical studies. With the introduction of x ray crystallography Goldschmidt and his associates undertook structural analysis of important minerals and worked out laws for the distribution of elements in crystalline rocks. His principal coworkers in the period were T. Barth, G. Lunde, I. Oftedahl, L. Thomassen, and W. H. Zachariasen. Goldschmidt moved to Göttingen in 1929 and began spectroscopic analyses for elements present only in traces, seeking thereby to develop the geochemistry of individual elements. He had to flee Germany in 1935 and, following a period in a concentration camp, he fled Norway in 1942 to settle in England.

V. I. Vernadsky developed an important center of geochemical research in the U.S.S.R. after 1917. While directed significantly toward the study of Soviet mineral resources, it has also contributed much to fundamental geochemistry. A. E. Fersman of this group published a four-volume treatise on geochemistry (1933–1939).

COSMIC CHEMISTRY

The nature and origin of the universe has held a peculiar fascination for chemists, as it did for the author of Genesis. With the accumulation of spectral data from stars and nebulae it is possible to estimate the cosmic abundance of the elements and bring forth hypotheses regarding cosmic history. With the recent developments in nuclear science, coupled with advances in theoretical and observational astronomy, a number of scientists have been encouraged to speculate about the origin of the elements.

Any attack on the problem requires a knowledge of the distribution of the elements, including the isotopic abundance. Geochemists have been collecting this data for some time for the elements of the earth's crust. Meteorites have contributed data for solid matter from the solar system, extensive analyses being made in the laboratories of Harrison Brown at California Institute of Technology and Harold Urey at Chicago. The composition of the stars can only be learned through spectral analysis. This, though useful, can give little information regarding isotopes and is only pertinent to the elements near the surface of the stars. A table of cosmic abundance, compiled in 1937 by Goldschmidt, was improved by Brown in 1949.[17] Such data are essential to cosmologists in their search for a theory of the origin and nature of the universe.

CONCLUSION

Modern inorganic chemistry has experienced a renaissance since the introduction of quantum concepts for dealing with bonding. Through application of such ideas it is possible to deal with the structure not only of conventional inorganic compounds but also of those representing varying degrees of uniqueness such as coordination complexes, chelates, clathrates, and π-complexes. In fact the twentieth century has seen the disappearance of the traditional barrier between organic and inorganic chemistry since

[17] H. Brown, *Rev. Mod. Physics*, **21**, 628 (1949). Also see H. E. Suess and H. C. Urey, *ibid.*, **28**, 53 (1956).

Fig. 22.8. HAROLD C. UREY.
(Courtesy of the University of Chicago.)

many of the most interesting compounds contain metal ions to bonded organic groups.

The periodic table has not only been filled, but also extended, since 1900. This has been achieved partly as the result of concerted efforts to discover the missing elements; but principally it has been the result of the success of transmutation associated with bombardment by high-velocity particles. In fact positions 43, 61, 85, and 87 of the periodic table were destined to remain empty until transmutation products could be found to fill them since half lives of these elements are all too short for any significant accumulation in nature.

The chemistry of all elements has been advanced since 1900, but particular progress has been made in understanding the rare earths, the rare gases, silicon, boron, fluorine, and the lesser-known transition elements such as zirconium and hafnium.

CHAPTER 23

ORGANIC CHEMISTRY V. GROWTH AND TRANSFORMATION

GENERAL PATTERNS

Organic chemistry has undergone great changes during the twentieth century, in both the nature of its achievements and its point of view. At the beginning of the century it was dominated by the Germans. To the extent that organic chemists outside Germany had any recognition, it was usually because they had studied in a German university. Young men who could not afford the two years of study in Germany for the doctorate generally contented themselves with local study under one of the German graduates.

The success of German organic chemistry from 1860 had been striking. The synthesis of long lists of compounds was an activity highly compatible with the German temperament, and the pages of the *Annalen* brought forth hundreds of such compounds. This is not to imply that the German chemists were mere hacks. Their knowledge of reactions was extensive and their application of reactions often showed deep insight into the subject. The interest of the German chemical industry was a constant stimulus to the preparation of new compounds. Working conditions were generally very satisfactory, and the steady flow of students made a high level of production possible. This pattern persisted until World War I and even beyond.

While the German organic chemist did not ignore theoretical matters,

Fig. 23.1. RICHARD WILLSTÄTTER.
(Courtesy of the Edgar Fahs Smith Collection.)

Fig. 23.2. ADOLF BUTENANDT.
(Courtesy of J. W. Williams.)

he was seldom deeply concerned with them. There were so many compounds to prepare that empirical approaches would generally suffice. The new breed of physical chemist might concern himself with such matters, but the organic chemist frequently bragged that he had known the answer for twenty years.

World War I did not greatly change organic chemistry in Germany, but it did initiate profound changes elsewhere. Because of the dominance of the German chemical industry at that time, particularly in the field of organic chemicals, the Allied powers found themselves without synthetic dyes, drugs, solvents, and many other necessary materials. They had to create a chemical industry with great speed, and the lesson learned was not forgotten when peace came. The prosperity of each nation's chemical industry was protected, and its expansion was encouraged. With it grew the strength of its own professional chemists. The dependence upon German training was broken. Until the thirties young men still welcomed an opportunity for study in Germany, but they no longer considered it as important as it had once been.

Following World War I leadership in organic chemistry began to shift to Switzerland, Britain, and the United States. Beside the German names of Willstätter, Kuhn, Windaus, and Butenandt stood the Swiss names of

Karrer, Ruzicka and Prelog, the English names of Robinson, Heilbron, Ingold and Haworth, the Scottish names of Todd, Purdie and Irvine, and the American names of Conant, Adams, Gilman, Fieser and Woodward.

Until 1925 organic chemistry was very much the chemistry of aromatic compounds. Coal tar was a readily available raw material in the days when gas lighting was commonplace. As electrical lighting grew in importance, the distillation of coal did not increase at a pace to keep up with the needs of the chemical industry, and industrialists turned to other sources of raw materials. It was quickly recognized that natural gas, petroleum, and fermentation products such as ethanol, butanol, and acetone provided good sources of organic chemicals. Thus aliphatic chemistry began to receive more attention than in the days when chemists preferred working with aromatic compounds which ordinarily gave crystalline, sharp-melting compounds and would discard aliphatic compounds which are more often oils or even tars.

Improvements in equipment since 1925 have made organic chemistry a less sloppy type of work. Sound borosilicate glassware, ground-glass joints, heating mantles, magnetic stirrers, efficient and convenient vacuum pumps, molecular stills, chromatography, and isotopic tracers are a few of the innovations which provide better control of synthesis and separation. The introduction of instrumentation, particularly infrared spectroscopy and nuclear magnetic resonance, has speeded up analysis and identification.

New techniques and new reagents and solvents make it possible to achieve reactions which were tedious or impossible before. High pressure hydrogenation, ozonization, and selective catalysis are cases in point.

Compounds for use as starting materials in synthesis also have become more readily available. In many laboratories it once was considered good training for a student to prepare his necessary compounds from common chemicals. This practice is hard to justify for the experienced chemist, however, and the availability of starting compounds of some complexity has been a real boon to organic progress. In 1916 Clarence G. Derick and his students at Illinois spent the summer synthesizing a large number of the less common but extensively used organic compounds. Two years later, C. E. K. Mees of the Eastman Kodak Company announced plans for the company to begin producing organic compounds widely used in synthesis, but not commercially available. This commercial venture has been very successful; the Organic Chemicals Division of the company listed more than 4,500 compounds in its 1960 catalog. Other companies also entered the field of specialty chemicals. The problem of synthesis of necessary compounds also is aided by the publication of *Organic Syntheses*, an annual volume of carefully tested methods for the synthesis of various compounds. This too grew out of the work at Illinois; the first volume appeared in 1921 under the editorship of Roger Adams (1889–1971).

The new theoretical outlook based upon electronic mechanisms permits better understanding of the reactive centers in organic compounds, the mechanism of reactions, catalytic effects, solvent effects, differences in reactivity, and similar matters. This point of view has made possible such synthetic achievements as Woodward's preparation of reserpine within a year after the proof of its structure.

Of course, the influence of the chemical industry on pure organic chemistry must not be ignored. By employing chemists in large numbers, both at the bachelors and doctoral levels, this branch of the industry stimulated the study of organic chemistry and the amount of research.

It becomes difficult at times to distinguish between advances in pure organic chemistry and industrial developments. Hence the reader will find that in certain cases the early phases of purely organic research will be treated in Chapters 25 and 26 since the substance was rapidly introduced into industrial chemistry. It should also be noted that the division between organic and biochemistry is a very nebulous one and in many cases the decision to discuss a subject in this or the next chapter is purely arbitrary.

THEORETICAL DEVELOPMENTS

The structural atom and the concept of the electron-pair bond, discussed in previous chapters, very naturally were incorporated into structural organic chemistry. The symbolic line representing the bond between atoms came to stand for the electron pair, although it was quite frequently convenient to use dots, crosses, or circles to represent electrons. The early ideas of Lewis and Sidgwick were very quickly expanded to fit the characteristics of organic compounds.

Since certain classes of compounds are more reactive than others, since certain functional groups are more reactive than others, and since solvents, reagents, and catalysts often have directing influences on reactions, it was evident that the electron-pair bond is responsive to a variety of intra- and intermolecular forces. Lewis recognized this when he pointed out that in compounds like $H_3C:CH_3$, $H_2N:NH_2$, and $Cl:Cl$ there is an equal sharing of the bond between the two atoms, but that in many compounds a permanent polarization is apparent on account of an unequal sharing of the bond—for example, $H_3C:NH_2$, $H_3C:OH$, $H_3C:Cl$—leading to a detectable dipole moment.

Robert Robinson[1] (1886–1975) and Christopher K. Ingold[2] (1893–1970) undertook the examination of the probable role played by electrons in organic reactions, particularly insofar as high and low electron densities figure in the creation of reactive centers. Ingold devoted considerable

[1] R. Robinson, *Outline of an Electrochemical (Electronic) Theory of the Course of Organic Reactions*, Institute of Chemistry of Great Britain and Ireland, London, 1932.
[2] C. K. Ingold, *Chem. Revs.*, **15**, 225 (1934).

Fig. 23.3.   ROBERT ROBINSON.
(Courtesy of A. L. Wilds.)

attention to the matter and created a classification scheme which has been widely adopted. Quantum mechanics also have been applied to arrive at statistical distribution of electron densities, quantized states, and resonance energies. Here Pauling played an important role, with many others also contributing. Naturally many chemists regretted the introduction of quantitative treatment, but even they have been able to gain some understanding with qualitative concepts like polarity, electron attraction, and relative electronegativity.

Ingold introduced such concepts as inductive and tautomeric effects and realized that each of these, in addition to a permanent effect in creating differences in electron densities, is responsive in a dynamic manner to environmental influences. The nature of nearby groups was studied for their influence on permanent and dynamic polarization (general inductive effects [——►]). Tautomeric effects [↷] were also evaluated in systems with double and triple bonds—where one atom tends to withdraw an electron from the unsaturated center for its own use or where an atom tends to release an electron pair with a resultant tendency toward multiple bond formation. In this manner such matters as the *ortho-* and *para-* directing character of hydroxy and amino groups on aromatic rings became explainable.

Reagents have been classified as to their relative character. The Lewis concept of acid and base has proved particularly useful in this work. A. Lapworth proposed that reagents be classified as *anionoid* and *cationoid* according to whether they resemble anions or cations. Robinson expanded upon this. Ingold introduced the term *nucleophilic* for reagents which give electrons to or share them with an acceptor nucleus, or *electrophilic* reagent.

These concepts have been extensively elaborated; they are modified and expanded as experimental and theoretical advances provide source material pertinent to the constructs. The development of ideas on chemical bonding has been particularly important in the extension of organic reaction theory. And the contributions of molecular orbital theory, π-bonding, and resonance have all been of particular significance.

Free Radicals

Various compounds considered to be free radicals during the mid-1800's were shown, after Cannizzaro clarified the molecular concept, to have double the expected molecular weight. Thus the so-called cyanogen, cacodyl, methyl, and ethyl radicals proved to be compounds with a formula double that which had been postulated, for example, $(CN)_2$, $(CH_3)_4As_2$, C_2H_6, C_4H_{10}. During the decades that followed it was generally assumed that free radicals could not exist, or at least could not be isolated or otherwise proven to exist. Arguing from his ideas on bivalent carbon, John U. Nef sought in 1897 to demonstrate the intermediary role of free radicals, but his views were at variance with those of most chemists. Only a year before, Ostwald had suggested that the nature of organic radicals precluded the possibility of isolating them.

A few years later Moses Gomberg (1866–1947) of Michigan synthesized tetraphenylmethane. He next attempted the preparation of hexaphenyl-ethane by the action of silver on triphenylmethylchloride. The product failed to show the expected properties, being unexpectedly reactive and giving colored solutions. All of Gomberg's studies were inconsistent with the formula $(C_6H_5)_3C$—$C(C_6H_5)_3$ but pointed toward the existence of triphenyl methyl $(C_6H_5)_3C$, a compound of trivalent carbon. The compound reacted readily with oxygen and halogens. Gomberg reasoned that it was not possible to place six phenyl groups around two carbon atoms without interference and hence the substance must exist as the free radical. Small groups like oxygen or iodine might easily combine with the triphenylmethyl to restore the tetravalence of the carbon atom.

Gomberg continued the study of compounds with various aryl groups on a carbon atom, three such groups to one carbon, and observed a

Fig. 23.4. WERNER E. BACHMANN (left) AND MOSES
GOMBERG.
(Courtesy of A. L. Wilds.)

consistent pattern of free radical formation. These compounds are always
highly reactive toward oxygen, halogens, and hydrogen, and always
colored. Hexaphenylethane was shown to exist, but on dissolving in a
polar solvent it dissociates into yellow triphenyl methyl. Nitrogen compounds were found to behave in similar fashion. For example, tetraphenyl
hydrazine dissociates to $(C_6H_5)_2N$ in solution.

With the introduction of the covalent bond it was realized that the
molecule breaks into two groups, each with a free electron $(C_6H_5)_3C\cdot$,
which would explain the color. These unpaired electrons readily pair with
unpaired electrons of the oxygen molecule or with halogen or hydrogen
to form a covalent bond, or accept an electron from a metal atom to form
an ionic compound. Solutions of hexaphenylethane in benzene are nonconductors, but in an ionizing solvent such as liquid sulfur dioxide conductivity is high, which suggests dissociation into positive and negative ions

$$(C_6H_5)_3C:C(C_6H_5)_3 \rightleftarrows (C_6H_5)_3C^+ + (C_6H_5)_3C:^-$$

The greater stability of the triphenyl methyl free radical compared with the methyl free radical has been explained on the basis of its large resonance energy.

The study of low-molecular-weight free radicals has received considerable attention. Since around 1950 the concept of free radicals has played a significant role in reaction mechanisms, particularly as the outgrowth of the researches of Morris S. Kharasch at Chicago, Werner Bachmann at Michigan, Francis Owen Rice at Johns Hopkins and Catholic University, and Karl Ziegler of Mülheim. Some of the implications have been considered in Chapter 20 and will not be pursued further here.

Mechanisms

Researchers have devoted considerable time to reaction mechanisms, but progress has been slow since organic reactions are seldom simple. Consequently the study of mechanisms traditionally has lagged behind structural theory.

Work on the rate of halogenation of ketones and similar studies by Lapworth at the beginning of the century revealed the role of enolization during acid and base catalysis of such reactions. Such projects have been extended into many areas with the result that studies of kinetics and mechanisms have significantly broadened our understanding of organic reactions. Photochemical techniques, isotopic tracers, polarography, nuclear magnetic resonance spectroscopy, and low-temperature techniques have become valuable tools in such studies.

STEREOCHEMISTRY

Optical Isomerism

The theory by which the asymmetric carbon atom is treated as a tetrahedron proved exceedingly fruitful and its extension to include mirror image structures in general was useful. Not only is the asymmetric carbon atom used as a basis for optical activity, but also any configuration where mirror images are formed is associated with such activity. Van't Hoff's theory includes the concept of free rotation around single bonds. He indicated that where rotation is restricted the possibility for enantiomer formation exists.

This prediction was validated by the resolution of compound I by the

I   II

younger Perkin, Pope, and Wallach in 1909. Compound II was resolved by W. H. Mills and A. M. Bain a year later. These resolutions illustrate asymmetry which occurs when molecules have their atoms in two perpendicular planes without opportunity for rotation. Many other examples of optical isomerism brought about by restricted rotation have been encountered, for example, in the allenes.

Van't Hoff predicted that optical activity should be expected in the allenes when four different groups occupy the end positions on the allenic group, III.

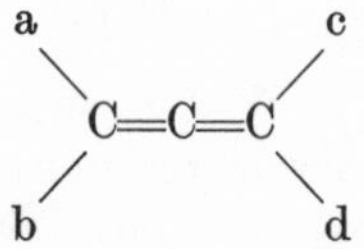

III 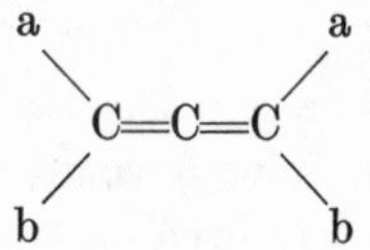 IV

Somewhat later it was recognized that activity might also be expected in the simpler allene of structure IV. Only in 1935 was experimental verification obtained when Mills and P. Maitland carried out the following asymmetric dehydration using active camphorsulfonic acids as a catalyst and obtained an active allene.

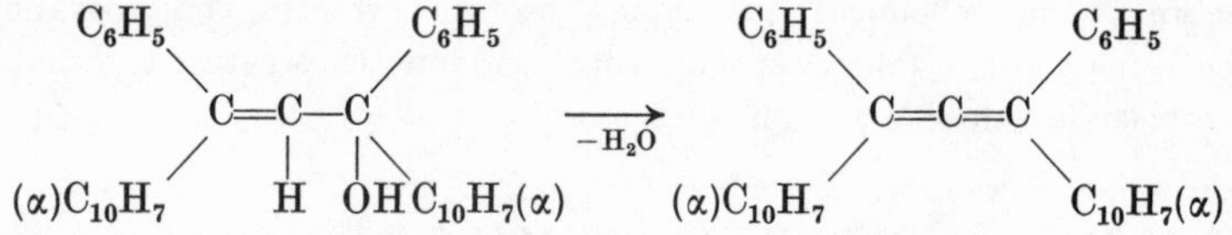

When an inactive catalyst was used the racemic mixture was formed. Further evidence for the correctness of van't Hoff's prediction was obtained about this time by Elmer P. Kohler and his students. A compound having a similar type of stereochemistry was resolved by Pope, Perkin, and Wallach in 1909, in this case the ring conferring the rigidity of a double bond (V).

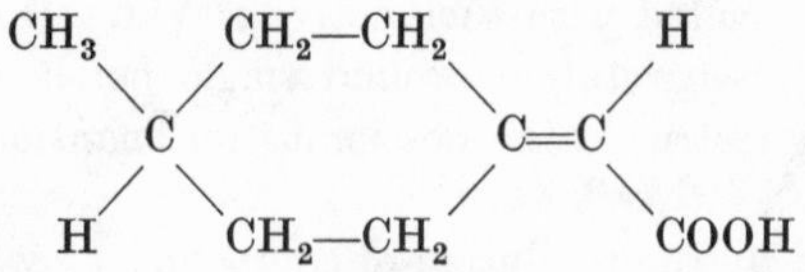

V

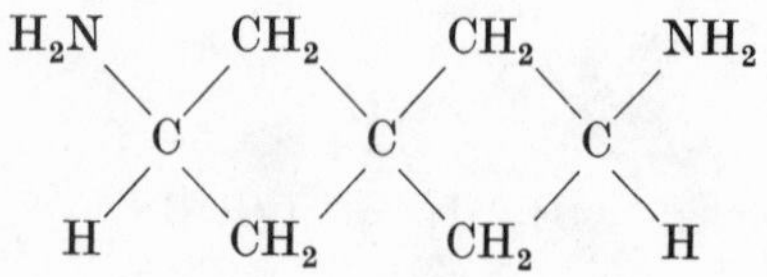

VI

A related type of compound, the spirane, was predicted by O. Aschan to be active. The first resolution of such a compound (VI) was accomplished by S. E. Janson and Pope in 1932. Other compounds of this sort were resolved by Mills, Leuchs, Boeseken, and others.

Van't Hoff's theory did not envisage the discovery made by James Kenner and his associates at Sheffield in the early twenties when they resolved a diphenyl with carboxyl and nitro groups on each of the ortho positions.

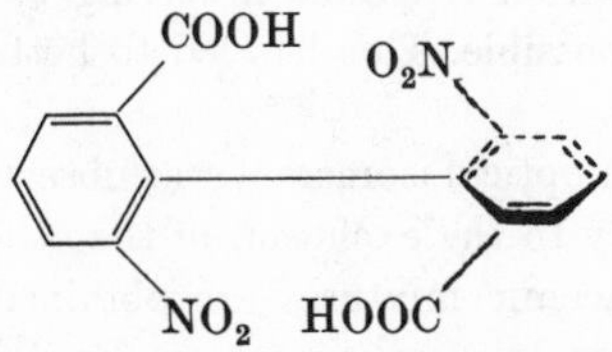

$$\text{COOH} \quad \text{O}_2\text{N}$$
$$\text{NO}_2 \quad \text{HOOC}$$

Here, the mirror images result from the apparent inability of the large groups on the ortho positions of the rings to move past one another. Kenner, E. E. Turner, and others resolved many similar compounds to show that steric hindrance to free rotation can be the basis for formation of enantiomers. It was learned that such isomerism will result wherever restricted rotation or intramolecular overcrowding is present.

Further work on optical activity established that the molecular rotation of compounds should be treated as the sum of the contributions of the individual asymmetric centers. Significant progress in this type of theoretical approach was made in the field of sugar chemistry by Claude S. Hudson (1881–1952) of the National Institutes of Health in the United States. Karl Freudenberg (b. 1886) of Heidelberg showed that comparable changes in structure bring about comparable changes in rotation.

The problem of absolute configuration remained a knotty one. Customarily configuration has been related directly or indirectly to that of D-(+)-glyceraldehyde.

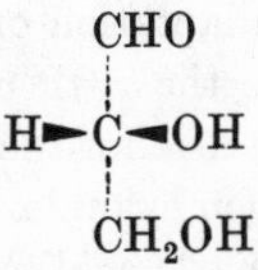

$$\text{CHO}$$
$$\text{H} \blacktriangleright \text{C} \blacktriangleleft \text{OH}$$
$$\text{CH}_2\text{OH}$$

This configuration was arbitrary until 1951 when J. M. Bijvoet, A. F. Peerdeman, and A. J. Van Bommel showed it to be correct on the basis of x ray studies of the crystal of sodium rubidium tartrate.[3] Chemical correlations frequently make it possible to relate the absolute configuration of the compound under study directly or indirectly to glyceraldehyde,

[3] J. M. Bijvoet, A. F. Peerdeman, and A. J. Van Bommel, *Nature*, **168**, 271 (1951).

provided no inversions at the asymmetric centers are involved. Even where inversions occur, it is sometimes possible to take them into account and by various devices arrive at absolute configurations for such complicated molecules as terpenoids and steroids.

The mechanism of the Walden inversion was a troublesome one for many years. Joseph Kenyon and his associates were able to show that inversion takes place at a specific stage of the reaction scheme for certain simple molecules. Ingold demonstrated that inversion is always involved in nucleophilic substitution reactions involving two steps (S_N2) except where structurally impossible. This has led to better understanding and control of the reaction.

Early in the study of optical isomers it was observed that in nature one isomer is found, usually to the exclusion of the other. In the laboratory, on the other hand, racemic mixtures are obtained during synthesis of compounds with asymmetric carbon atoms. Emil Fischer discussed the problem of synthesis of a particular isomer in 1894, and a few years later Willy Marckwald examined in detail the synthesis of active compounds. He considered an "asymmetric synthesis" to be one in which an optically active compound is produced from symmetrical molecules by the use of active intermediates without application of any resolution methods. Marckwald prepared methylethylacetic acid from methylethylmalonic acid by using active brucine and obtained a mixture containing 55 per cent of the *levo* acid.[4]

Alexander McKenzie (1869–1951) prepared *l*-mandelic acid from α-ketophenylacetic acid through the use of *l*-menthol. McKenzie subsequently carried out the synthesis of active lactic acid from pyruvic acid by esterifying with alcohols such as *l*-menthol, *l*-borneol, and *l*-amyl, reducing, and hydrolyzing. The product in these cases always contains both isomers, but one isomer will be present in excess. McKenzie and his collaborators studied a large number of asymmetric syntheses during the next quarter century.[5]

The use of enzymes to aid in asymmetric synthesis has been quite successful. This is, of course, an adaptation of the kind of syntheses carried out by living organisms where the path of the reaction is enzyme-controlled. L. Rosenthaler used emulsin for the preparation of *l*-mandelic acid from benzaldehyde and hydrogen cyanide. Using fumarase, Henry Dakin prepared *l*-malic acid from fumaric acid. Y. Sumiki used aspartase from yeast for the preparation of *l*-aspartic acid from fumaric acid and ammonia. G. Embden prepared *d*-arginine from pyruvic acid using liver, and C. Neuberg converted methylphenylketone to the corresponding *levo* alcohol with the aid of reductase. Enzymatic asymmetric syntheses are

---

[4] W. Marckwald, *Ber.*, **37**, 349 (1904).

[5] A. McKenzie, *J. Chem. Soc.*, **85**, 1294 (1904); and many later papers in this journal to 1932.

found to result in a strong preponderance of one of the optical isomers, sometimes approaching 100 per cent, in contrast to purely chemical asymmetric syntheses where the preponderance of one isomer over the other is generally only a few per cent.

Optical Rotatory Dispersion. Biot recognized in 1817 that the rotatory power of an optically active substance varies with the wavelength of light used. Measurements during the next forty years utilized various light sources but, following the introduction of the Bunsen burner, measurements were customarily made at the wavelengths of the D lines of sodium. Consequently after 1860 there was only sporadic investigation of change of specific rotation with change in wavelength. The small amount of work done came largely from laboratories of physical chemistry, particularly those of T. M. Lowry at Cambridge and W. Kuhn at Basel. These investigators were interested in the relation between physical properties and chemical constitution. The only organic chemists to make extensive use of optical rotatory dispersion before 1940 were L. Tschugaev, P. A. Levine, K. Freudenberg, and H. Rupe.

The principle of optical rotatory dispersion took on great significance after 1952 when O. C. Rudolph and Sons placed a photoelectric spectropolarimeter on the market. With this instrument it is possible as a routine matter to measure the optical rotation over a wide range of wavelengths including the ultraviolet. Carl Djerassi, at Wayne University, has been particularly active in developing the consequences of the measurement for organic chemistry. Change in rotatory power with wavelength can be related to the structural peculiarities of the compound and is therefore of diagnostic value in studying configuration and conformation.

Geometric Isomerism

Wislicenus' work in experimentally verifying van't Hoff's predictions regarding geometric isomerism was quickly extended to the nitrogen compounds. In 1890 Hantzsch and Werner showed that there should be a *cis-trans* isomerism in nitrogen compounds like the oximes if the three valences of the nitrogen atom are not in the same plane. Thus it was possible to explain Goldschmidt's observation that benzildioxime is isomerized by heating in ethanol and Beckmann's report of two forms of benzaldoxime. When it was understood later that the nitrogen behaves as a tetrahedron, this isomerism could be explained by the opposite corners being occupied by the OH group and an unshared electron pair.

The determination of configuration of the oximes was difficult, and until the twenties assignments in the literature were frequently incorrect. During this decade chemists began to relate the configuration to the ease of forming cyclic compounds. O. L. Brady and G. Bishop studied the behavior of the 2-chloro-5-nitrobenzaldoximes toward alkali. The ring closure was considered to be that of the *anti* isomer. Jacob Meisenheimer

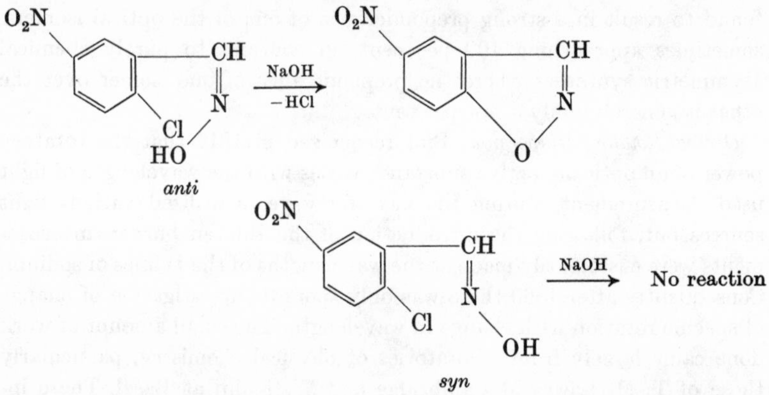

$$O_2N-C_6H_3(Cl)(OH)-CH=N \xrightarrow[-HCl]{NaOH} O_2N-C_6H_3-CH=N-O$$

anti

$$O_2N-C_6H_3(Cl)-CH=N-OH \xrightarrow{NaOH} \text{No reaction}$$

syn

(1876–1934) used the method extensively in proving the structure of similar compounds.

The Beckmann rearrangement also became a useful method for establishing configuration. This reaction was discovered in 1886 when Beckman treated benzophenone oxime with phosphorus pentachloride and obtained benzanilide.

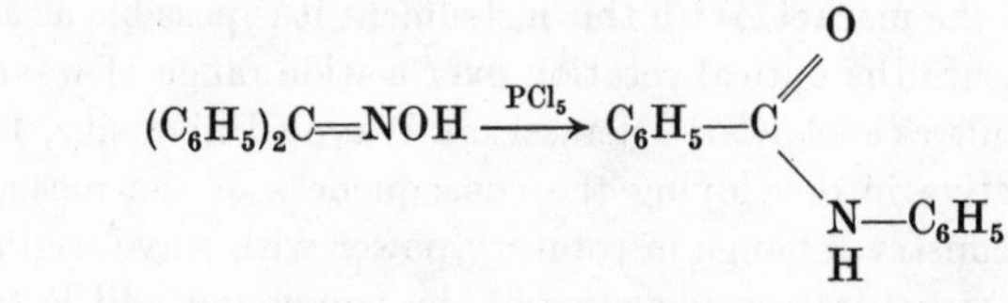

$$(C_6H_5)_2C{=}NOH \xrightarrow{PCl_5} C_6H_5-\underset{\underset{H}{N-C_6H_5}}{\overset{O}{C}}$$

The reaction was found to be a very general one brought about by such reagents as phosphorus oxychloride, acetyl chloride, sulfuric acid, certain metal chlorides, and various other reagents. Extensive studies by Meisenheimer not only clarified the nature of the rearrangement but also revealed that the character of the final products might serve as a clue to the configuration of the oxime used as a starting material.[6]

After 1890 it was also apparent that geometric isomers might be expected for many compounds containing double-bonded nitrogen atoms, and such isomers were soon reported in a number of laboratories. Hans von Euler and A. Hantzsch at Würzburg observed that colorless, water-soluble p-methoxybenzene diazonium cyanide isomerizes into two colored, water-insoluble compounds which are soluble in organic solvents and hydrolyze to amides. Many other cases of isomerism of compounds prepared from diazonium salts have been reported.

Azobenzene, as commonly prepared, always had the *trans* form, but in 1937 G. S. Hartley succeeded in isolating the *cis* form which has a dipole moment of 3.0 Debye units, whereas the *trans* form has no dipole moment.

[6] For reviews of the Beckmann rearrangement see A. H. Blatt, *Chem. Revs.*, **12**, 215 (1933); B. Jones, *ibid.*, **33**, 335 (1944).

The study of the configuration of systems with carbon-to-carbon double bonds also posed knotty problems. We have seen in Chapter 12 that the *cis* structure of maleic acid was established when it was observed that maleic anhydride forms easily, thus leaving the *trans* structure for fumaric acid. This approach is useful wherever the nature of the compound is such that cyclization is possible, for example, Fittig's preparation of coumarin from the *cis* form of o-hydroxycinnamic acid. The *trans* form of the same acid does not form coumarin. This method of structure determination is fraught with dangers, however, since some geometric isomers isomerize easily and may therefore give fallacious results. Alternatively, ring opening was utilized since this will yield the *cis* form, for example, Fittig's production of maleic acid by the oxidation of *p*-benzoquinone or of benzene.

The production of a compound of known configuration is helpful wherever it is reasonably certain that isomerization will not take place during the reaction. Thus the configuration of the solid and liquid (m.p. 15° C.) crotonic acids was established by Karl von Auwers[7] who showed that trichlorocrotonic acid is converted by reduction into solid crotonic acid, while hydrolysis (which converts the —CCl₃ group to —COOH) results in the formation of fumaric acid, known to be *trans*. Hence the solid isomer of crotonic acid must be the *trans* form, whereas the liquid acid must have the *cis* configuration.

Physical methods could be used to establish structures after it was learned that the *cis* form generally has the greater solubility in inert solvents, the lower melting point, the higher heat of combustion,[8] and, for acids, the higher ionization constant.[9] Werner pointed out that there is a close resemblance between *cis* and *trans* isomers in aliphatic compounds and the *ortho* and *para* isomers in aromatic compounds.[10] Dipole moments are also used for simple compounds, the more symmetrical *trans* form having the lower dipole moment. Debye used x ray measurements to determine the distance between chlorine atoms in the two dichloroethylenes, confirming the conclusions arrived at by other methods. Other physical constants, such as density, absorption spectra, Raman spectra, parachor, and refraction are helpful in certain cases; reaction rates are also utilized.

Studies on geometric isomers have revealed that one of the isomers is frequently more stable than the other. Heat, light, and reagents such as iodine, halogen acids, nitrous acid bring about isomerization in the direction of the more stable isomer. Theoretical studies of the reaction have

[7] K. von Auwers and H. Wissebach, *Ber.*, **56**, 715 (1923).

[8] Stohmann, *Z. physik. Chem.*, **10**, 416 (1892); W. Longuinine, *Ann. chim.*, [6] **23**, 189 (1891); W. A. Roth and R. Stoermer, *Ber.*, **46**, 260 (1913).

[9] W. Ostwald, *Z. physik. Chem.*, **3**, 242, 278, 380 (1889); *Ber.*, **24**, 1106 (1891). Also see H. Gilman, *Organic Chemistry*, Wiley, New York, 1938, vol. 2, p. 450.

[10] A. Werner, *Lehrbuch der Stereochemie*, Jena, 1904.

yielded indifferent results. Reagents and conditions have a great deal to do with paths of isomerization and, in fact, addition to unsaturated centers.

Geometric isomerism also exists in saturated rings of three, four, five and six members since the rigidity of the ring restricts free rotation around the bonds in the ring. In these compounds optical activity may also be encountered. For example, in disubstituted cyclopropanes, I is the *cis* form and is also *meso*, but the *trans* forms II and III are mirror images.

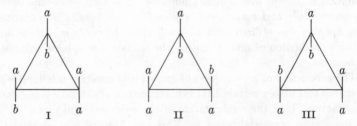

Similar relationships hold in larger rings. The configuration frequently is determined on the basis of ability to resolve the product into optical isomers, which points toward the *trans* form, whereas inability to resolve suggests the *cis*. Jacob Böeseken (1868–1948) of Delft successfully attacked the configuration of 1, 2-dihydroxy compounds by observing their reaction toward boric acid. The *cis* compounds form a cyclic complex showing strongly acid properties. The *trans* isomers do not form such a complex.

$$\left[\begin{array}{c} \underset{R}{\overset{R}{\underset{\displaystyle C-O}{|}}} \quad \underset{R}{\overset{R}{\underset{\displaystyle O-C}{|}}} \\ \underset{R}{\overset{\displaystyle C-O}{}}\quad B \quad \overset{\displaystyle O-C}{\underset{R}{}} \end{array}\right]^{-} H^{+}$$

In saturated condensed ring systems *cis-trans* isomerism may occur, provided the nature of the substituents does not result in undue strain. W. Hückel prepared the *cis* and *trans* form of decahydronaphthalene.

cis   *trans*

Böeseken has also identified such compounds. Paul D. Bartlett prepared cyclohexene oxide and obtained only the *cis* form, indicating that a three-membered ring is too strained for *trans* formation to take place.

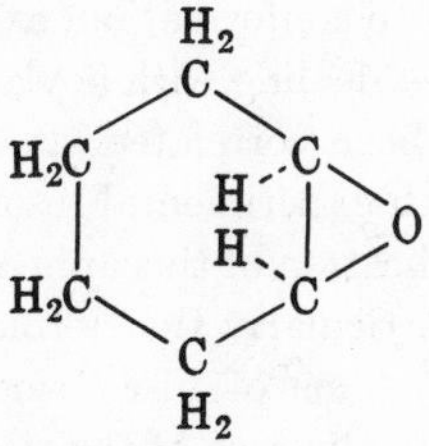

Conformational Analysis

The term *conformation* was introduced by Haworth[11] to apply to the arrangement of the atoms of a given molecule in space. Because of rotation around single bonds, strains, and similar factors, an almost infinite number of conformations is theoretically possible for a given molecule. Energy considerations, however, are such that a few conformations are preferred. The differences in energy states between such conformations are sufficiently low that conformational isomers ordinarily cannot separate. The application of conformational analysis to reacting systems, which is very largely a post-1945 development, represents the first major structural breakthrough since the introduction of the concepts of van't Hoff in 1874.

Conformational analysis of ethane reveals that the conformation of lowest energy is represented by the maximum staggering of C—H bonds (I), while the maximum energy state is observed when the front C—H bonds eclipse those in the rear (III). The basis for preferred conformations

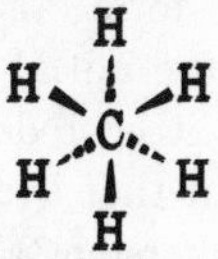

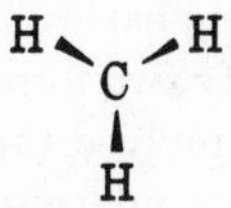

is associated with the energy interactions between nonbonded atoms. These interactions are weakly attractive, up to the sum of van der Waals radii, but become repulsive as atoms become closer. This tends to explain the greater stability of the fully staggered conformation. Of course, other factors such as hydrogen bonding, dipole effects, and integral charge interactions may modify the optimal conformation in actual instances.

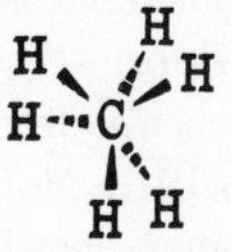

[11] W. N. Haworth, *The Constitution of Sugars*, Arnold, London, 1929, p. 90.

Non-bonded interactions can be calculated on theoretical grounds,[12] although empirical approaches[13] yield better results.

Various physical and chemical properties can be related with the preferred conformation. Certain absorption frequencies in the ultraviolet and infrared have been related to conformation, as have partition and adsorption phenomena. Equilibria dealing with acylation rates, ester hydrolysis, and oxidation rates have been correlated, as have the relative stabilities of epimers. Agreement with experimental results is good.

The most extensive application of this approach has been to compounds having saturated rings, particularly the steroids and triterpenoids. These compounds, with their systems of fused rings, present a multitude of stereochemical problems. In the case of the steroids, many of the answers were worked out on the basis of conventional approaches by 1945. The wealth of data accumulated during twenty years of study served as raw material for the development of conformational analysis. This in turn speeded up the solution of the stereochemistry of various triterpenoids.

In large part the structural problems derived from the stability of ring systems. Baeyer introduced his Strain Theory in 1885 to account for the abundance of five- and six-membered ring compounds and the lack of compounds with smaller and larger rings. He pointed out that since the four valences of the carbon atom should be directed toward the corners of a tetrahedron, the angle between bonds must normally be 109° 28′. In planar rings with five or six atoms the bond angle does not depart too seriously from this value (108° and 120°, respectively), but in smaller and larger rings the distortion is serious.

H. Sachse (1854–1911) pointed out in 1890 that the strains called for in cyclohexane and larger rings were based on the assumption that the carbon atoms lay on the same plane. If the ring took on a puckered configuration, the strains would be eliminated. Molecular models showed that cyclohexane should form two puckered isomers, the C-form (also termed "bed," or "bathtub") and the Z-form ("chair"). Because initial attempts to identify two forms failed, Sachse's ideas were given little credence until E. Mohr revived the concept in 1918. Mohr pointed out that cyclohexane might be a mixture of both isomers in which interconversion takes place with ease and suggested that decalin (decahydronaphthalene) should exist in *cis* and *trans* forms which could interconvert only as the result of very serious bond distortions. The postulated isomers were isolated by W. Hückel in 1925.

Physical evidence pointed to the greater stability of the chair form of

---

[12] See F. H. Westheimer and J. E. Mayer, *J. Chem. Phys.*, **14**, 733 (1946) and references therein; T. L. Hill, *ibid.*, **16**, 938 (1948) and references therein; D. H. R. Barton, *J. Chem. Soc.*, **1948**, 340.

[13] C. W. Beckett, K. S. Pitzer, R. Spitzer, *J. Am. Chem. Soc.*, **69**, 2488 (1947); R. B. Turner, *ibid.*, **74**, 2118 (1952); W. S. Johnson, *ibid.*, **75**, 1498 (1953).

six-membered rings when this was stereochemically possible. The work of Barton, Hassel, Pitzer, Spitzer, Prelog, Cookson, and their associates all pointed to the preponderance of the chair form in fused rings such as are present in steroids.[14]

The work of D. H. R. Barton (b. 1918) and his colleagues at London and Glasgow on the conformational analysis of the triterpenoids has advanced the stereochemistry of this difficult class of compounds very markedly. The concept has also proved useful in the study of the alkaloids and the pyranose sugars. In aliphatic chemistry the method is applied to such problems as the nature of addition to carbonyl groups in asymmetric synthesis.

Conformational analysis cannot by itself master the stereochemistry of molecules; other approaches such as x ray crystallography, absorption spectroscopy and, of course, ordinary chemical evidence are required. X ray crystallography was particularly important in establishing the stereochemistry of cholesterol, of vitamin B_{12}, and of other complex molecules.

SYNTHESIS

Until about 1940 synthetic activities in organic chemistry followed much the pattern prevalent at the beginning of the century, the primary difference being a greatly increased number of investigators, an increased number of synthetic methods, and improved facilities. The approach to synthetic problems was highly empirical and objectives were limited. Only rarely did anyone attempt a synthesis involving a large number of steps. Because of the need to be familiar with a large number of specific compounds and reactions, organic chemists tended to specialize in sugar chemistry, alkaloids, dyes, terpenes, proteins, fats, steroids, or some similar field and to be little interested in the whole field of chemistry.

Since 1940 the complexion of synthetic organic chemistry has been revolutionized by the application of theoretical principles to the programming of synthesis problems and the use of instrumentation as a guide at each step on the way. Ambitious projects, sometimes involving more than thirty distinct operations, have been successfully undertaken.

The chemistry of natural products has played a vital role in stimulating this change. Biochemists' interest in vitamins and enzymes, coupled with the drug industry's interest in antibiotics, hormones, and natural substances such as the rauwolfia alkaloids, spurred study of the synthesis of complex molecules with multiple reactive centers.

[14] R. S. Rasmussen, *J. Chem. Phys.*, **11**, 249 (1943) and citations therein; D. H. R. Barton in A. Todd, *Perspectives in Organic Chemistry*, Interscience, New York, 1956, pp. 82–86, 93.

The Early Period

Such well-established "name" reactions as those of Wurtz, Williamson, Perkin, Kolbe, Cannizzaro, Friedel-Crafts, Reimer-Tiemann, Lossen, Hofmann, Hell-Volhard-Zelinsky, Hantzsch, Skraup, Friedlander, Gabriel, Jacobsen, Knorr, Michael, Reformatsky, Willgerodt, Gattermann, Sandmeyer, Curtius, Nef, Chugaev, Stobbe, Claisen, and Knoevenagel continued to be used extensively.[15] Modifications which extended their applicability frequently were introduced. At the same time new reactions were discovered which made old objectives more easily attained or permitted new syntheses.

The Grignard reagent, discussed in Chapter 12, was introduced in 1899 but was not fully appreciated until the twentieth century. Grignard himself extended it to the preparation of a wide variety of compounds and it is used by inorganic chemists also. The reaction has been studied by various investigators, most notably perhaps by Henry Gilman (b. 1893) at Iowa State, and extended into areas beyond those suggested by Grignard.

Since the Grignard reagent is reactive toward replaceable, or active, hydrogen—for example, water, alcohol, ammonia, HCl—it may be applied analytically to the estimation of such replaceable hydrogen. This application was first suggested by Leo A. Tschugaev (1872–1922) of St. Petersburg; it was further developed by his student Th. Zerewitinov.

Other reactions introduced early in the century were the Bouveault aldehyde synthesis, the Bucherer reaction for converting aromatic phenols to amines, the Ullmann reaction for converting aromatic halides to the hydrocarbon with copper, and the Ullmann condensation for joining simple ring compounds into more complex fused rings. All of these reactions are applicable to aromatic compounds and reflect the intense interest in dye chemistry during the first decade of the century.

The Bouveault-Blanc reduction of this same period provided a method for converting acids to the corresponding alcohol. The reduction is brought about by the reaction of sodium and ethanol in the presence of the ester of the acid to be thus reduced. The Clemmensen reduction converts carbonyl groups to a methylene group by the use of amalgamated zinc in acid. The Dakin reaction uses hydrogen peroxide in alkaline solution to convert aromatic aldehydes to phenols.

During World War I there was little new activity in the field of synthetic chemistry except for the Rosenmund reduction. In this reaction, acid

---

[15] Short descriptions of these and other name reactions will be found in Alexander R. Surrey, *Name Reactions in Organic Chemistry*, Academic Press, New York, 1954, together with brief biographies of the chemists who introduced the reaction. Also see Kurt G. Wagner, *Autoren-Namen als chemische Begriffe*, Verlag Chemie, Weinheim, 1951.

chlorides are converted to aldehydes by introducing hydrogen into a solution containing palladium catalyst.

The Barbier-Wieland degradation for shortening the chain length of organic acids by one unit was introduced in 1913 by Barbier and improved in 1926 by Wieland. A method for moving in the opposite direction was introduced by Arndt and Eistert in 1935. They utilized diazomethane to lengthen acid chains by one unit.

Another reaction of marked significance was that discovered at Kiel by Otto Diels (1876–1954) and Kurt Alder (1902–1958) in 1928. They observed that butadiene reacts vigorously with maleic anhydride to give a six-membered ring compound, cis-Δ^4-tetrahydrophthalic anhydride, in quantitative yield.

The reaction, which involves 1,4 addition of the ethylenic compound to the diene with formation of a double bond at the 2,3 position, was found to be a very general one. Conjugated dienes react readily with a compound having a double or triple bond activated by carbonyl, carboxyl, nitrile, or nitro groups. The reaction is significant, not only for the compounds which can be produced and for the light shed upon the nature of addition reactions involving 1,4 addition, but in the analytical realm as well. H. P. Kaufmann of Münster introduced the Diene number as a means of measuring analytically the presence of conjugated unsaturation. This is particularly important in the fatty oil field since tung and oiticica oils are used in paints and varnishes. Both of these oils contain fatty acids which are unique for their conjugated unsaturation.

The Schmidt reaction for the conversion of carboxylic acids into amines was introduced by Karl Friedrich Schmidt (b. 1887) in 1923. The reagent, hydrazoic acid, was also found useful for converting aldehydes to nitriles and ketones to amides. The Stephen reaction, which utilizes hydrochloric acid and stannic chloride to convert nitriles to the corresponding aldehydes, was introduced in 1925 by Henry Stephen of Manchester.

The Meerwein condensation, introduced in 1939 by Hans L. Meerwein (1879–1965) of Marburg, results in combination between an aromatic diazonium halide and an α,β-unsaturated carbonyl compound. The versatility of the reaction was markedly extended by C. Frederick Koelsch who showed that the coupling can be directed to the β-position. Meerwein, at an earlier date, was one of the independent discoverers of the Meerwein-Ponndorf-Verley Reduction for reducing carbonyl compounds to alcohols in the presence of an aluminum alkoxide. Carried in the opposite direction, which is possible since it is ordinarily an equilibrium reaction, this reaction is known as an Oppenheimer oxidation. This works most satisfactorily for conversion of secondary alcohols to ketones, although it is used to some extent for the oxidation of primary alcohols.

Hydrogenation. Catalytic hydrogenation is a useful technique, both for synthetic work and for the interpretation of theoretical problems. The original developments of Sabatier and Senderens at the beginning of the century were soon adapted for industrial application. The need for suitable high-pressure equipment delayed widespread use of hydrogenation techniques in organic research until after World War I. Much important work was done in the thirties.

Suitable catalysts for hydrogenation reactions were also slow to be developed. Paal introduced a method for preparing platinum catalyst early in the century. Other finely divided metals, especially nickel, were utilized, but methods for preparing catalysts were poorly standardized and results were often disappointing. A nickel-aluminum alloy patented in 1927 by Murray Raney is widely used for the preparation of nickel catalyst by dissolving out the aluminum with sodium hydroxide. Roger Adams and his colleagues at Illinois reduced metal oxides for catalytic use. Homer Adkins (1892–1949) and his associates at Wisconsin pioneered in the development of copper chromite as an effective catalyst.

The Recent Period

The modern period in organic synthesis started in the forties, although some troublesome syntheses had been accomplished in the previous decade, for example, thiamine by R. R. Williams and J. K. Cline; ribo-flavin independently by Paul Karrer and Richard Kuhn; pyridoxine by S. A. Harris and Karl Folkers and independently by Kuhn; ascorbic acid by T. Reichstein and independently by Kuhn; α-tocopherol in three laboratories—those of Karrer, A. Todd, and Lee I. Smith; the anti-hemorrhagic vitamin K in the laboratories of E. A. Doisy and of Louis Fieser; equilenin by W. Bachman, J. W. Cole, and A. L. Wilds; panto-thenic acid by Folkers and also by Kuhn and H. Wieland. These successes were somewhat eclipsed by the successful total synthesis of quinine by R. B. Woodward and W. E. Doering; of cortisone by L. H. Sarett; of patulin and strychnine by Woodward; of morphine by Marshall Gates and by D. Ginsberg; of biotin in the Merck laboratories by Folkers, A. Grussner, and Y. SubbaRow; of folic acid by C. W. Waller; of cholesterol and vitamin D_3 independently by Woodward and R. Robinson; of β-carotene by Hans Inhoffen and by Karrer; of vitamin A by O. Isler; of insulin by F. Sanger; and of chlorophyll *a* by Woodward and by Martin Strell.

The remarkable aspect of these syntheses was the rapidity with which

Fig. 23.5. HOMER ADKINS.
(Courtesy of A. L. Wilds.)

Fig. 23.6. PAUL KARRER.
(Courtesy of P. Karrer.)

they followed the establishment of the structure of the compound. In some cases synthetic methods were even of importance in unsnarling questionable problems involving structure. These syntheses represented the power of the new point of view in organic chemistry, since the steps were frequently mapped out on theoretical grounds before any laboratory work was done. Experimental operations were continuously checked, infrared absorption playing an important role in this area. Work of this magnitude reflected the growth of team research. Hardly any of these syntheses were one-man jobs. In the university laboratories they represented the contributions of a series of graduate and postdoctorate students under the guidance of the laboratory director. In industrial laboratories there was perhaps an even greater concentration of talent in the teams which accomplished these breakthroughs.

These accomplishments reflect a characteristic of midtwentieth century science—its great dependence on the exchange of ideas. The era of the narrow specialist has given way to the era of the generalized approach to problems. This does not mean that there are still no specialists—the laboratories are full of them. The greatest work, however, is being done by scientists of broad outlook, and by specialists who operate as part of a team made up of a broad spectrum of specialties.

Microbiological Synthesis. A particularly interesting synthetic development, valuable both in organic research and in industrial production is the use of microorganisms. Molds and other organisms are used extensively in the production of antibiotics. Here the microorganism is responsible for the final product but little is known regarding intermediate stages. However, microorganisms also have been utilized to carry out one step in a series of synthetic operations. They are peculiarly suited for this application since they can carry out a stereospecific reaction which, in strictly chemical syntheses, would lead to a mixture of isomers. Ascorbic acid, *l*-ephedrine, pyridoxal, and pyridoxamine, certain anthraquinones, and certain penicillins are synthesized with the aid of suitable microorganisms. The method is also employed in the steroid field.

NATURAL PRODUCTS

The compounds found in natural organic substances continue to attract the attention of many leading organic chemists. The chemistry of the terpenes, proteins, fats, carbohydrates, and alkaloids presents structural and synthetic problems which tax the ingenuity of the best minds in the field. The biochemical activities in the fields of nutrition, physiology, and genetics continue to raise new problems in connection with the vitamins, hormones, enzymes, pigments, and nucleic acids. Some of these subjects will be discussed in the chapter on biochemistry. Others are very largely problems of organic chemistry and will be discussed here.

Carbohydrates

Although Emil Fischer correctly explained the existence of two methyl glucosides by assigning a ring structure to the compounds, he did not extend the ring structure to glucose itself, feeling that evidence for such an extension was inadequate. He failed to recognize that the problem was related to the phenomenon of mutarotation discovered in 1846 by Dubrunfaut who found that the optical rotation of freshly prepared glucose solutions falls steadily until a specific rotation of $+52.5°$ is reached. In 1895 Charles Tanret reported the preparation of two glucose isomers, one with a specific rotation of $+113°$, the other of $+19°$. After the isomers are dissolved in water, the rotation of each changes to an equilibrium value of $+52.5°$.

Edward F. Armstrong (1878–1945) related the two forms of glucose to Fischer's methyl glucosides in 1903 when he showed that the α-glucoside is hydrolyzed by emulsin to give the high-rotating form of glucose (α), while maltase emulsifies the β-glucoside to give the low-rotating form (β). During the next thirty years there was a concerted study of the structure of sugar molecules, with special attention being given the nature of the oxygen-containing ring. Particularly important contributions were made in the laboratories of Claude S. Hudson (1881–1952; United States Public Health Service); James C. Irvine (1877–1952; St. Andrews); W. Norman Haworth (1883–1950; Durham, Birmingham); and Edmund L. Hirst

(1898–1975; Birmingham, Bristol, and Edinburgh). Purdie and Irvine introduced the preparation of methyl ethers of the sugars. The ethers were used with great effectiveness by these workers and by Haworth in establishing the location of ring closure. The pentaacetates, first prepared by Franchimont in 1879 and 1892, were also useful as were the acetone addition compounds first prepared by Emil Fischer in 1895 and studied carefully by Irvine in 1913. About 1926 it became evident to Haworth and Hirst that methyl glucoside commonly occurs in the form of a pyranose ring. Haworth later established that the furanose ring structure is also possible, although the equilibrium in glucose results primarily in the pyranose ring.

Glucose	α-Methylgluco-	α-Methylgluco-	
Aldehyde structure	pyranoside	furanoside	
(Fischer)	(Haworth, Hirst)	(Haworth)	

Hudson carefully studied optical activity in the sugars. Using van't Hoff's principle of additivity (1898), he devised rules for dealing with optical effects at various asymmetric centers. Hudson was also instrumental in applying periodic acid to the solution of structural problems in carbohydrate chemistry. The reagent was introduced for the splitting of α-glycols by L. Malaprade in 1928. Hudson used it effectively to get information on terminal groups, branches in chains, and order of linkage in polysaccharides. Lead tetraacetate, introduced by Rudolf Criegee, also proved useful in the splitting of molecules where glycol linkages occur.

By the thirties the most pressing structural problems regarding monosaccharides had been solved and a fruitful attack was being launched on the glycosides and the polysaccharides. The latter presented an exceedingly complex problem but, because of the economic importance of starch and cellulose, work continues in many laboratories. The efforts of organic chemists in this area have been fruitfully supplemented by those of physical chemists.

Fats

The study of fats and waxes has been pursued actively since 1900. Several methodological contributions have helped simplify the problem of separating and identifying constituents. For many decades the main procedure for separating fatty acids was to take advantage of the difference in solubility of the lead salts of liquid (primarily unsaturated, soluble) and solid (saturated, insoluble) acids in ether, according to the method introduced by Gusserow and Varrentrapp. This technique was improved in the twenties when Ernst Twitchell (1863–1929) substituted alcohol for ether. Separation of individual fatty acids from the saturated fraction was achieved by fractional crystallization until fractionating columns became efficient enough to permit distillation of the methyl esters at low pressure. Unsaturated acids were usually brominated, crystallized, and identified. J. B. Brown introduced low-temperature crystallization methods about 1940, thereby greatly improving the separation of unsaturated acids. The phase studies of acid mixtures made by H. A. Schuette and his students at Wisconsin provided valuable data for analytical studies. The application of gas chromatography to fatty ester mixtures became a powerful tool in the fifties.

Proteins and Amino Acids

Because of their importance in living processes, the study of nitrogen compounds has been actively pursued. The proteins have received particular attention, but a variety of other types of nitrogen compounds have been encountered among the hormones, vitamins, coenzymes, nucleic acids, and drugs.

Few additional amino acids have been discovered since 1900. Emil Fischer, after developing the fractional distillation of ethyl esters as a separation technique, isolated valine, proline, and hydroxyproline in protein hydrolysates. William C. Rose (b. 1887) at Illinois made the important discovery of threonine in casein when he failed to get good growth in rats on a synthetic diet containing no protein. Edward C. Kendall (1886–1972) of the Mayo Institute isolated the iodine containing thyroxine from thyroid gland in 1915. Its structure was established by Charles R. Harington and George Barger at Manchester.

Except for several amino acids occurring in nonprotein sources, all such acids have an amino group on the α-carbon atom and all have the same optical configuration at this asymmetric center. Karl Freudenberg (b. 1886) of Heidelberg showed that natural alanine is related to L-(+)-lactic acid. Therefore all of the natural amino acids have the L-configuration (although most of them are dextrorotatory). Acid-base properties of the acids appear to be in accord with the zwitterion formula, a term coming from the German word for hermaphrodite.

$$\text{R}-\overset{\overset{\displaystyle \text{NH}_3^+}{|}}{\underset{\underset{\displaystyle \text{H}}{|}}{\text{C}}}-\text{COO}^-$$

Analysis is complicated by the dual character of the acids, although titration methods have been devised by L. J. Harris, J. B. Conant and N. F. Hall, and Sørensen. The Sørensen titration involves blockage of the amino group by formaldehyde. A widely adopted analytical procedure was introduced in 1911 by D. D. Van Slyke. When sodium nitrite is used on amino acids, nitrogen gas is liberated and measured volumetrically (manometrically).

Determination of individual amino acids has been enhanced in the last two decades by the introduction of chromatographic techniques. Martin and Synge introduced paper chromatography for protein hydrolysates in 1945, the position of individual acids being revealed by treatment with ninhydrin which forms a colored compound upon reacting with amino acids. Stanford Moore and William H. Stein at Rockefeller Institute for Medical Research used starch columns in 1948 and ion-exchange resins in 1951 to bring about quantitative separations.

Microbiological assay for arginine, introduced in 1946 by E. E. Snell at Wisconsin, has been extended to other acids. David Rittenberg at Columbia perfected the isotope dilution technique for amino acid analyses in 1940.

Of particular significance in protein chemistry is the development of end-group analysis whereby the sequence of amino acids in peptide chains may be studied. In 1945 Sanger introduced the use of 2,4-dinitrofluoro-benzene for labeling terminal amino groups. Upon hydrolysis the labeled acid is present as a yellow 2,4-dinitrobenzene derivative which can be identified by paper chromatography. Techniques of this sort have enabled Frederick Sanger (b. 1918) of Cambridge to map the distribution of amino acids in the insulin molecule. P. Edman of Sweden introduced the use of phenylthioisocyanate in 1950 as an agent for the removal of the terminal amino acid in a protein or polypeptide. These approaches, coupled with chromatographic analysis and physicochemical methods have proved fruitful in the study of naturally occurring peptides such as glutathione, oxytocin, and vasopressin and bacterial peptides such as the tyrocidines and gramicidins.

Steroid Structures

The widespread occurrence of cholesterol in animal tissues and the discovery of the bile acids stimulated interest in the structure of these compounds. Organic chemistry had progressed by the twenties to a point where such leaders as Adolf Windaus (1876–1959) at Göttingen and

Fig. 23.7. ADOLF WINDAUS.
(Courtesy of the Edgar Fahs Smith
Collection.)

Fig. 23.8. HEINRICH WIELAND.
(Courtesy of David Green.)

Heinrich O. Wieland (1877–1957) at Munich were willing to attack structural problems of this magnitude. Until 1932 the work was almost entirely on cholesterol (Windaus) and on the readily available bile acids, cholic and desoxycholic (Wieland). Cholesterol was recognized as a complex nucleus carrying a secondary alcohol and an isooctyl side chain. The bile acids were recognized as hydroxy derivatives of cholanic acid containing the same nucleus as cholesterol with a side chain of $-CH(CH_3) \cdot CH_2CH_2 \cdot COOH$. The nucleus was shown to be made up of four fused rings with a high degree of saturation. It was difficult to determine the relationship of the rings on the basis of the oxidative degradation reactions used. At the time Wieland and Windaus received their Nobel Prizes (1927 and 1928, respectively), the proposed structures were

Old Cholesterol Structure

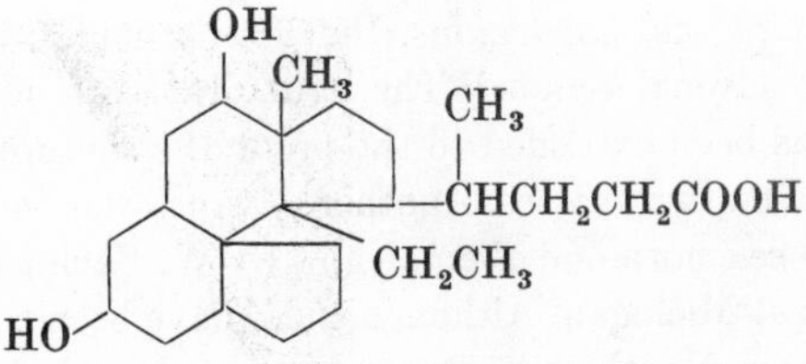

Old Cholic Acid Structure

The attachment of the two carbon atoms constituting the ethyl group was provisional, and Wieland soon was forced to conclude that they did not belong where they had been placed. Selenium dehydrogenation and x ray evidence finally led to a cyclopentanoperhydrophenanthrene nucleus and structures as follows:

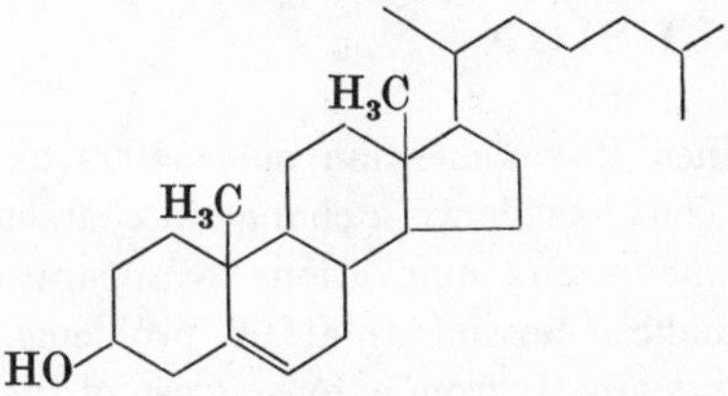

Cholesterol

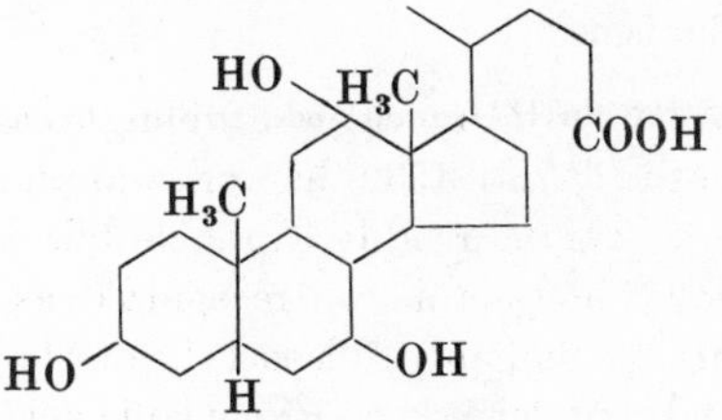

Cholic acid

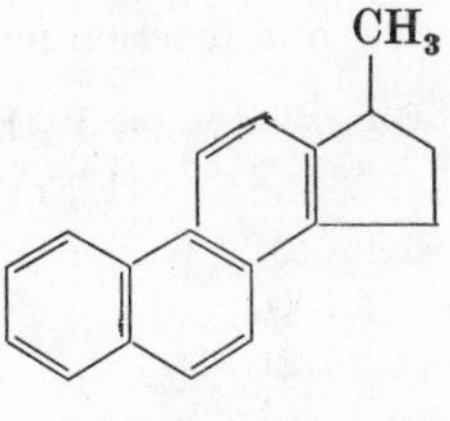

Diels' hydrocarbon

A large part of the thirties was consumed in resolving knotty problems regarding the formation of various degradation products. Diels' hydrocarbon, formed in small amounts from both compounds by selenium dehydrogenation, was the subject of much controversy. Diels and Leopold Ruzicka (1887–1976) each studied its degradation and synthesis very extensively. Many other workers were involved in the clarification of the chemistry of these compounds.[16]

The nucleus found in these compounds turned out to occur commonly in nature. Besides the sterols and bile acids, the same ring system was found in the cardiac glycosides, the toxins from the parotid glands of

---

[16] A. Windaus, *Z. physiol. Chem.*, **213**, 147 (1932); I. Heilbron, J. C. E. Simpson, and F. S. Spring, *J. Chem. Soc.*, **1933**, 626; O. Rosenheim and H. King, *Ann. Rev. Biochem.*, **3**, 87 (1934).

toads, the saponins and sapogenins, the sex hormones, and the cortical hormones of the adrenal glands.[17] The structural work on cholesterol and the bile acids has been extended to the proof of structure of many representatives of these other classes. Another advance was the total synthesis of a steroid, the sex hormone *dl*-equilenin, by W. Bachmann, J. W. Cole, and A. L. Wilds at Michigan. Although there have been total syntheses of this and other steroids, their work marked a milestone in the field since it not only represented a successful synthesis of great complexity, but also served to confirm the conclusions regarding the structure of these compounds previously based on degradation reactions. Another sex hormone, estrone, was synthesized from equilenin by R. E. Marker in 1936, but this synthesis was of no importance in clarifying structure.

CONCLUSION

Organic chemistry has maintained its momentum since 1900 as a vigorous field of chemical research, but has shown a change in character in the last decade or so. Improvements and innovations in apparatus, reagents, and analytical methods make it possible to attack problems of great complexity. This attack benefits greatly from extensive use of theoretical principles, particularly those dealing with electronic concepts in bonding and reaction mechanisms.

[17] For example, see W. H. Strain in H. Gilman, ed., *Organic Chemistry*, Wiley, New York, 2nd ed., 1943, vol. 2, p. 1341.

CHAPTER 24

BIOCHEMISTRY II.
THE DYNAMIC PERIOD

Biochemical research was often looked upon with contempt during the nineteenth century. It was considered imprecise since biological chemists rarely worked with pure materials and the experiments were often carried out without proper attention to suitable controls. This criticism was frequently invalid and the critics often failed to appreciate the problems confronting the investigators; nevertheless biochemistry seemed an unpromising field to many chemists in more precise areas of research.

Since 1900 biochemists have earned great respect. Of course their success is attributable in part to the groundwork laid in the earlier period, in large part to their more critical attitude and to the more advanced nature of their experimental methods. Biochemistry became an exact science only when it was able to draw upon improved techniques of analytical, organic, and physical chemistry.

The most spectacular advances have been made in the field of nutrition. At the beginning of the century the protein and mineral requirements of animals were only barely recognized and the vitamin concept was unknown. By 1950 the major nutritional problems had been solved and biochemists were turning to more subtle problems such as those dealing with cellular metabolism and hereditary mechanisms. The status of nutrition was now such that limitations on an adequate diet were no longer scientific, but principally economic. Rickets and scurvy had become medical curiosities in many parts of the world. Pellagra was no longer the scourge of the United States South that it once had been, and beriberi was largely under control in the Orient. To the extent that deficiency diseases exist today, they result from ignorance or poverty or both.

Significant attacks began to be made upon such troublesome problems as photosynthesis, metabolism of sugar, chemical control of body processes, and so on. Since 1940 biochemistry, it may be said, has been in its "dynamic" period, in contrast to the earlier "static" period when interest lay primarily in the nature of the specific compounds involved in physiological processes. Major interest is focused now on the nature of material and energy transfers during physiological changes. Isotopic tracers have proved invaluable in this work and biochemistry now draws heavily on physical chemistry in its search for understanding of the chemistry of life.

NUTRITION

At the beginning of the twentieth century concepts of adequate nutrition were somewhat more refined but actually little changed from Liebig's views a half-century earlier. The point of view was well expressed in 1898 by C. F. Langworthy,[1] an associate of Atwater, in his "laws of nutrition." Foods were thought to have two purposes: (1) to supply energy for heat and work and (2) to provide material for building and repair. Except for nitrogenous material needed for building and repair, Langworthy considered the three components—proteins, fats, and carbohydrates—as interconvertible, rating in the ratio of 1:2.5:1 as energy producers. This notion was consonant with the view of Voit and of Atwater. The latter published nutrient tables comparing the prices of the three classes of components in various foods. These tables stressed the low cost of cereals and placed water-rich foods such as milk, eggs, meat, seafoods, fruits, and vegetables at a disadvantage.[2]

Caloric values and a protein controversy which raged during the first decade of the new century were of major concern to food and feed experts. The traditional view that meat eating leads to vigorous health was challenged by Russell Chittenden of Yale who, on the basis of experiments on himself and on athletes, concluded that high-protein diets lead to various ailments arising from toxins produced by protein decomposition in the colon. Serious scientists as well as food faddists insisted that protein intake should be merely sufficient to replace nitrogen losses.

The controversy, which was characterized by a high degree of subjectivity and a minimum of controlled experimentation, gradually exhausted itself as it became evident that the matter was complicated by the presence or absence of trace nutrients, the existence of which was coming to be dimly recognized. In part these nutrients were associated with proteins themselves since amino acid composition was found to vary from protein to protein, but even more significant were indications that highly purified

[1] C. F. Langworthy, *Expt. Sta. Record*, **9**, 1003 (1898).
[2] W. O. Atwater, *Chemistry and Economy of Food*, U.S. Dept. Agr., Bull. 21 (1898).

fats, proteins, carbohydrates, and minerals cannot be formulated into adequate diets for experimental animals or human beings. The evidence had been accumulating since the Franco-Prussian War, but the implications were overlooked until 1906, when at least a dozen papers were gathering dust in the literature.[3]

Frederick Gowland Hopkins (1861–1947) of Cambridge recognized the significance of unknown food factors. When he and S. W. Cole isolated tryptophan in 1900, they observed that this amino acid is responsible for the protein test discovered by Adamkiewicz in 1875. Since the protein

Fig. 24.1. FREDERICK GOWLAND HOPKINS.
(Courtesy of David E. Green.)

from maize (zein) fails to give the test they concluded that zein contains no tryptophan. Hopkins and E. C. Willcock set up experiments with mice to find out if tryptophan is important in nutrition. Using a diet of zein, starch, sugar, ash, paper, different fats, and lecithin, they observed the average survival time to be fourteen days. When tryptophan was present in the basic diet, the survival time was extended to 28 days, indicating the need for tryptophan, but suggesting the general inadequacy of the diet. Supplementation with a small quantity of milk led to normal growth

[3] E. V. McCollum, *A History of Nutrition*, Houghton, Mifflin, Boston, 1957, pp. 207–209.

and survival. In a lecture in 1906 Hopkins[4] pointed out that synthetic diets are fundamentally inadequate and that natural foods contain countless other substances necessary for health. He pointed to rickets and scurvy as diseases that might be prevented or cured by proper diets. Obviously influenced by recent work on hormones, he suggested that these compounds must be synthesized from appropriate precursors such as digestion products of proteins furnished by the diet.

Evidence for the dietary origin of rickets and scurvy was available but was generally ignored by the medical profession which was just then overimpressed with the microbial origin of disease. Cod-liver oil was a folk remedy for rickets but since the etiology of the disease is complicated and the "remedy" is not always curative, it had little medical standing. The same was true of sunshine which was recognized as effective in certain quarters. Scurvy was even more clearly of dietary origin; many cases were on record of its cure by fresh fruits and vegetables. In 1795 the Scottish surgeon James Lind induced the British Navy to supply its seamen with lemon juice, but because of adulteration and improper preservation it was frequently ineffective.

Beriberi, which plagued the Orient after polished rice became popular, was also becoming recognized as dietary in origin. K. Takaki, Director-General of the Japanese Navy, reduced the incidence of the disease on shipboard by increasing the use of meat, vegetables, and condensed milk while cutting down on polished rice. Concurrently (1889) in the Dutch East Indies, Christiaan Eijkman (1858–1930) observed a neurosis in fowl which resembled the symptoms of beriberi in humans. From feeding experiments with chickens he learned that the neuritic symptoms occurred when polished rice was fed but not when unpolished rice was substituted. Eijkman also observed that the incidence of beriberi was high among prisoners fed polished rice but low among those receiving brown rice.

By 1912 further research indicated the inadequacy of highly purified diets for experimental animals. Particularly revealing were the studies on single-grain rations carried out at Wisconsin. S. M. Babcock had been skeptical about the feeding tables developed for animals on the basis of proximate analyses of feedstuffs, suspecting that there might be quality factors not revealed in conventional analyses for protein, carbohydrate, and fat. Four of his associates, Edwin B. Hart (1874–1953), Elmer V. McCollum (1879–1967), Harry Steenbock (1886–1967), and George C. Humphrey[5] (1875–1947) initiated an experiment using sixteen heifer calves. Four of the animals were fed a balanced ration derived entirely from the wheat plant. Three similar sets of calves were fed similarly balanced rations derived from the corn plant, the oats plant, and a mixture from all three

[4] F. G. Hopkins, *Analyst*, **31**, 385 (1906).
[5] E. B. Hart, E. V. McCollum, H. Steenbock, and G. C. Humphrey, Wis. Agr. Exp. Sta., *Research Bull.*, 17 (1911).

plants. The corn-fed animals grew well, went through normal pregnancies, and gave birth to healthy calves. The wheat-fed animals grew poorly, appeared sickly, and aborted or gave birth to calves which failed to survive. The other two sets of animals fell into an intermediate group, none of the animals being in good health. All of the wheat-fed cows died before termination of the four-year experiment, except for one who flourished when transferred to a corn diet during the last year. A healthy, corn-fed cow transferred to the wheat ration at the same time declined very rapidly. The experiment was somewhat different in character from those made with synthetic rations but served to emphasize marked differences even between foods of natural origin.

Such prominent authorities as Voit and Emil Abderhalden were convinced that synthetic diets should be adequate. Voit suggested that since they were unpalatable it should be expected that animals would not eat well and would therefore not appear healthy. McCollum sought to overcome this objection by using sugars, mixed proteins, cheese distillate, and freshly rendered bacon fat in the diets of rats, but without success.

In 1909 Thomas B. Osborne (1859–1929) and Lafayette B. Mendel[6] (1872–1935) at Yale began an extensive series of feeding experiments designed to study the nutritive importance of various amino acids. During his extensive studies on plant proteins, Osborne had accumulated a large number of highly purified compounds and determined the amino acid distribution, showing that proteins vary widely in their chemical composition. Using these proteins, he and Mendel were able to study the need for the different amino acids by supplying proteins in which a particular acid was missing. However it was necessary first to design a diet on which rats grew normally when a good protein like casein was present. This was accomplished only after a "protein-free milk," composed of lactose, minerals, yellowish pigments, and miscellaneous trace components, was developed to supplement the basic diet of lard, starch, and a nutritious protein such as casein. Osborne and Mendel also introduced the practice of raising rats on screens to prevent erratic results associated with coprophagy. (It was later learned that intestinal bacteria synthesized some of the missing dietary factors.)

The vitamin concept was introduced in 1912 by the Polish biochemist Casimir Funk (1884–1967) while he was working at the Lister Institute in London. Here he had become familiar with beriberi, which some of his associates considered to be caused by an amino acid deficiency. Funk had gained amino acid experience while serving as an assistant to Abderhalden in Emil Fischer's laboratory in Berlin. He isolated the proteins of rice and rice polishings, but it soon became evident that the cause of the disease was not an amino acid deficiency. In working with rice polishings,

---

[6] T. B. Osborne and L. B. Mendel, Carnegie Inst, of Washington, *Bull.*, 156, parts 1 & 2 (1911).

Fig. 24.2. ELMER V. McCOLLUM.
(Courtesy of E. V. McCollum.)

and later with yeast which was also curative for the disease, he carried out a fractionation of the nonamino acid nitrogen compounds, obtaining a highly curative concentrate which appeared to belong to the pyrimidine class of compounds and a crystalline substance which was identified as nicotinic acid. Since both of these substances are organic bases and are involved in vital processes, Funk proposed that the name *vitamine* be applied to such trace nutrients whose lack brought on such deficiency diseases as beriberi, scurvy, pellagra, and possibly rickets.[7] The name, despite some opposition, came into general use, but was shortened to *vitamin* when it became apparent that not all of these substances are amines.

Vitamin A and the Carotenes

In experiments on rats McCollum and M. Davis learned that addition of butterfat or the fat of egg yolk stimulated the growth of their animals, whereas they failed when lard or olive oil was used as a calorific source in the synthetic diets. McCollum and Davis were fortunate in unsuspectingly using lactose contaminated with whey constituents and permitting their rats access to their feces, since their diet would otherwise have still been deficient. They went on to show that the growth factor, termed "fat soluble A" (later vitamin A), was associated with the nonglyceride portion of fats and was also present in glandular tissues such as liver and kidney and in the leaves of plants but not in fatty tissues of animals or in vegetable oils.

The next two decades were very active ones in the study of the vitamin

[7] C. Funk, *J. State Med.*, **20**, 341 (1912); *Lancet*, **2**, 1266 (1911); *J. Physiol.*, **43**, 395 (1911), **45**, 489 (1912–13), **46**, 173 (1913); *Die Vitamine*, Wiesbaden, 1913. Also see Benjamin Harrow, *Casimir Funk, Pioneer in Vitamins and Hormones*, Dodd, Mead, New York, 1955.

content of foods, with results frequently being confused as a consequence of impure diets, species variability, and multiplicity of deficiency factors.

In 1919 Steenbock associated vitamin A activity with yellow foods which contain carotene, but other investigators demonstrated activity in sources lacking carotene such as liver and fish-liver oils. It was finally established that carotenes convert to the vitamin and thus afford protection against vitamin A deficiency. Extensive studies on carotenoid pigments of plants in the laboratories of Palmer, Willstätter, Kuhn, Zechmeister, and Karrer led to the separation of the closely related carotenes, xanthophylls, and lycopene and the study of their properties. Karrer observed the presence of the β-ionone ring in some of the compounds and went on to establish the structure of β-carotene and the related pigments. The structure of vitamin A was found to correspond to a half-molecule of β-carotene with an alcohol group at the end of the side chain. Synthesis was later accomplished in the laboratories of R. C. Fuson at Illinois and Kuhn at Heidelberg.

Vitamin D

For a time the prevention and cure of rickets was associated with fat-soluble A, but in 1919 Edward Mellanby showed that puppies still developed rickets on diets supplemented with foods rich in the growth factor. In 1922 McCollum destroyed vitamin A in cod-liver oil by oxidation and found that rats developed the eye soreness associated with vitamin A deficiency but did not become rachitic. He introduced the term *vitamin D* for the antirachitic factor.

Diets developed by Henry C. Sherman (1875–1955) at Columbia revealed the role of calcium and phosphorus in the disease. McCollum's group at Johns Hopkins studied the calcification of bones in animals on various diets.

For years there had been a vague correlation between rickets and lack of sunlight. In the early twenties several reports indicated that rats on a rachitic diet benefited by exposure to sunlight or ultraviolet or even to ultraviolet irradiation of rat jars and their inanimate contents. At this point Steenbock and Archie Black reported that irradiation of a rickets-producing diet prevented the disease. A similar observation was reported by Alfred F. Hess and M. Weinstock at Columbia. The Steenbock discovery was patented by a Wisconsin alumni group and licensed to food and drug processors for enrichment of yeast, milk, and other products, the proceeds being utilized for support of research at the university.

Steenbock and Hess both sought the source of the pro-vitamin and quickly associated it with the sterol portion of fats. The chemistry of vitamin D was finally unraveled in the thirties by groups associated with Windaus at Göttingen and F. A. Askew at the National Institutes of Health in England. Windaus' work on the structure of the sterols actually played an important part in solving the problem since ergosterol proved convertible to vitamin D_2 by irradiation. Still another active compound, named D_3, proved to come from 7-dehydrocholesterol.

The Antiscorbutic Vitamin

Work on vitamin C progressed slowly, although the guinea pig was shown to be susceptible to lack of the factor in 1907. The instability of the factor and the difficulty of designing an otherwise suitable diet impeded research until the twenties. In 1928 Albert Szent-Györgyi (b. 1893), working on biological oxidations in Hopkins' laboratory, isolated a strongly reducing substance from adrenal glands, orange juice, and cabbage which was named *hexuronic acid*. In 1931 Charles G. King at Pittsburgh isolated

Fig. 24.3. HARRY STEENBOCK.
(Courtesy of the University of Wisconsin.)

Fig. 24.4. ALBERT SZENT-GYÖRGYI.
(Courtesy of David E. Green.)

a crystalline substance which protects guinea pigs from scurvy at a level of 0.5 mg. per day. J. Svirbely and Szent-Györgyi established the identity of King's crystals with hexuronic acid and went on to prepare large quantities from Hungarian red peppers.

Structural studies were pursued in the laboratories of Hirst, Haworth, and Karrer. Hirst established a relationship to the sugar L-gulose, and the structural proof soon followed. Taddeus Reichstein's laboratory in Basel accomplished a synthesis before structural proof was final. The name *ascorbic acid* was suggested in 1933 by Szent-Györgyi and Haworth. Commercial syntheses soon made the vitamin a comparatively cheap chemical.

The Vitamin B Complex

When McCollum introduced the term *water-soluble B* in 1915 he had no reason to believe that it would ultimately include more than a half-dozen nutritive factors besides the one curative of beriberi (which came to be designated *vitamin B_1*). However work with different sources (rice polishings, yeast, liver), different basic diets, and different species slowly revealed that a complex of vitamins was included under the original term.

The antiberiberi factor was most clearly definable, and the early work was aimed at its isolation and identification. Potent concentrates were prepared from rice polishings as early as 1912, but further purification was difficult. Finally in 1926 the Dutch workers B. C. P. Jansen and W. P. Donath isolated active crystals. Other workers prepared similar crystals and reported the presence of sulfur. When Robert R. Williams (1886–1965), who had worked on the problem in the Philippines, became chemical director of the Bell Telephone Laboratories he developed a method for isolating the crystals in a yield of 5 mg. per ton of rice polishings. Collaborating with Merck chemists, he established that the active compound is a pyrimidine nucleus linked to a thiazole derivative. In 1936 Williams developed a commercial synthesis. The name *thiamin*, proposed by Williams, came into general use as *thiamine*. During World War II the addition of synthetic thiamine to white flour and other deficient cereal products was introduced.

By 1920 chemists had fairly well established that a growth factor is associated with the antiberiberi factor, and that in the absence of this growth factor rats develop dermatitis as well. As diets were improved the need for such a factor, variously termed B_2, G, or P-P (pellagra preventive), was established. In 1933 L. E. Booher established that rat dermatitis is cured by a yellow pigment prepared from whey powder. Shortly before this Otto Warburg and W. Christian had prepared a yellow oxidation enzyme from bottom yeasts. Kuhn showed the yellow pigment in Warburg's enzyme to be spectroscopically similar to Booher's whey pigment and to yellow pigments from yeast, liver, heart muscle, spinach, and eggs. Kuhn and Booher concluded that the pigment was vitamin B_2. Synthesis was achieved in 1935 in the laboratories of Kuhn and Karrer when it was learned that the vitamin, named *riboflavin*, contains an *iso*-alloxazine nucleus attached to the sugar D-ribose. In contrast to work on most other vitamins, the nutritional role was established only after the synthetic vitamin was available for study since few experimental diets had been free of the vitamin.

Pellagra, endemic in Spain, Italy, and the United States South, was finally recognized as a deficiency disease in the twenties largely as the result of the work of Carl Voegtlin (1879–1960) and Joseph Goldberger (1874–1929) of the United States Public Health Service. They showed that introducing milk, eggs, yeast, or meat into the high-maize, Southern type diets cures the disease. Experimental studies with animals were confusing, however, since rat dermatitis is not equivalent to human pellagra but represents a riboflavin deficiency. The dog ultimately proved the right species when it was recognized that blacktongue is equivalent to human pellagra.

In 1935 H. von Euler's laboratory established that cozymase, essential as a coenzyme in alcoholic fermentation, gives nicotinic acid on hydrolysis. In 1936 nicotinic acid was found to be a hydrolysis product of the coferment (triphosphopyridine nucelotide) isolated by Warburg and Christian. A year later Y. SubbaRow isolated the acid from liver. Conrad Elvehjem (1901–1962) and his Wisconsin associates R. J. Madden, Frank Strong, and D. W. Woolley soon reported nicotinic acid and its amide to be curative for blacktongue in dogs. This was confirmed in other laboratories and extended to the cure of human pellagra.

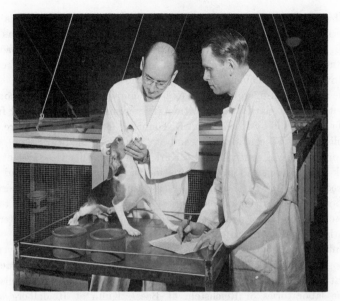

Fig. 24.5. Conrad A. Elvehjem (right) with S. N. Gershoff.
(Courtesy of the University of Wisconsin.)

Despite the effectiveness of nicotinic acid in curing pellagra, there still remained some anomalies regarding the disease. A pellagra-like deficiency in rats, discovered by Goldberger in 1926, failed to respond to nicotinic acid treatment but did respond to yeast or liver. Further, it was learned, human pellagra is found most frequently when the diet is rich in maize and rarely when the diet contains an equivalent proportion of wheat flour, rice, oats, or millet, all comparatively poor sources of nicotinic acid.

In 1938 H. Chick's laboratory in England and Elvehjem's at Wisconsin reported that the growth of rats on a diet of highly purified materials supplemented by maize grits can be improved by adding either nicotinic acid or tryptophan. Later studies showed that pellagra-susceptible species possess the ability to convert tryptophan to nicotinic acid. Since maize is lower in tryptophan than other cereals, there is little amino acid for conversion.

As pure vitamins became available for use in synthetic diets, it was realized that not all deficiency symptoms were being prevented. Rats, pigeons, and chicks on synthetic diets supplemented with all known vitamins frequently showed poor growth and skin and nerve disorders, which ultimately led to decline and death unless the diet was supplemented with extracts of yeast or liver. Students of bacterial growth were also finding vitamins important and were reporting unsatisfactory growth unless natural supplements were added to the nutrient medium. During the thirties there were many reports of such deficiencies in animals and bacteria; the responsible factors were designated by such terms as B_3, B_4, B_5, B_6, H, W, factors I and Y, filtrate factor, antiacrodynia factor, and antialopecia factor. A great deal of confusion surrounded these factors and investigators frequently failed to confirm each other's results.

The animal and microbiological experiments which led to the discovery of biotin helped clarify the situation. The compound was isolated from dried eggs in 1936 by F. Kögl and B. Tönnis at Utrecht. Its structure was established by Vincent du Vigneaud at the Cornell Medical School, and the compound was synthesized by Karl Folkers and his associates at the Merck Laboratories. About the same time *meso*-inositol, a compound discovered in 1850, was shown to be essential for the health of certain microorganisms and for rats and mice. Inositol and biotin proved to be the active components of "bios," the name given organic materials essential for yeast growth by E. Wildiers in 1901.

Panthothenic acid was first encountered as a growth factor for yeasts by Roger Williams at Oregon State in 1933; it was purified and characterized at the Merck Laboratories in 1940. Total synthesis was accomplished in the laboratories of Reichstein at Basel, Kuhn at Heidelberg, and Wieland at Munich. Several laboratories demonstrated its nutritive importance for animals.

Pyridoxine was uncovered in connection with studies of rat acrodynia, at one time confused with pellagra. The factor was isolated in crystalline form by five different laboratories in 1938; the structure proof and synthesis were carried out at the Merck Laboratories under S. A. Harris and Folkers.

Para-aminobenzoic acid (PABA), first detected in Elvehjem's laboratory in connection with gray hair in piebald rats, was found essential in the growth of various bacteria.

The pteroic acid factors involved in the prevention of pernicious anemia were originally encountered in studies of bacterial growth by Esmond E. Snell and William H. Peterson at Wisconsin. Later, at Texas, Snell and his associates isolated an active fraction from spinach leaves (termed *folic acid*, from the Latin *folium*, for leaf). Proof of structure and synthesis were completed in 1946 by a sixteen-member group led by Robert B. Angier at the Lederle Laboratories.

A final member of the B-complex, *cobalamin* or B_{12}, was isolated in 1948 at Merck and by E. Lester Smith of the Glaxo Laboratories in England. Its structure was attacked by Alexander Todd at Cambridge and by Folkers' group at Merck, but success might have been long delayed except for the x ray studies in the laboratories of Dorothy Crowfoot Hodgkin at Oxford and Kenneth N. Trueblood at UCLA. The compound, which has a complicated porphyrin structure complexing an atom of cobalt, has proved important in treatment of pernicious anemia.

Other Fat Soluble Vitamins

Around 1920 it was independently observed in the laboratories of Henry A. Mattill in Iowa and Herbert M. Evans in California that rats on certain experimental diets suffered reproductive disorders which were prevented if wheat germ, lettuce, or alfalfa leaves were added to the diet. An active substance, named *tocopherol*, was isolated from the unsaponifiable matter of wheat germ oil in 1936. The structure of the several active compounds was deduced in 1937 by E. Fernholz at Princeton, and a successful synthesis was made by Karrer.

Certain unsaturated fatty acids, particularly linoleic and arachidonic, were found by George O. and Margaret M. Burr at Minnesota to play a role in the nutrition of rats.

A fat-soluble vitamin involved in the coagulation of blood was encountered in 1929 by C. P. Henrik Dam of Copenhagen in connection with his studies on chicks. Isolation and identification was accomplished over a decade later in four laboratories, those headed by H. J. Almiqust at California, Edward Doisy at Washington University (St. Louis, Mo.), Louis Fieser at Harvard, and Paul Karrer at Zurich.

Somewhat parallel to the discovery of vitamin K was the isolation by Karl Paul Link and his associates at Wisconsin of dicoumarol, a substance responsible for hemmorhage in cattle eating spoiled sweet clover hay.[8] Dicoumarol and related compounds are useful in the prevention of blood coagulation.

Other Nutritive Essentials

Early and sporadic studies of the role of amino acids in nutrition revealed the essential nature of ten of these compounds. Osborne and Mendel's work at Yale early in the century, followed by that of Emil Abderhalden in Germany and H. H. Mitchell at Illinois, suggested that lysine, histidine, tryptophan, cystine, and tyrosine must be supplied in the diet of the rat. Further research in the thirties by William C. Rose at Illinois led to the discovery of threonine and revealed that lysine, tryptophan, histidine, leucine, isoleucine, phenylalanine, methionine, valine, arginine, and threonine are essential in the diet of the rat.[9] Other studies showed that man requires the same acids except for arginine and histidine.

Concurrently with the work on vitamins and amino acids, inorganic elements were studied. In some cases the essential nature of the element was easily demonstrated. In others the total need is low and it was difficult to design diets which contained none of the element under

[8] H. A. Campbell, W. K. Roberts, W. K. Smith, and K. P. Link, *J. Biol. Chem.*, **136**, 47 (1940); **138**, 1 (1941); Campbell and Link, *ibid.*, **138**, 21 (1941); M. A. Stahmann, C. F. Huebner, and Link, *ibid.*, **138**, 513 (1941).

[9] W. C. Rose, *Physiol. Revs.*, **18**, 109 (1938).

consideration. By 1940 it was quite clear that experimental and domestic species need sodium, potassium, calcium, magnesium, iron, zinc, manganese, cobalt, chlorine, iodine, fluorine, and phosphorus.[10]

PHOTOSYNTHESIS

Following de Saussure's work in 1804 there was little positive investigation of the photosynthetic reaction before 1900. Liebig speculated in 1843 that plant acids were the intermediates between carbon dioxide and sugars, basing his hypothesis on the sourness of unripe fruits: In 1870 Baeyer introduced the formaldehyde hypothesis whereby carbon dioxide was first believed to be reduced to formaldehyde which subsequently became condensed to sugar. Even before this Butlerov had converted paraformaldehyde into a sweet syrup with the aid of alkali. The syrup, a complex mixture, was not attacked by yeast. Similar results were obtained a quarter-century later by O. Loew, using formaldehyde. The formaldehyde hypothesis was a popular one, especially with Edward C. C. Baly (1871–1948) who published reports of experiments which tended to support it.

Other investigators failed to find evidence of even traces of formaldehyde in plant material where photosynthesis had been taking place, and it was well known that even small amounts of formaldehyde have a toxic effect on living cells. H. A. Spoehr examined the whole matter critically in his monograph *Photosynthesis* (1925) and concluded that the formaldehyde hypothesis was without foundation.

Blackman obtained evidence in 1905 that the reduction of carbon dioxide in photosynthesis takes place in several steps and demonstrated kinetically that one or more of the reactions takes place in the dark and is temperature-dependent. It was learned later that numerous steps in the photosynthetic process are not photochemical.

The source of the liberated oxygen was at first believed to be the carbon dioxide, particularly since equimolecular amounts of gas are involved. However several lines of evidence suggested water as the source, which was finally proven in 1941 by Samuel Ruben and Martin D. Kamen through the use of oxygen-18 as a tracer.

A new insight into the photosynthetic process was introduced late in the twenties through studies of photosynthesis in bacteria. In their studies of anaerobic fermentations, Kluyver and Doncker learned that hydrogen is transferred to acceptors other than oxygen. Cornelius B. van Niel of Stanford related photosynthesis in higher plants with photosynthesis in sulfur bacteria. In green sulfur bacteria, for example, he found CO_2

---

[10] E. V. McCollum, *A History of Nutrition*, Houghton, Mifflin, Boston, 1957 deals with the role of inorganic nutrients in Chapters 21, 22, and 26.

and hydrogen sulfide to be photochemically converted to organic material, water, and sulfur. In purple sulfur bacteria, carbon dioxide, water, and reduced forms of sulfur were converted to organic matter and sulfuric acid. Van Niel looked upon these reactions as fundamentally similar to photosynthesis in green plants. Such bacteria even contain a chlorophyll-like substance. In 1941 van Niel suggested a general equation in which a hydrogen donor H_2A is necessary for the photoreduction of carbon dioxide to an organic building block (CH_2O) with the release of A. In green plants the hydrogen donor is water, in green sulfur bacteria it is hydrogen sulfide, in other bacteria, an organic substrate is converted to a more oxidized form.

Further understanding of photosynthesis came from several sources. The clarification of intermediary carbohydrate metabolism through the work of Embden, Meyerhoff, Krebs, the Coris, and others established the products and processes by which sugars are utilized. In turn this knowledge was helpful in studying the synthetic process. Techniques involving the use of isotopic tracers have been of major importance since 1940.

Ruben, Hassid, and Kamen at the California Radiation Laboratory carried out photosynthetic studies on barley, using the 22-minute half-lived carbon isotope of mass 11. They found that the plant utilizes $^{11}CO_2$ in the light and also in the dark, provided the labeled carbon dioxide is added soon after illumination. The bulk of the carbon-11 is found in water-soluble compounds. These same investigators carried out similar studies with the green alga *Chlorella*, which is used extensively in photosynthetic studies. They learned that the ^{11}C is to be found in compounds containing carboxyl and hydroxyl groups. In 1941 Ruben and Kamen identified carbon-14 produced from nitrogen by an (n, p) reaction. This isotope of carbon, with its 5,600 year half life, is far more suitable for metabolic studies than carbon-11 but it was little used until after World War II because of the press of other problems.

The tracer study of photosynthesis was begun in earnest in the California Radiation Laboratory by Melvin Calvin (b. 1911). Preilluminated *Chlorella* was exposed to $^{14}CO_2$ in order to study the dark uptake of labeled carbon. Using techniques of solvent extraction, precipitation, and ion exchange it was possible to identify ^{14}C in such compounds as phosphoglyceric acid, malic acid, and alanine. Employing two-dimensional paper chromatography and radiochemical techniques Calvin's group was able to study the formation of labeled amino acids, carboxylic acids, triose phosphates, hexose phosphates, phosphoglyceric acid, ribulose diphosphate, sedoheptulose monophosphate, and other reaction products, thereby gradually causing the photosynthetic reaction to unfold its secrets.[11]

---

[11] M. Calvin, *et al.*, *Carbon Dioxide Fixation and Photosynthesis*, Symposia for the Society of Experimental Biology, no. 5, Academic Press, New York, 1951.

Chlorophyll

The nature of the chlorophyll molecule has received extensive study since 1900. The name was given to the green coloring matter of plants by Pelletier and Caventou in 1817, but almost from the beginning the pigment was suspected to be a mixture. In 1873 F. G. Sorby separated two green and two yellow pigments by partition between solvents. Chromatography was used by Tswett early in the new century for the same purpose. In 1906 Willstätter and his associates prepared relatively pure chlorophyll and subjected it to careful study. The green pigment of leaves proved to consist of two compounds, chlorophyll a and b, present in the proportion of 3 to 1. The formula of chlorophyll a was found to be $C_{55}H_{72}N_4O_5Mg$; the b variety differs in that it has an additional oxygen atom and two less hydrogen atoms. The formulas were verified in 1932 by Hans Fischer and by Arthur Stoll.

Nencki had shown in 1901 that heme of hemoglobin is readily degraded to substituted pyrroles and that chlorophyll also yields pyrroles. Willstätter went on to identify these pyrroles and thereby began the reconstruction of the porphyrin structure. He also showed that hydrolysis results in the splitting off of methyl and phytyl alcohols.

Willstätter's fundamental work was extended during the twenties and thirties, particularly in the laboratories of Hans Fischer (1881–1945) at Munich and James B. Conant (1893–1978) at Harvard. Fischer recognized a fundamental skeleton in hemin of four pyrrole molecules with methyl, ethyl, vinyl, and propionic acid fragments. Around 1930 Conant deduced that the phytyl alcohol group in chlorophyll must be esterified to the propionic acid side chain. Fischer synthesized the prophyrins derived from hemin and from chlorophyll. The synthesis of chlorophyll a itself was accomplished in 1960 by a group in Woodward's laboratory at Harvard and independently by M. Strell, A. Kolajanoff, and H. Killer of the Technische Hochschule in Munich.

INTERMEDIARY METABOLISM

The end products of animal metabolism were recognized fairly early in the study of biochemistry. Unraveling the pathway by which the carbohydrate, fat, and protein molecules are degraded to carbon dioxide, water, and urea was, however, a difficult piece of work. There are still unanswered questions.

While hypothetical mechanisms for the intermediate steps in metabolism could be postulated, it was not easy to identify intermediates since many of them ordinarily exist for only a fraction of a second. Thus experimental evidence favoring or nullifying a postulated mechanism was difficult to gather. Much of the data was accumulated piecemeal, a fragment coming from one laboratory, a fragment from another.

The study of enzymes was particularly fruitful. Once a pure enzyme was obtained its activity could be studied without interference from other enzyme systems. The study of fermentation processes also led to important clues which were extended to processes occurring in animal tissues. Certain chemicals were found to poison specific enzymes, thus causing an accumulation of the substance the enzyme ordinarily attacked.

In 1897 Edward Buchner showed that a cell-free extract of yeast can ferment sugar to alcohol. In 1878 Kühne used the term *enzyme* (Gr., *en*, in; *zyme*, yeast) to designate the responsible agent. And in 1904 Arthur Harden and W. J. Young of the Lister Institute reported the dialysis of yeast juice with the separation of a protein and nonprotein fraction. Neither fraction is active while alone, but fermentation occurs when the two portions are combined. The dialyzable fraction contains a substance, *Coenzyme I*, which serves as an adjunct, or prosthetic group, on the enzyme, thus making it functional. Harden and Young also discovered that phosphorylated intermediates are present during the fermentation process. They could identify fructose-1,6-diphosphate (Harden-Young ester).

Enzymes

The elucidation of enzyme chemistry played an important part in gaining an understanding of intermediary processes. Early studies of digestive enzymes were important only to the extent that they permitted pioneering in the field of organic catalysts. It was soon learned that the digestive enzymes are not a part of cellular systems and cellular metabolism is a truly complex phenomenon. It was suspected by 1900 that enzymes are proteins with a comparatively specific catalytic function, for example, hydrolytic (esterases, carbohydrases, proteinases, peptidases, phosphorylases, amidases) and oxidative (dehydrogenases, oxidases, catalase). The proteinaceous nature of enzymes seemed evident to Traube and other nineteenth-century students but was questioned during the 1920's when Willstätter purified yeast invertase and several other enzymes. He reported high catalytic activity but failed to obtain color reactions for tryptophan and concluded that the enzymes themselves are not proteins, although they are frequently associated with proteins.

Urease, which catalyzes the hydrolysis of urea to carbon dioxide and ammonia, was prepared in crystalline form from Jack-bean meal by James B. Sumner (1887–1955) in 1926. Several years later John H. Northrop (b. 1891) crystallized pepsin. By 1960 about 75 enzymes had been crystallized. These compounds all proved to be proteins, frequently with prosthetic groups, and it was shown that Willstätter was led astray by the fact that even at great dilutions enzyme activity is often so pronounced that the common protein color reactions are inoperative.

Fig. 24.6. OTTO WARBURG.
(Courtesy of Dean Burk, National Institutes of
Health.)

Prosthetic groups were encountered in enzyme studies from the beginning of the century when Harden and Young recognized Coenzyme I, or Cozymase. The identified prosthetic groups frequently were compounds of pyrimidines or purines, a pentose sugar (usually ribose or deoxyribose), phosphate units, and members of the vitamin B complex. They function in dehydrogenation reactions, racemization, decarboxylation, or in the transfer of phosphate, methyl, amino, or acetyl groups. Another type of prosthetic group reported was the iron-porphyrin type found not only in hemoglobin but also in catalases and carboxylases responsible for the decomposition of hydrogen peroxide and organic peroxides.

The study of the factors influencing rates of enzymatic reactions is based largely on the work of Lenor Michaelis and M. L. Menton in 1913. They developed a theory for the influence of substrate concentration on the velocity of the catalyzed reaction and worked out a quantitative treatment for it. Although refined and extended by later workers, it still provides the foundation for rate studies.

The use of manometric techniques in enzyme studies can be traced to the work of Joseph Barcroft (1872–1947) in 1902. His method was improved by Otto Warburg's respiration manometer in which the oxygen uptake of a metabolizing system is studied by observing changes in pressure as the reaction proceeds.

Carbohydrate Metabolism

Numerous laboratories contributed to the unraveling of carbohydrate breakdown. It became apparent fairly early that the conversion of hexoses to carbon dioxide takes place in the muscle by means of an anaerobic process which produces 3-carbon fragments followed by an aerobic conversion to the oxides. Carl Neuberg identified acetaldehyde in 1911 by a trapping technique, using sodium bisulfite to form an addition product which is not attacked. A few years later Otto F. Meyerhof (1884–1951) showed that the coenzymes present in alcoholic fermentation are also found in muscle cells.

Further progress was made by G. Emden, J. Kluyver, K. Lohmann, H. von Euler, J. K. Parnas, O. Warburg, and other investigators. Work on alcoholic fermentation made it possible to postulate the pathway of hexose breakdown through phosphorylated 3-carbon fragments to pyruvic acid, acetaldehyde, and ethyl alcohol. The pathway is known as the Embden-Meyerhof scheme, after the two investigators who clarified many of the steps. The completely anaerobic sequence provides for a significant energy release.

Several important advances were made in the thirties. Albert Szent-Györgyi studied the respiration of minced pigeon breast muscle, important for its high rate of respiration, and observed that while the rate of oxygen uptake falls off with time, the original rate can be restored by the addition of salts of such acids as succinic, fumaric, malic, or oxalacetic. The closely related malonic acid, however, inhibits the process. T. Thunberg showed that muscle contains such enzymes as succinic dehydrogenase, fumarase, and malic dehydrogenase. Hans A. Krebs (1900–1981) showed that respiratory rates are also increased by salts of citric acid, α-ketoglutaric acid, pyruvic acid, and such amino acids as L-glutamic and L-aspartic. Martins and Knoop studied the oxidation of citrate by muscle tissue, reporting the sequence: citric $\rightleftarrows$ *cis* aconitic $\rightleftarrows$ D-isocitric $\rightleftarrows$ oxalosuccinic $\rightleftarrows$ α-ketoglutaric.

Carl and Gerty Cori, at Washington University, St. Louis, developed techniques for preventing the next step in a sequence by blocking the activation of the necessary enzyme. By this means they were able to elucidate the steps in the synthesis and degradation of glycogen. They successfully synthesized glycogen *in vitro* from glucose in the presence of adenosine triphosphate (ATP) and the enzymes, hexokinase, phosphoglucomutase, and phosphorylase, the glucose being successively converted to glucose-6-phosphate, glucose-1-phosphate, and glycogen. ATP serves in the first step as a phosphate donor, being restored from adenosine diphosphate (ADP) in the terminal step.

Krebs, working at Sheffield in 1940, suggested a rather complicated cycle involving the decarboxylation of pyruvic acid and the transfer of

Fig. 24.7. CARL AND GERTY CORI.
(Courtesy of David E. Green.)

Fig. 24.8. HANS A. KREBS.
(Courtesy of David E. Green.)

the resulting acetate group to oxalacetic acid with the formation of citric acid, followed by a sequence of dehydrogenations and decarboxylations terminating in the restoration of oxalacetic acid and the complete oxidation of the acetic acid residue. His cycle involves the catalytic aid of a sequence of enzymes plus such cofactors as coenzyme A which is involved in the transfer of acetyl groups, adenosine di- and tri-phosphate participating in phosphate transfer, and tri-phosphopyridine nucleotide acting as a hydrogen transfer agent. The Krebs' cycle has undergone various refinements but has proved sound in its main essentials. The use of respiratory poisons such as malonic acid or sodium arsenite blocks certain steps in the cycle and causes the accumulation of identifiable acids. Furthermore the cycle accounts for the rapid intramuscular oxidation of glutamic acid, aspartic acid, and alanine. These three amino acids are interconvertible by transamination to α-keto acids, which are a part of the Krebs cycle.

Lipid Metabolism

The occurrence of natural fatty acids almost exclusively as straight chains with an even number of carbon atoms posed a problem which had to be taken into consideration in all attempts to deal with synthesis and degradation. It was known that fats can be formed from carbohydrates, and a hexose molecule might readily furnish two 3-carbon fragments for glycerol synthesis. The explanation of long hydrocarbon chains, such as those present in fatty acids, offered greater difficulty.

The useful hypothesis of β-oxidation came out of F. Knoop's studies in 1904. Knoop fed dogs a number of organic acids which contained a phenyl group at the end of the hydrocarbon chain. He knew that benzoic acid is detoxified in animals and eliminated as hippuric acid in the urine; that phenylacetic acid is excreted as phenaceturic acid. After feeding the dogs ω-phenyl acids of progressively longer chain length, Knoop reported the recovery of hippuric acid when the phenyl group was on an odd-numbered chain (for example, phenyl propionic acid, phenyl valeric acid) but phenaceturic acid when the chain was even-numbered (for example, phenyl butyric acid). This suggested that in fatty acid oxidation the attack occurs on the β-carbon atom and removes carbon atoms in units of two. Degradation of a chain from the acid end appeared to continue until the phenyl group blocked further attack at the β-position.

Further evidence favoring β-oxidation came from studies of diabetics, whose blood, breath, and urine contains "acetone bodies"—β-hydroxybutyric acid, acetoacetic acid, and acetone. Embden showed by liver perfusion experiments that these compounds derive from fatty acids, the oxidation of which fails to go beyond the 4-carbon acid stage in diabetics.

Hypotheses for the synthesis of fatty acids also sought to deal with the introduction of 2-carbon units. M. Nencki suggested an aldol condensation of acetaldehyde units, followed by reduction of the hydroxyl group and oxidation of the terminal aldehyde group. J. B. Leathes and H. S. Raper held a somewhat similar view, but postulated dehydration at the hydroxyl positions to form double bonds which might later become saturated. Ida Smedley introduced a synthetic mechanism starting from pyruvic acid and acetaldehyde which provided for a buildup of 2-carbon units.

Although the Smedley hypothesis was widely accepted for several decades, it failed to account for various factors which are involved and was superceded in the forties. Rittenberg and Block prepared acetic acid labeled with deuterium and carbon-13 and fed it to animals which were then found to contain large amounts of the labeled atoms in their fatty acids.

Following the discovery of coenzyme A (CoA) by Feodor Lynen in 1951 Fritz Lipmann and Severo Ochoa independently showed that the acetylated form of the coenzyme is involved in fatty acid synthesis. This scheme accounted for the large energy change involved, but later was challenged as the result of the work of Salih J. Wakil at Wisconsin's Enzyme Institute.

New light was thrown upon the storage and use of fats in 1935 when Rudolf Schoenheimer (1898–1941) began his studies with deuterium-labeled fats. Up to this time it had been supposed that an animal in energy balance oxidizes freshly consumed fats for its energy needs. Fat deposits were looked upon as reserve supplies used only when the diet was inadequate. Schoenheimer hydrogenated linseed oil with deuterium and fed the labeled fat to mice maintained at constant weight. Within four days 44 per cent of the labeled fatty acids were detectable in the fat deposits. The experiments showed that fatty acids were steadily being deposited and that the deposits were constantly drawn upon for metabolic needs. In other studies with labeled palmitic acid deuterium was soon detected in stearic, oleic, lauric, and myristic acids. Schoenheimer's studies with labeled carbohydrates and amino acids revealed further that all body constituents are in a highly dynamic state.

Another fruitful area of lipid study has been that dealing with fatty livers. In the earlier work on diabetes it was learned that while depancreatectomized dogs can be maintained in a fairly satisfactory state by injecting insulin, their livers become very high in fat content. Charles Best found that this can be alleviated by feeding lecithin or, better still, choline.

The role of lecithin and other phospholipids in metabolism was never established in the early studies. Lecithin was isolated from egg yolk in 1846 by N. T. Gobley and found to contain nitrogen and phosphorus besides fatty acids and glycerol. Strecker identified the nitrogenous base as choline, a compound which he had previously isolated from bile. Brain chemistry, extensively studied by J. L. W. Thudichum in the 1880's, revealed the presence not only of lecithin, but of cephalin, and the sphingolipids. The exact chemical nature of the latter compounds was not fully clarified until the 1940's when Ernest Klenk at Cologne and Herbert E. Carter at Illinois established the nature of the base sphingosine and its place in the structure of the sphingomyelins and cerebrosides.

Best's work on the prevention of fatty livers revealed that fats leave the liver as choline phospholipids. The capacity of the liver to synthesize lecithin was related to the choline supply through tracer studies with phosphorus-32 and nitrogen-15. In the absence of choline lecithin synthesis can be stimulated by analogs like arsenocholine $(CH_3)_3As^+CH_2CH_2OH$. It was further observed that fatty livers fail to develop on high-protein diets, methionine being the responsible component. Guided by these findings, Vincent du Vigneaud (1901–1978) formulated the concept of the labile methyl group and the process of transmethylation.

Du Vigneaud hypothesized that the role of methionine in preventing fatty livers might be attributed to its serving as a source of easily available methyl groups for the synthesis of choline. The hypothesis was verified by feeding methionine which had a deuterium-labeled methyl group. It also was demonstrated that animals can synthesize methionine from deuterium-labeled methyl groups in choline and from betaine. There remained, however, the question of the animal's capacity to synthesize labile methyl groups. This possibility was suggested by the anomalous behavior of individual animals in certain experiments. A small amount of

Fig. 24.9. VINCENT DU VIGNEAUD.
(Courtesy of David E. Green.)

synthesis was observed in experiments in which rats were supplied with deuterium-enriched water, the deuterium showing up in the methyl groups of choline. The possibility that the synthesis might be attributable to bacteria in the intestinal tract was disproved with the aid of James Reyniers (1908–1967) of Notre Dame University. Reyniers had developed techniques for rearing and studying animals in a germ-free environment. When two such rats were fed deuterium-enriched water they were found to contain deuterium-enriched labile methyl groups.

Nitrogen Metabolism

The intermediary metabolism of nitrogen compounds has received considerable attention, but many questions are yet to be answered. Oxidation transamination was observed by Krebs in 1933 and for a time was thought to be a general reaction. Later studies showed that L-amino acid oxidases occur commonly only in snake venoms and certain molds, although D-amino acid oxidases occur commonly in animal tissues, apparently serving to destroy the unnatural amino acids wherever encountered.

Transamination, which involves an exchange of α-amino and α-keto groups between two organic acids, has proved an important metabolic reaction. In 1934 R. M. Herbst and Engel observed the reaction in a model system involving α-amino phenylacetic acid and pyruvic acid;

alanine, benzaldehyde and carbon dioxide were detected as reaction products. The latter two products presumably resulted from the decarboxylation of phenyl-α-ketoacetic acid, $C_6H_5 \cdot CO \cdot COOH$. Evidence for transamination in pigeon breast muscle was obtained in 1937. Various amino acids react with α-ketoglutaric acid to form glutamic acid or with oxalacetic acid to form aspartic acid. Schoenheimer's isotopic tracer studies using ^{15}N for labeling glycine or leucine established that the nitrogen-15 is soon found in most amino acids. It was also observed about this time that the α-keto analogues of the essential amino acids serve as satisfactory substitutes in the diet.

In 1945 David E. Green reported the isolation from the heart muscle of swine of two enzymes which catalyze transamination reactions. The same year E. E. Snell observed that in cases of pyridoxine (B_6) deficiency, transamination reactions proceed poorly. A decade later he showed, employing model systems involving no enzymes, the reaction of pyridoxal phosphate and glutamic acid to form α-keto glutaric acid and pyroxamine phosphate and the reverse reaction. The isolated transaminases contained the pyridoxal phosphate-pyridoxamine phosphate system as a coenzyme.

Ornithine Cycle. Urea was isolated from urine by Rouelle in 1773. Its immediate source was not recognized until 1904 when A. Kossel and Dakin found the enzyme arginase in animal tissues and observed its role in the hydrolysis of arginine to urea and ornithine. The metabolic formation of the heavily nitrogenous end group of arginine from ammonia and carbon dioxide remained in doubt until 1932 when Krebs reported that either ornithine or citrulline stimulates the formation of urea in liver slices. Krebs postulated a cycle in which ornithine reacts with ammonia and carbon dioxide to form citrulline. On reaction with ammonia this gives arginine which in the presence of arginase is hydrolyzed to urea and ornithine.

Later studies revealed that the cycle is fundamentally correct, but not as simple as postulated. In 1954 Sarah Rathner showed that the ammonia in the citrulline-ornithine step has its origin in aspartic acid in an enzymatically controlled reaction involving transformation of ATP to the monophosphate and the simultaneous formation of arginino succinic acid. The latter is enzymatically cleaved to fumaric acid and arginine. The conversion of ornithine to citrulline was also found to be enzymatically controlled. Studies by Philip P. Cohen (b. 1908) and his associates at Wisconsin revealed a complex mechanism involving ATP, magnesium ion, carbanyl-glutamic acid, and at least two enzymes. The relationship between this cycle and the citric acid cycle was soon established; fumaric acid and aspartic acid represent bridging compounds.

Synthesis of Nitrogen Compounds. Contributions from many laboratories have served to clarify the interrelationships in synthesis of nitrogen compounds. The use of isotopic tracers has helped establish the precursors and pathways in amino acids, vitamins, and porphyrin. For example, tracer studies of porphyrin synthesis by David Shemin (b. 1911) of Columbia revealed glycine and succinic acids to be the principal precursors. Glycine also is a precursor in purine synthesis, with carbon dioxide, ammonia, and formaldehyde contributing fragments, not directly, but through aspartic and glutamic acids.

The source of the aromatic rings in certain amino acids was a puzzle until tracer studies revealed glucose to be the starting substance. The steps in the conversion of glucose to phenylalanine, tyrosine, and tryptophan were clarified through the use of mutant strains of microorganisms which had lost the capacity to synthesize certain enzymes necessary for normal metabolism. The principal contributions have come from the laboratories of Beadle and Tatum at Stanford who used *Neurospora* strains and of Bernard D. Davis at New York University who used strains of *E. coli*.

HORMONES

The term *hormone* (Gr., I rouse to activity) was introduced by William M. Bayliss (1860–1924) and Ernest Henry Starling (1866–1927), physiologists at University College, London, following their discovery of secretin in 1902. Isolated from the duodenal mucosa, this polypeptide stimulates the secretion of pancreatic juice. There had been earlier evidence of chemical secretions which affect the regulation of body chemistry.

Insulin. The role of the pancreas in carbohydrate metabolism was revealed in 1889 by von Mering and Minkowski who produced diabetes in dogs by removing the gland. Although the existence of the hormone was still hypothetical, the name *insulin* was proposed in 1916 by Edward A. Sharpey-Schafer, since he suspected that the secretion originated in the Islands of Langerhans, peculiar cell clusters found in the pancreas in 1869 by the German physician Paul Langerhans. Early efforts to isolate the hormone failed because the extracts were contaminated with trypsin, the protein-splitting enzyme also produced by the pancreas.

Frederick Banting (1891–1941), working in J. J. R. Macleod's physiological laboratory at the University of Toronto, together with Charles H. Best (1899–1978) and later James B. Collip (b. 1892), obtained a non-toxic extract in 1922, at first from dogs, later from fetal calves, and finally from beef pancreas. The hormone soon became available for treatment of diabetic patients. The chemistry of insulin was extensively studied at Johns Hopkins in the laboratory of J. J. Abel, at Columbia by Oskar Wintersteiner, and in Britain by D. A. Scott and C. R. Harington. The protein nature of the hormone was quickly established, and Scott identified zinc in association with the crystalline compound. As mentioned earlier, Sanger established the structure of the molecule and synthesized it.

Adrenalin. In 1895 G. Oliver and Sharpey-Schafer observed that an extract of the adrenal glands has a powerful effect in raising the blood pressure. Several years later Abel obtained the benzoyl derivative of the responsible compound from the medulla of the glands. This was synthesized by Stoltz in 1904 and also by Dakin. The compound was named *adrenalin*, or *epinephrin*. In amounts of 0.2 to 0.5 mg., it is effective in increasing the blood pressure, accelerating the heart beat, and counteracting the effect of excess insulin. The structure proved to be related to ephedrine, which is found in a Chinese herb, and to such synthetic compounds as benzedrine (Amphetamine), Propadrine, and Neosynephrine which are used in the treatment of bronchial asthma and nasal congestion and for contracting the capillaries.

Pituitary Hormones. The pituitary gland at the base of the brain proved to be a source of a number of hormones which have an important role in controlling the action of certain muscles and in stimulating various organs to the production of hormones. The hormones of the posterior lobe of the gland have a rapid action of short duration, those of the anterior lobe a slower but more prolonged effect. Two of the posterior hormones which have been studied, *oxytocin* and *vasopressin*, increase blood pressure, stimulate uterine contractions, and reduce urinary secretion. Both hormones are octapeptides; the formulas were established in 1953. Du Vigneaud prepared oxytocin by synthesis from the amino acids.

The anterior lobe contains at least six hormones, all of them polypeptides or proteins, sometimes with carbohydrate present in the molecule. One of the hormones stimulates growth; the others stimulate the thyroid, the adrenal cortex (ACTH), the mammary glands, the ovarian follicles, the corpus luteum, and the testes.

Sex Hormones. The effects of castration were observed very early in the history of agriculture. The differences between the ox and bull, the gelding and stallion, the capon and cock, the eunuch and man were too obvious to escape notice. It was long suspected that the effects of castration resulted from lack of a substance normally secreted by the testes, but the first formal statement of the idea of an internal secretion was put forth by Johannes Müller in 1833. In 1849 A. A. Berthold showed that transplantation of a cock's testes into a capon restored the latter to a normal-appearing cock.

In the twenties Fred C. Koch's (1876–1948) laboratory at the University of Chicago reported distinct progress in the extraction from bull testicles of a substance which, on injection into capons, produces restoration of secondary sexual characteristics, in particular the growth of the comb. Less active extracts were also obtained from urine. In 1931 Adolf Butenandt (b. 1903) reported the isolation of 15 mg. of a crystalline hormone, *androsterone*, from 15,000 l. of urine. His proposed structure was confirmed by L. Ruzicka who synthesized the compound from cholesterol.

A closely related steroid was also isolated and identified. These compounds, however, do not have the potency of testicular extracts. A highly active steroid, *testosterone*, was isolated in the Amsterdam laboratory of Ernst Laqueur (1880–1947) in 1935. Soon after this Butenandt and Ruzicka synthesized the identical compound, thus establishing its structure.

Studies on female sex hormones were pursued vigorously in the twenties, particularly by Butenandt at Göttingen and Edward A. Doisy (b. 1893) at St. Louis University. These workers independently isolated *estrone* in 1929; the search dated back to 1923 when Doisy showed the presence in follicular fluid of a substance which induces estrus in rats and developed a test for its estimation. The substance was at first concentrated from the urine of pregnant women. B. Zondek showed in 1930 that the urine of the pregnant mare is much richer and later found the best source to be stallion urine. The formula of estrone was readily established once the fundamental problem of the steroid nucleus was solved.

Reduction of the ketone group of estrone gave a compound, *estradiol*, which is more active than the parent compound. This finding led Doisy's group to undertake the isolation of the more active compound from a half ton of sows' ovaries. It was also isolated from the urine of pregnant mares as were the weakly estrogenic *estriol* and *equilenin*.

Estrone extracted from mare's and human pregnancy urine was soon introduced into human therapy. As more of the primary products have become available they have been chemically converted to more suitable compounds. Estradiol was at first prepared by the reduction of estrone, but now it is partially synthesized from natural steroids such as cholesterol. It has been found further, that certain synthetic products such as stilbestrol and hexestrol show marked estrogenic activity.

The natural female hormones formed in the ovary during the ripening of follicules were found to be responsible for the changes in the uterus associated with estrus. Following expulsion of the ovum, the ruptured follicle gradually fills with a yellowish material known as *corpus luteum*. The corpus luteum produces a hormone which completes the alteration of the uterus begun by the estrogen. This hormone, *progesterone*, prepares the uterine wall for implantation of the fertilized egg, inhibits the ripening of further follicules, and stimulates development of the mammary glands. If fertilization of the ovum does not take place, the corpus luteum degenerates and a new follicle begins to ripen. The isolation of progesterone was achieved in 1934 by Butenandt at Danzig, Karl Slotta at Breslau, Wintersteiner at Columbia, and Albert Wettstein in the CIBA laboratory in Basle. The structure was soon established and commercial production of the hormone soon thereafter achieved, using stigmasterol, cholesterol, cholic acid, or the sapogenin diosgenin.

Cortical Hormones. In 1855 the degeneration of the adrenal glands was related to Addison's disease. Late in the 1920's it was observed that the lives of adrenalectomized animals can be prolonged by the use of extracts of the adrenal cortex. This initiated a series of investigations, particularly in the laboratories of Reichstein in Switzerland, Kendall at the Mayo Clinic, and Wintersteiner at Columbia. Twenty-eight steroids were isolated within a few years; a twenty-ninth, aldosterone, was isolated in 1953. The structures were determined very largely in Reichstein's laboratory. Seven of the compounds, which show a hormonal activity, play a role in the control of electrolyte balance in body fluids, kidney function, and carbohydrate formation from proteins. Certain members are useful in the treatment of Addison's disease, and in 1949 Hench and Kendall observed the relief offered by cortisone for sufferers from rheumatoid arthritis. Synthetic studies have led to even more active compounds with further ring unsaturation or with fluorine atoms present.

CONCLUSION

Biochemistry has become a vigorous specialty since 1900 since it has been possible to move forward from the basic knowledge of chemistry and physiology provided by nineteenth-century investigators. Progress began to be made in the fields of hormones and nutrition by the beginning of the century. While the early studies were cumbersome in method and highly empirical they netted encouraging results. Later on the principles arising out of the fields of basic chemistry began to take on significance. However nearly all of the work on vitamins was completed during the period when analytical results were obtained by assay procedures utilizing living organisms.

Progress in understanding the vitamins was slow because the concept of trace nutrients was an unfamiliar one. Furthermore the diseases involved were often complicated by the problem of multiple nutrients. The unraveling of the story of the vitamin B complex was particularly difficult, being complicated by the fact that certain members of the complex are seldom encountered at levels which result in deficiency disease. Despite the difficulties, the vitamin problem was essentially solved by 1940.

Although some early progress was made in work on photosynthesis and intermediary metabolism, the attack on these problems has advanced markedly since 1940. The availability of isotopic tracers makes it possible to carry out studies which could not be designed earlier. The experience gained in dealing with trace amounts of vitamins and hormones has proved valuable in the work on metabolic intermediates. Unquestionably of great significance to success in biochemical studies has been the introduction of analytical methods utilizing chromatography.

CHAPTER 25

INDUSTRIAL
CHEMISTRY II.
CHEMICALS
FOR INDUSTRIAL USE

TWENTIETH-CENTURY TRENDS

The chemical industry became firmly established in a significant role in the Western economy during the nineteenth century. Since 1900 it has become one of the dominant industries and markedly influences techniques and products in other industrial fields. This phenomenal growth since 1900 was stimulated at first by wartime conditions and demands. As it resolved the problems raised by World War I, the industry acquired consciousness of the vital role of research which has led to the cycle: research begets new and improved products and techniques which in turn suggest new areas for research in addition to those explored in response to other needs.

At the beginning of the twentieth century Germany was supreme in the chemical field as the result of the conscious promotion of scientific education, the enthusiastic use of scientists in industry, and vigorous utilization of mutually advantageous economic devices within the industry. The British chemical industry was well-developed primarily in the field of heavy chemicals, but was still suffering from the futile efforts of the Leblanc soda producers to remain in business in the face of competition from the Solvay process.[1] The fine chemicals industry was insignificant in England except for the company of Ivan Levinstein. In the other European nations there was some activity, especially in heavy chemicals production. Fine chemicals were in large part controlled by the Germans. Only the Swiss firms were able to hold consistently markets for quality chemicals in the face of German competition.

The Allied chemical industry was woefully ill-prepared for World War I. The German industry was vigorous and healthy, that of the Allies was underdeveloped and moribund. The Germans were able not only to supply the conventional chemical needs of a nation at peace but also to provide the explosives and other chemicals required for warfare. The commercial development of the Haber process for fixation of atmospheric nitrogen came at an opportune time since otherwise the British navy would have been able to create a nitrate shortage by preventing import of Chilean saltpeter, until then essential for German agriculture and industry.

For a time the unavailability of German dyes, drugs, potash chemicals, photographic chemicals, optical glass, and chemical glassware was a severe blow to the Allied Powers and to sympathetic neutrals. The shortage, however, stimulated strenuous British and American efforts to meet the more critical needs, and by the end of the war vigorous chemical industries were developing in both countries. The Allies finally were able to meet explosive with explosive, and chemical poison with chemical poison. Germany was handicapped during the last months of the war by a rubber shortage which resulted from the Allied blockade. German chemists introduced a synthetic rubber, but it was of inferior quality.

After the war the Allied nations were determined that theirs would be healthy chemical industries. The strong industries developed in some nations during the war were in a position to meet peacetime needs. Other countries were not in so fortunate a position and government subsidies and other types of favored treatment were necessary to insure the development and growth of fledgling chemical industries. In the postwar climate this preferential treatment was usually forthcoming. The seizure and exploitation of German patents, favorable tariffs, and outright subsidies were common practices.

Germany never regained its dominant position, but its chemical industry remained vital. By developing new products and aggressive cartel arrangements, I. G. Farben, the voice of the German industry, was able to hold a significant share of the world market. In Britain, Imperial Chemical Industries, Ltd.—organized in the twenties out of United Alkali Co., the Brunner, Mond interests, British Dyestuffs Corp., Nobel, and two cyanide

---

[1] L. F. Haber, *The Chemical Industry during the Nineteenth Century*, Clarendon Press, Oxford, 1958, pp. 96 ff., 180.

companies—was dominant. The trend toward consolidation was very marked in postwar Europe and was actually encouraged by many of the governments.

The German tradition of government encouragement of industry can be traced back to the Prussia of Frederick the Great and even to many of the smaller principalities. This pattern was firmly fixed under Bismarck and Kaiser Wilhelm II. During the years of economic distress after World War I the government characteristically encouraged any activities which would strengthen Germany's position in the world chemical market. The formation of the Interessen Gemeinschaft für Farbenindustrie in 1925 met with strong governmental approval. As early as 1904 the German dye manufacturers were moving toward consolidation. Badische, Bayer, and AGFA joined in a cartel-type community-of-interest arrangement in which markets were divided and patents exchanged. At the same time Höchst obtained a 99 per cent interest in the stock of Kalle, and a 75 per cent interest in Casella. Later Badische-Bayer-AGFA and Höchst negotiated a joint understanding, and in 1916, faced with the prospect of thriving dye industries in Britain and America, they agreed to a more binding cartel. For the German chemical industry the period following the end of the war was chaotic. I. G. Farben, formed under the management of Carl Bosch of Badische, represented a trust controlling the major part of the chemical industry. It and Deutsche Gasolin combined to dominate the German petroleum market; with Glanzstoff-Bemberg it brought rayon under control; and through stock interests in Nobel and Rheinisch-West-fälische it "tied up" explosives. The synthetic ammonia operations at Oppau, where a devastating ammonium nitrate explosion had occurred in 1921, were expanded to the point where Germany became a major figure in the world fertilizer market. Badische's synthetic methanol wrecked havoc on foreign wood-distilling operations. The Bergius' process for hydrogenation of coal was pushed, as was research on synthetic rubber. These activities, of marginal importance during the twenties, were revived and expanded following the Nazi rise to power.

In France the Establissements Kuhlmann was a similar but much smaller scale consolidation. The parent firm traced back to 1825 when the small Kuhlmann sulfuric acid plant began operation. The Verein für Chemische und Metallurgische Production von Aussig represented a similar combine in Central Europe.

The large chemical combines not only sought to wield economic power within national boundaries, but also entered into cartel arrangements with one another in order to control world markets. A German-Swiss-French dye cartel was functioning even before its formal organization in 1928. Operating arrangements were kept secret, but it was apparent that the Germans controlled the rich market in the Orient and the French that of South America, while the Swiss operated in Southern Europe.

In the United States the trend toward bigness was also evident during the twenties. Because of the long-standing American opposition to monopoly, no one company rose to the dominant position of I. G. Farben in Germany or I.C.I. in Britain, but through internal expansion and merger such companies as du Pont, Union Carbide and Carbon, Monsanto, Standard Oil (several uncombined companies), Hercules, American Cyanamid, Allied Chemical and Dye, became powerful organizations.

During the Great Depression chemical companies suffered too, but tended to recover more rapidly than firms in other branches of industry. Research activities began to pay off as markets developed for new textiles, plastics, alloys, and other products. During this period coal-tar chemicals began to lose ground as sources of synthetic compounds to aliphatic chemicals derived particularly from petroleum and natural gas. Coal-tar, "king" during the nineteenth century, was deposed by petrochemicals.

World War II placed heavy demands upon the chemical industry, not only for traditional explosives and other chemicals, but also for light metals, synthetic rubber, high quality aviation fuels, synthetic oils and fats, medicinals, and purified isotopes for nuclear weapons. The industry of both warring camps was sufficiently vigorous and imaginative to meet the demands placed upon it. It is doubtful whether the German chemical industry could have produced nuclear weapons in view of the heavy and sustained bombing of German industrial centers toward the end of the war. But despite Allied bombing German industry was able to develop and place into production the dreaded V-2 missiles.

In the United States peace did not prove catastrophic to an industry grown to monstrous proportions in response to the demands of war. Pent-up consumer demands, reconstruction needs, new products, and new uses generally have maintained production at a high level. And, of course, the tensions of the Cold War have had a marked effect on the chemical industry which must meet the demands of nuclear research and work on jet propulsion systems and rocket fuels, for example. The space age has directed attention to many elements and compounds which once were mere curiosities in someone's laboratory. Metals such as titanium and zirconium and fuels such as hydrazine and the boron hydrides suddenly are in demand.

PRODUCTION OF STANDARD CHEMICALS

The major characteristics of the heavy chemical industry in 1900 were the demise of the Leblanc process, the growth of the contact process for manufacturing concentrated sulfuric acid, the rapid development of electrochemical industries, and the recognition that a practical method for the fixation of atmospheric nitrogen was needed.

Sulfuric Acid and Sulfur

Three developments dating from the late 1800's have influenced the sulfuric acid industry throughout this century—the production of acid from by-product sulfur dioxide, the contact process, and the introduction of Frasch process sulfur. For some time the metal smelting industry had been producing more and more by-product sulfur dioxide. The fumes given off during the roasting of sulfide ores are so objectionable that in the 1890's governments began imposing regulations to insure their control. This legislative trend spread in the new century, stimulating the introduction of the contact process for utilizing sulfur dioxide produced during smelting operations.

The impact of the contact process, which was brought into commercial operation during the 1890's, began to be felt in the early 1900's. Several different processes involving differences in catalysts and conditions of operation were developed. While platinum was the principal catalyst, iron and vanadium oxides also were employed.

Fig. 25.1. HERMAN FRASCH.
From a portrait by Robert Vos in the Chemists
Club, N.Y.
(Courtesy of the Freeport Sulfur Company.)

After a decade of development the Frasch process was first introduced commercially in 1901 in Louisiana. The German-born Herman Frasch (1851–1914) emigrated to the United States in 1868. After attending the Philadelphia College of Pharmacy, he worked in the petroleum industry in Cleveland, where he invented a process for desulfurizing crude oil and for reviving exhausted oil wells by hydrochloric acid treatment. Early in the nineties he became interested in the sulfur-bearing salt domes in Louisiana. Elemental sulfur was known to be present in these geologic formations in the limestone fissures 150 meters beneath the surface muck and quicksands which prevented conventional mining.

Frasch developed a technique for exploiting these deposits. Three concentric pipes were sunk into the formation; superheated water sent down the outer pipe melted the sulfur. Compressed air forced down the central pipe created pressure which forced the molten sulfur up the intermediate pipe from where it ran into large storage vats. Although sulfur was successfully pumped late in 1894, seven years elapsed before mechanical and financial difficulties were solved. The enterprise benefited by the opening of the nearby Spindletop oil field which provided a source of fuel for supplying the heavy demand for hot water. Commercial production expanded rapidly—from under 7,000 long tons in 1901 to 220,000 tons by 1905.[2] Frasch sulfur broke the sway of the Sicilian brimstone monopoly and reversed the trend away from the use of elemental sulfur in acid production. In addition the Frasch interests were able to supply sulfur in quantity for agricultural uses, for the rapidly expanding rubber industry, and for the paper industry which had introduced the bisulfite pulping process.

Other Acids

Although sulfuric acid remained the primary industrial acid, other inorganic acids began assuming more important roles and quantity production of organic acids was undertaken.

Nitric acid continued to be important as a nitrating agent in the production of explosives and a great variety of organic chemicals. Its history is inseparable from that of ammonia and had best be treated there.

In the twentieth century hydrochloric acid has become increasingly important. It is used in many organic reactions such as the preparation of vinyl chloride for plastics, chloroprene for synthetic rubber, and alkyl chlorides as intermediates for many syntheses; also in the delinting of cottonseed and the treatment of oil wells. While large amounts of the acid are still prepared from sodium chloride, by-product acid is produced as the result of chlorination of organic compounds, for example, the production of phenol by chlorination of benzene.

In 1927 Charles Pfizer & Co. introduced a commercial process for the production by fermentation of citric acid, which until then had been produced in Italy from citrus juices. The process was based on J. N. Currie's discovery that the mold *Aspergillis niger* produces a significant quantity of citric acid during the fermentation of molasses or starch at

[2] Williams Haynes, *American Chemical Industry, A History*, Van Nostrand, New York, 6 vols., 1944–1954, vol. 1, 1954, p. 266.

low pH. Production from cull citrus fruits and pineapples continues to satisfy a part of the demand. The food industry is the principal customer for citric acid.

The Pfizer company, founded in 1849 in Brooklyn, New York, has pioneered in fermentation chemicals since its success with the citric acid process. It has introduced gluconic acid for use in the pharmaceutical field, fumaric acid for varnish resins, and itaconic acid. Because of its experience in industrial fermentations Pfizer naturally was among the first to produce antibiotics.

Lactic acid obtained from the fermentation of whey was introduced by a subsidiary of the National Dairy Products Corporation in 1936. This acid is also produced by the fermentation of molasses and glucose. S. M. Weisberg of the National Dairy Laboratories was responsible for the development of the whey process. Besides its value in the food and drug industries, lactic acid is used in electroplating and is esterified to produce lacquer solvents.

Salts

While we cannot here deal with the production and uses of salts in detail, a few developments are worthy of notice. Aluminum salts, of course, are still important in chemical operations. Alum, essential in dyeing, tanning, and baking powder, is now in great demand for water purification. Anhydrous aluminum chloride came into use with the industrialization of Friedel-Crafts reactions. It is employed, for example, in the synthesis of anthraquinone from benzene and phthalic anhydride and in the condensation of phenols with phosphorus oxychloride to prepare tricresyl phosphates, used as plasticizers and gasolene additives. Activated aluminum compounds are in great demand as adsorbents and catalysts.

Chromium salts and chromic acid have become increasingly important since the perfection of chrome-plating processes by Colin G. Fink of Columbia University and others. A virtual nationwide fetish for chrome plate on automobiles, household products, and industrial decor has made the United States the world's heaviest user of the metal, even though it has almost no domestic sources.

The demand for fluorides, which rose when the Freons (CCl_2F_2, etc.) were introduced as refrigerant gases, increased still further when the Manhattan Project began separating uranium isotopes by gaseous diffusion of the hexafluorides. And research on the organic fluorides has resulted in the introduction of products, Teflon, for example, for which fluorine is required in quantity.

Alkalies and Chlorine

While soda ash still is important as an industrial alkali, with the electrolytic production of chlorine sodium hydroxide has become available in ever increasing amounts. In some areas it has displaced the carbonate as well as gaining new markets of its own.

The demand for chlorine has grown continually. The expansion of the paper industry has steadily strained the capacity of producers to supply chlorine for pulp bleaching. Further, more and more chlorine has been required for the treatment of municipal water supplies and for sanitation in the food industry and in hospitals.

For large-scale purposes chlorine is generally used directly and is shipped in liquid form in steel cylinders and even in tank cars. Bleaching powder, which can be converted to chlorine solution as needed is usually preferred for smaller scale applications, as in dairies and on farms. High-concentration powder was introduced in 1928 by the Mathieson Alkali Works as HTH (high test hypochlorite). This is essentially calcium hypochlorite $Ca(OCl)_2$, in contrast to ordinary bleaching powder $CaCl(OCl)$, where only half of the chlorine is available for oxidizing purposes.

In 1936 in Hopewell, Virginia, the Solvay Process Co. introduced the nitrosyl process for preparing chlorine and sodium nitrate. The reaction of nitric acid and sodium chloride makes it possible to produce chlorine without alkali. The demand for metallic sodium has risen since the metal became necessary in the production of tetraethyl lead for gasolene. The electrolytic cell developed by J. Cloyd Downs in 1924 is widely used in the production of sodium and chlorine from molten sodium chloride.

The Fixed Nitrogen Problem

In 1898 Sir William Crookes, fearful that a food shortage was imminent, urged that per acre yield of wheat be sharply increased by intensive cultivation. But he warned this would quickly deplete the world's resources of fixed nitrogen unless a commercial process for the fixation of atmospheric nitrogen was developed. (Fertilizer nitrogen then came almost entirely from Chilean saltpeter and by-product ammonia from gas plants.) Even as Crookes spoke, several possible approaches to the development of such a process were being considered.

In Norway Kristian Birkeland (1867–1917), professor of physics at the University of Christiana (Oslo), and Samuel Eyde, an engineer, brought a process into commercial operation in 1903. Their approach, based upon a technique utilized by Cavendish in 1784, had been examined and rejected in Britain and the United States. The high temperatures required to bring about even a moderately favorable equilibrium concentration of nitric oxide were difficult to attain without having to invest at prohibitive cost in a power source. Birkeland and Eyde succeeded primarily because of availability of cheap electricity in Norway. The arc was made more effective by spreading it out with a strong magnetic field. The nitric oxide formed by passing air through the arc was oxidized to the brown tetroxide upon cooling, after which it could be dissolved in water to form nitric acid. This was neutralized with limestone to give calcium nitrate, a product sold for fertilizer as "Norwegian saltpeter."

Because so much electricity was required, the process did not spread to other parts of the world and finally was displaced by the Haber process. The last Birkeland-Eyde plant terminated its operations in 1928. Nevertheless the direct combination of nitrogen and oxygen continued to have appeal, and various investigators sought economical means of attaining the necessary high temperatures. A serious effort was made during the

Fig. 25.2. WILLIAM CROOKES.
(Courtesy of the J. H. Walton Collection.)

forties by Farrington Daniels and his associates at Wisconsin. Their gas-fired furnaces and system of heat interchangers made the process sufficiently promising to persuade the Food Machinery Corporation to operate a pilot plant for a time.

The cyanamide process was also commercialized during the first decade of the century. F. Rother, assistant to Professor Adolf Frank (1834–1916), and Dr. Nikodem Caro of the technical high school at Charlottenburg prepared the compound by passing nitrogen over calcium carbide at a temperature of 1000° C. Patents were taken out by Frank and Caro in 1902. Commercial production began in Germany and later in Canada and the United States. Cyanamide, which slowly hydrolyzes to ammonia in the soil, was sold for direct use as a fertilizer. Treated with steam or hot caustic solution it hydrolyzes rapidly. It was an important source of synthetic ammonia for about a decade. Like the Birkeland-Eyde process, it was dependent on a cheap source of electric power since the necessary reactions for producing calcium carbide and converting it to cyanamide were carried out in electric furnaces.

The direct combination of nitrogen and hydrogen was also receiving much attention just prior to World War I. The reaction had been studied as early as 1840 by Regnault, but only when it was examined from the standpoint of thermodynamic principles was it found to be commercially practical. Nernst's early theoretical studies were followed up by Fritz Haber (1868–1934) who demonstrated by 1905 that it might be made a working process. Carl Bosch (1874–1940), an engineer with the Badische firm, was active in the industrialization of the process, which is often designated the Haber-Bosch process. The reaction is carried out at pressures of 200 atmospheres. This required pioneering in the field of pumps and reaction vessels. Since the equilibrium is favored by low temperatures, catalysts which would speed up the reaction to a practical degree had to be developed. Various metallic oxides were found suitable, usually in empirical mixtures. Temperatures of 550° C. are necessary in order to achieve reasonable reaction rates, despite the harmful effect on the equilibrium.

Hydrogen was produced by the action of steam on coke, which results in water gas. In order to obtain hydrogen free of carbon monoxide, the Badische group developed a process whereby steam and water gas were passed over a metallic catalyst which brought about oxidation of the carbon monoxide to dioxide while producing more hydrogen. Nitrogen was produced in pure form by fractionation of liquid air produced by the Linde process. By dissolving in water ammonia was extracted from the equilibrium mixture formed after nitrogen and hydrogen passed over the Haber catalysts. Badische rapidly expanded production during the war years, and in 1918 Germany produced 200,000 tons of synthetic ammonia. Germany thus did not need Chilean nitrates.

Since nitric acid is essential for the manufacture of modern explosives, Germany needed a method for converting ammonia into nitric acid. It was provided by the Ostwald process which was brought into commercial production before the Haber process. The catalytic oxidation of ammonia to nitric oxide over hot platinum had been studied as early as 1839 by Frédéric Kuhlmann of Lille. In the early 1900's Ostwald and Eberhard Brauer[3] undertook a systematic investigation of the reaction to see if it could not be made reliable. Patents were taken out in 1902, and by 1909 construction of a plant had begun. By 1914 Germany was producing 250 tons of nitric acid per day by oxidation of ammonia, much of it used for explosives manufacture. Some was combined with ammonia for ammonium nitrate fertilizer.

Other countries also sought to develop a nitrogen fixation process, especially during the war because submarine warfare made the supply of

[3] Wm. Ostwald and E. Brauer, *Chem. Zeit.*, **27**, 100 (1903).

Chilean saltpeter uncertain. There was considerable interest in the cyanamide process, particularly in the United States where it had been introduced in 1917. A large plant was set up by the United States government at Muscle Shoals, Alabama, but the war ended before it was brought into operation. In the meantime the Haber process was further developed, and ammonia plants were built in several countries. In time the process was modified and made more efficient. Pressures of 800 to 1000 atmospheres were attained through the work of the Frenchman Georges Claude (1870–1938) and the Italian Luigi Casale (1882–1927). As suitable alloy steels for reaction vessels became available, such pressures could be used.

The large-scale development of the Haber process naturally had ruinous effects upon the economy of Chile which for 75 years had been the world's principal source of fixed nitrogen. Producers of Chilean nitrate tried to retain their market but could not avoid a marginal position. Extracting the iodine and boron compounds in the crude mineral permitted them to maintain production.

INDUSTRIAL GASES AND THEIR USES

During the nineteenth century oxygen was produced on a very limited scale; it was used principally in the combustion of coal gas in the limelight and occasionally for medical purposes. Oxygen was produced by decomposition of potassium chlorate until the Brin process was perfected around 1886. This process, based upon the oxidation of red-hot barium monoxide by air at increased pressure followed by the decomposition of the dioxide to monoxide and oxygen at reduced pressure, reduced the price of oxygen to the point where its use in blowpipes became widespread.

During the 1890's there was considerable interest in the potential of acetylene as a fuel gas, but its instability under pressure made it dangerous to handle in the conventional manner. Generators were developed with which it could be formed from calcium carbide and water as needed. In 1897 Claude found that it is readily absorbed in acetone and can be safely handled in cylinders with the acetone solution absorbed on infusorial earth. In 1895 Le Chatelier called attention to the high temperature of the oxyacetylene flame. During the following years researchers sought to develop a torch which would effectively utilize the two gases. Early in the 1900's the Fouché brothers, Edmond and Davis, perfected such a torch and developed it for welding and metal cutting.

As the use of the oxyacetylene torch grew in the metal-working and glass-working trades after 1906, the demand for acetylene and oxygen increased rapidly. Where cheap electricity was available electrolysis of water became competitive with the Brin process, but liquefaction of air soon superseded both processes as a source of oxygen. During the 1890's, processes for air liquefaction were developed by William Hampson in England, Tripler in the United States, and Carl von Linde in Germany. In 1902 Linde patented a process for effectively separating the components of air by fractional distillation. His process was soon exploited commercially in Germany and, after 1906, in the United States. By 1914 compressed oxygen of good purity was available in large quantity. The nitrogen produced simultaneously became a major asset when the Haber process was introduced to industry.

Until 1900 hydrogen gas was used principally to fill balloons and to burn with oxygen to produce the limelight. Hydrogen had generally been produced by the action of sulfuric acid on zinc. With the rise of the electrochemical industries it was produced along with oxygen in the electrolysis of water and with chlorine and sodium hydroxide in the electrolysis of salt solution. Interest in the hydrogenation processes developed from the researches of Sabatier, Senderens, and others created an unprecedented demand for hydrogen.

Paul Sabatier (1854–1941) showed in 1897 that when hydrogen and the vapor of unsaturated organic compounds are passed over a finely divided metal such as nickel the molecule is saturated. In 1902 the German inventor K. P. W. T. Normann patented a process for bubbling hydrogen through hot oil containing a suspension of finely divided nickel. With this process oils could be transformed into solid fats suitable for margarine manufacture. The early prejudice against such fats was overcome during World War I in Germany which suffered a serious shortage of natural fats. The hydrogenation of oils became a large industry following the war. Fish and whale oils figured prominently in European margarine production; hydrogenation served to eliminate the fishy flavor normally present in such oils. In the United States hydrogenation was applied principally to cottonseed oil for the production of cooking fat.

As the demand for industrial hydrogen increased with the wider use of hydrogenation processes and the Haber process, electrolytic sources proved inadequate. Some hydrogen was prepared by passing steam over hot iron, but the principal source soon became water gas. The hydrogen is separated from the carbon monoxide either by liquefaction by the Linde-Frank-Caro process, or by catalytically converting the carbon monoxide to dioxide with steam and dissolving the dioxide in water.

Initially oils were hydrogenated at atmospheric pressures. About 1911 Vladimir Ipatieff (1867–1952) began his studies of high-pressure hydrogenation which revolutionized both research and industrial practice, and Friedrich Bergius (1884–1949) initiated his work on the hydrogenation of coal. Bergius sought to develop a method for the production of hydrocarbons suitable for use as motor fuel in Germany, which had no petroleum resources. Bergius obtained a heavy oil from powdered coal suspended in oil and heated with hydrogen and a catalyst. The process was not

Fig. 25.3. Vladimir N. Ipatieff.
(Courtesy of Universal Oil Products Company.)

competitive with petroleum motor fuels but did contribute to Germany's needs during World War II. Gasolene companies outside Germany also have indicated considerable interest in the process; it may prove significant as it becomes more costly to exploit petroleum supplies.

Perhaps of even greater interest in this connection is the Fischer-Tropsch process developed in the twenties by the Germans Franz Fischer (1877–1947) and Hans Tropsch (1889–1935). Water gas passed over suitable metal oxide catalysts at about 200° C. and moderate pressure is converted into hydrocarbons suitable for motor fuels. The process was ready for commercial production of hydrocarbons after 1935 and was an important factor in the German war effort, as evidenced by the attention given to synthetic oil plants by Allied bombers. The hydrocarbons thus produced also were oxidized to fatty acids to help Germany overcome a shortage of food fats.

During the years following World War I a large number of hydrogenation reactions were commercially developed, for example, naphthalene to the solvents tetralin and decalin; phenol to cyclohexanol for use in nylon.

The Badishe company began research on the conversion of water gas to methanol about 1913. The French chemist Patart patented a workable process in 1922 and commercial production was begun soon thereafter. The wood distilling industry, already hard hit by competition from acetone produced by the fermentation of starch, collapsed. Use of synthetic

methanol as a solvent, an antifreeze, and for the production of formaldehyde for plastics, expanded rapidly.

The inert gases also became important industrially. In 1914 Irving
Langmuir introduced the use of argon as a filler for incandescent light
bulbs. Neon was first used in advertising signs during the twenties. Both
of these gases are produced from liquid air. There was also interest in
helium because of its lifting power combined with its nonflammability.
Hamilton P. Cady (1874–1943) and D. F. McFarland of the University
of Kansas found more than 1 per cent of helium to be present in certain
samples of natural gas. The United States Navy immediately became
interested and discovered the richest natural gas resources to be those
near Amarillo, Texas. The federal government assumed control of the
resources and became the world's chief producer; the contract for extracting the helium was awarded the Linde Air Products Co. Besides its
use in lighter-than-air craft, helium is used in admixture with oxygen
for deep sea diving and caisson work and as an inert atmosphere in the
welding of certain metals.

SOLVENTS

Traditional solvents such as ethanol, methanol, naphtha, benzene,
acetone, and carbon tetrachloride still are important in industry, but not
nearly as important as earlier in the century. World War I drastically
stimulated acetone production and butanol, an associated by-product,
profoundly changed the solvent picture in the postwar era. At the same
time, a large variety of new solvents—for example, chlorinated hydrocarbons and hydrogenated aromatics—came into production for specialized uses.

The British propellant *cordite* required the use of acetone for the nitro-
cellulose-nitroglycerin mixture. Acetone was prepared from calcium
acetate, most of which came from the acetic acid present in "pyroligneous
acid," a by-product of charcoal production. During World War I the
supply was completely inadequate. Even acetic acid produced by fermentation (alcoholic and acetic) of saccharine materials did not meet the
demand for acetone.

A German process for the catalytic conversion of acetylene to acetic acid
was placed in operation in Canada and the United States. Calcium carbide,
needed for the production of acetylene, was readily available near centers
of hydroelectric power, but the conversion of acetylene to acetaldehyde, a
necessary step in the process, proved very troublesome. The German
patents on the process, which had been seized by the Alien Property
Custodian, were cleverly uninformative.

Another process was found to be more useful. With the acetone crisis at its height it was recalled that Russian-born Chaim Weizmann (1874–1952) at the University of Manchester was producing butyl alcohol by the fermentation of starch. Weizmann was interested in butyl alcohol as a source of butadiene, which he believed might be useful in the production of synthetic rubber. The fermentation had been discovered in 1910 by the French chemist Fernbach, but it was Weizmann who developed the process by isolating an organism particularly effective in fermenting starch to the desired acetone. The organism, *Clostridium acetobutylicum*, fermented starch to give two parts butyl alcohol and one part acetone, the

Fig. 25.4. CHAIM WEIZMANN.
(Courtesy of the Weizmann Institute of Science.)

latter of no interest to Weizmann but of extreme interest to the British military. The process was quickly brought into commercial production in government-operated distilleries in the United States (Terre Haute and Peoria), Canada, and India.

After the war the American plants were bought by the newly formed Commercial Solvents Corporation. Interest quickly turned from acetone to the butyl alcohol which had piled up as a comparatively useless by-product of the process. The alcohol and its derivatives proved extremely valuable solvents for fast-drying nitrocellulose lacquers which "over night" revolutionized the automobile industry. The slow drying of conventional finishes was a formidable bottleneck in that industry until 1920 when Edward M. Flaherty, a chemist with du Pont, developed a lacquer by dissolving nonexplosive nitrocellulose in amyl acetate along with suitable gums and plasticizers. Amyl acetate, prepared from amyl alcohol, a by-product of the whiskey industry, quickly became a rare chemical with the advent of Prohibition in the United States. It was found that butyl acetate is an even better solvent, so the success of the starch fermentation industry in peacetime was assured. In England the process was introduced com-mercially by the Distillers Company.

Weizmann was called before Prime Minister Lloyd George who wished to recommend him to the King for whatever honor he might request. Weizmann refused knighthood and a pension but asked for British spon-sorship of the Zionist movement to restore Palestine as a homeland for the Hebrews. Shortly thereafter the Cabinet published the Balfour Declaration. Through the long years before the wish became a fact Weizmann exerted his own efforts and used the income from his patents to further the cause. When the state of Israel was born it was only natural that he should become its first President.

ORGANIC SYNTHESIS CHEMICALS

In 1900 the bulk of organic raw materials was derived from coal tar, with alcoholic fermentation, wood distillation, and naval stores supplying subsidiary amounts. Today coal tar is far less significant although it is still an important source of aromatic chemicals. Fermentation of molasses and starches furnishes industrial ethanol in large quantities. The wood distillation industry succumbed to competition from methanol produced by synthesis and acetone from fermentation. Naval stores, of course, are still important. One American company, Hercules Chemical, bases a significant percentage of its business on products synthesized from turpentine, rosin, and pitch.

Acetylene very quickly gained recognition as a chemical synthesis gas. It began to figure prominently as an industrial raw material following Walter Reppe's (b. 1892) studies of its chemical behavior at the laborator-ies of the Badische firm at Ludwigshafen.

The major source of process chemicals now is natural gas and petroleum. Until 1930 chemical research in the petroleum industry was not extensive and was concentrated largely on immediately practical matters. There was not even a large body of knowledge regarding either the particular hydrocarbons present in various sources of petroleum, or the physical and chemical properties of individual hydrocarbons. Finally in 1931 a syste-matic study of petroleum hydrocarbons was undertaken by the U.S. Bureau of Standards and the American Petroleum Institute. Individual petroleum companies have come to appreciate the importance of large-scale chemical research. In part they were guided by the example of chemical companies like Sharples Solvent, Carbide and Carbon Chemicals, Dow, and Monsanto, which utilize hydrocarbon gases as raw materials for the manufacture of organic chemicals.

AGRICULTURAL SOURCES OF CHEMICALS

Products of the soil, for example, naval stores and cellulose, have always figured to some extent as chemical raw materials. Ethanol has long been produced by the fermentation of molasses and, after malting, of starch. Starch itself is an important industrial material; it is used as an adhesive, a sizing agent for paper, a finishing agent for textiles, a food ingredient as the starting material for the manufacture of commercial glucose, and in the production of butanol and acetone. Cellulose—in pure form from cotton and in less pure form from wood—is important in the paper industry and is used too in the production of explosives and plastic films and filaments. Sucrose is another industrially significant carbohydrate. Fatty materials have many new uses. Proteins too are used in industry as chemical materials, chiefly in plastics, where casein from milk and zein from corn are particularly important. Furfural, produced as a hydrolysis product of oat hulls, was introduced as a chemical raw material by the Quaker Oats Co. during the twenties. Utilized at first primarily as a solvent, it now is important also as a process chemical for plastics and nylon production.

In 1935 there was organized in the United States an enthusiastic group which called itself the Farm Chemurgic Council. Although the group had the blessing of Francis P. Garvan, a long-time propagandist for American self-sufficiency in chemicals, and Henry Ford, the leading figures in the movement were William Jay Hale (1876–1955) and Charles Holmes Herty (1867–1938). Hale was an organic chemist with Dow Chemical and the son-in-law of Herbert Henry Dow, the company's founder. Herty had served as a chemistry professor at the Universities of Georgia and North Carolina and in 1935 was research director of the Georgia Department of Forestry and Development. Herty had long been a promoter of forest products as a source of chemical materials.

The organizers of the Chemurgic Council, gathering at a time when agriculture everywhere was in a depressed state, saw in chemistry the salvation to the farm problem. Arguing that the capacity of the population to consume food was clearly limited, they suggested that the only way to dispose of agricultural surpluses was to use them in the processing of chemical products. The purpose of the council was to publicize such uses and encourage research which would reveal new uses for farm crops. Henry Ford widely publicized his company's intention to develop and use soybean products in its automobiles. The studies of George Washington

Carver (1864-1943) of the Tuskⁿgee Institute on materials which might be produced from peanuts, sweet potatoes, and other crops from the South attracted considerable attention. A concerted effort was made to win acceptance of fermentation alcohol as a motor fuel.

The results of the chemurgy movement were generally unspectacular. Many agricultural products were already utilized. Increasing such usage was principally an economic matter. Where agriculture was able to supply raw materials more cheaply than the coal mine or the oil well, agricultural products were used. During World War II, when the production of synthetic rubber had to be stimulated at any cost, a large quantity of alcohol was converted to butadiene. Normally such conversion is uneconomic.

NEW METALS AND NEW ALLOYS

Carbon steel dominated the metals picture in 1900, although traditional alloys such as bronze, brass, and solder were holding their own. The demand for pure copper was increasing with the rise of the electrical industry; the canned food industry was using more and more tin for tin plating; and the galvanizing of sheet steel required greater amounts of zinc. Aluminum was being produced, but there was yet no significant demand for the metal.

Iron and Steel

Ferrous metals were still the most important commercial metals, but products and operations were beginning to change significantly. Wrought iron has been largely supplanted by mild steel. The open hearth process now accounts for the largest quantity of steel, although sizable quantities of Bessemer steel are still made. The electric furnace is widely used in the production of high quality steel for special purposes. Chemical control of steel-making operations is commonplace. Steady progress in the understanding of operations has led to improved products. Almost no steel produced today is without alloying metals. With careful control of alloying ingredients and appropriate heat and mechanical treatment, a wide variety of steels with special properties can be produced.

The most significant development in blast furnace operation is the use of air enriched with oxygen. This innovation, introduced during World War II, significantly increases the productivity of furnaces.

Taconite as an Iron Ore

World War II brought about the near exhaustion of the rich iron ores of the famous Mesabi Range in Minnesota from which, since 1884, more than two billion tons of ore have been shipped through the Sault Ste. Marie locks to the smelting centers on the southern shores of Lakes

Michigan and Erie. Consequently American steel companies are interested in ore deposits in Venezuela, Brazil, Liberia, and Labrador as well as the extensive deposits of taconite in the Lake Superior area. For years taconite had been virtually ignored because of its hardness and low iron content.

After many years of research Edward Wilson Davis, a professor of metallurgical engineering at the University of Minnesota, developed a commercially practical method for processing of taconite. The ore is too hard for ordinary drilling and blasting methods. Davis devised a "jet piercer" which literally burns into the rock with a flame of kerosene and oxygen. The powdered ore is concentrated by magnetic and chemical processes and then pelletized to get it into a form suitable for shipping and smelting.

Alloys—Especially Alloy Steels

By 1880 there was considerable interest in alloys. The metallographic microscope and the phase rule permitted alloys to be studied from a less exclusively empirical point of view. In 1863 Henry Clifton Sorby (1826–1908) of Sheffield developed techniques for polishing and etching metal surfaces so they could be examined under a microscope. Crystalline structure began to be correlated with tensile strength, ductility, and other properties. The application of the phase rule to studies of alloys followed Roozeboom's pioneering work which began in 1887. Utilizing the two techniques researchers learned a great deal about alloys and developed several practical alloys. X ray techniques and the electron microscope also have been invaluable in work on alloys.

One of the earliest improvements in steel was the introduction of manganese (13 per cent) by Robert A. Hadfield in 1882. Hadfield found that manganese confers resistance to shock and wear. Manganese in small amounts had been used as a deoxidizer since Robert F. Mushet's work about a decade earlier. Spiegeleisen, an iron-carbon mixture containing 20 per cent manganese was widely used as an additive; the manganese reduced any residual iron oxides to iron and the carbon served to bring that element up to the desired level. With the introduction of high-manganese steels, spiegeleisen had too little manganese to serve as an additive. Ferromanganese, which contains around 80 per cent manganese displaced it. Manganese steels have been in heavy demand, especially because their toughness makes them valuable for armor plate for ships and tanks. In 1900 L. Aitchison showed that tungsten and molybdenum in steels cause the latter to hold their temper in high-speed cutting tools. In 1896 C. E. Guillaume found that steels containing about 40 per cent of nickel have a very low coefficient of expansion and hence are useful for surveyors tapes, clock pendulums, and metal-in-glass seals. Such alloys became known as invar and platinite.

Rust-resistant steels were introduced in 1912 by H. Brearly who used nickel and chromium as alloying ingredients in concentrations up to 8 and 18 per cent, respectively. These steels are of great importance in the food, pharmaceutical, and chemical industries. Because of their hardness and toughness chromium steels are ideal for armor and in armor-piercing projectiles. Chromium is introduced in the form of ferrochrome which is produced principally from chromite ore by electric furnace and blast furnace methods.

Extensive research into alloy steels has resulted in a large variety of specialized products with unique uses. Frequently, combining several alloying ingredients enhances the effect of each ingredient. Besides the alloying elements mentioned above, vanadium, silicon, nitrogen, boron, and other elements are utilized. Often new fabrication methods have been required.

Light Metals

Aluminum. Although some aluminum was produced late in the 1800's, the metal first achieved a degree of importance only after World War I. Before then it had been used primarily for cooking utensils and sundry light articles. Since World War I, there has been continuing progress in the design of strong alloys containing copper, manganese, magnesium, and other metals. Manufacturers were quick to use appropriate aluminum alloys in automobile engine parts, railroad coaches, and airplanes. The wood and cloth airplane of World War I gave way to aluminum alloy planes by 1930. During World War II aluminum production was a matter of critical concern among all belligerents. Since World War II aluminum suppliers have invested so heavily and imaginatively in research to develop new alloys and uses that the metal is outranked only by iron in importance.

Magnesium. Magnesium became important during World War II although it had been produced in small amounts since 1900 by Griesheim-Elektron and soon thereafter by Dow Chemical. The German firm used Stassfurt salts and magnesia as source materials and developed a process for electrolysis of magnesium oxide in a molten bath of fluorides of magnesium, barium, and sodium which resembled the Hall process, magnesium and oxygen being liberated at the electrodes.

The Dow company was founded in the nineties by Herbert H. Dow for the electrolytic production of bromine from brines pumped from underground in the vicinity of Midland, Michigan. Its operations were so successful that Dow was able to move into the British market for bromides in competition with the German bromine syndicate. The company countered German dumping of bromides by moving into the German market itself. Soon thereafter the Germans withdrew from the American market, one of the few cases in which they were unable to subdue foreign competition prior to World War I.

In the early years of its operations Dow confined its operations to

Fig. 25.5. HERBERT H. DOW.
(Courtesy of Dow Chemical Company.)

bromine and bleaching powder. However its brines produced a significant quantity of magnesium chloride. About 1916 a method was developed for the preparation of anhydrous magnesium chloride and a cell was designed for the electrolytic reduction of the molten chloride. For many years there was little demand for the light metal, but Dow persisted in research aimed at developing uses. By alloying it with aluminum, manganese, zinc, and other metals, useful light-weight alloys were produced. I. G. Farben in Germany was carrying on similar developmental studies.

World War II created a heavy demand for magnesium, particularly for aircraft parts and incendiary bombs in which the magnesium case was set afire by a thermite mixture inside. The burning magnesium was difficult to extinguish because of the ease with which it reacted with water. Dow anticipated the wartime demand and expanded its production from Michigan brine and in 1941 opened a plant in Freeport, Texas, for the extraction of magnesium and bromine from sea water. Although the amount of magnesium ion in sea water is around 0.13 per cent, this can be profitably extracted. Calcium hydroxide prepared from oyster shells is used to precipitate magnesium hydroxide which is converted to the anhydrous chloride with hydrogen chloride and electrolyzed.

Two other drastically different processes were placed in operation with the financial support of the United States government. The ferrosilicon process, developed originally in Canada by Lloyd M. Pidgeon, is based on the conversion of calcined dolomite to magnesium and calcium silicate. The carbothermic method is based upon a process developed by the

Austrian Fritz J. Hansgirg. It involves the calcination of magnesite, briquetting of magnesia with carbon, high temperature reduction—which gives carbon monoxide and magnesium vapor—quenching in hydrogen or natural gas to obtain magnesium powder, and distillation of the magnesium from attendant impurities in vacuo. This process was handicapped by explosion hazards and actually produced little pure magnesium. Variant electrolytic processes based on the Elektron and the Dow processes were also in use for a time. Dow, the principal American producer during the war, retains that position.

Other New Metals

Several other metals have become important. Of these, titanium is of special interest since it is the tenth most abundant element in the earth's crust. Despite this abundance, it was little more than a laboratory curiosity because of the difficulty of smelting it in a usable form. William J. Kroll introduced a process involving reduction of titanium chloride with magnesium-sodium alloy in an atmosphere of argon or helium. Kroll was a German refugee when he interested the U.S. Bureau of Mines in his process. It has been fully developed, and titanium production has reached full commercial status; in 1958 American capacity was 36,000 tons. Although as strong as steel, the metal has half the density of steel. It is resistant to corrosion in salt water and will prove to be a widely useful metal if cost can be lowered.

Zirconium is also produced by the Kroll process. Commercial capacity in 1958 was 3,000 tons per year in the United States. The metal ores are twice as abundant as copper ores and thirteen times as abundant as lead ores. Zirconium is used principally in the nuclear energy and the jet propulsion fields. High cost discourages other applications of the metal. Hafnium, a by-product of zirconium production, is used for control rods in nuclear reactors. Because of its high neutron absorptive capacity, it must be scrupulously removed from zirconium when the latter metal is used in nuclear reactors.

Gallium has been produced by Aluminum Company of America since 1949 but has not yet been exploited commercially. Beryllium came into commercial production about 1920 but initially was used almost exclusively as an alloying ingredient in certain copper alloys where high resiliency was desired. With the advent of the nuclear age, it is in greater demand. Lithium is used in the preparation of such chemicals as lithium hydride, valued as a transportable source of hydrogen, and lithium aluminum hydride $LiAlH_4$, a versatile reducing agent in organic reactions.

The rise of the transistor industry during the last decade has created a demand for semiconducting metals such as germanium and silicon in a high state of purity. Silicon has been employed in steelmaking for many years; but for this purpose a high state of purity is not required and ferrosilicon is generally used. This alloy is an electric furnace product utilized primarily as a deoxidizer in steel production. Large amounts (12 per cent) of silicon in steel confer resistance to acid, giving the alloy Duriron which is ideal for acid-handling equipment.

Boron production also has grown to meet the demand for boron hydrides for rocket fuels. Boron also is used in steelmaking and in the production of Borazon, a boron nitride with a hardness exceeding that of diamonds.

Cermets

With the onset of the space age, the demand for heat-resistant materials has increased markedly. Metals with high thermal resistance have been investigated regardless of cost. The most important outgrowth of nuclear and jet engineering has been the development of cermets, solid materials which are produced from resistant metals and ceramics and have some of the virtues of both.

Flotation

One of the major developments in ore processing has been the flotation process which makes it possible to work low-grade ores profitably. In 1890 a copper ore containing less than 30 per cent of copper was considered not worth smelting. Today ores containing 1 per cent of copper are profitably worked.

The process was developed for copper ores by English and Australian engineers (Ballot, Cattermole, Chapman, Higgins, Picard, and Sulman). Finely ground sulfide ores are placed in a cell with water and an oil such as pine oil, eucalyptus oil, or kerosene, and a chemical frothing agent is added. The metal sulfide particles are wetted by the oil and floated to the top by the froth formed by agitation in the cell. A high degree of recovery of valuable ore is possible since the silicate impurities are wetted by the water and remain at the bottom of the cell as a sludge.

The flotation process has drastically changed ore-dressing practices, particularly in the copper, zinc, and lead industries. Not only has it made possible the concentration of ores, but also by suitable chemical treatment, separation of one mineral from another is sometimes achieved. For example, in mixed zinc and lead ores the zinc is complexed with cyanide while the lead sulfide is floated. Then the zinc is released for flotation by adding copper salts.

CONCLUSION

Since 1900 the chemical industry has changed from one dominated by Germany to one of worldwide proportions, the transition taking place during World War I. There also has been a trend away from coal tar as a primary source of synthesis chemicals; petrochemicals and agriculture-derived compounds have taken their place beside coal tar derivatives.

Both wars stimulated expansion of the industry. New sources of raw materials were explored and new products were manufactured in order to supply the insatiable needs of a major war effort. With the coming of peace the new raw materials continued to be exploited and there were serious efforts to divert new products into useful peacetime fields.

Particularly noteworthy developments in the chemical industry have been the growth of the contact process, the Haber process, the spread of flotation techniques in the ore dressing industries, and the broad-scale application of electrochemical methods.

CHAPTER 26

INDUSTRIAL
CHEMISTRY III.
CHEMICALS FOR
CONSUMERS

It is not always possible to distinguish with finality between chemicals intended for further processing (chemical intermediates) and those intended for ultimate consumption (consumer products). Metals, acids, bases, salts, gases, solvents, and so forth, are sometimes incorporated into consumer products which are eaten, worn out, destroyed, or otherwise dissipated, but they more generally are employed in industrial operations which lead to consumer products. For example, consider explosives. The military is an ultimate consumer of explosives. Mining companies use explosives to make ores available for further processing. Use of explosives in construction of roads, canals, and tunnels represents an intermediate role but comes close to ultimate consumption. Hence, it is evident that distinctions between substances treated in this chapter and the preceding one are not clean cut. Generally the products treated in Chapter 25 are used industrially for further processing, those treated here are sold for use by the general public.

SYNTHETIC DRUGS

The search for chemicals with pharmaceutical value was pursued vigorously during the last part of the nineteenth century. One of the most enthusiastic workers in chemotherapeutics was Paul Ehrlich whose major contributions came in the present century although his research went back to the 1870's. Ehrlich (1854–1915), while still a student at Breslau and Leipzig, developed a passionate interest in dyes and their behavior toward living tissues. His cousin Carl Weigert (1845–1904) had taught him the technique of staining bacteria. Weigert was not the first to have

Fig. 26.1. PAUL EHRLICH.
(Courtesy of the National Library of
Medicine, Washington, D.C.)

Fig. 26.2. ERNEST F. A. FOURNEAU.
(Courtesy of the American Institute
for the History of Pharmacy.)

used colored compounds as biological stains; chromic acid, carmine, and hematoxylin had been used on tissues for more than a decade before Weigert used methyl violet to reveal bacteria in animal tissues (1875). Differential staining spread very rapidly among bacteriologists and histologists. Robert Koch used behavior toward stains as a diagnostic tool in identifying bacteria, H. C. J. Gram introduced his differential technique in 1884, and many others made further contributions.

Because of the selectivity of specific dyes for certain bacteria or certain kinds of tissues, Ehrlich arrived at the notion that it should be possible to treat diseases by administering the right dye. He showed in 1887 that methylene blue stains living nerve cells but not the adjacent tissues. Similarly it stains certain bacteria but not others. Should it not be possible to discover certain dyes which attach themselves to a specific organism, thereby killing it without damaging the cells of the host?

In 1889 Ehrlich became a member of Robert Koch's Institute for Infectious Diseases in Berlin. He was already closely associated with Emil von Behring in 1892 when Behring discovered his antitoxin for diphtheria; Ehrlich had much to do with the development of the serum. He later became director of the State Serum Institute at Frankfurt-am-Main. Even though he was occupied with the production and testing of serum, Ehrlich maintained his search for dyes with a high specificity for pathogens while being relatively nontoxic toward higher animals. He obtained the cooperation of the Casella chemical works which passed on to him samples of new compounds produced in their laboratories. In addition, with the establishment of the Georg Speyer-Haus in 1906, he was able to surround himself with a staff of assistants, both chemists and bacteriologists, for the synthesis and alteration of compounds and for the study of their effects on pathogens and animals.[1]

At an early stage Ehrlich had developed his side chain theory of germicidal action. According to this theory it should be possible to design a molecule having a side chain with a complementarity to a parasite. By means of the side chain, attachment of the molecule to the organism should handicap its activity and possibly kill it. Since the side chains should be specific for the pathogenic organisms but not the host cells, effective magic bullets could be designed. His thoughts were influenced in part by the success of *serum therapy*. Here the pathogens themselves stimulated the formation of specifically active substances lethal to the pathogen without harming the host. Since effective serums could not be produced for numerous diseases, particularly those due to animal parasites, it was necessary to develop *chemotherapy* by creating specifically lethal compounds.

Since a chemical toxic to one organism is almost certain to show toxicity toward other cells, Ehrlich developed the *therapeutic index* as a measure for the safety of a chemical for use as a drug, the therapeutic index being the ratio of the maximum dose tolerated by the host animal to the curative dose.

Early in the 1900's Ehrlich and Kiyoshi Shiga found that Trypan Red is particularly effective against diseases caused by trypanosomes. The related Trypan Blue was shown to be even more effective by F. E. P. Mesnil and Maurice Nicolle. These drugs had some success in the treatment of tropical diseases such as sleeping sickness and the *mal de caderas*. In 1906 Koch introduced the use of Atoxyl in the treatment of trypanosome diseases in humans. The compound had been prepared by Béchamp in 1863 and was believed to be the anilide of arsenic acid. It was reported

---

[1] Martha Marquardt, *Paul Ehrlich*, Heinemann, London, 1949, p. 121.

to have trypanocidal activity in 1905 by the Liverpool physicians H. W. Thomas and A. Breinl; the name Atoxyl was used since it appeared to be nontoxic to the host. Ehrlich, who was interested in arsenic compounds because of the periodic relation of arsenic to nitrogen, confirmed the activity of the compound toward trypanosomes, but found it to be too toxic for use because of its damaging effects on the optic nerve. He questioned Béchamp's structure and proposed the correct one since his experience with dyes suggested that a free amino group was present.

At about this time the causative organism for syphilis, *Treponema pallidum*, was discovered by Erich Hoffmann and Fritz Schaudinn. Schaudinn pointed out that the organism, a spirochete, has properties more nearly protozoan than bacterial, and it was hoped that trypanocidal drugs would be effective against it. Ehrlich designed new arsenic compounds which could be tested on diseased rabbits and mice. Reasoning that the azo group was an effective unit in such dyes as Trypan Red, he hypothesized that trivalent arsenic might be more effective than pentavalent arsenic such as found in Atoxyl. His chemist Alfred Bertheim prepared such compounds; Arsacetin (acetyl-*p*-amino-phenylarsinic acid) was found to be particularly effective against experimental trypanosomiasis. Compound number 418, arsenophenylglycine, proved even more effective. More compounds were synthesized in the laboratory and tested under the direction of his Japanese assistant Sahachiro Hata. Success was attained against syphilis in 1909 with the compound 606, which was later marketed as Salvarsan or Arsphenamine. Later (1912) a more convenient compound was prepared (914, Neosalvarsan). Treatment with these arsenicals was introduced very rapidly. While not without shortcomings,

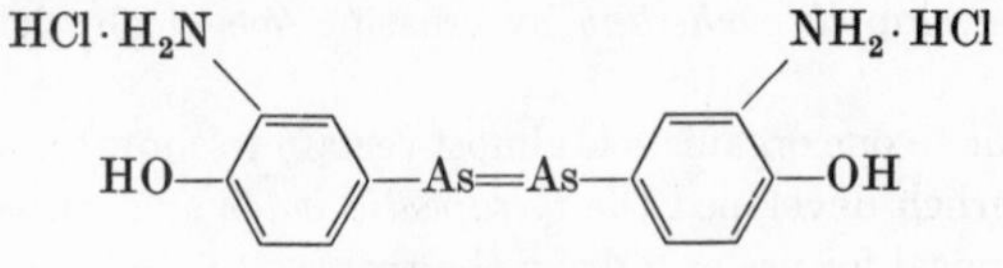

Salvarsan
Arsphenamine (606)

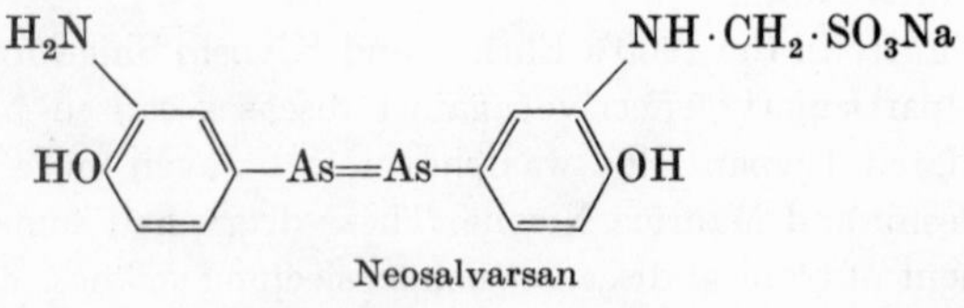

Neosalvarsan

Neosalvarsan remained a standard treatment for syphilis until effective antibiotics were introduced in the forties.

The success of the arsenicals as specifics led to great optimism that the chemical industry might create similar chemotherapeutic agents. Any enthusiasm was doomed to disappointment since study of effective compounds failed to suggest how molecules might be constructed so as to make them specific for certain diseases without being harmful to the host. Except for Bayer 205 and several other compounds, there was no real success in creating chemotherapeutic agents between 1910 and 1930.

Bayer 205, or Germanin, was introduced in 1920 as a specific against African sleeping sickness. The German Bayer Co. (the American unit was seized during World War I and forced to become an independent company) made the drug available under such rigid restrictions that it was charged that the drug was being used as a political weapon to bring about the return of Germany's lost colonies. Ernest Fourneau (1872–1949) of the Pasteur Institute identified the compound despite its not being patented and the Bayer company's refusal to make it available for research purposes. A team of scientists with the British Dyestuffs Corporation was also active in solving the problem. By examining the prewar German patent literature, Fourneau learned that a great deal of research was directed toward complex urea derivatives. There was further evidence in the patent literature of interest in amino-benzoyl units attached by amide linkages and in terminal naphthylamine sulfonic acid terminal groups as in the Trypan dyes. By a sort of detective process, hundreds of possible compounds were eliminated and the field was narrowed to 25. Each of these compounds was synthesized and tested biologically and chemically. One of them, Fourneau 309, proved to have the trypanosomal activity, the nontoxicity, and the chemical stability of Bayer 205. Compared with 56 mg. of the German product the Pasteur Institute had somehow obtained, it proved to be identical. The Germans refused to admit the identity; Fourneau 309 was patented in England and America. The compound is highly effective against the early stages of sleeping sickness but without effect on the later stages. Tryparsamide, prepared in 1919 by Walter Abraham Jacobs and Michael Heidelberger of the Rockefeller Institute of Medical Research, proved useful in the later stages in which involvement of the central nervous system occurs.

Antimalarials

Antimalarials received a great deal of attention during the late 1800's. Although quinoline was early identified as a part of the molecule of quinine, the remainder was referred to as "the second half" for many years. Skraup, Königs, and Rabe were active in identifying degradation products from "the second half" around the turn of the century; such products were designated by the use of *loipon* or *meros* in their names. The structure of this half was slowly clarified; a synthesis was achieved by Paul Rabe (1869–1952) in 1931. The total synthesis of quinine was finally accomplished by Woodward and Doering in 1944.

In the meantime the search for a synthetic substitute was carried on in various laboratories. I. G. Farben introduced Plasmochin in 1926. It kills the quinine-resistant sexual form of the malaria parasite, but is too toxic for general use. Atabrine was introduced about 1930 but did not come into widespread use until quinine supplies were cut off from Java during World War II. It attacks the schizont stage of the parasite, as does quinine, and prevents the destruction of red blood corpuscles, the cause of chills and fever. An acridine-type dye, it causes temporary yellow pigmentation. Because of its availability and effectiveness it is still used.

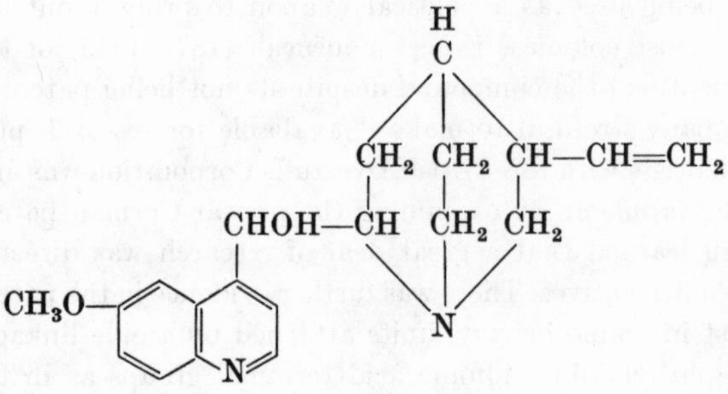

Quinine

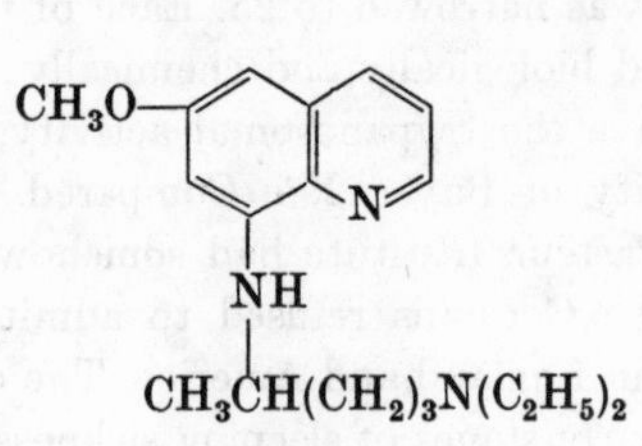

Plasmochin

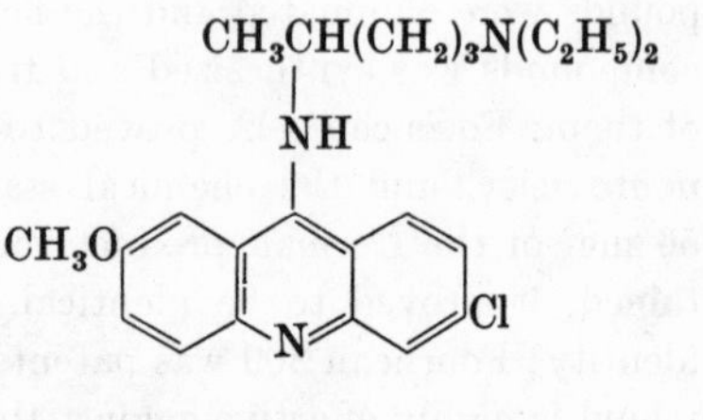

Atabrine

Because of the antimalarial shortage during the early years of the war, a vigorous research program was launched in the United States to synthesize and test all sorts of compounds as possible antimalarial agents. The tests revealed a number of promising compounds among the 14,000 considered, but none of these was an adequate substitute for Atabrine and quinine.

Sulfa Drugs

The sulfa drugs were introduced in the thirties; Prontosil, a product of I. G. Farben, was the first. It was patented in 1932 and tested against streptococcal and staphlococcal infections in experimental animals by Gerhard Domagk (1895–1964). Clinical tests proved it to be strikingly effective and it was soon brought into general medical use. An examination of the red dye by J. Tréfouël and associates at the Pasteur Institute revealed

Fig. 26.3. GERHARD DOMAGK.
(Courtesy of the National Library of
Medicine, Washington, D.C.)

that, while the compound is ineffective against bacteria *in vitro*, it breaks
down to sulfanilamide in the tissues. Fourneau found sulfanilamide to be
as effective as Prontosil itself and it quickly came into use since it was not

Prontosil

Sulfanilamide   PABA

patentable, having been prepared by P. Gelmo in 1908. It was charged
that I. G. Farben had learned of its effectiveness and then tacked on the
second ring in order to protect its knowledge with a patent.

The success of sulfanilamide led to a vigorous examination of other
compounds containing the sulfamide group. Within five years more than
a thousand compounds had been synthesized and tested by pharmaceutical companies. A small number of the compounds were found to be

therapeutically valuable and quickly found their way into medical practice.

The mode of action of the sulfa drugs was attacked by Paul Fildes and D. D. Woods, biochemists at Cambridge University. They observed that *in vitro* sulfanilamide is more effective against a particular organism in certain culture media and less effective in other media. Yeast extract appeared to contain a substance which checks the effectiveness of the drug. When yeast extract is present in the culture media, a larger quantity of sulfanilamide is required to kill the bacteria. This action resembles the tendency for malonic acid to inhibit the enzyme succinase in its role of dehydrogenating succinic acid to fumaric acid.

Woods found that the para-aminobenzoic acid (PABA) in yeast extract sufficiently resembles sulfanilamide that an antagonism occurs in organisms requiring PABA in their metabolism. A few years later studies on folic acid revealed that PABA is an integral part of the folic acid molecule. Bacterial species which synthesize their own folic acid find this function blocked when significant quantities of sulfanilamide are present. Higher animals, as well as certain harmless bacteria, require ready formed folic acid in the food supply and so are not harmed by sulfanilamide.

Fildes proposed that it should be possible to design drugs which resemble necessary metabolites. Various workers uncovered such antagonists.[2] Pyrathiamine inhibits growth of organisms which cannot synthesize thiamine while pantoyltaurine exhibits a similar antagonism in organisms requiring pantothenic acid. In general this approach to chemotherapy has not been strikingly successful since microorganisms are very versatile in obtaining their nutrients and we still know too little about their metabolism. A further problem is encountered in the immunity which so frequently develops in a bacterial species after use of a drug, a problem of which Ehrlich was aware. While susceptible strains are readily killed, resistant organisms survive to become dominant.

Antibiotics

Sulfa drugs represented a major breakthrough in chemotherapy but in the forties were forced to give way, to a significant degree, to the antibiotics. Although antibacterial effects of molds had been reported and even investigated at least several times, the history of this class of drugs really began with Alexander Fleming (1881–1955) in 1928. Returning to his laboratory after a short vacation, he observed a blue mold growing on a Petri dish which had originally been overgrown with colonies of staphlococci. Around the mold there was a halo within which no bacterial colonies were growing. The mold evidently had produced a lethal substance which diffused outward through the culture media, killing bacteria wherever they were present. Fleming learned that the media containing

[2] Paul Fildes, *Brit. J. Exptl. Path.*, **22**, 293 (1941).

the antibacterial substance, which he named *Penicillin* after the mold *Penicillium notatum*, was not toxic to laboratory animals.

The press of other duties at St. Mary's Hospital in London prevented his vigorously following up his observations, and the advent of sulfa drugs led to the general disinterest in his report. Only in 1936 did penicillin come under renewed examination at Oxford by the Australian-born Howard Walter Florey and the German refugee Ernst Boris Chain. They confirmed Fleming's results and went on to isolate an impure powder of uniquely potent killing power toward certain germs in test animals. During 1942 a pure yellow powder was produced by chemist Chain and successfully used by Fleming against meningitis. By this time Florey had visited the United States where plans were set up for industrial production by Merck, Pfizer, and Squibb, with aid from the federal government. The heavily subsidized research and production effort led to the solution of production problems in time for penicillin to become available in sufficient quantity to aid the war effort.

Considerable effort also was devoted to the chemistry of penicillin. Results of work in Britain and the United States soon revealed the existence of several kinds of penicillin. When the structure of penicillin was finally clarified (through the work of Dorothy Crowfoot Hodgkin[3] in England) it was found that the various penicillins have the same general structure but vary in a side chain. It soon became apparent that this molecule would not be easily synthesized. However biosynthetic procedures led to striking success in producing a large variety of penicillins. In 1957, following nine years' work, John C. Sheehan and K. R. Henery-Logan synthesized penicillin V.

Beginning around 1940 interest in antibiotics mounted rapidly. Basic research on bacterial inhibitors produced by microorganisms had been slowly gaining momentum in the preceding years. In 1939 René J. Dubos (b. 1901), who had been working at the Rockefeller Institute, announced the isolation of an antibacterial substance from *Bacillus brevis*. This was soon fractionated into tyrocidine and gramicidin. Although they were obtained in crystalline form, they were each further separated into related polypeptides. Gramicidin S, reported in the U.S.S.R. by Gause in 1944, is also a polypeptide. While these compounds are effective against gram-positive organisms, their toxicity restricts their development for use in medicine.

The studies of Waksman and his associates at Rutgers led to the discovery of streptomycin in 1943. Selman A. Waksman (1888–1973), a leading soil bacteriologist, isolated the substance from *Streptomyces griseus*, a soil organism. Streptomycin is an effective agent against kidney infections, tuberculosis, and several other ailments which fail to respond to penicillin. Considerable study was necessary to establish the complicated structure:

[3] Dorothy Crowfoot Hodgkin, *Advancement of Sci.*, **6**, 85 (1949).

a base (streptidine), a sugar (streptose), and an amino-glucose residue. The glucose residue is unique among natural sugar derivatives in showing the L-configuration.

Early in the forties pharmaceutical houses undertook the examination of products of soil microorganisms from all over the world in a frantic search for new and profitable antibiotics. Chloramphenicol (Chloromycetin), introduced by Parke, Davis, originated from a mold (*Streptomyces venezuelae*) obtained from a sample of Venezuelan soil. This compound has such a simple structure that commercial synthesis was quickly achieved.

Aureomycin was isolated by Benjamin Duggar (1872–1956) in 1948 from *Streptomyces aureofaciens* in the Lederle laboratories, and Pfizer scientists obtained terramycin from *Streptomyces rimosus*. In the fifties both were found to be derivatives of tetracycline which is itself isolated from natural sources.[4] These compounds are valuable in the control of a wide variety of bacterial, viral, and rickettsial diseases, and hence are termed *broad-spectrum* antibiotics.

Antibiotics are also used in the feeding of animals and the preservation of poultry, fish, and meats. In animal feeds the antibiotics apparently bring about an increased utilization of food as the result of control of bacterial flora in the digestive tract.

Corticoid Drugs

Cortisone and ACTH (*a*dreno-*c*ortico-*t*ropic-*h*ormone) were introduced by the pharmaceutical industry in 1949 as drugs for rheumatic fever and rheumatoid arthritis. Cortisone was first isolated by Edward C. Kendall of the Mayo Clinic from the cortex (bark) of adrenal glands of animals. The principal work on cortical hormones was carried out in the thirties at Mayo Clinic and at Reichstein's laboratory in Switzerland. Philip S. Hench, also of the Mayo Clinic, demonstrated its effectiveness in certain

Serine—tyrosine—serine—methionine—
phenylalanine—histidine—glutamic acid—
arginine—tryptophan—glycine—glycine—
valine—proline—lysine—lysine—lysine—
arginine—arginine—valine—lysine—valine—
proline—tyrosine (amide)

Synthetic ACTH

Cortisone

---

[4] A. C. Finlay, G. L. Hobby, S. Y. P'an, P. P. Regna, J. B. Routien, D. B. Seeley, G. M. Shull, B. A. Sobin, I. A. Solomons, J. W. Vinson, and J. H. Kane, *Science*, **111**, 85 (1950).

cases of arthritis and rheumatic fever. The adrenals of 180,000 sheep are required to produce one gram. The chemical composition was immediately examined. Woodward's group established the structure and went on to synthesize the compound in 1953. Since it is a steroid-class compound, raw material for synthesis is available and, although complicated, commercial synthesis is practical.

ACTH, obtained from the pituitary gland of hogs, proved to be a polypeptide. The study of its structure was pursued vigorously by teams at American Cyanamid and at the Universities of California and Pittsburgh. In 1960, after seven years' work, the Pittsburgh group directed by Klaus Hofmann synthesized a polypeptide having all of the biological activity of ACTH.

A great deal of research effort by pharmaceutical laboratories and also in the universities has resulted in steady progress in designing compounds with greater effectiveness. Steroid research, particularly, has contributed improvements on the cortisone molecule. Hydrocortisone is among them, but it has a number of undesirable side effects. Related compounds with increased unsaturation and added fluoro, hydroxyl, or methyl groups in certain positions give greater potency.

Other Drugs

Of particular interest is the development of compounds useful in the treatment of mental diseases. Tranquilizers have been found effective in calming highly disturbed patients and making them amenable to psychotherapy. Of a variety of drugs produced for this purpose, reserpine, derived from a Hindu folk remedy for insanity, has attracted most attention. Within a few years the structure was proven, and Woodward's group has synthesized the molecule. Psychic energizers, drugs useful in bringing patients from states of depression, also have been created.

EXPLOSIVES

By 1900 new military explosives were beginning to come into use. Black gunpowder, which had brought an end to feudalism, remained a standard propellant through the Spanish American War (1898). In the Boer War (1899–1902) and the Philippine Insurrection (1899–1902) smokeless powder began to supplant the traditional mixture of charcoal, sulfur, and saltpeter. Around 1889 *cordite*, a gelatinized mixture of nitrocellulose in nitroglycerine, was developed by Frederick Abel (1827–1902) of the British Royal Military Academy and James Dewar. It was found to be a superior propellant which does not give off a telltale cloud upon firing. Nobel, it should be noted, had been working on the problem even earlier.

During the Boer War the British also used a new type of munition, the "disruptive," or high-explosive, shell. The explosive in such a shell is not set off when the shell is hurled from the artillery piece. Upon impact a detonator sets off the explosive charge which hurls the shattered metal case in all directions. The British disruptive, named *lyddite* after the Lyde Proving Grounds, was primarily picric acid. The French had experimented with picric acid shells as early as 1885, using the name *melinite* for their product; the Japanese called a similar explosive *shimosite*. In Germany the laboratory of the Badische firm produced a variety of nitrated organic compounds which were tested by the German army. Trinitrotoluene (TNT) was found to be most satisfactory. It was then learned that nearly as effective a disruptive shell could be produced by using a mixture of TNT and ammonium nitrate.

World War I

During World War I TNT and picric acid explosives were brought into large-scale use. The Germans, with their flourishing coal tar chemicals industry, were prepared for the production of high explosives. The French and British were not nearly as well prepared. The British produced considerable coal tar (a significant part of which had been sold to German firms) and had established a world-wide trade in phenol and cresols. Because they were better prepared to produce these compounds, they and the French concentrated on picric acid explosives rather than turning to TNT which would have required the development of toluene production.

The United States, while not yet a belligerent, found itself deeply involved in the chemical phases of the war. French and British demands for naval stores were felt at once. Phenol was also in demand. American production had been limited to phenol as a by-product of coking plants. Synthesis from benzene was started almost at once. Acetone, essential as a solvent in the production of cordite, was also in great demand. At the same time America suffered because it was difficult to import German dyestuffs and medicinal products. Germany needed American cotton and foodstuffs, but trade was greatly reduced by British control of the seas. Because of shortage on the one hand and demand on the other, organic chemical production rose rapidly in the United States.

Despite a flourishing metals industry the United States had been recovering only a small proportion of the by-products from coke production. Beehive coke ovens, completely wasteful of coal gas and coal tar, were commonplace beside coalfields adjacent to metal smelting centers. Cheaply built and as readily abandoned, they were "monuments" to business leaders contemptuous of America's heritage of natural resources. These wasteful procedures had to be abandoned in favor of more expensive by-product ovens which would permit the recovery of the tar so rich in starting materials for organic syntheses.

World War II

When World War II began, TNT and picric acid were still the preferred military high explosives and they continued to play an important role in conventional military operations. Several other products which acquired specialized uses were developed during the war. The name RDX was given the product resulting from the nitration of hexamethylenetetramine. It was known as early as 1899, but because of its sensitivity and high cost had been rejected by most military groups. The British developed production and handling methods when the need for an explosive with greater power than TNT became crucial. James H. Ross of McGill University devised a production process utilizing formaldehyde and ammonium nitrate. Bachmann of Michigan further improved the Ross method, and large-scale production was begun by the Tennessee Eastman Co. Pentaerythritol tetranitrate (PETN), produced by nitration of pentaerythritol, is another highly sensitive explosive which, in admixture with TNT, is used in mines, bombs and torpedoes.

OTHER CHEMICALS FOR WARFARE

Toxic Gases

Toxic gases were introduced into warfare by the German army in the offensive at Ypres on April 22, 1915. Chlorine gas was released from cylinders in the German trenches and carried by the wind into the British trenches where heavy casualties resulted. Gas was used throughout the war, both sides trying to gain an advantage by employing new toxins. The development of gas masks utilizing adsorbent chemicals such as nut charcoal prevented such warfare from becoming extremely effective. Nevertheless the public reaction reached such proportions that an international agreement was signed after the war outlawing gas warfare. Toxic gases played no role in World War II, although belligerents maintained supplies of gases and gave a great deal of attention to gas defense. No doubt the high cost and comparative ineffectiveness of gas warfare coupled with fear of retaliation had more to do with failure to use gas than the moral influence of the treaty.

Although chlorine was the first gas used, it was superseded by phosgene as World War I progressed. Phosgene is colorless, heavy, and odorless and gives no warning of its presence. It has none of the choking effects of chlorine and does its damage slowly upon inhalation, being hydrolyzed to hydrochloric acid in the lungs. The liquid diphosgene was also used since it is persistent and slowly hydrolyzes to phosgene. The French introduced hydrogen cyanide gas, but it is too light to be persistent.

In addition to lung irritants vesicants, or blister gases, were used.

Strictly speaking, these are oily compounds which cause violent blistering upon contact with skin. The most widely used chemical of this sort was mustard gas, which will persist on vegetation for days under favorable weather conditions.

World War II research led to new discoveries in the vesicant field; the nitrogen mustards were extensively studied. Besides their vesicant action these compounds penetrate the skin and have a systemic action on living cells similar to exposure to x rays. These compounds are being investigated for their influence on hereditary mechanisms and for possible effects on cancerous diseases.

German research led to "nerve gases" which block certain enzyme-controlled reactions in nervous tissues. Subsequent research has uncovered interference with phosphorylation reactions and has led to an intensive search for a possible antidote.

Incendiaries

Incendiaries, which have figured prominently in warfare throughout history, were developed to a high level of effectiveness in World War II. Thermite magnesium bombs were dropped on London in great numbers by the Luftwaffe. Later the visits were returned in kind on German cities. Japan suffered heavily from fire raids because of the tinder-rich buildings in its cities.

The initial shortage of magnesium in the United States led to the development of substitute incendiary materials. A Harvard group, headed by Louis Fieser and reinforced by staff members of Arthur D. Little, Inc., studied the influence of aluminum naphthenate as a thickening agent for gasolene. The addition of aluminum soaps of coconut fatty acids led to Napalm which is used in bombs and flame throwers. The aluminum soap mixture of naphthenate, palm acids, and oleic acid added to gasolene gives a jellied product which is a superior incendiary material.

PETROLEUM PRODUCTS

Petroleum, known as "rock oil" since antiquity, only rose to industrial importance after 1859 when Edwin L. Drake developed the technique of drilling for oil rather than waiting for it to accumulate slowly at the surface. Within ten years oil had become a big business in the United States. Kerosene as an illuminant filled a growing need, since whale oil was in short supply after decades of vigorous whaling. Lubricating oils were also needed as a result of increasing use of machinery; lard oil and vegetable oils no longer were able to meet the demand. Petroleum also yielded solvents and fuel, and continued to be used in medicines.

Processing techniques for handling petroleum were available by the time it became abundant. Benjamin Silliman, Sr. at Yale had studied it sufficiently to understand its general composition. Illuminating oils had been produced from petroleum in England and the United States in the 1850's. An enterprising native of Pittsburgh, Samuel M. Kier, had not only sold bottled petroleum (Seneca Oil) as a cure for every known ailment, but had sought the advice of James Curtis Booth, a Philadelphia chemist, as to other uses. Booth advised distillation. Kier designed a crude still copied from the whiskey stills used in the locality and used it to produce lamp oil. This practice spread quickly since lamp oil met an obvious demand. The first distillate, naphtha, was usually discarded for it had not yet found use as a solvent. The final distillate was an oil rich in paraffin. The lamp oil was the valuable product. It was usually treated with surfuric acid, caustic soda, or other chemicals in order to decolorize and deodorize it. As new oil fields were opened, the problem of dealing with undesirable impurities such as sulfur compounds frequently confronted refiners. These problems were attacked in completely empirical fashion until 1900.

The development of the internal combustion engine and its use as the motive power for the automobile changed the character of petroleum production. Originally the kerosene lamp was the principal consumer of petroleum products, with the low-boiling naphtha fraction being an unwelcome by-product. Oil dealers were frequently tempted to increase the yield of kerosene by adding the low-boiling fractions. Since this was apt to explode when used in lamps, the "flash point" of kerosene had to be regulated by the federal and state governments. With the advent of the automobile and the growth of electrical lighting, the lower-boiling fraction, in the form of gasolene, became the desirable product. The adulteration of gasolene with kerosene created a need for government regulation.

Methods for obtaining a larger yield of gasolene from petroleum were sought. Researchers developed the cracking process whereby large hydrocarbon molecules are decomposed to gasolene-sized hydrocarbons through the use of high temperatures and pressures. The Standard Oil Co. of Indiana introduced a thermal cracking process developed by William M. Burton in 1913. Carbon Petroleum Dubbs (1881–1962) developed an alternative process. The Universal Oil Products Co. was set up as a licensing corporation to exploit the Dubbs patents and later became deeply involved in hydrocarbon research. A number of other cracking processes have been patented.

During World War I investigators became concerned with the quality of gasolene for airplane engines. It was learned that more power could be produced if the compression ratio of the engine were increased but that this would lead to severe knocking, even with good quality gasolene. Experimentation with other liquid fuels revealed that benzene is superior to gasolene and that cyclohexane is better still. Benzene was in short supply because of other urgent needs, and cyclohexane was a synthetic prepared by the hydrogenation of benzene. Researches made by Charles F. Kettering (1876–1958) and his associates of the Dayton Metal Products Co. failed to produce an aviation fuel before the war ended, but did reveal that engine knock is related to the chemical nature of the fuel.

On the basis of a Kettering hypothesis that a red dye in the fuel would improve heat absorption and thus vaporization in the engine cylinder, iodine was tried and found effective, but not because of its color. This touched off an Edisonian-type search for a less expensive antiknock agent. Aniline was found to be effective. Selenium and tellurium compounds proved even better. Investigation of metal-organics established that tetraethyl tin is highly effective. Thomas Midgley, Jr. (1889–1944) and Carol Hochwalt (b. 1899) then prepared tetraethyl lead which proved to be most suitable of all.

Quantity production of tetraethyl lead posed a problem. Through the work of Charles A. Kraus (1875–1967) then at Clark University, a commercial process was developed. General Motors, which had absorbed Kettering's Dayton laboratory in 1920, and Standard Oil Co. of New Jersey formed the Ethyl Gasolene Corporation to produce the necessary chemicals and license their use.[5]

This research led to the introduction of the octane number as a rating of gasolene quality. Midgley and his associates found that isooctane (2,2,4-trimethyl pentane) performed very well in a test engine, whereas *n*-heptane appeared to be the poorest burning hydrocarbon available. Gasolenes were then compared in engine performance with isooctane-heptane mixtures, a 75 octane gasolene corresponding in performance with a 75:25 octane-heptane mixture.

In order to prevent fouling of the engine cylinders and valves with lead oxide during combustion, it was necessary to introduce bromine into the gasolene. When this did not prove satisfactory, dibromoethylene was used. This led to heavy demands for bromine which persuaded Dow to begin extracting bromine from sea water in North Carolina.

Soon after manufacture of tetraethyl lead was begun at Bayway, New Jersey, there was a serious outbreak of lead poisoning at the plant. This led to an examination of the health aspects of the use of the fuel additive. A committee set up by the United States Surgeon-General reported that there were no good grounds for prohibiting use of Ethyl gasolene of a specified composition. Despite the use of tetraethyl lead in almost all motor fuel and despite the proliferation of automobiles there has never been another medical study of the effects on human health.

The burning properties of most gasolenes are improved by this antiknock compound and its use has spread widely, particularly in America where

[5] Williams Haynes, *American Chemical Industry*, *A History*, vol. 4, *The Merger Era*, Van Nostrand, New York, 1948, p. 402.

the driving of high-powered automobiles is a national obsession. Other antiknock compounds have been studied. I. G. Farben was developing iron carbonyl in 1924, but it proved to cause spark plug troubles and the German industry obtained the rights to the Ethyl Corporation patents. During the past decade research has led to great interest in sandwich-type metal-organics.

One of the major developments during World War II was the introduction of catalytic cracking on a large scale. This involves the use of metal alumino-silicates as catalysts to obtain a better distribution of hydrocarbon fragments than is possible in thermal cracking. The cracking fragments are also richer in the kind of hydrocarbons which burn well in an internal combustion engine. During the war the polymerization process by which gaseous hydrocarbons are united to give molecules in the gasolene range was also developed. This enables refiners to utilize the unsaturated gases produced in large quantity during cracking. Copper pyrophosphate-type catalysts are useful in this process. Alkylation was another process introduced just before the war to combine olefins and saturated hydrocarbons to give branched hydrocarbons, hydrogen fluoride or fuming sulfuric acid being the principal catalysts. This process made it commercially feasible to prepare isooctane for airplane fuel. Hydrocarbons of even higher octane number are produced by alkylation.

Catalytic reforming processes were developed to bring about the conversion of naphthenes and certain paraffins into aromatic hydrocarbons, thus improving the quality of the gasolene. These operations are carried out at high temperatures and pressures in the presence of molybdic acid or platinum catalysts deposited on alumina carriers.[6]

PLASTICS

Following his discovery of nitrocellulose Schönbein observed that it is soluble in an ether-alcohol mixture, giving a viscous liquid he termed *collodion*. A hard transparent film remains after evaporation of the solvent. Collodion quickly found use in medicine-and in the field of photography. In 1883 the English inventor Joseph W. Swan (1828–1914) developed a process for making nitrocellulose fibers. These were used in early electric lamps where, after being suitably charred, they served as carbon filaments. In France Hilare de Chardonnet (1839–1924) also developed a process for making fibers by extrusion. These were then hydrolyzed to cellulose to reduce flammability and marketed as "artificial silk" in 1891.

A plastic known as Xylonite was produced in 1865 by Alexander Parkes (1813–1890) of Birmingham, England. By mixing nitrocellulose, alcohol, camphor, and castor oil, a material was obtained which could be molded

[6] C. N. Kimberlin, Jr., *J. Chem. Educ.*, **34**, 574 (1957).

under pressure while warm. In America a somewhat improved product was produced by John Wesley Hyatt (1837–1920) from nitrocellulose and camphor. This was marketed commercially from 1872 as Celluloid. These plastics were used for photographic films, combs, men's collars, portable windows, and various small specialty articles. Despite their flammability Xylonite and Celluloid were widely used. Their success naturally led to a search for other plastic materials.

The casein plastics Galilith and Erinoid came into production after 1897 when W. Krische, a Hanoverian lithographer, and Adolf Spitteler, a Bavarian chemist, found a method for polymerizing casein with formaldehyde to a hard mass resembling ivory in physical characteristics. Since that time a variety of plastics have been prepared from the proteins of peanuts, soybeans, and other natural sources.

Phenol-formaldehyde plastics, under the name Bakelite, were developed in the first decade of the present century by Leo H. Baekeland (1863–1944). Baekeland was educated in chemistry in his native Belgium before emigrating to New York where he set up a laboratory as a consulting chemist. His first important invention, *Velox*, a photographic print paper which could be used under artificial illumination, netted him a sizable fortune when he sold it to Eastman Kodak. He then turned to the study of the reaction of phenol and formaldehyde.

Various chemists, following Baeyer in 1871, had reported the formation of dark-colored, tarry materials when these chemicals reacted. Since organic chemists were interested in pure crystalline compounds, these resinous products were considered unimportant and were invariably consigned to the trash container. Baekeland was able to bring about a controllable polymerization which led to a valuable thermosetting plastic. The General Bakelite Company began production in 1910.[7] The insulating properties, together with the ease of molding, led to wide usage in the electrical industry, particularly after World War I when phenol was no longer needed for the large-scale manufacture of picric acid. The newly developing radio industry made extensive use of phenol-formaldehyde resins.

During the next three decades the plastics industry expanded very rapidly with a wide variety of new types being brought into commercial production. Beetle-ware resulted from the polymerization of urea or thiourea with formaldehyde. This type of plastic was developed by Fritz Pollack of Vienna and was widely used in the production of tableware.

In the thirties the acrylic resins were developed by polymerizing acrylic acid $CH_2=CH\cdot COOH$ or related compounds. Methyl methacrylate $CH_2=C(CH_3)\cdot COOCH_3$ became the monomer used for the production of Perspex by Imperial Chemical Industries, Lucite by du Pont, and Plexiglas by Rohm and Haas. Methyl methacrylate polymers were observed

[7] W. Haynes, *op cit.*, vol. 6, *Company Histories*, 1949, 437.

as early as 1872, and the Swiss chemist G. W. A. Kahlbaum made an unbreakable beer glass of the polymer in 1888. Production of such a plastic was begun in Germany in 1927 and in England and the United States about a decade later.[8] Other commercial plastics are produced from vinyl chloride, styrene, ethylene, and tetrafluorethylene.

High Polymer Research

Initially plastics research was highly empirical. In 1926 Hermann Staudinger (1881–1965) showed that thermoplastic substances are made up of linear molecules having sizes of 1,000 Å and more. In thermoplastic materials these long molecules show little tendency to flow at ordinary temperatures, but on heating the rigidity is lost. In thermosetting plastics the original linear molecules still retain reactive centers where cross-linking is possible; hence a rigid three-dimensional molecule is formed during the application of heat, resulting in a permanently hard material.

The early contributions of Staudinger initiated vigorous exploration of the science of high polymers. Macromolecules, when studied systematically from the viewpoint of physical chemistry, were found to differ from small molecules only insofar as size introduces new patterns of behavior. Such research not only has been significant in the development and improvement of commercial plastics but also has contributed to the better understanding of natural high polymers such as rubber, cellulose, starch, silk, and wool. The theory of rubber-like elasticity developed through the work of Herman Mark (b. 1895) at Brooklyn, Paul J. Flory at Cornell University, and Frederick Wahl in Germany was useful in furthering progress in the field. Mark pioneered in the use of x rays in studying fibers while he was at the Kaiser Wilhelm Institut at Dahlem in the midtwenties. While he was director of high polymer research for I. G. Farben (1927–1932), his laboratory at Ludwigshafen laid the foundations for the commercial production of polystyrene plastics, the polyvinyls, the polyacrylics, and the buna-type rubbers. When the Nazis came to power he took a professorship at the University of Vienna while being retained as a consultant for I. G. Farben. Following the Anschluss he fled from Austria, briefly holding a position with a Canadian cellulose firm before taking a post at the Brooklyn Polytechnic Institute where he founded the Polymer Research Institute and served as a consultant to du Pont.

The work of Giulio Natta (b. 1903) of Milan on reaction mechanisms and behavior of catalyst systems was responsible for improved understanding of high polymers, particularly polypropylenes. By x ray crystallography Natta established the helical structure of the large molecules formed by polymerization of propylene and related hydrocarbons.

Karl Ziegler (1898–1973) of the Max Planck Institute for Coal Research

---

[8] F. Sherwood Taylor, *A History of Industrial Chemistry*, Heinemann, London, 1957, p. 263.

at Mülheim, through his extensive researches on organometallic catalysts, developed industrially important methods for the addition of alkali metal alkyls to unsaturated bonds. Through his work the synthesis of polyethylene polymers became possible.

TEXTILES

Although Réaumur[9] suggested in 1734 that man might imitate silk, it was not until after the discovery of nitrocellulose that any serious attempts were made in this direction. Early in the 1880's Swan in England and Chardonnet in France produced nitrocellulose filaments from collodion solutions. By 1891 Chardonnet had devised a way of hydrolyzing off the nitrate groups so as to make the product less flammable, and commercial production of "artificial silk" was begun at Besançon.

In 1897 Hermann Pauly began producing a cellulose fiber by the cuprammonium process in which cellulose was dissolved in ammoniacal copper hydroxide solution and precipitated with sulfuric acid. The solubility of cellulose in the reagent had been discovered by Eduard Schweitzer in 1857, or possibly by John Mercer even earlier.

A third process was developed in 1892 by Charles Frederick Cross and Edward John Bevan in England. They dissolved cellulose in alkali and carbon disulfide to obtain a thick solution (viscose) which could be regenerated into cellulose fibers or sheets (cellophane).

Cross and Bevan were also responsible for the development of cellulose acetate as a fiber (celanese) and as a film material. Paul Schützenberger carried out the first acetylation of cellulose in 1865, but the commercialization of the reaction developed very slowly.

These cellulose fibers were at first marketed as "artificial silk," even though they were cellulosic rather than proteinaceous in character and differed rather extensively in properties from natural silk. The term *rayon* gradually came into use for fabrics made from chemically treated cellulose. Through research the products have been improved to the point where they enjoy general consumer acceptance on their own merits. For some uses, as for the cord in automobile tires, rayon is found superior to natural cotton. Over the years the viscose process has proven the most economic and has driven the Chardonnet and cuprammonium processes into retirement. Cellulose acetate has properties different from viscose and has maintained a position in the marketplace.

Other Synthetic Textiles

The rise of purely synthetic fibers has been very dramatic. The first of these to become commercially important was 66-nylon, marketed in the

[9] R. A. F. de Réaumur, *Mémoires pour servir a l'Histoire des insectes*, Paris, 7 vols., 1734–1742.

form of women's stockings in 1940. During the war years, because of military uses for the plastic, it was not possible to meet consumer demand. Still nylon gained a popularity which virtually eliminated silk as a textile material in the West. The completely synthetic textiles developed during and since the war have had a devastating effect upon the market for natural fibers such as wool and silk.

Fig. 26.4. WALLACE H. CAROTHERS.
(Courtesy of the du Pont Company.)

Nylon was the outgrowth of research directed by Wallace Hume Carothers (1896–1937) of du Pont on the basic chemistry of large molecules. Such studies had been initiated in the twenties when Charles M. A. Stine (1882–1954) became director of du Pont's laboratory. It was felt that certain areas of basic research should be pursued by an industrial laboratory even though commercial results might not be evident for many years. Carothers left his position at Harvard to head the work on polymer chemistry. The synthetic rubber neoprene also was a direct outgrowth of these studies.

The researchers were interested in duplicating or at least approaching the characteristics of silk, a protein. Amino acids do not lend themselves easily to controlled condensation, so other monomers were examined. Adipic acid $HOOC(CH_2)_4COOH$ and hexamethylene diamine $H_2N(CH_2)_6 \cdot NH_2$, compounds which can easily be manufactured from coal tar chemicals, were condensed. Nylon was introduced to the market with great fanfare as a textile made from coal, air, and water. About the time that du Pont was developing nylon, I. G. Farben was commercializing Perlon, a related polymer produced from caprolactam.

Orlon is a synthetic fiber produced by du Pont by the polymerization of acrylonitrile $CH_2{=}CH \cdot CN$. Here, hydrogen bonding between adjacent polymethylene chains is responsible for a great deal of the strength. Somewhat similar synthetic fibers from acrylonitrile are produced by American Cyanamid (X-51), Chemstrand (Acrilon), and Union Carbide (Dynel).

The polyesters were examined for practical possibilities by Carothers, but commercial development first occurred in England. J. R. Whinfield of the Calico Printers Association developed a successful fiber in 1946 by polymerizing terephthalic acid and ethylene glycol. This was produced by Imperial Chemical Industries as Terylene. The du Pont company introduced a similar fiber under the name Dacron.

The synthetic fibers in no case duplicate a natural fiber; their properties are in part superior and in part inferior to standard fibers. Nearly all of the synthetics are highly moisture resistant and, therefore, dry rapidly. Some of them readily take a heat set which causes them to be highly wrinkleproof and to hold desired creases well. All of the synthetics have created serious dyeing problems, necessitating the creation of new dyes and new dyeing methods.

RUBBER, NATURAL AND SYNTHETIC

Except for the introduction of vulcanization in 1839 through the work of Thomas Hancock in England and Charles Goodyear in the United States, rubber received little attention from chemists before 1900. Of course, theirs can hardly be considered a chemical contribution since the investigations were pursued on an empirical rather than on a scientific basis. The rubber industry, to the extent that one existed in the nineteenth century, found chemistry of no real concern in its operations. The industry had its first real success with the introduction of the bicycle tire. When a practical automobile was introduced at the beginning of the twentieth century the rubber industry was faced with a demand that was ultimately to convert it into a large chemical industry. The pneumatic tire developed by John Boyd Dunlop in 1896 represented a masterpiece of mechanical handling of a difficult raw material. Chemistry was not involved, and the early problems of the industry were primarily problems of supply. Uncertainties of supply, even after the introduction of plantation rubber in 1907, spurred efforts to develop a satisfactory synthetic. And practical problems encountered in handling a heat- and oxygen-sensitive raw material also led manufacturers to seek scientific aid.

The earliest scientific contributions were in the field of accelerators, chemicals which speed up the vulcanization process and permit the use of lower temperatures. George Oenslager of the Diamond Rubber Company of Akron discovered that aniline speeds up the rate of vulcanization. Other organic bases also were found to have a similar effect. Aniline was secretly brought into use, giving Diamond Rubber a distinct advantage over competitors. Later the less toxic thiocarbanilide was introduced. All tires produced by Diamond Rubber contained one of these accelerators. The company was merged with the B. F. Goodrich Co. in 1912. Meanwhile other rubber fabricators were groping toward similar achievements. The first German patents for rubber accelerators were issued in 1914.

About the same time that accelerators were introduced it was found that other chemicals enhance the properties of rubber. Zinc oxide was at first added merely as a filler, but it turned out that a beneficial effect was exerted. Carbon black, it was learned, markedly increases the mileage of tire treads.

The deterioration of rubber from exposure to air was found to be associated with the formation of peroxides, with subsequent splitting of the rubber molecules to smaller size. Antioxidants in the form of aromatic amines, phenols, and quinones were found to be effective in retarding such deterioration. During the decade following World War I the scientific advances in rubber compounding so improved the quality of the automobile tire that the industry found itself in a state of economic depression brought on by its own efficiency.

Purely chemical studies of the nature of rubber began in 1860 when C. Greville Williams distilled the crude substance and identified isoprene as the principal decomposition product. William A. Tilden prepared isoprene from various terpenes around 1882 and established its structural formula. He found that a rubber-like gum forms when isoprene from turpentine is left standing in sunlight.

Work on rubber progressed very slowly because large molecules posed a dilemma most chemists preferred to avoid. Early investigations showed the presence of double bonds. Ozonolysis studies between 1905 and 1912 by Harries and later by Plummer (1931) resulted in the identification of levulinic acid, revealing a head-to-tail linkage of isoprene units. In 1925 Katz showed by x ray studies that stretched rubber has a crystalline structure. It was also learned that natural rubber molecules have a *cis* structure, whereas nonelastic *gutta-percha* consists of isoprenoid units with *trans* configuration in which blockage by methyl groups prevents elasticity.

Because Europe and North America were dependent upon the tropics for natural rubber, there was considerable interest in developing a synthetic rubber even before World War I began. Germany suffered greatly during the war years from a shortage of natural rubber and introduced a synthetic "Methyl rubber" obtained from the polymerization of 2,3-dimethyl butadiene which they were producing from acetone by hydrogenation to pinacol followed by dehydration. The resemblance to natural rubber was superficial and the product failed to survive once the war came to an end.

Research on synthetic rubber was continued in Germany and elsewhere. During the thirties commercial production of several kinds of synthetic polymers with rubber-like properties was begun.

Fig. 26.5. Julius A. Nieuwland.
(Courtesy of Notre Dame University.)

Du Pont's neoprene was an outgrowth of research on the reactions of acetylene carried out at Notre Dame University by Julius A. Nieuwland (1878–1936). Father Nieuwland passed acetylene through a solution of copper chloride and ammonium chloride to obtain a mixture of polymers in which divinylacetylene, a yellow oil, was predominant. This polymerized further to yield a poor rubber. Carothers further explored the polymerization of acetylene. Divinylacetylene proved unsatisfactory for rubber production but was developed into a synthetic drying oil which gave a hard resinous finish. Even more important, the du Pont group found that another polymerization product of acetone, monovinylacetylene, might be treated with hydrogen chloride to give chloroprene, an excellent monomer for synthetic rubber. Neoprene rubber proved to be superior to natural rubber in its resistance to sunlight, oxidation, and the attack of many chemicals.

The buna-type rubbers were first developed through research at I. G. Farben. Butadiene is the principal monomer for this type of rubber, but, as was learned early in the 1900's, butadiene does not polymerize to form a good rubber. By polymerization with a copolymer such as styrene or acrylonitrile a satisfactory rubber was obtained. The name was derived from *bu*tadiene and *na*trium, sodium being used as a polymerization catalyst. Buna-S, using styrene, and Buna-N, using acrylonitrile, became the principal synthetic rubbers during World War II both in Germany and in the British-United States sector, although other types of synthetic rubber were developed, for example, butyl rubber from isobutene and Thiokol from ethylene dichloride and sodium polysulfide.

AGRICULTURAL CHEMICALS

Western agriculture has been revolutionized by scientific research which has provided improved chemical fertilizers, insecticides, rodenticides, herbicides, pharmaceuticals, and soil improvers. Fertilizers are an outgrowth of nineteenth-century agricultural chemistry but the rest of these agricultural chemicals are twentieth-century products.

The major change in fertilizers resulted from the introduction of processes for nitrogen fixation and the exploitation of new reserves of phosphate and potash salts and steady improvement in methods for their processing. The importance of trace elements in plant nutrition is appreciated now, and fertilizer producers are concerned with the presence of boron, cobalt, manganese, copper, molybdenum, vanadium, tungsten, and zinc.

Insecticides

Insecticides, of course, have other than agricultural uses. Insects present esthetic and health hazards besides preying on farmers' crops and represent a major adversary for man because they are in many respects highly efficient biologically. Insects are found in a myriad of forms, they proliferate at a stupendous rate, they have voracious appetites, and they frequently acquire a genetic tolerance to an unfavorable environment. To make the problem of control more difficult, beneficial insects such as bees are as readily killed by poisons as the harmful insects. In addition, massive control measures are frequently deadly to higher animals (birds, amphibians, reptiles, fishes) which serve as a natural check on the insect population. During the twentieth century man, who so often has placed severe stresses on nature, many times has "helped" harmful insects. Mass production of many specific crops has prevented the establishment of an ecological balance which, under normal conditions, would hold the insect population in check.

The large-scale use of insecticides began in the 1870's when the potato beetle was spreading rapidly eastward across the United States. Paris Green proved an effective control chemical. Calcium arsenate came into use when the boll weevil began large-scale attacks on cotton fields, and lead arsenate gained favor among orchardists in their efforts to control the codling moth. Sulfur and copper sulfate were used for the control of fungus infestations in horticulture.

Inorganic insecticides dominated the field until World War II. Organic substances were limited to a few natural products such as rotenone, nicotine, and pyrethrum which were more costly than the inorganic poisons and hence were restricted in use. Experience with inorganics led to progressively increasing problems. Insects developed resistant strains so quickly that heavier and more frequent application became necessary. In some cases this resulted in direct damage to vegetation or led to a build-up of arsenic compounds in the soil which made it difficult to grow crops of any kind. Fruits and vegetables were marketed with significant residues of toxic compounds on them. While acid and alkaline washes reduced such residues to safe levels, some growers had to be forced to wash their produce by regulatory agencies. During the 1930's there was much quarreling in the United States between consumer and producer groups with respect to insecticide residues.

Synthetic organic insecticides became important after 1939 when Paul Müller (1900–1965) and associates of the Swiss firm J. R. Geigy observed the effectiveness of dichlorodiphenyltrichloroethane (DDT) against insects. The compound had been synthesized in 1874 by O. Zeidler. Geigy began to market the substance as Gesarol. Gesarol—and modifications of it— was widely used by the German army during the war. In 1943 the United States army began using it with great success to control malaria in Casablanca and typhus in Naples. Before the invasion of Leyte the island was dusted by planes. DDT proved highly effective in the control of mosquitos, flies, lice, and agriculturally important insects since it acts both as a contact and stomach poison. It has become immensely popular as an insecticide.

With increased use, certain shortcomings of DDT have become evident. It is ineffective against aphids, boll weevils, Mexican bean beetles, and certain other types of insects. Houseflies and certain other insects soon developed highly resistant strains which are not killed, even by heavy doses of DDT. Also it is absorbed by farm animals, the compound appearing in the fatty tissues and in the milk fat where the residues could cause injury to human consumers. Many DDT analogs have been prepared and tested in order to find superior insecticides. Methoxychlor is much less toxic to higher animals and is not concentrated in the fat and milk. Other useful modifications also have been developed.

Extensive investigations of halogen compounds has revealed insecticidal properties in many of them. Imperial Chemical Industries introduced hexachlorocyclohexane (666) in 1945. Of nine possible isomers, that termed *gamma* proved to be the active one. It is effective against the boll weevil and forage crop insects but because of the production of off-flavors is unsuitable for use on food crops. Lindane, the purified gamma isomer, is

Fig. 26.6. PAUL MÜLLER.
(Courtesy of the National Library of Medicine,
Washington, D.C.)

effective in smaller quantities and hence is not as objectionable. Other effective chlorinated insecticides are Chlordan, an octachloro derivative of substituted methanoindene; Toxaphene, a chlorinated derivative of camphene which is produced from pine-oil by Hercules; and two Diels-Alder condensation products, Dieldrin and Aldrin.

Organophosphorus insecticides resulted from the wartime research of Gerhard Schrader of the I. G. Farben laboratories. This work, reportedly aimed toward the development of nerve gases toxic to humans, was uncovered by American and British intelligence teams following the collapse of Germany. Of the more than 300 compounds studied by Schrader, Parathion has proved to be a particularly valuable insecticide. This and other organic phosphates are toxic to a great many insects in very low concentration since they inhibit cholinesterase, the enzyme in the nerve tissues of animals which hydrolyzes acetylcholine and thus plays an important role in the transmission of nerve impulses. Since cholinesterase is important in the nerve function of human beings as well as in insects, major hazards are involved in the manufacture and use of such phosphorus insecticides.

Rodenticides

Before World War II the chemical attack on rodents was waged primarily with such well-known poisons as strychnine, red squill, thallium sulfate, and zinc phosphide. These were never entirely satisfactory since the rat, being an intelligent animal, quickly learns to avoid poisoned food after being made ill or having observed disasters among his relatives. Wartime research, sponsored by the United States government, into more powerful or more subtle rodenticides led to the introduction of Compound 1080 and ANTU. Sodium fluoroacetate (1080) is relatively tasteless and is easily incorporated into liquid and solid bait. Alpha-naphthylthiourea (ANTU) is fatal to brown Norway rats, causing damage to the blood vessels of the lungs and causing them to literally drown from leakage of their own blood into their lungs. Both of these highly effective compounds are dangerous to use because of toxicity to dogs and to other beneficial animals.

Since compounds related to coumarin interfere with coagulation of the blood, K. P. Link investigated the effectiveness of such compounds as rodenticides. Warfarin, named after the *W*isconsin *A*lumini *R*esearch *F*oundation which controls the patent, was developed out of this research. The compound is apparently tasteless to rodents and causes a delayed death from slow internal hemorrhages. Since rats do not die in the vicinity of the poisoned bait, relatives do not become suspicious of it.

Herbicides

Before 1940 most weed killers were inorganic compounds like sodium chloride, sodium chlorate, ferrous sulfate, and sodium arsenate which kill all types of vegetation and leave the soil sterile until the chemical is leached out. Research on plant hormones led to a new class of organic herbicide, often with a somewhat selective action. During the war the United States Chemical Warfare Service and Department of Agriculture, the Boyce Thompson Institute, and the University of Chicago sought to develop chemicals which might be used to destroy enemy crops. Similar studies were made in England.

It was found that 2,4-dichlorophenoxyacetic acid (2,4-D) kills broad-leaved plants when used in small amounts but has very little effect on grasses. The compound is related to indole acetic acid which was shown by Went in 1928 to be a growth hormone for plants.

Further research led to organic herbicides which are effective against grasses without seriously damaging broad-leaved plants. The availability of chemical herbicides has had a significant effect on American agriculture. These compounds also are used to control undesirable plants in waterways,

Fig. 26.7.   KARL PAUL LINK.
(Courtesy of the University of Wisconsin.)

along highways, and in recreational areas. However, they are not always used with wisdom from the standpoint of conservation and ecology.

Plant control chemicals have also been developed which retard growth, speed ripening, intensify color, induce seedlessness, cause defoliation, and prevent dropping of fruit before picking time.

CONCLUSION

There has been a spectacular increase in the number and kinds of chemical products made available to ultimate consumers. Chemotherapy finally became significant with the production of sulfa drugs, antibiotics, and corticoid drugs. Petroleum research has led not only to greater diversion of distillates into gasolene but also to vastly improved quality of motor fuels. Plastics, synthetic textiles, and elastomers have grown to major importance, particularly since fundamental research on macro-molecules led to application of scientific knowledge to the production of high polymers with uniquely desirable properties. Agriculture has been revolutionized as a result of availability of synthetic insecticides, herbi-cides, and similar chemicals.

Vigorous industrial research development has been responsible for the revolution in consumer products. No company could maintain its market position without a thriving scientific department. The rapid development of new products frequently leads to rapid obsolescence of earlier products. One major company frequently boasts that the bulk of its sales are in products unknown thirty years before. The need for competitive advan-tages has forced the industries to actively search out new consumer products.

CHAPTER 27

GROWTH AND PROBLEMS

As chemistry moves into the last four decades of the twentieth century it is characterized by a vigor such as it never knew in any of its earlier periods. During the past century the Baconian dream that the study of natural philosophy could create wonders for the welfare of humankind has become real. The dye industry and the electrical industry paved the way a century ago by applying discoveries of basic chemistry and physics which had accumulated up to that time. Now, in the middle of the twentieth century, technology is placing ever greater demands upon basic science.

The discipline of chemistry is characterized by an enthusiasm and optimism among those in the field and is watched with a kind of wide-eyed wonder by most people who fail to understand the subject but are vaguely convinced that science has great powers to produce material wonders and devastating horrors.

At the same time that chemists enjoy considerable prestige, they are beset by a variety of problems, many of them without apparent solutions. Many of these problems are of chemists' own making. Not only in chemistry, but in virtually all fields of learning, men must digest a staggering mass of knowledge in order to master their own specialty. At the same time the consequences of their activities in their own specialties should not be completely ignored. The introduction of new chemicals into the environment may have unexpected repercussions. The introduction of new reactions has made established industries obsolete. New knowledge about nutrition and improvements in sanitation and chemotherapy have increased life-expectancy and triggered a population explosion. The development of nuclear weapons has thrown doubt upon the ability of human civilization to survive.

GROWTH OF THE CHEMICAL PROFESSION

The growth since 1900 of the chemical profession is reflected in membership changes in the American Chemical Society. In 1900 membership was under 2,000; in 1920 it had climbed to 15,000; in 1940 to 26,000; in 1950 to 63,000; and by 1960 there were over 85,000 members on the rolls.

Activity in different parts of the world is reflected in a recent study by Brooks[1] who examined five random issues of *Chemical Abstracts* published in 1957 and 1958 for national origins of papers in several chemical areas. Despite the smallness of the sample and the shortness of the time period covered by the study, it brings out quite well the location of greatest research activity and application. Table 27.1, based on data taken from

Table 27.1 RELATIVE DISTRIBUTION OF CHEMICAL ACTIVITY IN 1957–1958[a]

Country	General and Physical	Organic	Biological	Electronic and Nuclear	Metallurgy and Metallography	Total
England	204	227	54	211	55	751
France	58	70	33	119	38	318
Germany	149	175	73	117	71	585
Japan	110	168	77	146	75	576
Russia	376	139	38	294	254	1,101
United States	367	344	291	953	183	2,138
All others	232	248	147	311	79	1,017
Grand total	1,496	1,371	713	2,151	755	6,486

[a] As reflected in a study of *Chemical Abstracts* by B. T. Brooks, *J. Chem. Educ.*, **35**, 469 (1958). Figures represent the number of abstracts counted in five randomly selected issues.

Brooks, shows the United States with a substantial lead in total number of publications. The United States figure for electronic and nuclear chemistry looks as if it might be atypical, but even if it were reduced to a third of the tabulated figure, the United States is still left with a strong over-all record. This, however, is not unexpected for a populous and prosperous nation having abundant resources and well-developed industries. By contrast, although not shown in the table, the quality and quantity of chemical research coming out of much smaller nations such as Sweden, The Netherlands, and Switzerland is particularly impressive.

Britain, France, and Germany continue to maintain a strong position in chemical research but no longer provide the almost exclusive leadership

[1] B. T. Brooks, *J. Chem. Educ.*, **35**, 468–469 (1958).

they did during the eighteenth and nineteenth centuries. The rise of Russia and Japan to prominence is largely a twentieth-century phenomenon, with much of the growth occurring after World War I. A small number of Russian chemists made vitally important contributions during the nineteenth century, but growth of the science in that country became striking only after the Revolution. By 1950 Russia had clearly become a leading figure in the chemical world.

Chemical Education

The education of chemists has received a great deal of attention since 1900. At times there was grave fear that not enough young people were going into the field. This was particularly true during wartime, both hot and cold. During the depression there was a temporary oversupply.

Chemical curricula have undergone considerable change. Undergraduate training early in the century emphasized descriptive chemistry. A familiarity with the fields of general inorganic, analytical, organic, and physical chemistry was considered adequate, and physical chemistry was frequently studied without benefit of calculus. This pattern started to change in the twenties when a background in calculus began to be stressed for introductory courses in physical chemistry. Now most good curricula still stress knowledge of the same four basic areas, but the nature of the courses has drastically changed. There is far less emphasis on descriptive content and far more on theoretical material. Large amounts of physical chemistry have crept into the introductory course, into analytical chemistry, and even into introductory organic courses. Senior inorganic courses based on physical chemistry frequently are required, and a part of the analytical chemistry has been extended into the upperclass years when theoretical principles and instrumental methods can be stressed.

Graduate education has tended, in general, to follow the pattern established in Germany during the nineteenth century. In fact many of the leading graduate centers in the United States and other countries were developed by chemists educated in Germany. Research is heavily stressed.

The emphasis in graduate studies has shifted toward more extensive grounding in theoretical chemistry. Organic chemistry largely resisted the impact of physical chemistry until 1940, but then underwent extensive changes, particularly in the hands of the younger chemists entering the field. Even biochemistry, which had long remained highly descriptive, has turned more and more to theory.

Chemical engineering felt the impact of physical chemistry quite early. The discipline developed at the beginning of this century out of electrical engineering, when electrochemical industries were expanding rapidly. Initially chemical engineering was highly empirical, but it soon took advantage of thermodynamics and other theoretical tools as the chemical industry placed greater and greater demands upon its engineers. Chemical engineering was one of the first branches of the engineering profession to develop and encourage graduate study.

Postdoctoral study has become very widespread since World War II. Earlier, many wished to study under a renowned professor following receipt of the doctorate but few students could afford to do so. With the new emphasis upon science, fellowship funds and research contracts supported by industry and government agencies are increasingly available to postdoctoral students. Postdoctoral fellows enabled professors to turn out research at a faster pace since such fellows required less supervision than raw graduate students.

PROBLEMS OF THE CHEMICAL PROFESSION

Proliferation of Chemical Literature

With the rapid expansion of chemical research there has been a veritable explosion of chemical literature. It is virtually impossible to keep abreast of a major area of chemistry and it is growing more and more difficult even to keep in touch with the area of one's specialty. To cope with the problem there has arisen a group of literature specialists, particularly in the United States and Soviet Russia, who seek to design classification systems that are simple, unambiguous, and amenable to machine searching.

In part the proliferation of chemical literature is associated with the increase in the number of chemists, but it is also aggravated by certain publication pressures. The ever-present pressure for priority is accentuated by the larger number of workers in the field. Consequently chemists are inclined to write up results before the researches are complete. Long papers representing a comprehensive piece of research are becoming less and less common. Fragmentary researches are rushed into print as rapidly as results are accumulated. A further reason for fragmentary publication is the "publish or perish" atmosphere prevailing in many college administrations.

The production of papers not only has caused the standard chemical society journals to expand to heroic proportions, but also has brought about a succession of new journals, many of them dealing with specialized aspects of chemistry. *Chemical Abstracts* now reports the results of chemical work published in more than 8,000 journals.[2] While most of these are not strictly chemical journals, they frequently contain papers of a chemical nature.

The growth of journals sponsored by the American Chemical Society is illustrative of the growth of publications. The *Journal of the American Chemical Society* was started in 1879, remaining a modest publication to the end of the century even though it absorbed the six-year old *Journal*

[2] For the coverage of *Chemical Abstracts* see *List of Periodicals Abstracted by Chemical Abstracts*, Chem. Abs. Service, Ohio State University, Columbus, Ohio, 1956.

of Analytical and Applied Chemistry in 1893. Edward Hart, founder of the latter journal, took over the editorship of the *Journal of the American Chemical Society* at that time. In 1902 editorship passed on to William A. Noyes, chief chemist of the U.S. Bureau of Standards, who in 1907 became head of the chemistry department at the University of Illinois. Noyes' eighteen-year editorship saw great expansion of the society's publication activities. The *Journal* became the principal American publication in its field and in 1913 absorbed the competitive *American Chemical Journal* which had been founded by Ira Remsen in 1879. The *Journal* continued to grow in size and stature as the editorship passed on to Arthur B. Lamb (ed. 1918–1949) of Harvard, then to William A. Noyes, Jr. of the University of Rochester, the son of the earlier editor, and in 1963 to Marshall Gates of Rochester.

Chemical Abstracts was initiated by the Society in 1907 under the editorship of the elder Noyes. An antecedent *Review of American Chemical Research* had been started at Massachusetts Institute of Technology by Arthur A. Noyes, a distant relative of William A. Noyes. This review was gradually expanded in scope and incorporated into the *Journal.* W. A. Noyes took editorial responsibility in 1902. When *Chemical Abstracts* was founded as an independent journal its scope was expanded to include chemical work published by non-Americans, and it was unique in being the earliest abstract journal to cover both basic and applied chemistry. Noyes passed the editorship on to Austin M. Patterson in 1909, and, after poor health forced the latter to curtail his work on the journal, Evan J. Crane (b. 1889) began his editorship which terminated only with his retirement in 1959. *Chemical Abstracts* was in its early days no match for the long established *Chemisches Zentralblatt* in its coverage of basic chemistry. Through the years of Crane's editorship, however, it showed steady improvement in quality and coverage to become ultimately the leading journal in the field of scientific abstracting.

The needs of industrial chemists were recognized by the founding of the *Journal of Industrial and Engineering Chemistry* in 1909. This publication (which dropped the *Journal of* from its title in 1923) gave rise to a *News Edition* in 1923 and an *Analytical Edition* in 1929. The latter became an independent ACS publication in 1947 as *Analytical Chemistry.* The semimonthly *News Edition* was renamed *Chemical and Engineering News* in 1942 and made a weekly five years later. The *Journal of Chemical and Engineering Data* was started in 1959.

The *Journal of Physical Chemistry* was founded in 1896 by Wilder D. Bancroft of Cornell University, a former student of Ostwald who was promoting the growth of physical chemistry in America. Ownership of the *Journal* passed to the ACS in 1932, when Samuel C. Lind became editor. *Chemical Reviews* began publication in 1924, the *Journal of Organic Chemistry* in 1936, the *Journal of Agricultural and Food Chemistry* in 1953, *Inorganic Chemistry* in 1962, *Biochemistry* and *Medicinal Chemistry* in 1963.

Besides these journals sponsored by the Society, several Divisions sponsor journals of their own. The *Journal of Chemical Education* was founded by the Division of Chemical Education in 1924, under the editorship of Neil E. Gordon of Wayne University. Since 1928 *Rubber Chemistry and Technology* has translated and reprinted papers from worldwide journals which are of interest to members of the Division of Rubber Chemistry. *Chymia*, founded in 1948 under the editorship of Henry M. Leicester, is an annual sponsored by the Division of the History of Chemistry.

In most other countries at least one major journal is published by the chemical society. Specialized societies also tend to support their own journals. Additional journals have been founded by private individuals or by publishing houses.

Economic, political, and military events have been responsible for termination, suspension, and revision of certain journals. The depression of the thirties saw a number of lesser journals go out of existence or consolidate with other publications. World War II disrupted German journals badly. The *Berichte der deutschen chemischen Gesellschaft* suspended publication in 1945. It was revived in 1947 under the title *Chemisches Berichte*. The *Zeitschrift für physikalische Chemie* was also a wartime fatality. It was independently revived in both East and West Germany and comes out in different Leipzig and Frankfurt editions.

Reviews are especially important in modern science since only by reading them can the average scientist hope to keep somewhat abreast of his field and learn something about work being done in related areas. *Chemical Reviews* has held an important place in its field since 1924. Books presenting annual reviews of a specialized field are particularly valuable since they represent an effort to be systematically exhaustive for a particular area.

Although the *Berlinisches Jahrbuch für Pharmacie* (1795–1840) must be considered the earliest chemical annual, it was Berzelius' *Jahres-Bericht* which pioneered in the field between 1821 and 1847. Similar annuals figured prominently until the end of the century when abstracting journals began to take on special importance. The Chemical Society of London began sponsoring *Annual Reviews on the Progress of Chemistry* in 1904. However, it was no longer practical to cover all of chemistry in an annual review and few other such efforts were more than temporary. There was, however, a move toward annual reviews in specialized fields. Under the editorship of J. Murray Luck of Stanford, the *Annual Review of Biochemistry* pioneered in this sort of review. There are now a number of such annual reviews dealing with physiology, nuclear science, food research, and other specialized fields.[3]

Abstract journals are the lifeblood of the chemical literature, since only

[3] For a listing of such reviews see E. J. Crane, A. M. Patterson, and E. B. Marr, *A Guide to the Literature of Chemistry*, Wiley, New York, 2nd ed., 1957, pp. 207–212.

publications such as *Chemisches Zentralblatt, Chemical Abstracts, British Chemical and Physiological Abstracts,* and *Referativnyi Zurnal—Khimiya* are capable of keeping chemists abreast of the current literature. More than that, through their indexes they make it possible to carry out literature searches systematically and exhaustively.

Chemisches Zentralblatt, which has been published under a sequence of related titles since 1830, was taken over by the Deutschen Chemischen Gesellschaft in 1897. It was expanded in 1919 to include applied chemistry by absorbing the abstract section of *Zeitschrift für angewandte Chemie.* During World War II it appeared irregularly and its coverage was very incomplete. Publication ceased entirely in 1945. Then separate editions began appearing in the Eastern and Western Zones of Germany. These were both inadequate and also represented expensive duplication. They were ultimately combined and through the aid of *Chemical Abstracts* brought up to date.

British Chemical Abstracts was started in 1926, taking over the abstracting activities of the *Journal of the Chemical Society* in basic chemistry and of the *Journal of the Society of Chemical Industry* in the applied field. It was later expanded to include *Physiological Abstracts.* Both societies supported *British Chemical and Physiological Abstracts* until 1953 when it was considered too expensive a project to be continued.

The *Referativnyi Zhurnal—Khimiya* was started by the Soviet Institute of Scientific Information in 1953 for the purpose of abstracting pure and applied chemistry on a world-wide basis.

Comprehensive and informative abstracting is increasingly difficult and indexing seems to have gotten out of hand. The annual indexes of CA do not appear for one or even two years after completion of the abstracts portion of the volume. The Fifth Decennial Index for the years 1947–1956 only began to come off the printing presses late in 1960. The cost of decennial indexes is pricing everyone out of the market except for university, industrial, and governmental libraries. The cost of annual indexes has reached the point where many individual chemists are forced to give up their subscriptions. The cost of abstracting is met only by introduction of graded subscription rates bearing heavily on libraries and industrial subscribers.[4]

[4] In view of these costs, and in view of the widespread importance to the whole populace of the rapid spread of scientific information, it appears timely to look upon abstracting as an activity which should be financed on a national, or even better, an international basis. The abstracting of the scientific literature, or even better, scholarly literature as a whole, might logically be an activity of UNESCO. It seems more logical to have abstracts prepared once and distributed quickly than for each scientific society to carry out this expensive function independently. *Chemical Abstracts* has developed an excellent service during a half century. It might very well be selected to continue its work on an international basis with international support. Abstracts prepared by it might very well be published simultaneously in English, Russian, French, German, Spanish, Chinese, and possibly other languages.

Nomenclature

International communication would be well served if still greater agreement were possible between chemists of various nationalities regarding nomenclature. The International Union of Pure and Applied Chemistry has done excellent work in setting up rules of standard nomenclature in organic and inorganic chemistry. However there are still large areas where an international usage would benefit chemists generally.

While there is general agreement on symbols, there is still no general usage for names of the elements themselves. Elements known to the ancients have national names varying from country to country. The common elements like hydrogen, oxygen, carbon, and nitrogen have had

Table 27.2   COMPARISON OF TRADITIONAL AND STOCK NOMENCLATURE FOR $FeSO_4$[a]

Language	Traditional Name	Stock Name
Danish	Ferrosulfat	Jern-to-sulfat
Dutch	Ferrosulfat	Ijzer-twee-sulfaat
English	Ferrous sulfate	Iron-two-sulfate
Finnish	Ferrosulfaatti	Rauta-kaksi-sulfaatti
French	Sulfate ferreux	Sulfate de fer-deux
German	Ferrosulfat	Eisen-zwei-sulfat
Hungarian	Ferroszulfát	Vas-kettö-szulfát
Italian	Solfato ferroso	Solfato di ferro-due
Russian	Sulfat zheleza	Sernokisloe zhelezo-dva
Spanish	Sulfato ferroso	Sulfato de hierro-dos
Swedish	Ferrosulfat	Järn-twå-sulfat
Turkish	Ferrosulfat	Demir-ikki-sulfat

[a] From *Chemical Nomenclature*, Advances in Chemistry Series, No. 8, with permission of the American Chemical Society.

their names translated to the equivalent meaning in other languages, for example, in German—*Wasserstoff, Sauerstoff, Kohlenstoff, Stickstoff*. In still other cases the same element has two names in general use, for example, beryllium and glucinium, sodium and natrium, potassium and kalium, niobium and columbium, lutetium and cassiopeium, hafnium and celtum, antimony and stibium, tungsten and wolfram. The International Union has made provisional recommendations favoring use of the Latin name, where one exists, in all languages. In the case of the more recently discovered elements the Union favors beryllium, niobium, lutetium, hafnium, and wolfram.

The Union's recommendation of the Stock nomenclature for inorganic compounds introduces a problem, however, which could interfere with understanding among chemists. The Stock nomenclature proposes use of national names; in the area of well-established nomenclature it would have the effect of reducing familiarity with names, as revealed in table 27.2.[5]

NONPROFESSIONAL PROBLEMS CREATED BY CHEMISTRY

In the previous section we have examined several chemical problems which are chiefly and directly pertinent to the chemical profession but of little consequence outside the profession (although other scientific professions are plagued by the same types of problems). There are, however, certain other problems arising out of chemical activities which are of a more general nature and which affect the population as a whole. These problems demand the best wisdom of the world's leaders, and they will be resolved only very gradually, even where there is good will and a sincere desire for their solution. Chemists can help in their solution but will need the collaboration of the best minds in many other fields. Perhaps chemists can be of greatest service if they will become more conscious of the results of their activities and use their influence to delay the introduction of new products and new processes until they can be sure the advantages outweigh the disadvantages.

Nuclear Warfare

The threat to civilization arising out of the discovery of nuclear fission is clearly the most critical problem facing the world at the moment. Chemists and physicists have been criticized for their part in making such weapons possible. It is difficult, when one understands the sequence of the development, to determine which atomic scientist is most blameworthy —Democritus, Lucretius, Gassendi, Dalton, Berzelius, Avogadro, Stas, Cannizzaro, Mendeleev, Crookes, Becquerel, Thomson, Curie, Rutherford, Planck, Moseley, Bohr, Einstein, Sommerfeld, Heisenberg, Chadwick, Lawrence, Joliot, Fermi, Hahn, Seaborg, or Oppenheimer. They all had a hand in developing the knowledge behind the bomb! The basic discoveries had been made by 1939. Whatever moral reservations were held by workers on the Manhattan Project were tempered by the knowledge that German scientists were seeking to develop the weapon for their government.

The subsequent entry of other nations into the Nuclear Club has accentuated the danger, as has the successful development of fusion weapons. Testing of nuclear weapons creates a fallout hazard which alarms people throughout the world. A glimmer of hope arose in the summer of 1963 with the working out of a limited test ban agreement between Russia, Britain, and the United States.

[5] K. A. Jensen, in *Chemical Nomenclature*, Advances in Chemistry Series, No. 8, Am. Chem. Soc., Washington, 1953, p. 42.

The dangers to civilization are obvious. Although a nation striking a saturation blow with nuclear warheads might eliminate the defender's power to retaliate and thereby win world dominance, the radioactivity released by such a blow probably would cause such damage to plant and animal life that the victory would be a defeat for everyone. Hence it is imperative that the world's leaders continue to seek such rapport that nuclear weapons will be abandoned. However, the pattern of events makes it evident that no such rapport will soon be reached. World powers continue to be suspicious and selfish in the face of the threat to the future of civilization. It is evident that each of the nations in the Nuclear Club feels that it has the power to attack or retaliate if necessary and that the fallout hazard which might be loosed upon the world is less important than preservation of national sovereignty. Other important nations without nuclear weapons behave as if they discount the damage that would be done to their teeming populations or hope themselves to have such weapons in time to be a factor in large-scale international conflict.

Of course any kind of weapon is unpleasant, but this has not deterred human beings from fighting one another. Chemical and biological warfare can be highly destructive, but they lack the threat of total annihilation posed by nuclear weapons. It is their capacity to bring about total annihilation that sets nuclear weapons apart from earlier weapons and makes it imperative that human beings practice the art of living together even though their ideals and objectives differ radically.

Disposal of Nuclear Wastes

Also associated with nuclear processes is the problem of waste disposal. Fission fragments produced in nuclear reactors vary in their chemical nature and in their decay patterns. Most of them cannot be handled by conventional methods of waste disposal since some of the half lives are sufficiently long to cause radiation hazards if they enter ground water supplies. Dumping at sea is also intolerable since the radiation may have unforeseen effects on marine plants and animals. Hence, it has been necessary to go to great expense to treat reactor wastes so that they do not become future hazards. By holding them for a period of time the short-lived isotopes become largely spent. Long-lived residues can then be canned and buried or dropped into the depths of the sea. It has even been proposed that such wastes may be carried away from the earth and scattered into interstellar space. As the use of nuclear reactors for power production becomes widespread, the waste disposal problem will become more serious.

Industrial Wastes and Hazards

Chemicals have long had a reputation as noxious substances and those working with them frequently have been placed outside the center of respectable society. The ancient tanner carried on his business outside the city. The alchemist carried on his search for the philosopher's stone in dingy basements. Even today the budding young chemist in the family is relegated, with many misgivings, to the basement or attic.

Pollution of streams with toxic wastes or with decomposable organic materials is a problem of long standing. Pollution of the atmosphere with sulfur dioxide and other noxious gases is of particular concern in smelting districts. Discarding solid waste on surface soil not only takes land out of agricultural use but frequently leads to unpleasant odors and noxious leaching which causes deterioration of the neighboring community. Such pollution is undesirable not only because of the unpleasantness caused to human beings living nearby, but also because of the problems raised with respect to balance of nature. Streams overloaded with noxious wastes lose their capacity for self-purification and the natural equilibrium between the flora and fauna in such streams is upset.

When chemical operations were conducted on a small scale it was possible to regulate them or isolate them in such a manner that they did not become a public nuisance. Workmen in chemical establishments frequently labored under hazardous conditions, but the general public was not unduly exposed. With the growth of the chemical industry, both internal and external hazards were increased. During much of the nineteenth century, when businessmen customarily operated according to their own interpretation of Adam Smith's teachings, the chemical industry quite generally was indifferent to the consequences of its actions. Pollution of countryside and atmosphere caused such regions as Widnes, near Liverpool, to become barren wastelands. Considerable progress has been made since 1900. Adverse public opinion, legal action, and a more enlightened attitude on the part of industrialists have brought about much improvement, although all is still not sweetness and light.

The chemical industry is also characterized by direct hazards to its workers. The dangers associated with production of white lead were recognized in the eighteenth century as were the hazards of verdigris infections. The rise of the chemical industry in the nineteenth century was paralleled by the rise of industrial hazards. Acid and chlorine fumes were characteristic of the alkali and bleaching powder works. Bleaching powder itself was a highly irritant dust. The advent of the synthetic dye industry led to poisoning from vapors of benzene, aniline, and other coal tar derivatives. Such compounds also produced widespread skin disorders among workmen. Toward the end of the century it was recognized that certain compounds, particularly β-naphthylamine, were responsible for bladder cancer among workers.[6]

Although the chemical industries were notorious for bad working

[6] B. W. Richardson, *J. Royal Soc. Arts*, 24, 116, 146, 186, 203, 488 (1876); Thomas Oliver, ed., *Dangerous Trades*, Murray, London, 1902.

conditions during much of the nineteenth century, they underwent a period of reform toward the end of the century and are today, in most cases, models of enlightened concern for the welfare of the working man. The beginnings of change appeared first in Germany, during Bismarck's era when the government knocked the props from under the socialist movement by instituting the very reforms upon which socialists were basing their appeal. England more gradually followed in passage of regulations dealing with industrial health. In the United States such reforms came more slowly, but now management, as a consequence of its battle with labor, has reached a high level of responsibility where industrial safety is concerned. In fact safety consciousness in industry far surpasses that in academic institutions.

This stage has not been reached without horrible misfortunes. During the twenties women who painted the luminous numbers on watch dials pointed their brushes with their lips between strokes, thus ingesting radioactive materials which led to a slow and painful death. The case of the workers in trades dealing with granite, sand, and other silicate rocks is a more recent example. Inhalation of silica dust causes growth of fibrous tissues in the lungs, leading to loss of function and terminating in pulmonary tuberculosis. Asbestos dust leads to a similar lung degeneration. Work with white phosphorus, arsenicals, fluorine, mercury compounds, chromium compounds, and lead in its various forms has resulted in unfortunate poisoning cases. These are becoming more and more rare as proper attention is given to ventilation, suitable clothing, automatic operations, and other safety practices.

Environmental Chemical Hazards

People having no connection with the industry are increasingly exposed to chemical products which enter the environment in a variety of ways—in foods as additives, in drugs, in textiles as dyes, in the atmosphere and water as pollutants, in gardening supplies as pesticides, in household items as detergents, paints, deodorizers, and insecticides, in cosmetics as dyes and deodorants. Table 27.3 shows the growth in output of some of the products of the American chemical industry.[7] It has been estimated by the same source that 400 to 500 new chemical products—and new wastes —enter the market each year. Increased use of radioisotopes in industry and research presents an added environmental hazard. Much more must be learned about the hazards presented by these products and wastes. The more enlightened segments of the chemical industry are conscious of these problems and are carrying out research on them. However such research as is necessary to demonstrate probable safety under conditions of use is both expensive and time-consuming. For example, in the case

[7] *The Chemical Industry Facts Book*, Mfg. Chemists Assoc., Washington, 3rd ed., 1957, p. 18. Used with permission of the publisher.

of a food additive it should involve acute and chronic toxicity tests on at least two different species, one of them nonrodent. After varying time intervals animals should be sacrificed and examined for evidence of damage to internal organs and for evidence of abnormal growths. The U.S. Food and Drug Administration has been a leader in the development of such methods of pharmacological testing.

Table 27.3   U.S. OUTPUT OF VARIOUS CHEMICAL MATERIALS[a]

Product	1925	1937	1950	1955
Plastics	46	315	2,151	3,739
Fibers	58	329	1,496	1,718
Medicinal chemicals	3	17	49	79
Fertilizers	790	1,300	4,430	6,000
Surface-active agents	0	9	676	1,153
Antiknock agents	3	65	350	700
Dyes	86	122	196	168
Insecticides and other synthetic organic agricultural chemicals	27	102	286	506

[a] Quantities are in units of a million pounds except for fertilizers which are expressed in plant food thousand tons.

From *The Chemical Industry Facts Book*, 3rd edn., 1957, with permission of the Manufacturing Chemists Association.

It is important that such research be carried out by public agencies, although the primary cost burden should be met by the industries which will profit from sale or use of such chemicals. The industrial point of view, however, cannot always be depended upon to keep objectionable products from the market. An example is found in the case of weed killers placed upon the market several years ago by members of the French chemical industry. Roger Gautheret of the Sorbonne showed after extensive field experiments that the chemicals killed weeds satisfactorily but had damaging effects on the quality of crops. The industry, which had spent large sums on development and advertising, sought to have the report suppressed. Gautheret remained firm and the weed killers were removed from the market.[8]

Still another example of the need for attention to public welfare before marketing a new chemical is seen in the thalidomide tragedy which broke in 1962. The drug, very effective in inducing sleep without the dangers of barbiturates, was placed on the European market early in the 1960's. Efforts to have it cleared by the Food and Drug Administration for sale in the United States were delayed as a result of Marion Kelsey's suspicions that testing had not been adequate to show freedom from unfavorable

[8] Philip R. White, *Science*, **131**, 616 (1960).

side effects. Despite heavy pressures from the firm wishing to market thalidomide she held firm. Suddenly experience in Europe revealed births of grossly deformed children borne by mothers who had used the drug during the early months of pregnancy.

Ecological Hazards

Before 1900 imbalances brought about through the introduction of chemicals into the natural environment were seldom of more than local consequence. Today, with the magnitude of the chemical industry and the opportunity for large-scale operations, such chemical tampering with nature may easily result in imbalances which can never favorably readjust.

Chemical producers are constantly on the alert for new uses for their products. Extensive research has led to a host of new drugs, protective coatings, insecticides, herbicides, preservatives, rodenticides, etc. Frequently they have been placed into use without thought of ecological consequences, one of the most flagrant cases being the recent fire ant eradication program in southern United States. In order to eradicate the insect, large areas were dusted by airplane with potent insecticides, killing innumerable birds and wild animals in the process. Such a crash program, while checking one organism, can easily upset a region's ecology for long periods. Rachel Carson, in a best-seller, *Silent Spring*, dealt with these matters in 1962.

The time has now been reached where purveyors of chemicals must accept responsibility for the effects of their products. A product should not be marketed with the hope that it will not have adverse effects on innocent plants and animals which come into contact with it. It should not be reasoned that, since the amount used is small, it cannot possibly injure living species other than the ones for whom it is intended. The product must not only be thoroughly tested before it is introduced into commerce, but also there must be a concensus of opinion among experts in all affected fields that serious upsets will not be caused by proposed methods of use.

There must be a respect for the public welfare throughout the entire industry involved in the use of a particular product. The cranberry episode of 1959 was shocking, not because the aminotriazole residues found present were apt to cause cancer (the amounts likely to be ingested were hardly likely to produce cancers in human consumers although no one could be absolutely certain), but because the chemical had been used as a preharvest spray without having had a clearance for such use. Blame for the episode should have fallen, not on Arthur Flemming, who was carrying out his duties as Secretary of Health, Education, and Welfare, but upon growers who contemptuously disregarded label statements and used the chemical out of season. To the extent that salesmen encouraged such use, blame must fall on them too.

With the spread of modern science to all parts of the world it becomes more important than ever that environmental hazards be carefully evaluated. It is particularly important that final decisions be made by competent persons who have no personal interest in the matter. A company's own scientists, no matter how competent and upright they be, should not be the ultimate judges. Further, officials who make the decision must be kept free of the many pressures and conflict of interest situations which may influence them.

Natural Resources

The problem of dwindling natural resources can no longer be ignored. Population is growing at a more rapid rate than ever. Advances in sanitation, nutrition, and medicine have made it possible to cut infant and maternal mortality and increase life span. Indeed some consider the population "bomb" a more frightening explosive than the hydrogen bomb. Since 1650 the earth's population has increased from a half billion to over 2 2/3 billion.

Added to the increased population is the increasing industrialization of that population. Consumption of energy, water, mineral, soil, and wildlife resources is much more extensive at a high level of industrialization than in a peasant economy. As production of food, clothing, and material goods becomes more sophisticated, the per capita consumption of resources jumps out of proportion to the actual mass of finished product. In a handicraft economy little metal is used for tools and energy is largely provided by humans and animals. In the transition to a simple industrial economy the amount of metal needed for machinery is greatly increased and mineral fuels constitute the principal energy source. In an advanced industrial economy the use of metal and energy is further increased. The network of mines, factories, and transportation systems required to keep the economy functioning results in the consumption of large amounts of basic raw materials which are never handled by ultimate consumers. It is estimated that in the United States eight tons of steel are tied up in capital goods for every person in the nation. Annual per capita energy expenditure in America in 1950 amounted to the equivalent of eight tons of coal; in Western Europe the figure was about 2.5 tons; in Japan about one ton; in the rest of Asia about 1/20 ton. The needs will become larger as industrialization proceeds around the world. Already the pinch in resources is being felt.

Energy Resources. The industrial revolution was based on the substitution of energy from coal for that of human beings, animals, wind, and water. Petroleum and natural gas now figure prominently as energy sources. At one time there was little concern for future supplies—the Carboniferous Era had been a bountiful one and coal seemed available into the indefinite future. Petroleum also seemed available for the taking, and natural gas was simply a nuisance which was frequently set afire and burned at the wellhead.

All of this has changed. For several decades the known reserves of petroleum have been periodically judged ample for only about fifteen years. Fortunately, new reserves are being discovered with sufficient frequency to sustain this margin. However the convenience of the internal combustion engine for transportation purposes has caused petroleum consumption to rise at a prodigious rate. This, coupled with the value of petroleum products for heating purposes and for petrochemicals, creates a very serious problem. Chemistry has helped by creating methods for converting unimportant hydrocarbons into gasolene-size molecules and processes for altering the structure of molecules to create structures with improved combustion characteristics. Cracking and polymerization increases the yield of gasolene-type hydrocarbons. However these developments are merely useful in extending the date when the petroleum shortage becomes critical. By then it will be profitable to extract hydrocarbons from extensive oil shale deposits, and production of liquid fuels from coal will become commonplace, which in turn will speed the day when coal resources are exhausted.

Estimates of when the world's fossil fuels will be exhausted vary by a factor of ten, depending on whether one listens to the pessimists or the optimists. A really precise estimate is not possible. The magnitude of existing deposits is known in only a very approximate way. This is particularly true of petroleum because exploration continues to locate new pools. However these become progressively more expensive to exploit because of depth and inaccessibility. Also the rate of use continues to increase with increasing industrialization.

As attention turns to other energy sources, one finds shortcomings on every hand. Waterpower is available for much further development, particularly in Asia and Africa. The total prospects, however, are inadequate to supply more than a small fraction of total needs. Winds, tides, and the temperature differential between ocean layers are all potential energy sources of minor practical consequence.

Energy from nuclear reactions is sure to figure prominently in the future. The fission of one kilogram of uranium-235 produces energy equivalent to burning 6.6 million kg. of coal. The separation and use of uranium-235 is expensive and impractical, however, since it is better used as an intermediary in the conversion of uranium-238 to plutonium-239 and of thorium-232 to uranium-233, both fissionable isotopes. The richest uranium ores presently come from the Congo and Canada. India and Brazil are the major sources of monazite sands which are primary sources of thorium. However there are extensive deposits of poorer ores in many parts of the world and these are already being exploited. Even an average piece of granite contains uranium and thorium equivalent in energy content to 50 kilograms of coal per kg. of rock. Rich ores will be exhausted quickly once nuclear power plants become extensively operative, but low-grade sources will be available for a number of centuries of use. This source is exhaustible in perhaps 1,000 years at foreseeable rates of energy consumption.

If the maintenance of fusion under practical conditions is solved, and if capital costs are not prohibitive, the hydrogen of the world's oceans will be a bountiful source of energy for a long period into the future. Present-day experiments suggest that success is possible.

Solar energy is not an exhaustible resource. The earth has depended upon solar energy in an ultimate sense throughout history, since the photosynthesis carried on by plants has furnished men and animals with their energy needs. The sun also is responsible for the deposits of fossil fuels and for the evaporation of water which falls as rain and snow and thereby provides, when it collects at high elevations, the source of water-power.

There is considerable interest in more direct exploitation of solar energy. Very complex problems are involved, many of them involving large capital outlays. Simple and inexpensive solar cookers have been developed. The Bell Telephone Co. has introduced a solar battery used in earth satellites as well as in earthly installations. Work is also in progress on devices for heating homes, large-scale growth of algae, and distillation of sea water.

Science should be able to make the conversion of energy from one form to another somewhat more efficient; this would result in significant savings of resources. If the energy of nuclear reactions could be made directly available as electricity, the elimination of wasteful heat inter-changers might result in significant savings of uranium and thorium. It may also be possible to devise chemical systems related to photosynthesis whereby solar energy is directly transformed into chemical energy. Work on fuel cells suggests that the direct production of electrical energy by chemical combination of substances such as hydrogen and oxygen may be practical.

Water Resources. Except in arid lands water has been taken for granted throughout most of human history. Now this is no longer possible. Because of population growth and the increase in industrialization many localities are plagued by water shortage, which is becoming a grave problem in many parts of the earth.

Although average rainfall on land areas of the earth[9] amounts to 1.2 × 10^{17} liters per year (or 120,000 liters per day per inhabitant), it is so unevenly distributed that there is gross overabundance in some areas and great scarcity in others. Redistribution through irrigation is as old as civilization itself and the practice has spread as the world's population has grown. Practical considerations limit the extent to which irrigation

[9] B. F. Dodge, *Am. Scientist,* **48,** 514 (1960).

can be extended. Abundant sources of fresh water must be near the land to be irrigated and the power supply must be adequate to move the water to where it is needed. An alternative is to learn how to cause rain to fall when and where it is needed. Extensive experiments on cloud-seeding were carried out around 1950 by Irving Langmuir, Shafer, and associates. Dry ice or crystals of silver iodide were shown to be effective in initiating rainfall but erratic results caused the trials to be abandoned.

Urbanization coupled with growing industrialization has led to greatly increased consumption of water. In the United States per capita daily usage of water in the fifties amounted to about 600 liters for household purposes, 3,000 liters for industrial purposes, and 3,000 liters for irrigation. Typical industries used water at the rate of 275 tons for a ton of steel, 20 tons for a ton of sulfuric acid, 200 tons for a ton of paper, and 20 barrels for the refining of a barrel of petroleum. Air conditioning, which became very popular in the United States during the fifties, uses water in prodigious quantities.

The problem is complicated further by the fact that industrial waste waters frequently cause serious pollution when they are returned to rivers and lakes. Nature-lovers and industrialists have long argued about this. During the nineteenth century there was little effort to improve conditions. Now public opinion is bringing about a change in attitude—and practice. Remedial legislation is proving effective and industries are required to process waste materials so that gross damage to streams is avoided. Much more remains to be done but at least the major offenders have come to recognize their responsibility.

The shortage of fresh water in many populous places has spurred research on processes for desalting sea water. Some of these projects have reached the pilot-plant stage. Distillation is an obvious procedure but the heat costs make it so expensive that its use is limited to small-scale operations, as on shipboard. Various types of distillation show some promise, but by their very nature can never be made truly economical. Even solar energy, a relatively inexpensive source, is of limited potential value because of the large capitalization required to exploit an intermittent and low-intensity energy source. Freezing or crystallization processes also require massive amounts of energy. Separation of water in the form of hydrocarbon hydrates is one of the unique crystallization methods under study. Processes for separation of salt from brine include ion exchange and electrodialysis with semipermeable membranes.

Mineral Resources. The rapidly increasing use of metals in an industrialized world is leading to depletion of rich ores of nearly all metals. The reserves of iron ore in the United States have reached the point where the mining and processing of taconite has become profitable.

Copper, necessary in the electrical industry, is rapidly approaching a critical state. In the eighteenth century it was considered impractical to smelt ores containing less than 18 per cent of the metal. By 1900 the average ore being processed contained 5 per cent of copper; in 1950 the ore being smelted in the United States contained 0.9 per cent. Improvements in concentration methods and simultaneous extraction of other metals make it possible to meet needs for the metal. Eventually, it will be necessary to process ores with 0.1 per cent copper.

Other metals in short supply are lead, tin, zinc, and manganese. To some extent other metals may be substituted for them, but since properties determine uses, substitution is not always possible. For example, substitutes for lead in paints are becoming commonplace, but the use of lead in storage batteries consumes large quantities of the metal. It is possible that nickel or cadmium storage batteries may ultimately take the place of lead, particularly if large quantities of lead continue to be used in gasolene additives. The battery problem is further complicated by the scarcity of nickel and cadmium.

To some extent major substitutes will come into use. Aluminum and magnesium have already taken on great importance. To a significant degree they have not functioned merely as substitute metals but because of their lightness have made possible the production of aircraft and other materials where lightness is a special virtue. Even in the case of aluminum, which follows silicon and precedes iron in abundance in the earth's crust, the availability problem is upon us since high-grade deposits of bauxite are rapidly being depleted. Titanium, with the virtues of lightness, corrosion resistance, strength, toughness, and relative abundance, probably will become important as soon as production costs can be brought down to a reasonable level.

To some extent plastics, wood, ceramics, and glass can be substituted for metals, and such substitution will take place where suitability and economics so dictate. And recovery of metals from scrap will be extended beyond present practice, but a serious problem presents itself here since the practice of alloying complicates recovery in uncontaminated form. Perhaps the ultimate source for some of these metals will become sea water which is already being profitably processed for salt, magnesium, and bromine.

Resources of nonmetallic minerals also present a steadily worsening problem. The chemical industry draws heavily upon coal, petroleum, air, water, salt (NaCl), sulfur, and limestone. The prospects for coal, petroleum, and water have been discussed. Air, the source of nitrogen, oxygen, and inert gases, presents no problem. Salt and limestone are plentiful, although not always favorably located. The outlook for low-cost sulfur is not good since the known sources of native sulfur available through the Frasch process are nearing exhaustion. Soon we shall be more dependent on pyrites, natural gas, and smelter fumes for sulfur. Ultimately it may become necessary to extract sulfur from gypsum.

Soil Resources. Tied in with the resource problem is that of tillable soil. Bad conservation practices have reduced the size of population the soil can support. Chemistry has aided materially in extending soil resources through the availability of fertilizers, soil conditioners, and other agricultural chemicals. On the other hand the chemical industry has played a negative role when through pollution of atmosphere and streams it has laid waste to whole areas, or when through its mining operations it has covered large areas with waste materials. Population growth steadily converts tillable soil into urban communities, and the mobility of industrial populations causes significant amounts of agricultural land to be chopped up by superhighways.

Soilless growth of plants using water containing dissolved chemicals has been demonstrated. The practicality of such operations on a large scale is definitely limited, however, at least insofar as the immediate future is concerned.

Microbiological food sources (yeasts, algae) may prove important if the world's population becomes unduly large. Or chemical synthesis may ultimately supply the necessary sugars, proteins, and vitamins. Before this becomes essential, some agricultural economies may be effected by direct consumption of plant products as is done in the Orient, rather than feeding them to animals and consuming the eggs, milk, and meat.

RESPONSIBILITIES FOR THE FUTURE

Up to this point in human history chemistry has contributed much to satisfy the intellectual curiosity and the material welfare of mankind. The rate of progress, extremely slow until two centuries ago, remains rapid. The concepts which have arisen in the past two centuries have been generally fruitful, as witness the present vigor of chemistry. However chemists must pursue their work with a searching scepticism which will reveal the flaws in current hypotheses before they lead to a period of stagnation such as that represented by alchemy.

Intellectual success perhaps affords leaders in the field their greatest satisfaction. Frequently they have been content to permit lesser figures to work out the applications. It is the applications, though, which loom most important in the public mind. The contributions of Lavoisier, Dalton, Berzelius, Avogadro, Kekulé, Willard Gibbs, van't Hoff, Emil Fischer, Rutherford, Bohr, and G. N. Lewis have never been as comprehensible as those of Leblanc, Perkin, Solvay, Frasch, Haber, Ehrlich, and Fleming. The importance of basic research must never be forgotten since application can only proceed as far as fundamental knowledge permits.

The United States in particular, and the Western World in general, is in a favorable position to pursue fundamental studies vigorously. The

United States is faced with the fact of an affluent society. Constantly diminishing efforts are going into the production function, whereas ever larger efforts are centered on marketing. The public is exposed to the persuasive powers of a sales organization set up to convince consumers that they cannot exist without a particular product. While there is still poverty in the United States, it is in some part a poverty stemming from inability to manage personal finances in the face of the glut of consumer goods and the delusion of easy credit available for their purchase. Although there is perhaps not yet a glut of material goods in Europe, the potential is there. In the light of the economic conditions in the industrialized parts of the world, there is a surprising reluctance to divert larger portions of income into public services, the one area where there is not a surfeit. Education, consumer protection, conservation, recreation, the arts, and science are the victims of general neglect—although science has received more public financial support in recent years, but primarily for military reasons.

There must be even greater recognition of the role of science, particularly basic science, in dealing with the problems of the present and the future. Research must be encouraged but not forced. Miraculous results must not be expected of everyone. Many scientists are plodders who will never do more than gather unspectacular facts, but such workers are still making useful contributions. Only a few will be Faradays, Pasteurs, Emil Fischers, and Rutherfords, with an uncanny ability to foresee the next significant experiment. This must be understood by those responsible for the support of science.

At the same time there must be balance in the support of all areas of learning. The world will suffer if all of the brightest young minds are diverted into science. There must be some who better understand the field of human relations for it is here that the most threatening problems of the present day are found. These cannot be solved by more and better science, although science can help by developing less tedious ways to perform physical tasks and methods for utilizing the earth's resources more effectively. The scientist must acquire a better understanding of the consequences of his activities. And nonscientific leaders must at the same time acquire a better insight into the nature and the limitations of science.

The pursuit of science can still do much for mankind, but man's use of science can also be his undoing.

APPENDIX I

Discovery of the Elements (KNOWN TO THE ANCIENT WORLD: carbon, sulfur, iron, tin, lead, copper, mercury, silver, gold; KNOWN BEFORE A.D. 1600: arsenic, antimony, bismuth, zinc.)

Year	Element	Discoverer
1669	Phosphorus	Brand
1737	Cobalt	Brandt
1748	Platinum	de Ulloa (described metal which was known earlier)
1751	Nickel	Cronstedt
1766	Hydrogen	Cavendish
1772	Nitrogen	D. Rutherford
1774	Chlorine	Scheele
1774	Manganese	Gahn
1774	Oxygen	Priestley (Scheele, publ. 1777; Bayen, not identified)
1781	Molybdenum	Hjelm (paper not publ. until 1890)
1783	Tellurium	Müller von Reichstein
1783	Tungsten	de Elhuyar brothers
1789	Uranium	Klaproth (as the oxide; metal isolated in 1841 by Péligot)
1789	Zirconium	Klaproth (as the oxide; metal isolated in 1824 by Berzelius)
1791	Titanium	Gregor (as oxide; oxide rediscovered and given name by Klaproth; metal isolated by Berzelius in 1825)
1794	Yttrium	Gadolin (as the impure oxide of numerous rare earths)
1797	Beryllium	Vauquelin (recognized as oxide, Wöhler and, indep., Bussy isolated metal in 1828)
1798	Chromium	Vauquelin
1801	Niobium	Hatchett (originally named columbium)
1802	Tantalum	Ekeberg
1803	Palladium	Wollaston
1803	Rhodium	Wollaston
1803	Cerium	Klaproth, and Berzelius and Hisinger (as the impure oxide of numerous rare earths)
1804	Osmium	Tennant
1804	Iridium	Tennant
1807	Potassium	Davy
1807	Sodium	Davy

Discovery of Elements (continued)

Year	Element	Discoverer
1808	Barium	Davy
1808	Strontium	Davy
1808	Calcium	Davy
1808	Magnesium	Davy
1808	Boron	Gay-Lussac and Thenard (Davy independently)
1811	Iodine	Courtois
1817	Lithium	Arfvedson
1817	Cadmium	Stromeyer
1818	Selenium	Berzelius
1824	Silicon	Berzelius
1825	Aluminum	Oersted (Wöhler isolated better sample in 1827)
1826	Bromine	Balard (Löwig isolated in 1825 but published later)
1829	Thorium	Berzelius
1830	Vanadium	Sefström (recognized by del Rio in 1801 but confused with Cr; metal isolated by Roscoe in 1869)
1839	Lanthanum	Mosander (as impure oxide; freed from didymia in 1841)
1843	Terbium	Mosander (as impure oxide; further sepn. by Marignac in 1886)
1843	Erbium	Mosander (as impure oxide containing six other oxides)
1844	Ruthenium	Klaus
1860	Cesium	Bunsen and Kirchhoff
1861	Rubidium	Bunsen and Kirchhoff
1861	Thallium	Crookes (Lamy prepared metal in 1861)
1863	Indium	Reich and Richter
1875	Gallium	Boisbaudran (Eka-Al)
1879	Holmium	Cleve (as oxide contaminated with dysprosia; spectroscopic evidence for in 1878 by Soret)
1879	Thulium	Cleve (as oxide)
1879	Scandium	Nilson (as oxide; Eka-B)
1879	Ytterbium	Nilson (as oxide contaminated with lutetium)
1880	Samarium	Boisbaudran (as oxide contaminated with Gd and Eu)
1880	Gadolinium	Marignac
1885	Praseodymium	Auer von Welsbach
1885	Neodymium	Auer von Welsbach
1886	Germanium	Winkler (Eka-Si)

Discovery of Elements (APPENDIX I continued on page 824)

Year	Element	Discoverer
1886	Fluorine	Moissan (numerous earlier attempts at isolation)
1886	Dysprosium	Boisbaudran
1894	Argon	Ramsay and Rayleigh
1895	Helium	Ramsay and, indep., Cleve
1898	Krypton	Ramsay and Travers
1898	Neon	Ramsay and Travers
1898	Xenon	Ramsay and Travers
1898	Polonium	Marie and Pierre Curie
1898	Radium	Marie and Pierre Curie and Gustave Bémont
1899	Actinium	Debierne
1900	Radon	Dorn, as Thoron by Rutherford and Owens
1901	Europium	Demarçay
1907	Lutetium	Urbain
1917	Protactinium	Hahn and Meitner (Soddy and Cranston independently)
1923	Hafnium	Coster and Hevesy
1925	Rhenium	Noddack, Tacke, and Berg
1932	Neutron	Chadwick
° 1939	Francium (87)	Perey
1939	Technetium (43)	Perrier and Segrè
°° 1940	Neptunium (93)	McMillan and Abelson
1940	Astatine (85)	Corson, Mackenzie, and Segrè
1940	Plutonium (94)	Seaborg, McMillan, Kennedy, and Wahl
1944	Americium (95)	Seaborg, James, and Morgan
1944	Curium (96)	Seaborg, James, and Ghiorso
1945	Promethium (61)	Marinsky and Glendenin
1949	Berkelium (97)	Thompson, Ghiorso, and Seaborg
1949	Californium (98)	Thompson, Street, Ghiorso, and Seaborg
1954	Einsteinium (99)	Ghiorso, Thompson, Higgins, Seaborg, Studier, Fields, Fried, Diamond, Mech, Pyle, Huizenga,
1954	Fermium (100)	Hirsch, Manning, Browne, Smith, Spence

° Elements discovered after 1932 have the atomic number shown in parentheses after the name.

°° All elements with atomic numbers greater than 92 (uranium) were discovered as synthetic products of high energy processes (hurling high velocity ions at an appropriate target element) or as debris in nuclear weapons tests (einsteinium and fermium). Most synthetic elements are highly radioactive and decay quickly. Elements beyond californium have been synthesized in only small amounts, sometimes only a few atoms. They are identified by decay processes, with such identifications frequently questionable. Priority debates regarding credit for discovery (synthesis) are eventually adjudicated by the International Union of Pure and Applied Chemistry (IUPAC). Through 1982 the IUPAC has approved names for no elements beyond Lawrencium (103).

(APPENDIX I continued on page 824)

APPENDIX II

Discovery of Radioactive Isotopes

Original Symbol	Isotope	Discoverer	Year	Emission	Half Life	Series
U	Mixture	Klaproth	1789			
Th	Mixture	Berzelius	1829			
Po	$^{210}_{84}$Po	Marie Curie	1898	α	138.4*d*	4n + 2
Ra	$^{226}_{88}$Ra	P. & M. Curie	1898	α	1620*y*	4n + 2
Ac	$^{227}_{89}$Ac	Debierne	1899	β 98.8%	22*y*	4n + 3
		Giesel	1902	α 1.2%		
Tn	$^{220}_{86}$Rn	Owens & Rutherford	1899	α	52*s*	4n
UX₁	$^{234}_{90}$Th	Crookes	1900	β	24.1*d*	4n + 2
Nt	$^{222}_{86}$Rn	Dorn (Curies)	1900	α	3.825*d*	4n + 2
ThX	$^{224}_{88}$Ra	Rutherford & Soddy	1902	α	3.64*d*	4n
RaA	$^{218}_{84}$Po	Rutherford & Brooks	1902–1904	α 99.96% β 0.04%	3.05*m*	4n + 2
RaB	$^{214}_{82}$Pb	Rutherford & Brooks	1902–1904	β	26.8*m*	4n + 2
RaC	$^{214}_{83}$Bi	Rutherford & Brooks	1902–1904	β 99.96% α 0.04%	19.7*m*	4n + 2
RaD	$^{210}_{82}$Pb	Rutherford & Brooks	1904–1905	β	20*y*	4n + 2
RaE	$^{210}_{83}$Bi	Rutherford & Brooks	1904–1905	α β	2.6 × 10⁶*y* 5.0*d*	4n + 2
RaF	$^{210}_{84}$Po	Rutherford & Brooks; also Curies	1904–1905	α	138.4*d*	4n + 2
AcX	$^{223}_{88}$Ra	Giesel Gadlewski	1904–1905	α	11.6*d*	4n + 3
An	$^{219}_{86}$Rn	Giesel (Debierne)	1904	α	3.9*s*	4n + 3
RdTh	$^{228}_{90}$Th	Hahn	1905	α	1.9*y*	4n
ThC	$^{212}_{83}$Bi	Rutherford Slater	1905	α 66.3% β 33.7%	60.5*m*	4n
RdAc	$^{227}_{90}$Th	Hahn	1906	α	18.2*d*	4n + 3
AcC′	$^{211}_{84}$Po	Marsden & Barrett Hahn & Meitner	1906	α	25*s*	4n + 3
ThC″	$^{208}_{81}$Tl	Marsden & Barrett Hahn & Meitner	1906	β	3.1*m*	4n
UI	$^{238}_{92}$U	McCoy & Ross Boltwood	1907	α	4.5 × 10⁹*y*	4n + 2
UII	$^{234}_{92}$U	Moore & Schlundt	1907	α	2.5 × 10⁵*y*	4n + 2
Io	$^{230}_{90}$Th	Boltwood Hahn Marckwald	1907 1907 1908	α	8 × 10⁴*y*	4n + 2
MsTh₁	$^{228}_{88}$Ra	Hahn	1907	β	6.7*y*	4n
MsTh₂	$^{228}_{89}$Ac	Hahn	1907	β	6.13*h*	4n
RaC′	$^{211}_{84}$Po	Hahn & Meitner Fajans	1909 1911	α	0.52*s*	4n + 2
ThA	$^{216}_{84}$Po	Geiger & Marsden	1910	α	0.16*s*	4n
UY	$^{231}_{90}$Th	Antonoff	1911	β	25.6*h*	4n + 3
UX₂	$^{234}_{91}$Pa	Fajans & Göhring Hahn & Meitner Soddy, Cranston, Fleck	1917	β	1.18*m*	4n + z
UZ	$^{234}_{91}$Pa	Hahn	1921	β	6.66*h*	4n + 2
AcK	$^{223}_{87}$Fr	Perey	1939	β	22*m*	4n + 3

APPENDIX III

RADIOACTIVE DECAY SERIES

A. *Uranium-238 (4n + 2) Series*

Atomic number and element

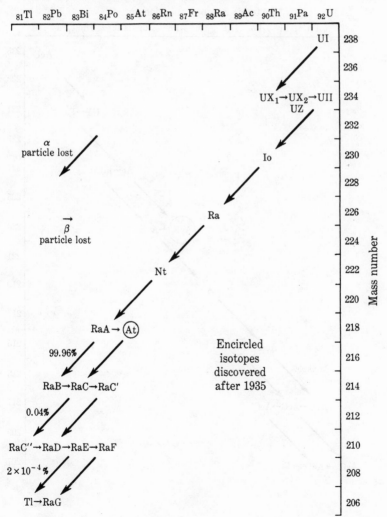

B. *Actinium-227 (4n + 3) Series, or Uranium-235 Series*

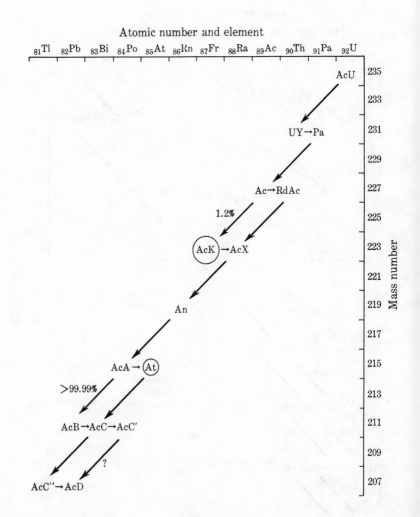

Thorium-232 (4n) Series

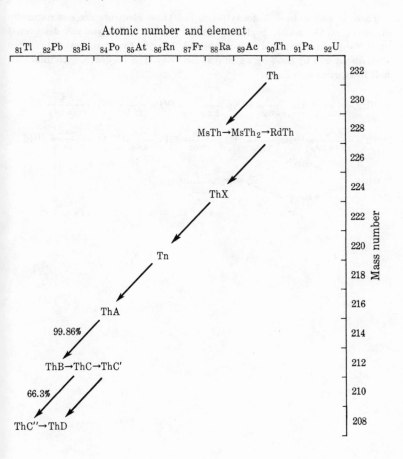

D. *Plutonium-241 (4n + 1) Series*

There is also a $4n + 1$ decay series but these elements are not naturally occurring so the series, beginning with Plutonium-241, was not discovered until Manhattan Project work commenced. The first five members of this series were discovered by Seaborg's group at Berkeley. The sequence, with indicated half lives, runs as follows:

$$^{241}\text{Pu} \xrightarrow[13.2y]{-\beta} \,^{241}\text{Am} \xrightarrow[462y]{-\alpha} \,^{237}\text{Np} \xrightarrow[2.2 \times 10^6 y]{-\alpha} \,^{233}\text{Pa} \xrightarrow[27.4d]{-\beta} \,^{233}\text{U} \xrightarrow[1.62 \times 10^5 y]{-\alpha} \,^{229}\text{Th}$$

$$^{229}\text{Th} \xrightarrow[7.34 \times 10^3 y]{-\alpha} \,^{225}\text{Ra} \xrightarrow[14.8d]{-\beta} \,^{225}\text{Ac} \xrightarrow[10.0d]{-\alpha} \,^{221}\text{Fr} \xrightarrow[4.8m]{-\alpha} \,^{217}\text{At} \xrightarrow[1.8 \times 10^{-2} y]{-\alpha} \,^{213}\text{Bi}$$

$$^{213}\text{Bi} \begin{array}{c} \nearrow \,^{213}\text{Po} \\ \searrow \,^{209}\text{Tl} \end{array} \searrow\nearrow \,^{209}\text{Pb} \xrightarrow[3.32h]{-\beta} \,^{209}\text{Bi}$$

$-\beta \qquad 47m \qquad -\alpha \qquad -\alpha \qquad 4.2 \times 10^{-6} s \qquad -\beta \qquad 2.2m$

APPENDIX IV

(Nationality of Nobelist is indicated in parentheses as: A, American; Ar, Argentine; 1, Austrian; Aus, Australian; B, British; Be, Belgian; C, Canadian; Ch, Chinese; Cz, zech; D, Dutch; Da, Danish; F, French; Fi, Finnish; G, German, pre 1945; Gw, West ermany; H, Hungarian; I, Italian; In, Indian; J, Japanese; K, Korean; La, Latvian; ι, Luxemburg; P, Polish; Pak, Pakistani; Po, Portuguese; R, Russian; Rum, Rumanian; Af, South African; Sp, Spanish; Swe, Swedish; Swi, Swiss; V, Venezuelan; Y, Yugoslav. hen born and educated in another country, this is indicated by "b." followed by the mbol for the country of birth.)

ear	Chemistry	Physics	Medicine and Physiology
01	J. H. van't Hoff (D)	W. K. Röntgen (G)	E. A. von Behring (G)
02	Emil Fischer (G)	H. A. Lorentz (D)	Ronald Ross (B)
		P. Zeeman (D)	
03	S. A. Arrhenius (Swe)	H. A. Becquerel (F)	Niels R. Finsen (Da)
		Pierre Curie (F)	
		Marie Curie (F b. P)	
04	William Ramsay (B)	J. W. Strutt (Rayleigh) (B)	Ivan Pavlov (R)
05	Adolf Baeyer (G)	Philipp Lenard (G)	Robert Koch (G)
06	Henri Moissan (F)	J. J. Thomson (B)	C. Golgi (I)
			S. Ramón y Cajal (Sp)
07	Eduard Buchner (G)	A. A. Michelson (A)	C. Laveran (F)
08	Ernest Rutherford (B)	G. Lippmann (F)	Paul Ehrlich (G)
			E. Metchnikov (F b. R)
09	Wilhelm Ostwald (G)	G. Marconi (I)	Theodor Kocher (Swi)
		Ferdinand Braun (G)	
10	Otto Wallach (G)	J. van der Waals (D)	Albrecht Kossel (G)
11	Marie Curie (F b. P)	Wilhelm Wien (G)	A. Gullstrand (Swe)
12	Victor Grignard (F)	Gustaf Dalén (Swe)	Alexis Carrel (A b. F)
	Paul Sabatier (F)		
13	Alfred Werner (Swi)	H. Kammerlingh-Onnes (D)	Charles Richet (F)
14	T. W. Richards (A)	Max von Laue (G)	R. Bárány (Au)
15	Richard Willstätter (G)	Wm. Henry Bragg (B)	No award
		Wm. Lawrence Bragg (B)	
17	No award	Charles G. Barkla (B)	No award
18	Fritz Haber (G)	Max Planck (G)	No award
19	No award	Johannes Stark (G)	Jules Bordet (Be)
20	Walther Nernst (G)	C. Guillaume (Swi)	A. Krogh (Da)
21	Frederick Soddy (B)	Albert Einstein (G)	No award
22	Francis W. Aston (B)	Niels Bohr (Da)	Archibald V. Hill (B)
			Otto Meyerhof (G)
23	Fritz Pregl (Au)	Robert A. Millikan (A)	Frederick G. Banting (C)
			John J. R. Macleod (C)
24	No award	Karl Siegbahn (Swe)	Willem Einthoven (D)
25	R. Zsigmondy (G b. Au)	James Franck (G)	No award
		Gustav Hertz (G)	
26	The Svedberg (Swe)	Jean B. Perrin (F)	Johan Fibiger (Da)
27	Heinrich Wieland (G)	Arthur H. Compton (A)	J. Wagner-Jauregg (Au)
		C. T. R. Wilson (B)	

Nobel Prize Winners in Chemistry, Physics, and Medicine (continued)

Year	Chemistry	Physics	Medicine and Physiology
1928	Adolf Windaus (G)	O. W. Richardson (B)	Charles Nicolle (F)
1929	Arthur Harden (B) H. v. Euler-Chelpin (Swe b. G)	Louis V. de Broglie (F)	F. G. Hopkins (B) Christiaan Eijkman (D)
1930	Hans Fischer (G)	C. V. Raman (In)	Karl Landsteiner (A b. Au
1931	Carl Bosch (G) Friedrich Bergius (G)	No award	Otto Warburg (G)
1932	Irving Langmuir (A)	Werner Heisenberg (G)	Charles Sherrington (B) E. D. Adrian (B)
1933	No award	P. A. M. Dirac (B) Erwin Schrödinger (Au)	Thomas H. Morgan (A)
1934	Harold C. Urey (A)	No award	G. H. Whipple (A) G. R. Minot (A) W. P. Murphy (A)
1935	Frédéric Joliot-Curie (F) Irène Joliot-Curie (F)	James Chadwick (B)	Hans Spemann (G)
1936	Peter J. W. Debye (G b. D)	Victor F. Hess (Au) Carl D. Anderson (A)	Henry H. Dale (B) Otto Loewi (Au)
1937	Walter N. Haworth (B) Paul Karrer (Swi)	Clinton J. Davisson (A) George P. Thomson (B)	Albert Szent-Györgyi (H)
1938	[a]Richard Kuhn (G)	Enrico Fermi (I)	Corneille Heymans (Be)
1939	[a]A. Butenandt (G) Leopold Ruzicka (Swi)	Ernest O. Lawrence (A)	[a]Gerhard Domagk (G)
1943	Georg von Hevesy (H)	Otto Stern (A b. G)	Edward Doisy (A) Henrik Dam (Da)
1944	Otto Hahn (G)	Isador Isaac Rabi (A)	Joseph Erlanger (A) Herbert Gasser (A)
1945	Arturi I. Virtanen (Fi)	Wolfgang Pauli (Swi b. Au)	A. Fleming (B) Ernst B. Chain (B b. G) Howard W. Florey (B)
1946	James B. Sumner (A) John H. Northrop (A) Wendell M. Stanley (A)	Percy W. Bridgman (A)	Herman J. Muller (A)
1947	Robert Robinson (B)	Edward Appleton (B)	Carl F. Cori (A b. Cz) Gerty T. Cori (A b. Cz) B. Houssay (Ar)
1948	Arne Tiselius (Swe)	P. M. S. Blackett (B)	Paul Müller (Swi)
1949	William F. Giauque (A)	Hideki Yukawa (J)	Walter R. Hess (Swi) Antonio E. Moniz (Po)
1950	Otto Diels (G) Kurt Alder (G)	Cecil F. Powell (B)	Philip S. Hench (A) E. C. Kendall (A) Tadeus Reichstein (Swi)
1951	Glenn T. Seaborg (A) Edwin M. McMillan (A)	John Cockroft (B) Ernest T. S. Walton (Ir)	Max Theiler (A b. SAf)
1952	A. J. P. Martin (B) R. L. M. Synge (B)	Felix Bloch (A) Edward M. Purcell (A)	Selman A. Waksman (A b. R)
1953	Herman Staudinger (G)	Fritz Zernike (D)	Fritz A. Lipmann (A. b. C Hans A. Krebs (B b. G)
1954	[b]Linus Pauling (A)	Max Born (B b. G) Walter Bothe (G)	J. F. Enders (A) F. C. Robbins (A) T. H. Weller (A)
1955	Vincent du Vigneaud (A)	Willis E. Lamb (A) Polykarp Kusch (A)	Hugo Theorell (Swe)

Nobel Prize Winners in Chemistry, Physics, and Medicine (continued)

Year	Chemistry	Physics	Medicine and Physiology
1956	Cyril N. Hinshelwood (B) Nikolai N. Semenov (R)	William Shockley (A) Walter A. Brattain (A) John Bardeen (A)	D. W. Richards, Jr (A) André F. Cournand (A b. F) Werner Forssmann (G)
1957	Alexander Todd (B)	Chen Nin Yang (Ch) Tsung Dao Lee (Ch)	Daniel Bovet (I b. Swi)
1958	Frederick Sanger (B)	Pavel A. Cherenkov (R) Igor E. Tamm (R) Ilya M. Frank (R)	George W. Beadle (A) Edward L. Tatum (A) Joshua Lederberg (A)
1959	Jaroslav Heyrovsky (Cz)	Emilio Segrè (A b. I) Owen Chamberlain (A)	Severo Ochoa (A b. Sp) Arthur Kornberg (A)
1960	Willard F. Libby (A)	Donald A. Glaser (A)	Frank McFarlane Burnet (Aus) Peter B. Medawar (B)
1961	Melvin Calvin (A)	Robert Hofstadter (A) Rudolf L. Mössbauer (G)	George von Békésy (H)
1962	John C. Kendrew (B) Max F. Perutz (B)	Lev D. Landau (R)	Francis H. C. Crick (B) James D. Watson (A) M. H. F. Wilkins (B)
1963[b]	Giulio Natta (I) Karl Ziegler (G)	J. Hans D. Jensen (G) Eugene P. Wigner (A b. H) Maria Goeppert Mayer (A b. G)	John C. Eccles (Aus) Allan L. Hodgkin (B) Andrew F. Huxley (B)
1964	Dorothy Crowfoot Hodgkin (B)	Charles H. Townes (A) Nikolai G. Basov (R) Alexander M. Prokhorov (R)	Konrad E. Bloch (A b. G) Feodor Lynen (G)
1965	Robert B. Woodward (A)	Richard P. Feynman (A) Julian S. Schwinger (A) Shinichero Tomonaga(J)	André Lwoff (F) François Jacob (F) Jacques Monod (F)
1966	Robert S. Mulliken (A)	Alfred Kastler (F)	Charles B. Huggins (A) Francis Peyton Rous (A)
1967	Manfred Eigen (G) Ronald G. W. Norrish (B) George Porter (B)	Hans A. Bethe (A b. F, ed. G)	Haldane K. Hartline (A) George Wald (A) Ragner Granit (Swe)
1968	Lars Onsager (A)	Luis W. Alvarez (A)	Robert W. Holley (A) H. Gobind Khorana (A b. In.) Marshall W. Nirenberg (A)
1969	Derek H. R. Barton (B) Odd Hassel (Nor.)	Murray Gell-Mann (A)	Max Delbrück (A b. G) Alfred D. Hershey (A) Salvador E. Luria (A b. I)
1970	Luis Leloir (Ar b. F)	Louis Neel (F) Hannes Alfven (Swe)	Julius Axelrod (A) Bernard Katz (B) Ulf von Euler (Swe)
1971	Gerhard Herzberg (C b. G)	Dennis Gabor (B)	Earl Sutherland, Jr. (A)
1972	Christian Anfinsen (A) Stanford Moore (A) William H. Stein (A)	John Bardeen (A) Leon Cooper (A) John Schrieffer (A)	Gerald Edelman (A) Rodney R. Porter (B)
1973	Ernst O. Fischer (Gw) Geoffrey Wilkinson (B)	Ivar Giaevar (A b. K) Leo Esaki (J) Brian Josephson (B)	Karl von Frisch (Gw) Konrad Lorenz (Gw b. Au) Nikolaas Tinbergen (B)

Nobel Prize Winners in Chemistry, Physics, and Medicine (continued)

Year	Chemistry	Physics	Medicine and Physiology
1974	Paul J. Flory (A)	Martin Ryle (B) Antony Hewish (B)	Albert Claude (A b. Lu) George Emil Palade (A b. Rum) Christian René de Duve (Be)
1975	John Cornforth (B b. Aus) Vladmir Prelog (Swi b. Y)	James Rainwater (A) Ben Mottelson (A b. Da) Aage Bohr (Da)	David Baltimore (A) Howard Temin (A) Renato Dulbecco (A b. I)
1976	Wm. L. Lipscomb (A)	Burton Richter (A) Samuel C. C. Ting (A)	Baruch S. Blumberg (A) Daniel C. Gajdusek (A)
1977	Ilya Prigogine (Be)	John Van Vleck (A) Philip Anderson (A) Nevill F. Mott (B)	Rosalyn S. Yalow (A) Roger Guillemin (A) Andrew V. Schally (A)
1978	Peter Mitchell (B)	Pyotr Kapitsa (R) Arno Penzias (A) Robert Wilson (A)	Daniel Nathans (A) Hamilton O. Smith (A) Werner Arber (Swi)
1979	Herbert C. Brown (A) George Wittig (Gw)	Steven Weinberg (A) Sheldon Glashow (A) Abdus Salam (Pak)	Allan M. Cormack (A) Geoffrey N. Hounsfield (B)
1980	Paul Berg (A) Walter Gilbert (A) Frederick Sanger (B)	James W. Cronin (A) Val L. Fitch (A)	Baruj Benacerraf (A b. V) Jean Dausset (F) George Snell (A)
1981	Kenichi Fukui (J) Roald Hoffmann (A b. P)	Nicolaas Bloembergen (A b. D) Arthur Schawlow (A) Kai M. Siegbahn (Swe)	David Hubel (A) Roger W. Sperry (A) Torsten Wiesel (A b. Swe
1982	Aaron Klug (B b. La)	Kenneth G. Wilson (A)	Sune K. Bergström (Swe) Bengt I. Samuelson (Swe) John R. Vane (B)
1983	Henry Taube (A b. C)	S. Chandrasekhar (A b. In) William A. Fowler (A)	Barbara McClintock (A)

a Indicates that the award had to be declined for political reasons. No awards were made in 1916, 1940, 1941, and 1942.

b The Peace Prize for 1963 was awarded to Linus Pauling, the 1954 chemistry laureate.

BIBLIOGRAPHIC NOTES

In order to avoid burdening the text with footnotes direct references to the literature have been omitted except for quotations and such other items where direct citation appears highly desirable. To a large extent this book is based upon an examination of original sources, critical studies, reviews, and compilations, many of which treat the subject in considerable detail and provide references to original publications. I am calling attention to such reviews and critical studies in the appropriate chapter bibliographies which follow. Anyone wishing to pursue a given subject more extensively may well consult such papers and be led by them to the original sources.

To trace an original paper by a particular author one may consult J. C. Poggendorff, *Biographisch-Literarisches Handwörterbuch zur Geschichte der exacten Wissenschaften* [Barth, Leipzig, 2 vols., 1863; supplements: vol. 3 for 1858–1883 (1898); vol. 4 for 1883–1903 (1904); vol. 5 for 1904–1922 (1926); vol. 6 for 1923–1931 (1936–1940); vol. 7 for 1932–1953 (1956–1962), the last volumes being published by Akademie-Verlag, Berlin].

The *Catalogue of Scientific Papers* published by the Royal Society of London (19 vols., 1867–1925) is a compilation of the scientific literature which appeared between 1800 and 1900. It is exceedingly valuable for tracing work done in the nineteenth century. The *International Catalogue of Scientific Literature* represents a continuation of the earlier *Catalogue*, but it fell victim to World War I and covers the literature of only the earlier years of the twentieth century.

Chemical Abstracts, published by the American Chemical Society since 1907, is the best current guide to the world's chemical literature. The Decennial Indexes are particularly valuable for rapid surveys of the work of a particular person or on a particular subject. The *Chemisches Centralblatt* is not as well indexed, but has the virtue of reaching back through antecedent publications to 1830.

Somewhat parallel in value to *Chemical Abstracts* are the Critical Bibliographies of the History and Philosophy of Science which have been published at frequent intervals in *Isis* since the journal was founded by the late George Sarton in 1913. Sarton had prepared 79 bibliographies by the time he relinquished the work in 1953. The bibliographies are now issued on an annual basis, with somewhat revised organization and less comprehensive scope. Their great value to historians of science is somewhat limited on account of the lack of a collective index.

A small publication which is very useful in tracking down important chemical papers is Paul Walden's *Chronologische Übersichtstabellen zur Geschichte der Chemie von ältesten Zeiten bis zur Gegenwart* (Springer, Berlin, 1952). It is a reasonably comprehensive compilation of significant discoveries, arranged chronologically. It has an author index but no subject index. Herbert Valentin's *Geschichte der Pharmacie und Chemie in Form von Zeittafeln* (Wissenschaftliche Verlagsgesellschaft, Stuttgart, 3rd ed., 1950) is a useful chronology which has an author and subject index but does not carry references to original sources.

Bibliographic works of special significance are Henry C. Bolton, *Select Bibliography of Chemistry, 1482–1892* (Smithsonian Institution, Washington, 1894–1904); John Ferguson, *Bibliotheca Chemica* (Maclehose, Glasgow, 1906, reprinted by Holland Press, London, 1954); the very rare *Catalogue of the Ferguson Collection* (Maclehose, Glasgow, 1943); and Denis I. Duveen, *Bibliotheca Alchemica et Chemica* (Weil, London, 1949). Bolton's bibliography and its two supplements (1899, 1904) are very comprehensive and place emphasis on contemporary chemical works. The other three bibliographies are catalogs, respectively, of the Young Collection in the Royal Technical College of Glasgow, the Ferguson Collection of the University of Glasgow, and the Duveen Collection now at the University of Wisconsin. All of these collections have their greatest strength in alchemical and early chemical works, although the Duveen collection also includes a large number of titles from the eighteenth and nineteenth centuries [see S. A. Ives and A. J. Ihde, "The Duveen Library," in *J. Chem. Educ.*, **29**, 244(1952)].

A small and very convenient general bibliographic work is George Sarton's *Horus, A Guide to the History of Science* (Chronica Botanica, Waltham, Mass., 1952) which contains very complete listings of works dealing with the history of the various sciences in all parts of the world.

Journals

The most useful journals in the history of chemistry are clearly *Chymia*, *Journal of Chemical Education*, and *Ambix*. *Chymia*, founded in 1948 as an annual for studies in the history of chemistry, has published many excellent papers, but because of financial problems has appeared irregularly. *Ambix* was founded in 1937 as the journal of the Society for the Study of Alchemy and Early Chemistry, but it too has published only intermittently. The *Journal of Chemical Education*, while not primarily a historical journal, has published many papers on historical subjects. The cumulative indexes covering the first 25 volumes and volumes 26–35 are useful tools. The *American Journal of Physics* fills a somewhat similar role in its field.

Important papers dealing with chemistry are frequently found in history of science journals such as *Isis*, *Osiris*, *Archives internationales d'histoire des sciences* (formerly *Archeion*), *Annals of Science*, *Lychnos*, *Centaurus*, and *Revue d'histoire des sciences et leurs applications*. This is also true of peripheral journals such as *Bulletin of the History of Medicine*, *Journal of the History of Medicine and Allied Sciences*, *Technology and Culture*, *Journal of the History of Ideas*, and *Cahiers d'histoire mondiale*. General scientific journals such as *Science, Nature,* and *Die Naturwissenschaften* often carry historical papers. *Endeavour*, published by Imperial Chemical Industries, has many good historical and review articles. *Science Progress* deals with recent developments. *Scientific American*, since its rejuvenation in 1948, carries many good articles. *American Scientist* is frequently useful, as was *Scientific Monthly* until its demise in 1957.

There have been other serial publications which dealt with history of science, medicine, and technology. Many of them were abandoned after a short time; some continue to appear at the present time. Many good historical papers lie comparatively buried in such journals. For a complete list, see Sarton's *Horus*, pp. 194–248.

Histories of Chemistry

General works dealing with the history of chemistry are numerous, though mostly limited in scope and of irregular quality. Several of the best date from the nineteenth century and therefore fail to deal with recent developments. However even the best twentieth-century imprints fail to bring the subject significantly into the present century, although there are numerous twentieth-century popularizations dealing with some phase of chemistry, usually from the human interest standpoint. While some of these books are shallow and full of errors of fact and interpretation, others have genuine value if used with an understanding of the author's background and purpose.

Thomas Thomson's *History of Chemistry* (Colburn & Bentley, London, 2 vols., 1830–1831) is frequently uncritical but is late enough to take account of the new understanding of chemistry which resulted from the work of Lavoisier, Klaproth, and Dalton. Further advance is evident in Ferdinand Hoefer's *Histoire de la chimie* (Didot, Paris, 2 vols., 1842–1843, particularly in the second edition, 1866–1869).

Unquestionably the most outstanding work of the century is Hermann Kopp, *Geschichte der Chemie* (Vieweg, Braunschweig, 4 vols., 1843–1847). The first volume presents a survey of chemical developments from antiquity to Kopp's time; the second is a study of the history of chemical operations, analytical chemistry, mineralogy, pharmacy, and alchemy and of the nature of chemical changes; volumes three and four are detailed histories of elements and compounds. The set is, even today, a valuable source of information regarding many phases of chemistry. Kopp later published a *Beiträge zur Geschichte der Chemie* (Vieweg, Braunschweig, 2 vols., 1869–1875) and *Die Entwicklung der Chemie in der neueren Zeit* (Oldenbourg, Munich, 1873). While these are works of some merit, neither is as scholarly, as well organized, or as readable as his earliest history.

A very good one-volume history is Albert Ladenburg, *Vorträge Über die Entwicklungsgeschichte der Chemie von Lavoisier bis zur Gegenwart* (Vieweg, Braunschweig, 1869). It was updated through new editions, the fourth appearing in 1907. There were two French translations of the late editions and an English translation of the second edition by L. Dobbin, *Lectures on the History of the Development of Chemistry since the time of Lavoisier* (Alembic Club, Edinburgh, 1900). The footnotes are rich in references to the journal literature. Another good German work is that of Ernst von Meyer, *Geschichte der Chemie* (Verlag von Veit, Leipzig, 1889); a fourth edition appeared in 1914. English translations by George M'Gowan, *History of Chemistry* (Macmillan, London & New York), appeared in 1891, 1898, and 1906.

C. A. Wurtz, *Dictionnaire de chimie pure et appliquée* (Hachette, Paris, 3 vols., in 5, 1868–1878), has a 94-page introductory essay, "Histoire des doctrines chimiques depuis Lavoisier." An English translation of the essay was published by Henry Watts, *History of Chemical Theory* (Macmillan, London, 1869). Wurtz's essay has considerable merit even though the opening sentences, "Chemistry is a French science. It was founded by Lavoisier, of immortal memory," caused a violent reaction in other countries, particularly Germany. M. E. Chevreul's *Histoire des connaissances chimiques* (Morgand, Paris, 1866) is not truly a history. As the first volume of a projected five-volume study which was never carried further, it is a rambling, speculative work dealing with the nature of chemistry. Raoul Jagnaux's *Histoire de la chimie* (Baudry, Paris, 2 vols., 1891) is a useful work with emphasis on chemical facts. The principal twentieth-century French work is Maurice Delacre, *Histoire de la chimie* (Gauthier-Villars, Paris, 1920).

Twentieth-century histories of chemistry in German are plentiful, but are of variable merit, none of them having achieved the eminence of Kopp's book of a century ago. W. Hertz's *Grundzuge der Geschichte der Chemie* (Enke, Stuttgart, 1916) is a small work of no particular distinction. Thor Ekecrantz's *Geschichte der Chemie* (Springer, Berlin, 1921) is a good short history. Eduard Färber's *Die Geschichtliche Entwicklung der Chemie* (Springer, Berlin, 1921) is a sound work which is not entirely superseded by the author's much later American book. Richard Meyer's *Vorlesung über die Geschichte der Chemie* (Akad. Verlagsgesellschaft, Leipzig, 1922) is particularly good on late nineteenth-century chemistry. Fritz Ferchl and A. Süssenguth's *Kurtzgeschichte der Chemie mit 200 Abbildungen* (Gesellschaft für Geschichte der Pharmacie, Mittenwald, 1936) is primarily a collection of illustrations. An English translation has appeared as *A Pictorial History of Chemistry* (Heinemann, London, 1939). H. E. Fiertz-David's *Die Entwicklungsgeschichte der Chemie* (Birkhäuser, Basel, 1945, 2nd ed., 1952) is a well-balanced survey giving considerable space to modern applied chemistry. Paul Walden's *Drei Jahrtausende Chemie* (Limpert, Berlin, 1944), as well as his very short *Geschichte der Chemie* (Athenäum-Verlag, Bonn, 2nd ed., 1950) are damaged by the author's pro-German, anti-British bias. Georg Lockemann's *Geschichte der Chemie* [Gruyter, Berlin, 2 vols., 1950–1955 (English translation as *The Story of Chemistry*, Philosophical Library, New York, 1959)] is a short, tightly written, fact-crammed compilation marred by numerous errors.

Twentieth-century English works are also plentiful and are generally of good quality. Edward Thorpe's *History of Chemistry* (Watts, London, 2 vols., 1909–1910) presents a good survey up to 1900. Wm. A. Tilden's *The Progress of Scientific Chemistry in our own Times* (Longmans, London, 1899, 2nd ed., 1913) deals with certain developments of the nineteenth century and contains biographical notes on a large number of chemists. James Campbell Brown's *History of Chemistry* (Churchill, London, 1913) has a strongly biographical tone. T. M. Lowry's *Historical Introduction to Chemistry* (Macmillan, London, 1915, 3rd ed., 1936) presents a good historical account of certain facts and theories without attempting to present a formal history of the subject. T. P. Hilditch's *A Concise History of Chemistry* (Methuen, London, 1911, 1922) is exactly that. Despite its conciseness, it contains a mass of factual information, with numerous tables serving as useful summaries. Eric John Holmyard's *Chemistry to the Time of Dalton* (Oxford Univ. Press, London, 1925) has a strongly biographical character.

Clearly first rate is J. R. Partington's *A Short History of Chemistry* (Macmillan, London, 1937), with slightly revised editions in 1948 and 1957. This same author is now publishing a four-volume comprehensive history of chemistry which fills a long-felt need, *A History of Chemistry* (Macmillan, London). Volume 2, dealing with the period between 1500 and 1750 was published in 1961; vol. 3, dealing with France from 1600 and Europe generally from 1750 bringing the subject through Dalton's work, appeared in 1962; vol. 4, published in 1964, deals with the period from 1800 into the twentieth century; vol. 1, announced as forthcoming, is to bring the subject from its origins to A.D. 1500. The volumes in print are organized biographically in an approximately chronological order. Work of each chemist is treated in great detail with full bibliographic listing of his works. The set will be invaluable for scholars in history of chemistry.

Alexander Findlay's *A Hundred Years of Chemistry* (Duckworth, London, 1937, 1948) deals in very able fashion with developments of the last century. A. J. Berry has published two small works dealing with selected topics which had their principal development in the last 75 years; the books are *Modern Chemistry* (Cambridge Univ. Press, Cambridge, 1946) and *From Classical to Modern Chemistry* (same publisher, 1954).

American works apparently begin with Francis P. Venable's *History of Chemistry* (Heath, Boston, 1894, last ed., 1922), a brief survey. F. J. Moore's *A History of Chemistry* (McGraw-Hill, New York, 1918) is strongly biographical in character. The second (1931) and third (1939) editions prepared by Wm. T. Hall are little changed except for added biographical sketches. John M. Stillman's *The Story of Chemistry* (Appleton, New York, 1924) is an excellent account of developments through Lavoisier. Although subsequent scholarship has thrown new light on several matters, it is still one of the best American books on the subject. A reprint edition under the title *The Story of Alchemy and Early Chemistry* (Dover, New York, 1960) is now available. Two other recent books of good quality are Eduard Farber, *Evolution of Chemistry* (Ronald, New York, 1952) and Henry M. Leicester, *The Historical Background of Chemistry* (Wiley, New York, 1956).

Reprints and Source Books

Reprints of important chemical papers and books are very limited in number. The best are those prepared by the Alembic Club of Edinburgh. This organization has published twenty-two titles which bring together some of the most significant papers in the rise of chemistry. Wm. Ostwald originated the *Klassiker der exakten Wissenschaften* (Engelmann, and later, Akademische Verlag, Leipzig, 246 vols., 1899–1956) which includes a number of chemical papers in the original German or in German translation. In some cases the whole work is reprinted, in others, only pertinent parts. Dover Publications has just initiated a Classics in Science series under the general editorship of Gerald Holton which may ultimately provide an extensive reprint series in English. The one volume published to date is O. Theodor Benfey, ed., *Classics in the Theory of Chemical Combination* (New York, 1963), a collection of pertinent papers by Wöhler, Laurent, Williamson, Frankland, Kekulé, Couper, van't Hoff, and Le Bel.

Henry M. Leicester and Herbert S. Klickstein's *A Source Book of Chemistry 1400–1900* (McGraw-Hill, New York, 1952) contains papers of 83 chemists, but most of the papers are short and frequently deletions have been made. Other source books in this series (Astronomy, Physics, Mathematics, Geology, Greek Science, Biology) sometimes contain papers on subjects related to chemistry. The chemistry source book has a useful bibliography of biographies.

The Harvard Case Histories in Experimental Science, under the editorships of James B. Conant and Leonard K. Nash, are not reprints in the strictest sense but contain selected excerpts along with analysis. Four of them deal with chemical subjects: Conant, *The Overthrow of the Phlogiston Theory* (no. 2, 1950); Nash, *The Atomic-Molecular Theory* (no. 4, 1950); Nash, *Plants and the Atmosphere* (no. 5, 1952); Conant, *Pasteur's Study of Fermentation* (no. 6, 1952). All numbers of the series were published by Harvard Univ. Press. They are also available in a collected set of two volumes, *Harvard Case Histories in Experimental Science* (1957). The case histories evolved out of the ideas on science teaching expressed by Conant in the Terry lectures given at Yale University and published in *On Understanding Science* (Yale Univ. Press, New Haven, 1947). These ideas were further expanded in *Science and Common Sense* (Yale Univ. Press, New Haven, 1951).

Biographical Works

Biographical works dealing with individuals will be mentioned in the chapter bibliographies where the work of these scientists is encountered. There are, however, a large number of collective biographies which give attention to the lives of chemists. These will be listed here and not mentioned again except for special reasons.

Of the many books dealing with the lives of great chemists, the most comprehensive, reliable, and useful has been Gunther Bugge, ed., *Das Buch der grossen Chemiker* (Verlag Chemie, Berlin, 2 vols., 1929–1930). It will be referred to hereafter simply as "Bugge." Eduard Farber, ed., *Great Chemists* (Interscience, New York, 1961) provides an English equivalent of even greater comprehensiveness. It will be cited hereafter as "Farber." Also see R. Sachtleben and A. Hermann, *Von der Alchemie zur Grossynthese, Grosse Chemiker* (Battenberg Verlag, Stuttgart, 1961) which has portraits and short biographical sketches of many famous chemists.

William A. Tilden's *Famous Chemists* (Routledge, London, 1921) is very good on a limited number of the more prominent figures. Wm. Ostwald's *Grosse Männer* (Akad. Verlagsgesellschaft, Leipzig, 1909) deals with the lives of several important chemists in connection with Ostwald's notion of "romantic" and "classical" types of genius. The original book is supplemented with a series of individual biographies published between 1910 and 1932.

E. J. Holmyard's *The Great Chemists* (Methuen, London, 1929) deals with the lives of fifteen chemists; his *Makers of Chemistry* (Clarendon Press, Oxford, 1931) contains a much larger number of shorter sketches. Benjamin Harrow's *Eminent Chemists of our Time* (Van Nostrand, New York, 1920) is a readable but somewhat worshipful series of biographical sketches. The same may be said of Bernard Jaffe's *Crucibles* (Simon & Schuster, New York, 1930 and later editions), although this is a more substantial work. A collection of about 250 portraits with brief biographical sketches will be found in Henry

M. Smith's *Torchbearers of Chemistry* (Academic Press, New York, 1949). This volume also contains a good bibliography of biographies.

Excellent biographies with critical evaluations of the subject's chemical work will be found in the *Memorial Lectures Delivered before the Chemical Society*. These lectures were originally published in the *Journal of the Chemical Society*, but four collected volumes have been published in London containing the lectures delivered in 1893–1900, 1901–1913, 1914–1932, and 1933–1943 (vols. I and II by Gurney and Jackson, vols. III and IV by the Chemical Society).

Mary Elvira Weeks' *Discovery of the Elements* (Journal of Chemical Education, Easton, Pa., 1933, 6th ed., 1956) is a rich source of biographical information on both prominent and obscure chemists. Philipp Lenard's *Grosse Naturforscher* (Lehmann, Munich, 1929) was translated into English by H. S. Hatfield under the title *Great Men of Science* (Macmillan, New York, 1934). It includes brief sketches of numerous scientists but is marred by a very noticeable pro-German bias.

Eduard Farber's *Nobel Prize Winners in Chemistry 1901–1950* (H. Schuman, New York, 1953) gives short biographies together with a discussion of the awardees' work. A second edition published in 1963 includes sketches of recipients between 1951 and 1960. The other volumes of the series are also of interest because of the interrelationship of the sciences. They are Niels Heathcote, *Nobel Prize Winners in Physics*, and Lloyd G. Stevenson, *Nobel Prize Winners in Medicine*.

There are also a number of collective biographies of some merit, although none of them adds significantly to the material in those already listed. They are still of some interest. The more satisfactory are the following:

M. M. P. Muir, *Heroes of Science, Chemists*, Soc. for Promoting Christian Knowledge, London, 1883.

T. E. Thorpe, *Essays in Historical Chemistry*, Macmillan, London, 1894.

Wm. Ramsay, *Essays Biographical and Chemical*, Constable, London, 1908.

E. Roberts, *Famous Chemists*, Unwin, London, 1911.

A. Griffiths, *Biographies of Scientific Men*, Sutton, London, 1912.

F. L. Darrow, *Masters of Science and Invention*, Harcourt, Brace, New York, 1923.

B. Harrow, *The Making of Chemistry*, John Day, New York, 1930.

J. N. Leonard, *Crusaders of Chemistry, Six Makers of the Modern World*, Doubleday, Doran, New York, 1930.

Henry Thomas and D. L. Thomas, *Living Biographies of Great Scientists*, Garden City Publ. Co., Garden City, N.Y., 1942.

Grove Wilson, *The Human Side of Science*, Cosmopolitan, New York, 1942.

D. B. Hammond, *Stories of Scientific Discovery*, Cambridge Univ. Press, Cambridge, 1933.

E. R. Trattner, *Architects of Ideas*, Carrick & Evans, New York, 1938.

S. K. Bolton, *Famous Men of Science*, Crowell, New York, 1938.

J. G. Crowther, *Men of Science*, Norton, New York, 1936.

J. G. Crowther, *Famous American Men of Science*, Norton, New York, 1937.

I. Nechaev, *Chemical Elements, the Fascinating Story of Their Discovery and of the Famous Scientists who Discovered Them,* Coward-McCann, New York, 1942.

B. Jaffe, *Men of Science in America,* Simon & Schuster, New York, 1944.

Charles G. Fraser, *Half-Hours with Great Scientists,* Univ. of Toronto Press, Toronto and Reinhold, New York, 1948.

A very useful bibliography of *Biographies of Engineers and Scientists* has been published by Thomas J. Higgins. It originated as a series of papers in various periodicals, the ones on chemists in *School Science and Math.,* **44,** 650(1944) and **48,** 438(1948), but these were published in collected form in *Research Publications* (Illinois Inst. of Technology, vol. 7, no. 1, pp. 1–62, 1949). References are all to books, including biographical collections as well as biographies of individuals.

Of course appropriate biographical information will be found in such standard references as *Dictionary of National Biography; Dictionary of American Biography; International Who's Who; Who's Who in America; Allgemeine Deutsche Biographie; American Men of Science;* Fritz Ferchl, ed., *Chemisch-Pharmaceutisches Bio- und Bibliographikon* (Neumaiern, Mittenwald, 2 vols., 1937–1938); A. V. Howard, *Chamber's Dictionary of Scientists* (Dutton, New York, 1951).

Festschrifts, Commemorative Volumes, Collected Papers, Symposia

It is ironical that good papers frequently lie buried in collected volumes brought together to honor a birthdate or report a symposium. Whereas papers published in traditional journals are encountered in the normal course of events, those in collected volumes are frequently missed. There have not been a large number of collected volumes dealing with the history of science, but there are a few which should be mentioned as repositories of papers dealing with the history of chemistry.

It is fitting that there should be a Festschrift honoring Edmund O. von Lippmann, since he was personally responsible for much work in the history of chemistry—J. Ruska, ed., *Studien zur der Geschichte der Chemie, Festgabe Edmund O. von Lippmann zum Siebzigsten Geburtstage* (Springer, Berlin, 1927). Lippmann's many papers have been published in collected volumes under the titles: *Abhandlung und Vorträge zur Geschichte der Naturwissenschaften* (Veit, Leipzig, 2 vols., 1906–1913) and *Beiträge zur Geschichte der Naturwissenschaften und der Technik* (Springer and Verlag Chemie, Berlin and Weinheim, 2 vols., 1923 and 1953).

Other Festschriften with chemical interest are: *Beiträge aus der Geschichte der Chemie dem Gedächtnis von Georg. W. A. Kahlbaum,* Paul Diergart, ed. (Deuticke, Leipzig & Vienna, 1909); *Essays on the History of Medicine, Presented to Karl Sudoff on the Occasion of his Seventieth Birthday,* C. Singer and H. E. Sigerist, eds. (Oxford Univ. Press, London, 1924); *Science, Medicine and History, Essays on the Evolution of Scientific Thought and Medical Practise, Written in Honor of Charles Singer* (Oxford Univ. Press, London and New York, 1953).

Critical Problems in the History of Science, Marshall Clagett, ed. (Univ. of Wisconsin Press, Madison, 1959) carries several excellent papers of chemical interest.

CHAPTER
BIBLIOGRAPHIC
NOTES

CHAPTER 1. PRELUDE TO CHEMISTRY

Of the various books dealing with technological knowledge in ancient times, the first volume of *A History of Technology*, edited by Charles Singer, E. J. Holmyard and A. R. Hall (Oxford Univ. Press, London, 1954), is particularly useful. In carrying the subject "From Early Times to Fall of Ancient Empires," the volume has good chronological tables and maps, as well as chapters on "Fire-making, Fuel and Lighting" (H. S. Harrison), "Discovery, Invention, and Diffusion" (Harrison), "Working Stone, Bone, and Wood" (L. S. B. Leakey), "Graphic and Plastic Arts" (Leakey), "Foraging, Hunting, and Fishing" (D. Forde), "Chemical, Culinary, and Cosmetic Arts" (R. J. Forbes), "Pottery" (L. Scott), "Extraction, Smelting, and Alloying" (Forbes), and "Measures and Weights" (G. F. Skinner).

J. R. Partington's *Origins and Development of Applied Chemistry* (Longmans, London, 1935) is a comprehensive study of the chemical arts in ancient times. A. Lucas' *Ancient Egyptian Materials and Industries* (Arnold, London, 3rd ed., 1948) is very useful, as are Reginald C. Thompson's *Dictionary of Assyrian Chemistry and Geology* (Clarendon Press, Oxford, 1936) and his *On the Chemistry of the Ancient Assyrians* (Oxford Univ. Press, London, 1925). A more recent study is Martin Levey's *Chemistry and Chemical Technology in Ancient Mesopotamia* (Elsevier, Amsterdam and New York, 1959).

A very helpful tool for the study of ancient technology is R. J. Forbes' *Bibliographia Antiqua. Philosophia Naturalis* (Nederlandische Institut voor het Nabije Oosten, Leyden, 1940–1950). Ten parts give references to journal articles and books dealing with all phases of ancient science and technology. Forbes is also the author of a valuable series of books published under the title *Studies in Ancient Technology* (Brill, Leiden, 6 vols., 1955–1958).

The best work on ancient metallurgy is Forbes' *Metallurgy in Antiquity* (Brill, Leiden, 1950). Also see J. A. N. Friend's *Iron in Antiquity* (Griffin, London, 1926) and T. A. Rickard's *Man and Metals* (McGraw-Hill, New York, 2 vols., 1932).

The best references on early medicine are the two volumes of Henry E. Sigerist, *History of Medicine* (Oxford Univ. Press, New York, vol. I, *Primitive and Archaic Medicine*, 1951; vol. II, *Early Greek and Indian Medicine*, 1961). These were the first of a projected six-volume treatise on the history of medicine, but the author's untimely illness and death prevented further work on the project. Max Neuberger's *Geschichte der Medicin* (Enke, Stuttgart, 2 vols., 1906–1911) was also intended to be a comprehensive history, but it was never finished; volume 2 ends with the Middle Ages. An English translation by Ernest Playfair, *History of Medicine* (Oxford Univ. Press, London, 2 vols., 1910–1925), is available and despite its age is still a valuable reference. Chapters on early medicine are to be found in such standard shorter works as: E. Ackerknecht, *A Short History of Medicine* (Ronald, New York, 1955); F. H. Garrison, *An Introduction to the History of Medicine* (Saunders, Philadelphia, 1913, 4th ed., 1929); A. Castiglioni, *History of Medicine* (Knopf, New York, 1941, 2nd ed., 1947); Douglas Guthrie, *History of Medicine* (Nelson, London, 1945); V. Robinson, *The Story of Medicine* (Tudor, New York, 1931); and C. J. Singer, *Short History of Medicine* (Oxford Univ. Press, Oxford, 1928, 2nd ed., with E. A. Underwood, 1962).

Knowledge of the material concepts of the early Greek philosophers is limited since most of the original writings have been lost, and the men are known chiefly through the commentaries of later philosophers. The standard source on early Greek philosophy is Hermann Diels' *Die Fragmente der Vorsokrater* (Wiedmannische Verlagsbuchhandlung, Berlin, 3 vols., 7th ed., 1951), which contains Greek and German texts of the extant fragments of the various philosophers, together with existing statements about each philosopher by the ancient writers. An extremely useful pair of English volumes for use with Diels has been prepared by Kathleen Freeman: *The Pre-Socratic Philosophers* (Blackwell, Oxford, 3rd ed., 1953) and *Ancilla to the Pre-Socratic Philosophers* (same publisher, 1952). Also see: E. Zeller, *Outlines of the History of Greek Philosophy*, transl. by L. R. Palmer from the thirteenth German ed. of 1931 (Meridian, New York, 1957); John Burnet, *Early Greek Philosophy* (Black, London, 4th ed., 1945); F. M. Cornford, *Principium Sapientiae, The Origins of Greek Philosophical Thought* (Cambridge Univ. Press, Cambridge, 1952); Werner Jaeger, *The Theology of the Early Greek Philosophers*, transl. from the German by L. Magnus (Scribners, New York, 3 vols., 1901); Bertrand Russell, *A History of Western Philosophy* (Simon & Schuster, New York, 1945); and R. A. Tsanoff, *The Great Philosophers* (Harper, New York, 2nd ed., 1964).

There are many printed editions of Plato's dialogs, the standard being the Greek text edited by Henricus Stephanus (Henri Estienne) with a Latin translation by Johannes Serranus [(Jean de Serres), Geneva, 1578]. The pagination of this edition has been followed in nearly all subsequent editions. The most famous English translation is that of Benjamin Jowett (Clarendon Press, Oxford, 5 vols., 1899–1906). There are numerous other translations of individual works, those published by the Loeb Classical Library being particularly useful since they are published in convenient size and carry the Greek text beside the English. The *Timaeus* has been translated into English by R. D. Archer-Hind (1888), A. E. Taylor (1909), and F. M. Cornford, the latter with a running commentary which makes it exceedingly valuable— *Plato's Cosmology* (Routledge & Kegan Paul, London, 1937).

The standard edition of Aristotle's works is the Greek text with Latin translation prepared by Immanuel Bekker for publication by the Berlin Academy (5 vols., 1831–1870). Most subsequent editions have retained the

CHAPTER 1 769

Bekker pagination. An English translation of the complete works was made under the editorship of W. D. Ross (Clarendon Press, Oxford, 1908–1931). Many of the Aristotelian works are available in the Loeb Library. T. E. Lones' *Aristotle's Researches in Natural Science* (West, Newman, London, 1912) and D'Arcy Wentworth Thompson's *Aristotle as a Biologist* (Clarendon, Oxford, 1916) deal with the scientific works, the emphasis being principally on biology. W. D. Ross' *Aristotle* (Methuen, London, 1923) is a general study of Aristotle and his works.

Works dealing with atomism are: W. J. Oates, *Stoic and Epicurean Philosophers* (Random House, New York, 1940); Cyril Bailey, *Greek Atomists and Epicurus* (Clarendon, Oxford, 1928); A. J. Festugière, *Epicurus and his Gods*, transl. from the French by C. W. Chilton (Blackwell, Oxford, 1955). C. Bailey has made a superb translation and study of Lucretius' poem in *Titi Lucreti Cari De Rerum Natura* (Oxford Univ. Press, London, 3 vols., 1949). Besides these detailed studies, Greek atomism is examined in Kurd Lasswitz, *Geschichte der Atomistik im Mittelalter bis Newton* (Voss, Hamburg, 2 vols., 1890); Joshua C. Gregory, *A Short History of Atomism* (Black, London, 1931); and Andrew G. Melsen, *From Atomos to Atom*, transl. from the Dutch by H. J. Koren (Duquesne Univ. Press, Pittsburgh, 1952).

Several general works deal nicely with Greek science and philosophy. René Taton, ed., *Histoire Générale des Sciences*, vol. 1, *La Science Antique et Mediévale (des origines à 1450)* (Presses Universitaires de France, Paris, 1957), has chapters giving good treatment of the period. An excellent short study is Marshall Clagett's *Greek Science in Antiquity* (Abelard-Schuman, New York, 1955). George Sarton finished two volumes of his projected *History of Science* before his death, the volumes bringing the subject just to the Christian Era: vol. 1, *Ancient Science through the Golden Age* (Harvard Univ. Press, Cambridge, 1952); vol. 2, *Hellenistic Science and Culture in the Last Three Centuries* B.C. (1959). Another good work is Pierre Brunet and Aldo Meili's *Histoire des Sciences: Antiquité* (Payot, Paris, 1935). A very extensive work is Abel Rey, *La science dans l'antiquité* (Michel, Paris, 4 vols., 1930–1948). Morris R. Cohen and I. E. Drabkin's *A Source Book in Greek Science* (McGraw-Hill, New York, 1948) contains a large amount of material from this period, but very little of it has chemical significance.

The *Historia Naturalis* of Pliny has been translated in its entirety by John Bostock and H. T. Riley, *The Natural History of Pliny* (Bohn's Classical Library, London, 6 vols., 1855–1857), but the chemical parts show certain weaknesses and errors of interpretation. A better partial translation is available in Kenneth Bailey, *The Elder Pliny's Chapters on Chemical Subjects* (Arnold, London, 2 vols., 1929–1932). There is an extensive chapter on Pliny in Lynn Thorndike, *A History of Magic and Experimental Science* (Macmillan, New York, vol. 1, 1923). This and subsequent volumes of Thorndike's monumental study have numerous chapters dealing with alchemists, natural philosophers, and physicians who contributed to the development of science before A.D. 1700 (Macmillan and Columbia Univ. Press, 8 vols., 1923–1958). There is also extensive information about persons and their works in G. Sarton, *Introduction to the History of Science* [(Homer to A.D. 1400), Williams & Wilkins, Baltimore, 3 vols. in 5, 1927–1948].

Alchemy

Alchemical scholarship has its beginnings with Marcelin Berthelot. In *Les origines de l'alchimie* (Steinheil, Paris, 1885) and *Introduction à l'étude de la chimie des anciens et du moyen âge* (Steinheil, Paris, 1889) he dealt with the general problems of alchemical studies. These works were supplemented with two additional works in which he published texts, French translations, and commentary: *Collections des anciens alchimistes Grecs* (Steinheil, Paris, 3 vols., 1887–1888); and *La chimie au moyen âge* (Steinheil, Paris, 1893).

It was not until 1919 that a new work of major importance was published: E. O. von Lippmann's *Entstehung und Ausbreitung der Alchemie* (Springer, Berlin, 1919). Two additional volumes in the form of encyclopedic addenda were published in 1931 and 1954 (Verlag Chemie, Weinheim). Lippmann's critical studies had the effect of showing that Berthelot had not settled all problems and paved the way for the studies of Holmyard and Stapleton in England and of Ruska and Kraus in Germany.

Julius Ruska threw doubt on the alleged antiquity of Arabic alchemy with his *Arabische Alchemisten* (Winter, Heidelberg, 2 vols., 1924). Paul Kraus placed new light on the historic character of Jabir in his reports in the *Dritte Jahresbericht des Forschungsinstituts für Geschichte der Naturwissenschaften in Berlin*, 1930 and in the 44th and 45th volumes of the *Memoirs de l'Institut d'Egypte*, 1943 and 1942, respectively.

As a consequence of the extensive studies of alchemy, several general works were produced which provide the reader with a more reliable picture of alchemy than did those which existed in the nineteenth century. However there are still being produced certain works on alchemy which are of little merit since the subject has an appeal for persons of strange religious bent. The best of the general works on alchemy is F. Sherwood Taylor's *The Alchemists* (H. Schuman, New York, 1949) which presents a clear and sympathetic survey. Also good is E. J. Holmyard, *Alchemy* (Penguin, Harmsworth, Middlesex, 1957). John Read's interest has led to four works of popular interest. *Prelude to Chemistry* (Macmillan, New York, 1937) deals very largely with literary works which had significance in European alchemy—and is the source of title for the first chapter of this book, for which my thanks to Mr. Read. *Humour and Humanism in Chemistry* (Bell, London, 1947) gives attention to the activities of certain late alchemists. *The Alchemist in Life, Literature and Art* (T. Nelson, London, 1947) deals with alchemy as a subject for writers and artists. All of these works are illustrated with plates from alchemical works and reproductions of paintings and sketches dealing with the subject. Read's most recent work is *Through Alchemy to Chemistry* (Bell, London, 1957). Other works of value are: Arthur John Hopkins' *Alchemy, Child of Greek Philosophy* (Columbia Univ. Press, New York, 1934) which deals with the role of color in alchemical thought; and Wilhelm Ganzenmüller's *Beiträge zur Geschichte der Technologie und der Alchemie* (Verlag Chemie. Weinheim, 1956), which is a collected volume of the author's papers dealing with various phases of alchemy, particularly as alchemy relates to the rise of technology.

Iatrochemistry

For a general view of medicine in the Middle Ages and Renaissance, see the histories of medicine mentioned earlier. On distillation there is an excellent study by R. J. Forbes, *Short History of the Art of Distillation from the Beginnings up to the Death of Cellier Blumenthal* (Brill, Leiden, 1948).

The literature on Paracelsus is exceedingly voluminous and much of it is downright bad. The best scholarship is that of Karl Sudhoff who prepared a critical edition of the works: *Sämtliche Werke* (Oldenbourg, Munich and Berlin, 14 vols., 1922–1933). There is no English translation of the complete works, the most extensive collection being that of Arthur E. Waite, *The Hermetic and Alchemical Writings of Paracelsus the Great* (Elliott, London, 1894). Among the more reliable biographies of Paracelsus are: Anna M. Stoddart, *The Life of Paracelsus* (Murray, London, 1911); Karl Sudoff, *Paracelsus, ein deutsches Lebensbild aus der Renaissancezeit* (Bibliogr. Inst., Leipzig, 1936); Henry M. Pachter, *Paracelsus* (H. Schuman, New York, 1951); and Walter Pagel, *Paracelsus, An Introduction to Philosophical Medicine in the Era of the Renaissance* (Karger, Basel, 1958). Also see A. J. Ihde, *Sixth Ann. Lecture Series* (College of Pharmacy, Univ. of Texas, Austin, 1963), p. 5.

Technology

The best comprehensive work dealing with technology during the Middle Ages and the Renaissance is the Singer, *et al.*, *A History of Technology*, volumes 2 and 3, which have chapters dealing with chemical matters. Shorter works of importance are: F. Sherwood Taylor, *A History of Industrial Chemistry* (Heinemann, London, 1957); Frank D. Adams, *The Birth and Development of the Geological Sciences* (Williams and Wilkins, Baltimore, 1938) R. J. Forbes, *Man the Maker* (H. Schuman, New York, 1950); Thomas A. Rickard, *Man and Metals* (McGraw-Hill, New York, 2 vols., 1932); and W. B. Parsons, *Engineers and Engineering in the Renaissance* (Williams and Wilkins, Baltimore, 1939).

Herbert Hoover and Lou Henry Hoover prepared an excellent English translation of Agricola's book on mining: *Georgius Agricola de re metallica* (The Mining Magazine, London, 1912; reprinted by Dover, New York, 1950). English translations of the other sixteenth-century metallurgical works have been prepared through the activities of Cyril S. Smith. With Martha Gundi he authored *The Pirotechnia of Vannoccio Biringuccio* (Am. Inst. of Min. and Met. Engineers, New York, 1942); with Annaliese G. Sisco, the *Bergwerk- und Probierbüchlein* (Am. Inst. Min. and Met. Engrs., New York, 1949) and *Lazarus Ercker's Treatise on Ores and Assaying* (Univ. of Chicago Press, Chicago, 1951). All of these works are useful in showing the state of assaying and metallurgical techniques in the Renaissance. They also reveal much regarding the state of other chemical technologies and of technology in general, the Agricola work being extensively illustrated with woodcuts showing technical operations.

For biographical material on the Renaissance technologists see: Hans Hartman, *Georg Agricola* (Wissensch. Verlag., Stuttgart, 1953); E. Darmstaedter, *Georg Agricola, 1494–1555, Leben und Werk* (Munich, 1926); C. L. Brightwell, *Palissy the Potter* (Carleton & Potter, New York, 1858); H. Morley,

Palissy the Potter (Chapman & Hall, London, 2 vols., 1852); and G. Tierie, *Cornelius Drebbel* (H. J. Paris, Amsterdam, 1932). Palissy's *Admirable Discourses* are available in an English translation by Aurèle La Rocque (Univ. of Illinois Press, Urbana, 1957).

The Seventeenth Century

For background on intellectual activities during the seventeenth century see: F. L. Nussbaum, *The Triumph of Science and Reason, 1660–1685* (Harper, New York, 1953); J. E. King, *Science and Rationalism in the Government of Louis XIV* (J. Hopkins Univ. Press, Baltimore, 1949); and Basil Willey, *The Seventeenth-Century Background* (Doubleday Anchor, New York, repr. ed., 1953). There are good sections on the century in John H. Randall, *Making of the Modern Mind* (Houghton, Mifflin, Boston, rev. ed., 1940) and in Crane Brinton, *Ideas and Men* (Prentice-Hall, New York, 1950).

The general histories of science all deal with the century in considerable detail. For extensive treatment see René Taton, ed., *Histoire Générale des Sciences*, vol. 2, *La Science Moderne (de 1450 à 1800)*. H. Butterfield's excellent *Origins of Modern Science* (Bell, London, 2nd ed., 1957) has a number of chapters dealing with the century, as has A. R. Hall, *The Scientific Revolution, 1500–1800* (Longmans, Green, London, 1954). Also see Marie Boas, *The Scientific Renaissance, 1450–1630* (Harper, New York, 1962). A very detailed and factual treatment will be found in A. Wolf, *A History of Science, Technology and Philosophy in the Sixteenth and Seventeenth Centuries* (Allen & Unwin, London, 1935, 2nd ed. prepared by D. McKie, 1950). There are numerous studies of Newton, two of the best being I. B. Cohen, *Newton and Franklin* (Amer. Phil. Soc., Philadelphia, 1956) and L. T. More, *Isaac Newton, A Biography, 1642–1727* (Scribners, London, 1934). Hélène Metzger made a superb study of the concepts of matter in *Newton, Stahl, Boerhaave* (Alcan, Paris, 1930).

On van Helmont see Walter Pagel, *Jo. Bapt. van Helmont* (Springer, Berlin, 1930). For Hooke there is a good biography by Margaret 'Espinasse, *Robert Hooke* (Heinemann, London, 1956). Also see R. T. Gunther, *The Life and Work of Robert Hooke*, vol. 6 of *Early Science in Oxford* (the author, Oxford, 1930). There are no books on John Mayow, but for short studies see H. Guerlac, *Isis*, **45**, 243(1954) and *Actes Seventh Int. Cong. Hist. Sci.*, **1953**, p. 332; J. R. Partington, *Isis*, **47**, 217, 405(1956); and D. McKie in vol. 1 of the Singer Commemorative Essays, p. 484.

There are numerous studies of Boyle, the best examination of his chemical work being that of Marie Boas, *Robert Boyle and Seventeenth-Century Chemistry* (Cambridge Univ. Press, Cambridge, 1958). The best biography is Louis T. More, *The Life and Works of the Honourable Robert Boyle* (Oxford Univ. Press, New York, 1944). Boyle's collected works are available in two standard editions, both edited by Thomas Birch, *The Works of the Honourable Robert Boyle* (Millar, London, 5 vols. folio, 1744 and 6 vols. quarto, 1772). Both editions contain a biography by Birch. Peter Shaw was the editor of an abridged edition, *The Philosophical Works of the Honourable Robert Boyle* (Innys & Manly, London, 2nd ed., 1738). For a bibliography see John Fulton, *A Bibliography of the Honourable Robert Boyle* (Clarendon, Oxford, 2nd ed.,

1961). *The Sceptical Chymist* is very rare in the original editions but a convenient reprint is available in the Everyman's Library ed. (Dent, London, 1911). A. J. Ihde, *Chymia*, **9**, 47(1963) deals with some of Boyle's thoughts on transmutation.

On seventeenth-century atomism the principal works are: Kurd Lasswitz, *Geschichte der Atomistik im Mittelalter bis Newton* (Voss, Hamburg, 2 vols., 1890); J. C. Gregory, *A Short History of Atomism* (Black, London, 1931); A. G. Van Melsen, *From Atomos to Atom*, transl. from the Dutch by H. J. Koren (Duquesne Univ. Press, Pittsburgh, 1952). Good papers dealing with the materialistic philosophy are: T. S. Kuhn, *Isis*, **43**, 12(1952) and Marie Boas, *Osiris*, **10**, 412(1952).

A fine study of the founding of scientific societies was made by Martha Ornstein, *The Role of Scientific Societies in the Seventeenth Century* (Univ. of Chicago Press, Chicago, 2nd ed., 1928). Also see: Dorothy Stimson, *Scientists and Amateurs, A History of the Royal Society* (Schuman, New York, 1948); and Harcourt Brown, *Scientific Organizations in Seventeenth-Century France* (Williams & Wilkins, Baltimore, 1934).

On the early history of the phlogiston theory see: J. H. White, *The History of the Phlogiston Theory* (Arnold, London, 1932); and Joshua C. Gregory, *Combustion from Heracleitos to Lavoisier* (Arnold, London, 1934).

CHAPTER 2. PNEUMATIC CHEMISTRY

There are no extensive works dealing exclusively with the developments of this period, but most works treating the "reign" of phlogiston and its rejection pay some attention to the pneumatic chemists' studies.

Stephen Hales life and scientific contributions are developed in A. E. Clark-Kennedy, *Stephen Hales, An Eighteenth-Century Biography* (Cambridge Univ. Press, Cambridge, 1929). This work is very meager on Hales' chemical ideas, and it is best to go directly to Hales' own works, *Vegetable Staticks* (Innys and Woodward, London, 1727) and *Haemastaticks* (1733) for his chemistry. Short studies of Hales and his scientific work are: H. Guerlac, *Archives Internationale Hist. Sci.*, **15**, 393(1951); R. Foregger, *Anaesthesia*, **11**, 235(1956); G. E. Burget, *Ann. Med. Hist.*, **7**, 109(1925); D. F. Harris, *Sci. Monthly*, **3**, 440(1904); and P. M. Dawson, *Bull. J. Hopkins Hosp.*, **15**, 185(1904).

Joseph Black's thesis is available in the English translation of A. Crum Brown, communicated by Leonard Dobbin to *J. Chem. Educ.*, **12**, 225, 268 (1935). The paper on magnesia alba was first published in *Essays and Observations, Physical and Literary. Read before a Society in Edinburgh, and Published by them*, **2**, 157(1756); the paper is most readily available as *Alembic Club Reprint*, No. 1, Edinburgh, 1898. Black's lectures were edited and published after his death by John Robison, *Lectures on the Elements of Chemistry Delivered in the University of Edinburgh; by the late Joseph Black* (Creech, Edinburgh, 2 vols., 1803). The only book-length biographies of Black are Wm. Ramsay, *The Life and Letters of Joseph Black* (Constable, London, 1918) and W. Ramsay, *Joseph Black, A Discourse* (Glasgow, 1904). Short biographical accounts include the following: Max Speter in G. Bugge, *Das Buch* *der grossen Chemiker* (Verlag Chemie, Berlin, 1929, vol. I, pp. 240–252); Wm. A. Tilden, *Famous Chemists* (Routledge, London, 1921, p. 22); D. McKie, *Annals of Sci.*, **1**, 101(1936); and E. W. J. Neave, *Isis*, **25**, 372(1936). An excellent study of Black's work on fixed air has been made by H. Guerlac, *Isis*, **48**, 124, 433(1957). R. Foregger, *Anaesthiology*, **18**, 257(1957) has also reported on the discovery of carbon dioxide.

The life of Cullen, Black's teacher, is described in John Thomson, *An Account of the Life, Lectures, and Writings of William Cullen, M.D.* (Blackwood, Edinburgh and London, 1859).

Some of Scheele's publications are available in English in *The Chemical Essays of Carl Wilhelm Scheele* (Murray, London, 1786); and C. W. Scheele, *Chemical Observations and Experiments on Air and Fire* (J. Johnson, London, 1780). Tilden (*op. cit.* (p. 53), has a biographical chapter on Scheele, as does G. Lockemann in Bugge (*op. cit.*, vol. 1, p. 274). Georg Urdang's *Pictorial Life History of the Apothecary Chemist Carl Wilhelm Scheele* (Am. Inst. of the Hist. of Pharmacy, Madison, Wis., 1942) is a short but informative brochure. O. Zekert's *C. W. Scheele, sein Leben und seine Werke* (Mittenwald, 3 vols., 1931–1933) and F. A. Flückinger's *Zur Erinnerung an Scheele, ein Jahrhundert nach seinem Ableben* (Halle, 1886) are useful German works.

Comprehensive biographies of Cavendish are George Wilson, *The Life of the Honourable Henry Cavendish* (Cavendish Society, London, 1851), and A. J. Berry, *Henry Cavendish, his Life and Scientific Work* (Hutchison, London, 1960). Georg Lockemann has a short biography in Bugge (*op. cit.*, vol. 1, p. 253), and B. Jaffe, *Crucibles* (Simon & Schuster, New York, 1930) has a popular treatment of his life. A more comprehensive picture of the man in his times can be seen in W. R. Aykroyd, *Three Philosophers, Lavoisier, Priestley and Cavendish* (Heinemann, London, 1935). An account of Cavendish's work on air will be found in *Alembic Club Reprint* No. 3.

Joseph Priestley has been the subject of extensive study. The earliest biographical work is John Corry's *The Life of Joseph Priestley* (Wilks, Grafton, Birmingham, 1804) which only barely anticipates the publication of Priestley's autobiographical notes, *Memoirs of Dr. Joseph Priestley, to the Year 1795, written by Himself; with a Continuation to the Time of His Decease, by his Son, Joseph Priestley, And Observations on His Writings, by Thomas Cooper . . . and the Rev. William Christie* (Allenson, London, 1806). John F. Fulton and Charlotte H. Peters have prepared an excellent bibliography of Priestley's educational and scientific works which is published in *Papers of the Bibliographical Soc. of America*, **30**, 150(1936), with a lithoprinted addendum of *Works* (Yale Univ. School of Medicine, New Haven, 1937). A large part of the scientific correspondence is available in H. C. Bolton, *Scientific Correspondence of Joseph Priestley* (Collins, Philadelphia, 1892). J. T. Rutt's *Life and Correspondence of Joseph Priestley* (Hunter, London, 2 vols., 1831–1832) deals principally with the theological activities and includes hundreds of letters. Other biographical works are: Thomas E. Thorpe, *Joseph Priestley* (Dent, London, 1906); D. H. Peacock, *Joseph Priestley* (Macmillan, New York, 1919); Anne Holt, *A Life of Joseph Priestley* (Oxford Univ. Press, London, 1931); R. M. Caven, *Joseph Priestley, 1733–1804* (Inst. of Chem. of Great Britain, London, 1933); E. F. Smith, *Priestley in America, 1794–1804* (Blakiston's,

Philadelphia, 1920); and J. G. Gillam, *The Crucible, The Story of Joseph Priestley* (Hall, London, 1954). G. Lockemann has a chapter on Priestley in Bugge (*op. cit.*, vol. 1, p. 263). C. A. Browne deals with Priestley's American activities in *J. Chem. Educ.*, **4**, 159, 184(1927). This issue of the *Journal of Chemical Education* has several other papers dealing with Priestley's American activities and influence. Bicentennial observations by P. J. Hartog, A. N. Meldrum, and H. Hartley are published in *J. Chem. Soc.*, **1933**, 896–920. An account of Wm. Cobbett's (Peter Porcupine) attack on Priestley was published by Lyman C. Newell, *J. Chem. Educ.*, **10**, 151(1933). Philip Hartog, *Ann. Sci.*, **5**, 25(1941), makes a good comparison of the views of Priestley and Lavoisier. Accounts of the Birmingham riots will be found in Priestley's *Memoirs* and in K. Loewenfeld, *Mem. & Proc., Manchester Lit. & Philos. Soc.*, **57**, Part 19 (1913), and F. R. Maddison, *Notes Royal Soc. London*, **12**, 98(1956). R. E. Schofield, *Ann. Sci.*, **13**, 148(1957), presents a good study of Priestley's scientific background. J. S. Hepburn, *J. Franklin Soc.*, **244**, 63, 95(1947) gives an account of Priestley's Pennsylvania associations. H. Guerlac, *J. Hist. Med.*, **12**, 1(1957) deals with the early work on gases.

CHAPTER 3. LAVOISIER AND THE CHEMICAL REVOLUTION

The literature is rich in works dealing with the period of Lavoisier. Indeed the Chemical Revolution has received perhaps more attention than any other period in the whole history of chemistry. Despite this, there are numerous unanswered questions and much investigation remains to be done.

An excellent bibliography of Lavoisier's works has been prepared by Denis I. Duveen and Herbert S. Klickstein, *A Bibliography of the Works of Antoine Laurent Lavoisier, 1743–1794* (Dawson, and Weil, London, 1954).

Three of the major works of Lavoisier were quickly translated into English: *Essays Physical and Chemical*, trans. by Thomas Henry (J. Johnson, London, 1776); *Method of Chymical Nomenclature*, trans. by J. St. John (Kearsley, London, 1788); and *Elements of Chemistry*, trans. by Robert Kerr (Creech, Edinburgh, 1790)—the latter being available in a 1945 reprint (Edwards Bros., Ann Arbor, Mich., for the St. John's College Great Books Program).

Madame Lavoisier brought together her husband's last researches as the *Mémoires de Chimie* [Madame Lavoisier] (Paris, 1805). Publication of an edition of collected works was undertaken by the French Ministry of Public Instruction, *Oeuvres de Lavoisier* (Académie des Science, Paris, 6 vols., 1862–1893). Volume 1 contains the *Opuscules* and the *Traité*; vol. 2, the *Mémoirs*; vols. 3 and 4 miscellaneous chemical papers and reports; vol. 5 studies on geology, mineralogy, and gunpowder; vol. 6 reports and studies on non-chemical subjects, most of them made in the later part of his life. Very recently publication of Lavoisier's correspondence has been undertaken by the Académie under the editorship of René Fric. The correspondence makes up vols. 7 and 8 of the *Oeuvres*. The first part appeared in 1955 (Michel, Paris); the second in 1957.

Books dealing with Lavoisier's life and work include: Edouard Grimaux's *Lavoisier, 1743–1794* (Alcan, Paris, 1888) is a standard work based upon correspondence, manuscripts, and family papers but is short on the scientific work and marred by many misprints and minor errors; Marcelin Berthelot's *La Révolution Chimique, Lavoisier* (Alcan, Paris, 1890) is based on a study of Lavoisier's notebooks, with extensive quotations; Douglas McKie, *Antoine Lavoisier, The Father of Modern Chemistry* (Lippincott, Philadelphia, 1935); McKie's *Antoine Lavoisier, Scientist, Economist, Social Reformer* (Schuman, New York, 1952) should not be confused with the previous title which deals much more extensively with the scientific work; J. A. Cochrane, *Lavoisier* (London, 1931); M. L. Foster, *Life of Lavoisier* (Smith College Monograph, Northampton, Mass., 1926); Sidney J. French, *Torch and Crucible, The Life and Death of Antoine Lavoisier* (Princeton Univ. Press, Princeton, 1941); Maurice Daumas, *Lavoisier* (Gallimard, Paris, 1941); Daumas, *Lavoisier, théoricien et expérimenteur* (Presses universitaires, Paris, 1955); Lucien Scheler, *Lavoisier et la Révolution française* (Hermann, Paris, 1956); René Dujarric de la Rivière, *Lavoisier, économiste* (Masson, Paris, 1949); de la Rivière and M. Chabrier, *La vie et l'oeuvre de Lavoisier d'apres ses écrits* (Michel, Paris, 1959).

Among the studies of Lavoisier's scientific activities and influence, the following are particularly noteworthy: Andrew N. Meldrum, *The Eighteenth-Century Revolution in Science—The First Phase* (Longmans, Green, Calcutta, 1930); Hélène Metzger, *La philosophie de la matière chez Lavoisier* (Hermann, Paris, 1935); James B. Conant, *The Overthrow of the Phlogiston Theory* (Harvard Univ. Press, Cambridge, 1950); Max Speter, *Lavoisier und seine Vorläufer* (Enke, Stuttgart, 1910). The influence of Lavoisier's work in Germany is carefully examined in Georg Kahlbaum and August Hoffmann, *Die Einführung der Lavoisier'schen Theorie im Besonderen in Deutschland* (Barth, Leipzig, 1897). Henry Guerlac's *Lavoisier—The Crucial Year* (Cornell Univ. Press, Ithaca, 1961) is a splendid study of the background to Lavoisier's work on combustion in 1772.

The best study of the decline of the phlogiston theory is that of J. R. Partington and D. McKie, *Annals of Science*, **2**, 361(1937); **3**, 1, 337(1938); and **4**, 113(1939). Other works giving considerable attention to the problem are J. H. White, *The History of the Phlogiston Theory* (Arnold, London, 1932); and Joshua C. Gregory, *Combustion from Heracleitos to Lavoisier* (Arnold, London, 1934). Guyton de Morveau's conversion to Lavoisier's doctrine is examined by D. I. Duveen and H. S. Klickstein, *Osiris*, **12**, 342(1956). These authors have made a number of other fine studies of the spread of Lavoisier's ideas: *Ann. Science*, **10**, 321(1954); *Isis*, **45**, 278, 368(1954); *Proc. Am. Philos. Soc.*, **98**, 466(1954); *Ann. Sci.*, **11**, 103, 271(1955) and **13**, 30(1957)—deals with the relations of Lavoisier and Benjamin Franklin; *J. Chem. Educ.*, **31**, 60(1954), **34**, 502(1957), **35**, 233, 470(1958)—deals with Lavoisier and the French Revolution; *Bull. Hist. Med.*, **29**, 164(1955)—deals with medical contributions. The following by Duveen are principally biographical: *Isis*, **42**, 233(1951); *Notes Royal Soc. London*, **13**, 56(1958); *Chymia*, **4**, 13(1953)—deals with Madame Lavoisier.

Henry Guerlac has a good study of Lavoisier's biographers in *Isis*, **45**, 51(1954) and deals with his education in *Isis*, **47**, 211(1956). W. A. Smeaton deals with the education of Lavoisier and Fourcroy in *Ann. Science*, **11**,

309(1955). Duveen traces the publication history of the *Traité* in *Isis*, **41**, 168(1950). The influence on pharmacy is examined by Georg Urdang, *Am. J. Pharm. Educ.*, **18**, 216(1954). D. McKie, *Notes Royal Soc. London*, **7**, 1(1949) publishes the text of sixteen letters seized at Lavoisier's home during the Terror. M. Daumas, *Arch. int. hist. sci.*, **38**, 19(1957) deals with the treatment of Lavoisier by historians.

Details of Lavoisier's scientific work are examined in the following: A. N. Meldrum, *Archeion*, **14**, 15, 246(1932); G. Bertrand, *Arch. int. hist. sci.*, **29**, 807(1950); J. R. Partington, *Ann. Science*, **9**, 96(1953) and *Nature*, **178**, 1360(1956). A somewhat broader examination of his viewpoint is to be found in the following: T. L. Davis, *Isis*, **16**, 82(1932); S. E. Toulmin, *J. Hist. Ideas*, **18**, 205(1957); S. J. French, *J. Chem. Educ.*, **27**, 83(1950); T. D. Phillips, *Isis*, **46**, 53(1955); Harold Hartley, *Proc. Royal Soc. London*, **134B**, 348(1947); M. Daumas, *Thales*, **6**, 69(1949–1950); Aldo Mieli, *Archeion*, **14**, 51(1932); H. Metzger, *Archeion*, **14**, 31(1932); and Pierre Lemay, *Prog. med.* (Paris), **85**, 228(1957).

There are numerous chapter-length biographies of Lavoisier; it usually is sufficient to consult one of the standard collections of biographies of scientists. For sketches of his principal contemporaries, see notes for Chapter 2. Besides those mentioned there, Bugge has chapters dealing with the lives of Berthollet, Fourcroy, and Klaproth. Several biographies of interest are: Georges Bouchard, *Guyton Morveau, chimiste et conventionnel 1737–1816* (Libraire Acad. Perrin, Paris, 1938); R. D. de la Rivière, *E. I. du Pont de Nemours, élève de Lavoisier* (Libraire de Champs-Elysées, Paris, 1954); P. Cap, *G. F. Rouelle* (Jannet, Paris, 1842); L. R. Gottschalk, *Jean Paul Marat* (Greenberg, New York, 1927); W. A. Smeaton, *Fourcroy, Chemist and Revolutionary, 1755–1809* (the author, London, 1962).

The development of the metric system is treated in Adrien Favre, *Les origines du système métrique* (Presses Universitaires, Paris, 1931); Edward Nicholson, *Men and Measures, A History of Weights and Measures* (Smith, Elder, London, 1912); and John Perry, *The Story of Standards* (Funk & Wagnalls, New York, 1955).

Charles Gillespie, *Behavorial Science*, **4**, 67(1959) has a good examination of science during the French Revolution. Henry Guerlac, *Chymia*, **5**, 73(1959) reports on some antecedents to the Chemical Revolution. There are several studies of scientific activities in the early days of the École polytechnique; A. Binet, *Histoire de l'École Polytechnique* (Paris, 1892); Jean Pierre Callot, *Histoire de l'École Polytechnique, ses legendes, ses traditions, sa gloire* (Paris, 1959); W. A. Smeaton, *Ann. Science*, **10**, 224(1954), deals with chemical instruction.

CHAPTER 4. CHEMICAL COMBINATION AND THE ATOMIC THEORY

For works dealing with the early phases of analytical chemistry, the reader is referred to the bibliographic notes for Chapter 11.

An excellent study of the growth of knowledge regarding chemical composition was made six decades ago by Ida Freund, *The Study of Chemical*

Composition, An Account of its Method and Historical Development (Cambridge Univ. Press, Cambridge, 1904). Although of ancient vintage, it represents a careful and extensive study of the whole problem of chemical combination as it was faced by chemists during the nineteenth century. Another useful work of the same period is M. M. Pattison Muir's *A History of Chemical Laws and Theories* (Wiley, New York, 1907). This is broader in its coverage of chemical concepts than Freund but not so thorough in its treatment. Muir's *Treatise on the Principles of Chemistry* (Cambridge Univ. Press, Cambridge, 1884) also retains some value.

A splendid study of the general problem has been published as a Harvard Case History (no. 4) in Experimental Science under the authorship of Leonard K. Nash: *The Atomic-Molecular Theory* (Harvard Univ. Press, Cambridge, 1950). Nash has extended his study in a paper in *Isis*, **47**, 101(1956). A. N. Meldrum made a useful set of studies published in the *Memoirs of the Literary and Philosophical Society of Manchester*, **55**, nos. 3, 4, 5, 6, 19, and 22 (1910–1911). Also see Meldrum's small books, *Avogadro and Dalton* (Clay, Edinburgh, 1904) and *The Development of the Atomic Theory* (Oxford Univ. Press, London, n. d.). H. E. Roscoe and A. Harden's *A New View of the Origin of Dalton's Atomic Theory* (Macmillan, London, 1896) is of special value since it contains extensive quotations and some photographic reproductions from Dalton's notebooks, the latter having been destroyed during bomber raids on Manchester during World War II. Other works bearing on Dalton's early work on atomic theory are: Thomas Thomson, *History of Chemistry* (Colburn and Bentley, London, 2 vols., 1830–1831, vol. 2, Chap. 6); William C. Henry, *Memoir of the Life and Scientific Researches of John Dalton* (Cavendish Soc., London, 1854); H. E. Roscoe, *John Dalton and the Rise of Modern Chemistry* (Macmillan, London, 1895); R. A. Smith, *Memoir of John Dalton and the History of Atomic Theory up to his Time* (Ballière, London, 1856); J. P. Millington, *John Dalton* (Dent, London, 1906); L. J. Neville-Polley, *John Dalton* (Macmillan, London, 1920); and E. M. Brockbank, *John Dalton and some unpublished letters of personal and scientific interest, with additional information about his colour-vision and atomic theory* (Manchester Univ. Press, Manchester, 1944). Numerous short papers deal with Daltonian atomism, several of importance being: J. R. Partington, *Ann. Sci.*, **4**, 245(1939) and *Scientia*, **49**, 221(1955); H. F. Coward, *J. Chem. Educ.*, **4**, 23(1927); and D. I. Duveen and H. S. Klickstein, *ibid.*, **32**, 333(1955); Frank Greenway, *Manchester. Memoirs*, **100**, 1(1959). Dalton's *New System of Chemical Philosophy* [vol. 1, part 1, Russell, Manchester, and Bickerstaff, London, 1808; part 2, Russell & Allen, Manchester, and Bickerstaff, London, 1810; vol. 2, part 1 (all publ.), Wilson, London, 1827] has been reproduced in facsimile edition by Dawson (London, 1953).

Higgins' atomic theory has been examined in the following publications: J. H. White, *Sci. Prog.*, **24**, 300(1929); E. R. Atkinson, *J. Chem. Educ.*, **17**, 3(1940); J. R. Partington, *Nature*, **167**, 120(1951); F. Soddy, J. R. Partington, *ibid.*, p. 734; T. S. Wheeler, *Endeavour*, **11**, 47(1952). Higgins' books are quite scarce, but the text on atomism is now available in T. S. Wheeler and J. R. Partington, *The Life and Work of William Higgins* (Pergamon, New York, 1960).

The only extensive study of Klaproth is that of Georg E. Dann, *Martin Heinrich Klaproth, Ein deutscher Apotheker und Chemiker* (Akademie Verlag, Berlin, 1958). Elsie G. Ferguson has a good paper on the analytical contributions of Bergman, Klaproth, Wollaston, and Vauquelin, *J. Chem. Educ.*, **17**, 554(1940) and **18**, 3(1941). For the contributions of Berzelius see the notes for Chapter 6. There are no extended studies of Berthollet or Proust. Freund (*op. cit.*) has two chapters dealing with their work and the controversy, and Berthollet generally receives considerable attention in works dealing with Lavoisier.

Avogadro is the subject of I. Guareschi, *Amadeo Avogadro und die Molekulartheorie* [Barth, Leipzig, 1903 (vol. 7 of Kahlbaum's Monographien aus der Geschichte der Chemie)] and of Meldrum's *Avogadro and Dalton* (cited above). Gay-Lussac is treated in Edmond Blanc and Léon Delhoume, *La vie émouvante et noble de Gay-Lussac* (Gauthier-Villars, Paris, 1950). Good biographical chapters of Geoffroy, Klaproth, Berthollet, Proust, Fourcroy and Vauquelin, Richter, Dalton, and Gay-Lussac and Thenard are found in Bugge, *Das Buch der grossen Chemiker* (vol. 1).

Alembic Club Reprint no. 2, *Foundations of Atomic Theory*, contains papers of Dalton, Wollaston, and T. Thomson. Reprint no. 4, *Foundations of Molecular Theory*, has pertinent papers of Dalton, Gay-Lussac, and Avogadro.

CHAPTER 5. ELECTROCHEMISTRY AND THE DUALISTIC THEORY

A number of books deal with the history of electricity, the earliest being Joseph Priestley's *History and Present State of Electricity, with Original Experiments* (J. Dodsley [etc.], London, 1767). This book, of course, deals with the subject before Galvani and Volta's experiments but furnishes a very detailed background up to that time. Edmund T. Whittaker's *History of the Theories of Aether and Electricity* (Nelson, London, 2 vols., 1951–1953) deals with the conceptual aspects of the subject to 1925. Several works treat the subject from a biographical point of view: M. F. O'Reilly (Potamian) and J. J. Walsh, *Makers of Electricity* (Fordham Univ. Press, New York, 1909); Dorothy M. Turner, *Makers of Science, Electricity and Magnetism* (Oxford Univ. Press, London, 1927); Rollo Appleyard, *Pioneers of Electrical Communication* (Macmillan, London, 1930); Dayton C. Miller, *Sparks, Lightning, Cosmic Rays. An Anecdotal History of Electricity* (Macmillan, New York, 1939). There are also sections on the history of electricity in such books as Henry Crew, *The Rise of Modern Physics* (Williams & Wilkins, Baltimore, 1928) and Max von Laue (trans. by Ralph E. Oesper), *History of Physics* (Academic Press, New York, 1950). Charles G. Fraser's *Half Hours with Great Scientists* (Reinhold, New York, 1948) contains excerpts from the writings of many of the great physicists, including the pioneers in electrical research.

Bern Dibner, *Galvani-Volta* (Burndy Library, Norwalk, Conn., 1952) has a good treatment of the researches of these two physicists. Also see C. J. Brockman's papers on the origin of voltaic electricity in *J. Chem. Educ.*, **5**, 549(1928), **6**, 1293, 1726(1929).

The lives of Davy and Faraday have received extensive study, and collected works of both have been published. John Davy, brother of Sir Humphry, served as editor for the *Collected Works of Humphry Davy* (Smith, Elder, London, 9 vols., 1839–1840). John Davy also published the *Memoirs of the Life of Sir Humphry Davy, Bart.* (Longman, Rees, Orme, Green & Longman, London, 2 vols., 1836) and *Fragmentary Remains, Literary and Scientific, of Sir Humphry Davy, with a sketch of his life and selections from his correspondence* (Churchill, London, 1858). The standard biography is that of J. A. Paris (*The Life of Humphry Davy*, Colburn & Bentley, London, 2 vols., 1831). Other biographies are: H. Mayhew, *The Wonders of Science, or, Young Humphry Davy* (Harper, New York, 1856); T. E. Thorpe, *Humphry Davy, Poet and Philosopher* (Macmillan, London, 1896); J. C. Gregory, *The Scientific Achievements of Sir Humphry Davy* (Oxford Univ. Press, London, 1930); and W. Prandtl, *Humphry Davy. Jöns Jacob Berzelius* (Wissenschaftliche Verlagsgesellschaft, Stuttgart, 1948).

Most of Faraday's papers were published by him in collected form in his later years: *Experimental Researches in Chemistry and Physics* (Taylor, London, 1859); *Experimental Researches in Electricity* (Taylor, London, 3 vols., 1839–1855). The notebooks, which remained in the possession of the Royal Institution, were published as *Faraday's Diary* (G. Bell, London, 7 vols., 1932–1936). Biographies include: H. Bence Jones, *The Life and Letters of Faraday* (Longmans, Green, London, 2 vols., 1870); J. H. Gladstone, *Michael Faraday* (Macmillan, London, 1872); W. Jerrold, *Michael Faraday, Man of Science* (Partridge, London, 1891); John Tyndall, *Faraday as a Discoverer* (Appleton, New York, 1894); Sylvanus P. Thompson, *Michael Faraday, His Life and Work* (Cassell, London, 1898); *Michael Faraday (1791–1867)* (Small, Maynard, Boston, 1925); R. A. Hadfield, *Faraday and his Metallurgical Researches* (Chapman & Hall, London, 1931); R. Appleyard, *A Tribute to Michael Faraday* (Constable, London, 1931); W. Cramp, *Michael Faraday and Some of his Contemporaries* (Pitman, London, 1931); T. Martin, *Faraday* (Duckworth, London, 1934); and T. Martin, *Faraday's Discovery of Electro-Magnetic Induction* (Arnold, London, 1949).

There are also numerous short treatments of the lives of Davy and Faraday. Wilhelm Ostwald wrote the chapters on these men for Bugge, *Das Buch der grossen Chemiker*. Robert Siegfried is author of the Davy chapter in E. Farber's *Great Chemists* (Interscience, New York, 1961); A. J. Ihde prepared the chapters on Faraday and Berzelius. Wm. A. Tilden has chapters on Davy, Dalton, Gay-Lussac, Proust, Berzelius, Faraday, and Avogadro in his *Famous Chemists* (Routledge, London, 1921). Davy's chemical philosophy is examined by R. Siegfried in *Chymia*, **5**, 193(1959). Faraday's chemical contributions have been examined by L. C. Newell, *J. Chem. Educ.*, **3**, 1248(1926) and **8**, 1493(1931) and his electrochemical laws by Rosemary G. Ehl and A. J. Ihde, *ibid.*, **31**, 226(1954). Also see the latter paper for information on Carlo Matteucci who has received little attention in the English literature although there is an Italian biography by N. Bianchi, *Carlo Matteucci e l'Italia del suo tempo* (Turin, 1874).

Berzelius' dualistic theory has not received any extensive modern study. References to Berzelius will be found primarily in the bibliographic notes for Chapter 6.

CHAPTER 6. THE PERIOD OF PROBLEMS

Much of the best material dealing with this period is to be found in the works of Freund, Muir, and Nash which are referred to in the bibliographic notes for Chapter 4.

Although Berzelius was clearly the key figure in the chemistry of this period, there has not been an adequate study of his role. The one fairly extensive biographical work is *Jöns Jacob Berzelius, Autobiographical Notes* edited by H. G. Söderbaum and translated into English by Olaf Larsell (Williams & Wilkins, Baltimore, 1934). A five-volume *Bibliografi över J. J. Berzelius* has been prepared by Arne Holmberg (Stockholm, 1933–1953). The famous textbook went through several editions, with considerable expansion with each edition, and was even more widely circulated in its German and French translations. Berzelius' monument to the chemistry of his times is, of course, the *Jahres-Bericht über die Fortschritte der physischen Wissenschaften* which first appeared in 1822 and was issued annually to the time of his death. It was published in Swedish but is better known in the German translation prepared for many years by Wöhler (1823–1841). For some remarks regarding the history of the *Jahres-Bericht* see W. Ostwald, *J. Chem. Educ.*, **32**, 373(1955). A number of Berzelius' early papers appeared originally or in reprint form in Thomson's *Annals of Philosophy*.

There are two published collections of Berzelius' letters: J. Carriere served as editor for *Berzelius und Liebig. Ihre Briefe von 1831–1845* (Lehmann, Munich and Leipzig, 1893); Otto Wallach edited the *Briefwechsel zwischen J. Berzelius und F. Wöhler* (Engelmann, Leipzig, 2 vols., 1902). Impressions of the Stockholm laboratory are to be found in Wöhler's recollections published in *Ber.*, **8**, 838(1875) and in *Am. Chemist*, **6**, 131(1875). H. Rose's recollections appear in translation in *Am. J. Science*, [2], **16**, 1, 173, 305(1853). Biographical chapters of value are: H. G. Söderbaum in Bugge, *Das Buch der grossen Chemiker*; W. Prandtl in his book, *Humphry Davy. Jöns Jacob Berzelius* (Wissenschaftliche Verlagsgesellschaft, Stuttgart, 1948); Wm. Tilden in his *Famous Chemists*; and A. J. Ihde in E. Farber, *Great Chemists*. Also see R. Winderlich, *J. Chem. Educ.*, **25**, 500(1948); H. Hartley, *Notes & Records, Royal Soc.*, **6**, 161(1949); Georg Kahlbaum, *Mitt. Gesch. Med. Naturw.*, **3**, 277(1904); P. Walden, *Z. angew. Chem.*, **43**, 325, 351, 366(1930).

The life and work of Dulong is examined by Pierre Lemay and Ralph E. Oesper, *Chymia*, **1**, 171(1948). For the life of Mitscherlich see R. Winderlich in *J. Chem. Educ.*, **26**, 358(1949); A. Williamson, *J. Chem. Soc.*, **17**, 440(1864); and G. Bugge in vol. 1 of *Das Buch der grossen Chemiker*. His works are collected in the *Gesammelte Schriften von Eilhard Mitscherlich* (Mittler, Berlin, 1896). On Gmelin see P. Walden, *J. Chem. Educ.*, **31**, 534(1954); and E. Pietsch and E. Beyer, *Ber.*, **72A**, 5(1939). For Prout see S. Glasstone, *J. Chem. Educ.*, **24**, 478(1947); O. T. Benfey, *ibid.*, **29**, 78(1952); and R. Siegfried, *ibid.*, **33**, 263(1956). The two papers of Prout in the *Annals of Philosophy*, **7**, 111, 343(1816) are more readily accessible in *Alembic Club Reprint* no. 20, which also carries related papers of Stas and Marignac. For Dumas see the notes for the next chapter. The work of Stas is treated by J. Mallet, *J. Chem. Soc.*,

782   CHAPTER BIBLIOGRAPHIC NOTES

63, 1(1893); E. Morley, *J. Am. Chem. Soc.*, **14**, 173(1892); and J. Timmermans, *J. Chem. Educ.*, **15**, 353(1935).

CHAPTER 7. ORGANIC CHEMISTRY I: RISE OF ORGANIC CHEMISTRY

Organic is the only area of chemistry which has attracted sufficient historical interest to have several books written on the subject. Except for Walden's, however, they fail to reach into the twentieth century. The earliest and least adequate of the histories is Carl Schorlemmer's *The Rise and Development of Organic Chemistry* (Macmillan, London, rev. ed., 1894). A much better history, but available only in German, is Carl Graebe's *Geschichte der organischen Chemie* (Springer, Berlin, 1920). This volume carries the story of organic chemistry to 1880 and was to have been followed by a second volume which failed to materialize before the author's death in 1927. A purported volume 2 was published in 1941 under the authorship of Paul Walden, *Geschichte der organischen Chemie seit 1880* (Springer, Berlin, 1941). It is in essence a new treatise which places primary stress on experimental discoveries rather than on interpretation. Still another book of considerable merit is Edvard Hjelt's *Geschichte der organischen Chemie von ältesten Zeit bis zur Gegenwart* (Vieweg, Braunschweig, 1916).

There are two very useful tabulations which are exceedingly helpful in searching out information regarding organic chemistry. Edmund O. von Lippmann's *Zeittafeln zur Geschichte der organischen Chemie* (Springer, Berlin, 1921) is a chronological tabulation of significant contributions to the subject, starting in 1500 and ending in 1890. It is reasonably inclusive, includes references to the original literature, and has good subject and author indexes. Paul Walden's *Chronologische Übersichtstabellen zur Geschichte der Chemie von den ältesten Zeiten bis zur Gegenwart* (Springer, Berlin, 1952) is more ambitious in that it attempts to do a similar job for the whole field of chemistry. It succeeds quite well in this objective as far as organic chemistry is concerned but is not all that might be desired with respect to other fields. Arranged chronologically, it is divided into time periods and after 1800 into the fields of organic and inorganic chemistry. References to the original literature are included, and there is an author index but no subject index.

There is only one book-length biography of Wöhler, J. Valentin's *Friedrich Wöhler* (Wissenschaftliche Verlagsgesellschaft, Stuttgart, 1949). A. W. Hofmann has a long biographical study in *Ber.*, **15**, 3127(1882), and there is a good chapter by R. Winderlich in Bugge, *Das Buch der grossen Chemiker* (vol. 2). Edgar Fahs Smith tells of his student days in Wöhler's laboratory in *J. Chem. Educ.*, **5**, 1554(1928). H. S. van Klooster discusses Wöhler's American pupils, *ibid.*, **21**, 158(1944). B. Lepsius has a short paper in *Ber.*, **65A**, 89(1932). Contemporary reactions to the synthesis of urea are examined by W. H. Warren, *J. Chem. Educ.*, **5**, 1539(1928), Douglas McKie, *Nature*, **153**, 608(1944), and Ernest Campaigne, *J. Chem. Educ.*, **32**, 403(1955). The original paper was published in *Ann. Phys.*, **12**, 253(1828) and is available in English translation in H. M. Leicester and H. S. Klickstein's *A Source Book in Chemistry* (McGraw-Hill, New York, 1952, p. 309). The work with Liebig on oil of bitter almonds was published in *Ann.*, **3**, 249(1832) and is translated in Leicester and Klickstein (*op. cit.*, p. 312). Wm. Prandtl's *Deutsche Chemiker in der ersten Hälfte des neunzehnten Jahrhunderts* (Verlag Chemie, Weinheim Bergstrasse, 1956) has good chapters on Wöhler and Liebig.

Liebig's life and work have attracted widespread interest among biographers. August Wm. von Hofmann's *The Life Work of Liebig in Experimental and Philosophic Chemistry* (Macmillan, London, 1875) is a short sketch by one of his most talented students. W. A. Shenstone's *Justus von Liebig; His Life and Work* (Cassell, London, Macmillan, New York, 1895) is the earliest of subsequent works which include: J. Volhard, *Justus von Liebig* (Barth, Leipzig, 1909); G. F. Knapp, *Justus von Liebig nach dem Leben gezeichnet* (Munich, 1903); R. Blunck, *Justus von Liebig* (Limpert, Berlin, 1938); and Herta von Dechend, *Justus von Liebig in eigenen Zeugnissen und Solchen seiner Zeitgenossen* (Verlag Chemie, Weinheim Bergstrasse, 1953). The latter is a compilation of excerpts from his letters and those of his correspondents. R. Winderlich is the author of a biographical chapter in Bugge (*op. cit.*, vol 2). For short biographical sketches, see A. W. Hofmann in *J. Chem. Soc.*, **28**, 1065(1875); G. F. Knapp, *Ber.*, **36**, 1315(1903); R. E. Oesper, *J. Chem. Educ.*, **4**, 1461(1927); E. R. Schiertz, *ibid.*, **8**, 223(1931); E. Berl, *ibid.*, **15**, 553(1938); and C. A. Browne in F. R. Moulton, ed., *Liebig and after Liebig* (Am. Assoc. Adv. Sci., Washington, 1942). Liebig's letters have been partially published: J. Carrière, ed., *Berzelius und Liebig, Ihre Briefe von 1831–1845* (Lehmann, Munich and Leipzig, 1893); A. W. von Hofmann, ed., with the assistance of Emilie Wöhler, *Aus Justus Liebigs und Friedrich Wöhlers Briefwechsel in dem Jahren 1829–1873* (Vieweg, Braunschweig, 2 vols., 1888). Selected letters from the latter have recently been reprinted under the editorship of Robert Schwarz by Verlag Chemie (Weinheim Bergstrasse, 1958).

Liebig's books appeared in numerous editions, both in German and in translation. For English editions of the books on organic and agricultural chemistry, see Browne's paper in *Liebig and after Liebig*, p. 1.

For the life and work of Chevreul see: Albert B. Costa, *Michel Eugene Chevreul, Pioneer in Organic Chemistry* (Wis. Hist. Soc., Madison, 1961); Mme. de Champ, *Michel-Eugène Chevreul, vie intimé, 1786–1879* (Paris, 1930); A. Bourgougnon, *J. Am. Chem. Soc.*, **11**, 71(1889); A. W. Hofmann, *Ber.*, **22**, 1163(1889); H. Metzger, *Archeion*, **14**, 6(1932); G. Sarton, *Bull. Hist. Med.*, **8**, 419(1940); M. E. Weeks and L. O. Amberg, *J. Am. Pharm. Assoc.*, **29**, 89(1940); and Pierre Lemay and R. E. Oesper, *J. Chem. Educ.*, **25**, 62(1948).

There is no lengthy study of Dumas, but a number of short biographies are available: F. Henrich has a good chapter in Bugge (*op. cit.*, vol. 2); F. Le Blanc, *Bull. Soc. Chim.* [2], **42**, 549(1884); E. Maindron, *ibid.* [2], **45**, a(1886); W. Perkin, *J. Chem. Soc.*, **47**, 310(1885); A. Hofmann, *Ber.*, **17**, 629(1884) and *Nature*, **21**, 1(1889); G. Urbain, *Bull. Soc. chim. France* [5], **1**, 1425(1934); and Jane W. Alsobrook, *J. Chem. Educ.*, **28**, 630(1951).

Bunsen's works have been published in the *Gesammelte Abhandlungen* (Engelmann, Leipzig, 1904, 3 vols.), edited by Wm. Ostwald and Max Bodenstein. For an excellent sketch of Bunsen's life and work see Henry Roscoe's memorial lecture in *J. Chem. Soc.*, **77**, 513(1900), reprinted in *Memorial Lectures of the Chemical Society, 1893–1900* (Gurney & Jackson, London, 1901, vol. 1, p. 513). O. Fuchs has a chapter in Bugge (*op. cit.*, vol. 2, p. 78). Other sketches are: W. Ostwald, *Z. Elektrochem.*, **7**, 608(1900); T. Curtius, *Ber.*, **41**, 4875(1908); R. E. Oesper, *J. Chem. Educ.*, **4**, 431(1927) and **18**, 253(1941); H. Reinboldt, *Chymia*, **3**, 223(1950), and G. Lockemann and R. E. Oesper, *J. Chem. Educ.*, **32**, 456(1955). For a lengthier work see G. Lockemann, *Robert Wilhelm Bunsen. Lebensbild eines Deutschen Naturforschers* (Wissenschaftliche Verlagsgesellschaft, Stuttgart, 1949).

For Stromeyer see G. Lockemann and R. Oesper, *J. Chem. Educ.*, **30**, 202(1953). On Tobias Lowitz see H. Leicester, *ibid.*, **22**, 149(1945). Fourcroy is treated by W. A. Smeaton in *Fourcroy, Chemist and Revolutionary* (author, London, 1961), in *Nature*, **175**, 1017(1955) and in *Ann. Sci.*, **11**, 257(1955); and by M. Bloch in Bugge (*op. cit.*, vol. 1, p. 356). Vauquelin is included in the same chapter. Thomas Graham is examined by T. E. Thorpe in his *Essays in Historical Chemistry* (Macmillan, London, 1894) which also has chapters on Berzelius, Avogadro, Liebig, Dumas, Wöhler, and Gay-Lussac. For Graham, also see: R. A. Smith. *The Life and Work of Thomas Graham* (J. Smith, Glasgow, 1884); A. W. Hofmann, *Ber.*, **2**, 753(1869) and *Proc. Roy. Soc.*, **18**, xvii(1870); R. A. Gortner, *J. Chem. Educ.*, **11**, 279(1934); A. Ruckstuhl, *ibid.*, **29**, 594(1951); and Max Speter in Bugge (*op. cit.*, vol. 2, p. 69). For Pelletier and Caventou, see Marcel Delapene in *J. Chem. Educ.*, **28**, 454(1951).

CHAPTER 8. ORGANIC CHEMISTY II: ORGANIZATION

The events of this chapter are treated to some extent in the works of Schorlemmer, Graebe, Hjelt, and Walden described in the notes for Chapter 7. The standard histories of chemistry also attempt to deal with the period, but usually too sketchily to do it justice. O. T. Benfey's *From Vital Force to Structural Formulas* (Harvard Univ. Press, Cambridge, 1959), a good case study of the period, only available to date in offset from mimeograph. Benfey has a collection of original papers on valence theory, *Classics in the Theory of Chemical Combination* (Dover, New York, 1963).

Laurent's contributions are set forth in two excellent papers by Clara de Milt, *J. Chem. Educ.*, **28**, 198(1951) and *Chymia*, **4**, 85(1953). Also see the unsigned obituary (probably by Williamson) in *J. Chem. Soc.*, **7**, 149(1855); A. Wurtz, *Le Moniteur Scientifique*, **4**, 482(1862); E. Grimaux, *Rev. Sci.* [4], **6**, 161(1896); J. Nicklès, *Am. J. Sci.*, **66**, 103(1853); and M. Bloch in Bugge, *Das Buch der grossen Chemiker* (vol. 2, p. 92). The life of Gerhardt has been written by E. Grimaux and C. Gerhardt, Jr., *Charles Gerhardt, sa vie, son oeuvre, sa correspondence* (Masson, Paris, 1900). C. Gerhardt, Jr., also edited the *Correspondence de Charles Gerhardt* (Paris, 1924). For short biographies see: M. Bloch in Bugge (vol. 2, p. 92); W. Ostwald, *Grosse Männer* (Akad. Verlagsgesellschaft, Leipzig, 1909); E. Riegel, *J. Chem. Educ.*, **3**, 1106(1926); and M. Tiffeneau, *Compt. rend.*, **215**, 214(1942).

Hofmann's life is examined by Jacob Volhard, *August Wilhelm von Hofmann, Ein Lebensbild* (Berlin, 1902). The Chemical Society Memorial Lectures include tributes by L. Playfair, F. Abel, W. Perkin, and H. Armstrong, *J. Chem. Soc.*, **69**, 575(1896). The lectures are reprinted in the first volume of collected lectures. Short biographies of Hofmann appear in: B. Lepsius, *Ber.*, **51**,

1(1918), and in Bugge (vol. 2, p. 136); L. Maquenne, *Bull. soc. chim. France* [3], **9**, Ia(1893); and Otto Witt, *Chem. Industrie*, **15**, 181(1892).

Runge's life is examined by B. Anft and R. E. Oesper in *J. Chem. Educ.*, **32**, 566(1955). H. Leicester has a biography of Zinin, *ibid.*, **17**, 303(1940). The life of Williamson is treated in W. A. Tilden, *Famous Chemists* (p. 228) and by G. C. Foster, *J. Chem. Soc.*, **87**, 605(1905). There are several biographical works on Frankland, the most extensive being the *Sketches from the Life of Edward Frankland; born January 18, 1825, died August 9, 1899*, edited by his daughter (Spottiswood, London, 1902). Others are: J. Wislicenus, *Ber.*, **33**, 3847(1900); H. McLeod, *J. Chem. Soc.*, **87**, 574(1905); and W. A. Tilden, *Famous Chemists* (p. 216). Kolbe is described by G. Lockemann in Bugge (vol. 2, p. 124); A. W. Hofmann in *Ber.*, **17**, 2809(1884); E. Meyer in *J. prakt. Chem.* [n. s.] **30**, 417(1884); and W. Perkin, *J. Chem. Soc.*, **47**, 323(1885). For Wurtz see C. Friedel, *Notice sur la vie et les travaux de Charles Adolphe Wurtz* (Paris, 1885); A. W. Hofmann, *Adolphe Wurtz, Ein Lebensbild* (Braunschweig, 1888); G. Bugge in Bugge (vol. 2, p. 115); Hofmann, *Ber.*, **17**, 1207(1884); Friedel, *Bull. soc. chim. France* [2] **43**, i(1885); W. Perkin, *J. Chem. Soc.*, **47**, 328(1885); G. Urbain, *Bull. soc. chim. France* [5], **1**, 1425(1934); and Florence E. Wall, *J. Chem. Educ.*, **28**, 355(1951). The latter paper also deals with Faraday and Hofmann.

Although Kekulé has received much attention from biographers and students of his works, the same is not true of Couper. He almost sank into oblivion until rescued by Richard Anschütz, the biographer of Kekulé. The principal sources of information on Couper are R. Anschütz, *Life and Chemical Work of Archibald Scott Couper*, transl. by A. Crum Brown (Grant, Edinburgh, 1909); a paper by Brown in *J. Chem. Educ.*, **11**, 331(1934); and a chapter by O. T. Benfey in Farber's *Great Chemists*. Anschütz is the author of the two-volume *August Kekulé, Leben und Worken* (Verlag Chemie, Berlin, 1929). F. R. Japp is the author of the excellent memorial lecture in *J. Chem. Soc.*, **73**, 97(1898); or see vol. 1 of the collected lectures. O. T. Benfey was the organizer of the Kekulé-Couper symposium held by the History of Chemistry Division, American Chemical Society, in 1958. The *Journal of Chemical Education* published the papers presented there in the March and July 1959 issues. The principal paper by Herbert C. Brown dealt with the foundations of structural theory. Erwin N. Hiebert examined the experimental basis of Kekulé's valence theory. The remaining papers by H. M. Leicester, A. J. Ihde, and E. Campaigne—dealt with structural problems subsequent to 1858. Miscellaneous papers dealing with Kekulé's life and work are: R. Winderlich in Bugge (vol. 2, p. 200); H. Landolt, *Ber.*, **29**, 1971(1896); G. Schultz, *Ber.*, **23**, 1265(1890); G. Richardson, *J. Am. Chem. Soc.*, **18**, 1107(1896); E. Rimbach, *Ber.*, **36**, 4613(1903); and L. Darmstaedter and R. E. Oesper, *J. Chem. Educ.*, **4**, 697(1927).

There are chapters on Cannizzaro's life by B. Vanzetti and Max Speter in Bugge (vol. 2, p. 173); Tilden in his *Famous Chemists* (p. 173); Marotta in Farber's *Great Chemists;* A. Gautier in *Bull. soc. chim. France* [4], **7**, I(1910); L. C. Newell in *J. Chem. Educ.*, **3**, 1361(1926); N. Parravano, *ibid.*, **4**, 836(1927); and W. A. Tilden in *J. Chem. Soc.*, **101**, 1677(1926). For details on the Karlsruhe Congress see Clara de Milt, *Chymia*, **1**, 165(1948) and A. J. Ihde, *J. Chem.*

Educ., **38**, 83(1961). The Cannizzaro paper circulated at the congress is available in English translation as *Alembic Club Reprint* no. 18.

CHAPTER 9. CLASSIFICATION OF THE ELEMENTS

The best work on the discovery of the elements is Mary Elvira Weeks' *Discovery of the Elements* (Journal of Chemical Education, Easton, Pa., 6th ed., 1956). The book had its inception in a series of papers published in the *Journal of Chemical Education* in 1932 and 1933. The papers were made available as a paperback reprint in 1933. Subsequent enlargements have led to a series of new editions, culminating in the sixth which is brought up to date with a chapter by H. Leicester on the discovery of synthetic elements. Another work on the same subject is E. Pilgrim, *Entdeckung der Elemente* (Munders-Verlag, Stuttgart, 1950).

For the early history of spectroscopy see: T. H. Pearson and A. J. Ihde, *J. Chem. Educ.*, **28**, 267(1951); and E. C. C. Baly, *Spectroscopy* (Longmans, Green, London, 3 vols., 3rd ed., 1924, vol. 1).

A number of books deal with the history of the periodic table. Francis P. Venable's *The Development of the Periodic Law*, (Chemical Publ. Co., Easton, Pa., 1896) is old but quite complete for the period up to its time. A. E. Garrett's *The Periodic Law* (Paul & Trubner, London, 1909) adds little that is new. An excellent study of the development of various types of tables was published in 1934 by G. N. and Mary B. Quam, *J. Chem. Educ.*, **11**, 27, 217, 288(1934). An extensive bibliography is included.

Although there are a number of books on Mendeleev in Russian, there is only one in English and it is written in the form of biographical fiction. The author, Daniel Q. Posin, handles *Mendeleyev* (McGraw-Hill, New York, 1948) quite faithfully as far as the larger aspects of the subject's life are concerned but invents detail to suit his stylistic needs. The book contains a bibliography of Mendeleev's publications and of works about him. The numerous biographical sketches include: P. Walden in Bugge (vol. 2, p. 241); W. A. Tilden, *J. Chem. Soc.*, **95**, 2077(1908) and in his *Famous Chemists* (p. 241); T. E. Thorpe in his *Essays in Historical Chemistry* (p. 350); B. Jaffe in his *Crucibles* (p. 199); B. Harrow in his *Eminent Chemists of our Time* (p. 19); P. Walden in *Chem. Ztg.*, **31**, 167(1907) and in *Ber.*, **41**, 4719(1908); H. Leicester in Farber's *Great Chemists* and in *Chymia*, **1**, 67(1948); and W. Winicov in *J. Chem. Educ.*, **14**, 372(1937).

J. Lothar Meyer has received attention from the following biographers: P. Walden in Bugge (vol. 2, p. 230); J. H. Long, *J. Am. Chem. Soc.*, **17**, 664 (1895); K. Seubert, *Ber.*, **26R**, 1109(1895); P. Phillips Bedson, *J. Chem. Soc.*, **69**, 1403(1896).

A good study of Newlands and the law of octaves has been made by W. H. Taylor, *J. Chem. Educ.*, **26**, 491(1949). On Prout see S. Glasstone, *ibid.*, **24**, 478(1948) and Robert Siegfried, *ibid.*, **33**, 263(1956). For Döbereiner see: F. Henrich, *Z. angew. Chem.*, **36**, 482(1923); E. Theis, *ibid.*, **50**, 46(1937); W. Prandtl, *J. Chem. Educ.*, **27**, 176(1950); Hugo Döbling, *Die Chemie in Jena zur Goethezeit* (G. Fischer, Jena, 1928); W. Prandtl, *Deutsche Chemiker in der ersten Halfte des Neunzehnten Jahrhunderts* (Verlag Chemie, Weinheim

Bergstrasse, 1956, p. 37); and Alwin Mittasch, *Döbereiner, Goethe und die Katalyse* (Hippokrates-Verlag, Stuttgart, 1951).

On Kirchhoff see A. W. Hofmann, *Ber.*, **20**, 2771(1887). On Roscoe see *The Life and Experiences of Sir Henry Roscoe. Autobiography* (Macmillan, New York, 1905); E. Thorpe, *J. Chem. Soc.*, **109**, 395(1916) and the book *The Right Honourable Henry E. Roscoe* (Longmans, Green, New York, 1916).

Boisbaudran's life is discussed by W. Ramsay, *J. Chem. Soc.*, **103**, 742(1913) and by G. Urbain, *Am. Chem. J.*, **48**, 381(1912). On Nilson see O. Petterson, *J. Chem. Soc.*, **77**, 1277(1900). For Winkler see O. Brunck, *Ber.*, **39**, 4491(1907) and also Bugge (vol. 2, p. 336).

CHAPTER 10. THE DIFFUSION OF CHEMICAL KNOWLEDGE

Information on early chemical instruction is meager. W. A. Smeaton has a good paper on chemistry at the early École Polytechnique in *Ann. Sci.*, **10**, 224(1954). Georg Lockemann and R. E. Oesper, *J. Chem. Educ.*, **30**, 202(1953), deal with Stromeyer and the beginnings of laboratory instruction. A. J. Ihde has a general review of the development of laboratories in *Science Teacher*, **23**, 325(1956). The teaching laboratories of Fuchs and of Döbereiner are discussed in Wm. Prandtl, *Deutsche Chemiker in der ersten Hälfte des Neunzehnhunderts* (Verlag Chemie, Weinheim Bergstrasse, 1956). This book also deals with the educational activities of von Kobell, Liebig, Wöhler, Schönbein, Mitscherlich, H. Rose, and Gustav Magnus. Also see Prandtl in *J. Chem. Educ.*, **28**, 136(1951) on Fuchs.

For references to the lives of Gay-Lussac, Berzelius, Liebig, Wöhler, Bunsen, and Hofmann see the bibliographic notes for the immediately preceding chapters. On Accum see C. A. Browne, *J. Chem. Educ.*, **2**, 829, 1008 (1925). Eaton is examined by E. M. McAllister in *Amos Eaton, Scientist and Educator* (Pennsylvania Univ. Press, Philadelphia, 1941); Lomonosov by Boris N. Menschutkin in *J. Chem. Educ.*, **4**, 1079(1927) and more extensively in *Russia's Lomonosov, Chemist, Courtier, Physicist, Poet*, transl. by J. E. Thal and E. J. Webster (Princeton Univ. Press, Princeton, 1952). There is also a short sketch of Lomonosov by A. Smith in *J. Am. Chem. Soc.*, **34**, 109(1912). For Thomas Thomson see J. C. Irvine, *J. Chem. Educ.*, **7**, 2809 (1930); J. R. Partington, *Ann. Sci.*, **6**, 115(1949); and Herbert S. Klickstein, *Chymia*, **1**, 37(1948).

For the early history of Liebig's laboratory see H. G. Good, *J. Chem. Educ.*, **13**, 557(1936). Liebig's own student days are described by E. R. Schiertz, *ibid.*, **8**, 223(1931). His American pupils are discussed by H. S. Van Klooster, *ibid.*, **33**, 493(1956). Van Klooster also deals with Wöhler's American pupils, *ibid.*, **21**, 158(1945) and the student days of Edgar Fahs Smith in Wöhler's laboratory are described by Wm. McPherson, *ibid.*, **5**, 1554(1928).

The founding of the Royal College of Chemistry is described by John J. Beer, *ibid.*, **37**, 248(1960). The role of chemistry at Glasgow University over two centuries is examined in Andrew Kent, ed., *An Eighteenth-Century Lectureship in Chemistry* (Jackson, Glasgow, 1950).

The growth of chemical education in the United States is examined by Lyman C. Newell for the period before 1820 in *J. Chem. Educ.*, **9**, 677(1932);

for 1820 to 1870 by C. A. Browne, *ibid.*, p. 696; by Harrison Hale for 1870 to 1914, *ibid.*, p. 729; by Frank B. Dains for 1914 to 1930, *ibid.*, p. 745. H. S. Van Klooster deals with the beginnings of laboratory instruction in *Chymia*, **2**, 1(1949); I. B. Cohen deals with early instruction at Harvard in *ibid.*, **3**, 17(1950); A. J. Ihde and H. A. Schuette examine the growth of chemical instruction at the University of Wisconsin in *J. Chem. Educ.*, **29**, 65(1952). The role of Silliman in American chemistry is described by John Fulton and E. H. Thomson, *Benjamin Silliman, 1779–1865, Pathfinder in American Science* (Schuman, New York, 1947).

The development of chemical publications receives attention in E. J. Crane, A. M. Patterson, and Eleanor B. Marr, *A Guide to the Literature of Chemistry* (Wiley, New York, 2nd ed., 1957). Rozier's journals are analyzed by D. McKie in *Ann. Sci.*, **13**, 73(1957) and by E. W. J. Neave in *ibid.*, **6**, 416(1950) and **7**, 101, 144, 284, 393(1951–1952). The history of Liebig's *Annalen der Chemie* is examined by H. S. Van Klooster in *J. Chem. Educ.*, **34**, 27(1957).

The growth of chemical societies is recorded principally in anniversary publications of the several societies. Tom S. Moore and James C. Philip are authors of *The Chemical Society 1841–1941, A Historical Review* (Chemical Society, London, 1947). A small amount of historical material about the French society is to be found in *Celebration du Centenaire de sa Fondation* (Soc. Chimique de France, Paris, 1958). The history of the Deutschen Chemischen Gesellschaft during its first 75 years is described by Paul Walden, *Ber.*, **75A**, 166(1942). This same issue contains other items dealing with the activities of the society. Walden also has a short paper in *Die Chemie*, **55**, 367(1942), and H. Stadlinger has a notice in *Chem. Ztg.*, **66**, 491(1942).

The history of the American Chemical Society has had excellent treatment by Charles A. Browne and Mary Elvira Weeks' *A History of the American Chemical Society, Seventy-five Eventful Years* (American Chemical Society, Washington, 1952). The society also published less comprehensive commemorative works at its twenty-fifth and fiftieth anniversaries: *Twenty-fifth Anniversary Number of the Proceedings of the American Chemical Society* (Washington, 1902); and "A Half Century of Chemistry in America" in the Golden Jubilee Number, *J. Am. Chem. Soc.*, **48**, 254 pp. (Aug. 20, 1926). A companion volume to the Browne-Weeks *History* is titled *Chemistry . . . key to Better Living* (Am. Chem. Soc., Washington, 1951). It is without a listed editor but consists of a collection of papers reprinted from *Chemical and Engineering News* and *Industrial and Engineering Chemistry* during the seventy-fifth Anniversary Year. Included are histories of the various divisions of the society.

CHAPTER 11. ANALYTICAL CHEMISTRY I: SYSTEMATIZATION

The history of analytical chemistry has received very little attention. Thomas Thomson gives it a chapter in the second volume of his *History of Chemistry* (Colburn & Bentley, London, 1831); Ernst von Meyer gives it a section in his *History of Chemistry* (Macmillan, London, 1891); and a chapter is included in A. J. Berry, *From Classical to Modern Chemistry* (Cambridge

Univ. Press, Cambridge, 1954). Other books on the history of chemistry give analysis only incidental treatment. The same is largely true of journal articles. Frank J. Welcher has a short history of qualitative analysis in *J. Chem. Educ.*, **34**, 389(1957); Mary O. Hillis treats the history of microanalysis, *ibid.*, **22**, 348(1945); D. Malinin and John Yoe look at the development of laws of colorimetry, *ibid.*, **38**, 129(1961); Frank Greenway deals with early developments in *Endeavour*, **21**, 91(1962); and F. Szabadváry has a review of the history of, analytical chemistry, in German, in *Periodica Polytechnia, Chem. Engr. Series*, **2**, 49(1957), and several papers on the subject in Spanish in *Rev. univ. ind. Santander*, **4**, 53, 273(1962), **5**, 511(1963). The early history of volumetric analysis has been carefully studied by E. Rancke Madsen in *The Development of Titrimetric Analysis Until 1806* (Gad, Copenhagen, 1958).

The activities of the Fresenius family in analytical chemistry are treated by Remegius Fresenius in his *Geschichte des chemischen Laboratoriums zu Wiesbaden* (Wiesbaden, 1873). A brief historical account by R. Fresenius, transl. by R. E. Oesper, may be found in the *Register of Phi Lambda Upsilon*, **33**, 69(1948). Also see H. Grünhurt, *Ber.*, **52A**, 33(1919).

On Emil Erlenmeyer see H. Kiliani, *Z. angew. Chem.*, **22**, 481(1909); M. Conrad, *Ber.*, **43**, 3045(1910); and W. Perkin, *J. Chem. Soc.*, **99**, 1645(1911). On Stas see E. Morley, *J. Am. Chem. Soc.*, **14**, 173(1892); J. Mallet, *J. Chem. Soc.*, **63**, 1(1893); and J. Timmermans, *J. Chem. Educ.*, **15**, 353(1935).

For the life of Morley see Howard R. Williams' *Edward Williams Morley* (Chem. Educ. Publ. Co., Easton, Pa., 1957) and short papers by F. Clarke, *J. Chem. Soc.*, **123**, 3435(1923); C. Thwing, *Science*, **73**, 276(1931); C. C. Kiplinger, *J. Chem. Educ.*, **1**, 129(1924); and H. Focke, *Chem. Engr. News*, **22**, 244(1944). William A. Noyes' work is sketched by B. S. Hopkins in *J. Am. Chem. Soc.*, **66**, 1045(1944); that of F. W. Clarke by A. R. Martin in *J. Chem. Educ.*, **30**, 566(1953). The work of T. W. Richards has had considerable attention: G. Baxter, *Science*, **68**, 333(1921); H. Hartley, *J. Chem. Soc.*, **133**, 1937(1930); George S. Forbes, *J. Chem. Educ.*, **9**, 453(1932); and A. J. Ihde in Farber's *Great Chemists*.

On Mohr see R. E. Oesper, *J. Chem. Educ.*, **4**, 1357(1927); R. Hasenclever, *Ber.*, **33**, 3827(1900); and John M. Scott, *Chymia*, **3**, 191(1950). For Hempel see F. Foerster, *Z. angew. Chem.*, **30**, 1(1917); E. Graefe, *Chem-Ztg.*, **41**, 85(1917); and H. Wichelhaus, *Ber.*, **50**, 1(1917). For Winkler see O. Witt, *Chim. Industrie*, **27**, 613(1904); O. Brunck, *Ber.*, **39**, 4491(1907) and also in Bugge (vol. 2, p. 336). Lunge's work is treated by E. Berl, *J. Chem. Educ.*, **16**, 453(1939) and *Chem. Met. Engr.*, **46**, 258(1939); by F. Haber, *Ber.*, **56A**, 31(1923); and by E. Bosshard in Bugge (vol. 2, p. 351). For Bunsen see the notes for Chapter 7; for Roscoe, the notes for Chapter 9. Volhard's work is treated by E. Renouf, *Am. J. Chem.*, **43**, 281(1910) and by O. Vorländer, *Ber.*, **45**, 1855(1912).

The use of the microscope in the study of metals receives extensive treatment by Cyril S. Smith in his *History of Metallography* (Univ. of Chicago Press, Chicago, 1960); Smith's Chapter 13 deals with the contributions of Henry Sorby.

On the work of Kjeldahl see W. Johannsen, *Ber.*, **20**, 2771(1900); R. E. Oesper, *J. Chem. Educ.*, **11**, 457(1934); and Stig Veibel, *ibid.*, **26**, 459(1949).

For a review of methods for determination of organic nitrogen, see H. A. Schuette and F. C. Oppen, *Trans. Wis. Acad. Sciences*, **29**, 355(1935).

For the development of analytical apparatus see Ernest Child, *The Tools of the Chemist* (Reinhold, New York, 1940). On the chemical balance see: I. W. Brandel and Edward Kremers, *The Balance*, Monograph no. 12 in the Pharmaceutical Science Series (Pharm. Rev. Publ. Co., Milwaukee, 1906) and Robert P. Multhauf, *Proc. Am. Phil. Soc.*, **106**, 210(1962).

CHAPTER 12. ORGANIC CHEMISTRY III: CONSOLIDATION

The period of organic chemistry represented by this and the next chapter receives considerable attention in the standard histories of organic chemistry by Schorlemmer, Graebe, and Hjeldt and in most of the standard histories of chemistry. There is also a good chapter in A. Findlay's *A Hundred Years of Chemistry* (Duckworth, London, 2nd ed., 1948). A. J. Berry has a chapter on stereochemistry in his *Modern Chemistry, Some Sketches of its Historical Development* (Cambridge Univ. Press, Cambridge, 1946).

Kekulé, in a speech at the commemoration of the twenty-fifth anniversary of the benzene ring, revealed some of the background for his ideas on structural chemistry. The speech was published in *Ber.*, **23**, 1302(1890) and is available in an English translation by O. Theodor Benfey, *J. Chem. Educ.*, **35**, 21(1958). Benfey is also responsible for the Kekulé-Couper Symposium held at an American Chemical Society meeting in 1958 at which the following pertinent papers were read: Herbert C. Brown on foundations of structural theory, *J. Chem. Educ.*, **36**, 104(1959); Henry M. Leicester on Butlerov's contributions to structural theory, *ibid.*, p. 328; and A. J. Ihde on the unraveling of geometric isomerism and tautomerism, *ibid.*, p. 330. For other works on Kekulé see the notes for Chapter 8.

A. R. Surrey's *Name Reactions in Organic Chemistry* (Academic Press, New York, 2nd ed., 1961) contains information about more than 100 reactions commonly identified with a chemist's name. In each case the author gives a brief biographical sketch and a short description of the reaction, together with pertinent literature references. Somewhat in the same vein is Kurt G. Wagner's *Autoren-Namen als chemische Begriffe* (Verlag Chemie, Weinheim Bergstrasse, 1951) which is less detailed but includes theories, laws, rules constants, reagents, and apparatus. Short definitions are given and most pertinent references are included, but there is no biographical information.

For Butlerov's life and work see Henry M. Leicester in *J. Chem. Educ.*, **17**, 203(1940) and **36**, 328(1959) and in Farber's *Great Chemists*. Butlerov was much in the news during the 1950's when Russian chemists objected to Pauling's resonance concepts and sought to establish priority for Butlerov's structural concepts. A report of the Soviet Academy of Sciences, authored by D. N. Kursanov and seven others, on structural theory has been translated by Irving S. Bengelsdorf and published in *J. Chem. Educ.*, **29**, 2(1952). A relevant extract from a Soviet philosophical journal is also included in *ibid.*, p. 13. An analytical paper on theoretical chemistry in Russia was published by I. Moyer Hunsberger, *ibid.*, **31**, 504(1954).

On Schützenberger see C. Friedel, *Bull. soc. chim. France* [3] **19**, i(1898);

and Tenney L. Davis, *J. Chem. Educ.*, **6**, 1403(1929). On Dewar see H. Armstrong, *J. Chem. Soc.*, **131**, 1066(1928); and A. Findlay in *British Chemists* (Chem. Soc., London, 1947), A. Findlay and W. H. Mills, eds. There is a chapter by E. H. Rodd in the later volume on Henry E. Armstrong; and also J. V. Eyre's *Henry Edward Armstrong* (Butterworth, London, 1958), which gives much insight into the shortcomings of scientific education in Britain during Armstrong's time.

On Ladenburg see W. Herz, *Ber.*, **45**, 3597(1911) and F. Kipping, *J. Chem. Soc.*, **103**, 1871(1913). On Graebe there is an anonymous paper in *Z. angew. Chem.*, **40**, 217(1927) and one by P. Duden and H. Decker, *Ber.*, **61**, 9(1928). There are sketches on Fittig by F. Fichter, *Ber.*, **44**, 1339(1911); R. Meldola, *J. Chem. Soc.*, **99**, 1651(1911); and Ira Remsen, *Am. Chem. J.*, **45**, 210(1911). For Körner see J. B. Cohen, *J. Chem. Soc.*, **127**, 2975(1925) and R. Anschütz, *Ber.*, **59**, 75(1926). For Skraup see H. Schrötter, *Ber.*, **43**, 3683(1910) and M. Kohn, *J. Chem. Educ.*, **20**, 471(1943). Loschmidt is discussed by M. Kohn, *ibid.*, **22**, 381(1945) and by S. E. Virgo, *Science Prog.*, **27**, 634(1933).

The life of Victor Meyer has been examined by R. Meyer in *Viktor Meyer, Leben und Wirken* (Leipzig, 1917) and in *Ber.*, **41**, 4505(1908); O. Witt, *Chem. Industrie*, **20**, 325(1897); by T. Thorpe, *J. Chem. Soc.*, **77**, 169(1900); B. Horowitz, *J. Franklin Inst.*, **182**, 363(1916); Gustav Schmidt, *J. Chem. Educ.*, **27**, 557(1950); F. Heinrich in Bugge (vol. 2, p. 374); and H. Goldschmidt, *Zur Errinerung an Victor Meyer* (Heidelberg, 1897). There is one book-length biography of Baeyer, K. Schmorl's *Adolf von Baeyer* (Wissenschaftliche Verlagsgesellschaft, Stuttgart, 1952). For short sketches see R. Willstätter in Bugge (vol. 2, p. 321) or the English translation in Farber's *Great Chemists*; W. H. Perkin, *Nature*, **100**, 188(1917) and *J. Chem. Soc.*, **123**, 1520(1923); J. Howe, *J. Am. Chem. Soc.*, **45**, Proc. 45(1923); F. Henrich, *J. Chem. Educ.*, **7**, 1231(1930); J. R. Partington, *Nature*, **136**, 669(1935); F. Richter, *Ber.*, **68A**, 175(1935).

It is not difficult to become enthusiastic about Pasteur, and biographers have found him virtually irresistible. As a consequence, there are numerous biographical studies, both short and long. The best of the books are René Vallery-Radot's *The Life of Pasteur*, transl. by R. S. Devonshire, (Doubleday, Page, Garden City, N.Y., 1923, with several reissues) and René Dubos' recent study *Louis Pasteur, Free Lance of Science* (Little, Brown, Boston, 1950). Vallery-Radot, the husband of Pasteur's daughter Marie-Louise, authored a biography of his father-in-law as early as 1885. This was translated into English by Lady C. Hamilton and appeared as *Louis Pasteur, His Life and Labours* (Longmans, Green, London and Appleton, New York, 1885). Vallery-Radot's much expanded *La Vie de Pasteur* (Hachette, Paris, 1922) became the basis for Devonshire's translation. Vallery-Radot also wrote *Madame Pasteur* (Flammarion, Paris, 1941). His son, Pasteur Vallery-Radot, is responsible for publication of the *Oeuvres de Pasteur* (Masson, Paris, 1933–1939). *Pasteur, Correspondance* (Grasset, Paris, 1940), and for a short biography, *Louis Pasteur*, transl. by Alfred Joseph (Knopf, New York, 1958). The Institut Pasteur's centennial Festschrift, *Centième Anniversaire de la Naissance de Pasteur* (Hachette, Paris, 1922) contains some good biographical papers. Pasteur's associate E. Duclaux is the author of a fine study, *Pasteur, Histoire d'un esprit* (Charaire et Cie, Sceaux, 1896) available in English translation, *Pasteur, The History of a Mind* (Saunders, Philadelphia, 1920). Of the numerous short sketches among the best are: R. Koch in Bugge (vol. 2, p. 154; and in an English translation by R. E. Oesper in Farber's *Great Chemists*; Emil Fischer, *Ber.*, **28**, 2336(1895); P. Frankland, *J. Chem. Soc.*, **71**, 683(1897); A. Fernbach, *Bull. soc. chim. France* [3] **15**, i(1896); G. Bertrand, *J. Chem. Educ.*, **11**, 614(1934); and H. W. Moseley, *ibid.*, **5**, 50(1928).

For Le Bel see Marcel Delépine, *Vie et oeuvres de J. A. Le Bel* (Masson, Paris, 1949); W. J. Pope, *J. Chem. Soc.*, **1930**, 2789; and E. Wedekind, *Z. angew. Chem.*, **43**, 985(1930). There is much more on van't Hoff. Ernst Cohen first authored *Jacobus Hendricus van't Hoff* (Akademische Verlagsgesellschaft, Leipzig, 1899) and then the much more extensive *J. H. van't Hoff, sein Leben und Wirken* (Akademische Verlagsgesellschaft, Leipzig, 1912). Cohen was also the author of the chapter in Bugge (vol. 2, p. 391; which appears in the English translation of R. E. Oesper in Farber's *Great Chemists*). Other short sketches are: W. Ostwald, *Ber.*, **44**, 2219(1911); J. Walker, *J. Chem. Soc.*, **103**, 1127 (1913); G. Bruni, *Smithsonian Rept.*, **1913**, 767; B. Harrow, *Eminent Chemists of our Time* (Van Nostrand, New York, 1920, p. 79); C. W. Faulk, *J. Chem. Educ.*, **11**, 355(1934); and O. T. Benfey, *ibid.*, **37**, 467(1960). For J. A. Wislicenus, see E. Beckmann, *Ber.*, **37**, 4861(1904) and W. Perkin, *J. Chem. Soc.*, **87**, 501(1905). The principal papers of Pasteur, van't Hoff, Le Bel, and Wislicenus are available in English translation in George M. Richardson, ed., *The Foundations of Stereochemistry* (American Book Co., New York, 1901).

Much has been written about Marcelin Berthelot. Books include C. Snyder, *The Rise of Synthetic Chemistry and its Founder* (New York, 1903); E. Lavasseur, *Marcellin Berthelot (1827–1907)* (Paris, 1907); A. M. A. Boutaric, *Marcellin Berthelot (1827–1907)* (Paris, 1927); and Albert Ranc, *La pensée de Marcelin Berthelot* (Bordas, Paris, 1948). For short biographies and studies see the excellent long paper of E. Jungfleisch, *Bull. soc. chim. France* [4] i–cclx(1913) which includes a complete bibliography of Berthelot's publications in organic, physical, history and education; H. B. Dixon, *J. Chem. Soc.*, **99**, 2353(1911); C. Graebe, *Ber.*, **41**, 4805(1908); P. Sabatier, *J. Chem. Educ.*, **3**, 1099(1926); A. A. Ashdown, *ibid.*, **4**, 1217(1927); Marcel Delépine, *ibid.*, **31**, 631(1954); H. S. Van Klooster, *ibid.*, **28**, 359(1951); H. E. Armstrong, *Nature*, **120**, 659(1927); E. Färber in Bugge (vol. 2, p. 190), or the translation of the same biography (by Dora Farber) in Farber's *Great Chemists*; and the chapter in W. Ramsay, *Essays Biographical and Chemical* (Constable, London, 1908).

For Friedel see the Memorial Lecture by J. M. Crafts, *J. Chem. Soc.*, **77**, 993(1900); A. Ladenburg, *Ber.*, **32**, 372(1899); M. Hauriot, *Bull. soc. chim. France* [3] **23**, i(1900); and A. Behal, *ibid.* [4] **51**, 1423(1932). On Crafts see A. A. Ashdown, *J. Chem. Educ.*, **5**, 911(1928) and *Ind. Eng. Chem.*, **19**, 1063 (1927). The life of Grignard is treated in C. Courtot, *Bull. soc. chim. France* [5] **3**, 1433(1936); H. Gilman, *J. Am. Chem. Soc.*, **59**, Proc. 17 (1937); and H. Rheinboldt, *J. Chem. Educ.*, **27**, 476(1950). For Curtius see H. Wieland, *Z. angew. Chem.*, **41**, 193(1928) and C. Duisberg, *ibid.*, **43**, 723(1930). On Beckmann see G. Lockemann, *Ber.*, **61A**, 87(1928) and R. E. Oesper, *J. Chem. Educ.*, **21**, 470(1944). On Paul Walden see the autobiographical paper translated by Oesper, *ibid.*, **28**, 160(1951) and G. Kerstein, *Nachr. deutsch. Gesellsch. Gesch. Med.*, **10**, 17(1958).

On Beilstein see F. Richter, *Ber.*, **71A**, 35(1938) and *J. Chem. Educ.*, **15**, 310(1938); E. Huntress, *ibid.*, **15**, 303(1938); and O. Witt, *J. Chem. Soc.*, **99**, 1646(1911). Also see F. Richter, *75 Jahre Beilsteins Handbuch der organischen Chemie* (Springer, Berlin, 1957).

CHAPTER 13. ORGANIC CHEMISTRY IV: NATURAL PRODUCTS

The general references mentioned in connection with Chapters 7, 8, and 12 are largely pertinent to this chapter. The principal new figures of importance are Emil Fischer and Otto Wallach.

Fischer left an autobiography, *Aus meinem Leben* (Springer, Berlin, 1922), which deals very well with his work and ideas. There is another biography by Kurt Hoesch, *Emil Fischer, Sein Leben und sein Werk* (Verlag Chemie, Berlin, 1921). For short sketches see A. Wohl, *Chem. Industrie*, **42**, 269(1919); K. Hoesch, *Ber.*, **54A**, 299(1921); M. Forster, *J. Chem. Soc.*, **117**, 1157(1920); P. Jacobson, *Chem. Ztg.*, **43**, 565(1919); M. Bergmann in Bugge (vol. 2, p. 408); B. Helferich in Farber's *Great Chemists*; L. Darmstaedter and R. E. Oesper, *J. Chem. Educ.*, **5**, 37(1928); P. A. Roussel, *ibid.*, **30**, 122(1953); C. V. Ratnam, *ibid.*, **38**, 93(1961); Claude S. Hudson, *ibid.*, **18**, 353(1941) and **30**, 120(1953). The last four citations deal primarily with Fischer's work on natural products. The first Hudson paper deals with the work on the configuration of glucose and carries a bibliography to all of the relevant literature.

For Wallach see L. Ruzicka, *J. Chem. Soc.*, **135**, 1582(1932); E. Farber, *Nobel Prize Winners in Chemistry* (Schuman, New York, 2nd ed., 1963, p. 42); and W. S. Partridge and E. R. Schierz, *J. Chem. Educ.*, **24**, 106(1947). On Kiliani there is an autobiographical paper in *J. Chem. Educ.*, **9**, 1908(1932). C. A. Browne has a biography of Tollens, *ibid.*, **19**, 253(1942) and O. Wallach an obituary in *Ber.*, **51**, 1539(1918).

CHAPTER 14. INORGANIC CHEMISTRY I: FUNDAMENTAL DEVELOPMENTS

The history of inorganic chemistry has been very badly neglected. There are no books dealing primarily with the subject and only a few journal articles. Perhaps the most comprehensive treatment of primarily inorganic material is that which appears in Mary Elvira Weeks' *Discovery of the Elements* (Chem. Educ. Publ. Co., Easton, Pa., 6th ed., 1956). There is a good section on inorganic chemistry in Ernst von Meyer, *History of Chemistry* (Macmillan, London, 1891) and a somewhat unorganized chapter, which is principally physical chemistry, appears in F. J. Moore and Wm. T. Hall, *A History of Chemistry* (McGraw-Hill, New York, 3rd ed., 1939).

For the life and work of Schönbein, see E. Färber in Bugge (vol. 1, p. 458); W. La Rue, *J. Chem. Soc.*, **22**, x(1869); R. E. Oesper, *J. Chem. Educ.*, **6**, 432, 677(1929). Georg W. A. Kahlbaum, in his *Monographien aus der Geschichte der*

Chemie (Barth, Leipzig), has brought together much material on Schönbein: with E. Schaer, *Christian Friedrich Schönbein, 1799–1868, Ein Blatt zur Geschichte des 19 Jahrhunderts* [2 vols., 1899, 1901 (vols. 4 and 6 of series)]; with E. Thon, *Liebig und Schönbein, Briefwechsel, 1858–1868* [1900 (vol. 5 of series)]. Kahlbaum was also the editor of *Schönbein und Berzelius, Zwanzig Briefe, 1836–1847* (Schwabe, Basel, 1898).

On Thenard see M. Flourens, *Elogue historique de L. J. Thenard* (Paris, 1860); Armand Paul Thenard, *Un grand Français, le chimiste Thenard* (Jobard, Dijon, 1950); anon., *Am. J. Sci.* [2] **24**, 408(1857), **25**, 430(1858); M. Bloch in Bugge (vol. 1, p. 386); and K. Webb, *Chemistry & Industry*, **64**, 2(1947). For Curtius see Heinrich Wieland, *Z. angew. Chem.*, **41**, 193(1928) and Carl Duisberg, *ibid.*, **43**, 723(1930).

The life of Gustav Magnus is treated by W. Prandtl in *Deutsche Chemiker in der ersten Hälfte des neunzehnten Jahrhunderts* (Verlag Chemie, Weinheim/ Bergstrasse, 1956, p. 303). On Caro see E. Darmstaedter in Bugge (vol. 2, p. 298); E. Renouf, *Am. Chem. J.*, **44**, 557(1910); and A. Bernthsen, *Ber.*, **45**, 1987(1912).

Moissan's work has received attention by Alfred Stock, *Ber.*, **40**, 5099(1907); P. Lebeau, *Bull. soc. chim. France* [4] **3**, i(1908); Wm. Ramsay, *J. Chem. Soc.*, **101**, 477(1912); and Benjamin Harrow, *Eminent Chemists of our Time* (Van Nostrand, New York, 1920, p. 138).

The discovery of the inert gases is described by Wm. Ramsay in his *The Gases of the Atmosphere, The History of their Discovery* (Macmillan, London, 1915) and by M. W. Travers in *The Discovery of the Rare Gases* (Arnold, London, 1928). On the life of Rayleigh see Robert John Strutt (Fourth Baron Rayleigh), *John William Strutt, Third Baron Rayleigh* (Arnold, London, 1924); R. Glazebrook, *Sci. Progress*, **14**, 286(1919); and various authors in *Nature*, **103**, 365(1919). For Ramsay see W. A. Tilden, *Sir William Ramsay, Memorials of his Life and Work* (Macmillan, London, 1918); M. W. Travers, *A Life of Sir William Ramsay* (Arnold, London, 1956); P. Walden in Bugge (vol. 2, p. 250); F. Sherwood Taylor, *Am. Scientist*, **41**, 449(1953); M. W. Travers, *Nature*, **135**, 615(1935); W. A. Tilden, *J. Chem. Soc.*, **111**, 369(1917); R. Moore, *J. Franklin Inst.*, **186**, 29(1918); and C. Moureu, *Bull. soc. chim. France* [4] **25**, 401(1919). On Lockyer there is T. Mary Lockyer and Winifred L. Lockyer, *Life and Work of Sir Norman Lockyer* (Macmillan, London, 1928). There is a chapter on Janssen in Macpherson, *Astronomers of Today* (Gall & Inglis, London, 1905). For Hillebrand see E. T. Allen, *J. Chem. Educ.*, **9**, 73(1932) and Virginia Bartow, *ibid.*, **26**, 367(1949).

The discovery of the rare earth elements is treated in Weeks (*op. cit.*) and to some extent in B. S. Hopkins, *Chemistry of the Rarer Elements* (Heath, Boston, 1923; expanded in the later *Chapters in the Chemistry of the Less Familiar Elements*, Stipes, Champaign, Ill., 1939). Hopkins' books also include brief historical remarks about the discovery of many of the elements which do not fall into the rare earth category. S. T. Levy's *The Rare Earths* (Longmans, Green, New York, 1915) has only very brief historical remarks. Another work which contains much material on the discovery of the elements is Helen Miles Davis' *The Chemical Elements* (Science Service and Ballantine Books, Washington and New York, 1959). This book, which contains many excerpts from the original papers reporting new elements, is a revision of the Summer 1952 issue of *Chemistry*.

For the life of Gadolin see T. E. T., *Nature*, **86**, 48(1911) and V. Ojala and E. R. Schierz, *J. Chem. Educ.*, **14**, 161(1937). Boisbaudran is treated by G. Urbain, *Am. Chem. J.*, **48**, 381(1912) and by W. Ramsay, *J. Chem. Soc.*, **103**, 742(1913). For Cleve see H. Euler and A. Euler, *Ber.*, **38**, 4221(1905) and E. Thorpe, *J. Chem. Soc.*, **88**, 1301(1906). M. A. Etard treats the life of Demarçay in *Bull. soc. chim. France* [3] **32**, 1(1904) and in *Chem. News*, **89**, 137(1904). For Marignac see E. Ador, *Bull. soc. chim. France* [3] **9**, i(1893) and *Ber.*, **27**, 979(1894); and P. Cleve, *J. Chem. Soc.*, **67**, 468(1895). For Auer von Welsbach see F. Sedlacek, *Auer von Welsbach* (Springer, Vienna, 1934); anon., *J. Chem. Educ.*, **6**, 2051(1929); Fritz Lieben, *ibid.*, **35**, 230(1958); and J. d'Ans, *Ber.*, **64**, 59 (1931). On Nilson see O. Petterson, *J. Chem. Soc.*, **77**, 1277(1900). On Urbain see P. Job, *Bull. soc. chim. France* [5] **6**, 745(1939); R. Oesper, *J. Chem. Educ.*, **15**, 201(1938); and G. Champetier and C. H. Boatner, *ibid.*, **17**, 103(1940). For Charles James see Melvin M. Smith, L. A. Pratt, and B. S. Hopkins, *The Life and Work of Charles James* (Northeastern Section of the Am. Chem. Soc., Cambridge, Mass., 1932); H. Iddles, *J. Chem. Educ.*, **7**, 812(1930); and B. S. Hopkins, *J. Wash. Acad. Sci.*, **22**, 21(1932).

There is a certain amount of historical background to the development of coordination theory in John C. Bailar, ed., *Chemistry of the Coordination Compounds* (Reinhold, New York, 1956). On Jørgensen's theoretical ideas see G. B. Kauffman, *J. Chem. Educ.*, **36**, 521(1959) and *Chymia*, **6**, 180(1960). On Werner see P. Pfeiffer, *J. Chem. Educ.*, **5**, 1090(1928); anon., *ibid.*, **7**, 1732(1930); E. Berl, *ibid.*, **19**, 153(1942); P. Karrer, *Helv. Chim. Acta*, **3**, 196(1920); K. Hofmann, *Ber.*, **53A**, 9(1920); and G. T. Morgan, *J. Chem. Soc.*, **117**, 639(1920).

CHAPTER 15. PHYSICAL CHEMISTRY I: ORIGINS

Physical chemistry, just as analytical and inorganic, has had little historical study, and no book on the subject is available. The standard histories of chemistry give physical chemistry some attention and histories of physics treat such subjects as kinetic theory, thermodynamics, and electrochemistry.

Alexander Findlay's *A Hundred Years of Chemistry* (Duckworth, London, 2nd ed., 1948) has two good chapters on nineteenth-century physical chemistry, and A. J. Berry's *Modern Chemistry, Some Sketches of its Historical Development* (Cambridge Univ. Press, Cambridge, 1946) has some good chapters on classical atomic theory, electrochemistry, studies of gases, solutions, and chemical change. Berry's *From Classical to Modern Chemistry* (Cambridge Univ. Press, 1954) has chapters on heat theory, electrochemistry, optics and chemistry, molecular magnitudes, and kinetics. T. W. Chalmers' *Historic Researches, Chapters in the History of Physical and Chemical Discovery* (Scribners, New York, 1952) has a good presentation of the growth of knowledge of heat and electricity.

On the relation between chemical constitution and properties see Samuel Smiles, *The Relations between Chemical Constitution and some Physical Properties* (Longmans, Green, London, 1910). This work, while primarily technical,

has a historical introduction and faithfully presents the point of view characteristic of the time. On Kopp see E. Thorpe, *J. Chem. Soc.*, **64**, 779(1893);
E. O. von Lippmann, *Archeion*, **14**, 1(1932), which deals with historical
activities; B. Bessmerthny, *ibid.*, **14**, 62(1932), which deals with chemical
work; J. Ruska, *J. Chem. Educ.*, **14**, 3(1937). For H. A. Lorentz see Louis de
Broglie, *Notice sur la vie et l'oeuvre de Hendrik Antoon Lorentz* (Institut de
France, Academie des Sciences, Paris, 1951); J. Larmor, *Nature*, **111**, 1(1923)
and **121**, 287(1928); A. Fokker, *Physica*, **8**, 1(1928); and G. L. de Haas-
Lorentz, ed., *H. A. Lorentz, Impressions of his Life and Work* (North-Holland,
Amsterdam, 1957). On Landolt see R. Pribram, *Ber.*, **44**, 3337(1911) and R. E.
Oesper, *J. Chem. Educ.*, **22**, 158(1945).

On the development of theories of heat there are numerous short papers
but no comprehensive studies. On the development of thermometers see C. B.
Boyer, *Sci. Monthly*, **57**, 442, 546(1943); H. C. Bolton, *Evolution of the Ther-
mometer* (Chem. Publ. Co., Easton, Pa., 1900); C. B. Boyer, *Am. J. Physics*,
10, 176(1942); N. E. Dorsey, *J. Wash. Acad. Sci.*, **36**, 361(1946); J. N. Friend,
Nature, **139**, 395, 586(1937); F. S. Taylor, *Ann. Sci.*, **5**, 129(1941); M. K.
Barnett, *Osiris*, **12**, 269(1956); J. A. Chaldecott, *Ann. Sci.*, **8**, 195(1952);
Sanborn C. Brown, *Am. J. Physics*, **22**, 13(1954). D. McKie and N. H. V.
Heathcote, *The Discovery of Specific and Latent Heats* (Arnold, London, 1935)
has a good treatment of the contributions of Black and Wilcke. On the
caloric theory see S. C. Brown, *Am. J. Physics*, **18**, 367(1950); M. K. Barnett,
Sci. Monthly, **62**, 165, 247(1946); A. Tilloch, *Phil. Mag.*, **9**, 158(1801)—a
critique of experiments on the weight of heat; Duane Roller, *The Early
Development of the Concepts of Heat and Temperature, The Rise and Decline
of the Caloric Theory* (Harvard Case Hist. in Exptl. Sci. no. 3; Harvard Univ.
Press, Cambridge, 1950); and Erwin N. Hiebert, *Historical Foundations of the
Law of Conservation of Energy* (Wis. Hist. Soc., Madison, 1961).

On the development of steam power see H. W. Dickinson and H. P. Vowles,
James Watt and the Industrial Revolution (Longmans, Green, London, 1943)
and I. B. Hart, *James Watt and the History of Steam Power* (Schuman, New
York, 1949). On the economic impact of steam power the following are
valuable: T. S. Ashton, *The Industrial Revolution, 1760–1830* (Oxford Univ.
Press, New York, 1948); W. Bowden, M. Karpovich, and A. P. Usher, *An
Economic History of Europe since 1750* (American Book Co., New York, 1937);
S. B. Clough and C. W. Cole, *Economic History of Europe* (Heath, Boston,
rev. ed., 1946); H. Heaton, *Economic History of Europe* (Harper, New York,
rev. ed., 1948); and L. Mumford, *Technics and Civilization* (Harcourt, Brace,
New York, 1934). Also see the fourth volume of *A History of Technology*
edited by C. Singer, *et al.* (Oxford Univ. Press, London and New York,
1958).

Count Rumford has had a special fascination for biographers. Among the
books are: G. E. Ellis, *Memoir of Sir Benjamin Thompson, Count Rumford,
with notices of his daughter* (Am. Acad. Arts & Sciences, Philadelphia, 1871);
J. A. Thompson, *Count Rumford of Massachusetts* (Farrar & Rinehart, New
York, 1935); Egon Larsen, *An American in Europe, The Life of Benjamin
Thompson* (Philosophical Library, New York, 1953). For shorter articles see
C. R. Adams, *Sci. Monthly*, **71**, 380(1950); L. C. Newell, *Science*, **68**, 67(1928);

S. C. Brown, *Am. J. Physics*, **20**, 331(1952), *Proc. Am. Phil. Soc.*, **93**, 316(1949) and *Am. Sci.*, **42**, 113(1954); T. Martin, *Nature*, **163**, 348(1949) and *Bull. Brit. Soc. Hist. Sci.*, **1**, 144(1951); C. H. Dwight, *Notes Roy. Soc. London*, **2**, 189(1955); and M. Watanabe, *Isis*, **50**, 141(1959). For the works of Rumford see *The Complete Works of Count Rumford* (Am. Acad. Arts & Sciences, Philadelphia, 4 vols., 1870–1875).

On Carnot and his work see Thomas S. Kuhn, *Am. J. Physics*, **23**, 387(1955) and *Isis*, **49**, 132(1958); V. K. La Mer, *Am. J. Physics*, **22**, 20(1954) and **23**, 95(1955); and Milton Kerker, *Sci. Monthly*, **85**, 143(1957) and *Isis*, **51**, 257 (1960). On Clapeyron see H. S. Van Klooster, *J. Chem. Educ.*, **33**, 299(1956). For J. Robert Mayer see R. E. Oesper, *ibid.*, **19**, 134(1942) and Otto Bluth, *Isis*, **43**, 211(1952). Helmholtz is treated in J. G. M'Kendrick, *Hermann Ludwig Ferdinand von Helmholtz* (Longmans, Green, London, 1899); L. Koenigsberger, transl. by F. A. Wilby, *Hermann von Helmholtz* (Clarendon Press, Oxford, 1906); Hermann Ebert, *Hermann von Helmholtz* (Wissensch. Verlag, Stuttgart, 1949); and A. C. Crombie, *Sci. Amer.*, **198**, no. 3, 94(1958). On Joule there is O. Reynolds, *Memoir of James Prescott Joule* (Manchester Lit. and Philos. Soc., Manchester, 1892) and A. Wood, *Joule and the Study of Energy* (Bell, London, 1925).

William Thomson (Kelvin) is examined in the following: *Lord Kelvin's early home; being the recollections of his sister, the late Mrs. Elizabeth King; together with some family letters and a supplementary chapter by the editor, Elizabeth Thomson King* (Macmillan, London, 1909); J. Munro, *Lord Kelvin* (Drane, London, 1902); Sylvanus P. Thompson, *The Life of William Thomson, Baron Kelvin of Largs* (Macmillan, London, 2 vols., 1910); A. Grey, *Lord Kelvin, An Account of his Life and Work* (Dent, London, 1908); A. G. King, *Kelvin, The Man, A Biographical Sketch* (Musson, London, 1925); H. N. Casson, *Kelvin* (Efficiency Mag., London, 1930); D. Wilson, *William Thomson, Lord Kelvin, His Way of Teaching Natural Philosophy* (J. Smith, London, 1910); A. Russell, *Lord Kelvin, His Life and Work* (Jack, London, 1912); A. Russell, *Lord Kelvin* (Blackie, London, 1938).

Thomas S. Kuhn has a fine paper on energy conservation as an example of simultaneous discovery in M. Clagett, ed., *Critical Problems in the History of Science* (Univ. of Wisconsin Press, Madison, 1959). Also see J. H. Keenan and A. Shapiro, *Mech. Eng.*, **69**, 915(1947); J. R. Partington, *Nature*, **170**, 730 (1952); J. F. Allen, *Sci. Monthly*, **69**, 29(1949); A. E. Bell, *Nature*, **151**, 519(1943); P. S. Epstein, *Textbook of Thermodynamics* (Wiley, New York, 1937); and Ernst Mach, *History and Root of the Principle of the Conservation of Energy* (Open Court, Chicago, 1911).

Maxwell's papers were edited by W. D. Niven, *Scientific Papers of James Clerk Maxwell* (Cambridge Univ. Press, Cambridge, 2 vols., 1890). For biographies see Lewis Campbell, *The Life of James Clerk Maxwell* (London, 1882); R. Glazebrook, *James Clerk Maxwell and Modern Physics* (London, 1896); R. L. Smith-Rose, *James Clerk Maxwell, F. R. S., 1831–1879* (Longmans, Green, London, 1948); J. J. Thomson, *et al.*, *James Clerk Maxwell—A Commemorative Volume* (Cambridge Univ. Press, Cambridge, 1931).

Several papers dealing with kinetic theory of gases are R. Hooykas, *Arch. int. Hist. Sci.*, **2**, 180(1948); S. G. Brush, *Ann. Sci.*, **13**, 188, 274(1957), **14**,

185(1958); R. Roseman and S. Katzoff, *J. Chem. Educ.*, **11**, 350(1934); and W. S. James, *Sci. Prog.*, **23**, 263(1928), **24**, 57(1929), and **25**, 232(1930). On van der Waals see R. E. Oesper, *J. Chem. Educ.*, **31**, 599(1954). On Hess see H. M. Leicester, *ibid.*, **29**, 581(1951) and T. W. Davis, *ibid.*, **29**, 584(1951). On Clausius see S. G. Brush, *Ann. Sci.*, **14**, 185(1958). Linde is treated by F. Pollitzer, *Z. Elektrochem.*, **41**, 1(1935), and by K. Peters, *Z. angew. Chem.*, **48**, 231(1935). For Kamerlingh-Onnes see E. Cohen, *J. Chem. Soc.*, **130**, 1193(1927), and R. E. Oesper, *J. Chem. Educ.*, **21**, 263(1944). Julius Thomsen is treated by E. Thorpe, *J. Chem. Soc.*, **97**, 161(1910) and by N. Bjerrum, *Ber.*, **42**, 4971(1909). On Horstmann see M. Trautz, *Ber.*, **63A**, 61(1930) and *Z. Elektrochem.*, **35**, 875(1929) and A. Mayer, *Naturwissenschaften*, **18**, 261(1930). Roozeboom and the phase rule are treated by H. S. van Klooster, *J. Chem. Educ.*, **31**, 594(1954).

J. Willard Gibbs was insufficiently appreciated during his lifetime but has come into his own since his death. An early study of his role in science is had in four papers by Fielding H. Garrison, *Popular Sci. Monthly*, **74**, 470, 551(1909), **75**, 42, 191(1909). Short papers worthy of examination are: F. Donnan, *J. Franklin Inst.*, **199**, 457(1925); J. Johnston, *J. Chem. Educ.*, **5**, 507(1928); E. B. Wilson, *Sci. Monthly*, **32**, 211(1931); C. A. Kraus, *Science*, **89**, 275(1939); R. E. Langer, *Am. Math. Monthly*, **46**, 75(1939); and R. E. Oesper, *J. Chem. Educ.*, **32**, 267(1955). Books are Muriel Rukeyser, *Willard Gibbs* (Doubleday, ·Doran, New York, 1942) and Lynde P. Wheeler, *Josiah Willard Gibbs, The History of a Great Mind* (Yale Univ. Press, New Haven, 1952).

Le Chatelier is examined by R. E. Oesper, *J. Chem. Educ.*, **8**, 442(1931); A. Silverman, *ibid.*, **14**, 555(1937) and reprinted in Farber's *Great Chemists*; W. J. Pope, *Nature*, **138**, 711(1936); C. H. Desch, *J. Chem. Soc.*, **1938**, 139; and M. Bodenstein, *Ber.*, **72A**, 122(1939).

For Ostwald see B. Harrow, *J. Chem. Educ.*, **7**, 2ᶠ97(1930); A. Findlay, *Nature*, **129**, 750(1932); E. Nernst, *Z. Elektrochem.*, **38**, 337(1932); P. Walden, *Ber.*, **65A**, 101(1932); F. G. Donnan, *J. Chem. Soc.*, **1933**, 316; W. D. Bancroft, *J. Chem. Educ.*, **10**, 539, 609(1933); E. P. Hillpern, *Chymia*, **2**, 57(1949); G. Jaffe, *J. Chem. Educ.*, **29**, 230(1952); E. A. Hauser, *ibid.*, **28**, 492(1951); E. Farber, *ibid.*, **30**, 600(1953); J. R. Partington, *Nature*, **172**, 380(1953); R. E. Oesper, *J. Chem. Educ.*, **31**, 398(1954); E. Fischer, *Naturw. Rundsch.*, **8**, 49(1955); and Walther Ostwald, *J. Chem. Educ.*, **34**, 328(1957). For books see Wm. Ostwald, *Lebenslinien eine Selbstbiographie* (Berlin, 1926); Paul Walden, *Wilhelm Ostwald* (Leipzig, 1933); and Grete Ostwald, *Wilhelm Ostwald, mein Vater* (Berliner Union, Stuttgart, 1953).

Arrhenius' life and work is examined by W. Palmaer in Bugge (vol. 2, p. 443), with a translation of R. E. Oesper in Farber's *Great Chemists*; B. Harrow has a chapter in his *Eminent Chemists of our Times* (Van Nostrand, New York, 1920); the Memorial Lecture of James Walker appears in *J. Chem. Soc.*, **131**, 1380(1928); E. Riesenfeld has a biography in *Ber.*, **63A**, 1(1930) and a volume in Ostwald's *Grosse Männer* series (vol. 11, Akad. Verlagsbuchhandlung, Leipzig, 1930). For Raoult see J. van't Hoff, *J. Chem. Soc.*, **81**, 969(1902); and F. Getman, *J. Chem. Educ.*, **13**, 153(1936). For Hittorf see H. Goldschmidt, *Z. angew. Chem.*, **27**, 657(1914); and anon., *J. Chem. Soc.*, **107**, 582(1915).

For Graham see the notes for Chapter 7. For Zsigmondy see H. Cohen, *Z.*

Elektrochem., **35**, 876(1929); A. Lettermoser, *Z. angew. Chem.*, **42**, 1069(1929); and H. Freundlich, *Ber.*, **63**, 1(1930). For Perrin see E. Eschangon, *Compt. rend.*, **214**, 725(1942) and R. Audubert, *J. chim. phys.*, **39**, 45(1942).

CHAPTER 16. BIOLOGICAL CHEMISTRY I: AGRICULTURAL, PHYSIOLOGICAL, AND FOOD STUDIES

The history of biological chemistry has had more attention than some of the other fields, but even so, the literature is meager. There is a comprehensive history, Fritz Lieben's *Geschichte der physiologischen Chemie* (Deuticke, Leipzig & Vienna, 1935), which deals very largely with the nineteenth-century developments. E. V. McCollum's *A History of Nutrition* (Houghton, Mifflin, Boston, 1957) also has considerable material on this century. Charles A. Browne's *A Source Book of Agricultural Chemistry* (Chronica Botanica, Waltham, Mass., 1944) is not primarily a source book but contains running commentary with interspersed quotations from writers on agricultural science from Thales to Liebig. The quotations, while short, are well chosen and the commentary reflects the author's deep understanding of agricultural chemistry.

The development of early knowledge of photosynthesis is ably handled by Leonard K. Nash in Harvard Case Book no. 5, *Plants and the Atmosphere* (Harvard Univ. Press, Cambridge, 1952). Also see Howard S. Reed, *Jan Ingenhousz, Plant Physiologist* (Chronica Botanica, Waltham, Mass., 1949) which contains a history of the discovery of photosynthesis and reprints all of the pertinent papers of Ingenhousz. On Ingenhousz also see H. S. Van Klőoster, *J. Chem. Educ.*, **29**, 353(1952).

The best source of information on the early development of soil science is Browne's *Source Book* mentioned above. There are also good chapters concerning Liebig's role in F. R. Moulton, ed., *Liebig and after Liebig* (Am. Assoc. Adv. Sci., Washington, 1942). On Boussingault see A. Fayol, *La Nature*, **1947**, 215 and Browne's *Source Book*. The latter work also has biographical material on Ingenhousz, Senebier, Chaptal, de Saussure, Sprengel, Mulder, and Liebig. On Winogradsky see S. A. Waksman, *Science*, **118**, 36(1953) and Waksman's book *Sergei N. Winogradsky, His Life and Work* (Rutgers Univ. Press, New Brunswick, N.J., 1954).

The various histories of medicine briefly mention chemical developments in physiology and medicine. Also see such works as Michael Foster, *Lectures on the History of Physiology* (Cambridge Univ. Press, Cambridge, 1901, 1924); and the Clio Medica series published in New York by Hober—John F. Fulton, *Physiology* (1931) and Graham Lusk, *Nutrition* (1933). For Beaumont see A. H. Smith, *J. Nutrition*, **44**, 1(1951) and the facsimile reprint of Beaumont's *Experiments and Observations on the Gastric Juice and the Physiology of Digestion* (Harvard Univ. Press, Cambridge, 1929) which contains a biographical essay by Wm. Osler. On Magendie there is a fine biography by James M. D. Olmsted, *François Magendie, Pioneer in Experimental Physiology and Scientific Medicine in Nineteenth-Century France* (Schuman, New York, 1944). Also see P. F. Fenton, *J. Nutrition*, **43**, 1(1951). Bernard has been treated biographically by J. D. M. Olmsted and E. Harris Olmsted, *Claude Bernard and the Experimental Method in Medicine* (Schuman, New York, 1952), and more extensively by J. Olmsted in *Claude Bernard, Physiologist* (Harper, New York, 1938). For short papers see Jean Mayer, *J. Nutrition*, **45**, 3(1951); Oswei Temkin, *Bull. Hist. Med.*, **20**, 10(1946); and F. G. Young, *Brit. Med. J.* (i), 1431(1957) and *Ann. Sci.*, **2**, 47(1937). Bernard's classic work on experimental medicine has been translated into English and has been reprinted several times.

The principal historical study of respiration and energy utilization in animals is that of M. Rubner, *Sitzungsber. Preuss. Akad. Wissensch., Phys. Math. Klasse*, **17**, 313(1931). Also see Chapter 10 in McCollum's *History of Nutrition*. For a biographical sketch of Rubner see H. C. Knowles, *Diabetes*, **6**, 369(1957). For Pettenkofer there is K. Kisskalt, *Max von Pettenkofer* (Wissensch. Verlagsgesellsch., Stuttgart, 1948).

The history of fermentation is treated in Wm. Bulloch, *History of Bacteriology* (Oxford Univ. Press, London, 1938) and in J. B. Conant, *Pasteur's Study of Fermentation* (Harvard Case Book no. 6, Harvard Univ. Press, Cambridge, 1952).

The growth of the public health movement is examined in George Rosen, *A History of Public Health* (MD Publications, New York, 1958). There is no comprehensive history of the pure foods and drugs movement. Mark Sullivan has a good description of the early movement in the United States in his *Our Times, The United States, 1900–1925* (Scribners, New York, 5 vols., vol. **2**, 1927, p. 471). Also see James H. Young, *The Toadstool Millionaires, A Social History of Patent Medicines in America before Federal Regulation* (Princeton Univ. Press, Princeton, 1961); Louis Filler, *Crusaders for American Liberalism, The Story of the Muckrakers* (Harcourt, Brace, New York, 1939); and Stewart H. Holbrook, *The Golden Age of Quackery* (Macmillan, New York, 1959). For the activities of Fredrick Accum see C. A. Browne, *J. Chem. Educ.*, **2**, 829, 1008(1925). Harvey W. Wiley's *The History of a Crime Against the Food Law* (H. W. Wiley, Washington, 1929) deals with the early years of enforcement in America. Wiley's *An Autobiography* (Bobbs-Merrill, Indianapolis, 1930) has much material on passage and enforcement of the law. There is a good biography by Oscar E. Anderson, Jr., *The Health of a Nation; Harvey W. Wiley and the Fight for Pure Food* (Univ. of Chicago Press, Chicago, 1958). Maurice Natenberg, *The Legacy of Doctor Wiley* (Regent House, Chicago, 1957) is oversensational and prone to minor errors. For short articles dealing with Wiley see W. D. Bigelow, *Ind. Engr. Chem.*, **15**, 88(1923); W. W. Skinner, C. A. Browne, W. G. Campbell, *et al.*, *J. Am. Assoc. Official Agr. Chemists*, **14**, iii(1931); E. J. Dies, *Titans of the Soil* (Univ. of North Carolina Press, Chapel Hill, 1949, Chap. 17); O. E. Anderson, *Am. Hist. Rev.*, **61**, 550(1956); and A. J. Ihde in Farber's *Great Chemists*.

CHAPTER 17. CHEMICAL INDUSTRY I: THE NINETEENTH CENTURY

A number of books deal with the history of chemical industry in one fashion or another. F. Sherwood Taylor's *A History of Industrial Chemistry* (Heinemann, London, 1957) surveys the subject from antiquity into the twentieth century. L. F. Haber's *The Chemical Industry during the Nineteenth*

Century (Clarendon, Oxford, 1958) is an excellent study of the century's activities, particularly with respect to the industry's economic growth. Archibald Clow and Nan L. Clow, *The Chemical Revolution* (Batchworth, London, 1952) is a fine study of the growth of the chemical industry in the British Isles from approximately 1750 until the early 1800's. Both Haber and Clow have extensive bibliographies of works dealing with the history of industrial chemistry.

The history of the chemical industry in America up to World War II is contained in a six-volume work by Williams Haynes, *American Chemical Industry, A History* (Van Nostrand, New York, 1945–1954). Volume 1 deals with the beginnings in America and takes the subject up to 1910. The emphasis is principally economic, but a great deal of scientific information is included. A chemical chronology, statistical appendixes, and extensive bibliography of books make this a very useful volume. Haynes' *Chemical Pioneers* (Van Nostrand, New York, 1939) has biographical chapters dealing with sixteen figures who were prominent in the development of the industry in America. C. J. S. Warrington and R. V. V. Nicholls, *A History of Chemistry in Canada* (Pitman, New York & London, 1949) has a number of chapters dealing with the chemical industry.

For the chemical industry in Britain see Stephen Miall, *A History of the British Chemical Industry* (E. Benn, London, 1931); D. W. F. Hardie, *A History of the Chemical Industry in Widnes* (Imp. Chem. Ind., Widnes, 1950); J. Fenwick Allen, *Some Founders of the Chemical Industry* (Sherratt & Hughes, Manchester, 2nd ed., 1907) which contains biographical chapters of British leaders such as Muspratt and Deacon; and Trevor I. Williams, *The Chemical Industry* (Penguin, London, 1953) which deals extensively with Britain but also has material on Continental and American developments.

The French industry has been examined by J. Garçon, *Histoire de la chimie en France* (the author, Paris, 1900) and by M. Fauque, *L'évolution de la grand industrie chimique en France* (Éditions Universitaires, Strasbourg, 1932). For Germany see R. Lorenz, *Die Enwicklung der deutschen chemischen Industrie* (Leipzig, 1919) and B. Lepsius, *Deutschlands chemische Industrie* (Stilke, Berlin, 1914).

Bleaching is treated in S. H. Higgins' *A History of Bleaching* (Longmans, Green, London, 1924).

Sulfuric acid manufacture is dealt with by H. W. Dickinson, *Trans. Newcomen Soc.*, **18**, 43(1937–8) and O. Guttmann, *J. Soc. Chem. Ind.*, **20**, 5(1901). The best history of the Leblanc process is P. Baud, *Compt. rend.*, **196**, 1498 (1933). Also see C. C. Gillispie, *Isis*, **48**, 152(1957) regarding the origins of the process. Desmond Reilly, *ibid.*, **42**, 287(1951) deals with acids, bases, and salts in the nineteenth century. Leblanc's life is treated in A. Anastasi, *Nicolas Leblanc, sa vie, ses travaux, et l'histoire de la soude artificielle* (Hachette, Paris, 1884) and in Bugge. For the life of Muspratt see D. W. F. Hardie, *Endeavour*, **9**, 29(1955) and D. Reilly, *J. Chem. Educ.*, **28**, 650(1951). The latter paper also deals with the Gambles. Glover is treated by N. F. Newbury, *Endeavour*, **1**, 42(1942). R. Étienne has a chapter on Solvay in Farber's *Great Chemists*, and there is a book-length biography, P. Héger and C. Lefébure, *La vie d'Ernest Solvay* (Lamertin, Brussels, 1919).

Not much has been written about the early development of the fertilizer industry except for the role of Liebig in soil science. The symposium volume edited by F. R. Moulton, *Liebig and after Liebig* (Am. Assoc. Adv. Science, Washington, 1942) has several chapters dealing with soil fertility and the introduction of commercial fertilizers. Also see A. R. Hall, *The Book of the Rothamsted Experiments* (Dutton, New York, 1905) and Samuel W. Johnson, *How Crops Feed* (Judd, New York, 1891). There is a review of the fertilizer industry by C. J. Brand, *Am. Fertilizer J.*, **93**, no. 3, 5(1940) and there is a good series of papers read at a Liebig Symposium in *ibid.*, no. 6, p. 5. Also see M. S. Anderson, *Ag. Food Chem.*, **8**, 84(1960) on the history of soil testing.

On the early days of the explosives industry see: Wm. S. Dutton, *One Thousand Years of Explosives* (Winston, Philadelphia, 1960); Melvin A. Cook, *The Amazing Story of Explosives* [Bull. Univ. of Utah, **42**, no. 7 (1952)]; and A. P. Van Gelder and H. Schlatter, *History of the Explosives Industries of America* (Columbia Univ. Press, New York, 1927). On Nobel see: M. Evlanoff, *Nobel-Prize Donor, Inventor of Dynamite—Advocate of Peace* (Revell, New York, 1944); H. E. Pauli, *Alfred Nobel, Dynamite King—Architect of Peace* (Fischer, New York, 1942); H. Schück and R. Sohlman, *The Life of Alfred Nobel*, transl. from the German by Bryan and Beatrix Lunn (Heinemann, London, 1929); and Nobel Foundation, ed., *Alfred Nobel, The Man and His Prizes* (Univ. of Oklahoma Press, Norman, 1951). On the du Ponts see: W. S. Dutton, *Du Pont, One Hundred and Forty Years* (Scribners, New York, 1942); M. James, *Alfred I. du Pont, The Family Rebel* (Bobbs-Merrill, Indianapolis, 1941); and J. H. Winkler, *The Du Pont Dynasty* (Reynal & Hitchcock, New York, 1935).

There are two books dealing with the development of the illuminating gas industry: Charles Hunt, *A History of the Introduction of Gas Lighting* (London, 1907); and Dean Chandler, *Outline of the History of Lighting by Gas* (London, 1936).

There is much on the dye industry. A very good study is John J. Beer's *The Emergence of the German Dye Industry* (Univ. of Illinois Press, Urbana, 1959). Beer also has papers dealing with the origins of industrial research, *Isis*, **49**, 123(1958) and with early theories of dyeing, *Isis*, **51**, 21(1960). There is a German work by H. Caro, *Über die Entwicklung der Theerfarbenindustrie* (Berlin, 1893). For Peter Griess see V. Heines, *J. Chem. Educ.*, **35**, 187(1958) and M. A. Smith and W. R. Gower, *ibid.*, **29**, 176(1952). There is a very fine paper dealing with the growth of the dyestuffs industry by R. E. Rose, *J. Chem. Educ.*, **3**, 973, 1132(1926) which contains swatches of cloth dyed with the dyes under discussion. On the decline of the dye industry in Britain see Walter M. Gardner, ed., *The British Coal Tar Industry, Its Origin, Development, and Decline* (Williams & Norgate, London, 1915).

There were two centennial symposia dealing with the discovery of mauve: Robert Robinson, *et al.*, *Perkin Centenary London, One Hundred Years of Synthetic Dyestuffs* (Pergamon, London, 1958); and Howard J. White, Jr., ed., *Proceedings of the Perkin Centennial* (Am. Assoc. Textile Chemists and Colorists, New York, n.d., ca. 1958). Also see: Robt. Robinson, *J. Chem. Educ.*, **34**, 54(1957) and *Endeavour*, **15**, 92(1956); Sidney M. Edelstein, *Am. Dyestuff Reporter*, **45**, 598(1956); John Read, *Sci. Am.*, **196**, no. 2, 110(1957);

R. Brightman, *Nature*, **177**, 815(1956); H. S. Van Klooster, *J. Chem. Educ.*, **28**, 359(1951); S. M. Edelstein in Farber's *Great Chemists*.

On the history of synthetic indigo see H. Brunck, *The History of the Development of the Manufacture of Synthetic Indigo* [Kuttroff, Pickhardt & Co. (for Badische), New York, 1901]. On Duisberg there is his autobiography, *Meine Lebenserinnerungen* (Reclam, Leipzig, 1933) and his papers, *Abhandlungen, Vorträge und Reden* (Verlag Chemie, Berlin, 1930). For a short sketch see H. E. Armstrong, *Nature*, **135**, 1021(1933).

For the synthesis of alizarin see: Louis F. Fieser, *J. Chem. Educ.*, **7**, 2609 (1930); C. Graebe and C. Liebermann, *Ber.*, **1**, 104, 106, 186(1868) and *Ann.* Supplement, **7**, 257(1870). On Graebe see: P. Duden and H. Decker, *Ber.*, **61A**, 9(1928) and C. Liebermann, *Ber.*, **44**, 551(1911). For Liebermann see O. Wallach and P. Jacobson, *Ber.*, **51**, 1136(1918). For Caro see: E. Renouf, *Am. Chem. J.*, **44**, 557(1910); A. Bernthsen, *Ber.*, **45**, 1987(1910); and E. Darmstaedter in vol. 2 of Bugge.

Except for two German works, not much has been written about the early days of the synthetic drug industry. The German works are: E. Hellig, *Pharmacie und chemische Grossindustrie, ihre Entwicklung und volkswirtschaftliche Beudenting* (Tübingen, 1922); and Wilhelm Vershofen, *Die Anfänge der chemisch-pharmazeutischen Industrie, Eine Wirtschaftshistorische Studie* (Berlin, 3 vols., 1949–58). Volume 3 of the latter deals with the period from 1870 to 1914. Also see G. Reddish, *Am. J. Pharm. Educ.*, **23**, 445(1959) on the early history of antiseptics and disinfectants; R. L. Ray, *J. Chem. Educ.*, **37**, 451(1960) on alkaloids; C. Hochwalt, *Chemistry*, **30**, no. 6, 10(1957) on aspirin; George Urdang, *Sci. Monthly*, **61**, 17(1945) on cinchona; W. R. Bett, *Alchemist*, **21**, 685(1958) on cocaine. For short treatments of early developments in synthetic drugs see: A. Findlay, *A Hundred Years of Chemistry* (Duckworth, London, 2nd ed., 1948, p. 144); Findlay, *Chemistry in the Service of Man* (Harper, New York, 8th ed., 1960); A. J. Ihde, *Cahiers Hist. Mond.*, **4**, 957 (1958) and *Sixth Ann. Lecture Series* (College of Pharmacy, Univ. of Texas, Austin, 1963), p. 16.

On photography there is Helmut and Alison Gernsheim, *The History of Photography* (Oxford Univ. Press, London, 1955); Lucia Moholy, *A Hundred Years of Photography* (Pelican, Harmondsworth, Middlesex, 1939); Beaumont Newhall, *Photography, A Short Critical History* (Museum of Modern Art, New York, 1937); Josef M. Eder, *History of Photography*, transl. from the German by Edward Epstean (Columbia Univ. Press, New York, 1945). There is a good chapter in C. Singer, *et al.*, *A History of Technology* (Oxford Univ. Press. London, 1958, vol. 5, p. 716).

On glassmaking in the nineteenth century see Singer (*op. cit.*, vol. 4, p. 358, and vol. 5, p. 671). On chemical glassware there is Ernest Child, *The Tools of the Chemist* (Reinhold, New York, 1940).

For the history of metallurgy in the nineteenth century see: Singer, *et al.* (*op. cit.*, vol. 4, Chap. 4 and vol. 5, Chaps. 3 and 4); Thomas A. Rickard, *Man and Metals* (McGraw-Hill, New York, 2 vols., 1932); O. Johannsen, *Geschichte des Eisens* (Stahleisen, Düsseldorf, 3rd ed., 1953); T. S. Ashton, *Iron and Steel in the Industrial Revolution* (University Press, Manchester, 2nd ed., 1951); W. O. Henderson, *Britain and Industrial Europe, 1750–1870* (University

Press, Liverpool, 1954); and L. Guillet, *L'evolution de la metallurgie* (Alcan, Paris, 1928). For Bessemer see *Sir Henry Bessemer F. R. S. An Autobiography* (Offices of *Engineering*, London, 1905). On Kelly there is J. N. Boucher, *William Kelly, A True History of the So-called Bessemer Process* (the author, Greensburg, Pa., 1924). On Thomas there are R. W. Burne, *Memoir and Letters of Sidney Gilchrist Thomas, Inventor* (Murray, London, 1891) and L. G. Thompson, *Sidney Gilchrist Thomas, An Invention and its Consequences* (Faber & Faber, London, 1940). Also see W. T. Jeans, *The Creators of the Age of Steel* (Chapman & Hall, London, 1884) which has chapters on Bessemer, Siemens, and Thomas.

For the early history of nonferrous metals see: L. Aitchison, *Sheet Metal Ind.*, **28**, 405, 519, 530(1951); and T. A. Rickard (*op. cit.*, vol. 2). On aluminum there is: C. J. Smithells, *J. Roy. Soc. Arts*, **98**, 822(1950); E. G. West, *Bull. Inst. Metall.*, **5**, 5(1955); and A. von Zeerleder, *Aluminium* (Akad. Verlag Gesellschaft, Leipzig, 1955). On nickel there is: A. C. Sturney, *The Story of Mond Nickel* (Mond Ni. Co., London, 1951); and A. J. Wadhams, *Metals & Alloys*, **2**, 166(1931). For Henri Deville see M. Daumas, *Revue hist. sci.*, **2**, 352(1949). For C. M. Hall see: H. Holmes, *J. Chem. Educ.*, **7**, 233(1930); K. Arndt, *Elektrotech. Z.*, **57**, 149(1936); and H. Amick, *Am. Mineral.*, **27**, 196(1942). For Mond see: John M. Cohen, *The Life of Ludwig Mond* (Methuen, London, 1956); H. Bolitho, *Alfred Mond* (Secker & Warburg, London, 1933); and H. Armstrong, *Nature*, **127**, 238(1931).

There is virtually nothing on the history of the electrochemical industries except for scattered remarks in the standard sources on the chemical industry. On Daniell see T. O. Sistrunk, *J. Chem. Educ.*, **29**, 26(1952). On Acheson see R. Szymanowitz, *ibid.*, **33**, 113(1956) and E. G. Acheson, *A Pathfinder, Discovery, Invention, and Industry* (The Press Scrapbook, New York, 1910).

CHAPTER 18. RADIOCHEMISTRY I: RADIOACTIVITY AND ATOMIC STRUCTURE

There is as yet no comprehensive historical treatment of radioactivity and its consequences, although many works treat sizable snatches of the subject. Because of its popular appeal much has been written on the subject, some of the works giving decent attention to the history and human interest side of the story. The situation is actually much better than is the historical record with respect to physical, analytical, and inorganic chemistry.

Some general works which have useful chapters on radioactivity and atomic structure are: A. Findlay, *A Hundred Years of Chemistry* (Duckworth, London, 2nd ed., 1948); William Wilson, *A Hundred Years of Physics* (Duckworth, London, 1950); Joshua C. Gregory, *A Short History of Atomism* (Black, London, 1931); A. G. Van Melsen, *From Atomos to Atom* (Duquesne Univ. Press, Pittsburgh, 1952); Mary E. Weeks, *Discovery of the Elements* (Chem. Educ. Publ. Co., Easton, Pa., 6th ed., 1956); and Arthur E. E. McKenzie, *The Major Achievements of Science* (Cambridge Univ. Press, Cambridge, 2 vols., 1960). Volume 2 of the latter work, which is made up of selections from the scientific literature, contains papers of J. J. Thomson, Rutherford, Soddy, Planck, Bohr, and Heisenberg. An interesting and well-written little book dealing with the period between 1895 and 1915 is Alfred Romer, *The Restless Atom* (Doubleday, Garden City, N.Y., 1960).

More strictly scientific in purpose, but with much useful historical background, are the following: Samuel Glasstone, *Source Book on Atomic Energy* (Van Nostrand, New York, 2nd ed., 1958); John A. Eldridge, *The Physical Basis of Things* (McGraw-Hill, New York, 1934); R. F. Humphrys and Robert Beringer, *First Principles of Atomic Physics* (Harper, New York, 1950); and Irving Kaplan, *Nuclear Physics* (Addison-Wesley, Reading, Mass., 1955). All of these books place their major emphasis on developments after 1920 but are still useful with respect to events of the earlier period. At a somewhat more elementary level, there is useful information in Lloyd Wm. Taylor, *Physics, The Pioneer Science* (Houghton, Mifflin, Boston, 1941); Herman T. Briscoe, *The Structure and Properties of Matter* (McGraw-Hill, New York, 1935); William H. Bragg, *Concerning the Nature of Things* (Harper, New York, 1925); Alexander Kolin, *Physics, Its Laws, Ideas, and Methods* (McGraw-Hill, New York, 1950); Gerald Holton, *Introduction to the Concepts and Theories in Physical Science* (Addison-Wesley, Reading, Mass., 1952); G. Holton and Duane H. D. Roller, *Foundations of Modern Physical Science* (Addison-Wesley, Reading, Mass., 1958); Konrad Krauskopf, *Fundamentals of Physical Science* (McGraw-Hill, New York, 4th ed., 1960); Konrad Krauskopf and Arthur Beiser, *The Physical Universe* (McGraw-Hill, New York, 1960); Ernest Grunwald and Russell H. Johnsen, *Atoms, Molecules, and Chemical Change* (Prentice-Hall, Englewood Cliffs, N.J., 1960); Charles Compton, *An Introduction to Chemistry* (Van Nostrand, New York, 1958); George Gamow, *Matter, Earth, and Sky* (Prentice-Hall, Englewood Cliffs, N.J., 1958); Herbert Priestley, *Introductory Physics, An Historical Approach* (Allyn & Bacon, Boston, 1958); G. S. Christiansen and Paul H. Garrett, *Structure and Change* (Freeman, San Francisco, 1960); T. A. Ashford, *From Atoms to Stars* (Holt, Rinehart & Winston, New York, 1960); Francis T. Bonner and Melba Phillips, *Principles of Physical Science* (Addison-Wesley, Reading, Mass., 1957); and Cecil J. Schneer, *The Search for Order* (Harper, New York, 1960). There are several good chapters dealing with the experimental work up to 1920 in T. W. Chalmers, *Historic Researches, Chapters in the History of Physical and Chemical Discovery* (Scribners, New York, 1952).

On Helmholtz see: J. G. M'Kendrick, *Hermann Ludwig Ferdinand von Helmholtz* (Longmans, Green, London, 1899); L. Koenigsberger, *Hermann von Helmholtz*, transl. by Wilby (Clarendon Press, Oxford, 1906); Hermann Ebert, *Hermann von Helmholtz* (Wissenschaftliche Verlag, Stuttgart, 1949); H. Gruber and V. Gruber, *Sci. Monthly*, **83**, 92(1956); P. Huber, *Isis*, **43**, 152(1952) and *Experimentia*, **7**, 356(1951); and A. C. Crombie, *Sci. Am.*, **198**, no. 3, 94(1958).

On Crookes there are: E. E. F. D'Albe, *The Life of Sir William Crookes* (London, 1923); W. Tilden, *J. Chem. Soc.*, **117**, 444(1920); J. Druce, *Sci. Progress*, **26**, 677(1932); A. Findlay and W. H. Mills, *British Chemists* (Chem. Soc., London, 1947, Chap. 1). For Röntgen there is Otto Glasser, *Dr. W. C. Roentgen* (Thomas, Springfield, Ill., 1945).

J. J. Thomson has an autobiography, *Recollections and Reflections* (Macmillan, New York, 1937). There is also a biography by the younger Lord

Rayleigh, *The Life of Sir J. J. Thomson* (Cambridge Univ. Press, Cambridge, 1943), and Rayleigh is responsible for the notice in *Obit. Notices, Fellows of the Royal Society*, **3**, 587(1941). George Thomson has published the following about his father: *Nature*, **178**, 1317(1956); *Physics Today*, **9**, no. 8, 19(1956); *Brit. J. Phys. Med.*, **20**, 102(1957); and with Joan Thomson, *Notes Rec. Roy. Soc.*, **12**, 201(1957). Also see D. J. Price, *Discovery*, **17**, 494(1956).

For C. T. R. Wilson see P. M. Blackett, *Memorial Biographies, Fellows of the Royal Society*, **6**, 269(1960). On Millikan see *The Autobiography of Robert A. Millikan* (Prentice-Hall, New York, 1950); and L. A. DuBridge, *Science*, **119**, 272(1954) and *Biog. Mem. Natl. Acad. Sci.*, **33**, 241(1959). On von Laue see P. P. Ewald, *Mem. Biog., Fellows Roy. Soc.*, **6**, 135(1960). On W. H. Bragg see E. N. da C. Andrade, *Obit. Not., Fellows Roy. Soc.*, **4**, 277(1942–44). On Moseley see E. Rutherford, *Nature*, **96**, 33(1915) and B. Jaffe, *Crucibles* (Simon & Schuster, New York, 1930, Chap. 15). For Becquerel see Oliver Lodge, *J. Chem. Soc.*, **101**, 2005(1912).

The best guide to the Curies is the excellent biography by Eve Curie, *Madame Curie*, transl. by Vincent Sheean (Doubleday, Doran, New York, 1938). There is also the earlier work by Marie Curie, *Pierre Curie*, transl. by C. and V. Kellogg (Macmillan, New York, 1923). A memorial meeting in honor of Marie Curie is reported by Marston T. Bogert in *Bull. Polish Inst. Arts, Sci. Am.*, **3**, 200(1945). Also see A. Russell, *J. Chem. Soc.*, **138**, 654(1935).

There are several biographies of Rutherford, but the definitive study still remains to be written. Of the existing biographies the best is that of A. S. Eve, *Rutherford, Being the Life and Letters of the Rt. Hon. Lord Rutherford* (Cambridge Univ. Press, Cambridge, 1939). Others are: I. B. N. Evans, *Man of Power, The Life Story of Baron Rutherford of Nelson* (Paul, London, 1939— reissued as *Rutherford of Nelson* by Penguin in 1943); N. Feather, *Lord Rutherford* (Blackie, London, 1940); and *Ernest Rutherford, Atom Pioneer* (Philosophical Library, New York, 1957). The first five Rutherford lectures are available as *Rutherford by those who knew him* (Physical Soc., London, 1956). For short sketches see Henry Tizard, *J. Chem. Soc.*, **1946**, 980 and A. S. Eve and J. Chadwick, *Obit. Not., Fellows Roy. Soc.*, **2**, 395(1936–1938). J. B. Birks, ed., *Rutherford at Manchester* (Heywood, London, 1962) is an excellent volume containing the addresses delivered at Manchester on the occasion of the fiftieth anniversary of the α-scattering experiments and reprints some of the important papers of the Manchester period.

For Geiger see the obituary notice in *Z. Physik*, **124**, 1(1947). On Soddy there is Muriel Howorth, *Pioneer Research on the Atom, The Life Story of Frederick Soddy* (New World Publ., London, 1958) and A. Fleck, *Obit. Not. Fellows Roy. Soc.*, **3**, 203(1957). On Aston see Eric H. Rideal, *Chem. & Ind.*, **1946**, 10.

Much has been published about Einstein. There are his own *Ideas and Opinions*, based on *Mein Weltbild*, Carl Seelig, ed. (Crown, New York, 1954); and *Out of My Later Years* (Philosophical Library, New York, 1950). An excellent biography is Philipp Frank, *Einstein, His Life and Times* (Knopf, New York, 1947). Other biographical works are: H. G. Garbedian, *Albert Einstein, Maker of Universes* (Funk & Wagnalls, New York, 1937); Paul A. Schlipp, ed., *Albert Einstein, Philosopher-Scientist* (Library of Living

Philosophers, Evanston, Ill., 1949); Leopold Infeld, *Albert Einstein, His Work and its Influence on our World* (Scribners, New York, 1950); Antonina Vallentin, *The Drama of Albert Einstein* (Doubleday, New York, 1954); and Otto Nathan and Heinz Russell, *Einstein on Peace* (Simon & Schuster, New York, 1960).

W. Pauli, ed., *Niels Bohr and the Development of Physics* (McGraw-Hill, New York, 1955) is a Festschrift made up largely of technical essays, but there is a sketch of Bohr's Manchester years. For Max Planck see his *Scientific Autobiography and Other Papers*, transl. by F. Gaynor (Philosophical Library, New York, 1949); J. C. O'Flaherty, *Am. Sci.*, **47**, 68(1959); J. Franck, *Science*, **107**, 534(1948); and M. Born, *Obit. Not. Fellows Roy. Soc.*, **6**, 161(1948–1949).

For G. N. Lewis see G. R. Robertson, *Chem. Engr. News*, **25**, 3290(1947). On Langmuir see W. R. Whitney, *Ind. Engr. Chem.*, **20**, 329(1928).

CHAPTER 19. RADIOCHEMISTRY II: THE NUCLEAR AGE

On this subject many of the references to the previous chapter are applicable. In addition, and particularly pertinent to the new viewpoints in physical thought, are the following: Max von Laue, *History of Physics*, transl. from the German by R. E. Oesper (Academic Press, New York, 1950); Carl T. Chase, *The Evolution of Modern Physics* (Van Nostrand, New York, 1947); Albert Einstein and Leopold Infeld, *The Evolution of Physics* (Simon & Schuster, New York, 1938); E. Rutherford, *The Newer Alchemy* (Cambridge Univ. Press, Cambridge, 1937); Louis de Broglie, *The Revolution in Physics*, transl. by R. W. Niemeyer (Noonday Press, New York, 1953); Harrie Massey, *The New Age in Physics* (Harper, New York, 1960); G. O. Jones, J. Rotblat, and G. J. Whitrow, *Atoms and the Universe* (Scribners, New York, 1956); Pascual Jordan, *Physics of the Twentieth Century*, transl. by E. Oshry (Philosophical Library, New York, 1944); John Eldridge, *The Physical Basis of Things* (McGraw-Hill, New York); and F. K. Richtmyer, *Introduction to Modern Physics* (McGraw-Hill, New York, 2nd ed., 1934). In a more popular vein are: Selig Hecht's *Explaining the Atom* (Viking, New York, 1949), an excellent introductory work by a professor of biophysics; Joseph G. Feinberg, *The Atom Story* (Philosophical Library, New York, 1953); Maurice Duquesne, *Matter and Antimatter* (Harper, New York, 1960); Donald J. Hughes, *The Neutron Story* (Doubleday, Garden City, N.Y., 1959); Robert R. Wilson and Raphael Littauer, *Accelerators, Machines of Nuclear Physics* (Doubleday, Garden City, N.Y., 1960); F. Soddy, *The Story of Atomic Energy* (New Atlantis, London, 1949); and K. Mendelssohn, *What is Atomic Energy?* (Sigma, London, 1946).

There are already a large number of books dealing with nuclear weapons and their implications. The official United States report on the development of the atomic bomb is by Henry DeWolf Smyth, *Atomic Energy for Military Purposes* (Princeton Univ. Press, Princeton, 1945). Samuel Goudsmit, *Alsos* (Schuman, New York, 1947) is a report on German activities during World War II. For other general facets of the problem one may consult the following: Leonard Bertin, *Atomic Harvest, A British View of Atomic Energy* (Freeman, San Francisco, 1957); Ralph E. Lapp, *Atoms and People* (Harper, New York, 1956); Michael Amrine, *The Great Decision, The Secret History of the Atomic*

Bomb (Putnam's, New York, 1959); Wm. L. Laurence, *Men and Atoms* (Simon & Schuster, New York, 1959); and R. E. Lapp, *Roads to Discovery* (Harper, New York, 1960). On national aspects, besides the above, see: Gordon E. Dean, *Report on the Atom, What You Should Know About the Atomic Energy Program of the United States* (Knopf, New York, 1953); Leslie R. Groves, *Now It Can Be Told* (Knopf, New York, 1961); Lewis Strauss, *Men and Decisions* (Doubleday, New York, 1962); Arnold Kramish, *Atomic Energy in the Soviet Union* (Stanford Univ. Press, Stanford, 1959); Melville J. Ruggles and A. Kramish, *The Soviet Union and the Atom, The Early Years* (Rand Corp., Santa Monica, Calif., 1956); George F. Kennan, *Russia, The Atom and the West* (Harper, New York, 1958); and (for a German viewpoint) Robert Jungk, *Brighter than a Thousand Suns, A Personal History of the Atomic Scientists*, transl. by J. Cleugh (Harcourt Brace, New York, 1958).

For the effects of atomic weapons there are the following: Samuel Glasstone, ed., *The Effects of Atomic Weapons* (U.S. Govt. Printing Office, Washington, 1957); John M. Fowler, ed., *Fallout, A Study of Superbombs, Strontium 90, and Survival* (Basic Books, New York, 1960); John R. Hersey, *Hiroshima* (Knopf, New York, 1946); R. E. Lapp, *The Voyage of the Lucky Dragon* (Harper, New York, 1958); David Bradley, *No Place to Hide* (Little Brown, Boston, 1948); P. M. S. Blackett, *Fear, War and the Bomb* (Whittlesey House, New York, 1956); and James F. Crow, *Effects of Radiation and Fallout* (Public Affairs Comm., Pamphlet no. 256, New York, 1957). On control of nuclear weapons there are: Harrison Brown, *Must Destruction be Our Destiny?* (Simon & Schuster, New York, 1946); Walter Gellhorn, *Security, Loyalty, and Science* (Cornell Univ. Press, Ithaca, 1950); Henry A. Kissinger, *Nuclear Weapons and Foreign Policy* (Doubleday, Garden City, N.Y., 1958); Linus C. Pauling, *No More War!* (Dodd Mead, New York, 1958); and Edward Teller and Albert A. Latter, *Our Nuclear Future, Facts, Dangers and Opportunities* (Criterion, New York, 1958).

On nonmilitary uses of nuclear energy and isotopes see: Jacob Sacks, *The Atom at Work* (Ronald, New York, 1951); Sam H. Schurr and Jacob Marshak, *Economic Aspects of Atomic Power* (Princeton Univ. Press, Princeton, 1950); Hans Thirrig, *Energy for Man, Windmills to Nuclear Power* (Indiana Univ. Press, Bloomington, 1958); Laura Fermi, *Atoms for the World, United States Participation in the Conference of the Peaceful Uses of Atomic Energy* (Univ. of Chicago Press, Chicago, 1957); Norman Landsdell, *The Atom and the Energy Revolution* (Penguin, Harmondsworth, Middlesex, 1958); and David O. Woodbury, *Atoms for Peace* (Dodd Mead, New York, 1955).

There are, to date, very few biographical works dealing with those who were deeply involved in the development of nuclear energy. Laura Fermi has written a very fine biography of her late husband, *Atoms in the Family, My Life with Enrico Fermi* (Univ. of Chicago Press, Chicago, 1954). For personal reminiscences see Arthur H. Compton, *Atomic Quest, A Personal Narrative* (Oxford Univ. Press, New York, 1956) and Otto Hahn, *New Atoms, Progress and Some Memories* (Elsevier, New York, 1950).

There is a very important study of attitudes and opinions of scientists, governmental leaders, and religious groups by Erwin N. Hiebert, *The Impact of Atomic Energy* (Faith and Life Press, Newton, Kansas, 1961). One should examine files of the *Bulletin of the Atomic Scientists* for reactions of many prominent scientists to the social and political problems raised by modern science.

CHAPTER 20. PHYSICAL CHEMISTRY II: MATURITY

As mentioned in the notes to Chapter 15, not much has been written on the history of physical chemistry. Many of the references listed there are also pertinent to this chapter, although few of them deal with developments since 1925. Perhaps the most comprehensive account of the rise of physical chemistry is Harry C. Jones' *A New Era in Chemistry* (Van Nostrand, New York, 1913). Of the various textbooks on the market, few pay attention to the history of the subject. The most notable exceptions are: A. Findlay, *Introduction to Physical Chemistry* (Longmans, Green, London, 1933); Walter J. Moore, *Physical Chemistry* (Prentice-Hall, Englewood Cliffs, N.J., 2nd ed., 1955); and Neil K. Adam, *Physical Chemistry* (Clarendon, Oxford, 1956).

Several articles deal with the development of physical chemistry in America; all were prepared for anniversary publications of the American Chemical Society. The earlier is by Wilder D. Bancroft, in *A Half Century of Chemistry in America*, C. A. Browne, ed., published as a special number (Aug. 20, 1926) of the *Journal of the American Chemical Society*. At the time of the seventy-fifth anniversary there was a series of surveys in *Industrial and Engineering Chemistry*, volume 43 (1951); physical chemistry was reviewed by Farrington Daniels and colloid chemistry by W. O. Milligan, J. W. Williams, and E. J. Miller. These, and other articles of the series, were reprinted in a commemorative volume, *Chemistry . . . Key to Better Living* (Am. Chem. Soc., Washington, 1951).

Several review publications deal with progress in physical chemistry. *Chemical Reviews*, which began publication in 1924 under the editorship of Wm. A. Noyes, frequently has extensive reviews of specialized topics in physical chemistry. More extensive are: *Annual Review of Physical Chemistry* (Ann. Revs., Stanford, vol. 1 in 1950 under the editorship of G. K. Rollefson and R. E. Powell); *Annual Review of Nuclear Science* (*idem.*, vol. 1 in 1951 under editorship of James G. Beckerley); *Advances in Colloid Science* [Interscience, New York, vol. 1 in 1942 (not publ. annually) under the editorship of E. O. Kraemer]; and *Advances in Catalysis and Related Subjects* (Academic Press, New York, vol. 1 in 1948 under the editorship of G. W. Frankenburg, V. I. Komarewsky, and E. K. Rideal).

On the development of structural atomic theory one should consult the works of Niels Bohr, particularly his *Theory of Spectra and Atomic Constitution* (Cambridge Univ. Press, Cambridge, 1922) which reprints three of his principal papers. Also see H. A. Kramers and Helge Holst, *The Atom and the Bohr Theory of its Structure* (Glydendal, Copenhagen, 1923). On Molecular theory see: Linus Pauling, *The Nature of the Chemical Bond* (Cornell Univ. Press, Ithaca, 1939); W. G. Palmer, *Valency Classical and Modern* (Cambridge Univ. Press, Cambridge, 1944); and Henry Margenau, *Am. J. Physics*, **12**, 119, 247(1944), **13**, 73(1945)—dealing with atomic and molecular theory since Bohr.

For biographical material the following are of interest. On Tamman see W. E. Gardner, *J. Chem. Soc.*, **1952**, 1959. On Henry Armstrong there is: J. V. Eyre, *Henry Edward Armstrong* (Butterworth, London, 1958); E. J. Rodd, *J. Chem. Soc.*, **1940**, 1418; F. W. Keeble, *Obit. Notices, Fellows Roy. Soc.*, **31**, 229(1939–1940); H. Hartley, *Chem. & Ind.*, **1945**, 398; and E. Armstrong, *ibid.*, **1941**, 80. There is an excellent sketch of Kahlenberg by Norris F. Hall, *Trans. Wis Acad. Sciences, Arts, Letters*, **39**, 83(1949) and **40**, part 1, 173(1950) and a shorter sketch by A. T. Lincoln, *News Edn., Ind. Engr. Chem.*, **16**, 336(1938). On Hückel see R. E. Oesper, *J. Chem. Educ.*, **27**, 674(1950). On Sørensen see E. K. Rideal, *Memorial Lectures Delivered before the Chemical Society* (Chem. Soc., London, vol. 4, 1951, p. 159). On E. C. Franklin see A. Findlay, *ibid.*, p. 99. For Brønsted there is R. P. Bell, *J. Chem. Soc.*, **1950**, 409, reprinted in Farber's *Great Chemists*.

For Nernst see: Albert Einstein, *Sci. Monthly*, **54**, 195(1942), reprinted in Farber's *Great Chemists*; T. Isuardi, *Chemia*, **12**, 189(1942); and J. Eggert, *Z. physik. Chem. Unterricht.*, **56**, 43(1943) and *Naturwissenschaften*, **31**, 412(1943).

For Gomberg see: C. Schoepfle and W. Bachman, *J. Am. Chem. Soc.*, **69**, 2921(1947), reprinted in Farber's *Great Chemists*, and A. White, *Ind. Engr. Chem.*, **23**, 10(1931). On Paneth see the obituary by G. R. Martin, *Nature*, **182**, 1274(1958) and the memorial by H. J. Emeléus, *Mem. Biog. Fellows Roy. Soc.*, **6**, 227(1960). On Sidgwick there is a chapter by H. T. Tizard in Farber's *Great Chemists*. For Pauli see E. I. Valko, *J. Chem. Educ.*, **27**, 2(1950). For Pauling see: *Chem. Engr. News*, **27**, 28(1949); and anon., *Bull. Atomic Sci.*, **16**, 382(1960).

CHAPTER 21. ANALYTICAL CHEMISTRY II: EXPANSION

There is virtually nothing on the history of analytical chemistry beyond the references mentioned in the notes for Chapter 11. There are a few reviews of progress in special fields—for example, see *Chemical Reviews; Analytical Chemistry*, and *Advances in Analytical Chemistry and Instrumentation* (Interscience, New York, vol. 1 in 1960 under the editorship of Charles N. Reilly). On some topics there is meager historical background in C. R. N. Strouts, J. H. Gilfillan, and H. N. Wilson, eds., *Analytical Chemistry* (Clarendon, Oxford, 2 vols., 1955). The second volume is quite good on instruments and chromatography.

On the history of indicators see: F. Szabadváry, *Acta Chimica, Acad. Scient. Hungaricae*, **20**, 253(1959) in German; and A. A. Baker, *Chymia*, **9**, 82(1963).

For more detail on the early history of chromatography see: J. Farradane, *Nature*, **167**, 120(1951); Herbert Weil and T. I. Williams, *ibid.*, p. 906; Gerhard Hesse and H. Weil, *Michael Tswett's First Paper on Chromatography* (Woelm, Eschwege, 1954); and Trevor I. Williams, *The Elements of Chromatography* (Blackie, London, 1953). On Tswett see L. Zechmeister, *Isis*, **36**, 108(1946) and T. Robinson, *J. Chem. Educ.*, **36**, 144(1959).

For biographical sketches of Emich, Pregl, and Feigl see G. Kainz, *J. Chem. Educ.*, **12**, 608(1958). On Chamot there is C. W. Mason, *Ind. Engr. Chem., Anal. Ed.*, **11**, 341(1939); For Tiselius see R. E. Oesper, *J. Chem. Educ.*, **28**,

538(1951). There is an excellent biographical sketch of Heyrovsky in P. Zurman and Philip Elving, *ibid.*, **37**, 562(1960). For a fine sketch of Emich see Nicholas D. Cheronis, *Microchem. J.*, **4**, 423(1960).

CHAPTER 22. INORGANIC CHEMISTRY II: DECLINE AND RISE

There is no comprehensive history of inorganic chemistry or any of its facets except for Mary Elvira Weeks' *Discovery of the Elements* (J. Chem. Educ., Easton, Pa., 6th ed., 1956). The Golden Jubilee issue of *J. Am. Chem. Soc.*, C. A. Browne, ed., *A Half-Century of Chemistry in America* (Aug. 20, 1926) has a chapter on mineral chemistry by Edgar Fahs Smith and one on inorganic chemistry by James L. Howe. The seventy-fifth anniversary volume, *Chemistry . . . Key to Better Living* has a section on inorganic chemistry by L. F. Audrieth, reprinted from *Ind. Engr. Chem.*, **43**, 269(1951). For specialized phases of the field some help may be had from: *Advances in Inorganic Chemistry and Radiochemistry* (Academic Press, New York, vol. 1 in 1959 under the editorship of H. J. Emeléus and A. G. Sharp); and *Progress in Inorganic Chemistry* (Interscience, New York, vol. 1 in 1959 under the editorship of F. Albert Cotton).

It is possible to pick up snatches of historical background from the various standard textbooks and reference books such as Gmelin, Mellor, and Friend. Also see: the multivolume *Comprehensive Inorganic Chemistry* (Van Nostrand, New York, 1953 ff.) which is under the editorship of M. Cannon Sneed, J. Lewis Maynard, and Robert C. Brasted; Therald Moeller, *Inorganic Chemistry* (Wiley, New York, 1952); H. J. Emeléus and J. S. Anderson, *Modern Aspects of Inorganic Chemistry* (Van Nostrand, Princeton, 3rd ed., 1960; J. R. Partington, *A Text-Book of Inorganic Chemistry* (Macmillan, London, 6th ed., 1950); G. D. Parkes, ed., *Mellor's Modern Inorganic Chemistry* (Longmans, Green, London, rev. ed., 1939); E. S. Gilreath, *Fundamental Concepts of Inorganic Chemistry* (McGraw-Hill, New York, 1958); N. V. Sidgwick, *The Chemical Elements and Their Compounds* (Oxford Univ. Press, London, 1950); E. S. Gould, *Inorganic Reactions and Structures* (Holt, New York, 1956); Walter Hückel, *Structural Chemistry of Inorganic Compounds* (Elsevier, New York, 2 vols., 1951–52); Charles Compton, *An Introduction to Chemistry* (Van Nostrand, Princeton, 1958)—elementary, but contains a significant amount of historical material; T. M. Lowry, *Historical Introduction to Chemistry* (Macmillan, London, 1915, rev. 1936); F. Basolo and R. G. Pearson, *Mechanisms of Inorganic Reactions* (Wiley, New York, 1958).

On special phases of inorganic chemistry see: John C. Bailar, Jr., ed., *The Chemistry of the Coordination Compounds* (Reinhold, New York, 1956); F. H. Spedding and A. H. Daane, *The Rare Earths* (Wiley, New York, 1961); Alfred Stock, *The Hydrides of Boron and Silicon* (Cornell Univ. Press, Ithaca, 1933); W. G. Palmer, *Valency, Classical and Modern* (Cambridge Univ. Press, Cambridge, 1944); L. Orgel, *An Introduction to Transition Metal Chemistry* (Wiley, New York, 1960); E. G. Rochow, *Introduction to the Chemistry of the Silicones* (Wiley, New York, 2nd ed., 1951); *The Chemistry of the Organometallic Compounds* (Wiley, New York, 1957); Dallas T. Hurd, *An Introduction to the*

Chemistry of the Hydrides (Wiley, New York, 1952); Linus Pauling, *The Nature of the Chemical Bond* (Cornell Univ. Press, Ithaca, 3rd ed., 1960); Glenn T. Seaborg, *The Transuranium Elements* (Yale Univ. Press, New Haven, 1958); Joseph J. Katz and G. T. Seaborg, *Chemistry of the Actinide Elements* (Wiley, New York, 1957).

For biographical references there are the following: on Alfred Stock see the sketch in his *Hydrides of Boron and Silicon* and Egon Wiberg in Farber's *Great Chemists*; on Moseley see B. Jaffe, *Crucibles* (Simon & Schuster, New York, 1930); on Charles James there is Walter C. Schumb, *Life and Work of Charles James* (Boston, 1932) and B. S. Hopkins, *J. Wash. Acad. Sci.*, **22**, 21(1932); on Sugden there is L. E. Sutton, *Obit. Not. Fellows Roy. Soc.*, **7**, 493(1950–51); on W. J. Pope there is a chapter by G. T. Moody and W. H. Mills in A. Findlay, ed., *British Chemists* (Chem. Soc., London, 1947); the same volume has a chapter on Lowry by C. B. Allsop and W. A. Waters, one on A. Lapworth by R. Robinson, and one on J. F. Thorpe by G. A. R. Kon and R. P. Linstead. F. Challenger has a biography of Kipping in *Obit. Not. Fellows Roy. Soc.*, **7**, 183(1950–51). For Schlesinger see *Chem. Engr. News*, **27**, 496(1949). For Swarts see G. B. Kauffman, *J. Chem. Educ.*, **32**, 301(1955) and J. Timmermans and R. E. Oesper, *ibid.*, **38**, 423(1961). On Goldschmidt see J. D. Bernal, *J. Chem. Soc.*, **1949**, 2108.

CHAPTER 23. ORGANIC CHEMISTRY V: GROWTH AND TRANSFORMATION

Of the general histories of organic chemistry mentioned in the notes for Chapter 7, only Walden's gives significant attention to twentieth-century developments. There is a chapter on organic chemistry in America by Treat B. Johnson in the Golden Jubilee number (Aug. 20, 1926) of *J. Am. Chem. Soc.* and a section by H. L. Fischer in *Chemistry . . . Key to Better Living*, reprinted from *Ind. Eng. Chem.*, **43**, 289(1951).

Organic textbooks of the past generally gave considerable attention to historical matters, but this becomes less true of contemporary texts. A few of the principal exceptions are: Louis F. Fieser and Mary Fieser, *Organic Chemistry* (Heath, Boston, 3rd ed., 1956) and *Advanced Organic Chemistry* (Heath, Boston, 1962); James B. Conant and Albert H. Blatt, *Chemistry of Organic Compounds* (Macmillan, New York, 5th ed., 1959); Carl R. Noller, *Textbook of Organic Chemistry* (Saunders, Philadelphia, 2nd ed., 1958); John Read, *Direct Entry to Organic Chemistry* (Methuen, London, 1948) and *A Textbook of Organic Chemistry, Historical, Structural and Economic* (Bell, London, 3rd ed., 1943); Edgar Wertheim and Harold Jeskey, *Introductory Organic Chemistry* (McGraw-Hill, New York, 3rd ed., 1956); and Paul Karrer, *Organic Chemistry*, transl. by A. J. Mee (Elsevier, New York, 4th Engl. ed. from the 11th German, 1950). The latter book contains an appendix listing a chronology of important discoveries in organic chemistry. Some of the chapters in Henry Gilman, ed., *Organic Chemistry, An Advanced Treatise* (Wiley, New York, 2nd ed., 4 vols., 1943–1953) give a reasonably good picture of historical matters. Some of the chapters in A. Todd, ed., *Perspectives in Organic Chemistry* (Interscience, New York, 1956) show good historical background. This volume is commemorative of Robert Robinson. Also see C. K. Ingold, *Introduction to Structure in Organic Chemistry* (Bell, London, 1956) and *Structure and Mechanism in Organic Chemistry* (Cornell Univ. Press, Ithaca, 1953); and Edwin S. Gould, *Structure and Mechanism in Organic Chemistry* (Holt, Rinehart & Winston, New York, 1959). For reviews see: *Progress in Chemistry of Organic Natural Products* (Walter J. Johnson, New York, vol. 18 in 1960 under the editorship of L. Zechmeister); *Advances in Organic Chemistry* (Interscience, New York, vol. 2 in 1960 under the editorship of R. A. Raphael, E. C. Taylor, and H. Wynberg); and *Advances in Carbohydrate Chemistry* (Academic Press, New York, vol. 15 in 1960 under the editorship of M. L. Wolfrom and R. S. Tipson). A very useful volume is Alexander R. Surrey, *Name Reactions in Organic Chemistry* (Academic Press, New York, 2nd ed., 1961), which not only gives a brief description of the reaction but also references to original publications and to reviews as well as a short biographical sketch of the chemist originally developing the reaction.

For biographical information on Nobel Prize winning organic chemists see E. Farber, *Nobel Prize Winners in Chemistry* (Abelard-Schuman, New York, 2nd ed., 1963) which has short sketches on Grignard, Sabatier, Willstätter, Wieland, Windaus, Hans Fischer, Haworth, Karrer, Kuhn, Butenandt, Ruzicka, Diels, Alder, Todd, and Sanger.

Willstätter left an autobiography, *Aus meinem Leben* (Verlag Chemie, Weinheim Bergstr., 1949). Also see: R. Kuhn, *Naturwissenschaften*, **36**, 1(1949); and Rolf Huisgen, *J. Chem. Educ.*, **38**, 10(1961). On Karrer see R. E. Oesper, *ibid.*, **23**, 392(1946). On Robinson see *Chem. Eng. News*, **31**, 3844(1953). On Haworth see E. L. Hirst, *Obit. Not. Fellows Roy. Soc.*, **7**, 373(1950–1951). For Conant see *Chem. Eng. News*, **25**, 2004(1947). For Gilman see *ibid.*, **29**, 5132(1951). There is a sketch of Woodward in *ibid.*, **25**, 2137(1947); of Nef by M. L. Wolfrom, *Biog. Mem. Natl. Acad. Sci.*, **34**, 204(1960). Gomberg's life is described by A. White, *Ind. Eng. Chem.*, **23**, 10(1931) and by C. Schoepfle and W. Bachmann, *J. Am. Chem. Soc.*, **69**, 2921(1947). For Sabatier see Eric K. Rideal, *Obit. Not. Fellows Roy. Soc.*, **4**, 63(1942–1944). On Edward Armstrong there is C. Gibson and T. P. Hilditch, *ibid.*, **5**, 635(1945–1948).

CHAPTER 24. BIOCHEMISTRY II: THE DYNAMIC PERIOD

Twentieth-century biochemistry has received only minimal attention. Fritz Lieben's *Geschichte der physiologischen Chemie* (Deuticke, Leipzig & Vienna, 1935) is the only comprehensive historical work and, while valuable for studies up to 1900, is sketchy on studies after the turn of the century. E. V. McCollum's *A History of Nutrition* (Houghton Mifflin, Boston, 1957) is very good on the nutritional side of biochemistry. It carries the story only to 1940, however, and is sketchy on those phases of the subject dealing with chemical identification of the vitamins and the proof of their structure.

For a running account of work in the biochemical field one may consult the *Annual Review of Biochemistry* (Annual Reviews, Stanford, vol. 1 in 1932 under the editorship of J. Murray Luck) which sets a high standard for reviews. Other reviews of value in the biochemical field are: *Annual Review of Physiology* (Annual Reviews, Stanford, vol. 1 in 1939 under the editorship of J. Murray Luck); *Annual Review of Microbiology* (Annual Reviews, Stanford, vol. 1 in 1947 under the editorship of Charles E. Clifton); *Advances in Carbohydrate Chemistry* (Interscience, New York, vol. 1 on 1945 under the editorship of W. W. Pigman and M. L. Wolfrom); *Advances in Protein Chemistry* (Academic Press, New York, vol. 1 in 1944 under the editorship of M. L. Anson and John T. Edsall); *Advances in Enzymology and Related Subjects* (Interscience, New York, vol. 1 in 1941 under the editorship of F. F. Nord and C. H. Werkman); *Advances in Food Research* (Academic Press, New York, vol. 1 in 1948 under the editorship of E. M. Mrak and George F. Stewart); and *Vitamins and Hormones, Advances in Research and Applications* (Academic Press, New York, vol. 1 in 1943 under the editorship of Robert S. Harris). Also consult the indexes of such review journals as *Chemical Reviews, Physiological Reviews,* and *Nutrition Abstracts and Reviews.*

There are reviews of biochemical developments in America in the anniversary publications of the American Chemical Society. *A Half Century of Chemistry in America* (Aug. 20, 1926 issue of *J. Am. Chem. Soc.*) has a review of the chemistry of physiology and nutrition by Graham Lusk, and one on agricultural chemistry by C. A. Browne. *Chemistry . . . Key to Better Living* (Am. Chem. Soc., Washington, 1951) has the following reviews, all reprinted from *Ind. Eng. Chem.*, vol. 43(1951): "Agricultural and Food Chemistry" by F. C. Blanck, *et al.*; "Medicinal Chemistry" by M. L. Moore; "Biological Chemistry" by M. X. Sullivan; "Water, Sewage, and Sanitation Chemistry" by A. M. Buswell, *et al.* There is also Russell H. Chittenden, *The Development of Physiological Chemistry in the United States* (Chemical Catalog Co., New York, 1930).

Textbooks do not often present a historical picture, but from some of them it is possible to obtain some information regarding sequence of certain developments. See: Joseph S. Fruton and Sofia Simmonds, *General Biochemistry* (Wiley, New York, 2nd ed., 1958); Edward S. West and Wilbert R. Todd, *Textbook of Biochemistry* (Macmillan, New York, 3rd ed., 1961); Benjamin Harrow and Abraham Mazur, *Textbook of Biochemistry* (Saunders, Philadelphia, 7th ed., 1958); Abraham White, Philip Handler, Emil Smith, and DeWitt Stetten, Jr., *Principles of Biochemistry* (McGraw-Hill, New York, 2nd ed., 1959); Felix Haurowitz, *Biochemistry, An Introductory Textbook* (Wiley, New York, 1955); and Roger J. Williams and Ernest Beerstecher, Jr., *Introduction to Biochemistry* (Van Nostrand, New York, 2nd ed., 1948).

Several books deal with developments in special areas: Felix Haurowitz, *Progress in Biochemistry since 1949* (Interscience, New York, 1959); David E. Green, ed., *Currents in Biochemical Research* (Interscience, New York, 1946) and *Currents in Biochemical Research 1956* (Interscience, New York, 1956); D. E. Green and W. E. Knox, eds., *Research in Medical Science* (Macmillan, New York, 1950); E. S. Guzman Barron, ed., *Modern Trends in Physiology and Biochemistry* (Academic Press, New York, 1952)—being the Woods Hole Lectures dedicated to the memory of Leonor Michaelis and containing a biography by Barron; G. Pincus, ed., *Recent Progress in Hormone Research* (Proc. Lawrentian Hormone Conf., Academic Press, New York, vol. 1, 1947); Gregory Pincus and K. V. Thimann, eds., *The Hormones, Physiology, Chemistry and Applications* (Academic Press, New York, 3 vols., 1948–1955).

Other works of some value are: E. V. McCollum, *The Newer Knowledge of Nutrition* (Macmillan, New York, 5 edns., 1918–1939, the latter with Elsa Orent-Keiles and Harry G. Day, each edition being much enlarged and rewritten); Leslie J. Harris, *Vitamins in Theory and Practice* (Cambridge Univ. Press, Cambridge, 4th ed., 1955); Morris B. Jacobs, ed., *Food and Food Products* (Interscience, New York, 2nd ed., 3 vols., 1951); Jacob Sacks, *Isotopic Tracers in Biochemistry and Physiology* (McGraw-Hill, New York, 1953); Martin Kamen, *Isotopic Tracers in Biology* (Academic Press, New York, 3rd ed., 1957); Jacob Sacks, *The Atom at Work* (Ronald, New York, 1951); For the study of photosynthesis by use of isotopic tracers see: Melvin Calvin, *et al.*, *Carbon Dioxide Fixation and Photosynthesis* (Symposia for the Soc. of Exptl. Biol., no. 5, Academic Press, New York, 1951) and J. A. Bassham and M. Calvin, *The Path of Carbon in Photosynthesis* (Prentice-Hall, Englewood Cliffs, N.J., 1957). Vincent du Vigneaud gives a good account of his work on sulfur metabolism in his *A Trail of Research* (Cornell Univ. Press, Ithaca, 1952).

For material on F. G. Hopkins there is a commemorative volume edited by Joseph Needham and Ernest Baldwin: *Hopkins and Biochemistry* (Heffer, Cambridge, 1949). Also see H. H. Dale, *Obit. Not. Fellows Roy. Soc.*, **6**, 113 (1948–1949) and R. A. Peters, *Biochem. J.*, **71**, 1(1959). On Lind see Louis H. Roddis, *James Lind, Founder of Nautical Medicine* (Schuman, New York, 1950). On Funk there is Benjamin Harrow, *Casimir Funk, Pioneer in Vitamins and Hormones* (Dodd Mead, New York, 1955). On Mellanby see J. B. Leathes, *Obit. Not. Fellows Roy. Soc.*, **31**, 173(1939–1940).

For Babcock see: Louis Kahlenberg, *Ind. Eng. Chem.*, **16**, 1087(1924); H. L. Russell, *Science*, **74**, 86(1931); Paul de Kruif, *Hunger Fighters* (Harcourt Brace, New York, 1928, Chap. 9); and A. J. Ihde in Farber's *Great Chemists*. On Steenbock there is de Kruif, *op. cit.*, Chap. 10. On Hart there is E. H. Harvey, *Chem. Eng. News*, **22**, 435(1944). On Sherman see E. C. Kendall, *J. Chem. Educ.*, **32**, 510(1955) and Paul L. Day, *J. Nutrition*, **61**, 1(1957). Some of Sherman's papers, addresses, and reviews are collected in *Selected Works of Henry C. Sherman* (Macmillan, New York, 1948). This book includes a biography and complete bibliography. On Goldberger see de Kruif, *op. cit.*, Chap. 11 and Mary Farrar Goldberger, *J. Am. Dietetic Assoc.*, **32**, 724(1956).

On Kluyver see A. F. Kamp, J. W. M. La Riviere, and W. Verhoeven, eds., *Albert Jan Kluyver, His Life and Work* (Interscience, New York, 1959). On Meyerhof there is D. Nachmanson, *et al.*, *Biog. Mem. Natl. Acad. Sci.*, **34**, 153(1960). For Harden see F. G. Hopkins and C. J. Martin, *Obit. Not. Fellows Roy. Soc.*, **4**, 3(1942–1944) and C. J. Martin and F. G. Hopkins, in A. Findlay and W. H. Mills, eds., *British Chemists* (Chem. Soc., London, 1947). For Hans Fischer see H. Wieland, *Angew. Chem.*, **62**, 1(1950) and R. E. Oesper, *J. Chem. Educ.*, **15**, 350(1938). There is a note on Warburg by T. L. Sourkes, *ibid.*, **29**, 383 (1952). On Barcroft see Kenneth J. Franklin, *Joseph Barcroft, 1872–1947* (Blackwell, Oxford, 1953) and E. J. Roughton, *Obit. Not. Fellows Roy. Soc.*, **6**, 315(1948–1949).

For Thudicum there is David L. Drabkin, *Thudicum, Chemist of the Brain* (Univ. of Pennsylvania Press, Philadelphia, 1958).

On Starling there are E. B. Verney, *Ann. Sci.*, **12**, 30(1956) and R. Colp,

Jr., *Sci. Amer.*, **185**, no. 4, 56(1951). On Banting there is I. E. Levine, *The Discoverer of Insulin, Dr. Frederick Banting* (Messmer, New York, 1959). For Abel see H. H. Dale, *Obit. Not. Fellows Roy. Soc.*, **2**, 577(1936–1938). For Kendall see *Chem. Eng. News*, **28**, 2078(1950).

There are short sketches of all Nobel Prizewinning biochemists between 1901 and 1950 in either E. Farber, *Nobel Prize Winners in Chemistry* (Abelard-Schuman, New York, 2nd ed., 1963) or Lloyd G. Stevenson, *Nobel Prize Winners in Medicine* (Schuman, New York, 1953). In the former volume this includes E. Fischer, Buchner, Wallach, Willstätter, Wieland, Windaus, Harden, Euler-Chelpin, H. Fischer, Haworth, Karrer, Kuhn, Butenandt, Ruzicka, Virtanen, Sumner, Northrup, Stanley, Robinson, Tiselius, du Vigneaud, Todd, and Sanger. In the latter volume there are included Kossel, Meyerhof, Banting, Macleod, Wagner-Jauregg, Hopkins, Eijkman, Warburg, Szent-Györgyi, Domagk, Doisy, Dam, Fleming, Chain, Florey, Müller, Kendall, Hench, Reichstein, Lippmann, Krebs, Enders, Theorell, Beadle, Tatum, Ochoa, and Kornberg.

CHAPTERS 25 and 26. INDUSTRIAL CHEMISTRY II and III: THE TWENTIETH CENTURY

The principal general work on the history of industrial chemistry is F. Sherwood Taylor's *A History of Industrial Chemistry* (Heinemann, London, 1957) which gives a broad overview. It is strong on the chemical influences underlying the development of industrial chemistry, weak on specifics of chemical industries, and fails to give much attention to the twentieth century.

There is a very comprehensive history of American chemical industry by Williams Haynes: *American Chemical Industry, A History* (Van Nostrand, New York, 6 vols., 1945–1954). Titles and period of the individual volumes are; I. *Background and Beginnings, 1609–1911* (1954); II. *World War I Period, 1912–1922* (1945); III. *World War I Period, 1912–1922* (1945); IV. *The Merger Era, 1923–1929* (1948); V. *Decade of New Products, 1930–1939* (1954); and VI. *The Chemical Companies* (1949). The latter volume contains historical sketches of 219 American companies, submitted by personnel of the respective companies. Each of the other volumes contains a detailed chronology and extensive appendixes giving information on production, prices, imports, exports, tariff rates, mergers, and other matters. The set is illustrated with many portraits. Emphasis is heavily economic and organizational, although there is considerable attention to the influence of chemical developments. For other works dealing with specific countries see the notes for Chapter 17.

Williams Haynes has also published a number of popular books dealing with modern chemical industries: *This Chemical Age, The Miracle of Man-Made Materials* (Knopf, New York, 1942); *The Chemical Front* (Knopf, New York, 1943); *Men, Money and Molecules, Popular Story of Chemical Business* (Doubleday, Doran, New York, 1936); *The Stone That Burns, A History of Sulfur* (Van Nostrand, New York, 1942); and *Cellulose, The Chemical That Grows* (Doubleday, Garden City, N.Y., 1953). Other popular works which give some attention to industrial chemistry include: A. Findlay, *Chemistry in the Service of Man* (Longmans, Green, London, 8th ed., 1957) and *The Spirit of Chemistry* (Longmans, Green, London, 1930); Wm. A. Tilden, *Chemical Discovery and Invention in the Twentieth Century* (Routledge, London, 1917); Harry N. Holmes, *Out of the Test Tube* (Emerson, New York, 1943); A Cressy Morrison, *Man in a Chemical World* (Scribners, New York, 1937); Sidney J. French, *The Drama of Chemistry* (University Society, New York, 1937); C. C. Furnas, *The Next Hundred Years* (Reynal & Hitchcock, New York, 1936); I. B. Cohen, *Science, Servant of Man* (Little, Brown, Boston, 1948); Bernard Jaffe, *Chemistry Creates a New World* (Crowell, New York, 1957); Lawrence P. Lessing, *Understanding Chemistry* (Interscience, New York, 1959); J. Newton Friend, *Man and the Chemical Elements* (Scribners, New York, 1953); Richard Clements, *Modern Chemical Discoveries* (Dutton, New York, 1954); Walter S. Landis, *Your Servant the Molecule* (Macmillan, New York, 1944); Jacob Rosen and Max Eastman, *Road to Abundance* (McGraw-Hill, New York, 1953); Ritchie Calder, *Profile of Science* (Allen & Unwin, London, 1951); G. G. Hawley, *Small Wonder, The New Science of Colloids* (Knopf, New York, 1947). In the twenties the Chemical Foundation, a New York organization set up to impress the public with the need for a strong American chemical industry, was responsible for the publication and widespread distribution of several volumes dealing with applied chemistry. These, like some of the volumes just listed, are badly dated but are still useful for certain types of information: H. E. Howe, ed., *Chemistry in Industry* (2 vols., 1924–1925); Joseph S. Chamberlain, ed., *Chemistry in Agriculture* (1926); and Julius Stieglitz, ed., *Chemistry in Medicine* (1928).

There is a chapter on Industrial Chemistry by Charles E. Munroe in the Golden Anniversary Issue (Aug. 20, 1926) of *J. Am. Chem. Soc.*, and the Seventy-fifth Anniversary publication, *Chemistry . . . Key to Better Living* (Am. Chem. Soc., Washington, 1951; being reprints from vol. 43 of *Ind. Eng. Chem.*) carries papers by W. A. Pardee, *et al.*, on "Industrial and Engineering Chemistry"; by F. S. Lodge on "Fertilizer Chemistry"; by R. P. Dinsmore on "Rubber Chemistry"; by W. D. Horne, S. M. Cantor, and R. W. Liggett on "Sugar Chemistry"; Gustav Egloff and Mary L. Alexander on "Petroleum Chemistry"; W. O. Kenyon on "Cellulose Chemistry"; A. C. Fieldner, A. W. Gauger, and G. R. Yohe on "Gas and Fuel Chemistry"; and C. R. Bragdon and M. M. Renfrew on "Paint, Varnish, and Plastics Chemistry." The American Institute of Chemical Engineers published three commemorative volumes at the time of their golden anniversary: F. J. Van Antwerpen and Sylvia Fourdrinier, eds., *First Fifty Years of the American Institute of Chemical Engineers* (The Inst., New York, 1958); Edgar L. Piret, ed., *Chemical Engineering Around the World* (The Inst., New York, 1958); and Wm. T. Dixon and A. W. Fischer, Jr., *Chemical Engineering in Industry* (The Inst., New York, 1958).

The Manufacturing Chemists Association, Washington, D.C., began in 1953 the publication of *The Chemical Industry Facts Book* which gives a large amount of information about the industry. The fifth edition, 1962, is much enlarged.

J. G. Glover and W. B. C. Cornell, eds., *The Development of American Industries* (Prentice-Hall, New York, rev. ed., 1951) devotes a number of chapters to various chemical industries. Also see the excellent *Encyclopedia of Chemical Technology*, R. E. Kirk and D. F. Othmer, eds. (Interscience, New York, 15 vols., 1947–1956); Jeffry R. Stewart, *Encyclopedia of the Chemical Process Industries* (Chemical Publ. Co., New York, 1956); and George L. Clark, ed., *The Encyclopedia of Chemistry* (Reinhold, New York, 1957).

Carl R. Theiler, *Men and Molecules*, transl. from the German by E. Osers (Harrap, London, 1960), which deals with various chemical products treating them from a historical standpoint. J. L. Enos, *Petroleum Progress and Profits, A History of Process Innovation* (MIT Press, Cambridge, 1962) is an excellent study of petroleum cracking.

Several standard reference works on industrial chemistry have brief historical sections in connection with various industries: R. Norris Shreve, *Chemical Process Industries* (McGraw-Hill, New York, 1945) and *Selected Process Industries* (McGraw-Hill, New York, 1950); and *Rogers Industrial Chemistry*, ed. C. C. Furnas, *et al.*, (Van Nostrand, New York, 2 vols., 6th ed., 1942—1st ed. by Allen Rogers in 1912).

There is a significant amount of historical material, usually in the form of news stories, economic information, reviews, symposia, biographical matter, and pictures, in various and sundry trade, industrial, and engineering journals. Of particular importance are: *Chemical and Engineering News* which before 1942 was the *News Edition of Industrial and Engineering Chemistry*; *Chemical Week* which was founded in 1926 as *Chemical Industries*; *Chemical Engineering* which was founded in 1902 as *Electrochemical Industry* and was long published as *Chemical and Metallurgical Engineering*; *Reports on the Progress of Applied Chemistry*; *Chemisch Weekblad*; *Chemische Industrie*; *Chemistry and Industry*; originally a part of the British *Journal of the Society of Chemical Industry*; and *Chimie et Industrie*. The magazine *Fortune* frequently has good articles on chemical industries and on individual companies.

Several books relate chemistry and twentieth-century warfare. Victor Lefebure's *The Riddle of the Rhine* (Chemical Foundation, New York, 1923) relates, in a somewhat propagandistic fashion, the German rise to chemical dominance and the effect on the Allied Powers in their pursuit of the World War I. With respect to the second war there are several works of interest. A series of histories of activities growing out of work of the U.S. Office of Scientific Research and Development was published. The introductory volume by James Phinney Baxter 3rd, *Scientists Against Time* (Little, Brown, Boston, 1946) gives an overview of organization, weapons development, chemical warfare, medical developments, and the atomic bomb. More specifically chemical in the same series is Wm. A. Noyes, Jr., ed., *Chemistry, A History of the Chemistry Components of the NDRC* (Little, Brown, Boston, 1948). Development of the atomic bomb is described in Henry D. Smyth, *Atomic Energy for Military Purposes* (Princeton Univ. Press, Princeton, 1945) and in greater detail in R. G. Hewlett and O. E. Anderson, Jr., *A History of the United States Atomic Energy Commission*, vol. 1, *The New World, 1939/1946* (Penn. State Univ. Press, University Park, Pa., 1962). Also see Williams Haynes, *The Chemical Front* (Knopf, New York, 1943) and James R. Newman, *The Tools of War* (Doubleday, Doran, Garden City, N.Y., 1943). On the U.S. Chemical Corps there are Chem. Corps. Assoc., *The Chemical Warfare Service in World War II* (Reinhold, New York, 1948); and the official histories, published in 1959, Leo P. Brophy and George J. B. Fischer, *The Chemical Warfare Service: Organizing for War* and Leo P. Brophy, Wyndham Miles, and Rexmond C. Cochrane, *The Chemical Warfare Service: From Laboratory to Field.* Both are a part of the series "U.S. Army in World War II, The Technical Services" published by the Office of Military History, Dept. of the Army, Washington, D.C. A projected third volume is to deal with *Chemicals in Combat.* Still another volume of interest is Leslie E. Simon, *German Research in World War II* (Wiley, New York, 1947).

In the field of collected biographies Williams Haynes, *Chemical Pioneers* (Van Nostrand, New York, 1939) deals with some American industrialists. The sketches dealing with figures whose activities moved significantly into the twentieth century are Mallinckrodt, Klipstein, Dennis, Hasslacher, Queeny, Washburn, Dow, and Warner. Farber's *Great Chemists* includes chapters dealing with Carothers, Midgley, Langmuir, Bosch, Haber, Ipatieff, Dow, Little, and Baekeland.

On Dow see Murray Campbell and Harrison Hatton, *Herbert H. Dow, Pioneer in Creative Chemistry* (Appleton-Century-Crofts, New York, 1951). For Bosch there is K. Holdemann, *Naturwissenschaften*, **36**, 161(1949). For Haber see E. Berl, *J. Chem. Educ.*, **14**, 203(1937) and J. Coates, *J. Chem. Soc.*, **1939**, 1642. On Weizmann there is *Trial and Error, The Autobiography of Chaim Weizmann* (Harper, New York, 1949) and Ritchie Calder, *The Hand of Life* (Widenfeld & Nicolson, London, 1959). For Herty see Florence E. Wall, *The Chemist*, **9**, 124(1932). For Cottrell there is Frank Cameron, *Cottrell, Samaritan of Science* (Doubleday, Garden City, N.Y., 1952).

On Ehrlich there are two books and several articles: Martha Marquardt, *Paul Ehrlich* (Heinemann, London, 1949); Hans Loewe, *Paul Ehrlich, Schopfer der Chemotherapie* (Wissensch. Verlags., Stuttgart, 1950); Glenn Sonnedecker, *Wis. Pharmacist*, **28**, no. 7, 8(1960); Iago Galdston,*Sci. Monthly*, **79**, 395(1954); Georg Urdang, *Pharm. J.*, **172**, 211(1954); Paul Karrer, *J. Chem. Educ.*, **35**, 392(1958), and A. J. Ihde, *Sixth Ann. Lecture Series* (College of Pharmacy, Univ. of Texas, Austin, 1963), p. 19.

On Domagk see Ralph E. Oesper, *J. Chem. Educ.*, **31**, 188(1954) and anon., *MD med. Newsmag.*, **1**, no. 5, 26(1957). On Fleming see: K. Surrey, *Sir Alexander Fleming* (Cassell, London, 1959); Andre Maurois, *The Life of Alexander Fleming* (Dutton, New York, 1959); and L. Colebrook, *Obit. Not. Fellows Roy. Soc.*, **2**, 117(1956). On Duggar see E. C. Stakman, *Science*, **126**, 690(1957) and Farrington Daniels, *Year Book Am. Philos. Soc.*, **1957**, 117. On Waksman there is an autobiography, *My Life With the Microbes* (Simon & Schuster, New York, 1954). Waksman has also published a long article on the isolation and utilization of streptomycin. On the general history of antibiotics and sulfa drugs see A. J. Ihde, *op. cit.*, p. 25.

For Kettering see T. A. Boyd, *Professional Amateur, The Biography of Charles Franklin Kettering* (Dutton, New York, 1957) and Zay Jeffries, *Biog. Mem. Natl. Acad. Sci.*, **34**, 106(1960). On Kraus see *Chem. Eng. News*, **28**, 2704(1950). On Midgley see *ibid.*, **22**, 1896(1944).

On Baekeland see W. P. Cohoe, *Chem. Eng. News*, **23**, 228(1945) and L. V. Redman, *Ind. Eng. Chem.*, **20**, 1274 (1928). On Staudinger see Willem Quarles, *J. Chem. Educ.*, **28**, 120(1951) and *J. Polymer Sci.*, **19**, 387(1956). For Mark there is Gerald Oster, *J. Chem. Educ.*, **29**, 544(1952) and Morton M. Hunt, *The New Yorker* (Sept. 13, 1958, p. 48 and Sept. 20, 1958, p. 46).

There is a chapter on the chemical engineers in James Kip Finch, *The Story of Engineering* (Doubleday, Garden City, N.Y., 1960). Also see: Friedrich Klemm, *A History of Western Technology* (Scribners, New York, 1959); L. Aitchison, *A History of Metals* (Heffer, Cambridge, 1959); Courtney Robert Hall, *History of American Industrial Science* (Library Publishers, New York, 1954); and Simon Marcson, *The Scientist in American Industry* (Harper, New York, 1960).

The September 1957 issue of *Scientific American* is devoted to the subject of plastics in their various ramifications, the lead article being by Herman Mark.

CHAPTER 27. GROWTH AND PROBLEMS

On the growth of the chemical profession see Charles A. Browne and Mary Elvira Weeks' *A History of the American Chemical Society, Seventy-five Eventful Years* (Am. Chem. Soc., Washington, 1952) which deals very effectively with the activities of this American society. A companion volume, *Chemistry . . . Key to Better Living* (Am. Chem. Soc., Washington, 1951), brings together numerous articles published in volume 43 of *Industrial and Engineering Chemistry* and volume 29 of *Chemical and Engineering News* during the Diamond Jubilee year. These deal with division histories and many other matters. The earlier Golden Jubilee issue (Aug. 20, 1926) of the *Journal of the American Chemical Society* also contains much useful information.

For the British picture see Tom S. Moore and James C. Philip, *The Chemical Society 1841–1941, A Historica. Review* (Chemical Soc., London, 1947). For the German society see Paul Walden, *Ber.*, **75A**, 166(1942). For France see *Celebration du Centennaire de sa Fondation* (Soc. chim. de France, Paris, 1958).

There is also a great deal of scattered but useful material in various chemical publications. The Annual Reports of the American Chemical Society, readily available in *Chem. Eng. News*, give extensive evidence of the growth of the profession.

On the extent of the chemical literature see: E. J. Crane, Austin M. Patterson, and Eleanor B. Marr, *A Guide to the Literature of Chemistry* [Wiley, New York, 2nd ed., 1957 (the 1st ed. published by the two senior authors in 1927 is useful in giving more information about some of the publications)]; G. M. Dyson, *A Short Guide to the Chemical Literature* (Longmans, Green, London, 1951); Byron A. Soule, *Library Guide for the Chemist* (McGraw-Hill, New York, 1938); M. G. Mellon, *Chemical Publications* (McGraw-Hill, New York, 3rd ed., 1958); Charles H. Brown, *Scientific Serials* (Assoc. College & Reference Libraries, Chicago, 1956). An almost complete listing of journals dealing with the literature of chemistry will be found in the *List of Periodicals Abstracted by Chemical Abstracts* (Chem. Abs. Service, Ohio State Univ., Columbus, Ohio, 1956). For an analysis of over 200 of the most extensively used journals and handbooks see James van Luik and Associates, *Searching the Chemical and Chemical Engr. Literature* (Purdue Univ., Lafayette, Ind., 2nd ed., 1957). There is much information regarding the scope of university research in chemistry in the United States in the biennial *Directory of Graduate Research* prepared by the Committee on Professional Training of the American Chemical Society since 1953. This publication lists faculties, publications, and doctoral theses in departments of chemistry, biochemistry, and chemical engineering in United States schools offering graduate work in these fields.

On chemical nomenclature and its problems see: M. P. Crosland, *Historical Studies in the Language of Chemistry* (Harvard Univ. Press, Cambridge, 1962); G. M. Dyson, *A New Notation and Enumeration for Organic Compounds* (Longmans, Green, London, 2nd ed., 1949); Wm. J. Wiswesser, *A Line-Formula Chemical Notation* (Crowell, New York, 1954); A. M. Patterson, L. T. Capell, and D. F. Walker, *The Ring Index of Organic Compounds* (Am. Chem. Soc., Washington, 2nd ed., 1960); W. C. Fernelius, "Chemical Nomenclature," *Advances in Chemistry*, no. 8, 1953; W. C. Fernelius, *Chem. Eng. News*, **26**, 161(1948) and with E. M. Larsen, L. E. Marchi, and C. L. Rollinson, *ibid.*, 540; and Walter E. Flood, *The Origins of Chemical Names* (Oldbourne, London, 1963).

On the problems faced by civilization as a consequence of scientific activity there are a number of books and numerous journal articles. In particular, the reader is referred to the files of *Bulletin of the Atomic Scientists* and to various general periodicals such as *Atlantic, Harpers, Saturday Review, The Reporter,* and *Fortune*. General scientific journals such as *Science,* and *Nature* give some attention to the problems. Particularly important in the field of books is Harrison Brown, *The Challenge of Man's Future* (Viking, New York, 1954). Also see: Stringfellow Barr, *Let's Join the Human Race* (Univ. of Chicago Press, Chicago, 1950) and *Citizens of the World* (Doubleday, Garden City, N.Y., 1952); Max Otto, *Science and the Moral Life* [New American Library (Mentor), New York, 1949]; Alfred N. Whitehead, *Science and the Modern World* (Macmillan, New York, 1925) and *The Aims of Education* (Macmillan, New York, 1929); Bertrand Russell, *The Impact of Science on Society* (Simon & Schuster, New York, 1953); Julian S. Huxley, *Man Stands Alone* (Harper, New York, 1941) and *On Living in a Revolution* (Harper, New York, 1944; reprinted in part as *Man in the Modern World*, Mentor, New York, 1948); Fred Hoyle, *Man and Materialism* (Harper, New York, 1956); James B. Conant, *Modern Science and Modern Man* (Columbia Univ. Press, New York, 1952); J. Robert Oppenheimer, *The Open Mind* (Simon & Schuster, New York, 1955); Robert L. Heilbroner, *The Future as History* (Harper, New York, 1959); Roger J. Williams, *Free and Unequal, The Biological Basis of Individual Liberty* (Univ. of Texas Press, Austin, 1953); Martin Gardner, *In the Name of Science* (Putnam's, New York, 1952); and C. P. Snow, *Science and Government* (Harvard Univ. Press, Cambridge, 1961).

The problems of education in a scientific age are brilliantly discussed by C. P. Snow in *The Two Cultures and the Scientific Revolution* (Cambridge Univ. Press, New York, 1959). For other works dealing with educational problems see: James B. Conant, *On Understanding Science* (Yale Univ. Press, New Haven, 1947) and *Science and Common Sense* (Yale Univ. Press, New Haven, 1951); Philipp Frank, *Modern Science and its Philosophy* (Harvard Univ. Press, Cambridge, 1950); I. B. Cohen and Fletcher G. Watson, eds., *General*

Education in Science (Harvard Univ. Press, Cambridge, 1952); Robert Ulich, *Crisis and Hope in American Education* (Beacon, Boston, 1951); Earl J. McGrath, ed., *Science in General Education* (Brown, Dubuque, Ia., 1948); Robert R. Haun, ed., *Science in General Education* (Brown, Dubuque, Ia., 1961); Eric Ashby, *Technology and the Academics* (St Martin's Press, London, 1959).

On energy resources see: Eugene Ayres and Charles A. Scarlott, *Energy Sources, The Wealth of the World* (McGraw-Hill, New York, 1952); M. King Hubbert, *Science*, **109**, 103(1949); Farrington Daniels, *Am. Scientist*, **38**, 521(1950); and U.S. Dept. of State, *Energy Resources of the World* (Publ. no. 3428, Washington, 1949). For other resources there are: Erich W. Zimmerman, *World Resources and Industries* (Harper, New York, 1951); Paul K. Hatt, *World Population and Future Resources* (American Book Co., New York, 1952); the President's Materials Policy Commission, *Resources for Freedom* (U.S. Govt. Printing Office, Washington, 5 vols., 1952). On water resources the reader is referred to: B. Frank and A. Netboy, *Water, Land and People* (Knopf, New York, 1950); Harold E. Thomas, *The Conservation of Ground Water* (McGraw-Hill, New York, 1951); and A. H. Carhart, *Water in Your Life* (Lippincott, Philadelphia, 1951). Soil conservation problems are treated in: H. H. Bennett, *Soil Conservation* (McGraw-Hill, New York, 1939); Paul B. Sears, *Deserts on the March* (University of Oklahoma Press, Tulsa, 1935); E. H. Graham, *Natural Principles of Land Use* (Oxford Univ. Press, London, 1944); R. O. Whyte and G. V. Jacks, *Vanishing Lands* (Doubleday, Doran, New York, 1939); L. Dudley Stamp, *Land for Tomorrow* (Indiana Univ. Press, Bloomington, 1952); Charles E. Kellog, *The Soils That Support Us* (Macmillan, New York, 1941).

For information relating directly to the food supply see: United Nations Food and Agriculture Organization, *World Food Survey* (FAO, Washington, 1946) and *Food and Agriculture, World Conditions and Prospects* (FAO, Washington, 1949); R. N. Salaman, *The History and Social Significance of the Potato* (Cambridge Univ. Press, Cambridge, 1949); Frank A. Pearson and F. A. Harper, *The World's Hunger* (Cornell Univ. Press, Ithaca, 1945); E. Parmalee Prentice, *Food, War and the Future* (Harper, New York, 1944); Josue de Castro, *Geography of Hunger* (Little, Brown, Boston, 1952); T. W. Schulz, ed., *Food for the World* (Univ. of Chicago Press, Chicago, 1945); and Robert Brittain, *Let There Be Bread* (Simon & Schuster, New York, 1952). For a good short history of modern medical and public health techniques see B. J. Stern, *Society and Medical Progress* (Princeton Univ. Press, Princeton, 1941). C. Curwen's *Plough and Pasture* (Corbett, London, 1947) is a good short history of agriculture. For an account of changing food habits see Jack Drummond and A. Wilbraham, *The Englishman's Food* (Cape, London, 1939).

On the broad subject of waste of the world's resources one may consult: Kirtley F. Mather, *Enough and to Spare* (Harper, New York, 1944); Fairfield Osborn, *Our Plundered Planet* (Little, Brown, Boston, 1948); and William Vogt, *Road to Survival* (Wm. Sloan, New York, 1948) as well as various titles cited above.

The growth of the world's population is examined in the following works: W. S. and E. S. Woytinsky, *World Population and Production* (Twentieth

Century Fund, New York, 1953); Raymond Pearl, *Natural History of Population* (Oxford Univ. Press, London, 1939); Karl Sax, *Standing Room Only, The Challenge of Overpopulation* (Beacon, Boston, 1955); Elmer Pendell, *Population on the Loose* (Wilfred Funk, New York, 1951); Robert C. Cook, *Human Fertility, The Modern Dilemma* (Wm. Sloane, New York, 1951). The classical work of Thomas Malthus, *An Essay on the Principles of Population* (1798) is reprinted in numerous editions. Mentor recently brought out a reprint (MD295) under the title, *Three Essays on Population*, including Malthus' essay together with recent essays by Julian Huxley and Frederick Osborn.

On the subject of nuclear weapons and their implications the reader is referred to the notes for Chapter 19.

Discovery of Elements APPENDIX I (continued from page 749)

Year	Element	Discoverer
1955	Mendelevium (101)	Ghiorso, Harvey, Choppin, Thompson, and Seaborg
1958	Nobelium (102)	Synthesis by Ghiorso, Sikkeland, Walton, and Seaborg at the Lawrence Radiation Laboratory, California Berkeley, and in U.S.S.R. at the Joint Institute of Nuclear Research at Dubna by Flerov, et al. Berkeley claim confirmed by IUPAC, which also rejected an earlier claim by the Nobel Institute of Physics.
1961	Lawrencium (103)	Ghiorso, Sikkeland, Larsh, and Latimer
1968	Rutherfordium (104) or	Ghiorso and Berkeley group
1964	Kurchatovium	Flerov and Dubna group
1970	Hahnium (105)	Ghiorso and Berkeley group
1967	or Nielsbohrium	Flerov and Dubna group
1974	Element 106	Ghiorso and Berkeley group Also claimed by Flerov and Dubna group
1981	Element 107	Münzenberg, et al. (German Institute for Heavy Ion Research [GSI])
1982	Element 109 (single atom)	Münzenberg, et al. (GSI)

INDEX OF NAMES

INDEX OF SUBJECTS

A CATALOGUE OF
SELECTED DOVER BOOKS
IN ALL FIELDS OF INTEREST

A CATALOGUE OF SELECTED DOVER BOOKS IN ALL FIELDS OF INTEREST

CELESTIAL OBJECTS FOR COMMON TELESCOPES, T. W. Webb. The most used book in amateur astronomy: inestimable aid for locating and identifying nearly 4,000 celestial objects. Edited, updated by Margaret W. Mayall. 77 illustrations. Total of 645pp. 5⅜ x 8½.
20917-2, 20918-0 Pa., Two-vol. set $10.00

HISTORICAL STUDIES IN THE LANGUAGE OF CHEMISTRY, M. P. Crosland. The important part language has played in the development of chemistry from the symbolism of alchemy to the adoption of systematic nomenclature in 1892. ". . . wholeheartedly recommended,"—Science. 15 illustrations. 416pp. of text. 5⅝ x 8¼. 63702-6 Pa. $7.50

BURNHAM'S CELESTIAL HANDBOOK, Robert Burnham, Jr. Thorough, readable guide to the stars beyond our solar system. Exhaustive treatment, fully illustrated. Breakdown is alphabetical by constellation: Andromeda to Cetus in Vol. 1; Chamaeleon to Orion in Vol. 2; and Pavo to Vulpecula in Vol. 3. Hundreds of illustrations. Total of about 2000pp. 6⅛ x 9¼.
23567-X, 23568-8, 23673-0 Pa., Three-vol. set $32.85

THEORY OF WING SECTIONS: INCLUDING A SUMMARY OF AIRFOIL DATA, Ira H. Abbott and A. E. von Doenhoff. Concise compilation of subatomic aerodynamic characteristics of modern NASA wing sections, plus description of theory. 350pp. of tables. 693pp. 5⅜ x 8½.
60586-8 Pa. $9.95

DE RE METALLICA, Georgius Agricola. Translated by Herbert C. Hoover and Lou H. Hoover. The famous Hoover translation of greatest treatise on technological chemistry, engineering, geology, mining of early modern times (1556). All 289 original woodcuts. 638pp. 6¾ x 11.
60006-8 Clothbd. $19.95

THE ORIGIN OF CONTINENTS AND OCEANS, Alfred Wegener. One of the most influential, most controversial books in science, the classic statement for continental drift. Full 1966 translation of Wegener's final (1929) version. 64 illustrations. 246pp. 5⅜ x 8½.(EBE)61708-4 Pa. $5.00

THE PRINCIPLES OF PSYCHOLOGY, William James. Famous long course complete, unabridged. Stream of thought, time perception, memory, experimental methods; great work decades ahead of its time. Still valid, useful; read in many classes. 94 figures. Total of 1391pp. 5⅜ x 8½.
20381-6, 20382-4 Pa., Two-vol. set $17.90

YUCATAN BEFORE AND AFTER THE CONQUEST, Diego de Landa. First English translation of basic book in Maya studies, the only significant account of Yucatan written in the early post-Conquest era. Translated by distinguished Maya scholar William Gates. Appendices, introduction, 4 maps and over 120 illustrations added by translator. 162pp. 5⅜ x 8½.
23622-6 Pa. $3.00

THE MALAY ARCHIPELAGO, Alfred R. Wallace. Spirited travel account by one of founders of modern biology. Touches on zoology, botany, ethnography, geography, and geology. 62 illustrations, maps. 515pp. 5⅜ x 8½.
20187-2 Pa. $6.95

THE DISCOVERY OF THE TOMB OF TUTANKHAMEN, Howard Carter, A. C. Mace. Accompany Carter in the thrill of discovery, as ruined passage suddenly reveals unique, untouched, fabulously rich tomb. Fascinating account, with 106 illustrations. New introduction by J. M. White. Total of 382pp. 5⅜ x 8½. (Available in U.S. only) 23500-9 Pa. $5.50

THE WORLD'S GREATEST SPEECHES, edited by Lewis Copeland and Lawrence W. Lamm. Vast collection of 278 speeches from Greeks up to present. Powerful and effective models; unique look at history. Revised to 1970. Indices. 842pp. 5⅜ x 8½. 20468-5 Pa. $9.95

THE 100 GREATEST ADVERTISEMENTS, Julian Watkins. The priceless ingredient; His master's voice; 99 44/100% pure; over 100 others. How they were written, their impact, etc. Remarkable record. 130 illustrations. 233pp. 7⅞ x 10 3/5. 20540-1 Pa. $6.95

CRUICKSHANK PRINTS FOR HAND COLORING, George Cruickshank. 18 illustrations, one side of a page, on fine-quality paper suitable for watercolors. Caricatures of people in society (c. 1820) full of trenchant wit. Very large format. 32pp. 11 x 16. 23684-6 Pa. $6.00

THIRTY-TWO COLOR POSTCARDS OF TWENTIETH-CENTURY AMERICAN ART, Whitney Museum of American Art. Reproduced in full color in postcard form are 31 art works and one shot of the museum. Calder, Hopper, Rauschenberg, others. Detachable. 16pp. 8¼ x 11.
23629-3 Pa. $3.50

MUSIC OF THE SPHERES: THE MATERIAL UNIVERSE FROM ATOM TO QUASAR SIMPLY EXPLAINED, Guy Murchie. Planets, stars, geology, atoms, radiation, relativity, quantum theory, light, antimatter, similar topics. 319 figures. 664pp. 5⅜ x 8½.
21809-0, 21810-4 Pa., Two-vol. set $11.00

EINSTEIN'S THEORY OF RELATIVITY, Max Born. Finest semi-technical account; covers Einstein, Lorentz, Minkowski, and others, with much detail, much explanation of ideas and math not readily available elsewhere on this level. For student, non-specialist. 376pp. 5⅜ x 8½.
60769-0 Pa. $5.00

THE SENSE OF BEAUTY, George Santayana. Masterfully written discussion of nature of beauty, materials of beauty, form, expression; art, literature, social sciences all involved. 168pp. 5⅜ x 8½. 20238-0 Pa. $3.50

ON THE IMPROVEMENT OF THE UNDERSTANDING, Benedict Spinoza. Also contains *Ethics, Correspondence*, all in excellent R. Elwes translation. Basic works on entry to philosophy, pantheism, exchange of ideas with great contemporaries. 402pp. 5⅜ x 8½. 20250-X Pa. $5.95

THE TRAGIC SENSE OF LIFE, Miguel de Unamuno. Acknowledged masterpiece of existential literature, one of most important books of 20th century. Introduction by Madariaga. 367pp. 5⅜ x 8½.
20257-7 Pa. $6.00

THE GUIDE FOR THE PERPLEXED, Moses Maimonides. Great classic of medieval Judaism attempts to reconcile revealed religion (Pentateuch, commentaries) with Aristotelian philosophy. Important historically, still relevant in problems. Unabridged Friedlander translation. Total of 473pp. 5⅜ x 8½. 20351-4 Pa. $6.95

THE I CHING (THE BOOK OF CHANGES), translated by James Legge. Complete translation of basic text plus appendices by Confucius, and Chinese commentary of most penetrating divination manual ever prepared. Indispensable to study of early Oriental civilizations, to modern inquiring reader. 448pp. 5⅜ x 8½. 21062-6 Pa. $6.00

THE EGYPTIAN BOOK OF THE DEAD, E. A. Wallis Budge. Complete reproduction of Ani's papyrus, finest ever found. Full hieroglyphic text, interlinear transliteration, word for word translation, smooth translation. Basic work, for Egyptology, for modern study of psychic matters. Total of 533pp. 6½ x 9¼. (USCO) 21866-X Pa. $8.50

THE GODS OF THE EGYPTIANS, E. A. Wallis Budge. Never excelled for richness, fullness: all gods, goddesses, demons, mythical figures of Ancient Egypt; their legends, rites, incarnations, variations, powers, etc. Many hieroglyphic texts cited. Over 225 illustrations, plus 6 color plates. Total of 988pp. 6⅛ x 9¼. (EBE)
22055-9, 22056-7 Pa., Two-vol. set $20.00

THE STANDARD BOOK OF QUILT MAKING AND COLLECTING, Marguerite Ickis. Full information, full-sized patterns for making 46 traditional quilts, also 150 other patterns. Quilted cloths, lame, satin quilts, etc. 483 illustrations. 273pp. 6⅞ x 9⅝. 20582-7 Pa. $5.95

CORAL GARDENS AND THEIR MAGIC, Bronsilaw Malinowski. Classic study of the methods of tilling the soil and of agricultural rites in the Trobriand Islands of Melanesia. Author is one of the most important figures in the field of modern social anthropology. 143 illustrations. Indexes. Total of 911pp. of text. 5⅝ x 8¼. (Available in U.S. only)
23597-1 Pa. $12.95

CATALOGUE OF DOVER BOOKS

THE PHILOSOPHY OF HISTORY, Georg W. Hegel. Great classic of Western thought develops concept that history is not chance but a rational process, the evolution of freedom. 457pp. 5⅜ x 8½. 20112-0 Pa. $6.00

LANGUAGE, TRUTH AND LOGIC, Alfred J. Ayer. Famous, clear introduction to Vienna, Cambridge schools of Logical Positivism. Role of philosophy, elimination of metaphysics, nature of analysis, etc. 160pp. 5⅜ x 8½. (USCO) 20010-8 Pa. $2.50

A PREFACE TO LOGIC, Morris R. Cohen. Great City College teacher in renowned, easily followed exposition of formal logic, probability, values, logic and world order and similar topics; no previous background needed. 209pp. 5⅜ x 8½. 23517-3 Pa. $4.95

REASON AND NATURE, Morris R. Cohen. Brilliant analysis of reason and its multitudinous ramifications by charismatic teacher. Interdisciplinary, synthesizing work widely praised when it first appeared in 1931. Second (1953) edition. Indexes. 496pp. 5⅜ x 8½. 23633-1 Pa. $7.50

AN ESSAY CONCERNING HUMAN UNDERSTANDING, John Locke. The only complete edition of enormously important classic, with authoritative editorial material by A. C. Fraser. Total of 1176pp. 5⅜ x 8½.
20530-4, 20531-2 Pa., Two-vol. set $16.00

HANDBOOK OF MATHEMATICAL FUNCTIONS WITH FORMULAS, GRAPHS, AND MATHEMATICAL TABLES, edited by Milton Abramowitz and Irene A. Stegun. Vast compendium: 29 sets of tables, some to as high as 20 places. 1,046pp. 8 x 10½. 61272-4 Pa. $17.95

MATHEMATICS FOR THE PHYSICAL SCIENCES, Herbert S. Wilf. Highly acclaimed work offers clear presentations of vector spaces and matrices, orthogonal functions, roots of polynomial equations, conformal mapping, calculus of variations, etc. Knowledge of theory of. functions of real and complex variables is assumed. Exercises and solutions. Index. 284pp. 5⅝ x 8¼. 63635-6 Pa. $5.00

THE PRINCIPLE OF RELATIVITY, Albert Einstein et al. Eleven most important original papers on special and general theories. Seven by Einstein, two by Lorentz, one each by Minkowski and Weyl. All translated, unabridged. 216pp. 5⅜ x 8½. 60081-5 Pa. $3.50

THERMODYNAMICS, Enrico Fermi. A classic of modern science. Clear, organized treatment of systems, first and second laws, entropy, thermodynamic potentials, gaseous reactions, dilute solutions, entropy constant. No math beyond calculus required. Problems. 160pp. 5⅜ x 8½.
60361-X Pa. $4.00

ELEMENTARY MECHANICS OF FLUIDS, Hunter Rouse. Classic undergraduate text widely considered to be far better than many later books. Ranges from fluid velocity and acceleration to role of compressibility in fluid motion. Numerous examples, questions, problems. 224 illustrations. 376pp. 5⅝ x 8¼. 63699-2 Pa. $7.00

THE AMERICAN SENATOR, Anthony Trollope. Little known, long unavailable Trollope novel on a grand scale. Here are humorous comment on American vs. English culture, and stunning portrayal of a heroine/villainess. Superb evocation of Victorian village life. 561pp. 5⅜ x 8½.
23801-6 Pa. $7.95

WAS IT MURDER? James Hilton. The author of *Lost Horizon* and *Goodbye, Mr. Chips* wrote one detective novel (under a pen-name) which was quickly forgotten and virtually lost, even at the height of Hilton's fame. This edition brings it back—a finely crafted public school puzzle resplendent with Hilton's stylish atmosphere. A thoroughly English thriller by the creator of Shangri-la. 252pp. 5⅜ x 8. (Available in U.S. only)
23774-5 Pa. $3.00

CENTRAL PARK: A PHOTOGRAPHIC GUIDE, Victor Laredo and Henry Hope Reed. 121 superb photographs show dramatic views of Central Park: Bethesda Fountain, Cleopatra's Needle, Sheep Meadow, the Blockhouse, plus people engaged in many park activities: ice skating, bike riding, etc. Captions by former Curator of Central Park, Henry Hope Reed, provide historical view, changes, etc. Also photos of N.Y. landmarks on park's periphery. 96pp. 8½ x 11. 23750-8 Pa. $4.50

NANTUCKET IN THE NINETEENTH CENTURY, Clay Lancaster. 180 rare photographs, stereographs, maps, drawings and floor plans recreate unique American island society. Authentic scenes of shipwreck, lighthouses, streets, homes are arranged in geographic sequence to provide walking-tour guide to old Nantucket existing today. Introduction, captions. 160pp. 8⅞ x 11¾. 23747-8 Pa. $7.95

STONE AND MAN: A PHOTOGRAPHIC EXPLORATION, Andreas Feininger. 106 photographs by *Life* photographer Feininger portray man's deep passion for stone through the ages. Stonehenge-like megaliths, fortified towns, sculpted marble and crumbling tenements show textures, beauties, fascination. 128pp. 9¼ x 10¾. 23756-7 Pa. $5.95

CIRCLES, A MATHEMATICAL VIEW, D. Pedoe. Fundamental aspects of college geometry, non-Euclidean geometry, and other branches of mathematics: representing circle by point. Poincare model, isoperimetric property, etc. Stimulating recreational reading. 66 figures. 96pp. 5⅝ x 8¼.
63698-4 Pa. $3.50

THE DISCOVERY OF NEPTUNE, Morton Grosser. Dramatic scientific history of the investigations leading up to the actual discovery of the eighth planet of our solar system. Lucid, well-researched book by well-known historian of science. 172pp. 5⅜ x 8½. 23726-5 Pa. $3.50

THE DEVIL'S DICTIONARY. Ambrose Bierce. Barbed, bitter, brilliant witticisms in the form of a dictionary. Best, most ferocious satire America has produced. 145pp. 5⅜ x 8½. 20487-1 Pa. $2.50

CATALOGUE OF DOVER BOOKS

HISTORY OF BACTERIOLOGY, William Bulloch. The only comprehensive history of bacteriology from the beginnings through the 19th century. Special emphasis is given to biography-Leeuwenhoek, etc. Brief accounts of 350 bacteriologists form a separate section. No clearer, fuller study, suitable to scientists and general readers, has yet been written. 52 illustrations. 448pp. 5⅝ x 8¼.   23761-3 Pa. $6.50

THE COMPLETE NONSENSE OF EDWARD LEAR, Edward Lear. All nonsense limericks, zany alphabets, Owl and Pussycat, songs, nonsense botany, etc., illustrated by Lear. Total of 321pp. 5⅜ x 8½. (Available in U.S. only)   20167-8 Pa. $4.50

INGENIOUS MATHEMATICAL PROBLEMS AND METHODS, Louis A. Graham. Sophisticated material from Graham *Dial*, applied and pure; stresses solution methods. Logic, number theory, networks, inversions, etc. 237pp. 5⅜ x 8½.   20545-2 Pa. $4.50

BEST MATHEMATICAL PUZZLES OF SAM LOYD, edited by Martin Gardner. Bizarre, original, whimsical puzzles by America's greatest puzzler. From fabulously rare *Cyclopedia*, including famous 14-15 puzzles, the Horse of a Different Color, 115 more. Elementary math. 150 illustrations. 167pp. 5⅜ x 8½.   20498-7 Pa. $3.50

THE BASIS OF COMBINATION IN CHESS, J. du Mont. Easy-to-follow, instructive book on elements of combination play, with chapters on each piece and every powerful combination team—two knights, bishop and knight, rook and bishop, etc. 250 diagrams. 218pp. 5⅜ x 8½. (Available in U.S. only)   23644-7 Pa. $4.50

MODERN CHESS STRATEGY, Ludek Pachman. The use of the queen, the active king, exchanges, pawn play, the center, weak squares, etc. Section on rook alone worth price of the book. Stress on the moderns. Often considered the most important book on strategy. 314pp. 5⅜ x 8½.
                                                                20290-9 Pa. $5.00

LASKER'S MANUAL OF CHESS, Dr. Emanuel Lasker. Great world champion offers very thorough coverage of all aspects of chess. Combinations, position play, openings, end game, aesthetics of chess, philosophy of struggle, much more. Filled with analyzed games. 390pp. 5⅜ x 8½.
                                                                20640-8 Pa. $5.95

500 MASTER GAMES OF CHESS, S. Tartakower, J. du Mont. Vast collection of great chess games from 1798-1938, with much material nowhere else readily available. Fully annotated, arranged by opening for easier study. 664pp. 5⅜ x 8½.   23208-5 Pa. $8.50

A GUIDE TO CHESS ENDINGS, Dr. Max Euwe, David Hooper. One of the finest modern works on chess endings. Thorough analysis of the most frequently encountered endings by former world champion. 331 examples, each with diagram. 248pp. 5⅜ x 8½.   23332-4 Pa. $3.95

CATALOGUE OF DOVER BOOKS

THE COMPLETE BOOK OF DOLL MAKING AND COLLECTING, Catherine Christopher. Instructions, patterns for dozens of dolls, from rag doll on up to elaborate, historically accurate figures. Mould faces, sew clothing, make doll houses, etc. Also collecting information. Many illustrations. 288pp. 6 x 9. 22066-4 Pa. $4.95

THE DAGUERREOTYPE IN AMERICA, Beaumont Newhall. Wonderful portraits, 1850's townscapes, landscapes; full text plus 104 photographs. The basic book. Enlarged 1976 edition. 272pp. 8¼ x 11¼. 23322-7 Pa. $7.95

CRAFTSMAN HOMES, Gustav Stickley. 296 architectural drawings, floor plans, and photographs illustrate 40 different kinds of "Mission-style" homes from *The Craftsman* (1901-16), voice of American style of simplicity and organic harmony. Thorough coverage of Craftsman idea in text and picture, now collector's item. 224pp. 8⅛ x 11. 23791-5 Pa. $6.50

PEWTER-WORKING: INSTRUCTIONS AND PROJECTS, Burl N. Osborn. & Gordon O. Wilber. Introduction to pewter-working for amateur craftsman. History and characteristics of pewter; tools, materials, step-by-step instructions. Photos, line drawings, diagrams. Total of 160pp. 7⅞ x 10¾. 23786-9 Pa. $3.50

THE GREAT CHICAGO FIRE, edited by David Lowe. 10 dramatic, eye-witness accounts of the 1871 disaster, including one of the aftermath and rebuilding, plus 70 contemporary photographs and illustrations of the ruins—courthouse, Palmer House, Great Central Depot, etc. Introduction by David Lowe. 87pp. 8¼ x 11. 23771-0 Pa. $4.00

SILHOUETTES: A PICTORIAL ARCHIVE OF VARIED ILLUSTRATIONS, edited by Carol Belanger Grafton. Over 600 silhouettes from the 18th to 20th centuries include profiles and full figures of men and women, children, birds and animals, groups and scenes, nature, ships, an alphabet. Dozens of uses for commercial artists and craftspeople. 144pp. 8⅜ x 11¼. 23781-8 Pa. $4.50

ANIMALS: 1,419 COPYRIGHT-FREE ILLUSTRATIONS OF MAMMALS, BIRDS, FISH, INSECTS, ETC., edited by Jim Harter. Clear wood engravings present, in extremely lifelike poses, over 1,000 species of animals. One of the most extensive copyright-free pictorial sourcebooks of its kind. Captions. Index. 284pp. 9 x 12. 23766-4 Pa. $8.95

INDIAN DESIGNS FROM ANCIENT ECUADOR, Frederick W. Shaffer. 282 original designs by pre-Columbian Indians of Ecuador (500-1500 A.D.). Designs include people, mammals, birds, reptiles, fish, plants, heads, geometric designs. Use as is or alter for advertising, textiles, leathercraft, etc. Introduction. 95pp. 8¾ x 11¼. 23764-8 Pa. $4.50

SZIGETI ON THE VIOLIN, Joseph Szigeti. Genial, loosely structured tour by premier violinist, featuring a pleasant mixture of reminiscenes, insights into great music and musicians, innumerable tips for practicing violinists. 385 musical passages. 256pp. 5⅝ x 8¼. 23763-X Pa. $4.00

CATALOGUE OF DOVER BOOKS

TONE POEMS, SERIES II: TILL EULENSPIEGELS LUSTIGE STREICHE, ALSO SPRACH ZARATHUSTRA, AND EIN HELDEN-LEBEN, Richard Strauss. Three important orchestral works, including very popular *Till Eulenspiegel's Marry Pranks,* reproduced in full score from original editions. Study score. 315pp. 9⅜ x 12¼. (Available in U.S. only)
23755-9 Pa. $8.95

TONE POEMS, SERIES I: DON JUAN, TOD UND VERKLARUNG AND DON QUIXOTE, Richard Strauss. Three of the most often performed and recorded works in entire orchestral repertoire, reproduced in full score from original editions. Study score. 286pp. 9⅜ x 12¼. (Available in U.S. only)
23754-0 Pa. $8.95

11 LATE STRING QUARTETS, Franz Joseph Haydn. The form which Haydn defined and "brought to perfection." (*Grove's*). 11 string quartets in complete score, his last and his best. The first in a projected series of the complete Haydn string quartets. Reliable modern Eulenberg edition, otherwise difficult to obtain. 320pp. 8⅜ x 11¼. (Available in U.S. only)
23753-2 Pa. $8.95

FOURTH, FIFTH AND SIXTH SYMPHONIES IN FULL SCORE, Peter Ilyitch Tchaikovsky. Complete orchestral scores of Symphony No. 4 in F Minor, Op. 36; Symphony No. 5 in E Minor, Op. 64; Symphony No. 6 in B Minor, "Pathetique," Op. 74. Bretikopf & Hartel eds. Study score. 480pp. 9⅜ x 12¼. 23861-X Pa. $10.95

THE MARRIAGE OF FIGARO: COMPLETE SCORE, Wolfgang A. Mozart. Finest comic opera ever written. Full score, not to be confused with piano renderings. Peters edition. Study score. 448pp. 9⅜ x 12¼. (Available in U.S. only)
23751-6 Pa. $12.95

"IMAGE" ON THE ART AND EVOLUTION OF THE FILM, edited by Marshall Deutelbaum. Pioneering book brings together for first time 38 groundbreaking articles on early silent films from *Image* and 263 illustrations newly shot from rare prints in the collection of the International Museum of Photography. A landmark work. Index. 256pp. 8¼ x 11.
23777-X Pa. $8.95

AROUND-THE-WORLD COOKY BOOK, Lois Lintner Sumption and Marguerite Lintner Ashbrook. 373 cooky and frosting recipes from 28 countries (America, Austria, China, Russia, Italy, etc.) include Viennese kisses, rice wafers, London strips, lady fingers, hony, sugar spice, maple cookies, etc. Clear instructions. All tested. 38 drawings. 182pp. 5⅜ x 8.
23802-4 Pa. $2.75

THE ART NOUVEAU STYLE, edited by Roberta Waddell. 579 rare photographs, not available elsewhere, of works in jewelry, metalwork, glass, ceramics, textiles, architecture and furniture by 175 artists—Mucha, Seguy, Lalique, Tiffany, Gaudin, Hohlwein, Saarinen, and many others. 288pp. 8⅜ x 11¼. 23515-7 Pa. $8.95

THE CURVES OF LIFE, Theodore A. Cook. Examination of shells, leaves, horns, human body, art, etc., in *"the* classic reference on how the golden ratio applies to spirals and helices in nature . . . . "—Martin Gardner. 426 illustrations. Total of 512pp. 5⅜ x 8½. 23701-X Pa. $6.95

AN ILLUSTRATED FLORA OF THE NORTHERN UNITED STATES AND CANADA, Nathaniel L. Britton, Addison Brown. Encyclopedic work covers 4666 species, ferns on up. Everything. Full botanical information, illustration for each. This earlier edition is preferred by many to more recent revisions. 1913 edition. Over 4000 illustrations, total of 2087pp. 6⅛ x 9¼. 22642-5, 22643-3, 22644-1 Pa., Three-vol. set $28.50

MANUAL OF THE GRASSES OF THE UNITED STATES, A. S. Hitchcock, U.S. Dept. of Agriculture. The basic study of American grasses, both indigenous and escapes, cultivated and wild. Over 1400 species. Full descriptions, information. Over 1100 maps, illustrations. Total of 1051pp. 5⅜ x 8½. 22717-0, 22718-9 Pa., Two-vol. set $17.00

THE CACTACEAE,, Nathaniel L. Britton, John N. Rose. Exhaustive, definitive. Every cactus in the world. Full botanical descriptions. Thorough statement of nomenclatures, habitat, detailed finding keys. The one book needed by every cactus enthusiast. Over 1275 illustrations. Total of 1080pp. 8 x 10¼. 21191-6, 21192-4 Clothbd., Two-vol. set $50.00

AMERICAN MEDICINAL PLANTS, Charles F. Millspaugh. Full descriptions, 180 plants covered: history; physical description; methods of preparation with all chemical constituents extracted; all claimed curative or adverse effects. 180 full-page plates. Classification table. 804pp. 6½ x 9¼. 23034-1 Pa. $13.95

A MODERN HERBAL, Margaret Grieve. Much the fullest, most exact, most useful compilation of herbal material. Gigantic alphabetical encyclopedia, from aconite to zedoary, gives botanical information, medical properties, folklore, economic uses, and much else. Indispensable to serious reader. 161 illustrations. 888pp. 6½ x 9¼. (Available in U.S. only) 22798-7, 22799-5 Pa., Two-vol. set $15.00

THE HERBAL or GENERAL HISTORY OF PLANTS, John Gerard. The 1633 edition revised and enlarged by Thomas Johnson. Containing almost 2850 plant descriptions and 2705 superb illustrations, Gerard's *Herbal* is a monumental work, the book all modern English herbals are derived from, the one herbal every serious enthusiast should have in its entirety. Original editions are worth perhaps $750. 1678pp. 8½ x 12¼. 23147-X Clothbd. $75.00

MANUAL OF THE TREES OF NORTH AMERICA, Charles S. Sargent. The basic survey of every native tree and tree-like shrub, 717 species in all. Extremely full descriptions, information on habitat, growth, locales, economics, etc. Necessary to every serious tree lover. Over 100 finding keys. 783 illustrations. Total of 986pp. 5⅜ x 8½. 20277-1, 20278-X Pa., Two-vol. set $12.00

GREAT NEWS PHOTOS AND THE STORIES BEHIND THEM, John Faber. Dramatic volume of 140 great news photos, 1855 through 1976, and revealing stories behind them, with both historical and technical information. Hindenburg disaster, shooting of Oswald, nomination of Jimmy Carter, etc. 160pp. 8¼ x 11. 23667-6 Pa. $6.00

CRUICKSHANK'S PHOTOGRAPHS OF BIRDS OF AMERICA, Allan D. Cruickshank. Great ornithologist, photographer presents 177 closeups, groupings, panoramas, flightings, etc., of about 150 different birds. Expanded *Wings in the Wilderness*. Introduction by Helen G. Cruickshank. 191pp. 8¼ x 11. 23497-5 Pa. $7.95

AMERICAN WILDLIFE AND PLANTS, A. C. Martin, et al. Describes food habits of more than 1000 species of mammals, birds, fish. Special treatment of important food plants. Over 300 illustrations. 500pp. 5⅜ x 8½. 20793-5 Pa. $6.50

THE PEOPLE CALLED SHAKERS, Edward D. Andrews. Lifetime of research, definitive study of Shakers: origins, beliefs, practices, dances, social organization, furniture and crafts, impact on 19th-century USA, present heritage. Indispensable to student of American history, collector. 33 illustrations. 351pp. 5⅜ x 8½. 21081-2 Pa. $4.50

OLD NEW YORK IN EARLY PHOTOGRAPHS, Mary Black. New York City as it was in 1853-1901, through 196 wonderful photographs from N.-Y. Historical Society. Great Blizzard, Lincoln's funeral procession, great buildings. 228pp. 9 x 12. 22907-6 Pa. $8.95

MR. LINCOLN'S CAMERA MAN: MATHEW BRADY, Roy Meredith. Over 300 Brady photos reproduced directly from original negatives, photos. Jackson, Webster, Grant, Lee, Carnegie, Barnum; Lincoln; Battle Smoke, Death of Rebel Sniper, Atlanta Just After Capture. Lively commentary. 368pp. 8⅜ x 11¼. 23021-X Pa. $11.95

TRAVELS OF WILLIAM BARTRAM, William Bartram. From 1773-8, Bartram explored Northern Florida, Georgia, Carolinas, and reported on wild life, plants, Indians, early settlers. Basic account for period, entertaining reading. Edited by Mark Van Doren. 13 illustrations. 141pp. 5⅜ x 8½. 20013-2 Pa. $6.00

THE GENTLEMAN AND CABINET MAKER'S DIRECTOR, Thomas Chippendale. Full reprint, 1762 style book, most influential of all time; chairs, tables, sofas, mirrors, cabinets, etc. 200 plates, plus 24 photographs of surviving pieces. 249pp. 9⅞ x 12¾. 21601-2 Pa. $8.95

AMERICAN CARRIAGES, SLEIGHS, SULKIES AND CARTS, edited by Don H. Berkebile. 168 Victorian illustrations from catalogues, trade journals, fully captioned. Useful for artists. Author is Assoc. Curator, Div. of Transportation of Smithsonian Institution. 168pp. 8½ x 9½.
23328-6 Pa. $5.00

SECOND PIATIGORSKY CUP, edited by Isaac Kashdan. One of the greatest tournament books ever produced in the English language. All 90 games of the 1966 tournament, annotated by players, most annotated by both players. Features Petrosian, Spassky, Fischer, Larsen, six others. 228pp. 5⅜ x 8½.   23572-6 Pa. $3.50

ENCYCLOPEDIA OF CARD TRICKS, revised and edited by Jean Hugard. How to perform over 600 card tricks, devised by the world's greatest magicians: impromptus, spelling tricks, key cards, using special packs, much, much more. Additional chapter on card technique. 66 illustrations. 402pp. 5⅜ x 8½. (Available in U.S. only)   21252-1 Pa. $5.95

MAGIC: STAGE ILLUSIONS, SPECIAL EFFECTS AND TRICK PHOTOGRAPHY, Albert A. Hopkins, Henry R. Evans. One of the great classics; fullest, most authorative explanation of vanishing lady, levitations, scores of other great stage effects. Also small magic, automata, stunts. 446 illustrations. 556pp. 5⅜ x 8½.   23344-8 Pa. $6.95

THE SECRETS OF HOUDINI, J. C. Cannell. Classic study of Houdini's incredible magic, exposing closely-kept professional secrets and revealing, in general terms, the whole art of stage magic. 67 illustrations. 279pp. 5⅜ x 8½.   22913-0 Pa. $4.00

HOFFMANN'S MODERN MAGIC, Professor Hoffmann. One of the best, and best-known, magicians' manuals of the past century. Hundreds of tricks from card tricks and simple sleight of hand to elaborate illusions involving construction of complicated machinery. 332 illustrations. 563pp. 5⅜ x 8½.   23623-4 Pa. $6.95

THOMAS NAST'S CHRISTMAS DRAWINGS, Thomas Nast. Almost all Christmas drawings by creator of image of Santa Claus as we know it, and one of America's foremost illustrators and political cartoonists. 66 illustrations. 3 illustrations in color on covers. 96pp. 8⅜ x 11¼.
23660-9 Pa. $3.50

FRENCH COUNTRY COOKING FOR AMERICANS, Louis Diat. 500 easy-to-make, authentic provincial recipes compiled by former head chef at New York's Fitz-Carlton Hotel: onion soup, lamb stew, potato pie, more. 309pp. 5⅜ x 8½.   23665-X Pa. $3.95

SAUCES, FRENCH AND FAMOUS, Louis Diat. Complete book gives over 200 specific recipes: bechamel, Bordelaise, hollandaise, Cumberland, apricot, etc. Author was one of this century's finest chefs, originator of vichyssoise and many other dishes. Index. 156pp. 5⅜ x 8.
23663-3 Pa. $2.75

TOLL HOUSE TRIED AND TRUE RECIPES, Ruth Graves Wakefield. Authentic recipes from the famous Mass. restaurant: popovers, veal and ham loaf, Toll House baked beans, chocolate cake crumb pudding, much more. Many helpful hints. Nearly 700 recipes. Index. 376pp. 5⅜ x 8½.
23560-2 Pa. $4.95

CATALOGUE OF DOVER BOOKS

ILLUSTRATED GUIDE TO SHAKER FURNITURE, Robert Meader. Director, Shaker Museum, Old Chatham, presents up-to-date coverage of all furniture and appurtenances, with much on local styles not available elsewhere. 235 photos. 146pp. 9 x 12. 22819-3 Pa. $6.95

COOKING WITH BEER, Carole Fahy. Beer has as superb an effect on food as wine, and at fraction of cost. Over 250 recipes for appetizers, soups, main dishes, desserts, breads, etc. Index. 144pp. 5⅜ x 8½. (Available in U.S. only) 23661-7 Pa. $3.00

STEWS AND RAGOUTS, Kay Shaw Nelson. This international cookbook offers wide range of 108 recipes perfect for everyday, special occasions, meals-in-themselves, main dishes. Economical, nutritious, easy-to-prepare: goulash, Irish stew, boeuf bourguignon, etc. Index. 134pp. 5⅜ x 8½. 23662-5 Pa. $3.95

DELICIOUS MAIN COURSE DISHES, Marian Tracy. Main courses are the most important part of any meal. These 200 nutritious, economical recipes from around the world make every meal a delight. "I . . . have found it so useful in my own household,"—N.Y. Times. Index. 219pp. 5⅜ x 8½. 23664-1 Pa. $3.95

FIVE ACRES AND INDEPENDENCE, Maurice G. Kains. Great back-to-the-land classic explains basics of self-sufficient farming: economics, plants, crops, animals, orchards, soils, land selection, host of other necessary things. Do not confuse with skimpy faddist literature; Kains was one of America's greatest agriculturalists. 95 illustrations. 397pp. 5⅜ x 8½. 20974-1 Pa. $4.95

A PRACTICAL GUIDE FOR THE BEGINNING FARMER, Herbert Jacobs. Basic, extremely useful first book for anyone thinking about moving to the country and starting a farm. Simpler than Kains, with greater emphasis on country living in general. 246pp. 5⅜ x 8½. 23675-7 Pa. $3.95

PAPERMAKING, Dard Hunter. Definitive book on the subject by the foremost authority in the field. Chapters dealing with every aspect of history of craft in every part of the world. Over 320 illustrations. 2nd, revised and enlarged (1947) edition. 672pp. 5⅜ x 8½. 23619-6 Pa. $8.95

THE ART DECO STYLE, edited by Theodore Menten. Furniture, jewelry, metalwork, ceramics, fabrics, lighting fixtures, interior decors, exteriors, graphics from pure French sources. Best sampling around. Over 400 photographs. 183pp. 8⅜ x 11¼. 22824-X Pa. $6.95

ACKERMANN'S COSTUME PLATES, Rudolph Ackermann. Selection of 96 plates from the Repository of Arts, best published source of costume for English fashion during the early 19th century. 12 plates also in color. Captions, glossary and introduction by editor Stella Blum. Total of 120pp. 8⅜ x 11¼. 23690-0 Pa. $5.00

THE ANATOMY OF THE HORSE, George Stubbs. Often considered the great masterpiece of animal anatomy. Full reproduction of 1766 edition, plus prospectus; original text and modernized text. 36 plates. Introduction by Eleanor Garvey. 121pp. 11 x 14¾. 23402-9 Pa. $8.95

BRIDGMAN'S LIFE DRAWING, George B. Bridgman. More than 500 illustrative drawings and text teach you to abstract the body into its major masses, use light and shade, proportion; as well as specific areas of anatomy, of which Bridgman is master. 192pp. 6½ x 9¼. (Available in U.S. only) 22710-3 Pa. $4.50

ART NOUVEAU DESIGNS IN COLOR, Alphonse Mucha, Maurice Verneuil, Georges Auriol. Full-color reproduction of *Combinaisons ornementales* (c. 1900) by Art Nouveau masters. Floral, animal, geometric, interlacings, swashes—borders, frames, spots—all incredibly beautiful. 60 plates, hundreds of designs. 9⅜ x 8-1/16. 22885-1 Pa. $4.50

FULL-COLOR FLORAL DESIGNS IN THE ART NOUVEAU STYLE, E. A. Seguy. 166 motifs, on 40 plates, from *Les fleurs et leurs applications decoratives* (1902): borders, circular designs, repeats, allovers, "spots." All in authentic Art Nouveau colors. 48pp. 9⅜ x 12¼.
23439-8 Pa. $6.00

A DIDEROT PICTORIAL ENCYCLOPEDIA OF TRADES AND IN-DUSTRY, edited by Charles C. Gillispie. 485 most interesting plates from the great French Encyclopedia of the 18th century show hundreds of working figures, artifacts, process, land and cityscapes; glassmaking, paper-making, metal extraction, construction, weaving, making furniture, clothing, wigs, dozens. of other activities. Plates fully explained. 920pp. 9 x 12.
22284-5, 22285-3 Clothbd., Two-vol. set $50.00

HANDBOOK OF EARLY ADVERTISING ART, Clarence P. Hornung. Largest collection of copyright-free early and antique advertising art ever compiled. Over 6,000 illustrations, from Franklin's time to the 1890's for special effects, novelty. Valuable source, almost inexhaustible.
Pictorial Volume. Agriculture, the zodiac, animals, autos, birds, Christmas, fire engines, flowers, trees, musical instruments, ships, games and sports, much more. Arranged by subject matter and use. 237 plates. 288pp. 9 x 12.
20122-8 Clothbd. $15.00

Typographical Volume. Roman and Gothic faces ranging from 10 point to 300 point, "Barnum," German and Old English faces, script, logotypes, scrolls and flourishes, 1115 ornamental initials, 67 complete alphabets, more. 310 plates. 320pp. 9 x 12. 20123-6 Clothbd. $15.00

CALLIGRAPHY (CALLIGRAPHIA LATINA), J. G. Schwandner. High point of 18th-century ornamental calligraphy. Very ornate initials, scrolls, borders, cherubs, birds, lettered examples. 172pp. 9 x 13.
20475-8 Pa. $7.95

GEOMETRY, RELATIVITY AND THE FOURTH DIMENSION, Rudolf Rucker. Exposition of fourth dimension, means of visualization, concepts of relativity as Flatland characters continue adventures. Popular, easily followed yet accurate, profound. 141 illustrations. 133pp. 5⅜ x 8½.
23400-2 Pa. $2.75

THE ORIGIN OF LIFE, A. I. Oparin. Modern classic in biochemistry, the first rigorous examination of possible evolution of life from nitrocarbon compounds. Non-technical, easily followed. Total of 295pp. 5⅜ x 8½.
60213-3 Pa. $5.95

PLANETS, STARS AND GALAXIES, A. E. Fanning. Comprehensive introductory survey: the sun, solar system, stars, galaxies, universe, cosmology; quasars, radio stars, etc. 24pp. of photographs. 189pp. 5⅜ x 8½. (Available in U.S. only)
21680-2 Pa. $3.75

THE THIRTEEN BOOKS OF EUCLID'S ELEMENTS, translated with introduction and commentary by Sir Thomas L. Heath. Definitive edition. Textual and linguistic, notes, mathematical analysis, 2500 years of critical commentary. Do not confuse with abridged school editions. Total of 1414pp. 5⅜ x 8½.
60088-2, 60089-0, 60090-4 Pa., Three-vol. set $19.50